AF874817

Handbook of the Mammals of Europe

Series Editors

Klaus Hackländer, Department of Integrative Biology and Biodiversity Research, Institute of Wildlife Biology and Game Management, University of Natural Resources and Life Sciences, Vienna (BOKU), Vienna, Austria

Frank E. Zachos, Natural History Museum Vienna, Mammal Collection, Vienna, Austria
Department of Evolutionary Biology, University of Vienna, Vienna, Austria
Department of Genetics, University of the Free State, Bloemfontein, South Africa

This handbook offers a unique collection of information on all wild mammal species living in Europe and will serve as a standard reference guide for all mammalogists and readers interested in research on mammals. The introductory volume provides an overview of mammalian phylogeny, the history and current status of European mammals and their management as well as their habitats and the history of mammalogy in Europe. The remaining volumes present comprehensive species-specific chapters covering all aspects of mammalian biology, including palaeontology, physiology, genetics, reproduction and development, ecology, habitat, diet, mortality and behaviour. The economic significance and management of mammals and future challenges for research and conservation are addressed as well. Each chapter includes a distribution map, a photograph of the animal, and a list of key literature. This authoritative handbook provides current and detailed descriptions of all European mammals; it will appeal to academics and students in mammal research alike, as well as to practitioners whose work involves mammal management, control, use and conservation.

Luca Corlatti • Frank E. Zachos
Editors

Terrestrial Cetartiodactyla

With 71 Figures and 28 Tables

Springer

Editors
Luca Corlatti
ERSAF - Stelvio National Park
Office of Scientific Research
and Conservation
Bormio (SO), Italy

Chair of Wildlife Ecology and
Management
University of Freiburg
Freiburg, Germany

Frank E. Zachos
Natural History Museum Vienna,
Mammal Collection
Vienna, Austria

Department of Evolutionary Biology
University of Vienna
Vienna, Austria

Department of Genetics
University of the Free State
Bloemfontein, South Africa

ISSN 2730-7387 ISSN 2730-7395 (electronic)
Handbook of the Mammals of Europe
ISBN 978-3-030-24474-3 ISBN 978-3-030-24475-0 (eBook)
https://doi.org/10.1007/978-3-030-24475-0

© Springer Nature Switzerland AG 2022
This work is subject to copyright. All rights are reserved by the Publisher, whether the whole or part of the material is concerned, specifically the rights of translation, reprinting, reuse of illustrations, recitation, broadcasting, reproduction on microfilms or in any other physical way, and transmission or information storage and retrieval, electronic adaptation, computer software, or by similar or dissimilar methodology now known or hereafter developed.
The use of general descriptive names, registered names, trademarks, service marks, etc. in this publication does not imply, even in the absence of a specific statement, that such names are exempt from the relevant protective laws and regulations and therefore free for general use.
The publisher, the authors, and the editors are safe to assume that the advice and information in this book are believed to be true and accurate at the date of publication. Neither the publisher nor the authors or the editors give a warranty, expressed or implied, with respect to the material contained herein or for any errors or omissions that may have been made. The publisher remains neutral with regard to jurisdictional claims in published maps and institutional affiliations.

Cover illustration: Cover photos with kind permission from: Kurt Kracher (Red deer), Frank E. Zachos (European bison), Denis Bertanzetti (Alpine chamois)

This Springer imprint is published by the registered company Springer Nature Switzerland AG.
The registered company address is: Gewerbestrasse 11, 6330 Cham, Switzerland

Preface and Acknowledgments

Ungulates are special among European mammals. They comprise the largest (terrestrial) European mammals, they are important ecologically as well as economically, and they are fun to look at or watch. Humans have been fascinated by deer and wild cattle for millennia and, going back to cave paintings, have found artistic and spiritual inspiration in them. For all these reasons and more, they are also among the best-studied of all mammals. The present volume of the *Handbook of the Mammals of Europe* contains chapters on each European ungulate species, depicting and celebrating their diversity. A diversity which has been reduced considerably compared to former times and ages when recently and not so recently extinct species like the auerochs, the woolly rhino, or wild horses roamed the European plains and forests. Today, only even-toed ungulates (Cetartiodactyla) remain in Europe (not counting feral horses), and it is these that this volume covers. The somewhat awkward title "Terrestrial Cetartiodactyla" is due to one of the most interesting (although not necessarily surprising) insights into the evolutionary relationships of ungulates: whales are derived even-toed ungulates, and their closest living relatives are the two extant species of hippopotamus. Because whales are therefore nested within the other artiodactyls, the name Artiodactyla has been changed to Cetartiodactyla, and since the whales (Cetacea) are dealt with in a volume of their own, we are left with the terrestrial cetartiodactyls in this volume.

The question of where the boundaries of Europe are is an eternal one that has no definitive answer. Although Europeans might not like to hear this, Europe is a political construct, not a geographical continent, and therefore any delimitation will be arbitrary to a degree. We discuss this briefly in the introductory volume to this handbook, but the consequence is that some readers might miss certain species that are not covered here.

Chapters cover taxonomy, systematics and paleontology, distribution, morphology, physiology, genetics, life history traits, habitat and diet, behavior, parasites and diseases, population ecology, management and conservation, and future challenges for research and management for each species. We strove to maintain this structure as strictly as possible across chapters, with few exceptions, mainly those covering non-native species.

This handbook is a great opportunity for mammalogy and mammalogists alike, and we are grateful to Springer Publishers for inviting us as editors. In particular, we would like to thank Barbara Wolf, Lars Koerner, and Veronika

Mang for their constant support and enthusiasm in making this handbook happen. Further, a big thank you to all chapter authors for their time and effort and, last but not least, to the external referees of the species chapters who shared their time, knowledge, and expertise with us and greatly increased the quality of this volume.

Bormio, Italy Luca Corlatti
Vienna, Austria Frank E. Zachos
October 2022

Contents

About the Series Editors

Klaus Hackländer
Department of Integrative Biology and Biodiversity Research, Institute of Wildlife Biology and Game Management, University of Natural Resources and Life Sciences, Vienna (BOKU),
Vienna, Austria

Frank E. Zachos
Natural History Museum Vienna
Mammal Collection
Vienna, Austria

Department of Evolutionary Biology
University of Vienna
Vienna, Austria

Department of Genetics
University of the Free State
Bloemfontein, South Africa

About the Volume Editors

Luca Corlatti is Senior Wildlife Ecologist at the Stelvio National Park, Italy, and Affiliate Professor of Wildlife Ecology at the Albert Ludwig University of Freiburg, Germany. He obtained a MSc in Forest and Environmental Sciences at the University of Padua in 2006. After research experiences in Italy, Denmark, Sweden, Austria, and Scotland, in 2013 he obtained a PhD in Evolutionary Biology at the University of Siena. In 2016, he was recipient of a Marie Curie-Skłodowska Postdoctoral Fellowship in Behavioral Ecology at the Freiburg Institute for Advanced Studies; between 2017 and 2021, he held a Postdoc position with the Chair of Wildlife Ecology and Management at the University of Freiburg. His research interests primarily focus on the ecology, behavior, and management of mountain-dwelling ungulates. He is Associate Editor of the journals *Mammalian Biology* and *Wildlife Biology* and member of the Caprinae Specialist Group of the International Union for the Conservation of Nature.

Frank E. Zachos is head of the Mammal Collection at the Natural History Museum in Vienna, Austria, an Affiliated Lecturer at the University of Vienna, and an Affiliated Professor at the Department of Genetics, University of the Free State in Bloemfontein, South Africa. He studied biology, the history of science, and philosophy in Kiel and Jena, Germany, and received his PhD in Zoology in 2005. In 2009, he completed his habilitation (postdoctoral degree) in Zoology and Evolutionary Biology. His main research interests are intraspecific biodiversity, phylogeography, population genetics, and conservation of mammals and birds, with a particular focus on red deer and other ungulates. He also has a long-standing interest in species concepts and the species problem, as well as other theoretical and philosophical issues in evolutionary biology and systematics. He was the editor-in-chief of the journal *Mammalian Biology* from 2007 to 2021 and is currently the editor of the Mammalia series within the *Handbook of Zoology*.

Contributors

Pelayo Acevedo Instituto de Investigación en Recursos Cinegéticos, IREC (UCLM-CSIC-JCCM), Ciudad Real, Spain

Pia Anderwald Swiss National Park, Zernez, Switzerland

Marco Apollonio Department of Veterinary Medicine, University of Sassari, Sassari, Italy

Mari Cruz Arnal Departamento de Patología Animal, Facultad de Veterinaria, Universidad de Zaragoza, Zaragoza, Spain

Jon M. Arnemo Department of Wildlife, Fish, and Environmental Studies, Faculty of Forest Sciences, Swedish University of Agricultural Sciences (SLU), Umea, Sweden

Department of Forestry and Wildlife, Management, Faculty of Applied Ecology and Agricultural Sciences, Inland Norway University of Applied Sciences, Elverum, Norway

Miroslava Barančeková Department of Infectious Diseases, Veterinary University Brno, Palackeho trida, Brno, Czech Republic

Bruno Bassano Gran Paradiso National Park, Alpine Wildlife Research Center, Torino, Italy

Iris Biebach Department of Evolutionary Biology and Environmental Studies, University of Zurich, Zurich, Switzerland

Kurt Bollmann Biodiversity and Conservation Biology, Swiss Federal Research Institute WSL, Birmensdorf, Switzerland

Tomasz Borowik Mammal Research Institute, Polish Academy of Sciences, Białowieża, Poland

Gilles Bourgoin CNRS, Laboratoire de Biométrie et Biologie Evolutive UMR 5558, Université de Lyon, Université Lyon 1, Villeurbanne, France

VetAgro Sup, Université de Lyon, Marcy l'Etoile, France

Alice Brambilla Department of Evolutionary Biology and Environmental Studies, University of Zurich, Zurich, Switzerland

Gran Paradiso National Park, Alpine Wildlife Research Center, Torino, Italy

Jon E. Brommer Department of Biology, University of Turku, Turku, Finland

Jorge Cassinello Estación Experimental de Zonas Áridas (EEZA-CSIC), Almería, Spain

Norma G. Chapman Bury St. Edmunds, UK

Roberta Chirichella Department of Veterinary Medicine, University of Sassari, Sassari, Italy

Arnold S. Cooke Huntingdon, UK

Luca Corlatti ERSAF - Stelvio National Park, Office of Scientific Research and Conservation, Bormio (SO), Italy

Chair of Wildlife Ecology and Management, University of Freiburg, Freiburg, Germany

Jean-Marc Cugnasse Vignes Basses, Brousse le Château, France

Anna Maria De Marinis Italian Institute for Environmental Protection and Research (ISPRA), Ozzano dell'Emilia (BO), Italy

Göran Ericsson Department of Wildlife, Fish, and Environmental Studies, Faculty of Forest Sciences, Swedish University of Agricultural Sciences (SLU), Umea, Sweden

Daniel Fernández de Luco Departamento de Patología Animal, Facultad de Veterinaria, Universidad de Zaragoza, Zaragoza, Spain

Francesco Ferretti Research Unit of Behavioural Ecology, Ethology and Wildlife Management, Department of Life Sciences, University of Siena, Siena, Italy

Jean-Michel Gaillard CNRS, Laboratoire de Biométrie et Biologie Evolutive, UMR5558, Université de Lyon, Université Lyon 1, Villeurbanne, France

Pavao Gančević Department of Fisheries, Apiculture, Wildlife Management and Special Zoology, Faculty of Agriculture, University of Zagreb, Zagreb, Croatia

Ricardo García-González Instituto Pirenaico de Ecología (IPE-CSIC), Jaca, Spain

Mathieu Garel Office Français de la Biodiversité, Direction de la Recherche et de l'Appui Scientifique, Service Anthropisation et Fonctionnement des Écosystèmes Terrestres, Gières, France

Luisa Garofalo Istituto Zooprofilattico Sperimentale del Lazio e della Toscana M. Aleandri, Rome, Italy

Eleftherios Hadjisterkotis Ministry of Agriculture, Rural Development and Environment, Agricultural Research Institute, Nicosia, Cyprus

Sabine E. Hammer Institute of Immunology, University of Veterinary Medicine, Vienna, Austria

Achaz von Hardenberg Conservation Biology Research Group, Department of Biological Sciences, University of Chester, Chester, UK

Juan Herrero Technical School, University of Zaragoza, Huesca, Spain

Mark Hewison CEFS, Université de Toulouse, INRA, Castanet Tolosan, France

Øystein Holand Norwegian University of Life Sciences, Aas, Norway

Laura Iacolina Department of Chemistry and Bioscience, Aalborg University, Aalborg, Denmark

Department of Biodiversity, University of Primorska, Koper, Slovenia

Klemen Jerina Department of Forestry, Biotechnical Faculty, University of Ljubljana, Ljubljana, Slovenia

Lukas Keller Department of Evolutionary Biology and Environmental Studies, University of Zurich, Zurich, Switzerland

Marta Kołodziej-Sobocińska Mammal Research Institute, Polish Academy of Sciences, Białowieża, Poland

Rafał Kowalczyk Mammal Research Institute, Polish Academy of Sciences, Białowieża, Poland

Jarmila Krojerová-Prokešová Institute of Vertebrate Biology, Czech Academy of Sciences, Brno, Czech Republic

Adrian M. Lister Natural History Museum, London, UK

Rita Lorenzini Istituto Zooprofilattico Sperimentale del Lazio e della Toscana M. Aleandri, Centro di Referenza Nazionale per la Medicina Forense Veterinaria, Grosseto, Italy

Bejan Lortkipanidze NACRES – Centre for Biodiversity Conservation & Research, Tbilisi, Georgia

Sandro Lovari Research Unit of Behavioural Ecology, Ethology and Wildlife Management, Department of Life Sciences, University of Siena, Siena, Italy

Jonas Malmsten Department of Wildlife, Fish, and Environmental Studies, Faculty of Forest Sciences, Swedish University of Agricultural Sciences (SLU), Umea, Sweden

Pascal Marchand Office Français de la Biodiversité, Direction de la Recherche et de l'Appui Scientifique, Service Anthropisation et Fonctionnement des Écosystèmes Terrestres, Juvignac, France

Stefano Mattioli Research Unit of Behavioural Ecology, Ethology and Wildlife Management, Department of Life Sciences, University of Siena, Siena, Italy

Ivan Mizin Russian Arctic National Park, Arkhangelsk, Russia

Nicolas Morellet CEFS, Université de Toulouse, INRA, Castanet Tolosan, France

Wiebke Neumann Department of Wildlife, Fish, and Environmental Studies, Faculty of Forest Sciences, Swedish University of Agricultural Sciences (SLU), Umea, Sweden

Magdalena Niedziałkowska Mammal Research Institute, Polish Academy of Sciences, Białowieża, Poland

Carlos Nores INDUROT, University of Oviedo, Mieres, Asturias, Spain

Haritakis Papaioannou Department of Biological Applications and Technology, School of Health Studies, University of Ioannina, Ioannina, Greece

Holger Piegert Thale, Germany

Glenn Plumb International Union for Conservation of Nature, Species Survival Commission, Bison Specialist Group, Livingston, MT, USA

Tomasz Podgórski Mammal Research Institute, Polish Academy of Sciences, Białowieża, Poland

Department of Game Management and Wildlife Biology, Faculty of Forestry and Wood Sciences, Czech University of Life Sciences, Praha, Czech Republic

Elodie Portanier CNRS, Laboratoire de Biométrie et Biologie Evolutive UMR 5558, Université de Lyon, Université Lyon 1, Villeurbanne, France

CNRS, Station Biologique de Roscoff, UMR 7144, Sorbonne Université, Roscoff, France

Jenni Poutanen Department of Biology, University of Turku, Turku, Finland

Rory Putman Institute of Biodiversity, Animal Health and Comparative Medicine, University of Glasgow, Glasgow, UK

Luca Rossi Department of Veterinary Sciences, University of Turin, Turin, Italy

Toni Safner Department of Plant Breeding Genetics, Biometrics and Experimentation, Faculty of Agriculture, University of Zagreb, Zagreb, Croatia

Julián Santiago-Moreno Department of Animal Reproduction, INIA-CSIC, Madrid, Spain

Massimo Scandura Department of Veterinary Medicine, University of Sassari, Sassari, Italy

Niels Martin Schmidt Aarhus University, Roskilde, Denmark

Nikica Šprem Department of Fisheries, Apiculture, Wildlife Management and Special Zoology, Faculty of Agriculture, University of Zagreb, Zagreb, Croatia

Mikkel Stelvig Copenhagen Zoo, Frederiksberg, Denmark

Carole Toïgo Office Français de la Biodiversité, Unité Ongulés Sauvages, Gières, France

Hélène Verheyden CEFS, Université de Toulouse, INRA, Castanet Tolosan, France

Joaquin Vicente SaBio, Instituto de Investigación en Recursos Cinegéticos - IREC (CSIC-UCLM-JCCM), Ciudad Real, Spain

Paul Weinberg North Ossetian State Nature Reserve, Alagir, Russian Federation

Robert B. Weladji Concordia University, Montreal, QC, Canada

Mikael Wikström Finnish Wildlife Agency, Helsinki, Finland

Frank E. Zachos Natural History Museum Vienna, Mammal Collection, Vienna, Austria

Department of Evolutionary Biology, University of Vienna, Vienna, Austria

Department of Genetics, University of the Free State, Bloemfontein, South Africa

Wild Boar *Sus scrofa* Linnaeus, 1758 1

Massimo Scandura, Tomasz Podgórski, Joaquin Vicente, and Laura Iacolina

Contents

M. Scandura (✉)
Department of Veterinary Medicine, University of Sassari, Sassari, Italy
e-mail: scandura@uniss.it

T. Podgórski
Mammal Research Institute, Polish Academy of Sciences, Białowieża, Poland

Department of Game Management and Wildlife Biology, Faculty of Forestry and Wood Sciences, Czech University of Life Sciences, Praha, Czech Republic
e-mail: t_podgorski@ibs.bialowieza.pl

J. Vicente
SaBio, Instituto de Investigación en Recursos Cinegéticos - IREC (CSIC-UCLM-JCCM), Ciudad Real, Spain
e-mail: joaquin.vicente@uclm.es

L. Iacolina (✉)
Department of Chemistry and Bioscience, Aalborg University, Aalborg, Denmark

Department of Biodiversity, University of Primorska, Koper, Slovenia
e-mail: lauraiacolina@gmail.com

© Springer Nature Switzerland AG 2022

L. Corlatti, F. E. Zachos (eds.), *Terrestrial Cetartiodactyla*, Handbook of the Mammals of Europe,
https://doi.org/10.1007/978-3-030-24475-0_17

Common Names

English	(Eurasian) wild boar
German	Wildschwein
French	Sanglier
Spanish	Jabalí
Italian	Cinghiale
Russian	Кабан

Taxonomy, Systematics and Paleontology

The origin of *Sus scrofa* (Fig. 1) is in Asia, where a radiation of the genus *Sus* into several different taxa has occurred during the last 5 million years (Frantz et al. 2013a). In particular, an evolutionary engine for such suids (like the Rift Valley for hominids) was represented by the islands of South East Asia, where a vast number of related taxa occur nowadays. According to genomic data, the speciation of the Eurasian wild boar started during the Pliocene (4–4.5 million years ago (mya)), accompanied by its spread across the Asian continent and the radiation into regionally diverging populations (Frantz et al. 2013a). The spread of the species to the west was belated and its appearance in Europe, according to the first fossil records, goes back to the late Early Pleistocene around 900-800 thousand years ago (kya) (Cherin et al. 2020) during a period of great woodland expansion. During the Late Pleistocene, the wild boar was severely affected by cooling periods, restricting its range to southern regions. Converging paleontological and genetic data revealed refugial areas in southern Europe where the species survived during the last glaciation (Vilaça et al. 2014; Veličković et al. 2015). The following postglacial recolonization restored the species in almost the whole continent. An important role in the species' history was played by pig domestication, a long-lasting process that started in Asia about 10 kya and received a tremendous boost during the last centuries (White 2011). The relevance of pig domestication resides in the parallel evolution of the wild ancestor with its domesticated form, manipulated by humans. Their persistent contact and crossbreeding have shaped populations of the two forms over time (Frantz et al. 2013a) and still represent a powerful evolutionary force for the Eurasian wild boar.

High levels of intraspecific variation led to a repeatedly revised taxonomy. The first comprehensive assessment distinguished 16 subspecies, clustered into four groupings based on geographical and morphological criteria: Western, Eastern,

Fig. 1 Wild boar (photograph by and courtesy of I. I. Serval)

Indian, and Indonesian "races," the first one occurring in Europe (Groves 1981). Later on, Genov (1999) reviewed the variation in cranial morphology and confirmed the wild boar as a single polytypic species. This view was then questioned (Groves 2007) and more recently revised, elevating most of the 16 originally identified subspecies to species rank (Groves and Grubb 2011), although this splitting approach has received severe criticism. Under this classification, *Sus scrofa* represents a narrower taxon ranging from the Iberian Peninsula and Maghreb to Central Asia and is the only wild pig occurring in Europe. Here, four different subspecies were proposed on the basis of morphological and karyological data (Groves 2007):

S. s. scrofa Linnaeus, 1758: western subspecies (from Iberia to Belarus); variable size from south (smaller) to north (larger)

S. s. attila Thomas, 1912: from central Belarus and the Carpathians to Western Russia; a large subspecies

S. s. meridionalis Major, 1882: endemic to the islands of Sardinia and Corsica, formerly believed to include the wild boar inhabiting the south of Spain; a small subspecies

S. s. lybicus Gray, 1868: from the Balkans to the Near East; a small subspecies

Recent phylogeographic studies weakly supported this partition, confirming the genomic peculiarity of *S. s. meridionalis* from Sardinia (Iacolina et al. 2016) and suggesting some levels of differentiation for the populations inhabiting mainland Italy (where the endemic subspecies *S. s. majori* De Beaux and Festa, 1927 had been proposed in the past, see Scandura et al. 2008) and the southern Balkans (where several endemic lineages occur, Alexandri et al. 2012). On the other hand, no sharp genetic discontinuity emerged in Eastern Europe between putative *S. s. scrofa* and *S. s. attila* (Vilaça et al. 2014).

Current Distribution

The wild boar, also thanks to human action, has one of the widest distributions among terrestrial mammals and is abundant in many parts of its range. Such a broad range relies on the presence of individuals at different domestication stages, from wild to domestic, going through feral and hybrid forms. The species is now present on all continents but Antarctica, including many oceanic islands. In Europe, wild boar are present in most continental areas and on many islands (see Fig. 2). Being a thermophilic species, the northern boundary of its distribution is limited by snow depth and winter temperature, and in the last decades, it has notably expanded northwards in consequence of climate change (Danilov and Panchenko 2012). Island populations have been strongly influenced by humans. Historical populations occur in Corsica and Sardinia, where the species was introduced during the Neolithic. Once extinct, wild boar were recently reintroduced in Sicily and reappeared in Great Britain and Ireland, where the local populations proved to be represented by hybrid individuals, likely escaped from farms or illegally released (Frantz et al. 2012; McDevitt et al. 2013). In Cyprus, the species was illegally reintroduced in 1994, but went extinct again in 2004 (Hadjisterkotis and Heise-Pavlov 2006). Many other minor islands host populations of doubtful or admixed origin. In continental Europe, after a reduction in both distribution and abundance till the beginning of the twentieth century, especially after World War II the species experienced a strong recovery. Also thanks to milder climatic conditions, it has naturally recolonized regions like Estonia and Denmark, despite the attempts to prevent it in the latter country (Apollonio et al. 2010). A hybrid stock escaped from farms is the source of the present population in Sweden (Lemel et al. 2003), whose spread has recently reached southern Norway (Østfold county, VKM et al. 2018).

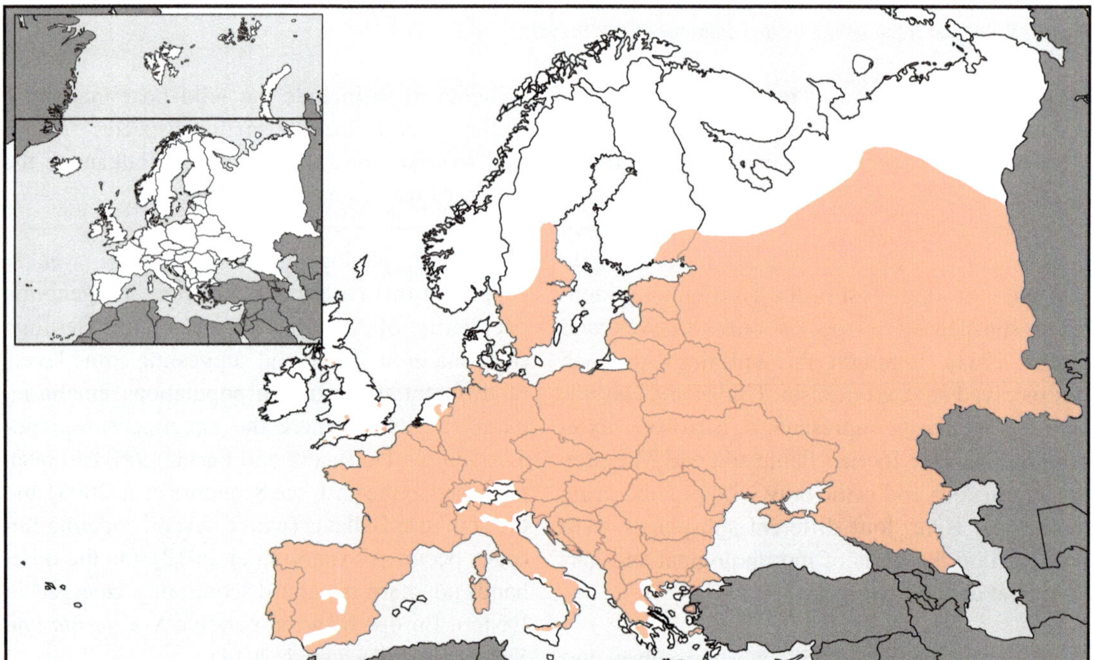

Map template: © Getty Images/iStockphoto

Fig. 2 Distribution map of the European wild boar, based on the IUCN Red List of Threatened Species. Version 2018–1 with kind permission. (Map template: © Copyright Getty Images/iStockphoto)

Description

Wild boar distribution covers a wide range of habitats and significant variation in coat color and size can be observed (Fig. 3). The general trend sees the smallest animals (maximum 70–90 kg) in the south and on Mediterranean islands, while larger animals (maximum *c.* 300 kg) are present in the northern and north-eastern parts of its range. Accordingly, other body measurements also show wide variation. One of the most striking is the head. Skull size has been a key trait for taxonomy, ranging in length between 30 and 47 cm across European wild boar (Keuling et al. 2018). Body length goes from 90 to 200 cm, but tail length (15–40 cm) and shoulder height (55–110 cm) are also highly variable (Keuling et al. 2018). The species is strongly built, with moderately short tail and legs, with the anterior legs longer than posterior ones, and the frontal part of the skull is more developed than the frontal area. Additionally, females are smaller than males, on average around 40% (Keuling et al. 2018). Coat color varies with age. Piglets are striped up to about 4 months, then they turn into a reddish color and, around one year of age, they display the adult pelage. The latter goes from brown to almost black and, with age, it can turn into grey.

Fig. 3 Wild boar from different areas in Europe. (**a**) Sardinia, Italy (photograph by M. Scandura); (**b**) Coto Doñana, Spain (photograph by J. Vicente); (**c**) Bulgaria (photograph by and courtesy of F. Morimando)

Table 1 Estimated tooth eruption. Teeth in parentheses might be absent

Age	Incisors	Canines	Premolars	Molars
Birth	$\mathrm{i}\,\frac{3}{3}$	$\mathrm{c}\,\frac{1}{1}$		
2 months	$\mathrm{i}\,\frac{1}{1}\,\frac{3}{3}$	$\mathrm{c}\,\frac{1}{1}$	$\mathrm{p}\,\frac{3}{3}\,\frac{4}{4}$	
2–4 months	$\mathrm{i}\,\frac{1}{1}\,\frac{2}{2}\,\frac{3}{3}$	$\mathrm{c}\,\frac{1}{1}$	$\mathrm{p}\,\frac{2}{2}\,\frac{3}{3}\,\frac{4}{4}$	
5–6 months	$\mathrm{i}\,\frac{1}{1}\,\frac{2}{2}\,\frac{3}{3}$	$\mathrm{c}\,\frac{1}{1}$	$\mathrm{p}\,\frac{2}{2}\,\frac{3}{3}\,\frac{4}{4}$	$\mathrm{M}\,\frac{1}{1}$
7–8 months	$\mathrm{i}\,\frac{1}{1}\,\frac{2}{2}\,\frac{3}{3}$	$\mathrm{c}\,\frac{1}{1}$	$\mathrm{p}\,\frac{2}{2}\,\frac{3}{3}\,\frac{4}{4},\,(\mathrm{P}\,\frac{1}{1})$	$\mathrm{M}\,\frac{1}{1}$
8–9 months	$\mathrm{i}\,\frac{1}{1}\,\frac{2}{2},\mathrm{I}\,\frac{3}{3}$	$\mathrm{c}\,\frac{1}{1}$	$\mathrm{p}\,\frac{2}{2}\,\frac{3}{3}\,\frac{4}{4},\,(\mathrm{P}\,\frac{1}{1})$	$\mathrm{M}\,\frac{1}{1}$
10–12 months	$\mathrm{i}\,\frac{1}{1}\,\frac{2}{2},\mathrm{I}\,\frac{3}{3}$	$\mathrm{C}\,\frac{1}{1}$	$\mathrm{p}\,\frac{2}{2}\,\frac{3}{3}\,\frac{4}{4},\,(\mathrm{P}\,\frac{1}{1})$	$\mathrm{M}\,\frac{1}{1}\,(\frac{2}{2})$
12–15 months	$\mathrm{i}\,\frac{2}{2},\mathrm{I}\,\frac{1}{1}\,\frac{3}{3}$	$\mathrm{C}\,\frac{1}{1}$	$\mathrm{p}\,\frac{2}{2}\,\frac{3}{3}\,\frac{4}{4},\,(\mathrm{P}\,\frac{1}{1})$	$\mathrm{M}\,\frac{1}{1}\,\frac{2}{2}$
15–18 months	$\mathrm{i}\,\frac{2}{2},\mathrm{I}\,\frac{1}{1}\,\frac{3}{3}$	$\mathrm{C}\,\frac{1}{1}$	$\mathrm{p}\,\frac{2}{2}\,\frac{3}{3},\;\mathrm{P}\,(\frac{1}{1})\,\frac{4}{4}$	$\mathrm{M}\,\frac{1}{1}\,\frac{2}{2}$
20 months *c.*	$\mathrm{I}\,\frac{1}{1}\,\frac{2}{2}\,\frac{3}{3}$	$\mathrm{C}\,\frac{1}{1}$	$\mathrm{P}\,(\frac{1}{1})\,\frac{2}{2}\,\frac{3}{3}\,\frac{4}{4}$	$\mathrm{M}\,\frac{1}{1}\,\frac{2}{2}$
22–28 months	$\mathrm{I}\,\frac{1}{1}\,\frac{2}{2}\,\frac{3}{3}$	$\mathrm{C}\,\frac{1}{1}$	$\mathrm{P}\,(\frac{1}{1})\,\frac{2}{2}\,\frac{3}{3}\,\frac{4}{4}$	$\mathrm{M}\,\frac{1}{1}\,\frac{2}{2}\,\frac{3}{3}$

Winter pelage, which includes a conspicuous dorsal mane in adult males, starts growing in the summer, the change is complete by the autumn, and it is then lost in late winter early spring. Upper and lower canines are well developed in the form of tusks. In males, upper canines curve out and upward and lower canines are long, protruding from the mouth and kept sharp by rubbing against the upper ones; they have a predominant role of defense. At birth, wild boar have eight primary teeth, while adults have 44 permanent teeth. Changes in dentition are commonly used to estimate age of individuals (Table 1).

Physiology

The wide distribution of the species highlights its ecological and physiological plasticity, enabling it to adapt and exploit the opportunities offered by a wide range of environments. Physiologically it is, therefore, a "generalist" animal. The wild boar is the ancestor of domestic pig (*Sus scrofa domesticus*, the most important animal species used for meat production worldwide), which has been often used as a model to biomedical research. This species has highly developed auditory and olfactory senses for the detection of predators, communication, and foraging. Wild boar do not differentiate all colors (Fulgione et al. 2017) and do not have a *tapetum lucidum* (residual light amplifier), and therefore, the ability to see in the dark is less relevant than the olfactory and auditory senses to perceive information about the environment. The tactile sense is well developed, especially in the oral and lip regions in order to detect food during rooting. A variety of scent glands secrete odorous compounds: preputial, anal, metacarpal, mandibular and salivary, tusk, lip, Harderian, and eyelid. Steroid pheromones in the saliva and preputial secretions inform on the reproductive status of males and may induce females in estrus to stand for copulation (Vandenbergh 1988). Digestion is rapid, adapted to the fast transition of food, and efficient for a large variety of nutrients (can be classified as dietary generalists), although the fermentation of cellulose, occurring in the *caecum*, is only partial. During late pregnancy and lactation, females experience higher protein and energy requirements, which may affect the survival of piglets (Vetter et al. 2015).

Wild boar prefer warm temperatures and do not tolerate extreme cold and hot dry environments. This species shows a limited metabolic capacity to produce heat without shivering and has very few sweat glands. Therefore, cold winters may impact survival, especially among piglets, since mass-specific metabolic rate is low and thermoregulatory costs cannot be compensated by the available resources when energetic food is scarce. In southern latitudes, hot summers, together with seasonal scarcity of food, can also impact survival. However, wild boar have developed behavioral thermoregulation strategies to cope with cold and hot conditions (Vetter et al. 2015). Climate change may favor wild boar population growth by buffering the negative effect of cold winters on survival and reproduction and increasing food availability (Vetter et al. 2015). Wild boar may store large amounts of fat, which helps to survive when food is scarce (Merta et al. 2014). Body condition (the amount of energy stored in organs and tissues) is especially relevant to health, reproductive performance, and population dynamics of the species (see section "Life History"). Consequently, kidney fat index and other measures of fat deposits (e.g., brisket, rump fat thickness) and different biometrical procedures (including regression approaches based on multiple biometric measures) have been used to assess it (e.g., Risco et al. 2018). Hematological and biochemical parameters in wild boar sera can be used to obtain insight into its metabolism and physiology. Nonetheless, reference values often differ among studies and with that of domestic pigs, evidencing a wide range of factors may affect them, such as environment, season, diet, age, and stressors (Casas-Díaz et al. 2015).

Genetics

Chromosomes

$2n = 36–38$, shared with domestic pigs. The number of chromosomes is variable because of a Robertsonian translocation involving chromosomes 15 and 17 (McFee et al. 1966). $2n = 36$ is the basic condition of western populations, whereas $2n = 38$ is typical of Asian wild boar (from South-East Asia to Turkey and Russia). Crossings between the two groups produce fertile hybrid individuals with $2n = 37$, which are common in admixed European populations (see Table S1 in Scandura et al. 2011a and references therein).

Phylogeny and Phylogeography

Phylogenetic relationships within *Sus scrofa* have been reconstructed by the analysis of mitochondrial DNA (mtDNA) and, more recently, by genome-wide analysis, contributing to disclose complex evolutionary interactions, including long-lasting introgressive hybridization with domesticated pigs and interspecific gene flow with other related suids in South-East Asia (Groenen 2016).

Whole-genome sequencing data provided a high-resolution phylogeny of the genus *Sus*, giving insights into the chronology of divergence between European and Asian populations, dated back to around 1 mya (Groenen et al. 2012; Frantz et al. 2013a). Genomic data are also informative on the demographic history of the European population, which reached a minimum during the last glacial maximum, around 20 kya, in parallel with the species' retreat to southern refugia (Groenen et al. 2012).

Different mtDNA clades are observed in the Eurasian wild boar, most of which occur in southern Asia (Larson et al. 2005), according to the abovementioned evolutionary history of the species. Only a few lineages are found in Europe, namely a pan-European clade (E1), an endemic Italian clade (E2), and an East Asian clade (A), whose occurrence in European wild boar is commonly attributed to genetic introgression from domestic pigs belonging to international commercial breeds (Scandura et al. 2011a; Vilaça et al. 2014). In fact, pig breed amelioration in past centuries involved the intentional crossbreeding of European pigs with Asian breeds (White 2011). Some Near Eastern (NE) haplotypes are also occasionally found in Eastern Europe and seem the result of natural gene flow (Alexandri et al. 2012). The overall phylogeographic pattern of the species in Europe is consistent with a major impact by Quaternary peri-glacial dynamics rather than by recent human-induced events (Scandura et al. 2008; Vilaça et al. 2014). Accordingly, a higher genetic diversity in southern peninsulas (i.e., glacial refugia) and a gradient of decreasing diversity northwards are observed (Alexandri et al. 2012; Vilaça et al. 2014; Veličković et al. 2015), as well as the signal of a postglacial population expansion (Scandura et al. 2008).

Genetic Diversity and Structuring

The genetic diversity of European wild boar is low compared to that of the Asian wild boar, mostly as a consequence of the ancient colonization history and of the bottlenecks undergone during the Quaternary glaciation (Groenen et al. 2012). Nonetheless, the genetic diversity of local populations can vary remarkably as a consequence of demographic fluctuations, degree of isolation, hybridization with domestic pigs, and human-mediated gene flow (Scandura et al. 2011a). Only a few studies have assessed genetic variation at a continental scale, while a number of investigations have explored local situations that are not directly comparable because of the use of different molecular markers and sampling designs. As expected on the basis of glacial/interglacial dynamics, southern peninsulas (Iberia, Italy, and Balkans) host a large amount of the overall diversity observed in European wild boar, showing endemic mtDNA lineages and haplotypes (Scandura et al. 2008; Alexandri et al. 2012; Vilaça et al. 2014; Veličković et al. 2015). High genetic variation in some wild boar populations may be affected by the local degree of anthropogenic introgression (see section "Hybridization"). At mtDNA, the effect of introgression from domestic pigs is reflected by the occurrence of Asian haplotypes, whereas at autosomal markers (i.e., microsatellites and single nucleotide polymorphisms (SNPs)), the assessment of the impact of introgression is more challenging. Therefore, genetic variation in wild boar populations should be evaluated with caution and only the richness of its endemic component can be interpreted as really informative under an evolutionary and conservation perspective. For example, populations like Castelporziano, Maremma, and Sardinia in Italy (Scandura et al. 2008; Iacolina et al. 2016), central Greece and the Dinaric region in the Balkans (Alexandri et al. 2012; Veličković et al. 2015), and Southern Iberia (Alves et al. 2010) show a remarkable proportion of endemic genetic variation.

Though the present status of the species depicts it as overabundant and almost continuously distributed in Europe (see section "Current Distribution"), the wild boar underwent a period of strong range fragmentation and local bottlenecks in the past centuries. This, along with the impact of land use modifications and human infrastructures, has left a detectable genetic signature in many European populations. As a result, many populations appear genetically structured (Scandura et al. 2011b; Goedbloed et al. 2013; Renner et al. 2016). Nonetheless, factors determining genetic discontinuities are not always easy to interpret and contrasting patterns of gene flow are observed in different areas (Renner et al. 2016).

Island populations deserve a separate mention, as most of them arose from one or multiple introductions in historical or more recent times. Consequently, their status varies greatly, as does their genetic variation, quite often diverging from the closest continental populations and possibly affected by hybridization with domestic pigs (McDevitt et al. 2013; Canu et al. 2018).

Hybridization

Hybridization, between the wild boar and its domestic counterpart, has been assessed in several European countries, using a variety of molecular and morphological markers. Wild x domestic hybridization levels vary greatly across Europe, from absent (Iberia, using mtDNA; Alves et al. 2003) to very high (England, using microsatellite and mtDNA; Frantz et al. 2012) and so does the geographic distribution of the phenomenon. For example, it was reported to be widespread in the Netherlands (Goedbloed et al. 2013) and Luxembourg (Frantz et al. 2013b), whereas in Greece and in Sardinia, it appeared to be limited to a few areas (Koutsogiannouli et al. 2010; Scandura et al. 2011b). Gene flow between the two forms is usually related to human practices, be it release of admixed individuals or free-ranging farming practices (e.g., McDevitt et al. 2013). Consequences of hybridization are not fully understood: Canu et al. (2016) argued that coat color changes due to crossbreeding can lead to a lack of camouflage that might increase the chances of being spotted by hunters or natural predators; Goedbloed et al. (2015) reported decreased resistance to pathogens, whereas Fulgione et al. (2016) noticed a fitness increase. Additionally, an improvement in meat quality was observed after introgression of wild genes into the domestic population (Matiuti et al. 2010). Nonetheless, the most commonly reported effects were alterations to the local gene pool, possibly leading to a loss of adaptation, increased population size or invasiveness, and morphological changes (Koutsogiannouli et al. 2010; Fulgione et al. 2016). An issue in assessing hybridization in natural populations is represented by the choice of diagnostic markers. Although mtDNA has been largely used, it can only enable the detection of alleles introgressed in the matriline, while autosomal markers (like microsatellites and SNPs) are needed to identify hybrid individuals and to correctly estimate their prevalence in the population. Furthermore, some quantitative trait loci (e.g., melanocortin 1 receptor, nuclear receptor subfamily 6, group A, member 1) can be useful to track the introgression of nonneutral domestic alleles (Frantz et al. 2013b; Canu et al. 2016). In addition to the wild x domestic hybridization, introgression between different putative wild boar subspecies has resulted from animal translocations, mostly carried out for hunting purposes (e.g., in Sardinia, Scandura et al. 2011b).

Life History

Growth

The two sexes exhibit a similar growth rate during their first year of life (reaching approximately 30–45 kg), after which males grow faster than females; they reach 90% of their asymptotical mass, respectively, at around 3 and 2 years of age (Spitz et al. 1998; Brogi et al. 2021). Growth continues throughout their lifetime, with local differences depending on variations in food availability. Such differences are less pronounced in males, which adopt compensatory strategies according to the environment, and affect more strongly females whose energy reserves are used to support pregnancy and lactation more than growth (Spitz et al. 1998).

Reproduction

The wild boar life history strategy is very uncommon among similarly sized ungulates, with a very high reproductive potential (Table 2). The reproductive biology of wild boar is a very complex process that depends on intrinsic and environmental factors, and involves highly plastic breeding tactics and allocation strategies in different ecological scenarios. Female wild boar have a higher reproductive effort than most other ungulate species. Therefore, they are highly dependent on food availability to compensate the energetic investment and ensure survival of both mother and litter. In hunted populations, most females normally do not live for longer than two or three mating seasons, reaching early sexual maturity (Gamelon et al. 2011).

Although changes in climatic conditions influence the reproductive pattern, the wild boar generally shows several estrus cycles per season (seasonally polyestrous) and normally does not mate during the summer to avoid giving birth in winter when low temperatures impair the survival of the piglets. The main breeding period typically occurs in autumn-early winter (between October and December) in temperate regions (Fonseca et al. 2004; Ježek et al. 2011), with peaks in November–December when most of the reproductive females come into estrus and males show increased testes size, testosterone levels, and semen quality (Kozdrowski and Dubiel 2004). Piglets are born in spring, but births may occur throughout the year (Maillard and Fournier 2004). In northern areas, the mating season is shifted, starting in November and may continue until January. In years with high availability of food resources, births are earlier and significantly more synchronized than in poor years (Maillard and Fournier 2004). In areas with a stronger climatic seasonality, with a short period of high food abundance, births are highly synchronous compared to areas with high food diversity all year round (Santos et al. 2006).

Female sexual maturity depends on age and body mass (Gethöffer et al. 2007): a female must reach a threshold body mass to be able to reproduce (typically between 25 and 35 kg,

Table 2 Wild boar life history traits

Trait	Parameter	Value
Gestation (days)	Mean (min-max)	115 (112 to 130)
Weaning (months)	Range	3–4
Litter size (nr)	Mean range (min-max)	3.05–6.91 (1 to 14)
Litters per year (nr)	Min-max	1
Mass at birth (g)	Min-max	670–1090
Proportion of breeding females (%)		
Juvenile (<1 year)	Min-max	0–90
Yearling (1–2 years)	Min-max	35–100
Adult (>2 years)	Min-max	65–100
Female age at 1st reproduction (months)	Mean range (min)	8–22 (4)
Female body mass at 1st reproduction		
(kg)	Mean range (min)	24.6–33 (17)
(% adult body mass)	Mean range	33.3–40
Age at natal dispersal (months)		
Female	Min-max	7–22
Male	Min-max	9–20
Dispersal distance (km)		
Female	Max	20
Male	Max	250
Longevity (yr)		
In captivity	Max	27
In the wild	Max	13

exceptionally 20 kg or less, Servanty et al. 2009; Table 2). Males may physiologically be able to mate at 10 months of age, possibly related to body mass, but normally gain access to females later. Once females become sexually active, they try to reproduce every year under any environmental conditions.

Fecundity ranges between 1 and 14 embryos, with average litter size highly variable among areas (e.g., Servanty et al. 2007; Bywater et al. 2010 and references therein). The mean litter size for adult boar in Europe is 6.28 (Bywater et al. 2010), the largest litters occurring in Central Europe, where their average size varies between 4 and 7 (Frauendorf et al. 2016; Náhlik and Sandor 2003; Servanty et al. 2007), compared to average litters between 3 and 5 in Southern Europe (Fonseca et al. 2004). Females produce one litter per year. Though hypothesized, there is no confirmation that under certain circumstances they can produce two litters (Bieber and Ruf 2005). The number of embryos increases with age, body size, and body condition (e.g., Náhlik and Sandor 2003; Fonseca et al. 2004; Frauendorf et al. 2016). Litter sizes and the proportion of pregnant females are higher in good mast years, which also result in earlier start of estrus and a higher proportion of females breeding during their first year of life (Groot Bruinderink et al. 1994). Furthermore, predictability of seasonal resources may relate to litter size increases with latitude (Bywater et al. 2010), whereas in Mediterranean ecosystems, rainfall positively affects breeding parameters (Fernández-Llario and Mateos-Quesada 2005).

Gestation is about 115 days, and piglets, which can see immediately after birth, remain close to the nest for 4–6 days. Sex ratio of foeti seems not to be biased (Keuling et al. 2013). Maternal condition (i.e., body mass) and resource availability appear to relate to litter size but, as known so far, not to the sex ratio in the litter (Servanty et al. 2007). Contrary to juvenile females, adult ones can adjust their relative allocation to littermates, according to the amount of available food resources. In mast years, a high variance in offspring weights is observed within litters that is matched by the variation in milk production among teats, leading to a lower rivalry among siblings. This way more piglets can be raised, maximizing female reproductive success (Gamelon et al. 2013).

Survival

Although wild boar over 10 years were reported living in nature, the average life expectancy is far lower (Gamelon et al. 2014). In fact, in hunted populations, the average life span may not extend longer than 24 months. Sexes display similar senescence rates (Gamelon et al. 2014). Yearly survival rates under different environmental and management conditions are variable (juveniles: 0.06 to 1.00 with an average in hunted populations of 0.46; yearlings: 0.11 to 1.00, on average 0.41 in hunted populations; adults: 0.03 to 1.00, on average 0.64 in hunted populations; see Keuling et al. 2013). Under good environmental conditions (namely tree seeding, access to crops and mild climate), yearly survival of juveniles may double (Bieber and Ruf 2005). Recruitment of piglets to the female population has been reported to be low; and only less than half of the piglets may survive till the end of September (Náhlik and Sandor 2003). Survival differs between areas, sex, and age (Keuling et al. 2013). Based on telemetry data, piglet survival averaged 0.5 (with 181 days reference period), and overall survival in hunted populations was similar over 1-year period (specifically 0.47 for female piglets, 0.44 for male piglets, 0.46 for yearling females, 0.29 for yearling males, 0.66 for adult females, and 0.59 for adult males; Keuling et al. 2013). In Spain, survival rates for adults ranged from 0.44 in hunting grounds to 0.66 in protected areas (Barasona et al. 2016).

Habitat and Diet

Habitat Selection

Wild boar are flexible in their habitat use, and their ecological plasticity explains the broad distribution and wide range of occupied habitats (Segura et al. 2014). Primary habitats of wild boar are characterized by well-developed vegetation and include forests, shrublands, marshes, and river valleys. Food and shelter availability are the main factors shaping wild boar occurrence (Segura et al. 2014) and highest densities are observed in highly productive areas dominated by rich deciduous forests and agricultural areas (Melis et al. 2006). Agricultural landscapes, which provide abundant shelter and food, have become important secondary habitats for this species in the last decades. Standing crops of maize, rapeseed, and cereals can provide optimal habitats utilized partially or exclusively during the cultivation period (Dardaillon 1986; Keuling et al. 2009; Thurfjell et al. 2009). Open and exposed farmlands outside of cultivation season are generally avoided, but linear vegetation elements, such as rows of trees/shrubs, within the fields can be utilized for movement all year round (Thurfjell et al. 2009). Grasslands and pastures shared with livestock provide attractive foraging habitats (Dardaillon 1986). Finally, wild boar have become increasingly present in urban and peri-urban areas of most European cities (Podgórski et al. 2013). Wild boar use natural corridors (river valleys, tree- and bush-covered areas) to enter and move within cities, while permanent presence is often observed in city parks and woodlands (Stillfried et al. 2017; Castillo-Contreras et al. 2018).

The effect of natural predators on the habitat use of wild boar is poorly understood. Presence of wolves (*Canis lupus*), the species' main natural predator, does not seem to be perceived as a high predation risk and evokes few behavioral responses in wild boar (Kuijper et al. 2014). Human hunting can have stronger impact on habitat use and can lead to home range shifts from exposed to refuge areas (Tolon et al. 2009), dispersion of resting sites (Scillitani et al. 2010), and greater randomness in habitat use as compared to nonhunting period (Saïd et al. 2012). Females tend to seek safe habitats away from disturbance, while males are more risk-tolerant and can remain hidden close to hunting activities (Saïd et al. 2012). Habitat use patterns may vary according to seasonally changing availability of resources, such as water, food, and shelter (Singer et al. 1981; Keuling et al. 2009). For example, dry season in the Mediterranean can drive wild boar from dried-up marshes into cultivated areas (Dardaillon 1986).

Movement Ecology

Foraging and social interactions are usually performed in a relatively small area (approx. 25 ha) where animals move short distances at low speed, rarely exceeding 1 km/h (Spitz and Janeau 1990). When travelling between habitat patches (foraging spots, resting sites), wild boar move directionally and at faster pace, that is, trotting at 1–10 km/h (Spitz and Janeau 1990; Briedermann 2009; Morelle et al. 2015). When fleeing, they can gallop in short burst of up to 40 km/h. Daily distances travelled are usually shorter than 10 km (Podgórski et al. 2013). Longer daily distances were observed in fragmented environments, where between-patch movements are frequent (e.g., urban areas; Podgórski et al. 2013), and during intensive hunts (Scillitani et al. 2010). Over a 24-h period, wild boar can cover 45–90% of its annual range (Podgórski et al. 2013). This indicates that home range size of wild boar is relatively small given the movement capacity of the species, which can be thus considered sedentary.

Wild boar exhibit remarkable intraspecific variation in home range size across a wide range of habitats. Size of annual home range varies between 400 ha to 6000 ha with an average size of about 800 ha (Boitani et al. 1994; Keuling et al. 2008; Podgórski et al. 2013). The smallest ranges are observed in urban areas and in rich habitats, while the biggest ranges occur in mountainous areas and poor habitats (Singer et al. 1981; Podgórski et al. 2013). Range shifts between habitats were observed in heterogeneous landscapes offering seasonally changing resources (e.g., mountains, field-forest mosaic) (Dardaillon 1986; Keuling et al. 2009; Thurfjell et al. 2009). Sexual differences in home range size are ambiguous, some studies reported larger home ranges in males (Morini et al. 1995), whereas others observed no sex-related differences (Boitani et al. 1994). During the rut, adult males roam widely in search of receptive females and may temporarily extend their home ranges (Singer et al. 1981), whereas movements of pregnant females decrease around parturition (Morelle et al. 2015).

The majority of young wild boar (70–80%) do not disperse further than 5 km away from their natal ranges (Truvé and Lemel 2003; Podgórski et al. 2014a). Longer movements (5–30 km) are observed less frequently and are undertaken more often by dispersing males than females. Natal dispersal is most frequent during the second year of age (Podgórski et al. 2014b). Occasionally, long distance movements of 50–250 km in straight line are performed by young animals, adult males, and adult females with offspring (Andrzejewski and Jezierski 1978; Truvé and Lemel 2003). Hunting disturbance, particularly intensive methods such as frequent drive hunts, may induce escape movements resulting in greater distances travelled, larger ranges, and dispersion from resting sites (Scillitani et al. 2010).

Diet

Wild boar are omnivorous and opportunistic in their food preferences and their diet reflects local and seasonal food availability. Plant matter constitutes over 90% of the diet on the annual scale and dominates in terms of frequency and volume over other food sources (Briedermann 2009; Barrios-Garcia et al. 2012). Plant food in the wild boar diet is very diverse and includes seeds, fruits, leaves, stems, shoots, bulbs, and roots (Schley and Roper 2003). Agricultural crops are heavily used when available, particularly during the summer and autumn when their nutritional value is at its peak. Consumed agricultural food items include cereals, vegetables, legumes, fruits, and others (Genov 1981; Herrero et al. 2006). Maize is one of the preferred crops (Herrero et al. 2006; Schley et al. 2008) and is commonly used as a bait by hunters (Schley and Roper 2003). When availability of agricultural crops or supplementary food is low, natural forage, such as herbaceous plants, browse, roots, and tree seeds (e.g., acorns, beechnuts, chestnuts), becomes an important dietary component (Groot Bruinderink et al. 1994; Herrero et al. 2005; Merta et al. 2014).

Diet composition is dominated by agricultural crops (>70% of stomach content volume) in wild boar living in the mosaic landscape of woodlands and farmlands (Genov 1981; Herrero et al. 2006; Merta et al. 2014) and, together with plant roots, constituted over 70% of the wild boar diet in the Mediterranean wetlands (Giménez-Anaya et al. 2008). Herbaceous plants, browse, roots, and tree seeds (up to 40% of the diet during mast years) make up most of the diet in mixed lowland forests and mountainous areas (Groot Bruinderink et al. 1994; Herrero et al. 2005; Merta et al. 2014).

Animal material is consumed by wild boar all year round and includes at least 40 animal species and genera. Wild boar consume animal matter frequently (occurs in 90% of analyzed stomach contents), but at low total volume (about 3% of the stomach content with >2% invertebrates and the rest vertebrates; Schley and Roper 2003; Herrero et al. 2006). Invertebrate prey includes mainly earthworms, insects, and snails, whereas consumed vertebrates are small mammals (rodents, shrews), fish, amphibians, reptiles, and birds (Schley and Roper 2003). Animal food can be obtained by wild boar by scavenging or predation (Herrero et al. 2006; Giménez-Anaya et al. 2008; Barrios-Garcia et al. 2012). Large mammals are consumed as carrion, whereas small mammals (rodents, shrews, hares, rabbits) are also taken directly as prey (Schley and Roper 2003). Wild boar eat eggs and chicks of ground nesting birds (Barrios-Garcia et al. 2012). The composition of animal items in the diet varies greatly among seasons. In the case of small mammals, it is higher in autumn and winter, whereas earthworms are mainly consumed in spring and summer (Schley and Roper 2003). The use of carrion can increase during autumn and winter due to the greater availability of carcasses (Briedermann 2009).

Behavior

Social Behavior

Wild boar societies are centered around family groups composed of one to several adult females and their offspring from the last or second last breeding season (Dardaillon 1988; Podgórski et al. 2014a). Most members of the groups are genetically related to each other at the level of first- or second-order relatives (Kaminski et al. 2005; Poteaux et al. 2009; Podgórski et al. 2014a). Size of social groups usually ranges between 5 and 10 individuals (Gabor et al. 1999; Poteaux et al. 2009; Podgórski et al. 2014a). Social groups are generally stable and coherent but may temporarily merge to form larger units (up to 30 animals) and single individuals may occasionally shift between groups (Poteaux et al. 2009; Podgórski et al. 2014a, b). Wild boar are not territorial and undefended home ranges of neighboring groups partly overlap, with individuals from different groups interacting regularly (Boitani et al. 1994; Podgórski et al. 2014b).

Male offspring leave maternal groups early in their life, usually around one year of age, and become solitary boars, rarely found within groups outside of the breeding season (Hirotani and Nakatani 1987; Dardaillon 1988). Adult males engage in dynamic and short-lived intraspecific relationships (Podgórski et al. 2014b), involving interactions with mating competitors or assessment of females' reproductive status and mating. During the rut, which takes place in late autumn and early winter, males temporarily join female groups for mating (Graves 1984; Dardaillon 1988). Female offspring show stronger fidelity to maternal groups and most of them do not disperse far (Hirotani and Nakatani 1987; Dardaillon 1988; Kaminski et al. 2005; Podgórski et al. 2014a). New groups can be formed following the permanent separation of yearling females from their maternal group or the split of a larger social unit

(Kaminski et al. 2005; Poteaux et al. 2009). Adult females maintain stable, long-lasting relationships and rarely shift between groups for long periods of time (Gabor et al. 1999; Poteaux et al. 2009; Podgórski et al. 2014b).

Mating Behavior

Breeding activity of wild boar is seasonal and the mating system is polygynandrous (Pérez-González et al. 2014). During the mating season, boars actively compete for access to estrus sows, which can mate with more than one boar within the 2–3 days of estrus, resulting in litters with multiple paternity (Poteaux et al. 2009; Gayet et al. 2016). Similarly, a single boar can fertilize several sows, which tend to synchronize their estrus locally (Canu et al. 2015). Farrowing takes place mainly in March and April (Gethöffer et al. 2007; Ježek et al. 2011; Rosell et al. 2012). Prior to parturition, pregnant females temporarily separate from their group and choose secluded sites for building a farrowing nest, in which piglets remain for a few days after parturition and then join the maternal group. Piglets are weaned at around 4 months of age but already at 4 weeks they start rooting and processing solid food (Špinka 2009). Due to synchronized estrus of sows within a group, multiple litters of similar age may be present in a group at the same time, and females may participate in cooperative nursing (Graves 1984).

Senses

Wild boar have a well-developed olfactory sense which is used in foraging, communication, navigation, and predator avoidance. Scent signals are used to locate food items at close distance (Suselbeek et al. 2014), assess predation risk (Kuijper et al. 2014), locate familiar individuals (Kittawornrat and Zimmerman 2011), stimulate reproductive activity (Kirkwood et al. 1983), and navigate within the home range. Wild boar also have a good auditory capacity and a rich repertoire of vocal signals used in social communication. There are around 20 types of calls, such as grunts, squeals, and trumpets, which may vary in amplitude, frequency, and modulation depending on the behavioral context (Špinka 2009). Vision is poorly developed and its role in communication is limited. Visual signals, usually displayed by competing or threatened animals, include ears and body positioning, erection of the dorsal mane, tail wiggling, bristle rising, and back arching (Graves 1984).

Activity

Wild boar activity typically lasts between 6 and 12 h a day. In natural and undisturbed conditions, wild boar are active during day and night, with alternating periods of activity and rest (Podgórski et al. 2013; Brivio et al. 2017). In human-dominated landscapes, wild boar have become largely nocturnal (Lemel et al. 2003; Keuling et al. 2008; Brivio et al. 2017). In urban environments, wild boar are mostly nocturnal, independently of the seasonal changes in day length, in order to minimize interference with humans (Podgórski et al. 2013). In rural areas, activity usually peaks around dawn and dusk and drops in the middle of the night. Seasonal variation in the activity patterns is generally low (Lemel et al. 2003; Keuling et al. 2008), but daily adjustments are observed in response to changes in temperature, precipitation, and humidity (Brivio et al. 2017).

Parasites and Diseases

Pathogens and Parasites

Their demography, ability to cross-breed with pigs, wide distribution, adaptability to a variety of habitats and to suburban areas, feeding habits, sociability, and high contact rates with many species, all expose wild boar to a plethora of pathogens. Their infectious and parasitic diseases have extensively been studied, mainly descriptively, because they are shared with: (1) humans (e.g., *Trichinella* spp.), (2) livestock-domestic pigs are susceptible to the same pathogens but other species might be involved (e.g., cattle – *Bos taurus* – Barasona et al. 2014); (3) endangered species (e.g., Aujeszky's disease with Iberian lynx – *Lynx pardinus* – Masot et al. 2016), and also because of pathogens' impact on its population dynamics (Barasona et al. 2016). The knowledge on wild boar diseases (based on long-term studies) has increased markedly during the last two decades. A recent review ranked the most frequently studied pathogens and host species in long-term studies on wildlife, and with respect to Europe and Asia wild boar was the most studied host (Barroso et al. 2021). Wild boar are thus relevant in the One Health context, due to their role as a true reservoir host for pathogens shared among wildlife, livestock, and humans (Fig. 4 provides details on the number of studies including wild boar as host species.

The list of infectious and parasitic diseases of wild boar is long and includes several zoonoses (see Ruiz-Fons et al. 2008; Jori et al. 2017 for a review). The most significant infectious diseases involving wild boar over the last two decades have been highly host-specific viruses: African swine fever (ASF) and Classical swine fever (CSF). There are also some remarkable multi-host pathogens such as tuberculosis (TB), foot-and-mouth disease virus, and zoonotic nematodes like *Trichinella* spp. Parasitic diseases, including ectoparasites, are normally of lower concern because many of them are usually adapted to one host species and rarely transmit to others. Wiethoelter et al. (2015) reported the top 10 diseases at the wildlife–livestock interface of which wild boar may host nine: two viruses (avian influenza and rabies), four bacteria (salmonellosis, TB, brucellosis, and

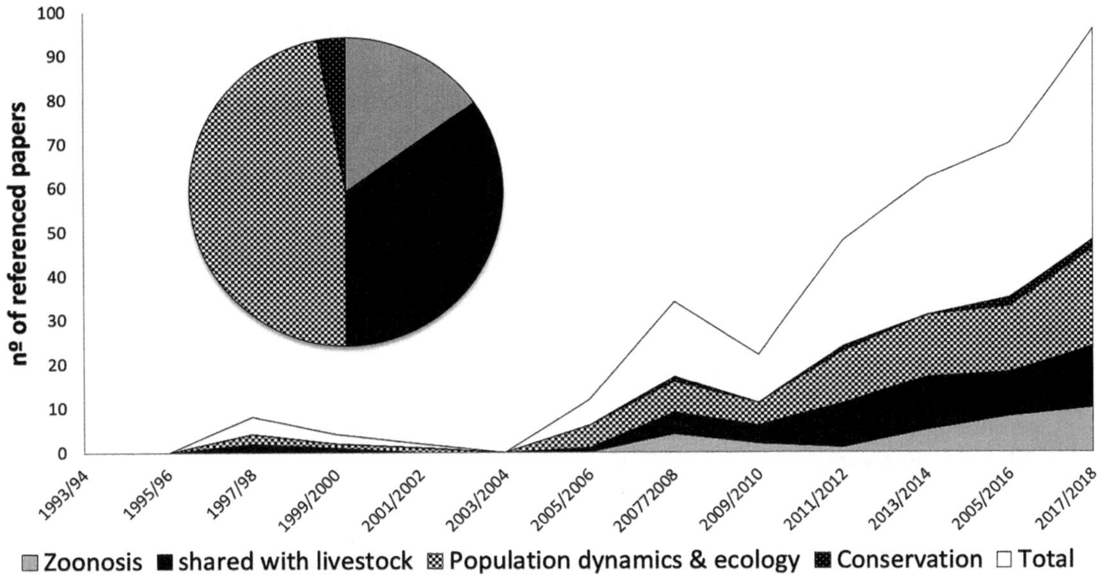

Fig. 4 Bi-annual number of georeferenced papers on long-term epidemiological studies including wild boar as host species in Scopus + Medline + Pubmed from 1993 to 2018. Databases were accessed on April 15, 2018. The keyword used was "wildlife diseases." We initially retrieved 6541 references, which resulted in 535 papers once duplicates and spurious results were removed and the following conditions applied: study longer than three consecutive years, annual sampling minimum of 10 individuals, same study area and populations over time, and wild animals in their natural environment (excluded lab and captive animals as well as clinical trials). Studies exclusively focusing on passive surveillance were also excluded. Finally, we filtered the papers which included, at least, wild boar as host species, and classified records according to the nature of the main conclusions: 1) zoonosis-related, 2) diseases shared with livestock, 3) population dynamics and/or ecology, and 4) conservation. Totals over the study period (proportions) are indicated in a pie chart

leptospirosis), one protozoon (toxoplasmosis), and two helminths (echinococcosis and trichinellosis).

Wild boar may function as a disease reservoir when they are able to maintain an infection in a given area in the absence of transmission from other hosts. In some cases, wild boar just maintain the infection secondarily to the main reservoir or are accidentally infected. The epidemiological role of wild boar is not easy to determine and requires compiling sound evidence about epidemiologic associations between reservoirs, genetic characterization of pathogens, and intervention studies (Naranjo et al. 2008). The possible transmission routes from and to other hosts are highly variable and can happen through both direct (contact with infected animals or carcasses, consumption of meat, oral, respiratory, conjunctival and transdermal routes, skin wounds) and indirect exposure (there is an indirect step or media: aerosols, consumption of contaminated food or water, through bites of arthropod vectors). Foodborne pathogens and antimicrobial resistance in indicator bacteria have been reported in urban wild boar, causing concerns for public health (Navarro-Gonzalez et al. 2013).

Pathogens can have direct or indirect (e.g., body condition mediated) impacts on the reproductive performance of wild boar (Ruiz-Fons et al. 2006), which can be partially compensated by an earlier return to estrus. Co-infections with multiple pathogens with different characteristics are frequent in wild boar, resulting in complex effects. For instance, some viral infections (e.g., porcine circovirus type 2) may impair the ability of wild boar to respond to other infections, including TB (Díez-Delgado et al. 2014). Population effects can also be relevant, as some diseases can lead to high mortality (e.g., CSF and ASF, Lange et al. 2012; Cortiñas Abrahantes et al. 2017). TB causes 30% of deaths in adult wild boar in endemic areas of Southern Spain (Barasona et al. 2016), which contrasts with a total natural death rate of 3% in Central Europe (Keuling et al. 2013).

Epidemiology

The factors involved in the maintenance and spread of pathogens by wild boar are varied and interdependent. Individual factors include sex, age, body condition, reproductive and immunological status, and genetics (e.g., genetic mechanisms are involved in susceptibility to TB, Queirós et al. 2018). Pathogen prevalence in wild boar is driven by changes in population densities and aggregation, sometimes caused by implementation of management practices (e.g., supplementary feeding, Vicente et al. 2013 for TB, Oja et al. 2017b for helminths and intestinal protozoa). Assessing how wild boar use their environment and how this affects interspecific interactions with wildlife and humans is therefore essential to estimate the risks for disease transmission and maintenance (e.g., Barasona et al. 2014, for scavenging: Carrasco-Garcia et al. 2018). The parallel growth of urban areas and wild boar populations in recent decades has contributed to increased interactions between wild boar, humans, and other animals alike. The removal of predators, recreational hunting (often under non-sustainable managements schemes resulting in overabundance; Gortázar et al. 2006), translocations, consumption and movement of wild boar meat and meat products without previous sanitary inspection, all increase the chances of spreading and sharing wild boar pathogens.

Population Ecology

Population Dynamics

Population dynamics of wild boar are driven by both natural and anthropogenic factors, and the most important natural drivers include mast of deciduous trees (such as acorns, beechnuts, and chestnuts), winter severity, and predation. Long-term data on wild boar population numbers in central and eastern Europe show that abundance of mast has a dominating positive effect on the population growth rates (Jędrzejewska et al. 1997; Bieber and Ruf 2005; Briedermann 2009; Vetter et al. 2015; Frauendorf et al. 2016). During the mast seeding years, when trees synchronously produce large seed crops, wild boar take advantage of the abundant food to accumulate energy reserves which enhances overwinter survival and subsequent reproduction (Jędrzejewska and Jędrzejewski 1998; Servanty et al. 2009; Canu et al. 2015; Frauendorf et al. 2016). Another natural factor strongly limiting wild boar numbers is winter severity, that is, the combination of average winter temperature and snow cover duration and depth. Deep snow and frozen soil make it difficult for wild boar to root and forage on vegetation and invertebrates. Harsh winters can cause marked declines in wild boar populations as a result of starvation and diseases which are responsible for most (73%) of natural mortality (Jędrzejewska et al. 1997; Jędrzejewska and Jędrzejewski 1998). This weather stochasticity mechanism shapes wild boar densities and dynamics across Europe, resulting in higher densities and faster population growth rates where winters are milder. Food abundance, however, can offset or even outweigh the negative effects of cold winters. Thus, winter severity will have weaker limiting effect on wild boar populations in highly productive regions or during mast seeding years (Melis et al. 2006; Vetter et al. 2015). The third and least important natural factor shaping wild boar numbers is predation. Wolves are the main natural predators of wild boar, contributing 50–100% to the predator-caused mortality across the species range, depending on the presence of other large carnivores that prey upon wild boar, such as brown bear (*Ursus arctos*) or lynx (*Lynx lynx*). However, impact of wolves on wild boar populations appears limited. Where wild boar and wolves co-occur, wolf predation makes up on average 16% (maximum 30%) of the natural mortality (Okarma 1995). Impact of wolves varies locally depending on the composition of ungulate communities and wild boar abundance. Wolf preferences may also change and wild boar can either represent the selected prey in some areas (Mattioli et al. 2011) or an auxiliary prey in others (Okarma 1995). The impact of natural predation is heavier on juveniles and yearlings (over 70% of kills), while adults can defend themselves effectively and are rarely attacked (Okarma 1995; Bassi et al. 2012).

Anthropogenic Impacts

Anthropogenic impacts on wild boar populations include direct effects of management and indirect effects related to climate change. Hunting is the main cause of wild boar mortality across Europe (Toïgo et al. 2008; Keuling et al. 2013). Hunting pressure can significantly affect life-history traits, for example, earlier age and timing of reproduction (Gamelon et al. 2011), and demographic structure of the populations by targeting mostly adults, in contrast to predation by wolves (Okarma 1995; Toïgo et al. 2008; Keuling et al. 2013). However, hunting does not seem to limit the currently observed growth of wild boar populations in Europe (Massei et al. 2015; Vetter et al. 2015). Supplementary feeding, which is common management practice across most European countries, can also shape population dynamics (Andrzejewski and Jezierski 1978). Food availability contributes to increased overwinter survival and buffers the negative effect of cold winters (Vetter et al. 2015). This effect will be greater at northern latitudes, where the limiting effect of cold winters is stronger. For example, wild boar densities in Estonia, where supplementary feeding was common (before the arrival of ASF), were much higher than in Finland, where climatic conditions were similar but supplementary food was never provided. Wild boar abundance in Estonia was strongly related to the number of supplementary feeding sites (Oja et al. 2014). Climate changes influence two major factors limiting wild boar population growth: winter severity and food abundance. First, winter temperatures increased throughout the twentieth century and these changes are associated with increased size of wild boar populations across central Europe (Vetter et al. 2015). The most likely mechanism driving this relationship is the increased survival during mild winters. Second, rising temperatures have also been shown to increase the frequency of mast seeding years (Bieber and Ruf 2005; Vetter et al. 2015), which have a positive effect on growth rates of wild boar populations (Bieber and Ruf 2005). Finally, changes in agricultural practices led to increased availability of energy-rich crops, such as maize, which is associated with higher reproductive output of wild boar and, often coupled with supplementary feeding in winter, contributes to population growth (Bieber and Ruf 2005; Servanty et al. 2009; Rosell et al. 2012).

Population Trends

Wild boar populations all over Europe have grown considerably during the last decades, despite large variation in climatic conditions and management across the continent (Fig. 5; Sáez-Royuela and Tellería 1986; Apollonio et al. 2010; Massei et al. 2015; Vetter et al. 2015). The increasing trend in wild boar numbers in Europe started in the 1960s and is continuing today (Sáez-Royuela and Tellería 1986; Massei et al. 2015). From 1982 to 2012, the average five-year population growth index, based on the hunting bag statistics from 18 European countries, varied between 1.4 and 1.7 (with 1 = no growth; Massei et al. 2015). This increase in population numbers is accompanied by geographical expansion towards the north (Apollonio et al. 2010). Other indices of wild boar abundance, such as crop damage, vehicle collisions, and environmental impacts, also show an increasing trend and confirm real growth in wild boar populations (Massei and Genov 2004; Schley et al. 2008; Apollonio et al. 2010; Morelle et al. 2013). Despite an increase in hunting bags (+150% from 1992 to 2012) and over 3 million wild boar annually

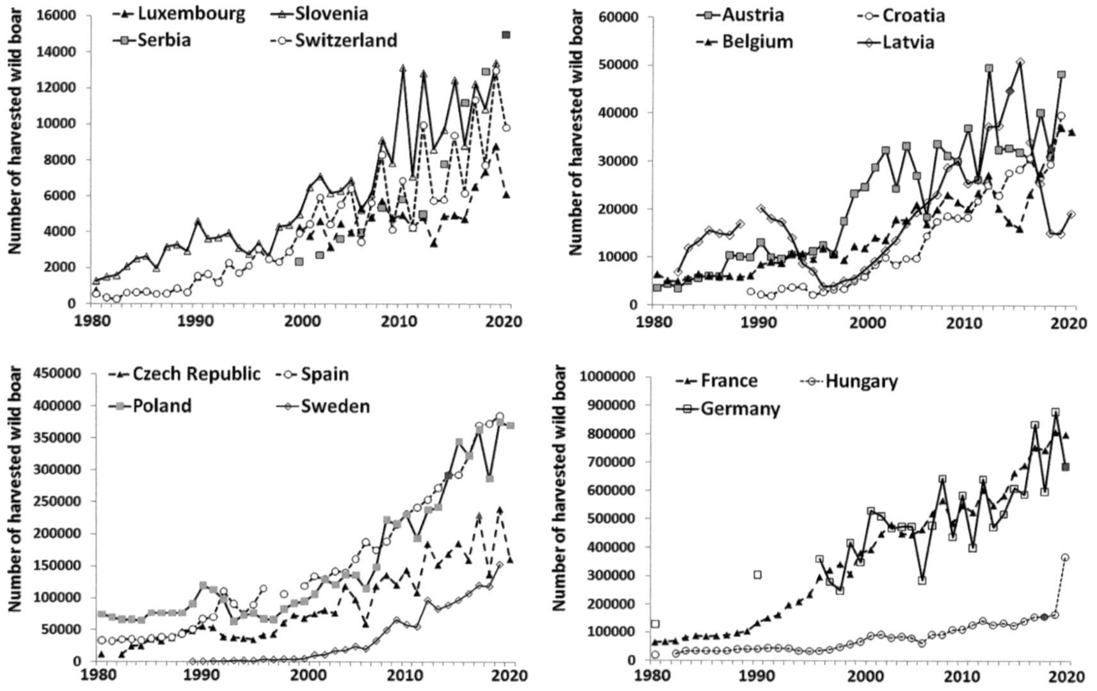

Fig. 5 Trends in wild boar hunting bags from selected European countries (1980–2020). Data provided from national/regional administrations to the Enetwild consortium (www.enetwild.com). Symbols in red indicate the year of the first African swine fever outbreak in the country

harvested in Europe, hunting seems to be not sufficient to limit wild boar population growth, which is thus expected to continue (Massei et al. 2015; Vetter et al. 2015). This is partly due to a declining number of hunters (−18% from 1992 to 2012) and their general unwillingness to reduce wild boar densities (Keuling et al. 2016). Other factors discussed above, such as increasingly milder winters and greater availability of natural and anthropogenic forage, are likely to continue to boost the growth of wild boar populations.

Conservation Status

The wild boar is the most abundant and widespread suid species in the world. Accordingly, it is classified as Least Concern by the IUCN (Oliver and Leus 2008). If the species, as a whole, is overabundant and does not raise any conservation concern, different situations may emerge at a local scale. The only threatened subspecies (or species, according to Groves and Grubb 2011) is *S. s. riukiuanus* Kuroda, 1924, living in Ryukyu Islands, Japan. The main, probably underestimated, issue in Europe is represented by the extensive anthropogenic hybridization leading to genetic homogenization and to the erosion of local genetic diversity. The current lack of sharp taxonomical units (see section "Taxonomy, Systematics and Paleontology") is likely to have been enhanced by human-mediated gene flow. Even where long-lasting isolation has favored genetic divergence, as in the case of the Sardinian wild boar, the introgression from commercial pig breeds and introduced non-native wild boar has impacted the local gene pool, jeopardizing the status of the *S. s. meridionalis* subspecies (Iacolina et al. 2016). At a regional scale, the genetic structure observed today often arises from different histories of releases by humans, hybridization, and human exploitation, artificially leading to diverging allele frequencies among local stocks (Goedbloed et al. 2013).

It is likely that the positive trend of the species has concealed the loss of native adaptive genetic variation across its European range. The ban of animal translocations and of the release of captive stocks, yet mainly associated to sanitary risks, will also help preventing further loss of adaptive potential.

Management

Introductions, Reintroductions, and Restocking

The Eurasian wild boar represents one of the most managed ungulate species in the world. Its present distribution is the result of introductions outside its native range, local extinctions due to overexploitation, restocking with animals translocated from other areas, farming, and escapes or releases of captive stocks. In Europe, the establishment of new populations by intentional introductions has only affected some minor islands (e.g., Canu et al. 2018), while previously extinct populations have been restored by reintroductions (e.g., in the Netherlands, Denmark, Serbia, Italy, and Greece; Apollonio et al. 2014) or by escapes from captivity (e.g., in Great Britain, Sweden, Ireland, and Slovenia; Apollonio et al. 2014). Restocking of depleted populations has been also a common practice in some areas (Apollonio et al. 2010) and has contributed to the recovery of the species across the continent. In many cases, human actions were not documented, and both the origin and number of released animals are unknown. Introductions and reintroductions were mostly successful, an exception being the wild boar in Cyprus, where the species was illegally restored in 1994, after a previous extinction, and died out again in 2004 (Hadjisterkotis and Heise-Pavlov 2006).

Impact, Conflicts with Humans, and Damage Control

Sus scrofa is listed by IUCN among the 100 worst invasive species in the world, because of its biology (i.e., rapid population growth rate) and overall impact on human activities and on the environment. Conflicts with humans are mainly associated with damages from feeding activity and the number of traffic accidents involving the species. The main impact is represented by crop damages that mostly affect cereals, especially maize, and are due not only to consumption but also to trampling (Schley and Roper 2003). In warm regions, heavy losses are also caused to vineyards and rice paddies (Calenge et al. 2004) and in continental regions to vegetables and grasslands (Schley et al. 2008). The main factors affecting the amount of damages are density of wild boar, distance of crops from natural refuges (e.g., woodlands), species cultivated, availability of natural food, and crop ripening period (Calenge et al. 2004; Schley et al. 2008; Thurfjell et al. 2009). Furthermore, damages are seasonally distributed according to geography and crop type (Schley et al. 2008). In Mediterranean areas, they peak in summer, induced by the shortage of water, whereas in temperate climates a maximum occurs in late winter, when food resources are scarce (Licoppe et al. 2013). Together with roe deer, the wild boar is responsible for most of the wildlife-vehicle collisions in Europe, especially in highly urbanized areas (Morelle et al. 2013). An annual peak in wild boar-caused accidents is observed in autumn-early winter, likely induced by an increased mobility during the rut and the hunting season (Morelle et al. 2013). Additional conflicts with humans arise from the habit to dig up the ground, looking for hypogeal food. Recreational areas like city gardens and parks or golf courses can be severely impacted by rooting, as can grazing areas for livestock (Licoppe et al. 2013). Besides, wild boar can occasionally affect livestock farming directly, by predation on lambs or new-born calves (Seward et al. 2004). Finally, a serious concern is represented by the transmission of diseases and zoonoses, which can affect livestock, pets, endangered wildlife, and humans (see "Parasites and Diseases").

In addition to the impact on human activities, at high densities, wild boar can represent a threat to local ecosystems, because of its trampling, rooting activity, and opportunistic feeding behavior (Massei and Genov 2004). Feeding on whole plants, fruits, bulbs, and tubers can alter the abundance and richness of plant species (Cuevas et al. 2012), and feeding on seeds and seedlings of forest trees can impact forest regeneration (Bongi et al. 2017). Predation on eggs and chicks can compromise the reproduction of ground-nesting birds (Oja et al. 2017a); grubbing and predation on earthworms, grubs, and small ground-dwelling mammals can modify animal communities and soil properties (Laznik and Trdan 2014).

Several methods are used to mitigate such impacts, differing in effectiveness, feasibility, costs, and social acceptance (Massei et al. 2011). Although not always true, a high impact by wild boar is often interpreted as a consequence of an overabundant population, so actions are undertaken to reduce their number. Traditional control methods include culling, as the main option, but also trapping (followed by suppression or translocation). In addition, methods of fertility control have been developed, based on immuno- or oral contraception (Massei et al. 2012). Other mitigation measures are intended to limit wild boar access to sensitive sites by metal fencing, electric fencing, diversionary feeding, and the use of chemical repellents and acoustic scarers. No eradication program has been successful in Europe, but experiences in the Americas suggest that a combination of different methods is more effective.

Hunting and Hunting Regulation

Besides being considered a pest, the wild boar is an important game species. It has been estimated that more than 3 million wild boar were harvested in 2012 in Europe (Massei et al. 2015). Traditional cooperative forms of hunting are based on drive hunts and are practiced especially in southern Europe (like the "braccata" in Italy and the "monteria" in Spain), while individual hunting (e.g., stalking, standing or high-seats) is more common in continental Europe. In some countries, harvest quotas are imposed by local authorities. Baits (mostly maize or other vegetables) are used in many areas as attractants during the hunting season. Hunting is generally allowed to licensed hunters from summer to early winter, with huge differences among countries; in some of them (e.g., in Portugal, Austria, Croatia, and

Estonia), wild boar hunting is allowed all year round, with possible restrictions to specific sex/age classes (Apollonio et al. 2010).

Although hunting is the main cause of wild boar mortality (around 85% of deaths, Keuling et al. 2013), it appears insufficient to counteract the positive trend of wild boar populations. Moreover, this population growth is paralleled by a general negative trend in the number of hunters which poses serious doubts on the capacity to effectively manage this species in the future (Massei et al. 2015).

Economic Value

Due to its size, abundance, current distribution, and the high level of interaction with human activities, the wild boar has enormous economic repercussions. In Poland alone, over a 5-year period, the compensation for damages amounted to 34.2 million €, whereas the revenue from the sale of the meat of shot animals amounted to 9.5 million € (Frackowiak et al. 2013). In Italy and France, the wild boar was reported as responsible for 90% of damages to crops and forests, causing an estimated annual loss exceeding 30 million € (Apollonio et al. 2010). A recent study, measuring the willingness-to-pay by hunters in Sweden, estimated at 113–529 SEK (10–50 €) the value of a wild boar, with large differences among hunter categories (Engelman et al. 2018). However, the value attributed to this game is context-dependent and may vary a lot among countries. Maximum values are reached by trophy hunting which is practiced in several countries, especially in central-eastern Europe. According to the International Council for Game and Wildlife Conservation (CIC) evaluation system, tusk size is the reference parameter to assess the quality of boar trophies. On this basis, a single harvested boar can be worth up to more than 1500 €.

Health Management

The control of wild boar diseases is a major challenge, especially for those shared with livestock, and takes advantage of the establishment of surveillance and monitoring schemes, together with health surveillance in domestic pigs. Suitable diagnostic tools, designed for pigs, are available. However, biosecurity measures should be implemented to prevent pathogen transmission, which can be bidirectional at the wild boar-livestock interface (Carrasco-Garcia et al. 2016). Additionally, wild boar management constitutes an essential aspect to prevent risk factors for many pathogens, since excessive densities and aggregation favor disease spread and maintenance (Gortázar et al. 2006; Cano-Terriza et al. 2018). Effective disease management requires tools from several fields which should be combined in an integrated control strategy. Different options can be applied and combined; however, a proper surveillance and monitoring scheme (for both disease and population; Sonnenburg et al. 2017) is always required to make the best decisions. Disease control can be achieved by different means, including (1) preventive actions (especially at the wildlife-livestock interface), (2) arthropod vector control, (3) host population control through random or selective culling, habitat management, or reproductive control (Massei et al. 2012), and (4) vaccination (Rossi et al. 2015; Díez-Delgado et al. 2018). Wild boar population control is performed through random or selective culling. However, despite a decline in population size of approximately 50% during the period 2014–2017, it did not prevent ASF spread in the Baltic States and Poland during the first years after detection (Cortiñas Abrahantes et al. 2017). Reproductive control is being researched (Massei et al. 2012), and field vaccination against certain pathogens has proved to be a potentially effective tool in some cases (for CSF, Rossi et al. 2015; for TB, Díez-Delgado et al. 2018) which should be integrated in control strategies. After a cost/benefit assessment, the alternative options of zoning or no-action should also be considered. Finally, the success of any disease control strategy in this species, which is part of the European hunting culture, also depends on stakeholders' collaboration and attitudes.

Future Challenges for Research and Management

The population growth and the spread of wild boar across Europe have brought the species to increase its interaction with humans, leading to a combination of worrying issues:

1. Urban wild boar: Once absent, wild boar are now a regular presence in the periphery and in green areas of cities like Berlin, Barcelona, and Rome. The management of these (peri-) urban populations is a real challenge, encompassing aspects like public education, species monitoring and control, public safety, and disease surveillance.
2. Population monitoring: An effective management of wild boar impact on human activities and natural systems, as well as the prevention of disease transmission, would require a reliable estimation of local population abundances and trends. However, estimating wild boar numbers is difficult to achieve because of their clumped distribution, social structure, use of resting sites in dense vegetation, and nocturnal activity. Although several methods have been proposed (see ENETWILD consortium et al. 2018), no standard exists so far and, in the common practice, science-based approaches give way to "guesstimates" or, quite often, to the use of (biased) hunting bag statistics. Nonetheless, great advances have been recently achieved by the ENETWILD consortium (https://enetwild.com/), which has produced suitability maps of wild boar occurrence and relative abundance in Europe by harmonizing hunting bag data. This project has also pointed out that hunting statistics can be suitable to determine wild boar density estimates, if a calibration with accepted rigorous methods is performed. This, however, deserves further research in a variety of contexts throughout Europe.
3. Hunting effectiveness: though hunting is recognized as a fundamental tool of population control, its effectiveness turned out to be constrained by social and legal aspects. In order to counteract more effectively the demographic trend and growing impacts of wild boar, new generations of specialized hunters and modifications to the current regulations are invoked (Apollonio et al. 2010; Massei et al. 2015). Professional hunting can also be of help in the future, especially in specific situations (e.g., in urban areas).

Given the present status and invasiveness of the species, the role of research on wild boar biology and management will be of utmost importance. Response to climate change, biological and ecological effects of different hunting regimes, biological consequences of the introgression of domestic pig genes, the genomic basis of the species' plasticity are among the most stimulating topics. Furthermore, a special effort should be devoted to technical aspects, like the development of more suitable and cost-effective monitoring procedures, the refinement of methods of population control (e.g., sterilization), or the development of vaccines against the most dangerous transmissible diseases (e.g., ASF). Finally, an important goal would be the dissemination of good practices and standards to reduce the current discrepancies among regions and countries in the management of the species.

References

Alexandri P, Triantafyllidis A, Papakostas S et al (2012) The Balkans and the colonization of Europe: the post-glacial range expansion of the wild boar, *Sus scrofa*. J Biogeogr 39:713–723

Alves E, Óvilo C, Rodríguez MC, Silió L (2003) Mitochondrial DNA sequence variation and phylogenetic relationships among Iberian pigs and other domestic and wild pig populations. Anim Genet 34:319–324

Alves PC, Pinheiro I, Godinho R et al (2010) Genetic diversity of wild boar populations and domestic pig breeds (*Sus scrofa*) in South-Western Europe. Biol J Linn Soc 101:797–822

Andrzejewski R, Jezierski W (1978) Management of a wild boar population and its effects on commercial land. Acta Theriol (Warsz) 23:309–339

Apollonio M, Andersen R, Putman R (2010) European ungulates and their management in the 21st century. Cambridge University Press, Cambridge, UK

Apollonio M, Scandura M, Šprem N (2014) Reintroductions as a management tool for European Ungulates. In: Putman R, Apollonio M (eds) Behavior and management of European Ungulates. Whittles Publishing, Dunbeath, pp 46–77

Barasona JA, Latham MC, Acevedo P et al (2014) Spatio-temporal interactions between wild boar and cattle: implications for cross-species disease transmission. Vet Res 45:122

Barasona JA, Acevedo P, Diez-Delgado I et al (2016) Tuberculosis-associated death among adult wild boars, Spain, 2009–2014. Emerg Infect Dis 22:2178–2180

Barrios-Garcia MN, Ballari SA, Barrios-García MN, Ballari SA (2012) Impact of wild boar (*Sus scrofa*) in its introduced and native range: a review. Biol Invasions 14: 2283–2300

Barroso P, Acevedo P, Vicente J (2021) The importance of long-term studies on wildlife diseases and their interfaces with humans and domestic animals: a review. Trans Emerg Dis 68:1895–1909

Bassi E, Donaggio E, Marcon A et al (2012) Trophic niche overlap and wild ungulate consumption by red fox and wolf in a mountain area in Italy. Mamm Biol 77:369–376

Bieber C, Ruf T (2005) Population dynamics in wild boar *Sus scrofa*: ecology, elasticity of growth rate and implications for the management of pulsed resource consumers. J Appl Ecol 42:1203–1213

Boitani L, Mattei L, Nonis D, Corsi F (1994) Spatial and activity patterns of wild boars in Tuscany, Italy. J Mammal 75:600–612

Bongi P, Tomaselli M, Petraglia A et al (2017) Wild boar impact on forest regeneration in the northern Apennines (Italy). For Ecol Manag 391:230–238

Briedermann L (2009) Schwarzwild, 2nd edn. Franckh-Kosmos Verlags-GmbH & Co. KG, Stuttgart

Brivio F, Grignolio S, Brogi R et al (2017) An analysis of intrinsic and extrinsic factors affecting the activity of a nocturnal species: the wild boar. Mamm Biol – Z Säugetierkd 84:73–81

Brogi R, Chirichella R, Brivio F, Merli E, Bottero E, Apollonio M (2021) Capital-income breeding in wild boar: a comparison between two sexes. Sci Rep 11:4579

Bywater KA, Apollonio M, Cappai N, Stephens PA (2010) Litter size and latitude in a large mammal: the wild boar *Sus scrofa*. Mamm Rev 40:212–220

Calenge C, Maillard D, Fournier P, Fouque C (2004) Efficiency of spreading maize in the garrigues to reduce wild boar (*Sus scrofa*) damage to Mediterranean vineyards. Eur J Wildl Res 50:112–120

Cano-Terriza D, Risalde MA, Jiménez-Ruiz S et al (2018) Management of hunting waste as control measure for tuberculosis in wild ungulates in south-central Spain. Transbound Emerg Dis 65:1190

Canu A, Scandura M, Merli E et al (2015) Reproductive phenology and conception synchrony in a natural wild boar population. Hystrix It J Mamm 26:77–84

Canu A, Vilaça ST, Iacolina L et al (2016) Lack of polymorphism at the MC1R wild-type allele and evidence of domestic allele introgression across European wild boar populations. Mamm Biol 81:477–479

Canu A, Apollonio M, Scandura M (2018) Unmasking the invader: genetic identity of invasive wild boar from three minor islands off Sardinia (Italy). Mamm Biol 93:29–37

Carrasco-Garcia R, Barasona JA, Gortázar C et al (2016) Wildlife and livestock use of extensive farm resources in South Central Spain: implications for disease transmission. Eur J Wildl Res 62:65–78

Carrasco-Garcia R, Barroso P, Perez-Olivares J et al (2018) Consumption of big game remains by scavengers: a potential risk as regards disease transmission in Central Spain. Front Vet Sci 5:4

Casas-Díaz E, Closa-Sebastià F, Marco I et al (2015) Hematologic and biochemical reference intervals for wild boar (*Sus scrofa*) captured by cage trap. Vet Clin Pathol 44:215–222

Castillo-Contreras R, Carvalho J, Serrano E et al (2018) Urban wild boars prefer fragmented areas with food resources near natural corridors. Sci Total Environ 615: 282–288

Cherin M, Alba DM, Crotti M, Menconero S, Moullé P-E, Sorbelli L, Madurell-Malapeira J (2020) The post-Jaramillo persistence of *Sus strozzii* (Suidae, Mammalia) in Europe: new evidence from the Vallparadís section (NE Iberian Peninsula) and other coeval sites. Quat Sci Rev 233:106234

Cortiñas Abrahantes J, Gogin A, Richardson J, Gervelmeyer A (2017) Epidemiological analyses on African swine fever in the Baltic countries and Poland. EFSA J 15:4732

Cuevas MF, Mastrantonio L, Ojeda RA, Jaksic FM (2012) Effects of wild boar disturbance on vegetation and soil properties in the Monte Desert, Argentina. Mamm Biol – Z Säugetierkd 77:299–306

Danilov PI, Panchenko DV (2012) Expansion and some ecological features of the wild boar beyond the northern boundary of its historical range in European Russia. Russ J Ecol 43:45–51

Dardaillon M (1986) Seasonal variations in habitat selection and spatial distribution of wild boar (*Sus scrofa*) in the Camargue, Southern France. Behav Processes 13:251–268

Dardaillon M (1988) Wild boar social groupings and seasonal changes in the Camargue, southern France. Z Saugetierkd 53:22

Díez-Delgado I, Boadella M, Martín-Hernando M et al (2014) Complex links between natural tuberculosis and porcine circovirus type 2 infection in wild boar. Biomed Res Int 2014:765715

Díez-Delgado I, Sevilla IA, Romero B et al (2018) Impact of piglet oral vaccination against tuberculosis in endemic free-ranging wild boar populations. Prev Vet Med 155:11–20

ENETWILD consortium, Keuling O, Sange M, et al (2018) Guidance on estimation of wild boar population abundance and density: methods, challenges, possibilities. EFSA Support Publ 2018:EN-1449. 48pp

Engelman M, Lagerkvist C-J, Gren I-M (2018) Hunters' trade-off in valuation of different game animals in Sweden. Forest Policy Econ 92:73–81

Fernández-Llario P, Mateos-Quesada P (2005) Influence of rainfall on the breeding biology of wild boar (*Sus scrofa*) in a Mediterranean ecosystem. Folia Zool 54:240–248

Fonseca C, Santos P, Monzón A et al (2004) Reproduction in the wild boar (*Sus scrofa* Linnaeus, 1758) populations of Portugal. Zoology 16:53–65
Frackowiak W, Gorczyca S, Merta D, Wojciuch-Ploskonka M (2013) Factors affecting the level of damage by wild boar in farmland in north-eastern Poland. Pest Manag Sci 69:362–366
Frantz AC, Massei G, Burke T (2012) Genetic evidence for past hybridisation between domestic pigs and English wild boars. Conserv Genet 13:1355–1364
Frantz LAF, Schraiber JG, Madsen O et al (2013a) Genome sequencing reveals fine scale diversification and reticulation history during speciation in *Sus*. Genome Biol 14:R107
Frantz AC, Zachos FE, Kirschning J et al (2013b) Genetic evidence for introgression between domestic pigs and wild boars (*Sus scrofa*) in Belgium and Luxembourg: a comparative approach with multiple marker systems. Biol J Linn Soc 110:104–115
Frauendorf M, Gethöffer F, Siebert U, Keuling O (2016) The influence of environmental and physiological factors on the litter size of wild boar (*Sus scrofa*) in an agriculture dominated area in Germany. Sci Total Environ 541:877–882
Fulgione D, Rippa D, Buglione M et al (2016) Unexpected but welcome. Artificially selected traits may increase fitness in wild boar. Evol Appl 9:769–776
Fulgione D, Trapanese M, Buglione M et al (2017) Pre-birth sense of smell in the wild boar: the ontogeny of the olfactory mucosa. Zoology 123:11–15
Gabor TM, Hellgren EC, Van den Bussche RA, Silvy NJ (1999) Demography, sociospatial behaviour and genetics of feral pigs (*Sus scrofa*) in a semi-arid environment. J Zool 247:311–322
Gamelon M, Besnard A, Gaillard JM et al (2011) High hunting pressure selects for earlier birth date: wild boar as a case study. Evolution 65:3100–3112
Gamelon M, Douhard M, Baubet E et al (2013) Fluctuating food resources influence developmental plasticity in wild boar. Biol Lett 9:20130419
Gamelon M, Focardi S, Gaillard JM et al (2014) Do age-specific survival patterns of wild boar fit current evolutionary theories of senescence? Evolution (N Y) 68:3636–3643
Gayet T, Devillard S, Gamelon M et al (2016) On the evolutionary consequences of increasing litter size with multiple paternity in wild boar (*Sus scrofa scrofa*). Evolution 70:1386–1397
Genov P (1981) The food composition of wild boar in northeastern and western Poland. Acta Theriol 26:185–205
Genov PV (1999) A review of the cranial characteristics of the Wild Boar (*Sus scrofa* Linnaeus 1758), with systematic conclusions. Mamm Rev 29(4):205–38
Gethöffer F, Sodeikat G, Pohlmeyer K (2007) Reproductive parameters of wild boar (*Sus scrofa*) in three different parts of Germany. Eur J Wildl Res 53: 287–297
Giménez-Anaya A, Herrero J, Rosell C et al (2008) Food habits of wild boars (*Sus scrofa*) in a mediterranean coastal wetland. Wetlands 28:197–203
Goedbloed DJ, van Hooft P, Megens HJ et al (2013) Reintroductions and genetic introgression from domestic pigs have shaped the genetic population structure of Northwest European wild boar. BMC Genet 14:2–10
Goedbloed DJ, van Hooft P, Lutz W et al (2015) Increased *Mycoplasma hyopneumoniae* disease prevalence in domestic hybrids among free-living wild boar. EcoHealth 12:571–579
Gortázar C, Acevedo P, Ruiz-Fons F, Vicente J (2006) Disease risks and overabundance of game species. Eur J Wildl Res 52:81
Graves HB (1984) Behavior and ecology of wild and feral swine (*Sus scrofa*). J Anim Sci 58:482–492
Groenen MAM (2016) A decade of pig genome sequencing: a window on pig domestication and evolution. Genet Sel Evol 48:1
Groenen MAM, Archibald AL, Uenishi H et al (2012) Analyses of pig genomes provide insight into porcine demography and evolution. Nature 491:393–398
Groot Bruinderink GWTA, Hazebroek E, van der Voot H (1994) Diet and condition of wild boar, *Sus scrofa scrofa*, without supplementary feeding. J Zool 233:631–648
Groves CP (1981) Ancestors for the pigs: taxonomy and phylogeny of the genus Sus. Dep Prehist Res Sch Pac Stud Aust Nat Univ Tech Bull 3:1–96
Groves C (2007) Current views on taxonomy and zoogeography of the genus *Sus*. In: Pigs and humans: 10,000 years of interaction. Oxford University Press, pp 15–29
Groves CP, Grubb P (2011) Ungulate taxonomy. Johns Hopkins University Press, Baltimore
Hadjisterkotis E, Heise-Pavlov PM (2006) The failure of the introduction of wild boar *Sus scrofa* in the island of Cyprus: a case study. Eur J Wildl Res 52:213–215
Herrero J, Irizar I, Laskurain NA et al (2005) Fruits and roots: wild boar foods during the cold season in the southwestern Pyrenees. Ital J Zool 72:49–52
Herrero J, García-Serrano A, Couto S et al (2006) Diet of wild boar *Sus scrofa* L. and crop damage in an intensive agroecosystem. Eur J Wildl Res 52:245–250
Hirotani A, Nakatani J (1987) Grouping-patterns and intergroup relationships of Japanese wild boars (*Sus scrofa leucomystax*) in the Rokko mountain area. Ecol Res 2:77--84
Iacolina L, Scandura M, Goedbloed DJ et al (2016) Genomic diversity and differentiation of a managed island wild boar population. Heredity (Edinb) 116:60–67
Jędrzejewska B, Jędrzejewski W (1998) Predation in vertebrate communities: the Białowieża Primeval Forest as a case study. Springer, Berlin
Jędrzejewska B, Jędrzejewski W, Bunevich A et al (1997) Factors shaping population densities and increase rates of ungulates in Bialowieza Primeval Forest (Poland and Belarus) in the 19th and 20th century. Acta Theriol 42:399–451
Ježek M, Štípek K, Kušta T et al (2011) Reproductive and morphometric characteristics of wild boar (*Sus scrofa*) in the Czech Republic. J For Sci 57:285–292
Jori F, Payne A, Kock R et al (2017) Disease transmission at the interface between wild and domestic suiform species in the old and new worlds. In: Meletti M, Meijaard E (eds) Ecology, conservation and

management of wild pigs and peccaries. Cambridge University Press, Cambridge, pp 388–403
Kaminski G, Brandt S, Baubet E, Baudoin C (2005) Life-history patterns in female wild boars (*Sus scrofa*): mother–daughter postweaning associations. Can J Zool 83:474–480
Keuling O, Stier N, Roth M (2008) How does hunting influence activity and spatial usage in wild boar *Sus scrofa* L.? Eur J Wildl Res 54:729–737
Keuling O, Stier N, Roth M (2009) Commuting, shifting or remaining?: different spatial utilisation patterns of wild boar *Sus scrofa* L. in forest and field crops during summer. Mamm Biol – Z Säugetierkd 74:145–152
Keuling O, Baubet E, Duscher A et al (2013) Mortality rates of wild boar *Sus scrofa* L. in central Europe. Eur J Wildl Res 59:805–814
Keuling O, Strauß E, Siebert U (2016) Regulating wild boar populations is "somebody else's problem"! – human dimension in wild boar management. Sci Total Environ 554:311–319
Keuling O, Podgórski T, Monaco A et al (2018) Eurasian wild boar *Sus scrofa* (Linnaeus, 1758). In: Melletti M, Meijaard E (eds) Ecology, conservation and management of wild pigs and peccaries, 1st edn. Cambridge University Press, Cambridge, UK, pp 202–233
Kirkwood RN, Hughes PE, Booth WD (1983) The influence of boar-related odours on puberty attainment in gilts. Anim Sci 36:131–136
Kittawornrat A, Zimmerman JJ (2011) Toward a better understanding of pig behavior and pig welfare. Anim Health Res Rev 12:25
Koutsogiannouli EA, Moutou KA, Sarafidou T et al (2010) Detection of hybrids between wild boars (*Sus scrofa scrofa*) and domestic pigs (*Sus scrofa* f. *domestica*) in Greece, using the PCR-RFLP method on melanocortin-1 receptor (MC1R) mutations. Mamm Biol 75:69–73
Kozdrowski R, Dubiel A (2004) The effect of season on the properties of wild boar (*Sus scrofa* L.) semen. Anim Reprod Sci 80:281–289
Kuijper DPJ, Verwijmeren M, Churski M et al (2014) What cues do ungulates use to assess predation risk in dense temperate forests? PLoS One 9:e84607
Lange M, Kramer-Schadt S, Blome S et al (2012) Disease severity declines over time after a wild boar population has been affected by classical swine fever - legend or actual epidemiological process? Prev Vet Med 106:185–195
Larson G, Dobney K, Albarella U et al (2005) Worldwide Phylogeography of wild boar reveals multiple centers of pig domestication. Science (80-) 307:1618–1621
Laznik Ž, Trdan S (2014) Evaluation of different soil parameters and wild boar (*Sus scrofa* [L.]) grassland damage. Ital J Anim Sci 13:3434
Lemel J, Truvé J, Söderberg B (2003) Variation in ranging and activity behaviour of European wild boar *Sus scrofa* in Sweden. Wildlife Biol 9:29–36
Licoppe A, Prévot C, Heymans M et al (2013) Wild boar / feral pig in (peri-)urban areas. In: International Union of Game Biologists – Congress IUGB 2013 – Brussels – Belgium
Maillard D, Fournier P (2004) Timing and synchrony of births in the wild boar (*Sus scrofa* Linnaeus, 1758) in a Mediterranean habitat: the effect of food availability. Galemys 16:67–74
Masot AJ, Gil M, Risco D et al (2016) Pseudorabies virus infection (Aujeszky's disease) in an Iberian lynx (*Lynx pardinus*) in Spain: a case report. BMC Vet Res 13:1–8
Massei G, Genov PV (2004) The environmental impact of wild boar. Galemys 16:135–145
Massei G, Roy S, Bunting R (2011) Too many hogs? A review of methods to mitigate impact by wild boar and feral hogs. Hum Wildl Interact 5:79–99
Massei G, Cowan DP, Coats J et al (2012) Long-term effects of immunocontraception on wild boar fertility, physiology and behaviour. Wildl Res 39:378
Massei G, Kindberg J, Licoppe A et al (2015) Wild boar populations up, numbers of hunters down? A review of trends and implications for Europe. Pest Manag Sci 71: 492–500
Matiuti M, Bogdan AT, Crainiceanu E, Matiuti C (2010) Research regarding the hybrids resulted from the domestic pig and the wild boar. Lucr Stiint – Zooteh Biotehnol Univ Stiint Agric Med Vet Banat Timisoara 43:188–191
Mattioli L, Capitani C, Gazzola A et al (2011) Prey selection and dietary response by wolves in a high-density multi-species ungulate community. Eur J Wildl Res 57: 909–922
McDevitt AD, Carden RF, Coscia I, Frantz AC (2013) Are wild boars roaming Ireland once more? Eur J Wildl Res 59:761–764
McFee AF, Banner MW, Rary JM (1966) Variation in chromosome number among European wild pigs. Cytogenetics 5:75–81
Melis C, Szafrańska PA, Jędrzejewska B, Bartoń K (2006) Biogeographical variation in the population density of wild boar (*Sus scrofa*) in western Eurasia. J Biogeogr 33:803–811
Merta D, Mocała P, Pomykacz M, Frackowiak W (2014) Autumn-winter diet and fat reserves of wild boars (*Sus scrofa*) inhabiting forest and forest-farmland environment in south-western Poland. Folia Zool 63:95
Morelle K, Lehaire F, Lejeune P (2013) Spatio-temporal patterns of wildlife-vehicle collisions in a region with a high-density road network. Nat Conserv 5:53–73
Morelle K, Podgórski T, Prévot C et al (2015) Towards understanding wild boar *Sus scrofa* movement: a synthetic movement ecology approach. Mamm Rev 45: 15–29
Morini P, Boitani L, Mattei L, Zagarese B (1995) Space use by pen-raised wild boars (*Sus scrofa*) released in Tuscany (Central Italy) – II: home range. IBEX J Mt Ecol 3:112–116
Náhlik A, Sandor G (2003) Birth rate and offspring survival in a free-ranging wild boar *Sus scrofa* population. Wildlife Biol 9:37
Naranjo V, Gortazar C, Vicente J, de la Fuente J (2008) Evidence of the role of European wild boar as a reservoir of *Mycobacterium tuberculosis* complex. Vet Microbiol 127:1
Navarro-Gonzalez N, Casas-Díaz E, Porrero CM et al (2013) Food-borne zoonotic pathogens and

antimicrobial resistance of indicator bacteria in urban wild boars in Barcelona, Spain. Vet Microbiol 167:686–689
Oja R, Kaasik A, Valdmann H (2014) Winter severity or supplementary feeding – which matters more for wild boar? Acta Theriol (Warsz) 59:553–559
Oja R, Soe E, Valdmann H, Saarma U (2017a) Non-invasive genetics outperforms morphological methods in faecal dietary analysis, revealing wild boar as a considerable conservation concern for ground-nesting birds. PLoS One 12:e0179463
Oja R, Velström K, Moks E et al (2017b) How does supplementary feeding affect endoparasite infection in wild boar? Parasitol Res 116:2131–2137
Okarma H (1995) The trophic ecology of wolves and their predatory role in ungulate communities of forest ecosystems in Europe. Acta Theriol 40:335–386
Oliver W, Leus K (2008) *Sus scrofa*. IUCN red list threat species 2008 e.T41775A10559847
Pérez-González J, Costa V, Santos P et al (2014) Males and females contribute unequally to offspring genetic diversity in the polygynandrous mating system of wild boar. PLoS One 9:1–22
Podgórski T, Baś G, Jędrzejewska B et al (2013) Spatiotemporal behavioral plasticity of wild boar (*Sus scrofa*) under contrasting conditions of human pressure: primeval forest and metropolitan area. J Mammal 94:109–119
Podgórski T, Scandura M, Jędrzejewska B (2014a) Next of kin next door – philopatry and socio-genetic population structure in wild boar. J Zool 294:190–197
Podgórski T, Lusseau D, Scandura M et al (2014b) Long-lasting, kin-directed female interactions in a spatially structured wild boar social network. PLoS One 9:1–11
Poteaux C, Baubet E, Kaminski G et al (2009) Socio-genetic structure and mating system of a wild boar population. J Zool 278:116–125
Queirós J, Alves PC, Vicente J et al (2018) Genome-wide associations identify novel candidate loci associated with genetic susceptibility to tuberculosis in wild boar. Sci Rep 8:1980
Renner SC, Suarez-Rubio M, Wiesner KR et al (2016) Using multiple landscape genetic approaches to test the validity of genetic clusters in a species characterized by an isolation-by-distance pattern. Biol J Linn Soc 118:292–303
Risco D, Gonçalves P, Mentaberre G et al (2018) Biometrical measurements as efficient indicators to assess wild boar body condition. Ecol Indic 88:43–50
Rosell C, Navàs F, Romero S (2012) Reproduction of wild boar in a cropland and coastal wetland area: implications for management. Anim Biodivers Conserv 35:209
Rossi S, Staubach C, Blome S et al (2015) Controlling of CSFV in European wild boar using oral vaccination: a review. Front Microbiol 6:1141
Ruiz-Fons F, Vicente J, Vidal D et al (2006) Seroprevalence of six reproductive pathogens in European wild boar (*Sus scrofa*) from Spain: the effect on wild boar female reproductive performance. Theriogenology 65:731–743
Ruiz-Fons F, Segalés J, Gortázar C (2008) A review of viral diseases of the European wild boar: effects of population dynamics and reservoir role. Vet J 176:158–169
Sáez-Royuela C, Tellería JL (1986) The increased population of the wild boar (*Sus scrofa* L.) in Europe. Mamm Rev 16:97–101
Saïd S, Tolon V, Brandt S, Baubet E (2012) Sex effect on habitat selection in response to hunting disturbance: the study of wild boar. Eur J Wildl Res 58:107–115
Santos P, Fernández-Llario P, Fonseca C et al (2006) Habitat and reproductive phenology of wild boar (*Sus scrofa*) in the western Iberian Peninsula. Eur J Wildl Res 52:207–212
Scandura M, Iacolina L, Crestanello B et al (2008) Ancient vs. recent processes as factors shaping the genetic variation of the European wild boar: are the effects of the last glaciation still detectable? Mol Ecol 17:1745–1762
Scandura M, Iacolina L, Apollonio M (2011a) Genetic diversity in the European wild boar *Sus scrofa*: phylogeography, population structure and wild x domestic hybridization: genetic variation in European wild boar. Mamm Rev 41:125–137
Scandura M, Iacolina L, Cossu A, Apollonio M (2011b) Effects of human perturbation on the genetic make-up of an island population: the case of the Sardinian wild boar. Heredity (Edinb) 106:1012–1020
Schley L, Roper TJ (2003) Diet of wild boar *Sus scrofa* in Western Europe, with particular reference to consumption of agricultural crops. Mamm Rev 33:43–56
Schley L, Dufrêne M, Krier A, Frantz AC (2008) Patterns of crop damage by wild boar (*Sus scrofa*) in Luxembourg over a 10-year period. Eur J Wildl Res 54:589–599
Scillitani L, Monaco A, Toso S (2010) Do intensive drive hunts affect wild boar (*Sus scrofa*) spatial behaviour in Italy? Some evidences and management implications. Eur J Wildl Res 56:307–318
Segura A, Acevedo P, Rodríguez O et al (2014) Biotic and abiotic factors modulating wild boar relative abundance in Atlantic Spain. Eur J Wildl Res 60:469–476
Servanty S, Gaillard JM, Allainé D et al (2007) Litter size and fetal sex ratio adjustment in a highly polytocous species: the wild boar. Behav Ecol 18:427–432
Servanty S, Gaillard J-M, Toïgo C et al (2009) Pulsed resources and climate-induced variation in the reproductive traits of wild boar under high hunting pressure. J Anim Ecol 78:1278–1290
Seward NW, VerCauteren KC, Witmer GW, Engeman RM (2004) Feral swine impacts on agriculture and the environment. Sheep Goat Res J 19:34
Singer FJ, Otto DK, Tipton AR, Hable CP (1981) Home ranges, movements and habitat use of European wild boar in Tennessee. J Wildl Manag 45:343–353
Sonnenburg J, Ryser-Degiorgis MP, Kuiken T et al (2017) Harmonizing methods for wildlife abundance estimation and pathogen detection in Europe – a questionnaire

survey on three selected host-pathogen combinations. BMC Vet Res 13:53

Špinka M (2009) Behaviour of pigs. In: Jensen P (ed) The ethology of domestic animals: an introductory text. CABI Publishing, Wallingford, pp 177–191

Spitz F, Janeau G (1990) Spatial strategies-an attempt to classify daily movements of wild boar. Acta Theriol 35:129–149

Spitz F, Valet G, Lehr Brisbin I (1998) Variation in body mass of wild boars from Southern France. J Mammal 79:251–259

Stillfried M, Fickel J, Börner K et al (2017) Do cities represent sources, sinks or isolated islands for urban wild boar population structure? J Appl Ecol 54:272–281

Suselbeek L, Adamczyk VMAP, Bongers F et al (2014) Scatter hoarding and cache pilferage by superior competitors: an experiment with wild boar, *Sus scrofa*. Anim Behav 96:107–115

Thurfjell H, Ball JP, Åhlén P-A et al (2009) Habitat use and spatial patterns of wild boar *Sus scrofa* (L.): agricultural fields and edges. Eur J Wildl Res 55:517–523

Toïgo C, Servanty S, Gaillard JM et al (2008) Disentangling natural from hunting mortality in an intensively hunted wild boar population. J Wildl Manag 72:1532–1539

Tolon V, Dray S, Loison A et al (2009) Responding to spatial and temporal variations in predation risk: space use of a game species in a changing landscape of fear. Can J Zool 87:1129–1137

Truvé J, Lemel J (2003) Timing and distance of natal dispersal for wild boar *Sus scrofa* in Sweden. Wildlife Biol 9:51–57

Vandenbergh JG (1988) Pheromones and mammalian reproduction. In: Knobil E, Neill JD (eds) The physiology of reproduction. Raven Press, New York, pp 1679–1696

Veličković N, Djan M, Ferreira E et al (2015) From north to south and back: the role of the Balkans and other southern peninsulas in the recolonization of Europe by wild boar. J Biogeogr 42:716–728

Vetter SG, Ruf T, Bieber C, Arnold W (2015) What is a mild winter? Regional differences in within-species responses to climate change. PLoS One 10:e0132178

Vicente J, Barasona JA, Acevedo P et al (2013) Temporal trend of tuberculosis in wild ungulates from Mediterranean Spain. Transbound Emerg Dis 60:92–103

Vilaça ST, Biosa D, Zachos F et al (2014) Mitochondrial phylogeography of the European wild boar: the effect of climate on genetic diversity and spatial lineage sorting across Europe. J Biogeogr 41:987–998

VKM, Skjerve E, Thurfjell H et al (2018) Wild boar population growth and expansion – implications for biodiversity, food safety, and animal health in Norway. Opinion of the Norwegian Scientific Committee for Food and Environment, Oslo

White S (2011) From globalized pig breeds to capitalist pigs: a study in animal cultures and evolutionary history. Environ Hist Durh N C 16:94–120

Wiethoelter AK, Beltrán-Alcrudo D, Kock R, Mor SM (2015) Global trends in infectious diseases at the wildlife–livestock interface. Proc Natl Acad Sci 112:9662–9667

Reeves' Muntjac *Muntiacus reevesi* (Ogilby, 1839)

2

Norma G. Chapman

Contents

N. G. Chapman (✉)
Bury St. Edmunds, UK
e-mail: ngchapman@btopenworld.com

© Springer Nature Switzerland AG 2022
L. Corlatti, F. E. Zachos (eds.), *Terrestrial Cetartiodactyla*, Handbook of the Mammals of Europe,
https://doi.org/10.1007/978-3-030-24475-0_18

Common Names

English	Reeves' muntjac or Chinese muntjac
German	Chinesischer Muntjak
French	Muntjac de Chine, muntjac de Reeves
Spanish	Muntíaco de Reeves
Italian	Muntjak della Cina
Russian	Китайский олень мунтжак

In England and in France, this deer is often referred to as Chinese muntjac, but that is confusing as several species of this genus occur in China. It has also been called Barking deer, which again could apply to all species in this genus. Named after John Russell Reeves (Flower 1929), the correct name is Reeves' muntjac (Ellerman and Morrison-Scott 1951); Reeves' muntjac is also used.

Taxonomy, Systematics and Paleontology

All muntjacs are native to southeast Asia, China, or India. Nine species are recognized, three having been discovered since the late 1980s (Heckeberg 2020; Lister 2020; Smith-Jones 2022). *M. reevesi* dates back to the Early Pleistocene (Ma et al. 1986). Reeves' muntjac *Muntiacus reevesi* (Fig. 1) is split into two subspecies, *M. r. reevesi* Ogilby, 1839, from China, and the slightly smaller and darker *M. r. micrurus* Sclater, 1875 (Formosan muntjac) from Taiwan. They share almost identical results of cytogenetic analyses, but Southern blot and Fluorescence in situ hybridization studies revealed some sequence divergence of satellite 1 DNA between these two, supporting the classification of the Formosan muntjac as a separate subspecies (Chiang et al. 2004). A new subspecies, *M. r. jiangkouensis* (Gu and Xu 1998), from Guizhou Province, China, was proposed from seven animals on minor phenotypic differences, but further information is lacking.

The origin of the lineage of *Muntiacus* can be traced to the Late Miocene in China, and it has been suggested that *Eostylocerus* from Early Pleistocene deposits in Yunnan, China, had a morphology approaching that of *Muntiacus* (Ma et al. 1986). The earliest known muntjac is *Muntiacus leilaoensis* from southwestern China (Croitor

Fig. 1 Mature buck, third head of antlers, molt to summer pelage beginning May 13 (photograph by N. Chapman)

2018). Traditionally, *Muntiacus* was considered to be a primitive forerunner of modern deer, but Croitor (2018) regarded it as a quite specialized taxon of the modern evolutionary radiation of the subfamily Cervinae. There are bone fragments possibly of *Muntiacus* from Poland, Ukraine, and France, so early forms may have extended into Europe (Czyzewska 1968).

Eight paleontological taxa were recognized by Groves (2016) who considered *M. atherodes* (Bornean yellow muntjac) to be the oldest extant species, followed by *M. reevesi*, but Geist (1998) regarded *reevesi* as the older, based on analysis of mitochrondrial and ribosomal DNA. These two forms occupied the edges of the range from which Groves and Grubb (1990) believed that a centrifugal geographic pattern of evolution occurred, with *M. crinifrons* and *M. rooseveltorum* evolving from the *reevesi* line. *M. reevesi* attained a wide distribution, mostly south of the River Yangste and reached Taiwan when there was a land bridge from the mainland in the Late Pleistocene.

The rapid and parallel chromosome number reductions, mostly resulting from tandem fusion, were analyzed by Wang and Lan (2000) for seven species. They concluded that the Giant muntjac, *M. vuquangensis,* discovered in 1994 (Schaller, Vrba 1996) is rightly placed in this genus. Heckeberg (2020), from mitochondrial and combined molecular analyses, placed this species in the same clade as *M. reevesi* together with *M. putaoensis, M. truongsonensis,* and *M. rooseveltorum*. A separate clade consists of *M. feae, M. muntjak, and M. crinifrons.* The latter is considered by some to be the same species as *M. gonshanensis* (Amato et al. 2000). *M. atherodes* was placed in a polytomy with these clades (Heckeberg 2020).

Current Distribution

Native Populations

In China, Reeves' muntjac occurs mostly within the subtropical southeast, extending from approximately 102° longitude to the east coast and from latitude 32° to the south coast taking in the Provinces of Shanxi, Gansu, Hubei, Anhui, Jiangxi, Fujian, Zhejiang, Hunan, Guizhou, Guangxi, Sichuan, and Yunnan. Snow cover limits its dispersal further north (Sheng 1991). Muntjac are heavily hunted, about 650,000 being taken annually (Ohtaishi and Gao 1990; Sheng 1991). Under such high pressure and degradation of the habitat, the population has decreased. The species is also present on three of the islands in the Zhoushan archipelago including the largest, Putuo. In Hong Kong, Reeves' muntjac is listed under Schedule 2 of the Wild Animals Protection Ordinance Cap 170 (2012) and the IUCN Red List 2016 (Timmins and Chan 2016) which suggests that this deer may still be extant here, but information is lacking.

M. r. micrurus is endemic to Taiwan occupying nearly all the broad-leaved forest in the mountainous zone above the intensive agricultural land and below the winter frost line. Much of their habitat is tall grassland (*Miscanthus sinensis*) and dense bamboo, shrubs, and herbs (McCullough et al. 2000). Formerly a captive colony of Formosan muntjac was maintained at Bielefeld University, Germany (Laurien 1997).

Introduced Populations

Reeves' Muntjac is very widely distributed in the United Kingdom, mainly in England (Fig. 2). Their history here began as imports from 1894 to 1906 to Woburn Park, Bedfordshire, subsequent releases, and numerous later translocations (Chapman et al. 1994b; Chapman 2021). In the 1960s–1970s, they were still regarded as a benign introduction (Dansie 1970). Release of muntjac in the UK was not completely illegal until the Invasive Alien Species (Enforcement and Permitting) Order 2019 was enacted (Moore 2021). An intensive survey in 1994 revealed the presence of muntjac in 745 10-km squares, almost half of the National Grid squares in England and Wales (Chapman et al. 1994b). Since 2002, the British Deer Society has organized surveys to monitor all deer species at roughly 5-year intervals. Range expansion seemed to peak at 12% per year between 2002 and 2005 (Ward et al. 2008, 2021). By 2016, reports came

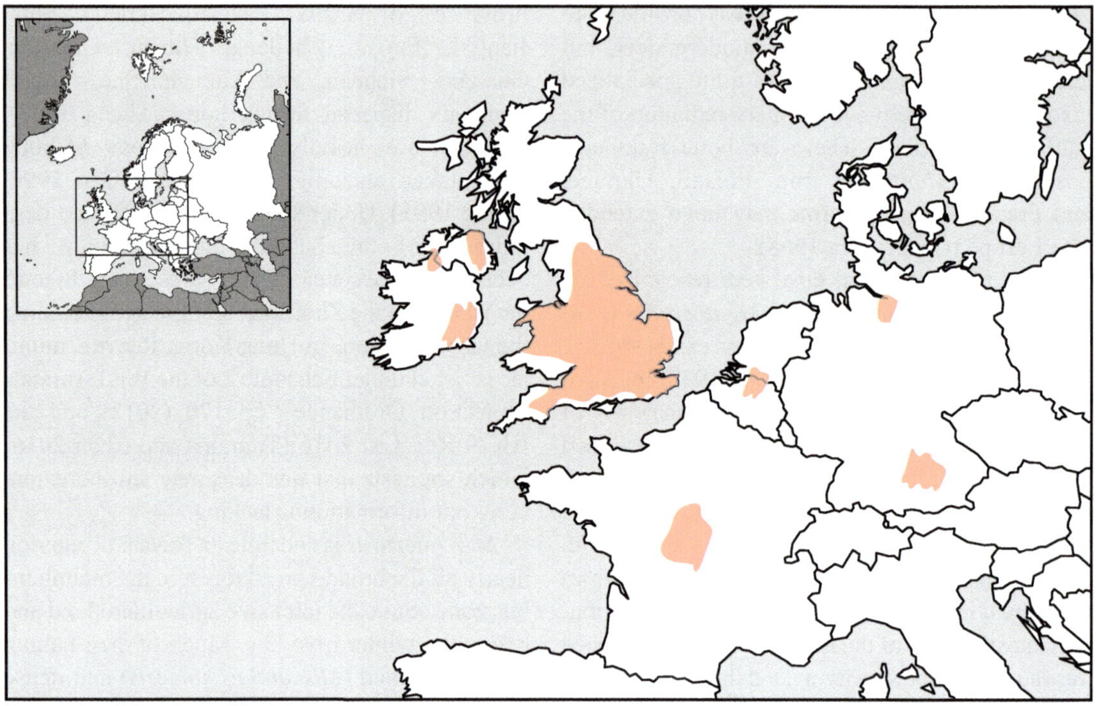

Fig. 2 Distribution of muntjak in the British Isles. See text for localized scattered reports in some other western European countries. (Map template: © Copyright Getty Images/iStockphoto)

from 887 10 km squares, an expansion of 8.7% since 2007 (Smith-Jones 2017). Almost all records were from the southern half of England with a gap only at the southeast tip, and few confirmed reports from the south-west peninsula. In northern England, definite records were sparse. In Wales, apart from along the border with England, reports were unconfirmed. From Scotland, confirmed records were received from only three widely scattered 10 km squares, fewer than in previous surveys which suggested that liberated or escaped animals did not prosper. For Northern Ireland, a road traffic casualty in 2009 was the first record, not far from a known captive population in County Down. Appropriate surveillance was put into force as the authorities were aware of the potential seriousness should the species become established (Dick et al. 2009, 2010; Hogg et al. 2013). East of Belfast in the forested vicinity of Mount Stewart, on the Ards Peninsula, one muntjac was shot in 2011 and another 2 years later. Camera trap photographs showed that the deer were present along at least a 3 km stretch of suitable habitat (Hogg et al. 2013). By 2016, records were received from six more 10 km squares at locations spread east to west across the southern part of Northern Ireland as far as County Fermanagh: These can be explained only by human agency as can isolated reports from the north, near the coast of County Coleraine (Smith-Jones 2017). By 2020, detections remained low and population sizes appeared to be small (Freeman and Hogg 2020). The UK muntjac population in 2018 was estimated at 115,000–147,000 (Matthews et al. 2018), but assessing population size is an imprecise science (Smart et al. 2004). The National Game Bag Census indicated an increase in culling of between 152% and 325% in the two decades up to 2015, but that only records animals shot on large private estates. In the Republic of Ireland, the first record was a male shot in 2008 in County Wicklow, south of Dublin. Further reports followed in that area, then from four other counties also in the east or south and from County Donegal in the northwest of the island. By 2014, almost

30 records were reported, rising to over 130 within 35 10 km squares by 2020 (National Biodiversity Data Centre, Ireland).

For continental Europe, in 2000 the Invasive Species Specialist Group of the IUCN prepared Guidelines for the Prevention of Biodiversity loss caused by Alien Invasive Species. When the European Regulation 1143/2014 came into force in 2019, the aim was to ensure that the establishment of A.I.S. is avoided throughout Europe. Muntjac was one of eleven mammalian species listed. Possession, exchange, trade, and release are all prohibited, and where they are present eradication is the aim. Before this Regulation came into force, muntjac were already present in several countries, having been imported from England decades before such translocations became illegal. Ward et al. (2021) mapped the range of land cover that muntjac might achieve across western Europe although extremely low-winter temperatures would likely be a limiting factor in northern areas: In unusually severe winters in the UK, heavy mortality was reported (Southern 1964). However, predicted climate warming (Metzger et al. 2008) could make northern parts of Europe more favorable, but some southern parts may become too arid. In the Netherlands, the first confirmed sightings, dating from 1997 to 1998, were from the Veluwe (Province of Gelderland) and from east of the river Ijssel (Achterhoek, Province of Overijssel). In, the estimated number reported to the Conseil de la Chasse was 50–100. The muntjac had very probably been imported from the U.K. and released with the intention of establishing a shootable population: Dutch and other European hunters had been coming to England for many years to shoot muntjac. Hollander (2013) had prepared a risk assessment and documented available information in 2015. By then, 153 reports had been received. Field research in 2016 indicated that a few muntjac remained at the Veluwe near Apeldoorn, and in the Province of Noord-Brabant there was a population, at Landgoed de Utrecht, suspected of having immigrated from Belgium. However, as some reports were from the middle of the country, it seems that the law of 2000 forbidding the trade and possession of muntjac was still being ignored. As in most countries, zoos need a license to keep them (Hollander 2015, La Haye pers. comm. 2019). Other sightings were from Hilvarenbeek, possibly Ossendrecht near Eindhoven, and the Maashorst region. In the eastern Zeeuws-Vlaanderen, a few muntjac were reported from the Westdorpe, Heikant area (Hollander 2016). In the 12 months up to June 2020, 42 observations of muntjac were reported from the provinces of Gelderland, Limburg, and Noord-Brabant (waarneming.nl website) indicating an established, growing population. Muntjac from Hertfordshire, England, were exported to Belgium in the 1980s (F. Marshall, pers. comm.), and in 1989 a prospective buyer, living half-way between Ghent and Antwerpen, was trying to source muntjac for captive breeding (M. Clark pers. comm.). Isolated sightings of muntjac were increasingly reported between 2008 and 2013 from localities near Brugge, Mol-Neerpelt, Hasselt, and also Braschaat where the deer came from a captive population, and some movement across the border to and from the Netherlands was postulated. In 2013, when muntjac were not considered as truly established, a very comprehensive Risk Analysis Report was prepared, drawing on the decades of experience from England (Baiwy et al. 2013). A best practice document followed (Casaer et al. 2015), and in 2019 the Feasibility of Eradication and Spread Limitation for Species of Union Concern sensu the EU IAS regulation (EU 1143/2014) (Adriaens et al.). Up to May 2019, nearly 400 observations had been submitted, mostly from East Flanders and Antwerp (www.waarnemingen.be/species/7700), and from 2018 to 2020 more than 220 muntjacs were confiscated or destroyed by nature inspection services in Flanders. From 2005, newspapers began to report muntjac in Denmark, near lake Røbæk in central Jutland and later from the island of Læsø, 19 km off the coast of Jutland (www.netnatur.dk/muntjac). Here, eight had been kept in captivity until the owner discovered that was illegal: He said all had been shot. In 2020 in Fyn (between Odense and Middlefart), one was shot. No specific risk assessment has been made, but the policy is to shoot as soon as a report is received (Environmental Protection Agency 2020). The presence of muntjac at Rambouillet in France in 1872 was mentioned by Dansie (1970). Introductions around 1891 to some private estates and forests (probably Chambord and Rambouillet) did not succeed in becoming established in the wild (Whitehead 1993), but colonies have been maintained in some animal parks including Clères in Normandy since before 1940 (Dubost 1970, 1971). In the late 1950s, escapees from Clères established a localized population which was present in 1982 (F. de Beaufort pers. comm.). No further escapes occurred from that park between 1966 and 1995 (P. Ciarpaglini pers. comm. 1995). Three muntjac from Woburn Park, England, were taken in about 1975 to an estate on the Seine, half-way between Paris and Le Havre, owned by a shooting enthusiast (editor of *La Chasse*) where they were to be placed in a large enclosure which previously had held hares (*Lepus europeaus*). Progeny from here were later passed to someone else, but the deer moved onto a neighboring estate (F. Marshall, pers. comm.). Richard (1982–1983) described the skull of a Reeves' muntjac killed in the woods at Bonnmare which may have escaped from a small zoo at Pitres in l'Eure, south of the Seine. In 1995, to the knowledge of the spokesman for the Association Nationale des Chasseurs de Grand Gibier, no muntjac were present in France, except possibly in a few zoos (A.J.H. de Boislambert, pers. comm.). Either this statement was in error, or further introductions occurred. In 1998, someone in France was sourcing muntjac from one or more wildlife parks in England (A. Bullimore pers. comm.). One muntjac was killed on a road in Brittany in 2012 (Belloy 2013) and another far away in Moselle in 2017. From 2000 to 2019, there had been sightings and a number shot in localized areas within three central Departéments southwest from Paris (Loir et Cher, Indre et Loire, Indre) (INPN 2020). The origin of at least some was from a captive collection which had held 30 muntjac, descendants of a male and four females imported from Magdeburg Zoo, Germany. Sightings 80 km away indicated an additional liberation/escape point (Hurel et al. 2019). Some muntjac from a wildlife park in eastern England were taken to Germany in the 1980s, but no details are available. Free-living muntjac were first reported in Germany in 2004, then followed mostly single sightings within five states, some of which were shot (Nehring and Skowronek 2017). Later reports came from widely scattered areas within Rheinland-Pfalz, including Bad Kreuznach, Kusel, Birkenfeld, Mayen, and near the towns of Trier and Koblenz (Hofmann 2018), but may have persisted only briefly. In Lower Saxony, in a forest near Hildesheim, a skull was found (U. Kierdorf. pers. comm.) In Northern Friesland in July 2020, one was shot on the order of the Federal State of Schleswig-Holstein within 2 days of being reported. A private deer park at Altenfelden in Oberosterreich, Austria, received some muntjac from Woburn Park, England, in the latter part of the nineteenth century (F. Marshall, pers. comm.). About 90 km to the southeast around 2018, there were reports of sightings of muntjac close to Mondsee, not far from where some had been held privately (pers. comm. anon). No established population was reported in 2020.

Many thousands of Reeves' muntjac are at liberty on two islands of Japan, following escapes from zoos after typhoons in 1960 and 1970 (Asada 2009). Control measures have been implemented (Tokyo Municipal Office (2017).

Abundance

No estimates of the population sizes in its native range are available. The UK muntjac population in 2018 was estimated at 115,000–147,000 (Matthews et al. 2018), but assessing population size is an imprecise science (Smart et al. 2004). Elsewhere in Europe, where the aim must be to eradicate the species, reports (in 2020) are mostly of scattered individuals or a few hundred sightings over a period. Freeman et al. (2016) showed that the UK populations are descendants of a small number of founding females.

Description

Reeves' muntjac is the smallest cervid free-living in Europe. Summer pelage is a rich, glossy red-brown over most of the body, but buff ventrally with a variable amount of white on the chin and inner aspect of the thighs. Neck, ears, and crown of males are often golden brown. Ears are broad, rounded, and 8.5 cm in length and almost black on some individuals. The rhinarium is black. Over the nasal bones, the hairs are short and dark. Males (bucks) have conspicuous black stripes over the frontal ridges, forming a V usually extending up the pedicles. Females (does) have a dark/black kite-shaped pattern on the forehead. The preorbital (suborbital) gland below each eye is a distinctive feature in both sexes.

Dorsal and flank hairs are 25–28 mm, but hairs on the edge of the rump are >40 mm. The tail is rich chestnut dorsally, white ventrally, and is very conspicuous when an alarmed deer flees, with tail erect. Forelegs are often almost black on the front, especially on males. The cleaves of the hooves are about 23 mm long: 66% of 262 animals assessed had at least one pair of unequal cleaves, almost always the outer being slightly longer (author's own data).

A black nuchal stripe may be present or absent in both sexes. Winter coat (achieved in September–October in England) is duller, darker brown. In England, molting occurs in April–June, beginning at the head and shoulders. At birth, the pelage is heavily spotted with buff spots which gradually fade away by about 8 weeks. Males develop their facial stripes by about 9 months.

Measurements

The values of the main measurements are reported in Table 1 (Author's own data). Very similar measurements of body and skull for animals in China are given by Sheng (1991) and Ma et al. (1986) who compared them with four other species of muntjac. The Formosan subspecies is slightly smaller, body mass given as 12 kg for males, 8 kg for females (McCullough 2000). The lateral (second and fifth) metacarpals are short (<30 mm) thin splints. Of the tarsal bones, the cubonavicular and external and median cuneiform are fused into a single bone (Hershkovitz 1982).

Skull

They are immediately distinguishable from any other genus by the very large, deep preorbital fossae and, on males, the frontal ridges which merge on to the long pedicles (Fig. 3). Antlers are streamlined, following the slope of the head. For animals in England aged at least 2 years, dimensions are given in Table 2 (Chapman 2008).

Dentition

Deciduous teeth 0.1.3.0 / 3.1.3.0, permanent dentition 0.1.3.3 / 3.1.3.3. The lower canine is incisiform and abuts the third incisor. The last deciduous premolar has three cusps, but its permanent replacement has only two, a useful guide when assessing age. Deciduous premolars two and three and permanent premolars two, three, and four are lophodont but all permanent premolars and molars are selenodont. Order of eruption of permanent mandibular teeth: molars, first and

Table 1 Measurements for muntjac at least 2 years old, shot or killed on roads from King's Forest. Suffolk, England

	Males			Females		
	Mean	Range	(n)	Mean	Range	(n)
Whole body mass (kg)	14.9	12.3–17.0	(41)	13.3	10–16.0	(30)
Carcass mass (kg)[a]	9.8	8.5–12.25	(32)	8.5	6.75–10.25	(25)
Head and body length (cm)	84	78–91	(34)	82	77–91	(23)
Tail to end vertebra (cm)	13	10.3–17.0	(36)	12	10–14.5	(25)
Tail incl. terminal hairs (cm)	16	13–17.5	(31)	15	13–18	(22)
Ear length (cm)	8.2	7.5–9.0	(43)	8.1	7–8.5	(29)
Hind foot length (cm)	22.7	21.5–24.5	(40)	22.4	20–23.5	(28)
Shoulder ht. (cm)	48.5	46–52	(27)	47	45–50	(20)
Girth (cm)	58	29–65	(29)	55	50–61	(20)

[a]carcass mass = head off, feet off at proximal ends of cannon bones, all viscera removed

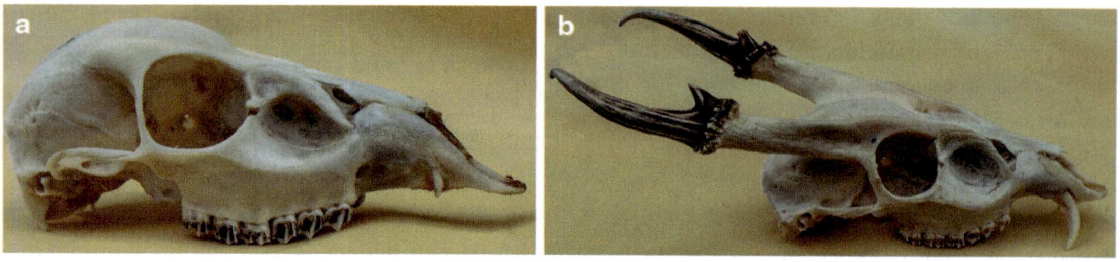

Fig. 3 (**a**) Skull of adult female; (**b**) skull of adult male (photographs by B. Soper)

Table 2 Skull length and width ($n = 23$ in males, $n = 19$ in females)

	Males		Females	
	Range	Mean	Range	Mean
Greatest length (mm)	162–176	167	154–169	162
Zygomatic width (mm)	73–82	78	71–78	74
Mandible length	123–135	129	122–133	127

second incisors, premolars, third incisor and canine. For maxillary teeth: first molar, canine in males *c.* 21 weeks, second and third molars, canine in females *c.* 53–57 weeks, premolars. A full set of permanent teeth is achieved by 83–92 weeks (Chapman et al. 1985a).

Sexual dimorphism of the upper canine is extreme. In the male, it is a curved tusk with a sharp point and up to 60 mm long, of which a third is within the alveolus: It is slightly mobile with fore and aft movement of about 7 mm and laterally 4 mm. The root closes by about 5 years. The outermost layer is enamel, as in Chinese water deer, *Hydropotes,* but unlike the rudimentary upper canines in other cervids. The tusks are very efficient weapons, capable of slashing another buck or potential predator, e.g., dog. Of 83 free-ranging bucks aged 3–5 years, 51% had one or both canines broken. In females, the canine is insignificant, only 17 mm of which two-thirds are within its alveolus: One or both were congenitally absent in a few cases (0.02%, n = 247) (Chapman 1997).

Antlers

Detailed antler data are available from daily observations over many years of a captive colony of muntjac kept under seminatural conditions in southeast England. Pedicles are first recognizable from 20–32 weeks. By 32–46 weeks of age, antlers appear and remain in velvet until 46–76 weeks.

The first antlers are small knobs or short spikes lacking a coronet. Young males subsequently synchronize with the antler cycle of older bucks by casting their first antlers in May or June (median date May 26) when they may be 51–112 weeks old. Thereafter, a regular pattern is established with new antlers growing during the summer (79–130 days, mean 106), velvet is cleaned from August to October (median date September 14), and hard antlers are retained until cast in late April to mid-July (median date 27 May) (Chapman and Chapman 1982). Second and subsequent antlers have a coronet and typically a short brow tine although its presence in 1 year does not necessarily mean it will be present the next year. Casting of both antlers on the same day occurred in 13.2% of 121 cycles: The longest intervals were 9–13 days (Chapman and Bartos 2014).

From the captive bucks, the heaviest pair of antlers (each 20 g) was an 11th pair, each just over 11 cm and had a brow tine <1 cm. Longer and heavier antlers occur in some free-ranging populations, e.g., >13 cm. The antlers curve backward and slightly inward, the span between the tips mostly 90 to 120 mm on mature bucks. The pedicles of the first antlers are slender (30-70 mm long), but subsequently remodeling of bone results in shorter, thicker structures, but even in old age it represents 19–43% of the total length of the cranial appendage so is long in comparison with pedicles of other deer.

Various morphological parameters of antler velvet in eight species were investigated by Bubenik (1993). The hairs of muntjac and roe velvet closely resembled skin hairs, were the most pigmented, and had the least complex sebaceous glands. Hair density was lowest in muntjac: The hairs were 0.04–0.010 in diameter.

Scent Glands

On the head, the position of the pair of preorbital sacs is very obvious. The opening is a curved slit below the corner of each eye, leading to a sac with strong semicircular sphincter musculature (Barrette 1976) which lies within a large fossa. The sac is composed of two pockets: The posterior contains a cream-colored paste of sloughed epidermal cells and lipids; the anterior pocket has fewer sebaceous and apocrine sweat glands and lacks paste. There is some sexual dimorphism – males having larger multilobed sebaceous glands and thicker hairs in the posterior pocket (Rehorek et al. 2005). During urination, defecation, and courtship, the slits are opened wide or the sac everted. Often at the same time, the long, very mobile tongue (14.5 cm) is flicked in and out of the mouth, so its tip (5.5 cm) passes over or into the open gland. The gland is active from an early age: Young fawns sometimes open the sacs while being groomed by the dam. Analysis of volatiles from the preorbital gland of muntjac showed the secretions to be individually distinct and with the potential for conspecifics to identify sex and population origin (Lawson et al. 2000, 2001).

The frontal glands, present in no other cervid, are shallow, almost hairless grooves 3–4 cm long on the forehead, within the black stripes of males/ at edge of dark marking of females, which are frequently wiped on the ground or vegetation, especially by bucks. Males also fray the bark of young trees with their incisors and rub the area with their pedicles. Scent-marking, mating, and social interactions are well illustrated by Dubost (1971) and also discussed by Barrette (1977c).

A third pair of scent glands, the interdigital, lies in a hair-lined cleft (about 3 cm long) between the cleaves of the hind feet. The secretion produced here may be deposited when walking. A chin gland (15-28 mm) is listed in a comparison of characteristics of five extant species of *Muntiacus* (Ma and Wang 1991).

Anterior orbital gland. Within the anterior aspects of the orbit, closely associated with the nictating membrane, lie the red-brown Harderian gland and the white nicitans gland now shown to be one bilobed gland, the anterior orbital gland. In males, it is significantly larger than in females (mean mass 7.62 g males; 1.42 g in females, $n = 25$). The white lobe secretes mucus: The other lobe produces serous and lipid secretions (Rehorek et al. 2005, 2007).

Feces also serve as scent markers. Fecal pellets are shiny, black, cylindrical, or nearly spherical, 5-16 mm long, 82% in 8-11 mm range; decomposition periods for 82 samples voided in February: 100% by 122 days (Chapman 2004a). Typically, 20–120 pellets are voided per defecation. The same spot may be used repeatedly, creating a latrine heap and leaving scent signals about the depositor(s).

Both sexes have paired paraurethral glands, respectively, alongside the vagina and proximal urethra. In males, their secretion may contribute to marking territory. Vulval licking by males is frequent: The secretion from the female may have a role in recognition of bonding between the sexes (Dansie and Williams 1973).

Age Determination

For field observations, the following categories were found useful: Fawns: up to 8 weeks old, coat spotted, or spots fading. Juveniles: indeterminate sex, 2–5 months, and up to 3/4 grown. Immature females: 5–8 months, 3/4 to full-grown. Mature females: >8 months, full size, and breeding age. Subadult males: 5–26 months, antler status varies from having pedicle bumps to hard first antlers. Adult males: enter this class between 14 and 26 months, after casting the first set of antlers (Claydon et al. 1986). The sequence of eruption of the teeth (Chapman et al. 1985a) and, for older deer, the pattern of wear on the molars can be a useful guide to age (Chapman et al. 2005).

Physiology

Muntjac are classed as concentrate selectors, having frequent feeding bouts, taking easily digestible plant parts, low in fiber, which are rapidly broken down by bacteria (Hoffman 1985). They have larger salivary glands relative to body mass than red deer or Chinese water deer (Kay 1987). The rumen has just two blind sacs and measures approximately 22 cm × 20 cm. Density of ruminal papillae varies with the seasons and position on the mucosa, but overall average density, from 113 rumina from free-ranging muntjac in England, was 66.96/cm^2. Average papilla length was 2.53 mm and width 0.86 mm (Pfeiffer 1993).

The thymus consists of two cervical and two thoracic lobes. Fetal development is rapid, and its mass does not change significantly between birth and 1 year. From then, it declines and by 6 years is very small. It seems less responsive to sexual changes than in fallow deer *Dama dama,* perhaps because of the year-round breeding. Involution occurs in old age or with trauma (Chapman and Twigg 1990).

From the pituitary gland, removed immediately after death, the ultrastructure of the adenohypophysis was observed (Young and Chaplin 1975). Thyroid ultrastructure was also described by Young (1976).

Genetics

M. reevesi has the highest number of chromosomes within the genus, $2n = 46$ for both sexes: All are acrocentric except for the small submetacentric Y. Chromosome studies on feral muntjac in England confirmed their specific identity as *M. reevesi* at a time when their status was in doubt (Chapman et al. 1983).

Hybridization in captivity between *M. reevesi* and *M. m. vaginalis* gave $2n = 26$ for a female and 27 for a male (Shi et al. 1980). Old records from London Zoo reported that such hybrids were fertile (Gray 1971), but Shi and Pathak (1981) found that in a hybrid male spermatogenesis was arrested at an early stage.

All the other species have much lower diploid numbers, including those discovered in the 1980s and 1990s (i.e., *M. putaoensis* in Myanmar (and more recently in Tibet Li et al. 2017), *M. truongsonensis* in Vietnam, and *M. vuquangensis* in Laos, Vietnam, and Cambodia) but the lowest is in the Indian muntjac *M. muntjak* where $2n$ is 7 for a male and 6 for female – the lowest known for any mammal (Wang and Lan 2000).

Life History

Growth

The mean body mass within 24 h of birth was 1209 g (range 900–1500 g), $n = 53$ (male 27, female 26), in a captive colony maintained in grass paddocks and fed, according to season, on carrots, potatoes, acorns, horse chestnuts, ivy *Hedera helix*, browse cut from deciduous trees, and large umbellifers (*Anthriscus sylvestris* and *Heracleum sphondylium*), sometimes supplemented with flaked maize. By 29 weeks and 33 weeks, two male fawns attained the mass of 7 kg and shoulder height of 40.5 cm. Some males before the age of 1 year achieved a body mass and shoulder height within the range for males two or more years old (see Table 1). Year-round breeding means that not all young are born in the same season, a factor which may influence growth rates in feral populations. For Formosan Reeves' muntjac, Pei (1996) gave information for postnatal growth using data from animals trapped in Taiwan. Age was assessed for 383 deer from eviscerated body mass, diastema length, and in males, upper canine length.

Reproduction

Unlike cervids native to Europe, muntjacs are aseasonal breeders with a postpartum estrus. Although adult muntjac have a synchronous antler cycle, there is little seasonal variation in the size and activity of testes, epididymides, or accessory reproductive glands. Spermatogenesis continued when antlers were in velvet, and year-round fertility was achieved (Chapman and Harris 1991) as it was for *M. r. micrurus* in Taiwan (Pei and Liu 1994; Pei et al. 1995). There is a disseminate prostate gland, and paired seminal vesicles.

If a doe does not conceive within a few days of giving birth, she will come into estrus at intervals of 14–15 days until mated successfully. A singleton is the norm, but there are rare reports of twin fetuses and a doe seen with two fawns. Implantation is almost always in the right horn of the uterus as first reported by Chapman and Dansie (1969) who also described the genital tract and fetal development (1970). The placenta is polycotyledonary.

First conceptions in the captive colony in the south of England occurred at 5–6 months, but more usually at 7–10 months with the potential to give birth at intervals of about 7 months. The minimum inter-birth period recorded in the same captive population was 211 days: mean for 12 does from 47 births was 219 days (author's own data). This fecundity can be maintained over many years, e.g., 22 fawns by the time one doe was 14 years old. Productivity is likely to be 1.6 fawns/year. Birth occurs in all months (Chapman 2020), and the sex ratio is close to parity ($n = 53$ captive births, and 200 uteri with sexable fetuses: author's own data): parity also reported from Berlin Zoo (Frädrich 1997). Neonates within 24 h of birth weighed 900-1500 g ($n = 30$), mean 1209 g, and have a length of about 42 cm, shoulder height 26 cm, girth 24 cm, hind foot 13 cm, and ear 5 cm.

Survival

In the aforementioned captive colony, both sexes attained ages in the midteens and elsewhere a doe died at 19 years 11 months. The oldest record in a zoo is given as 23.2 years (Müller et al. 2010). In a free-ranging population in the King's Forest, England, the oldest known survivor (female, ear tagged when young) lived until 13 years, but in a sample of 85 females killed on roads adjacent to the area, only 20% were estimated to be 5 or more years old (author's data).

Habitat and Diet

Habitat Selection

In many locations in the U.K., muntjac are sympatric with one or more other species of deer, most usually roe (*Capreolus capreolus*) and/or fallow (*Dama dama)* but in some areas red (*Cervus elaphus*), sika (*Cervus nippon*), or water deer (*Hydropotes inermis*) (Cooke 1998). Their preferred habitat includes a diversity of plant types, especially highly nutritious herbs, shrubs, and nut-producing trees (Chapman et al. 1985a; Chapman and Tutchener 2017). However, muntjac also utilize pine forests with less diversity where they favor older stands and areas with greater cover (Hemani et al. 2004). In England now, the range of habitats occupied by muntjac extends beyond woodland to fenland, arable landscapes, and series of gardens within villages and towns (Cooke 2019).

Diet

In England, but not on the continent, there have been many studies of the effect muntjac have on vegetation. However, many of the herb and tree species are the same or similar in western Europe, so the impacts would be expected to be similar. Direct observations (Cooke 1997, 2001; Diaz and Burton 1996), identification of plant epidermal cells from fecal pellets (Harris and Forde 1986), and inspection of rumen contents have all shown that muntjac browse on wide range of shrubs and graze on herbs. Grasses such as *Agrostis tenuis*, but not coarse grasses, can be important in winter and when sprouting early in spring (Jackson et al. 1977). Ivy (*Hedera helix*) is especially taken in winter, and bramble, *Rubus fructicosus* agg., is taken all-year, but some of its many species are not eaten (Chapman 2015/16). Where muntjac lived in a mosaic of deciduous woods and arable land, their diet included leaves from a wide range of trees, berries, and fungi. Wheat grains ingested in four winter months had been taken from feeders provided for pheasants (Chapman and Tutchener 2017).

Spatial Movements

The home range of an individual deer must vary according to the quality and diversity of the habitat. During a 5-year-long radio-tracking study in a mainly coniferous forest with about 12% broad-leaved trees (the King's Forest, Suffolk, England), home ranges of adults (calculated as minimum convex polygons) were 20–28 ha for males and 11–14 ha for females (Chapman et al. 1993). Subsequent to a serious gale, the habitat became more diverse, the muntjac population doubled, and ranges halved (Keeling 1995). Ranges of males encompassed ranges of several females with no or very little overlap with that of another male. Ranges of females did overlap. The same pattern of virtual separation of ranges of males but considerable overlap of female ranges has been reported in semicaptive situations (Dubost 1970; Miura 1984; Harding 1986).

In contrast, in southern Taiwan, where *M. r. micrurus* is heavily exploited by local hunters, studies of radio-collared muntjac were undertaken within 4.2 km^2 in the Central Mountain Range. Muntjac used all the microhabitats, but most activity was within the densely vegetated gaps in the forest which are created by wind throws caused by typhoons. The home range for four males ranged from 63 to 161 ha, and for three females the ranges overlapped, 68 to 168 ha (McCullough et al. 2000).

The ranging behavior and activity pattern of radio-collared adult muntjac (11,000 radio fixes) and adult roe (3000 fixes) were analyzed by Forde (1989). There was no significant seasonal change in home range size nor diurnal or seasonal shifts. On average, muntjac were active for 69% of each 24-h period. Activity peaked around dawn and dusk but was reduced at – 4 °C. Distances traveled were 0.7–1.1 km/day.

Behavior

Social Organization

Muntjac do not aggregate into groups. Most sightings are of a lone animal, or a doe with her fawn, buck and doe together, two females together, or both sexes with a fawn. An adult buck defends his home range which is likely to encompass the ranges of several females. Some overlap of their ranges is permitted by does although they have exclusive core areas. When moving unhurriedly within its home range, a muntjac potters along, often with nose toward the ground. When two are together, the second follows behind, on the exact route of the other. Muntjac are competent swimmers and have been seen crossing lakes.

Age of dispersal from the natal area is variable. In the King's Forest, some males moved on before growing pedicles, some much later. Some females established their mother's range overlapping. Most known relocations within the Forest were less than 4 km, but the longest known dispersal was of a tagged adult killed on a road 21 km away.

Foraging

Feeding periods peaked early in the morning, middle of the day, and dusk but could be at any time. Five active periods/24 h were typical (Forde 1989). Camera traps have shown that much feeding takes place at night (Cooke 2019). A feeding bout is followed by ruminating while lying up in cover, such as a tunnel within dense bramble.

Mating Behavior

A buck defends his home range within which he has the opportunity to copulate with any doe, irrespective of the stage of his antler cycle or time of year. In low density populations, a buck and doe may form a bond, frequently seen together whatever her reproductive state. An estrus doe often barks as she walks with her tail raised. Chases by a buck can be relentless, with neck stretched low with nose to her vulval area, licking her urine, exhibiting flehmen, and mounting her frequently: These activities are illustrated in the comprehensive account of social behavior by Barrette (1977a). The buck may lift the rump of the doe with his forehead or antlers and self-groom between these activities. Afterward, the female walks away or feeds (Yahner 1979).

Parental Care

A concealed spot is selected to give birth, in a dense patch of vegetation. The hiding phase lasts about 4 weeks. Thereafter, activity bouts by the fawn become more frequent (Yahner 1978). When a fawn is being suckled (there are two pairs of nipples), the dam frequently licks the anal area to stimulate defecation and thoroughly grooms other parts of the body. Yahner (1978) reported weaning to occur at 8 weeks although lactation has been observed in captivity up to 118 days after giving birth (author's own data).

Aggression

A dominance display precedes fighting between two bucks. Shoving, twisting, and clashing of antlers leads to a position whereby one male can rotate his head sideways and upward to deliver a downward blow with a tusk, inflicting cuts to the neck, side of the face, or ears of the other buck (Barrette 1977b). In England, a mature buck that was shot had 16 mm of antler protruding from close to the right ear, a further 11 mm was embedded in the skin. The thickness of the dermal shield in the neck area had prevented damage to the underlying muscle (Chapman 1996). Sparring, a harmless form of combat, has been observed in captivity. Two bucks make nose to nose contact before antler contact: A bout of allogrooming sometimes follows (Barrette 1977b).

Senses and Communication

As befits a solitary species whose natural habitat is densely vegetated, with very limited visibility, vocalization and olfaction are both of the utmost importance for communication. Barks of a muntjac are less gruff than those of roe deer but might be mistaken for those of a fox (*Vulpes vulpes*). These vocalizations are very loud, often repeated in rapid succession for many, even hundreds of times, by either sex. In Taiwan, the duration of 1467 bouts of barking by *M. r. micrurus* lasted from <1 s (one bark) to 27 min (262 barks) (Chen and Wang 1994). Yahner's (1980a) observations of a captive population concluded barks were mostly solicited when a strange object or potential threat was detected. Females bark when in estrus, utter a submissive squeak during courtship, and whimper when pursued by an over-amorous male. A fawn caught by a predator squeals and bleats. Both sexes scream extremely loudly when, for example, trapped in a fence or gate.

Scent Marking

Observations during 4 months of four males and four females in a large enclosure showed that scent marking, with open preorbital glands and wiping with frontal glands, was performed more by males than females, and dominants more than subordinates. Deer repeatedly marked over their own marking places and feces but sometimes over the spots where conspecifics had left urine or feces, so communal latrines developed (Barrette 1977a). The odor profile is different for individual deer so acts as social communication, informing conspecifics who is where (Lawson et al. 2000).

When facing a potential threat, the deer may stamp a fore foot with sufficient force to produce an audible thump. Grinding of teeth as part of a threat display by captive adult male Formosan muntjac was noted by Stadler and Hendrichs (1987), and Clark (1981) described this sound as clicking, usually produced by a buck when anxious.

Parasites and Diseases

In England, one species of biting louse *Damalinia indica* (= *Cervicola* or *Tricholipeurus indica*) is more common than the sucking louse *Selenopotes muntiacus,* but the burden is usually very light, occurring mainly in the groin and on the chin. Muntjac are also host to the tick *Ixodes ricinus*: The burden on muntjac was observed to be very much lighter than on roe deer (*Capreolus capreolus*) in King's Forest in England. The orange larvae of the mite *Neotrombicula autumnalis* have also been recorded (author's own data).

Lungworms have not been reported from muntjac in England although 120 lungs and trachea from five counties have been examined. Neither liverfluke nor bladderworm cysts have been recorded. Abomasal washings from 40 muntjac yielded small numbers (maximum 25) or no nematodes: *Spiculopteragia asymmetrica* and *Trichostrongylus* spp were identified (author's own data). In Japan, three species of exotic nematodes have been identified in a feral population (Setsuda et al. 2020).

Reports of infectious disease in muntjac in England are extremely rare. One case of *Mycobacterium bovis* was confirmed in southwest England (Delahay et al. 2001). In a survey of wild mammals in that region, 3 of 55 muntjac were culture positive for bovine tuberculosis (Delahay et al. 2005). At densities as low as $<6/\text{km}^2$, muntjac could be spill-out hosts and above $56/\text{km}^2$ could be maintenance hosts (Ward and Smith 2012).

Blood samples from 196 muntjac from Thetford Forest, England, were tested for pathogens. Two were positive for *Anaplasma phagocytophilum* (Duscher et al. 2020). In Northern Ireland, a novel gammaherpes virus (genus *Rhadinovirus*) was detected in wild muntjac (McKillen et al. 2017).

Muntjac are susceptible to foot-and-mouth disease (Gibbs et al. 1975). Pneumonia appeared to be the final cause of death for 46 starved muntjac in a dense, unmanaged population in England (Cooke et al. 1996). Hypertrophic osteopathy (Marie's disease) on lower limb bones has been observed (Chapman 2004a), and diseases of limb joints have also been reported from free-living muntjac (Green and Chapman 1993; Middleton 1975). Examples of scoliosis, spondylosis, mandibular and maxillary bone lesions, intervertebral disc disease, and carcinomas have been seen (author's own data). A possible poisoning by oxalic acid from eating leaves of sugar beet (*Beta vulgaris*) was reported (Chapman 1988). Hydromyelia was diagnosed in a 24-day-old captive fawn in the USA (Dutton et al. 2002). A novel gammaherpes virus, genus *Rhadinovirus*, was detected in a wild muntjac in Northern Ireland (McKillen et al. 2017).

If a captive muntjac requires veterinary procedure or a field study necessitates attaching a radio-transmitter on a collar, immobilization will be required. This can be achieved by injection of xylazine or methohexitone. The latter was more appropriate for use in the field but xylazine, giving a slower but calmer recovery period, and is suitable for a captive animal recovering in a small enclosure (Cooper et al. 1986). Designs for a suitable holding crate and examination crate have been described (Chapman et al. 1987). Deaths from exertional myopathy have been reported (Wallace et al. 1987). A veterinary review of keeping Chinese deer was given by Seidel (1993). Erythrocytes of many cervids exhibit sickling, forming a variety of bizarre shapes, burr-shaped in the case of Reeves' muntjac. Sickling is a hereditary physiological characteristic of the blood but does not result in hemolytic anemia or vascular occlusion as it does in man (Chapman 1977).

Population Ecology

Competition

Any competition for resources for muntjac in England is most likely to be with roe deer, and this would be expected in much of Europe. Both species are concentrate selectors seeking a wide range of herbaceous plants, shrubs, and fruits, but in winter both species especially exploit bramble (*Rubus*) (Forde 1989; Hemani et al. 2004). In the King's Forest, great spatial overlap, but low temporal overlap, was observed between these two species (Chapman et al. 1993; Forde 1989; Wray 1992) as was the case in Thetford Forest where there was apparent displacement of roe when muntjac densities increased (Hemani et al. 2005).

When muntjac and roe deer are sympatric, some reduction in roe density is to be expected. That was the conclusion of Forde (1989) whose research was in the King's Forest when muntjac density in his study area was believed to be 15/km^2 compared with 11.4 roe/km^2. In a similar habitat, Hemani et al. (2005) found areas where muntjac density was three times that of roe. His data also showed substantial reduction in body mass across the age classes for roe and a reduction in fertility (Hemani 2003; Dolman and Waeber 2008). Muntjac were more active by day and roe by night, so this reduced any interference competition (Forde 1989).

A high degree of dietary overlap between Brown hares (*Lepus europaeus*) and muntjac was seen in winter within the King's Forest as shown by the identification of plant-epidermal fragments in fecal pellets (Wray 1994), but in spring/summer hares moved to surrounding arable land.

Predators

The only wild predator of muntjac in the U.K. is the red fox (*Vulpes vulpes*). In the King's Forest study area, it was believed that about 50% of fawns were taken by foxes before reaching 2 months of age (Chapman and Harris 1996), and when foxes became resident in Monks Wood, Cooke (2014) observed a decline in the recruitment to the muntjac population. The level of predation will depend upon the density of both the muntjac and fox populations and the availability of alternative prey including roe deer fawns, small mammals, and rabbits. Muntjac would be suitable prey for wolves if the European distribution of the two overlapped. Dogs chase and maim or kill some muntjac and may receive serious injuries from the upper canines or antlers of a buck trying to defend itself.

Conservation Status

The IUCN Red List 2018 categorized Reeves' muntjac as Least Concerned although in its native range it is decreasing because of high hunting pressure and degradation of habitat (Sheng and Xu 1990).

Management

Where muntjac are present in continental Europe, the aim is to eradicate them, but in the U.K. that would be impossible because they are so long-established, numerous, and widely distributed. Small size and concealing habits mean that muntjac may pass unnoticed while their population increases and their reproductive potential enables a population to build up rapidly. Early control is essential to prevent the establishment of new populations.

Since 2002, the British Deer Society has organized surveys to monitor all deer species at roughly 5-year intervals. Between 1972 and 2002, muntjac range, expressed in 10 km^2, was estimated to have expanded at a compound rate of 8.2% a year (Ward 2005) and peaked at approximately 12% between 2002 and 2005 (Ward 2021). The 2016 survey reported muntjac in 887 10 km squares, an increase of 8.7% since 2007 (Smith-Jones 2017).

Population density depends on availability of diverse food sources year-round, cover, presence of other herbivores especially other deer species, level of predation, and disturbance factors such as forestry work and shooting. Several methods of assessing density have been tried. Assessing numbers by methods used for herding species is not appropriate for small, secretive muntjac. To estimate approximate densities in several deciduous woods in eastern England, Cooke (2019) scored a suite of field signs indicating the presence of muntjac: foot prints, paths, fecal pellets, scrapes, resting places, frayed woody stems (20-60 cm from the ground), and woody stems bitten through at 70–200 cm above ground level. His highest score suggested a density of 110 per km^2. For 206 ha, in which no culling took place, within the King's forest, a census figure for the total population (excluding fawns) was derived from intensive observations by experienced observers traversing the blocks and watching the perimeters. Over three consecutive winters, the population remained remarkably constant, with an average of 15 muntjac per km^2 but then increased to 27 per km^2 (Blakeley et al. 1997). Estimates of density can be misleading if no account is taken of deer which commute to feed in areas beyond their known woodland home base. Density has also been estimated by counting the standing crop of dung along transects (Hemani and Dolman 2005) or by thermal imaging (Hemani et al. 2007). In 2001 for one mainly coniferous block (13 km^2) of Thetford Forest, density was estimated by thermal imaging to be 20 muntjac per km^2. By this method in 2002, for the whole of that Forest (185 km^2) the estimate was 64 per km^2. On an adjacent 102 km^2 of heathland used for military training and sheep grazing, thermal imaging indicated a density of about six muntjac per km^2 (Waeber et al. 2013). Camera traps may be deployed to confirm identify and record behaviors and proved useful where muntjac were suspected of colonizing a new area in Northern Ireland (Dick 2017).

Snow impedes movement, and severe winters with persistent deep snow covering all available forage would be a limiting factor for muntjac to establish in some parts of Europe. Such winters are rare in England but occurred in 1962–1963 when many muntjac starved to death. Climate warming may enable muntjac to extend their range. When windchill was less than 0 °C Yahner (1980b), observing muntjac in large enclosures at the National Zoological Park, Virginia, noted that the deer stayed in a shelter or adjacent to tall vegetation.

Impact on Conservation Habitats

Impacts in conservation woodland can be serious where muntjac densities are high (Cooke 2004) with direct effects seen on woody vegetation (Tabor 1993; Cooke and Farrell 2001) and ground flora (Cooke 1994, 1997, 2006, 2021; Diaz and Burton 1996; Tabor 1999).

Rackham (1975) and Tabor (1993) highlight damage to woodland ground flora (especially oxslip *Primula elatior*) which they attribute to high grazing pressure from fallow deer and muntjac. Cooke has also reported comprehensively on the effects of muntjac at high densities on other elements of the ground flora (primroses, *Primula vulgaris*; bluebells, *Hyacinthoides nonscripta*; dog's mercury, *Mercurialis perennis*; and common spotted orchid, *Dactylorhiza fuchsia*), within Monks Wood (Cooke 1994, 1995, 2006) although these impacts were recorded at extremely high-population levels. The extent to which these heavy impacts are more widely representative is uncertain.

Heavy impacts on the field and ground layer in woodlands may have a pronounced effect on diversity of plant species present, and with indirect effects on other fauna including invertebrates (e.g., Pollard and Cooke 1994; Cooke and Farrell 2001; Flowerdew and Ellwood 2001; Fuller et al. 2005).

After many years with little management in many areas of the U.K., muntjac are now controlled to some extent by culling (Smith-Jones 2004; Downing 2014). They have no Closed Season because they breed at any time of year. In managed populations in England, for humane reasons, shooting a heavily pregnant female is good practice because her previous fawn will have become independent. Each European country has its own legislation relating to shooting (rifle caliber, etc.). Since 2019 under the Invasive Alien Species (Enforcement and Permitting) Order, any injured muntjac taken into care must not be released.

Fencing, at least 1.5 m high, may be erected round new plantations or areas needing special protection (Cooke and Lakhani 1996; Putman 1996a), and individual saplings can be protected by tree shelters (1.2 m tall). Removal of muntjac from residential areas remains a challenge.

Economic Value

In England, some land owners gain income from letting shooting rights or accompanying guest stalkers. The venison is gradually gaining in popularity and can be purchased from some game dealers. In the UK, little use is made of the pelts, but in China they are processed for high-precision polishing leathers.

Conflicts with Humans

Damage to forestry is localized though muntjac may seriously compromise establishment and may cause significant damage to coppice regrowth through browsing and stem breakage (Tabor 1993; Cooke 1994, 1998, 2006; Cooke and Farrell 2001). In the U.K., reports of damage to horticultural and agricultural interests are relatively few although some farmers have experienced localized damage (Putman and Moore 1998). Retail garden centers and market gardens need very secure perimeter fencing. In recent years, the presence of muntjac in villages and towns in the U.K. has brought them into conflict with residents. Their impact on garden plants and vegetable plots can be severe, distressing, and expensive for the owners (Chapman et al. 1994a). Access may be gained under or through a hedge or fence, over a wall (an adult doe jumped a 1.52 m wall adjacent to a road when disturbed) or through a gate. There are examples of immature and adult deer becoming stuck between vertical bars (8 cm apart) of metal gates (Chapman 2010). In these locations, control by shooting is rarely an option, and exclusion is often difficult to achieve. Deer vehicle collisions are a serious problem in the U.K., and those involving muntjac have been estimated to be around 18,600/year (The Deer Initiative).

Future Challenges for Research and Management

In the UK, constant surveillance and control are required to prevent further increase of this alien species which is so well established. In other countries, action needs to be taken immediately if a feral muntjac is reported: A population can build rapidly. Any country in a temperate or warmer zone where muntjac are present in captivity should be very vigilant regarding biosecurity and aware of the potential serious problems should any escape or deliberate liberations occur. Nentwig et al. (2018) regarded muntjac as one of the "100 worst" alien species in Europe. Under the Convention on Biological Diversity (1992), the Bern Convention (1979), and the European Habitats Directive (EEC 43/920), there are international obligations to address issues regarding Alien Invasive Species.

Acknowledgments Many of the studies referred to here were instigated by my late husband, Donald (1935-1982). Field work in the King's Forest continued with the support of many people under the direction of Prof. S. Harris. Others provided dead muntjac or organs or collaborated in various investigations. Correspondents abroad supplied distribution data. I am immensely grateful to all of them. My thanks are also offered to Dr. Arnold Cooke for helpful discussions over many years and to Prof. Rory Putman.

References

Adriaens T, Branquart E, Gosse D, Reniers J, Vanderhoeven S (2019) Feasibility of eradication and spread limitation for species of Union concern sensu the EU IAS regulation (EU 1143/2014) in Belgium

Amato G, Egan M, Schaller GB (2000) Mitochondrial variation in muntjac, evidence for discovery, rediscovery and phylogenetic relationships. In: Vrba E, Schaller GB (eds) Antelopes, deer and relatives. Yale University Press, New Heaven, pp 285–295

Asada N (2009) Distribution and population estimation for Reeves' muntjac in 2008 in Chiba prefecture, Japan. Rep Chiba Biodiv Centre 1:21–26

Baiwy E, Schockert V, Branquart E (2013) Risk analysis of Reeves' muntjac *Muntiacus reevesi*. Cellule interdépartmental sur les Espèces invasives (CiEi) DG03, SPW editions/Université de Liège, Gemloux

Barrette C (1976) Musculature of facial scent glands in the muntjac. J Anat 122:61–66

Barrette C (1977a) The social behaviour of captive muntjacs *Muntiacus reevesi* (Ogilby 1839). Z Tierpsychol 43:188–213

Barrette C (1977b) Fighting behaviour of muntjac and the evolution of antlers. Evolution 31:169–176

Barrette C (1977c) Scent-marking in captive muntjacs, *Muntiacus reevesi*. Anim Behav 25:536–541

Belloy A (2013) Réseau Ongulés sauvages. Lettre d'information no.17 Janvier 2013

Blakeley D, Chapman N, Claydon K, Claydon M, Harris S, Wakelam J (1997) Studying muntjac in the King's Forest, Suffolk. Deer 10:156–161

Bubenik GA (1993) Morphological differences in the antler velvet of Cervidae. In: Ohtaishi N, Sheng H-I (eds) Deer in China. Elsevier Science Publishers, Amsterdam, pp 56–64

Casaer J, Boone N, Devisscher S, Vercammen J, Adraiens T (2015) Best practice voor beheer van Chinese muntjak *Muntiacus reevesi* in Vlaanderen. Institut Natur-en Bosonderzoek, Brussels, p 60

Chapman DI (1977) Haematology of the deer. In: Archer RK, Jeffcott LB (eds) Comparative clinical haematology. Blackwell Scientific Publications, Oxford, pp 345–364

Chapman N (1988) Oxalic acid in the urine of wild muntjac deer (*Muntiacus reevesi*). Publ Veter Deer Soc 3(1):18–22

Chapman NG (1996) Muntjac in the wars. Deer 9:562–563

Chapman NG (1997) Upper canine teeth of *Muntiacus* (Cervidae) with particular reference to *M. reevesi*. Proceedings of the First International symposium on physiology and ethology of Wild and Zoo animals. Berlin, Sept. 18–21 1996. Supplementum zum Band 62 der Zeitschrift fur Säugetierkunde, pp 32–36

Chapman N (2004a) Bone of contention: Marie's disease. BBC Wildlife December: 44

Chapman NG (2004b) Faecal pellets of Reeves'muntjac, *Muntiacus reevesi*: defecation rate, decomposition period, size and weight. Eur J Wildl Res 50:141–145

Chapman NG (2008) Reeves' muntjac. In: Harris S, Yalden DW (eds) Mammals of the British Isles: handbook, 4th edn. The Mammal Society Southampton, London, pp 564–571

Chapman NG (2010) Trapped muntjac. Deer 15(5):57

Chapman N (2015/16) The prickly problem of brambles. Deer 17 (8):27

Chapman N (2020) Reproductive potential of muntjac. Deer 19(5):16–17

Chapman N (2021) History of introduction of Reeves' muntjac in Great Britain and ecological characteristics facilitating expansion of feral populations. Eur J Wildl Res 67:33

Chapman NG, Bartos L (2014) The asynchronous timing of antler casting of Reeves' muntjac (*Muntiacus reevesi*). Eur J Wildl Res 60:563–565

Chapman DI, Chapman NG (1982) The antler cycle of adult Reeves' muntjac. Acta Theriol 27:107–114

Chapman DI, Dansie O (1969) Unilateral implantation in muntjac deer. J Zool (Lond) 159:534–536

Chapman NG, Harris S (1991) Evidence that the seasonal antler cycle of adult Reeves' muntjac (*Muntiacus reevesi*) is not associated with reproductive quiescence. J Reprod Fertil 92:361–369

Chapman N, Harris S (1996) Muntjac. Mammal Society and British Deer Society, Fordingbridge

Chapman N, Tutchener J (2017) The diversity of diet of muntjac *Muntiacus reevesi* in woodlands in the Hitchin area. Trans Hertfordshire Natl Hist Soc 49:55–58

Chapman N, Twigg GI (1990) Studies on the thymus gland of British Cervidae, particularly muntjac, *Muntiacus reevesi*, and fallow, *Dama dama*, deer. J Zool (Lond) 222:653–675

Chapman DI, Chapman N, Mathews JG, Wurster-Hill DM (1983) Chromosome studies of feral muntjac deer (*Muntiacus* sp.) in England. J Zool (Lond) 201:557–588

Chapman DI, Chapman NG, Colles CM (1985a) Tooth eruption in Reeves' muntjac (*Muntiacus reevesi*) and its use as a method of age estimation(Mammalia: Cervidae). J Zool (Lond) 205:205–221

Chapman NG, Claydon K, Claydon M, Harris S (1985b) Distribution and habitat selection by muntjac and other species of deer in a coniferous forest. Acta Theriologca 30:287–303

Chapman NG, Claydon K, Claydon M, Harris S (1987) Techniques for the safe and humane handling of free-living muntjac deer (*Muntiacus reevesi*). Br Vet J 143: 35–43

Chapman NG, Claydon K, Claydon M, Forde PG, Harris S (1993) Sympatric populations of muntjac (*Muntiacus reevesi*) and roe deer (*Capreolus capreolus*): a comparative analysis of their ranging behaviour, social organisation and activity. J Zool (Lond) 229:623–640

Chapman N, Harris A, Harris S (1994a) What gardeners say about Muntjac. Deer 9(5):302–306

Chapman N, Harris S, Stanford A (1994b) Reeves' muntjac *Muntiacus reevesi* in Britain: their history, spread, habitat selection, and role of human intervention in accelerating their dispersal. Mammal Rev 24(3):113–160

Chapman NG, Claydon K, Claydon M, Claydon FP, Harris S, Keeling J (1997a) History and habitat preferences of muntjac in the King's Forest, Suffolk. Deer 10(5):289–294

Chapman NG, Furlong M, Harris S (1997b) Reproductive strategies and the influence of date of birth on growth and sexual development of an aseasonally-breeding ungulate: Reeves' muntjac (*Muntiacus reevesi*). J Zool (Lond) 241:551–570

Chapman N, Brown WAB, Rothery P (2005) Assessing the age of Reeves' muntjac (*Muntiacus reevesi*) by scoring wear of the mandibular molars. J Zool (Lond) 267: 233–247

Chen Y-C, Wang Y (1994) Weather and seasonal effects on Muntjac (*Muntiacus reevesi micrurus*) barking. Proceedings of 3rd International Congress on the Biology of Deer, Edinburgh: 272

Chiang P-Y, Lin CC, Liao S-J, Hsieh L-J, Li S-Y, Chao M-C, Li Y-C (2004) Genetic analysis of two subspecies of Reeves' muntjac by karyotyping and satellite DNA analyses. Zool Stud 43(4):749–758

Clark M (1981) Mammal watching. Severn House Naturalist's Library, London

Claydon K, Claydon M, Harris S (1986) Estimating the number of muntjac deer (*Muntiacus reevesi*) in a commercial coniferous forest. Bull Br Ecol Soc 17:185–189

Cooke AS (1994) Colonisation by muntjac deer *Muntiacus reevesi* and their impact on vegetation. In: Massey ME, Welch RC (eds) Monks Wood Nature Reserve: the experience of 40 years 1953–93. English Nature, Peterborough

Cooke AS (1995) Muntjac damage in woodland. Enact 3(3):12–14

Cooke AS (1997) Effects of grazing by muntjac (*Muntiacus reevesi*) on bluebells (*Hyacinthoides non--scripta*) and a field technique for assessing feeding activity. J Zool (Lond) 242:365–369

Cooke AS (1998) Survival and regrowth performance of coppiced ash (*Fraxinus excelsior*) in relation to browsing damage by muntjac deer (*Muntiacus reevesi*). Q J For 92:286–290

Cooke AS (2001) Information on muntjac from studying ivy (*Hedera helix*). Deer 11(9):498–500

Cooke AS (2004) Muntjac and conservation woodlands. In: Quine CP, Shore RF, Trout RC (eds) Managing woodlands and their mammals. Proc. Joint Mammal society/Forestry Commission symposium. Forestry Commission, Edinburgh, pp 65–69

Cooke AS (2006) Monitoring Muntjac Deer and their impacts in Monks Wood National Nature Reserve. English Nature Research Report 681. Peterborough; English Nature

Cooke AS (2014) Outfoxing the muntjac. Deer 17(2):30–35

Cooke AS (2019) Muntjac and water deer: natural history, environmental impact and management. Pelagic Press, Exeter. 374 pages

Cooke AS (2021) Colonisation, impacts in conservation woodland and management of Reeves' muntjac (*Muntiacus reevesi*) in an English county. Eur J Wildl Res 67:35

Cooke AS, Farrell L (2001) Impact of muntjac deer (*Muntiacus reevesi*) at Monks Wood Nature Reserve, Cambridgeshire, eastern England. Forestry 74:241–250

Cooke AS, Lakhani KH (1996) Damage to coppice regrowth by muntjac deer *Muntiacus reevesi* and protection with electric fencing. Biol Conserv 75:231–238

Cooke AS, Green P, Chapman NG (1996) Mortality in a feral population on muntjac *Muntiacus reevesi* in England. Acta Theriol 41:277–286

Cooper JE, Harris S, Forbes A, Chapman NG, Chapman DI (1986) A comparison of xylazine and methohexitone for the chemical immobilization of Reeves' muntjac (*Muntiacus reevesi*). Br Vet J 142:350–357

Croitor R (2018) Plio-Pleistocene deer of western palearctic: taxonomy, systematics, phylogeny. Institute of Zoology of the Academy of Sciences of Moldova, Chisinau

Czyzewska T (1968) Deer from Weze and their relationship with the Pliocene and recent Eurasiatic Cervidae. Acta Palaeontologica Polonica X111(4):537–604

Dansie O (1970) Muntjac. British Deer Society, 22 pages

Dansie O, Williams J (1973) Paraurethral glands in Reeves' muntjac deer, *Muntiacus reevesi*. J Zool (Lond) 171:469–471

Delahay RJ, De Leeuw ANS, Claridge M, Harris A, Cheeseman CL, Barlow AM, Millar MF, Clifton-Hadley RS (2001) First report of *Mycobacterium bovis* in a muntjac deer. Vet Rec 149:95–96

Delahay RJ, Smith GC, Barlow AM, Walker N, Harris A, Clifton-Hadley RS, Cheeseman CL (2005) Bovine tuberculosis infection in wild mammals in South-West England: a survey of prevalence and semi-quantitative assessment of the relative risks to cattle. Vet J 173: 287–301

Diaz A, Burton R (1996) The impact of predation by Muntjac deer *Muntiacus reevesi* on sexual reproduction of the woodland herb *Arum maculatum*. Deer 10(1): 14–19

Dick JTA (2017) Research update from the emerald isle. Deer 18(3):12–13

Dick JTA, Freeman M, Provan J, Reid N (2009) Muntjac knowledge transfer: ecology of introduced muntjac and appraisal of control procedures. *Quercus* project QU08–11. N.Ireland Environment Agency

Dick JTA, Freeman M, Provan J, Reid N (2010) First record of free-living Reeves' muntjac deer (*Muntiacus reevesi*) in Northern Ireland. Irish Nat J 31(2):152

Dolman P, Waeber K (2008) Ecosystem and competition impacts of introduced deer. Wildl Res 35:202–214

Downing G (2014) Stalking Muntjac: a complete guide. Quiller, Shrewsbury, p 192

Dubost G (1970) L'organisation spatiale et sociale de *Muntiacus reevesi* en semi-liberté. Extrait de Mammalia 34:331–355

Dubost G (1971) Observations éthologiques sûr le Muntjak (*Muntiacus muntjak* Zimmermann 1780 et *M. reevesi* Ogilby 1839) en captivité et semi-liberté. Z Tierpsychol 28:387–427

Duscher GG, Battisti E, Hodzic A, Waeber K, Steinbach P, Stubbe M, Heddergott M (2020) First detection and molecular identification of *Anaplasma phagocytophilum* in an introduced population of Reeves' muntjac (*Muntiacus reevesi*) in United Kingdom. Mol Cell Probes

Dutton CJ, Duncan M, Price HI (2002) Hydromyelia in a Reeves' muntjac (*Muntiacus reevesi*). J Zoo Wildl Med 33:256–262

Ellerman JR, Morrison-Scott TCS (1951) Checklist of Palaearctic and Indian mammals 1758 to 1946. British Museum. p 336

Flower SS (1929) List of the Vertebrated animals exhibited in the gardens of the Zoological Gardens of London 1828–1927. In: Mammals, vol 1. The Zoological Society of London, London

Flowerdew JR, Ellwood SA (2001) Impacts of woodland deer on small mammal ecology. Forestry 74(3):277–287

Forde P (1989) Comparative ecology of muntjac *Muntiacus reevesi* and roe deer *Capreolus capreolus* in a commercial coniferous forest. Ph.D. thesis, University of Bristol

Frädrich H (1997) Zur Haltung und Zucht von Muntjaks im Zoo Berlin. Der Zoologische Garten 67:28–33

Freeman M, Hogg K (2020) A decade of muntjac research in Northern Ireland. Deer 19(5):18–19

Freeman MS, Beatty GE, Dick JTA, Reid N, Provan J (2016) The paradox of invasion: Reeves' muntjac deer invade the British Isles from a limited number of founding females. J Zool 298:54–63

Fuller RJ, Noble DG, Smith KW, Vanhinsbergh D (2005) Recent declines in populations of woodland birds in Britain. Br Birds 98:116–143

Geist V (1998) Deer of the world: their evolution, behaviour and ecology. USA Stackpole Books, Mechanicsburg
Gibbs EPJ, Herniman KAJ, Lawman MJP, Sellers RF (1975) Foot-and-mouth disease in British deer: transmission of virus to cattle, sheep and deer. The Veterinary Record: June 28
Gray AP (1971) Mammalian hybrids. Commonwealth Agricultural Bureaux, Slough
Green P, Chapman N (1993) Joint disease in wild deer. Deer 9(1):30–31
Groves C (2016) Systematics of the Artiodactyla of China in the 21st century. Zool Res 37(3):119–125
Groves CP, Grubb P (1990) Muntiacidae. In: Bubenik GA, Bubenik AB (eds) Horns, pronghorns and antlers. Springer, New York, pp 134–168
Gu Y-He, Xu L-H (1998) *Muntiacus reevesi jiangkouensis*-a new subspecies from Guizhou, China. Acta Zool Sin 44:271–276
Harding SP (1986) The social behaviour of Reeves' muntjac in southern England. Appl Anim Behav Sci 29:512
Harris S, Forde P (1986) The annual diet of muntjac (*Muntiacus reevesi*) in the King's Forest, Suffolk. Bull Br Ecol Soc 17:19–22
Heckeberg NS (2020) The systematics of the Cervidae: a total evidence approach. Peer J 8:e8114
Hemani MR (2003) The ecology of roe deer (*Capreolus capreolus*) and muntjac (*Muntiacus reevesi*) in a forested landscape in eastern England. Ph.D. thesis, University of East Anglia
Hemani MR, Dolman PM (2005) The disappearance of muntjac (*Muntiacus reevesi*) and roe deer (*Capreolus capreolus*) pellet groups in a pine forest of lowland England. Eur J Wildl Res 51:19–24
Hemani MR, Watkinson AR, Dolman PM (2004) Habitat selection by sympatric muntjac (*Muntiacus reevesi*) and roe deer (*Capreolus capreolus*) in a lowland commercial pine forest. For Ecol Manag 194:49–60
Hemani MR, Watkinson AR, Dolman PM (2005) Population densities and habitat associations of introduced muntjac *Muntiacus reevesi* and native roe deer *Capreolus capreolus* in lowland pine forest. For Ecol Manag 215:224–238
Hemani MR, Watkinson AR, Gill RMA, Dolman PM (2007) Estimating abundance of introduced Chinese muntjac *Muntiacus reevesi* and native roe deer *Capreolus capreolus* using portable thermal imaging equipment. Mammal Rev 37:246–254
Hershkovitz P (1982) Neotropical deer. Part 1. Fieldiana Zoology, N.S.no.11
Hofmann RR (1985) Digestive physiology of the deer: their morphological specialisation and adaptation. Bull R Soc New Zealand 22:393–407
Hofmann M (2018) Allgemeinverfügung für den Abschuss des Chinesischen Muntjaks, einer invasiven Art von unionsweiter Bedeutung. Staatsanzeiger für Rheinland Pfalz.no.29: 769–700
Hogg K, Freeman M, Reid N, Dick J. 2013. Detecting invasive muntjac deer (*Muntiacus reevesi*) at low densities post-establishment with camera traps. Pilot summary report. Queens University Belfast, Belfast
Hollander H (2013) Risico-analyse muntjak (*Muntiacus reevesi*). Rapport 2013.09. Zoogdiervereniging, Nijmegen, the Netherland
Hollander H (2015) Reeves' muntjac (*Muntiacus reevesi*) and sika deer (*Cervus nippon*) in the Netherlands. Lutra 58(1):45–50
Hollander H (2016) Verspreidingsonderzoek muntjak *Muntiacus reevesi* in Nederland. Rapport no.2016.05 Bureau van de Zoodiervereniging, Nijmegen (with English summary)
Hurel P,Terlin M,Derré F,Barboiron A (2019) Le muntjac de Reeves, un nouveau cervidé exotique envahissant nécessitant une reaction rapide. Faune Sauvage ONCFS N321
Jackson JE, Chapman DI, Dansie O (1977) A note on the food of muntjac deer (*Muntiacus reevesi*). J Zool (Lond) 183:546–548
Kay RNB (1987) Weights of salivary glands in some ruminant animals. J Zool (Lond) 211:431–436
Keeling JGM (1995) Ecological determinants of muntjac deer *Muntiacus reevesi* behaviour. Ph.D. Thesis, University of Bristol
Laurien U (1997) Untersuchungen zur sozialen Ontogenese des Formosa Muntjaks (*Muntiacus reevesi micurus* Sclater 1875) in Gehegehaltung. Dissertation, Universität Bielefeld
Lawson RE, Putman RJ, Fielding AH (2000) Individual signatures in scent gland secretions of Eurasian deer. J Zool (Lond) 251:399–410
Lawson RE, Putman RJ, Fielding AH (2001) Chemical communication in Eurasian deer (Cervidae): do individual odours also code for attributes? J Zool (Lond) 253:91–99
Li G-G, Zhang M-X, Swa K, Maung K-W, Quan R-C (2017) Complete mitochondrial genome of the leaf muntjac (*Muntiacus putaoensis)* and phylogenetics of the genus *Muntiacus*. Zool Res 38(5):310–316
Lister A (2020) New results on the systematic of deer. Deer 19(7):38–41
Matthews F, Kubasiewicz LM, Gurnell J, Harrower CA, McDonald RA, Shore RF (2018) A review of the population and conservation status of British mammals. Technical summary. Natural England, Peterborough
McCullough DR, Pei KC-J, Wang Y (2000) Home range, activity patterns, and habitat relations of Reeves' muntjacs in Taiwan. J Wildl Manag 64(2):430–441
McKillen J, Hogg K, Lagan P, Ball C, Doherty RN, Collins L, Dick JTA (2017) Detection of a novel gammaherpesvirus (genus *Rhadinovirus*) in wild muntjac deer in Northern Ireland. Arch Virol 162:1737
Metzger MJ, Bunce RGH, Leemans R, Viner D (2008) Projected environmental shifts under climate change. European trends and regional impacts. Environ Conserv 35(1):64–75
Middleton HS (1975) Two pathological conditions in the Muntjac deer (*Muntiacus reevesi*). J Zool (Lond) 176: 271–274

Miura S (1984) Dominance hierarchy and space use pattern in male captive muntjacs, *Muntiacus reevesi*. J Ethol 2: 69–75

Moore N (2021) Invasive non-native species in Great Britain- policy and delivery, with specific reference to Reeves' muntjac. Eur J Wildl Res 67:36

Müller DWH, Lackey LB, Streich WJ, Hatt J-M, Clauss M (2010) Relevance of management and feeding regimens on life expectancy in captive deer. Am J Vet Res 71:275–280

Nehring S. Skowronek S (2017) Die invasiven gebietsfremden Arten der Unionsliste der Verordnung (EU) Nr.1143/2014. Bundesamt für Naturschutz

Nentwig W, Bacher S, Kumschick S, Pysek P, Vila M (2018) More than "100 worst" alien species in Europe. Biol Invasions 20:161–1621

Odend'hal S, Miller KV, Demaris S (1996) Preputial glands in Artiodactyla. J Mammal 77:417–421

Ohtaishi N, Gao Y (1990) A review of the distribution of all species of deer (Tragulidae, Moschidae and Cervidae) in China. Mammal Rev 20:125–144

Pei K (1996) Postnatal growth of the Formosan Reeves' Muntjac *Muntiacus reevesi micurus*. Zool Stud 35(2): 111–117

Pei K, Liu H-W (1994) Reproductive biology of male Formosan Reeves' muntjac (*Muntiacus reevesi micrurus*). J Zool (Lond) 233:293–306

Pei K, Taber RD, O'Gara WY (1995) Breeding cycle of the Formosan Reeves' muntjac (*Muntiacus reevesi micrurus*) in northern Taiwan, Republic of China. Mammalia 59:223–228

Pfeiffer C (1993) Vergleichend-morphologische Untersuchungen am Verdauungstrakt des Chinesischen Muntjak (*Muntiacus reevesi*) einschließlich saisonaler Veränderungen an der Vormagenschleimhaut. Dissertation Justus-Liebig- Universität, Gießen

Pollard E, Cooke AS (1994) Impact of muntjac deer *Muntiacus reevesi* on egg-laying sites of the white admiral butterfly *Ladoga camilla* in a Cambridgeshire wood. Biol Conserv 70:189–191

Putman R (1996a) Deer management on National Nature Reserves: problems and practices. English Nature research reports no. 173. English Nature, Peterborough

Putman RJ (1996b) Competition and resource partitioning in temperate ungulate assemblies. Chapman and Hall, London

Putman RJ, Moore NP (1998) Impact of deer in lowland Britain on agriculture, forestry and conservation habitats. Mammal Rev 28:141–163

Rackham O (1975) Hayley wood: its history and ecology. Cambridgeshire and Isle of Ely Naturalists' Trust, Cambridge

Rehorek SJ, Hillenius WJ, Kennaugh J, Chapman NG (2005) The gland and the sac-the preorbital apparatus of muntjacs. In: Mason RT, Le Master MP, Muller-Schwarz D (eds) Chemical signals in vertebrates, vol 10, pp 152–158. Springer Science and Business, New York

Rehorek SJ, Hillenius WJ, Sanjur J, Chapman NG (2007) One gland, two lobes: organogenesis of the "Harderian" and "nictitans" glands of the Chinese muntjac (*Muntiacus reevesi*) and fallow deer (*Dama dama*). Ann Anat 189:434–446

Richard L (1983) Note sûr un crane de petit cervidé Asiatique tue dans L'Eure. Actes du Museum de Rouen 46–57. ISSN 0180-9784

Seidel B (1993) Keeping Chinese deer in captivity- a veterinary review. In: Ohtaishi N, Sheng H-I (eds) Deer in China. Elsevier Science Publishers, Amsterdam, pp 390–400

Setsuda A, Kato E, Sakaguchi S, Tamemasa S (2020) *Chabaudstrongylus ninhae* and *Oesophagostomum muntiacum* in feral alien Reeves's muntjacs on Izu-Oshima Island. Jpn J Helminthol

Sheng H (1991) The deer in China. Reeves' muntjac. 126–148 (in Chinese and English). East China Normal University Press, Shanghai

Sheng H, Xu DN (1990) A preliminary study of the Reeves' muntjac (*Muntiacus reevesi*) of Jiangxi, Zhejiang and Anhui Provinces. J East China Normal Univ: Mammalian Ecology Supplement, pages 48 and 53 in English

Shi L, Pathak S (1981) Gametogenesis in a male Indian muntjac x Chinese muntjac hybrid. Cytogenet Cell Genet 30:152–156

Shi L, Ye Y, Duan X (1980) Comparative cytogenetic studies on the red muntjac, Chinese muntjac, and their F_1 hybrids. Cytogenet Cell Genet 26:22–27

Smart JCR, Ward A, White PCL (2004) Monitoring woodland deer populations in the U.K: an imprecise science. Mammal Rev 34(1):99–114

Smith-Jones C (2004) Muntjac: managing an alien species. Coch-Y-Bonddu Books, Machynlleth, p 216

Smith-Jones C (2017) The 2016 BDS deer distribution survey. Deer 18(5):15–18

Smith-Jones C (2022) A guide to the deer of the world. Quiller publishing, Stroud

Southern HN (1964) The handbook of British mammals. Blackwell Scientific Publications, Oxford

Stadler SG, Hendrichs H (1987) Zum Lautrepertoire eines ursprünglichen Cerviden, *Muntiacus reevesi micrurus* (Sclater, 1875). Sonderdruck aus Z f Säugetierkunde 52:372–382

Tabor RCC (1993) Control of deer in a managed coppice. Q J For 87:308–313

Tabor RCC (1999) The effects of muntjac deer, *Muntiacus reevesi,* and fallow deer, *Dama dama*, on the oxslip, *Primula elatior*. Deer 11:14–19

The Deer Initiative. http://www.thedeerinitiative.co.uk

Timmins J, Chan B (2016) *Muntiacus reevesi*. IUCB Red List of Threatened Species. 2016

Tokyo Municipal Office (2017) Control plan of feral alien Reeves's muntjac on Oshima Island, Tokyo in 2017

Waeber K, Spencer J, Dolman PM (2013) Achieving landscape-scale deer management for biodiversity conservation: the need to consider sources and sinks. J Wildl Manag 77:726–736

Wallace RS, Bush M, Montali RJ (1987) Deaths from exertional myopathy at the National Zoological Park from 1975 to 1985. J Wildl Dis 23:454–462

Wang W, Lan H (2000) Rapid and parallel chromosomal number reductions in muntjac deer inferred from mitochondrial DNA phylogeny. Mol Biol Evol 17(9):1326–1333

Ward A (2005) Expanding ranges of wild and feral deer in Great Britain. Mammal Rev 35(2):165–173

Ward AI, Smith GC (2012) Predicting the status of wild deer as hosts of *Mycobacterium bovis* infection in England. Eur J Wildl Res 58:127–135

Ward AI, Etherington T, Ewald J (2008) Five years of change. Deer 14(8):17–20

Ward AI, Richardson S, Mergeay J (2021) Reeves' muntjac populations continue to grow and spread across Great Britain and are invading continental Europe. Eur J Wildl Res 67:33

Whitehead GK (1993) The Whitehead encyclopedia of deer. Swan Hill Press, Shrewsbury

Wray S (1992) The ecology and management of European hares (*Lepus europeus*) in a commercial coniferous forest. Ph.D. thesis, University of Bristol

Wray S (1994) Competition between muntjac and other herbivores in a commercial coniferous Forest. Deer 9(4):237–242

Yahner RH (1978) Some features of mother-young relationships in Reeves muntjac (*Muntiacus reevesi*). Appl Anim Ethol 4:379–388

Yahner RH (1979) Temporal patterns in male mating behaviour of captive Reeves muntjac (*Muntiacus reevesi*). J Mammal 60:560–567

Yahner RH (1980a) Barking in a primitive ungulate *Muntiacus reevesi*: function and adaptiveness. Am Nat 116:157–177

Yahner RH (1980b) Activity patterns of captive Reeves' muntjacs. J Mammal 61(2):368–371

Young BA (1976) Some observation on the ultrastructure of the thyroid of certain Cervidae. J Zool (Lond) 180: 175–183

Young BA, Chaplin RE (1975) Some observations on the ultrastructure of the adenohypophysis of certain Cervidae. J Zool (Lond) 175:493–508

Further Reading

Ohtaishi N, Sheng H-I (1993) Deer of China: biology and management. In: Proceedings of a conference in Shanghai 1992. Elsevier, Amsterdam

Wemmer CM, ed. (1987) Biology and Management of the Cervidae. Smithsonian Institution Press, Washington, DC, p 517

Red Deer *Cervus elaphus* Linnaeus, 1758

3

Stefano Mattioli, Frank E. Zachos, Luca Rossi, Adrian M. Lister, and Luca Corlatti

Contents

S. Mattioli (✉)
Research Unit of Behavioural Ecology, Ethology and Wildlife Management, Department of Life Sciences, University of Siena, Siena, Italy
e-mail: steven.mattioli@libero.it

F. E. Zachos
Natural History Museum Vienna, Mammal Collection, Vienna, Austria

Department of Evolutionary Biology, University of Vienna, Vienna, Austria

Department of Genetics, University of the Free State, Bloemfontein, South Africa
e-mail: frank.zachos@nhm-wien.ac.at

L. Rossi
Department of Veterinary Sciences, University of Turin, Turin, Italy
e-mail: luca.rossi@unito.it

A. M. Lister
Natural History Museum, London, UK
e-mail: a.lister@nhm.ac.uk

L. Corlatti
ERSAF - Stelvio National Park, Office of Scientific Research and Conservation, Bormio (SO), Italy

Chair of Wildlife Ecology and Management, University of Freiburg, Freiburg, Germany
e-mail: luca.corlatti@ersaf.lombardia.it; luca.corlatti@wildlife.uni-freiburg.de

© Springer Nature Switzerland AG 2022

L. Corlatti, F. E. Zachos (eds.), *Terrestrial Cetartiodactyla*, Handbook of the Mammals of Europe,
https://doi.org/10.1007/978-3-030-24475-0_19

Common Names

English	Red deer
German	Rothirsch, Edelhirsch
French	Cerf élaphe
Spanish	Ciervo común, ciervo rojo
Italian	Cervo rosso, cervo nobile
Russian	Благородный олень

Now that the Central Asian red deer have been classified as a distinct species (see section "Taxonomy and Systematics"), the name red deer is, on a global scale, somewhat ambiguous. Therefore, the species dealt with here is sometimes called Western or European red deer, but in a European context, red deer will suffice.

Taxonomy and Systematics

Eurasian red deer (Fig. 1), Asian sika, and East Asian/North American wapiti or elk are very closely related taxa that are able to interbreed freely, which they do in captivity or where they co-occur naturally or through human introductions (see ▶ Chap. 4, "Sika *Cervus nippon* Temminck, 1838," in this volume). There is as yet no final consensus on species boundaries and taxonomy. For a long time, red deer and wapiti were combined into *Cervus elaphus* with various subspecies, to the exclusion of sika. A general problem, apart from interbreeding, was that in many analyses based on mitochondrial DNA sika and wapiti are sister taxa to the exclusion of

Fig. 1 Red deer herd: a stag with females and calves (photograph by Andrea Dal Pian)

the Eurasian red deer (for recent studies, see Lorenzini and Garofalo (2015), Doan et al. (2018) and Meiri et al. (2018)). This is in clear contradiction to the morphology of the three taxa, and the first comprehensive nuclear genetic study has indeed found a sister group relationship of wapiti/elk and red deer to the exclusion of sika (Hu et al. 2019). There also seems to be a consensus today to treat wapiti/elk as a distinct species (*Cervus canadensis*), leaving the name red deer for the European, West and Central Asian taxa. Even this latter group shows evidence of comprising two divergent taxa that could be and have been classified as distinct species: Western red deer *Cervus elaphus*, (Fig. 1) and Central Asian red deer *Cervus hanglu*. The latter, according to the most recent genetic datasets (Lorenzini and Garofalo 2015; Meiri et al. 2018), comprises the subspecies (formerly classified under *C. elaphus*) *bactrianus*, *yarkandensis* and the critically endangered *hanglu*.

The position of *C. hanglu* relative to *C. elaphus* and *C. canadensis* also depends on the molecular marker system used. Mitochondrial DNA yields a sister group relationship between *C. hanglu* and *C. elaphus* to the exclusion of wapiti (Lorenzini and Garofalo 2015; Doan et al. 2018; Meiri et al. 2018), while nuclear markers favor *C. hanglu* as sister to *C. canadensis* to the exclusion of *C. elaphus* (Hu et al. 2019).

For Europe, this does not change much – all red deer still belong to *Cervus elaphus*. The intraspecific taxonomy of the red deer in Europe, as would be expected for an animal as iconic and widespread, has a long history, and a large number of subspecies have been described (see Grubb (2005) for a detailed list). A thorough revision based on all available evidence and, importantly, comprehensive quantitative morphological data is long overdue. For Europe as defined here the following subspecies are often listed (for details and descriptions of their morphology, ecology, and behavior, see Dolan (1988), Geist (1998) and O'Gara (2002)):

C. e. elaphus Linnaeus, 1758 – Southern and Central Sweden

C. e. atlanticus Lönnberg, 1906 – Southwestern Norway

C. e. scoticus Lönnberg, 1906 – British Isles

C. e. hippelaphus Erxleben, 1777 – continental Europe
C. e. hispanicus Hilzheimer, 1909 – Iberian Peninsula
C. e. italicus Zachos et al., 2014 – Nature Reserve "Bosco della Mesola," Ferrara province of Italy, at the southern edge of the Po river delta
C. e. corsicanus Erxleben, 1777 – Tyrrhenian islands (Sardinia and Corsica)
C. e. maral Gray, 1850 – Anatolia, Caucasus, Northwestern Iran

Depending on taxonomic predilections, sometimes *C. e. elaphus* is taken to include all European red deer except for the subspecies *corsicanus*, *italicus*, and *maral*; or the *hippelaphus* subspecies only includes Western European mainland deer, while the Eastern populations are classified as *C. e. pannoniensis* Banwell, 1997 (sometimes called *montanus*, which is a preoccupied name); or *scoticus* is included within *hippelaphus*; and so on. The Barbary stag of Algeria and Tunisia, *C. e. barbarus* Bennett, 1833, occurs outside Europe but is closely related to *C. e. corsicanus* (see "Genetics" section).

Traditionally, subspecies classification was based on size, antlers, and pelage characteristics (including the rump patch and the neck mane) but recently bioacoustic analyses have also found differences between subspecies, particularly confirming the phenotypic distinctness of the Tyrrhenian and the Mesola red deer (Kidjo et al. (2008), Della Libera et al. (2015)) as well as the Iberian red deer (Volodin et al. 2013; Passilongo et al. 2013) that also shows a unique behavior of tongue protrusion while roaring (Frey et al. 2012). At least some of the populations classified as subspecies have also been shown to be genetically distinct (see "Genetics").

Paleontology

European red deer have a highly abundant and widespread fossil record, from their first occurrences around 800 kya or possibly as early as one million years ago (around the Early to Middle Pleistocene boundary) to the Holocene. The genus *Cervus* is believed to derive from *Pliocervus* of the Pliocene (*c.* 5.0–2.6 mya), the earliest representatives of the *Cervus* genus dating toward the end of that interval (Vislobokova 2008). The direct ancestor or sister group of the *Cervus elaphus/canadensis* clade is uncertain, but several species of the Late Pliocene to Early Pleistocene (3–2 mya) have been proposed, including *C. nestii,* well known from the Early Pleistocene of Italy and Britain, a fallow-deer-sized species with four-point antlers resembling those of small modern red deer subspecies such as *C. e. barbarus* (Croitor 2018). The earliest representatives of the *C. elaphus/canadensis* clade, with their characteristic pair of lower tines (brow and bez), occurred widely in Europe between 800 and 500 kya and have been named *C. e. acoronatus* as they lack the antler "crown" and resemble extant Bactrian deer in possessing a simple transverse terminal fork. They are replaced in Europe by coronate red deer from 400 kya onward (Lister 1986).

Throughout the sequence of Middle and Late Pleistocene interglacials in Europe, red deer were a consistently widespread and abundant element of the large mammal fauna. The species' range at different times extended from the Iberian Peninsula to the western flanks of the Urals (Meiri et al. 2018). In the intervening glacial phases, its range, like that of many temperate mammals, was reduced; this is best understood for the last glacial period (*c.* 100–12 kya) when the European range was largely restricted to the south, before expanding rapidly northward in the early Holocene (Sommer et al. 2008). Nonetheless, there are clear indications of more northerly populations, perhaps during brief warm periods (interstadials), represented, for example, by the deer fossils from Kent's Cavern (*c.* 40–30 kya) and Gough's Cave (*c.* 14 kya) in SW England. Niedziałkowska et al. (2021) suggest that winter cold was the main limiting factor to their distribution. Recent studies based on dental wear have shown the broad dietary flexibility of red deer through its Pleistocene record, with mixed feeding as the dominant mode but populations varying from browse dominated to graze dominated at different times dependent largely on the vegetational environment (Rivals and Lister 2016). The species also shows very significant variation in mean body mass, from 150 kg in some interglacial populations to 300 kg in the last glaciation (Saarinen et al. 2016). Nonetheless, a remarkable recent discovery is that the European range of *C. elaphus* was invaded during the last glacial phase by Siberian wapiti (*C. canadensis*) extending its range westward at least as far as the Crimea and Romania (Doan et al. 2018; Meiri et al. 2018). There is as yet no evidence of hybridization between the forms and they were probably separated ecologically and geographically, but more evidence is needed on this question.

Red deer, like other large mammal species in the Pleistocene, showed morphological divergence and often body size reduction when isolated on islands. On Jersey in the English Channel during the last interglacial (*c.* 120 kya), red deer dwarfed to approximately 20% of mainland body mass (Lister 1995). A somewhat reduced form on Sicily in the last glacial phase had remarkably long, gracile antlers, while on Malta size reduction went further, to individuals with estimated body mass of only 25 kg (d'Souza and Lister 2016).

Current Distribution

Red deer have a patchy but wide distribution in the Western Palearctic, occurring in most of Europe, North Africa, part of the Near East (Turkey), and Northern Iran (Caspian coast). In Levant red deer were exterminated in Middle Age (Uerpmann 1987, Tsahar et al. 2009). In Europe, the species is present from Great Britain and Ireland to the Iberian Peninsula through Central Europe and Scandinavia (up to 65° N) to the Balkans, European Russia, Crimea, and the Caucasus (Fig. 2). It is present also in Corsica (where it went extinct in 1970 and was reintroduced in 1985) and Sardinia. Red deer are extinct in Albania and are not present in Iceland and Finland.

Between the sixteenth and early twentieth centuries red deer declined or disappeared from many parts of their historical range due to overhunting, deforestation, and competition with livestock. Especially since the second half of the twentieth century red deer have recovered in most of the continent due to law enforcement, changes in land use, reduction of predators and domestic competitors, reintroductions, restocking operations, escapes from farms, and recolonization from adjacent countries. In the early 1980s, the total population size in Europe (USSR countries excluded) was estimated at around 1.1 million individuals (Gill 1990; Burbaitė and Csányi 2010), with a yearly total harvest of at least 0.275 million. In 2005 the global population estimate grew to some 2.4 million individuals, with an annual hunting bag of about 0.43 million (Belarus, Ukraine, Moldova, and W Russia excluded; Burbaitė and Csányi 2010; Apollonio et al. 2010; Deinet et al. 2013). The yearly total harvest grew to at least 0.73 million in 2017 (Linnell et al. 2020). The highest numbers occur in Spain, the UK, Germany, and Austria.

Densities typically range from 2–10 deer/km^2, but may reach values of 10–20 deer/km^2 in Scottish moorlands (Clutton-Brock and Albon 1989; Staines et al. 2008), 25–30 deer/km^2 in Sardinian maquis scrub (Lovari et al. 2007), and as high as 30 deer/km^2 on average in hunting estates in Mediterranean habitats of the Iberian Peninsula (Torres-Porras et al. 2014). In protected areas of the Central Italian Alps, minimum densities are recorded in summer, when animals are more dispersed, while densities up to 31 deer/km^2 may occur in winter-early spring, when the deer are concentrated at lower elevations in the valley floors (Corlatti et al. 2016). Though higher densities are locally possible, higher values can be typically reached only with intensive supplementary feeding.

Description

General Appearance

European red deer are large-sized deer, with a slender build and long legs (Fig. 1). The head is large and long; the eyes are relatively large; the ears and the neck are long. The tail is relatively short. Red deer have been classified as a saltatorial

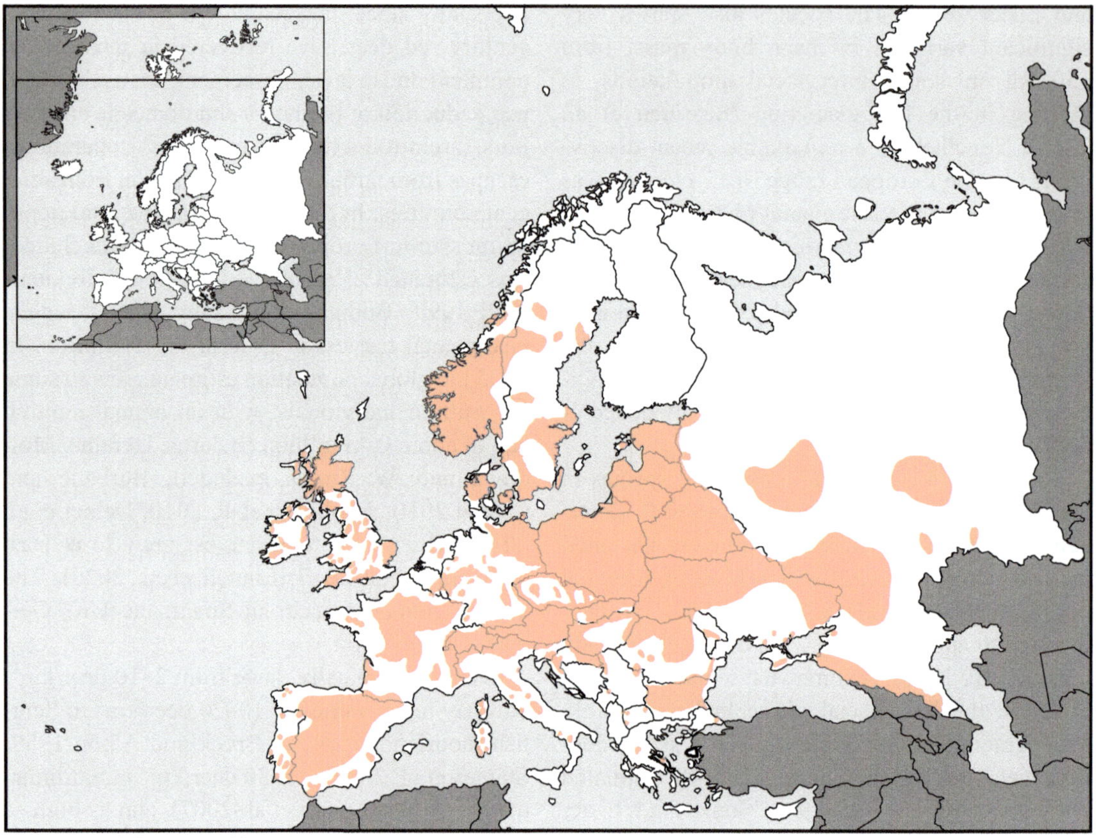

Map template: © Getty Images/iStockphoto

Fig. 2 Distribution range of red deer in Europe, based on data from the authors. (Map template: © Copyright Getty Images/iStockphoto)

and cursorial runner typical of wooded grasslands, an endurance runner only slightly inferior to the wapiti of the open and wooded grassland (Geist 1998). Sexual size dimorphism is pronounced in adults, with males quite markedly heavier and larger than females (Fig. 1).

The antlers are long, cylindrical, and well branched, typically with two basal points, the hook-like brow tine and the bez tine, a middle point called trez tine, and a terminal cup-like crown or royal of three or more tines. Between the trez tine and the crown an additional point can be present, the dagger (*Wolfssprosse* in German). It is also known - less properly - as surroyal, an old and slightly ambiguous term often used to indicate points of the terminal crown or simply an above-standard trophy.

Skull

The skull is elongated with a narrow, well-developed rostral region and rudimentary, pear-shaped upper canines. The preorbital cavity is large, the short premaxillary bone in the upper part extends caudally, the nasal ends far from the interorbital region and articulates with the maxillary without indentation (Bützler 1986).

Pelage

The lower lip and chin are whitish, and a black stripe (labial spot) runs down from the back of the mouth. The creamy rump patch is bordered on either side by a band of dark brown hair. There are some regional variations in color patterns of the rump patch: in the Swedish red deer it is yellowish without dark lateral border, in the Norwegian one it is entirely framed by a dark brown marking. The summer coat is typically reddish brown, with a whitish belly. The winter coat is prevailingly grayish brown, and adult males have a thick neck mane and a black-stained belly. The neck mane is less developed or even absent in Iberian and Pannonian stags. Moults occur in late April to May and in September–October. Newborn calves have a brown coat with many whitish spots which disappear at the age of two–three months. In some populations, for example the Mesola red deer (*C. e. italicus*), adults also show yellowish spots in their summer coat. Very pale to white coat color is occasionally observed and has been associated with a mutation in the tyrosinase gene (Reiner et al. 2020).

Glands

Metatarsal glands are present on the external side of the hind legs. Interdigital glands are rudimentary or absent. Antorbital or preorbital glands are present below the eyes. The ventral side of the tail has two modified subcaudal sweat glands. Antorbital glands are markedly open in adult stags at the peak of the rut (Bützler 1974), while in calves their secretion seems to be involved in establishing bonds with the mother (Hatlapa 1977). Open antorbital glands in calves signal stress or hunger (Bartoševa et al. 2012). Scent secretions of antorbital and metatarsal glands seem to convey information on sex, age, and individual (Lawson et al. 2000, 2001). The black belly spot or dark ventral patch of adult stags in autumn is an important scent area, with many volatile compounds originated by secretion of sebaceous glands and by the oxidation of urine (Martín et al. 2014). Urine sprayed during the rut contains very high concentrations of a catecholamine, and one of its metabolites, oxidized by air, forms dark pigments staining the belly pelage (Galván et al. 2019). This ventral patch has been shown to be a sexual signal during the rut (de la Peña et al. 2021).

Antlers

Pedicles generally begin to develop in male calves by the age of 8–10 months, in January–March. The start of pedicle growth is related to the attainment of a threshold body mass of *c.* 50–60 kg (Fennessy and Suttie 1985; Gaspar-López et al. 2008). In low-performance populations of the Scottish Highlands, Mesola Wood and Sardinia (Italy), the pedicle may sometimes not be visible until 14–20 months of age (cf. Mattioli 1993). Typically, at 9–12 months of age the first antlers start to grow, mostly simple, unbranched spikes of 20–40 cm, which are cleaned off velvet in September–October and cast from late April to June. The initiation of the first antler set is associated with reaching a threshold body mass of some 60–70 kg and a pedicle length of about 5–6 cm (Fennessy and Suttie 1985; Gaspar-López et al. 2008). In scarcely productive habitats some stags do not develop their first antler set until 3 or 4 years of age. Stags with very small rudimentary pedicles may not develop antlers for all their life (these animals are known as *hummels* in the Scottish Highlands, and as *Mönche*, literally “monks,” in German-speaking countries). Antlerless and small-antlered males in Spain have been found to be associated with high levels of homozygosity (Pérez-González et al. 2010).

When close to their highest potential, antlers of Western European red deer tend to be slightly shorter than those of Carpathian and Pannonian ones, but with a richer distal branching; Eastern European red deer tend to have longer basal tines (brow and bez) and a well-developed dagger (Geist 1998). In Mesola red deer antlers have a simplified conformation; a crown is rare, three-tined, cup-like, or fan-like (Mattioli and Ferretti 2014). Tyrrhenian red deer of Corsica and Sardinia also show simplified antlers, with a tendency to palmation (Caboni et al. 2006). Adult stags usually cast their antlers from late February to early April. After one week the new set begins to grow, generally reaching the final length in late June or early July. Velvet shedding occurs in July–August, 130–160 days after the previous casting. Older stags tend to cast antlers and clean velvet slightly earlier (Raesfeld and

Reulecke 1988). Increase in length follows a typical sigmoid curve, with the most rapid increase (on average 0.67 cm per day) occurring between 28 and 112 days after the previous antler casting; most of the mineralization occurs between days 91 and 112 (Muir et al. 1987). Antler production is costly in terms of tooth wear and expected longevity (Carranza et al. 2008) and stags can modulate their investment in antlers according to the social environment, i.e., the proportion of rivals and potential mates (Carranza et al. 2020). Antler size and antler branching complexity are associated with testicle size and the speed of spermatozoa (Malo et al. 2005). Antlers are therefore a clear signal not only of fighting ability but also of fertility.

Dentition and Age Determination

The dental formula for milk teeth is 0.1.3.0/ 3.1.3.0; for the permanent dentition, it is 0.1.3.3/ 3.1.3.3. The permanent bicuspid third premolar typically erupts at around 25–30 months of age. Patterns of eruption and replacement can differ among individuals and populations (Wagenknecht 1984; Azorit et al. 2002a). For animals older than 2–2.5 years, age can be estimated by examining the tooth wear of the lower cheek teeth (Wagenknecht 1984; De Marinis 2015). At about 11–12 years of age, the infundibulum of the first molar tends to disappear. Males often show a higher wear rate than females (Loe et al. 2003; Carranza et al. 2004), which is explained not only by different strategies of tooth depletion (Carranza et al. 2008) but also because sexual selection favored body size more than teeth size (hence underprovisioning teeth) in males of sexually dimorphic ungulates (Carranza et al. 2004; Carranza and Pérez-Barberia 2007). This sex difference in tooth wear was not found on the Isle of Rum (cf. Nussey et al. 2007). Tooth wear may vary markedly among areas, with consequences on senescence and longevity (Pérez-Barbería et al. 2015). Exact age determination is possible by sectioning incisors or molars (Mitchell 1967; Azorit et al. 2004; Veiberg et al. 2020).

Measurements

Mean whole body mass of adult males in Europe ranges between *c.* 100 and 220 kg (pre- and post-rut figures pooled, see Table 1). The heaviest recorded stags can reach a mass of 350–425 kg in August to early September (pre-rut) and of 270–320 kg in late October (post-rut, or lean mass) (Szunyoghy 1963). On average, the rut-induced body mass loss is around 15–25%. Mean body mass of adult females normally ranges from 55–130 kg (Table 1), with upper extremes of 160–190 kg. Adult males are generally 40–70% heavier than adult females. Some three or four centuries ago red deer were apparently larger, with Central European stags reaching summer body masses of 450 kg, and possibly also 500 kg (Mager 1941; Geist 1998). The greatest body masses are now attained in the Carpathian Mountains and in the Pannonian lowland (Table 1). Eviscerated body mass (without thoracic and abdominal organs) with the head represents around 72–79% of live mass in adult stags and around 69–73% in adult hinds (Dzięciołowski 1970; Mattioli and De Marinis 2009).

Mean head-trunk length ranges from 180–210 cm in adult stags and from 160–195 cm in adult females. Mean height at the withers ranges from 105–140 cm in males and 90–120 cm in females; mean tail length is 13–15 cm and ear length 18–21 cm (cf. Langvatn 1986; Raesfeld and Reulecke 1988; Mattioli and Ferretti 2014). Mean mandible length ranges from 260–330 mm in males and 235–306 mm in females. Mean condylo-basal skull length ranges from 320–435 mm in males (with records of 450–466 mm; Szunyoghy 1963; Szaniawski 1966) and from 290–385 mm in females (with a record of 410 mm). Body mass and measurements of skeletal traits, however, greatly depend on habitat cover and climatic conditions experienced by deer in specific areas (cf. Vannini et al. 2021)

Mean antler length of adult European stags ranges from *c.* 60–100 cm, with the longest recorded antlers reaching 130–144 cm (S. Csányi, pers. comm.) (Table 2). Antler mass tends to be related to body mass (Huxley 1931). Antler

Table 1 Examples of mean whole body mass of adult male (≥5 years old) and female (≥2 years old) red deer across Europe (in kg) (for males mostly pre-rut and post-rut mass pooled, for females mostly winter mass)

	Males	Females	Source
Sardinia (Italy)	98.0[a]	52.2[a]	Mandas L., unpublished
Mesola Wood (Italy)	108.7	76.2	Mattioli and Ferretti 2014
Rum, Scotland (United Kingdom)	112.0	78.6	Mitchell et al. 1976
SW Spain	116.4	81.4	Carranza J., unpublished
E Alps, Tarvisio (Italy)	148.4	86.2	Colombi D., unpublished
S Norway[b]	149.7	107.9	Mysterud A., unpublished
W Alps, Piedmont (Italy)[c]	164.9	105.7	Meneguz P. G. and Tizzani P, unpublished
N Apennine, Prato (Italy)	168.1	100.2	Becciolini et al. 2016
Massif Central, Charente (France)	170.1	107.7	Gervais P. and Chaland C., unpublished
Baranja, Slavonia (Croatia)	176.4	101.1	Tucak 1997
N Apennine, Casentino (Italy)	182.0	102.2	Mattioli L., unpublished
Backa (Serbia)	182.0	105.8	Tucak et al. 1999
Poland, lowland	199.9	116.7	Dzięciołowski 1970
Poland, Carpathians[c]	206.7	123.7	Tomek 2002

[a]Mean figures from only seven animals for both sexes, possibly not fully representative of the population (see Beccu 1989, with a range of around 95–120 kg for stags and 50–75 kg for hinds)
[b]Winter mass only
[c]Estimated from eviscerated mass (conversion factors 1.29 for males and 1.38 for females; see Mattioli and De Marinis 2009)

Table 2 Mean antler length (in cm) in adult (≥5 years old) red deer across Europe. n: sample size

	Mean	n	Source
Sardinia (Italy)	63.0	157	Caboni et al. 2006
Rum, Scotland (United Kingdom)	68.4	1142	Peters L., unpublished
Mesola Wood (Italy)	70.2	271	Mattioli and Ferretti 2014
Massif Central, Charente (France)	72.0	146	Gervais P., unpublished
S Norway	72.5	62	Røskaft 1978
Sierra Morena (Spain)	75.6	333	Azorit et al. 2002b
SW Spain	80.6	1092	Carranza J, unpublished
Harz Mountains, Lower Saxony (Germany)	81.3	4388	Drechsler 1980
Western Alps, Piedmont (Italy)	82.7	2127	Meneguz P. G. and Tizzani P., unpubl. 2018
Hohenbucko, Brandenburg (Germany)	83.7	40	Neumann 1968
Masuria (Poland)	84.6	423	Zalewski D., unpublished
Carpathian mountains (Slovakia)	86.5	712	Šmehyl P., unpublished
N Apennine, Prato (Italy)	88.0	76	Becciolini et al. 2016
Carpathian mountains (Poland)	88.6	185	Wierzbowska 1999
Eastern Alps (Slovenia)	89.5	285	Hafner 2011
Bükk Mountains (Hungary)	91.3	4024	Csányi S., unpublished
N Apennine, Pistoia and Bologna (Italy)	92.0	171	Mattioli 1996
Baranja (Croatia)	96.8	1595	Bečejac et al. 1984
Baranya (Hungary)	97.5	4959	Csányi S., unpublished
Baranja (Croatia)	98.2	382	Degmečić and Florijančić 2014

investment is generally high, with a mean production of around 3.6–4.8 kg of bony tissue for adult males with a mass of 165–185 kg after the rut (corresponding to a mass 4.5–6.2 kg of both antlers and the whole skull combined). Net antler mass (mass of both antlers without the skull) of medium-sized stags represents on average 2–2.6% of post-rut body mass (cf. Mattioli et al. 2021). The heaviest antlers of top trophy stags can reach 16–18 kg (Botev 1990, S. Csányi, pers. comm.), representing 5–6% of the body mass. From the sixteenth and early eighteenth centuries, antlers weighing 18–19.5 kg are known (Geist 1998). In contrast, adult males of maintenance ecotypes from Sardinia and the Scottish moorlands on average allocate only 1–1.2% of their body mass to antlers (Mattioli et al. 2021). Mean number of tines per antler pair ranges from 6–14 (cf. Mattioli and Ferretti 2014), with records of 32–35 under natural conditions (Mattioli 1996, S. Csányi, pers. comm.). Selective breeding and supplementary feeding including minerals in semi-captivity can strongly affect antler branching, with records of 47–50 tines in total (Warnham Park, the UK).

Physiology

Red deer exhibit consistent seasonal fluctuations in reproduction, metabolic rate, appetite cycles, fat deposition and depletion cycles, somatic growth, antler growth, and moult, regulated by endogenous signals related to the photoperiod (cf. Arnold 2020). In spring and summer, food intake increases in both sexes. Functionally, this allows males to accumulate fat resources before the rut and females to support late pregnancy and lactation. The peak in body mass is attained by stags in August to early September and by hinds in October–November (Mitchell et al. 1976). During the rut, adult males strongly reduce their food intake (hypophagia). In winter the energy requirements of both sexes decrease, thanks to the reduction of locomotor activity and of the endogenous heat production. The subcutaneous temperature of the peripheral parts of the body decreases during the night and in the early morning in winter ("nocturnal hypometabolism") (Arnold et al. 2004). Mean heart rate declines from 50–70 beats/min in the warm season to 40 beats in the cold season (Turbill et al. 2011; Arnold 2020). With the winter decline in food intake, the gastrointestinal tract decreases in size but the extraction of nutrients becomes more efficient (Arnold et al. 2015). Despite a lower energy expenditure and a better nutrient assimilation, by the end of winter animals often show a severe depletion of fat reserves and generally reach their lowest levels of body mass (Mitchell et al. 1976). As a consequence of their faster growth, higher food requirements, and the effects of androgen catabolism, males appear more sensitive to resource limitation than females (Clutton-Brock et al. 1982). Hormones strongly control breeding cycles in both sexes (Lincoln 1971; García et al. 2002; Gaspar-López et al. 2010). Estimates of daily metabolizable energy requirements for maintenance in free-ranging animals in summer are around 940 kJ per kg of metabolic weight ($BW^{0.75}$) (Jiang and Hudson 1992), which means 42–47 MJ per day for males of 160–185 kg, and 27–32 MJ for nonlactating females of 90–110 kg. Lactating females approximately double their energy expenditure (cf. Mitchell et al. 1977; Clutton-Brock et al. 1982).

Genetics

The red deer is among the European mammal species that have been studied in greatest detail, and only a short summary of our knowledge pertaining to its phylogeography and genetic structuring at various geographic scales, as well as its genetic diversity and the anthropogenic impacts on its gene pool(s) is possible in the scope of a handbook chapter. We can therefore only cite a small fraction of the available studies and refer the reader to the original publications we draw information from and to the literature cited therein.

Chromosomes

The karyotype of the red deer is the one typical of cervine deer: $2n = 68$ with a fundamental number of NF = 70. Only two of the 66 autosomes are metacentric, all others are acrocentric as is the X chromosome; the small Y chromosome is submetacentric (Fontana and Rubini 1990).

Phylogeography

A large number of studies have addressed the distribution history of red deer by means of molecular approaches, from early studies based on allozymes and restriction fragment length polymorphisms (RFLPs) to microsatellites and DNA sequences, including several studies based on ancient DNA sequencing. Together with detailed analyses and dating of fossil and subfossil material and its spatiotemporal distribution these have yielded results of a resolution normally not available for other European mammal species. The most comprehensive genetic studies are based on mitochondrial DNA sequences (Ludt et al. 2004; Skog et al. 2009; Niedziałkowska et al. 2011; Meiri et al. 2013, 2018; Doan et al. 2018, 2022; Queirós et al. 2019), and the major publications on distribution history in Europe based on the fossil record are Sommer et al. (2008) and Niedziałkowska et al. (2021). The following summary is based on these references unless stated otherwise.

Extant European red deer belong to one of five mitochondrial lineages (haplogroups) whose geographic ranges are largely as follows:

A – Western and Northern Europe from Iberia through France and the British Isles to Scandinavia and Central Europe

B – with a few exceptions (in all likelihood going back to translocations), this haplogroup is confined to the Tyrrhenian islands (Corsica and Sardinia) and to the red deer in Tunisia and Algeria

C – Eastern Central, Eastern, and Southeastern Europe

D – Po delta region in Italy; also found in Southeastern Poland (Borowski et al. 2016).

E – Caucasus, Turkey, Iran and formerly also W Russia (see Baskin and Danell 2003, p. 53, who cite other authors according to which red deer in European Russia were all exterminated by 1750; so present populations in that part would go back to reintroductions, mainly from Central Europe)

While there is no precise match between haplogroups and the subspecies mentioned above, it is nonetheless obvious that at least some of the subspecies are reflected by the phylogeographic results. Lineages D and E correspond, at least partly, to *C. e. italicus* and *C. e. maral*, respectively, and lineage B comprises *C. e. corsicanus* and *C. e. barbarus*. The other European subspecies are mostly part of lineage A, but if the Eastern European populations are acknowledged as *C. e. pannoniensis*, there is a certain coextension between that subspecies and lineage C. The two main lineages A and C co-occur in admixed populations in Central and Eastern Central Europe (for example, in Austria, Czechia, and Poland), but to what extent that is a natural mitochondrial suture zone or due to anthropogenic reintroductions after local and regional extirpation is not always clear. Within lineage A, Iberian red deer *C. e. hispanicus* have been shown to include at least two genetically distinct lineages, one of them more related to the rest of lineage A possibly due to northward recolonization after the Last Glacial Maximum (LGM) (Carranza et al. 2016). Analyses of past red deer distribution yielded the classical southern glacial refugia in Europe (Iberia, Italy, and Balkans), but they also occurred further north during the LGM, in Southern France in the west but increasingly also more northerly the further east one goes. Lineage A contributed most to the recolonization of the European range (probably from Southern France and, *pace* Queirós et al. 2019, Iberia), possibly preventing a further expansion of lineage C.

Lineage B today is confined to Sardinia, Corsica (where Sardinian red deer were reintroduced after extinction on Corsica in the second half of the twentieth century), and North Africa. These three populations are known to go back to human introductions several millennia ago. Recent ancient DNA analyses have shown the historical Italian mainland population to be the geographic origin of lineage B (Doan et al. 2017). This population became extinct in historical times when all red deer except for the Mesola population were extirpated, making the Tyrrhenian islands and North Africa "genetic museums" of the lost Italian glacial refugial lineage. Like the refugial lineages of many other Italian taxa, it probably never managed to cross the Alpine barrier after the LGM, contributing to the well-known high occurrence of endemism in Italy.

It is important to note that this is a very short and superficial summary of red deer phylogeography and recolonization after the LGM and that the past distribution pattern of the various lineages was much more complex than it is today. For example, lineage D occurred as far west as England, and both D and E and an extinct lineage were present in Crimea, where today only A is found. Wapitis were distributed as far west as Romania, and Western red deer as far east as the Urals. Also, in light of a tradition of translocating red deer across Europe for centuries, the fact that there is a relatively clear-cut geographic structure in their mitochondrial genome is perhaps somewhat surprising. This is particularly true given that the available translocation records do not suggest a strong bias toward stags and that the nuclear genetic analysis based on biparental microsatellites confirms the overall mtDNA structure in European red deer (in particular with respect to haplogroups A, B, and C, Zachos et al. 2016).

Genetic Diversity and Population Structure

Population genetic studies have been performed on red deer throughout the continent, from the British Isles and Scandinavia in the north to the Mediterranean in the south and from Iberia and the British Isles in the west to European Russia in the east. Arguably, we know more about the population genetics of red deer than of most (if not all) European large mammals. Even more than for the taxonomy and phylogeography of the species, the reader should keep in mind that the following is a very short and superficial overview of our knowledge on red deer genetics.

One of the problems with these studies from the last almost 40 years is that the choice of molecular markers varies, from early allozyme studies to studies based on RFLPs and then DNA sequences (mostly mitochondrial), nuclear microsatellites, and, most recently, genome-wide approaches. Diversity values are therefore usually not directly comparable, although the relative diversity when comparing different populations often holds across different marker systems. Zachos and Hartl (2011) give an overview of studies (including diversity values) until 2010. The only continent-wide population genetic study yielding directly comparable diversity values (including estimates of effective population sizes N_e) based on highly variable microsatellite data is the one by Zachos et al. (2016). Expectedly, they found a wide range of genetic diversity across Europe, but the two heavily bottlenecked subspecies that only narrowly escaped extinction – the Tyrrhenian *C. e. corsicanus* and the Mesola red deer *C. e. italicus* – consistently showed the lowest values for heterozygosity and allelic richness (a measure of allelic diversity corrected for differences in sample size), followed by populations in Scandinavia (representing the Norwegian and Swedish subspecies), which are also known to have experienced bottlenecks (e.g., Haanes et al. 2011). Similar results were also found for estimates of effective population sizes, with Sardinia and Mesola exhibiting critically low values of between 2 and 8, populations in Sweden, Norway, and parts of Germany showing low values of between 20 and 50, and the rest of Europe having larger values. Populations in Central and Eastern Europe seem to harbor higher levels of genetic diversity, which may be due to the admixture of regional gene pools in the contact area of the glacial refugial lineages A and C, human introductions of allochthonous red deer, larger stretches of natural habitat and thus lower fragmentation, or a combination of these.

Given that in large parts of Europe red deer occur in human-dominated landscapes, anthropogenic impacts on local and regional gene pools have been a frequent target of population genetic analyses. Habitat fragmentation is one of the most important drivers of population structuring in red deer, and a number of studies have dealt with it at various levels and testing for effects of highways and human infrastructure in general (see, for example, Frantz et al. (2012), Edelhoff et al. (2020)). Often clear signals of anthropogenic isolation have been found, sometimes in combination with signs of inbreeding depression such as shortened lower jaws (brachygnathia inferior) as commonly found in a long-term isolated population in Northern Germany (Zachos et al. 2007). Translocations are another frequent impact on gene pools and have been addressed by a number of studies, usually when trying to estimate the genetic signature of these translocations or (re-) introductions or when trying to identify autochthonous gene pools (e.g., Niedziałkowska et al. 2012; Pérez-Espona et al. 2013; Frantz et al. 2017; de Jong et al. 2020). Overall, in many parts of Europe, signals of genetic introgression through translocations can be found, but this did not blur the underlying phylogeographic structure (see above), suggesting that most translocations occurred within the respective ranges of the major genetic lineages rather than among them. In a large-scale study on more than 2000 red deer from Belgium and other European countries as potential sources for translocations, Frantz et al. (2017) found that *c.* 3.7% of all Belgian red deer were nonautochthonous animals or their descendants. Whether that is a representative value also for other countries remains unknown due to the lack of comparable studies elsewhere.

Apart from the type of studies mentioned so far, red deer genetics has also been studied extensively in at least two additional contexts. The change in allele frequencies at loci putatively linked to those determining antler traits as an inadvertent by-product of selective trophy hunting regimes has been demonstrated in French red deer (Hartl et al. (1995) and references therein). The famous long-term study of red deer on the Isle of Rum in Scotland includes high-resolution population genetic studies through space and time in an evolutionary ecology research framework, addressing small-scale genetic structuring, inbreeding depression, and heterozygosity-fitness correlations in which heterozygosity was positively correlated with lifetime breeding success (Slate et al. (2000), Nussey et al. (2005), Huisman et al. (2016), references therein and other publications by the group).

Hybridization

Red deer freely hybridize with several other *Cervus* species. Since sika deer *C. nippon* have been introduced in many European countries, this is the main species of concern for the dilution of red deer gene pools, and in several countries hybridization between the two has been found, sometimes to an extent that threatens the integrity of the native species. For details on red deer/sika hybridization, see ▶ Chap. 4, "Sika *Cervus nippon* Temminck, 1838." Another species that has repeatedly been introduced to Europe, mostly with the aim of increasing body and trophy size in local populations, is the wapiti. However, none of the phylogeographic or population genetic studies have found wapiti sequences, and where potential wapiti introgression was analyzed in detail, the level of admixture was found to be zero or close to zero e.g., in Scotland, Pérez-Espona et al. (2013).

Life History

Growth

Newborn calves weigh on average 8.0–8.8% of their mother's body mass (cf. Landete-Castillejos et al. 2003; Ceacero et al. 2018): For example, a hind of 80 kg in the Scottish Highlands or in Mesola Wood would give birth to a calf of *c.* 6–7 kg, while a 100 kg hind in the Polish lowlands or in the Italian Apennine would produce calves of *c.* 8–9 kg. In Sardinia the mean birth mass is 4.2–4.5 kg (Beccu 1989). Male newborn calves are on average 5–10% heavier than females. Weaning usually occurs at 6–8 months of age.

As expected in a highly dimorphic species, body mass growth greatly differs in the two sexes. Females generally attain their full mass at 3–5 years of age, while males continue to grow up to 6–8 years. A clear decline in body mass is noticeable in females very late, sometimes not before 20 years of age, while in males it can be already observed when they are 12–14 years old (Mysterud et al. 2001b). Local variations in the pattern of body mass growth can nonetheless occur, possibly in response to site productivity (Putman et al. 2019). In males, antler size usually peaks at *c.* 10–12 years of age (Drechsler 1980; Raesfeld and Reulecke 1988). A decline in antler size (mass and number of tines) is often evident from the age of 14–15 years onward (Raesfeld and Reulecke 1988; Drechsler 1988), sometimes slightly earlier (Mysterud et al. 2005). In red deer populations living in poorly productive habitats, where antler investment is less pronounced, a clear regression of antlers can sometimes be scarcely noticeable (Lemaître et al. 2014). Habitat-specific patterns in the expression of antler investment may also occur at a fine, intrapopulation scale: In Apennine red deer, for example, antler investment of males increased in areas characterized by relatively higher habitat heterogeneity and greater availability of open habitats (Mattioli et al. 2021).

Reproduction

Physiological sexual maturity is generally reached in both sexes at an age of *c.* 16–18 months. Females are polyestrous, with a mean estrous cycle of 18–20 days (Guinness et al. 1971; García et al. 2003). Estrus lasts *c.* 12–24 h and tends to be synchronized within the same female group (Iason and Guinness 1985). Over 80% of all conceptions occur within a time frame of 3–4 weeks, mostly in September–October. Small yearlings and females older than 13 years tend to ovulate later (Langvatn et al. 2004; Mysterud et al. 2008). In conditions of low population density and favorable habitat, most females will ovulate for the first time as yearlings and give birth as two-year-olds (Kröning and Vorreyer 1957; Ratcliffe 1984; Borowik et al. 2016). With high population density and low habitat productivity (e.g., Scottish Highlands, Alps), the age of primiparity may be delayed by a year or more. Body mass and fat reserves generally have positive effects on hind pregnancy probability in Scottish populations (Albon et al. (1986), but see Borowik et al. (2016) for Central Europe). Consequently, puberty is normally attained only when the female reaches a threshold body mass of around 70–75% of the final adult mass (Albon et al. 1983, 1986; Langvatn et al. 1996; Bertouille and de Crombrugghe 2002; Bonenfant et al. 2002). Pregnancy rate among prime-aged females is normally higher than 95%. Mean gestation length is 231–236 days (Guinness et al. 1971; Clutton-Brock et al. 1982). Prenatal growth follows a common pattern (Valentinčić 1958; Wenham et al. 1986). At low densities and in high-quality habitats, adult females tend to breed (almost) every year. At high densities and in poor habitats, females may struggle to recover from pregnancy and lactation and may therefore reproduce every other year (Mitchell et al. 1977; Clutton-Brock et al. 1982). Pregnancy probability is largely affected by density dependence in British red deer (Putman et al. 1996), while no effects of density-independent factors such as winter and summer temperature were found in Poland (Borowik et al. 2016). In a population of the Italian Alps, however, higher spring-summer temperature had negative, indirect effects on the probability of being pregnant, mediated by decreasing values of kidney fat index, a proxy of body condition (Corlatti et al. 2018). In females, a decline in fecundity is apparent only from the age of 14 years onward (Langvatn et al. 2004; Nussey et al. 2009). On the Isle of Rum in Scotland, female lifetime reproductive success varied between 0–9 (mean = 5.03, variance = 9.09) and was mainly influenced by winter calf survival, which in turn was influenced by maternal condition (Clutton-Brock et al. 1988). Red deer usually give birth to a single calf; twins are rare, less than 1% (cf. Sadleir 1987).

Males only have realistic chances to mate for the first time when they are able to hold a harem, usually not earlier than at an age of 4–5 years, though harems can be attended by subadult males who try to mate with estrus females when the harem-holder is distracted. In adult males there is strong competition for the access to mating and high individual variation of breeding success, with males of 8–10 years having the highest mating probability. On the Isle of Rum, some stags failed to breed during their complete lifetime, while others fathered more than 30–80 offspring surviving to one year of age; and most of the harem-holders are able to successfully mate for no more than four seasons (Clutton-Brock et al. 1988; Pemberton et al. 1992). Breeding success in males is related to body size and condition, which affect their fighting ability, while in females it is mostly connected with access to high-quality resources (Clutton-Brock et al. 1982). Males show a rapid reduction in breeding success from 11–12 years of age onward (Clutton-Brock et al. 1982; Nussey et al. 2009). On the Isle of Rum, male lifetime reproductive success varied between 0–32 (mean = 5.41, variance = 41.9) and was largely influenced by the stags' fighting abilities, i.e., by their body size (Clutton-Brock et al. 1988). Though variation in reproductive success is greater in males than in females, sex-related differences tend to decrease when offspring survival is accounted for (Clutton-Brock et al. 1988).

Survival

Sex ratio at birth is close to parity, but at low densities heavier mothers produce more sons than daughters, while with increasing population densities and weather severity the proportion of males born each year declines (Kruuk et al. 1999; Borowik and Jędrzejewska 2017). In unmanaged populations, the overall sex ratio is slightly female-biased, around 1:1.2–1.5. Highly skewed sex ratios are mainly the consequence of high density, poorly planned hunting, or trophy-oriented poaching (cf. Langvatn and Loison 1999; Corlatti et al. 2019a). Higher mortality and emigration rates of males contribute to a bias in adult sex ratio of natural populations toward females. Information on mortality patterns is still fragmentary. For both sexes, the most critical stage is the juvenile phase (up to 2 years of age) when high density, harsh weather, or natural predators can lower survival rates (Clutton-Brock et al. 1982; Kamler et al. 2007). Adult survival rates are generally high, but with differences between sexes: As expected for a strongly polygynous species, males tend to be more sensitive than females to food shortage and intense competition for food. In Scottish deer, after reaching a plateau in middle-aged animals, at 9–10 years of age survival probabilities rapidly decline in males, while they tend to have a slower decline in females (Clutton-Brock et al. 1988; Catchpole et al. 2004). A similar pattern occurs in Alpine-dwelling deer in Italy (Corlatti et al. 2019a) (Fig. 3).

Red deer can normally live up to 16–18 years of age, with records of 22 and 26 years, respectively, for males and females (cf. Mysterud et al. 2001b). Very old females frequently have an emaciated head and a grizzled facial area. Very old males often have a long saggy dewlap and simplified antlers and they move keeping their head and neck almost horizontal. Longevity (and associated sex-specific difference) is variable among populations; for instance, longevity for males and females is, respectively, 14 versus 15 years in Scotland and 10 versus 14 years in Spain (Pérez-Barbería et al. 2015).

Habitat and Diet

Habitat Selection

Red deer are classically considered an ecotone species, associated with open woodlands, woodland edges, or the interface between woodland and grassland (Mitchell et al. 1977). They rarely occupy large tracts of dense and unbroken forest. Red deer, however, are surprisingly adaptable and may be associated with many climatic and vegetation types from semiarid Mediterranean areas, such as the maquis scrubs of Andalusia and

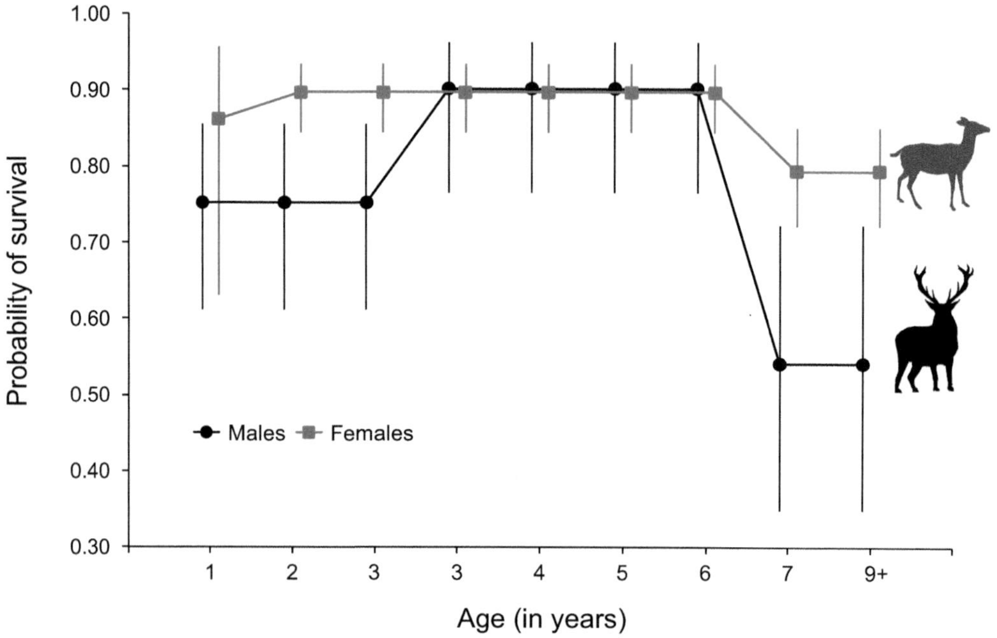

Fig. 3 Sex- and age-specific survival probabilities of 141 marked red deer in the Stelvio National Park in Italy between 2008 and 2017 (data from Corlatti et al. 2019a). Vertical lines represent 95% confidence intervals of the estimates

Sardinia, to the rich floodplains of the Danube, the summit meadows and high snowfall areas of the Alps and Scandinavia, the treeless moorlands of Scotland, and the steppes of Southern Ukraine. Red deer move easily in fresh snow of 20–30 cm, but prolonged snow cover can negatively affect calf mortality (Loison et al. 1999). Red deer tend to avoid terrains with snow cover of >50–60 cm, threshold levels that limit its northern distribution (cf. Baskin and Danell 2003). In the Alps, red deer can reach elevations of over 3000 m a.s.l. in summer and autumn (Büntgen et al. 2017), while in the Caucasus they have been observed at altitudes up to 5000 m a.s.l. (Danilkin 1999). In the last 50,000 years, a major limiting factor the distribution range has been the mean temperature in January (particularly when it was below −10 °C) (Niedziałkowska et al. 2021).

Movement and Space Use

To meet their seasonal and annual requirements, red deer need relatively large areas. Animals can be sedentary when their seasonal home ranges overlap, or migratory when winter and summer ranges are distinct in which case they can be close to one another or further apart. In some populations, animals are all residents (Kamler et al. 2008), while in others both tactics may occur. In mountainous habitats a variable proportion of deer are short-distance switchers or long-distance migrants, moving seasonally between low-elevation winter ranges and high-elevation summer ranges (Staines et al. 2008; Mysterud et al. 2011; Bocci et al. 2012; Kropil et al. 2015). Spring movements toward higher altitudes follow the forage maturation, while autumn movements toward the valley floors appear to be triggered by early snowfalls, decreasing temperature, forage senescence, and sometimes by the onset of the hunting season (Rivrud et al. 2016). In the Swiss National Park (central Alps), some adult males during winter remain at high elevations (up to 2800 m a.s.l.) on steep and snow-free slopes (Haller 2002). In Norway migratory stags move on average *c.* 30 km and hinds *c.* 20 km, with records of *c.* 100 and *c.* 75 km, respectively; the mean annual home range is 150 km^2 for migratory stags and 115 km^2 for migratory hinds (Rivrud et al. 2016; Meisingset et al. 2018). Mean annual home ranges of resident individuals are smaller, generally varying between 5–35 km^2 in stags and 2–8 km^2 in hinds (Kamler et al. 2008; Meisingset et al. 2018). Sedentary animals are particularly efficient in finding and exploiting variations in food resources at fine spatial scales, which allows them to avoid long-range movements, while switchers and migratory animals seek for large-scale spatial heterogeneity and predictability of resource productivity (Martin et al. 2018).

Diet

Red deer are classified as "intermediate feeders" (*sensu* Hofmann 1985) based on their digestive anatomy and physiology as well as their feeding style. Diet composition varies markedly, depending on the availability of plants in different habitats. On a continental scale, a literature review (Gebert and Verheyden-Tixier 2001) showed that red deer have a mixed diet of grass and sedges and concentrate food (forbs, leaves of shrubs and trees, fruits, and seeds). Deer can also adapt to seasonal changes in the quantity and quality of available food, e.g., by feeding alternatively on grasses or on browse. In winter, when most of the browse tends to be rich in indigestible lignine, red deer can shift to grasses when they are richer in soluble sugars and proteins (Verheyden-Tixier et al. 2008). In floodplain forests in Moravia (E Czechia), despite the relative abundance of grass and forbs, red deer prefer to eat woody plants from shrubs and trees the whole year (Prokešová 2004). In an Alpine Swiss area, red deer largely rely on graminoids throughout the year, while in winter they depend on good quality grass of farmed meadows in the valley floors (Zweifel-Schielly et al. 2012). In the Mediterranean maquis scrubs and woods, because of the scarcity of grasses and forbs and the rapid decrease of their nutritional value during the hot and dry summer, red deer primarily act as browsers, selecting shoots of woody plants with lower phenolic content (Bugalho and Milne 2003; Casula and Murgia 2017). In Great Britain, throughout the year grasses are the most important item in diet, together with heather and other dwarf shrubs such as bilberry, at least for upland deer populations. In winter, the contribution of dwarf shrubs to the diet increases and both sexes also take brambles, holly, and ivy when available. Ferns, lichens, and both deciduous and coniferous browse may also form a more significant part of the diet in autumn and winter.

Partly as a consequence of spatial segregation (Bonenfant et al. 2004a), but also as a result of differences in body size and nutritional requirement, significant differences between the sexes in diet composition may occur. The larger body size of males, with associated higher food requirements, leads to males feeding on different forage species or shifting their foraging style toward bulk even when not spatially segregated from females; at the same time, a larger ruminal volume allows males to digest bulk forages of poorer quality (Clutton-Brock et al. 1982). These differences may be intensified by competition from smaller-bodied females, which have significantly narrower muzzles, enabling them to crop preferred swards so short that males are unable to forage effectively. On open moorlands in Scotland, stags eat more heather and fewer grasses than hinds, especially in winter (Clutton-Brock et al. 1982; Staines and Crisp 1978; Staines et al. 1982).

Behavior

Social Behavior

The red deer is a gregarious species. The strongest and most stable social unit is the matrilinear group, usually formed by a grandmother, her daughters, and granddaughters. Stags either form unstable small groups with unrelated males, or are solitary. For most of the year females and males are socially and spatially segregated (Mitchell et al. 1977; Bonenfant et al. 2004a; Alves et al. 2013a). The degree of social and sexual segregation, however, is highly variable depending on population density and adult sex ratio. Mixed sex groups occur almost exclusively in early spring, during the vegetation green-up, and in early autumn, during the rutting season (Bonenfant et al. 2004a). In woodland habitats, red deer are generally solitary or in small groups (Staines et al. 2008; Dzięciołowski 1979, 1991); different groups may aggregate in open areas, but split up again into family groups when returning to forested areas. In open habitats, where larger groups may persist over long periods, complex social structures may establish within kin-related groups of hinds and among stag groups. Among females, a dominance hierarchy may develop, which appears to be primarily but not exclusively related to age (Clutton-Brock et al. 1982). In the first few days of their lives, newborn calves are hiders, i.e., they are often left alone by the mother, and if disturbed they "freeze" in a crouched posture. At 7–10 days they begin to follow their mother and join the group. Calves seem to inherit from their mothers at least part of their dominance status as adults (Veiberg et al. 2004).

Daughters' home ranges overlap with those of their mothers, while sons, when one or two years old, abandon their family group, wander widely and eventually settle in some area far from their birth place (Clutton-Brock et al. 1982). Pre- or postweaning orphaning increases mortality in both sexes; orphaning after weaning can also affect the physical development of juvenile males (Andres et al. 2013).

Mating Behavior

At the end of summer adult stags leave the bachelor groups and move to the traditional rutting grounds, and in early autumn they begin to compete for access to mates. The timing of the rut differs widely across Europe. For example, the peak of the rut occurs at around 5–15 September in Sardinia, 10–15 September in Southern Spain, 10–20 September in Southwestern Hungary, 20–25 September in the Northern Apennine (Italy), 20–30 September in Northeastern Hungary, 3–10 October in the central Alps, and 5–25 October in Scotland. In France the peak of conceptions is around 20–30 September, while in Norway it is around 15–30 October. Red deer are highly polygynous, with male mating success strongly dependent on fighting ability (Clutton-Brock et al. 1988). During the rut adult stags show several behavioral patterns (Bützler 1974; Clutton-Brock et al. 1982). They advertise their strength, challenge and evaluate potential contenders by means of roars, and exhibit direct and indirect threats such as canine display, ground pawing, vegetation thrashing and soil raking with antlers, wallowing, and urine spraying. They also wipe the urine-soaked neck and the antorbital gland against trees and rocks, herd females which attempt to leave the harem, and chase yearlings and subadults that try to approach the harem. Two rivals of comparable size often perform the so-called parallel walk, a slow march side by side, 5–20 m apart, maintaining the posture for a considerable distance, in order to assess the opponent's strength and avoid physical aggression. Challenges can escalate to a semi-ritualized fight as the two males, with antlers interlocked, push, twist, and wrestle trying to put the opponent off balance (Fig. 4). The dominant stag tests female receptivity by sniffing and lip curling (*flehmen*). The courtship approach or chivying consists of an outstretched neck posture, with slightly lifted antlers and frequent tongue flicking.

The most frequent male mating tactic appears to be the group guarding (i.e., "harem" defense). Harem size varies according to habitat; small family groups are common in woodlands (Bobek et al. 1992; Bonenfant et al. 2004b), 10–15 hinds are typical in open areas (Clutton-Brock et al. 1982; Staines et al. 2008). Behavioral variations within mating systems, however, are common and largely dependent on environmental structure, resource distribution, and density of males and females (Carranza et al. 1990). In Southern Spain, for example, the traditional mating tactic coexists with an alternative mating tactic, i.e., resource-based territoriality, where stags guard the scarce and patchily distributed food resources sought by hinds after the summer drought (Carranza et al. 1990, 1996).

Fig. 4 From top to bottom: sparring yearling stags, sparring subadults, and fighting adults (photographs by Andrea Dal Pian)

Communication

For most of the year males are almost mute, uttering only some nasal sounds or soft grunts during sparring fights or when in proximity to other deer. But from August, under the influence of testosterone, vocal folds grow and thicken, and adult stags begin to emit loud low-pitched calls, generally bouts of 3–8 roars, which at the peak of the rut can reach mean rates of 2 per minute in harem-holders (Bützler 1974; Clutton-Brock and Albon 1979). During the rut, males engage in very intense vocal activity as a means of attracting females and threatening opponents. Roaring rate is related to body size, fighting ability, and reproductive success, and roar characteristics appear to reflect male quality (Clutton-Brock and Albon 1979; Reby and McComb 2003). Red deer have a descended and highly mobile larynx which, during the roaring bouts, can be pushed further down toward the sternum, thus extending the vocal tract. The substantial extension of the head and neck region and the descent of the larynx create an elongated resonance chamber (Frey et al. 2021). Larger stags with longer vocal tracts tend to produce stronger calls. European red deer stags exhibit roaring variants, for example, in Spain (Passilongo et al. 2013), Corsica (Kidjo et al. 2008), Mesola Wood, Italy (Della Libera et al. 2015), and in the Pannonian Lowland (Volodin et al. 2019). Spanish stags extend their tongue during the roaring bout, possibly a cooling strategy in hot and arid ruts (Frey et al. 2012). Roars can influence female reproductive status, synchronizing and advancing the estrus (McComb 1987). Roars can be heard until January or early February. During the rut, adult stags frequently also use the chase bark (*Sprengruf*, or explosive call, in German), a threatening vocalization toward young males which try to intrude the harem, and toward hinds during herding behavior (Bützler 1974). Hinds utter a gruff bark or series of barks when alarmed and may also produce a low "mooing" sound when trying to locate young calves. Calves emit at least seven types of vocalizations, two nasal bleats (when asking for milk and when receiving maternal care) and five different high-pitched squeals for alarm, fear, and need of contact (Wölfel 1984).

Foraging Behavior and Activity Budget

Red deer are active by day and night. In undisturbed habitats such as the primeval forest of Białowieża, they have 12 bouts of activity of about 42 min throughout day and night, for a total of 8 h daily (Kamler et al. 2007). In Mediterranean habitats in Spain, radio-collared red deer showed some decrease in activity in the middle of the night (*c.* 2:00 h) and a much more marked decrease during the day (from *c.* 8:00 to 17:00 h; Carranza et al. 1991).

In the hunting-free areas of the Isle of Rum, red deer spend a total of 10–12 h per day feeding, usually divided into 6–10 bouts. In habitats with some human disturbance, they show between 5 and 9 feeding cycles daily, with the longest activity periods at dusk and dawn (Bubenik and Bubenikova 1967; Georgii and Schröder 1978). When frequently disturbed, red deer may become increasingly nocturnal. In open-hill country in Scotland, they are often found at high elevations during the day, descend to lower areas at dusk and return to higher areas by dawn. In forested habitats they tend to remain in, or close to, woodland cover during the day, venturing into open areas from dusk till dawn.

Parasites and Diseases

The majority of studies have been carried out on farmed deer, because of the economic significance of infections. Free-ranging red deer are generally thought to be relatively free of major diseases, partly because they usually occur at comparatively lower densities. However, risks of infection are not necessarily directly related to actual population size since aggregation of animals, usually due to human activities (e.g., artificial feeding), can increase the risk of pathogen transmission even in low-density deer populations (Milner et al. 2014).

Ectoparasites

Hard ticks (e.g., *Ixodes ricinus*, *Hyalomma lusitanicum*, *Dermacentor* spp., and *Haemaphysalis* spp.) are common (Valcárcel et al. 2016) as are deer ked *Lipoptena cervi*, nasopharyngeal bot fly larvae *Cephenemyia auribarbis* and *Pharingomyia cervi*, a sucking louse *Solenopotes burmeisteri*, the biting lice *Bovicola longicornis* and *B. tibialis*, as well as deer warble maggots *Hypoderma actaeon* and (much less frequently) *Hypoderma diana* (Boch and Schneidawind 1988). Their effects on individual condition and performance can be important if linked to other stressful events such as winter starvation (Charleston 1980; Fletcher 1982; Vicente et al. 2004). Red deer may also suffer from sarcoptic mange by the burrowing mite *Sarcoptes scabiei*. Usually, infection derives from direct or indirect contact with scabietic domestic goats or mountain-dwelling Caprinae (*Capra* spp. and *Rupicapra* spp.). Deadly outbreaks in red deer have been reported in Spain since the last decade of the past century, whereas cases and clusters are rare events in the Alps (Iacopelli et al. 2020; Oleaga et al. 2008). The protozoan *Besnoitia besnoiti* causing a severe chronic dermatitis in cattle, has also been reported in asymptomatic red deer (Gutiérrez-Expósito et al. 2016). Midge-transmitted species-specific nematodes of the genus *Onchocerca* are well known to produce large subcutaneous nodules on the dorsal parts and the external thigh of infected individuals (Boch and Schneidawind 1988).

Endoparasites

The digestive tract of red deer is home to ten species of protozoa (*Giardia duodenalis*, *Cryptosporidium parvum*, plus eight host-specific taxa of the genus *Eimeria*) and a number of helminths, including the generalist tapeworms *Paramphistomum cervi*, *Moniezia expansa*, and *M. benedeni*, and approximately two dozens of roundworms (Boch and Schneidawind 1988; Hamnes et al. 2006). Among the latter, the cervid-specific *Ostertagia leptospicularis* and *Spiculopteragia spiculoptera*, with their respective minor morphs *O. kolchida* and *S. mathevossiani*, are the dominant abomasal species, while *Nematodirus roscidus* is prevalent in the small intestine (Rehbein et al. 2002; Zaffaroni et al. 2000). Most protozoa and helminths of the gastrointestinal tract are only mildly pathogenic and rarely cause overt disease, but type II ostertagiasis-like episodes have been recorded in farmed deer (Connan 1991). The generalist liver flukes *Fasciola hepatica* and *Dicrocoelium dendriticum* may be present in red deer co-occurring with domestic ruminants, but the level of parasitism is usually low (French et al. 2016). The respiratory system frequently harbors deer-specific nematodes, *Dictyocaulus cervi* n.sp. (Pyziel et al. 2017) and the snail-transmitted *Varestrongylus sagittatus*, located in the large and small bronchi and in the lung parenchyma, respectively. The former is recognized as the single most important nematode parasite in farmed red deer, causing growth retardation, bronchitis, and respiratory distress in calves and yearlings (Alexander and Buxton 1994; Charleston 1980). The mosquito-transmitted nonpathogenic filarial worm *Setaria cervi* may be observed in the abdominal cavity during dressing. The highly prevalent snail-transmitted tissue nematode *Elaphostrongylus cervi* occasionally produces a clinical cerebrospinal disease in red deer and in sympatric small domestic ruminants (Alberti et al. 2011). Larvae of *Echinococcus granulosus*, a zoonotic tapeworm with a canid-ruminant life cycle, are occasionally found in the liver or lungs of red deer in Europe. A sylvatic cycle involving wolves, large cervids (though apparently not red deer), and the G8 and G10 strains of *E. granulosus* occurs in circumpolar countries (Davidson et al. 2016). Larvae of the canid-transmitted *Taenia hydatigena* are found on the serosae of the thoracic and abdominal viscera, though less frequently than in other wild ruminants (e.g., Caprinae). In the heart and muscles, 12 species of Protozoa (*Toxoplasma gondii* and 11 species of *Sarcocystis*, the latter occasionally causing a grey-greenish discoloration of carcasses) and the cystic larval stage of the canid-transmitted tapeworm *Taenia ovis krabbei* have been reported (Basso et al. 2020; Formenti et al. 2015; Letková et al. 2008). Blood protozoa include three species of *Babesia* and two of *Theileria* (Zanet et al. 2014).

Translocations of exotic deer species into Europe have incidentally caused the introduction of alien pathogens. In particular, wapiti and white-tailed deer *Odocoileus virginianus* have introduced the large American liver fluke *Fascioloides magna*, which has proved deadly in red deer, fallow deer *Dama dama*, roe deer *Capreolus capreolus*, and small domestic ruminants. Severe recurrent die-offs have been reported in a fenced hunting reserve in Northern Italy since the second half of the nineteenth century after the release of wapiti imported from the Western USA. Decades of attempts to control this alien parasite in free-ranging deer have not impeded the (still ongoing) spread of *F. magna* from the Czech Republic, where white-tailed deer were initially introduced, to Poland, Germany, and several countries along the Danube (Balbo et al. 1989; Malcicka 2015).

Infectious Diseases

While antibodies to a range of livestock and wildlife pathogens have been revealed in several serosurveys throughout Europe, only a handful of outbreaks of transmissible infectious diseases are known in red deer. Ocular disease caused by bovine herpesvirus-1 has been recorded in farmed calves (Nettleton et al. 1986). Farmed deer are also particularly susceptible to malignant catarrhal fever caused by ovine herpesvirus-2, which is thought to be carried by sheep (Heuschele and Reid 2001). A deer-specific parapoxvirus is the agent of a contagious stomatitis, producing proliferative lesions on the lips and erosions, vesicles, and ulcers in the mouth (Scagliarini et al. 2011). Hemorrhagic septicemia, an acute highly fatal disease caused by *Pasteurella multocida*, results in considerable (though sporadic) deer losses in Southern Europe. Climatic triggers have been called into question to explain the sudden virulentation of this opportunistic bacterial pathogen (Robinson et al. 2019). Chronic proliferative enteritis and mass loss (up to starvation) characterize Johne's disease caused by *Mycobacterium avium paratuberculosis*, affecting a wide range of ruminant and non-ruminant hosts, including red deer and livestock. Increased mortality in young stock, occasional deaths among adults, reduced growth, and lower pregnancy rates have been reported (Carta et al. 2013). Red deer are occasional victims of anthrax, a lethal bacterial infection caused by exotoxins of *Bacillus anthracis* (Fasanella et al. 2007). Red deer have been marginally affected by chronic wasting disease (CWD), a transmissible spongiform encephalopathy (TSE) caused by prions, affecting cervid hosts in North America and, recently, in Northern Europe (Vikøren et al. 2019). However, the closely related wapiti is among the most frequently infected deer in CWD outbreak areas in North America.

No significant role has been attributed to red deer in the medium- to long-term maintenance of major livestock diseases in Europe such as foot-and-mouth disease, brucellosis and, more recently, bluetongue disease (Rossi et al. 2019). However, there are examples of bovine tuberculosis foci maintained by red deer in the absence of infected livestock. The spillback of *Mycobacterium bovis* and *Mycobacterium caprae* from these deer to Tb-free livestock is also well documented (Delahay et al. 2007; Nigsch et al. 2019; Zanella et al. 2008, 2012).

Zoonoses

No major risks to human health derive from handling red deer or dressing their carcasses, although, as in other ungulates, the zoonotic agent *T. gondii* has been reported. Accordingly, pregnant women and immune-compromised patients should prudently refrain from consumption of unfrozen raw or rare venison. Deer keds *Lipoptena cervi* have been occasionally reported as the agent of a localized pruritic dermatitis in humans (Härkönen et al. 2009). Red deer are noncompetent hosts for the causative organism of Lyme disease, *Borrelia burgdorferi*, with evidence that the bacterium may actually be killed in infected ticks feeding on deer (Kurtenbach et al. 1998). Red deer are thus not directly involved in the transmission of Lyme disease to humans. However, red deer may act as a reservoir for adult ticks and may help maintain their abundance in the environment (Gilbert et al. 2012).

Population Ecology

Population Dynamics

Red deer population growth rate is strongly affected by both density-dependent and density-independent factors (Putman et al. 1996; Coulson et al. 2004). At low density, in productive habitats with no predators, red deer populations can increase up to 30–35% yearly. Low quality of food resources and high density tend to negatively affect traits such as age at first reproduction, calf survival, fecundity, and adult survival, although these parameters respond differently to different values of density, as predicted by Eberhardt's model (Clutton-Brock et al. 1985; Albon et al. 2000; Coulson et al. 2004). At high density post-winter recruitment can drop from 65–75 to 30–35 calves per 100 hinds (Mitchell et al. 1977) or even less (17 per 100 hinds in Mesola Wood, Mattioli et al. 2003). Red deer dynamics generally appears to be severely limited by adverse winter conditions, at least in more extreme environments, e.g. in Norway or in the Alps, especially at high densities (Forchhammer et al. 1998; Bonardi et al. 2017). The synergistic effect of density and weather conditions impacts red deer dynamics largely through the offspring female segment of the population, e.g., by lowering juvenile mass at birth and the subsequent reproductive success as adults (Albon et al. 1987), or decreasing calf survival and female reproductive performance (Albon and Clutton-Brock 1988; Clutton-Brock et al. 1988).

In most of the intact European forest ecosystems, the wolf (*Canis lupus*) is the main predator (Okarma 1995; Jędrzejewski et al. 2000). In Foreste Casentinesi National Park (Central Italy), red deer make up only one-third of wolf prey, with wild boar (*Sus scrofa*) and roe deer being the other most important prey species (Mattioli et al. 1995). Calves and sometimes hinds can be killed by the Eurasian lynx (*Lynx lynx*) (Jędrzejewski et al. 1993; Belotti et al. 2015; Heurich et al. 2016). Golden eagles (*Aquila chrysaetos*) and foxes (*Vulpes vulpes*) are known to prey on young calves. Humans can also severely impact red deer population dynamics through hunting (Coulson et al. 2004; Putman et al. 2019), although in sustainably managed populations, density and winter weather conditions still appear to be the major determinants of numerical fluctuations (Bonardi et al. 2017). The role of poaching in limiting red deer dynamics is still unclear and likely depends on the target of poachers, e.g., whether they focus on the male or the female segment of the population (Corlatti et al. 2019a).

Competition with Other Ungulates

Red deer are superior competitors to roe deer and chamois (*Rupicapra* sp.). For roe deer, interspecific competition has been suggested based on inversely related densities (Latham et al. 1997), habitat displacement (Danilkin and Hewison 1996), and lower body mass of roe deer fawns (Richard et al. 2010). For Apennine chamois (*Rupicapra pyrenaica ornata*), the exploitation of summit meadows by an increasing red deer population has negatively affected the diet of female chamois with a subsequent decline in kid survival (Lovari et al. 2014; Ferretti et al. 2015). In the Swiss National Park, higher red deer density has caused spatial displacement of Alpine chamois (*Rupicapra rupicapra rupicapra*) to less productive meadows and steeper slopes, with a clear decline in horn growth of kids and yearlings (Anderwald et al. 2016). In the Stelvio National Park (central-eastern Italian Alps), the increasing densities of red deer, alongside the synergistic effect of winter weather conditions, caused a strong numerical decline in Alpine chamois (Corlatti et al. 2019b), mainly acting on the female-kid segment of the chamois population (Donini et al. 2021a).

In several areas of Europe, there is habitat and diet overlap between red and fallow deer, with no direct evidence of competitive effects on behavior or population dynamics. In the enclosure of Žehušice (Czechia), fallow deer are more aggressive than red deer and may displace them from the feeding sites (Bartoš et al. 1996). In Mesola Wood (NE Italy), fallow deer proved more flexible and less density-dependent than red deer, increasing to levels that put the native red deer population at risk (Ferretti and Mattioli 2012), which led to the culling of fallow deer as part of the conservation measures to protect the Mesola red deer. Red deer are thought to be adversely affected by high numbers of introduced sika with which they easily hybridize (Ratcliffe 1987).

Red deer compete with sheep (Clutton-Brock and Albon 1989), while cattle appear to be complementary grazers with evidence of active facilitation of grazing, in particular for hinds, by the removal of coarser vegetation (Gordon 1988).

Effects of Climate Change

In some parts of Europe, climate warming translates into a lengthening of the vegetative period and an increase of food availability in spring and early summer for red deer. On the Isle of Rum, during a 38-year period from 1980–2007, higher temperatures in spring influenced the timing of reproduction and the antler cycle: mean parturition date advanced by 12 days, estrus date by 7 days, beginning of the rut by 6 days, end of the rut by 12 days, antler casting by 6 days, and velvet shedding by 7 days (Moyes et al. 2011). Climate change may also decrease the digestibility of plants in late summer and autumn, with negative consequences on body growth of juveniles and physical conditions of adults.

In mountainous habitats, increasing temperatures have caused an upslope shift in plant species, and subsequent behavioral adaptations in wildlife, including red deer, which significantly increased the elevations occupied over the past decades (Büntgen et al. 2017). The consequences of this upslope shift are difficult to predict: In the Italian Alps, the increase of spring-summer temperature tended to worsen body conditions and in turn reduced pregnancy probability, although the impact on population dynamics may be negligible (Corlatti et al. 2018). Conversely, the interactive effect of increased deer numbers at higher elevations and higher temperatures may accelerate the depletion of resources for chamois, e.g., in the Apennines (Ferretti et al. 2018). With global warming, habitat suitability of Western Norway will increase with an expansion of summer and winter ranges for both male and female migratory red deer (Rivrud et al. 2019). Climatic changes could also trigger resource-mediated effects on body mass or skeletal traits, resulting in smaller-sized deer with increasing temperature and drought (Vannini et al. 2021).

Conservation Status

At a continental scale, red deer are still increasing and are classified by the IUCN as "Least Concern" (Lovari et al. 2019). However, some populations are at risk. The Tyrrhenian red deer (*C. e. corsicanus*) is strictly protected under Appendix II of the Bern Convention and Annexes II and IV of the EU Habitats and Species Directive. In 2017, there were about 1400 red deer in Corsica and about 7000 in Sardinia (Murgia et al. 2017). Mesola red deer (*C. e. italicus*) are in danger of extinction, being restricted to a single small natural reserve. In Northern Macedonia and in Greece, red deer continue to be rare due to poaching, predation by feral dogs, and habitat degradation.

Management

A close relationship between humans and red deer has existed since prehistoric times. In the wake of their migrations, Neolithic farmers brought deer to new regions as a source of venison, skin, bones, and antlers. For example, red deer were translocated from peninsular Italy to Sardinia at least 4000–5000 years ago (Vigne 1992; Doan et al. 2017, 2022), from the Scottish mainland to the Hebrides and Orkney around 5000 ya (Stanton et al. 2016), and from Britain to Ireland in the same period (Carden et al. 2012). From the Middle Ages to the end of the eighteenth century, red deer were considered "high game" (*Hochwild*) in all of Europe, i.e., they were the exclusive quarry of kings, princes, high aristocracy, and high clergy. For centuries, royal and noble families continued to exchange deer, in the firm belief of improving the stock, a practice which has never completely disappeared even in modern times (Beninde 1940). Extensive deforestation, expansion of livestock farming, and poaching caused a dramatic decline of the species throughout Europe, and the species started to recover only at the beginning of the twentieth century or after the second world war (Deinet et al. 2013). Today, red deer have reached high numbers in much of Europe and pose serious management issues.

Large body size, consistent impact on agriculture and forestry, great mobility, sensitivity to disturbance, and hunting errors make correct management particularly challenging. Present management regimes differ considerably among countries, but models with high levels of artificiality (highly unnatural densities, fencing, and supplementary feeding) tend to prevail (Mysterud 2010). In some countries, for example, in Spain and Scotland, hunting can provide important revenue or boost property values and, consequently, extremely high densities are encouraged. In some regions many deer live in fenced areas (for example, Spain), and in some mountainous regions of central Europe, deer are kept in large enclosures during winter due to the lack of available wintering areas in the valleys, i.e., areas that are neither urbanized nor sensitive to potential damage. In many parts of the continent, supplementary feeding is the rule, e.g., during winter in Central and Northern Europe, and during summer in Spain. Born as an emergency measure to buffer climate extremes or to limit animal movements, expensive supplementary feeding has become essential for maintaining very high densities, although its effectiveness has often been questioned (Katona et al. 2014). In some countries and regions, red deer occurrence is legally permitted only in a few large forested areas, in others it is allowed almost everywhere without precise planning. The spatial scale of hunting management can be vast: In Germany, the special red deer management units cover around 300 km^2 (Wotschikowsky 2010); in the Northern Apennines, Italy, a red deer management unit includes a whole population, encompassing two regions and four provinces, covering 1700 km^2 and consisting of seven coordinated macro-districts (Mattioli unpubl.). In other countries, deer counts and shooting plans are independently made for small management units (small ownerships, small landowner consortia, and municipal hunting reserves) (Meisingset et al. 2018). Hunting bags are sometimes still male-biased, and excessive hunting pressure can dangerously lower the population mean age (Langvatn and Loison 1999; Milner et al. 2006; Skonhoft et al. 2013; Torres-Porras et al. 2014). In some countries, hunting halts in correspondence to the peak of the rut, in others it does not (Apollonio et al. 2011). In many countries, counting deer is not mandatory and shooting plans are based on rough estimates, on damage levels or simply ruled by local customs.

Most of the current management regimes should be extensively reformed. Given the potentially high impact on crops and forest regeneration, it would be beneficial to have a large-scale planning process (for example, at regional level) and use habitat suitability models to help decide where, and at what densities, the occurrence of the species can be tolerated, which will usually be far from urbanized areas in regions with large forest tracts. Given the high mobility of red deer, large enough management units (based on, e.g., monitoring of GPS-tagged animals) should be established and close cooperation between adjacent districts should be ensured (Meisingset et al. 2018). Counts should be mandatory, planned on a large scale and undertaken simultaneously to avoid multiple counts of the same animals, and a unique general shooting plan should be prepared for the whole large management unit, then subdivided among the different hunting grounds. Supplementary feeding should be discouraged because of the negative consequences on the populations (reduced natural selection, semi-domestication; at high concentrations of animals, increased aggression and stress, enhanced transmission of parasites and diseases, heavy local browsing, *etc.*) (Schmidt and Hoi 2002; Putman and Staines 2004; Milner et al. 2014). For counting, researchers have proposed or adopted several techniques (Mayle et al. 1999; Daniels 2006; Morellet et al. 2011; Grignolio et al. 2020), but not all appear sustainable in the long run. In the Scottish treeless moorlands, direct counts covering discrete blocks of land are regularly used (Clutton-Brock and Albon 1989). In the open woodlands of Central Europe, drive counts would be appropriate, but only at relatively high deer densities and at least on 10% of the total area (Pucek et al. 1975; Borkowski et al. 2011). In mountain regions with a sufficient presence of open areas (meadows), spotlight counts are advisable (Garel et al. 2010; Corlatti et al. 2016). In scrubwoods and in dense mountainous forests with scarce accessibility and low visibility, the use of pellet group counts, infrared camera distance sampling (Acevedo et al. 2008; Alves et al. 2013b), or roaring surveys (Mazzarone et al. 1991) should be preferred. Despite some criticism (Ciucci et al. 2009; Douhard et al. 2013), the nocturnal count of roaring stags with the reconstruction of the population structure by observation sessions, when compared with other counting methods (Dzięciołowski et al. 1996; Lovari et al. 2007), has provided consistent results and is currently used in Sardinia and the Apennines, Italy. Simple, deterministic cohort analysis also proved useful to reconstruct population size (Donini et al. 2021b). Besides knowing the absolute number of deer, culling plans can be improved by periodic estimates of demographic indices such as ovulation rate and calf winter recruitment. Shooting plans for selective hunting should be prepared according to technically sound methods, trying to maintain the natural structure of the population, with lower pressure on adult stags and slightly higher on hinds and calves. At least four age classes should be preferably recognized for males (calves, yearlings, subadults of 2–4 years of age, and adults of 5+ years) and three for females (calves, yearlings, and adults of 2+

years) (Drechsler 1988). Some simple biometric measurements on culled animals should be mandatory, with at least body mass and foot length for all animals, and mandible length for representative samples (Buchli 1979; Suttie and Mitchell 1983). Body size measures are good ecological indicators of change in demographic parameters, abundance, and habitat quality (Morellet et al. 2007).

Since bark stripping and heavy browsing not only depend on population density but also on forest type and structure and on silvicultural practices, the best solution for decreasing damage to forests cannot simply come from a control of red deer density levels, but also from a substantial change in silvicultural systems, moving from a timber production perspective with the prevalence of even-aged monospecific plantations to a more nature-oriented forest, able to guarantee a better distribution and a higher abundance of trophic resources for herbivores (Kuijper 2011).

The value of pure indigenous populations should be emphasized, also promoting their expansion at the expense of non-native or admixed stocks (see, for example, Spain, Greece, and Sweden). Reintroductions and restocking should be carefully planned. Finally, to minimize the risks of genetic erosion, landscape connectivity between isolated management units should be guaranteed (Edelhoff et al. 2020; Reiner et al. 2021).

Future Challenges for Research and Management

The red deer is certainly one of the best-studied mammals in Europe. Most of the research has been carried out in marginal habitats such as moorlands, or in somewhat artificial situations (fenced areas, forest plantations with supplementary feeding), often in unnatural high-density conditions. Furthermore, thus far most of the studies have been performed in Western and Central Europe. Treeless areas have been fundamental for behavioral observations, and overabundance has enabled research into how food restrictions can affect body growth and fertility. A more balanced and complete picture of red deer biology is desirable and requires to plan studies in more natural situations, also more frequently involving Eastern and Southeastern European countries. The recent expansion of large predators will offer exciting opportunities to investigate their impact on deer populations, their behavioral adaptations, and how their pattern of habitat selection will affect abundance and distribution of other components of European ecosystems. A regular standard collection of biometric data across the continent could improve our limited knowledge of the high plasticity of this species. Finally, a sharing system of up-to-date, open data on distribution, population abundance, and hunting bags collected in a standardized manner within each country should be promoted at a continental scale.

Acknowledgments We would like to thank Juan Carranza (University of Cordoba) for helpful comments on an earlier version of this chapter. For making unpublished data available we would like to thank Carranza J. (University of Córdoba, ES), Colombi D. (Regional Administration Friuli, Udine, I), Csányi S. (Szent István University, Gödöllő, H), Gaudry W. (Office National de la Chasse et de la Faune Sauvage, Paris, F), Gervais P. and Chaland C. (Fédération de chasseur de la Charente, Puymoyen, F), Lissovsky A.A. (Lomonosov State University, Moscow, RUS), Mandas L. (Agenzia Forestas, Cagliari, I), Meisingset E. (University of Oslo, N), Mattioli L. (Regional Administration Toscana, Arezzo, I), Meneguz P.G. (University of Turin, I), Murgia A. (Agenzia Forestas, Cagliari, I), Mysterud A. (University of Oslo, N), Peters L. (University of Edinburgh, UK), Šmehyl P. (University of Agriculture, Nitra, SK), Zagorodniuk I. (National Museum of Natural History, Kyiv, UA), and Zalewski D. (University of Olsztyn, PL). Gill R.M.A. (Forestry Commission, Farnham, UK), Ratcliffe P. (Ecological Land Use Consultancy, UK), and Lovari S. (University of Siena, I) kindly commented on the paragraph on management.

References

Acevedo P, Ferreres J, Jaroso R, Durán M, Escudero MA, Marco J, Gortázar C (2008) Estimating roe deer abundance from pellet group counts in Spain: an assessment of methods suitable for Mediterranean woodlands. Ecol Indic 10:1226–1230

Alberti EG, Gioia G, Sironi G, Zanzani S, Riccaboni P, Magrini M, Manfredi MT (2011) *Elaphostrongylus cervi* in a population of red deer (*Cervus elaphus*) and evidence of cerebrospinal nematodiasis in small ruminants in the province of Varese, Italy. J Helminthol 85:313–318

Albon SD, Clutton-Brock TH (1988) Climate and the population dynamics of red deer in Scotland. In: Usher MB, Thompson DBA (eds) Ecological changes in the uplands. Blackwell, Oxford, pp 93–107

Albon SD, Mitchell B, Staines BW (1983) Fertility and body weight in female red deer: a density dependent relationship. J Anim Ecol 52:969–980

Albon SD, Mitchell B, Huby BJ, Brown D (1986) Fertility in female red deer (*Cervus elaphus*)— the effects of body composition, age and reproduc- tive status. J Zool 209:447–460.

Albon SD, Clutton-Brock TH, Guinness FE (1987) Early development and population dynamics in red deer. II Density-independent effects and cohort variation. J Anim Ecol 56:69–81

Albon SD, Coulson TN, Brown D, Guinness FE, Pemberton JM, Clutton-Brock TH (2000) Temporal changes in key factors and key age groups influencing the population dynamics of female red deer. J Anim Ecol 69:1099–1110

Alexander TL, Buxton D (eds) (1994) Management and diseases of deer – a handbook for the veterinary surgeon. Veterinary Deer Society, London

Alves J, da Silva AA, Soares AMVM, Fonseca. (2013a) Sexual segregation in red deer: is social behaviour more important than habitat preferences? Anim Behav 85: 501–509

Alves J, da Silva AA, Soares AMVM, Fonseca C (2013b) Pellet group count methods to estimate red deer densities: precision, potential accuracy and efficiency. Mamm Biol 78:134–141

Anderwald P, Haller R, Filli F (2016) Heterogeneity on primary productivity influences competitive interactions between red deer and alpine chamois. PLoS One 11(1):e0146458

Andres D, Clutton-Brock TH, Kruuk LEB, Pemberton JM, Stopher KV, Ruckstuhl K (2013) Sex differences in the consequences of maternal loss in a long-lived mammal, the red deer (*Cervus elaphus*). Behav Ecol Sociobiol 67: 1249–1258

Apollonio M, Andersen R, Putman R (eds) (2010) European ungulates and their management in the 21st century. Cambridge University Press, Cambridge, UK

Apollonio M, Putman R, Grignolio S, Bartoš L (2011) Hunting seasons in relation to biological breeding seasons and the implications for the control or regulation of Ungulate populations. In: Putman R, Apollonio M, Andersen R (eds) Ungulate management in Europe: problems and practices. Cambridge University Press, Cambridge, pp 80–105

Arnold W (2020) Seasonal differences in the physiology of wild northern ruminants. Animal 14(S1):s124–s132

Arnold W, Ruf T, Reimoser S, Tataruch F, Onderscheka K, Schober F (2004) Nocturnal hypometabolism as an overwintering strategy of red deer (*Cervus elaphus*). Am J Physiol Regul Integr Comp Physiol 286:R174–R181

Arnold W, Beiglböck C, Burmester M, Guschlbauer M, Lengauer A, Schröder B, Wilkens M, Breves G (2015) Contrary seasonal changes of rates of nutrient uptake, organ mass, and voluntary food intake in red deer (*Cervus elaphus*). Am J Physiol Regul Integr Comp Physiol 309:R277–R285

Azorit C, Analla M, Calvo JA, Muñoz-Cobo J, Carrasco R (2002a) Teeth eruption pattern in red deer (*Cervus elaphus hispanicus*) in southern Spain. An Biol (Murcia) 24:107–114

Azorit C, Analla M, Carrasco R, Muñoz-Cobo J (2002b) Influence of age and environment on antler traits in Spanish red deer (*Cervus elaphus hispanicus*). Z Jagdwiss 48:137–144

Azorit C, Muñoz-Cobo J, Hervás J, Analla M (2004) Age through growth marks in teeth of Spanish red deer. Wildl Soc Bull 32:702–710

Balbo T, Rossi L, Meneguz PG (1989) Integrated control of *Fascioloides magna* infection in Northern Italy. Parassitologia 31:137–144

Bartoš L, Vankova D, Šiler J, Losos S (1996) Fallow deer tactic to compete over food with red deer. Aggress Behav 22:375–385

Bartoševa J, Ceacero F, Bartoš L (2012) Pre-orbital gland opening: part of sucking behavior in red deer (*Cervus elaphus*) calves. J Anim Sci 90:3207–3212

Baskin L, Danell K (2003) Ecology of ungulates: a handbook of species in eastern Europe and northern and central Asia. Springer, Berlin/Heidelberg

Basso W, Alvarez Rojas CA, Buob D, Ruetten M, Deplazes P (2020) *Sarcocystis* infection in red deer *(Cervus elaphus)* with eosinophilic myositis/fasciitis in Switzerland and involvement of red foxes (*Vulpes vulpes*) and hunting dogs in the transmission. Int J Parasitol Parasites Wildl 13:130–141

Becciolini V, Bozzi R, Viliani M, Biffani S, Ponzetta MP (2016) Body measurements from selective hunting: biometric features of red deer (*Cervus elaphus*) from Northern Apennine, Italy. Ital J Anim Sci 15:461–472

Beccu E (1989) Il cervo sardo. Delfino Ed., Sassari (I)

Bečejac B, Brna J, Mikuška J, Valter J (1984) Veličina rogovlja jelena obinog (*Cervus elaphus* L., 1758) na području Baranje i sjeverozapadne Bačke s obzirom na starosnu dob (Size of red deer antlers in the area of Baranja and northwestern Bačka according to age). RAD Jugoslavenske akademije znanosti i umjetnosti. Razred za Prirodne Znanosti 20:121–149. (in Croatian)

Belotti E, Weder N, Bufka L, Kaldusdal A, Küchenhoff H, Seibold H, Woelfing B, Heurich M (2015) Patterns of lynx predation at the interface between protected areas and multi-use landscapes in Central Europe. PLoS One 10(9):e0138139

Beninde J (1940) Die Fremdblutkreuzung (sog. Blutauffrischung) beim deutschen Rotwild. Zeitschrift für Jagdkunde (Sonderheft) 3:7–112

Bertouille SB, de Crombrugghe SA (2002) Fertility of red deer in relation to area, age, body mass, and mandible length. Z Jagdwiss 48:87–98.

Bobek B, Kosobucka M, Perzanowski K, Rebisz S (1992) Seasonal changes of the group size and sex ratio in various populations of red deer in southern Poland. Transactions of the 18th IUGB Congress, Krakow, pp 185–192

Bocci A, Angelini I, Brambilla P, Monaco A, Lovari S (2012) Shifter and resident red deer: intrapopulational and intersexual behavioural diversities in a predator-free area. Wildl Res 39:573–582

Boch J, Schneidawind H (1988) Krankheiten des jagdbaren Wildes. Verlag Paul Parey, Hamburg/Berlin

Bonardi A, Corlatti L, Bragalanti N, Pedrotti L (2017) The role of weather and density dependence on population dynamics of Alpine-dwelling red deer. Integr Zool 12: 61–76

Bonenfant C, Gaillard J-M, Loison A, Klein F (2002) Sex- and age-dependent effects of population density on life history traits of red deer *Cervus elaphus* in a temperate forest. Ecography 25:446–458

Bonenfant C, Loe LE, Mysterud A, Langvatn R, Stenseth NC, Gaillard J-M, Klein F (2004a) Multiple causes of sexual segregation in European red deer: enlightenments from varying breeding phenology at high and low latitude. Proc R Soc B Biol Sci 271:883–892

Bonenfant C, Gaillard J-M, Klein F, Maillard D (2004b) Variation in harem size of red deer (*Cervus elaphus* L.): the effects of adult sex ratio and age-structure. J Zool 264:77–85

Borkowski J, Palmer SCF, Borowski Z (2011) Drive counts as a method of estimating ungulate density in forests: mission impossible? Acta Theriol 56:239–253

Borowik T, Jędrzejewska B (2017) Heavier females produce more sons in a low-density population of red deer. J Zool 302:57–62

Borowik T, Wawrzyniak P, Jędrzejewska B (2016) Red deer (*Cervus elaphus*) fertility and survival of young in a low density population subject to predation and hunting. J Mammal 97:1671–1681

Borowski Z, Świsłocka M, Matosiuk M, Mirski P, Krysiuk K, Czajkowska M, Borkowska A, Ratkiewicz M (2016) Purifying selection, density blocking and unnoticed mitochondrial DNA diversity in the red deer, *Cervus elaphus*. PLoS One 11(9):e0163191

Botev N (1990) Development of the antlers of red deer in north-eastern Bulgaria. Transactions 19th I.U.G.B. Congress, Trondheim 1, pp 309–314

Bubenik AB, Bubenikova JM (1967) Twenty-four hour periodicity in red deer (*Cervus elaphus* L). In: Isaković I (ed) Transactions of the VII IUGB Congress, Beograd-Ljubljiana, pp 343–349

Buchli C (1979) Zur Populationsdynamik, Kondition und Konstitution des Rothirsches im und um den Schweizerischen Nationalpark. Hab. Thesis, Zurich, pp 99

Bugalho MN, Milne JA (2003) The composition of the diet of red deer (*Cervus elaphus*) in a Mediterranean environment: a case of summer nutritional constraint? For Ecol Manag 181:23–29

Büntgen U, Grenter L, Bolmann K, Jenny H, Liebhold A, Galván JD, Stenseth NC, Andrew C, Mysterud A (2017) Elevational range swift in four mountain ungulate species from the Swiss Alps. Ecosphere 8(4): e01761

Burbaitė L, Csányi S (2010) Red deer population and harvest changes in Europe. Acta Zool Lituanica 20: 179–188

Bützler W (1974) Kampf- und Paarungsverhalten, soziale Rangordnung und Aktivitätsperiodik beim Rothirsch. Fortschr Verhaltenforsch 16:1–76

Bützler W (1986) *Cervus elaphus* – Rothirsch. In: Niethammer J, Krapp F (eds) Säugetiere Europas. Vol. 2/II Paarhufer. AULA-Verlag, Wiesbaden, pp 107–139

Caboni A, Murgia C, Mattioli S (2006) Antler characteristics of the Sardinian red deer (*Cervus elaphus corsicanus*): a preliminary analysis. In: Bartoš L, Dušek A, Kotrba R, Bartošová-Víchová J (eds) Advances in deer biology. Deer in a changing world. Proceedings of the 6th Deer Biology Congress, Prague, p 266

Carden RF, McDevitt AD, Zachos FE, Woodman PC, O'Toole P, Rose H, Monaghan NT, Campana MG, Bradley DG, Edwards CJ (2012) Phylogenetic, ancient DNA, fossil and morphometric analyses reveal ancient and modern introductions of a large mammal: the complex case of red deer (*Cervus elaphus*) in Ireland. Quat Sci Rev 42:74–84

Carranza J, Pérez-Barberia FJ (2007) Sexual selection and senescence: male size-dimorphic ungulates evolved relatively smaller molars than females. Am Nat 170: 370–380

Carranza J, Alvarez F, Redondo T (1990) Territoriality as a mating strategy in red deer. Anim Behav 40:79–88

Carranza J, Hidalgo de Trucios SJ, Medina R, Valencia J, Delgado J (1991) Space use by red deer in a Mediterranean ecosystem as determined by radio-tracking. Appl Anim Behav Sci 30:363–371

Carranza J, Fernández-Llario P, Gomendio M (1996) Correlates of territoriality in rutting red deer. Ethology 102: 793–805

Carranza J, Alarcos S, Sánchez-Prieto CB, Valencia J, Mateos C (2004) Disposable-soma senescence mediated by sexual selection in an ungulate. Nature 432: 215–218

Carranza J, Mateos C, Alarcos S, Sánchez-Prieto CB, Valencia J (2008) Sex-specific strategies of dentine depletion in red deer. Biol J Linn Soc 93:487–497

Carranza J, Salinas M, de Andrés D, Pérez-González J (2016) Iberian red deer: paraphyletic nature at mtDNA but nuclear markers support its genetic identity. Ecol Evol 6(4):905–922

Carranza J, Pérez-Barberia FJ, Mateos C, Alarcos S, Torres-Porras J et al (2020) Social environment modulates investment in sex traits versus lifespan: red deer produce bigger antlers when facing more rivalry. Sci Rep 10:9234

Carta T, Álvarez J, Pérez de la Lastra JM, Gortázar C (2013) Wildlife and paratuberculosis: a review. Res Vet Sci 94:191–197

Casula P, Murgia A (2017) Selectivity and context dependence of Corsican red deer browsing in a Mediterranean coppice system. Hystrix 28:157–164

Catchpole EA, Fan Y, Morgan BJT, Clutton-Brock TH, Coulson T (2004) Sexual dimorphism, survival and dispersal in Red Deer. J Agric Biol Env Stat 9:1–26.

Ceacero F, Komárkova M, García AJ, Gallego L (2018) Different maternal strategies for male and female calves in a polygynous mammal. Curr Zool 65:269–277

Charleston WAG (1980) Lungworm and lice of the red deer (*Cervus elaphus*) and the fallow deer (*Dama dama*) – a review. N Z Vet J 28:150–152

Ciucci P, Catullo G, Boitani L (2009) Pitfalls in using counts of roaring stags to index red deer (*Cervus elaphus*) population size. Wildl Res 36:126–133

Clutton-Brock TH, Albon SD (1979) The roaring of red deer and the evolution of honest advertisement. Behaviour 69:145–169

Clutton-Brock TH, Albon SD (1989) Red deer in the highlands. BSP Professional Books, Oxford, UK

Clutton-Brock TH, Guinness FE, Albon SD (1982) Red deer: behavior and ecology of two sexes. University of Chicago Press, Chicago

Clutton-Brock TH, Major M, Guinness FE (1985) Population regulation in male and female red deer. J Anim Ecol 54:831–846

Clutton-Brock TH, Albon SD, Guinness FE (1988) Reproductive success in male and female red deer. In: Clutton-Brock TH (ed) Reproductive success. University of Chicago Press, Chicago, pp 325–343

Connan RM (1991) Type II ostertagiosis in farmed red deer. Vet Rec 128:233–235

Corlatti L, Gugiatti A, Pedrotti L (2016) Spring spotlight counts provide reliable indices to track changes in population size of mountain-dwelling red deer *Cervus elaphus*. Wildl Biol 22:268–276

Corlatti L, Gugiatti A, Ferrari N, Formenti N, Trogu T, Pedrotti L (2018) The cooler the better? Indirect effects of spring-summer temperature on fecundity in a capital breeder. Ecosphere 9:e02326

Corlatti L, Sanz-Aguilar A, Tavecchia G, Gugiatti A, Pedrotti L (2019a) Unravelling the sex- and age-specific impact of poaching mortality with multi-event modelling. Front Zool 16:20

Corlatti L, Bonardi A, Bragalanti N, Pedrotti L (2019b) Long-term dynamics of Alpine ungulates suggest interspecific competition. J Zool 309:241–249

Coulson TN, Guinness FE, Pemberton JM, Clutton-Brock TH (2004) The demographic consequences of releasing a population of red deer from culling. Ecology 85: 411–422

Croitor R (2018) Plio-Pleistocene deer of western Palearctic: taxonomy, systematics, phylogeny. Institute of Zoology of the Academy of Sciences of Moldova, Chişinău, p 140

D'Souza L, Lister AM (2016) An endemic radiation of deer in the Late Pleistocene of Malta. Abstracts, XIV Annual Meeting of the European Association of Vertebrate Palaeontologists Haarlem, Netherlands, p 140

Daniels MJ (2006) Estimating red deer *Cervus elaphus* populations: an analysis of variation and cost-effectiveness of counting methods. Mammal Rev 36: 235–247

Danilkin AA (1999) Oleny (Cervidae). GEOS, Moskva (in Russian)

Danilkin A, Hewison AJM (1996) Behavioural Ecology of Siberian and European roe deer. Chapman & Hall, London

Davidson RK, Lavikainen A, Konyaev S, Schurer J, Miller AL, Oksanen A, Skírnisson K, Jenkins E (2016) *Echinococcus* across the north: current knowledge, future challenges. Food Waterborne Parasitol 4:39–53

de Crombrugghe SA, Bertouille SB (1995) Body mass and lower jaw development of the female red deer as indices of habitat quality in the Ardennes. Acta Theriol 40: 145–162

de Jong JF, van Hooft P, Megens H-J, Crooijmans RPMA, de Groot GA, Pemberton JM, Huisman J, Bartoš L, Iacolina L, van Wieren SE, Ydenberg RC, Prins HHT (2020) Fragmentation and translocation distort the genetic landscape of ungulates: red deer in the Netherlands. Front Ecol Evol 8:535715

de la Peña E, Pérez-González J, Martin J, Vedel G, Carranza J (2021) The dark-ventral-patch of male red deer, a sexual signal that conveys the degree of involvement in rutting behavior. BMC Zool 6:18

De Marinis AM (2015) Valutazione dell'età nei Cervidi tramite esame della dentatura. Guida pratica all'identificazione delle classi d'età del Cervo. Manuali e Linee Guida 90.2/2013:1–136, ISPRA (in Italian)

Degmečić D, Florijančić T (2014) Utjecaj klimatiskih hidroloških čimbenika na razvoj rogovlja jelena običnog (*Cervus elaphus* L.) u hrvatskom dijelu Baranj skog Podinavlja (The impact of climate and hydrological factors on development of red deer (*Cervus elaphus* L.) antlers in the Croatian part of Baranja Danube region). Šumarski List 9–10:451–461. (in Croatian)

Deinet S, Ieronymidou C, McRae L, Burfield IJ, Foppen RP, Collen B, Böhm M (2013) Wildlife comeback in Europe: the recovery of selected mammal and bird species. Final report to Rewilding Europe by ZSL, BirdLife International and the European Bird Census Council. London

Delahay RJ, Smith GC, Barlow AM, Walker N, Harris A, Clifton-Hadley RS, Cheeseman CL (2007) Bovine tuberculosis infection in wild mammals in the South-West region of England: a survey of prevalence and a semi-quantitative assessment of the relative risks to cattle. Vet J 173:287–301

Della Libera M, Passilongo D, Reby D (2015) The acoustics of male rutting roars in the endangered population of Mesola red deer *Cervus elaphus italicus*. Mamm Biol 80:395–400

Doan K, Zachos FE, Wilkens B, Vigne J-D, Piotrowska N, Stanković A, Jędrzejewska B, Stefaniak K, Niedziałkowska M (2017) Phylogeography of the Tyrrhenian red deer (*Cervus elaphus corsicanus*) resolved using ancient DNA of radiocarbon-dated subfossils. Sci Rep 7(1):2331

Doan K, Mackiewicz P, Sandoval-Castellanos E, Stefaniak K, Ridush B, Dalén L, Weglenski P, Stankovic A (2018) The history of Crimean red deer population and *Cervus* phylogeography in Eurasia. Zool J Linnean Soc 183:208–225

Doan K, Niedziałkowska M, Stefaniak K, Sykut M, Jędrzejewska B et al (2022) Phylogenetics and phylogeography of red deer mtDNA lineages during the last 50 000 years in Eurasia. Zool J Linnean Soc 194:431–456

Dolan JM (1988) A deer of many lands – a guide to the subspecies of the Red Deer *Cervus elaphus* L. Zoonooz 62(10):4–34

Donini V, Pedrotti L, Ferretti F, Corlatti L (2021a) Disentangling demographic effects of red deer on chamois population dynamics. Ecol Evol 11: 8264–8280

Donini V, Corlatti L, Pedrotti L (2021b) Tracking red deer population size using cohort analysis. Mamm Biol 101: 675–680

Douhard M, Bonenfant C, Gaillard J-M, Hamann J-L, Garel M, Michallet J, Klein F (2013) Roaring counts are not suitable for the monitoring of red deer *Cervus elaphus* population abundance. Wildl Biol 19:94–101

Drechsler H (1980) Ueber die Geweihbildung bei Rothirschen im 'Rotwildring Harz' in den Jahren 1959–1978. Z Jagdwiss 26:207–219

Drechsler H (1988) Altersentwicklung und Altersansprache beim Rotwild: Gestalten, Gesichter, Gehabe, Geweihe, Gebisse. P. Parey Verlag, Hamburg/Berlin

Dzięciołowski R (1970) Relations between the age and size of red deer in Poland. Acta Theriol 15:253–268

Dzięciołowski R (1979) Structure and spatial organization of deer populations. Acta Theriol 24:3–21

Dzięciołowski R (1991) Ecological niches of five big ungulates in a forest tract. Folia Forestalia Polonica 33:56–69

Dzięciołowski R, Kossak S, Wasilewski M, Przypasniak J, Wawrzyniak P (1996) Estimating numbers and structure of red deer population in the Białowieza Forest. In: Paschalis P, Zajączkowski S (eds) Biodiversity protection in Primeval Białowieza Forest. FR SGGW, Warsaw, pp 77–90

Edelhoff H, Zachos FE, Fickel J, Epps CW, Balkenhol N (2020) Genetic analysis of red deer (*Cervus elaphus*) administrative management units in a human-dominated landscape. Conserv Genet 21:261–276

Fasanella A, Palazzo L, Petrella A, Quaranta V, Romanelli B, Garofolo G (2007) Anthrax in red deer (*Cervus elaphus*), Italy. Emerg Infect Dis 13:1118–1119

Fennessy PF, Suttie JM (1985) Antler growth: nutritional and endocrine factors. R Soc N Z Bull 22:239–250

Ferretti F, Mattioli S (2012) The Mesola red deer: present numbers and conservation perspectives. Hystrix 23: 35–43

Ferretti F, Corazza M, Campana I, Pietrocini V, Brunetti C, Scornavacca D, Lovari S (2015) Competition between wild herbivores: reintroduced red deer and Apennine chamois. Behav Ecol 26:550–559

Ferretti F, Lovari S, Stephens PA (2018) Joint effects of weather and interspecific competition on foraging behaviour and survival of a mountain herbivore. Curr Zool 65:165–175

Fletcher TJ (1982) Management problems and disease in farmed deer. Vet Rec 111:219–223

Fontana F, Rubini M (1990) Chromosomal evolution in Cervidae. Biosystems 24:157–174

Forchhammer MC, Stenseth NC, Post E, Langvatn R (1998) Population dynamics of Norwegian red deer: density-dependence and climatic variation. Proc R Soc Lond Ser B 265:341–350

Formenti N, Trogu T, Pedrotti L, Gaffuri A, Lanfranchi P, Ferrari N (2015) *Toxoplasma gondii* infection in Alpine red deer (*Cervus elaphus*): its spread and effects on fertility. PLoS One 10:e0138472

Frantz AC, Bertouille S, Eloy MC, Licoppe A, Chaumont F, Flamand MC (2012) Comparative landscape genetic analyses show a Belgian motorway to be a gene flow barrier for red deer (*Cervus elaphus*), but not wild boars (*Sus scrofa*). Mol Ecol 21:3445–3457

Frantz AC, Zachos FE, Bertouille S, Eloy M-C, Colyn M, Flamand M-C (2017) Using genetic tools to estimate the prevalence of non-native red deer (*Cervus elaphus*) in a Western European population. Ecol Evol 7:7650–7660

French AS, Zadoks RN, Skuce PJ et al (2016) Prevalence of Liver Fluke (*Fasciola hepatica*) in Wild Red Deer (*Cervus elaphus*): Coproantigen ELISA is a practicable alternative to faecal egg counting for surveillance in remote populations. PLoS One 11:e0162420

Frey R, Volodin I, Volodina E, Carranza J, Torres-Porras J (2012) Vocal anatomy, tongue protrusion behaviour and the acoustics of rutting roars in free-ranging Iberian red deer stags (*Cervus elaphus hispanicus*). J Anat 220: 271–292

Frey R, Wyman MT, Johnston M, Schofield M, Locatelli Y, Reby D (2021) Roars, groans and moans: anatomical correlates of vocal diversity in polygynous deer. J Anat 239:1336–1369

Galván I, Solano F, Zougagh M, de Andrés F, Murtada K, Ríos A, de la Peña E, Carranza J (2019) Unprecedented high catecholamine production causing hair pigmentation excretion in red deer. Cell Mol Life Sci 76:379–409

García AJ, Landete-Castillejos T, Garde JJ, Gallego L (2002) Reproductive seasonality in female Iberian red deer (*Cervus elaphus hispanicus*). Theriogenology 58: 1553–1562

García AJ, Landete-Castillejos T, Gomez-Brunet A, Garde JJ, Gallego L (2003) Characteristics of the oestrous cycle of Iberian red deer (*Cervus elaphus hispanicus*) assessed by progesterone profiles. J Exp Zool A Comp Exp Biol 298:143–149.

Garel M, Bonenfant C, Hamann J-L, Klein F, Gaillard J-M (2010) Are abundance indices derived from spotlight counts reliable to monitor red deer *Cervus elaphus* populations? Wildl Biol 16:77–84

Gaspar-López E, García AJ, Landete-Castillejos T, Carrión D, Estévez JA, Gallego L (2008) Growth of the first antler in Iberian red deer (*Cervus elaphus hispanicus*). Eur J Wildl Res 54:1–5

Gaspar-López E, Landete-Castillejos T, Estévez JA, Ceacero F, Gallego L, García AJ (2010) Biometrics, testosterone, cortisol and antler growth cycle in Iberian red deer. Reprod Domest Anim 45:243–249

Gebert C, Verheyden-Tixier H (2001) Variation of diet composition of red deer (*Cervus elaphus* L.) in Europe. Mammal Rev 31:189–201

Geist V (1998) Deer of the world: their evolution, behavior, and ecology. Stackpole Books, Mechanicsburg

Georgii B, Schröder W (1978) Radiotelemetrisch gemessene Aktivität weiblichen Rotwildes (*Cervus elaphus* L.). Z Jagdwiss 24:9–23

Gilbert L, Maffey GL, Ramsay SL, Hester AJ (2012) The effect of deer management on the abundance of *Ixodes ricinus* in Scotland. Ecol Appl 22:658–667

Gill RMA (1990) Monitoring the status of European and North American Cervids. Global environmental monitoring system. UNEP, Nairobi

Gordon I (1988) Facilitation of red deer grazing by cattle and its impact on performance. J Appl Ecol 25:1–10

Grignolio S, Apollonio M, Brivio F, Vicente J, Acevedo P, Palencia P, Petrovic K, Keuling O (2020) Guidance on estimation of abundance and density data of wild ruminant population: methods, challenges, possibilities. EFSA Support Publ 17(6):1876E, 1–54

Grubb P (2005) Order Artiodactyla. In: Wilson DE, Reeder DM (eds) Mammal species of the world. A taxonomic and geographic reference, 3rd edn. The Johns Hopkins University Press, Baltimore, pp 637–722

Guinness FE, Lincoln GA, Short RV (1971) The reproductive cycle of the female red deer, *Cervus elaphus* L. J Reprod Fertil 27:427–438

Gutiérrez-Expósito D, Arnal MC, Martínez-Durán D, Regidor-Cerrillo J, Revilla M, Fernández L, de Luco D, Jiménez-Meléndez A, Calero-Bernal R, Habela MA, García-Bocanegra I, Arenas-Montes A, Ortega-Mora LM, Álvarez-García G (2016) The role of wild ruminants as reservoirs of *Besnoitia besnoiti* infection in cattle. Vet Parasitol 223:7–13

Haanes H, Røed KH, Perez-Espona S, Rosef O (2011) Low genetic variation supports bottlenecks in Scandinavian red deer. Eur J Wildl Res 57:1137–1150

Hafner M (2011) Morfološke značilnosti rogovja navadnega jelena v vzhodnih Karavankah in Kamniško-Savinjskih Alpah (Morphological characteristics of antlers in the red deer population of eastern Karavanke and Kamnik Savinja Alps). Lovec 94:71–74

Haller H (ed) (2002) Der Rothirsch im Schweizerischen Nationalpark und dessen Umgebung. Nationalpark-Forschung in der Schweiz 91:1–144

Hamnes IS, Gjerde B, Robertson L, Vikøren T, Handeland K (2006) Prevalence of *Cryptosporidium* and *Giardia* in free-ranging wild cervids in Norway. Vet Parasitol 141:30–41

Härkönen S, Laine M, Vornanen M, Reunala T (2009) Deer ked (*Lipoptena cervi*) dermatitis in humans – an increasing nuisance in Finland. Alces 45:73–79

Hartl GB, Klein F, Willing R, Apollonio M, Lang G (1995) Allozymes and the genetics of antler development in red deer (*Cervus elaphus*). J Zool 237:83–100

Hatlapa HH (1977) Zur biologischen Bedeutung des Präorbitalorgan beim Rotwild: Prägung, Individualgeruch, Orientierung. Berl Münch Tierärztl Wochenschr 90:100–104

Heurich M, Zeis K, Küchenhoff H, Müller J, Belotti E, Bufka L, Woelfing B (2016) Selective predation of a stalking predator on ungulate prey. PLoS One 11(8): e0158449

Heuschele WP, Reid HW (2001) Malignant catarrhal fever. In: Williams ES, Baker IK (eds) Infectious diseases of wild mammals, 3rd edn. State University Press, Ames, pp 157–164

Hofmann RR (1985) Digestive physiology of the deer: their morphophysiological specialisation and adaptation. In: Drew K, Fennessy P (eds) Biology of deer Production. Roy Soc New Zeal Bull 22:393–407

Hu P, Shao Y, Xu J, Wang T, Li Y, Liu H, Rong M, Su W, Chen B, Cui S, Cui X, Yang F, Tamate H, Xing X (2019) Genome-wide study on genetic diversity and phylogeny of five species in the genus *Cervus*. BMC Genomics 20:384

Huisman J, Kruuk LEB, Ellis PA, Clutton-Brock T, Pemberton JM (2016) Inbreeding depression across the lifespan in a wild mammal population. Proc Natl Acad Sci U S A 113:3585–3590

Huxley JS (1931) The relative size of antlers in deer. J Zool 101:819–864

Iacopelli F, Fanelli A, Tizzani P, Berriatua E, Prieto P, Martínez-Carrasco C, León L, Rossi L, Candela MG (2020) Spatio-temporal patterns of sarcoptic mange in red deer and Iberian ibex in a multi-host natural park. Res Vet Sci 128:224–229

Iason GR, Guinness FE (1985) Synchrony of oestrus and conception in red deer (*Cervus elaphus* L.). Anim Behav 33:1169–1174

Jędrzejewski W, Schmidt K, Miłkowski L, Jędrzejewska B, Okarma H (1993) Foraging by lynx and its role in ungulate mortality: the local (Białowieża Forest) and the Palaearctic viewpoints. Acta Theriol 38:385–403

Jędrzejewski W, Jędrzejewska B, Okarma H, Schmidt K, Zub K, Musiani M (2000) Prey selection and predation by wolves in Białowieża Primeval Forest, Poland. J Mammal 81:197–212

Jiang Z, Hudson RJ (1992) Estimating forage intake and energy requirements of free-ranging wapiti (*Cervus elaphus*). Can J Zool 70:675–679

Kamler JF, Jędrzejewska B, Jędrzejewski W (2007) Activity patterns of red deer in Białowieża National Park, Poland. J Mammal 88:508–514

Kamler J, Jędrzejewski W, Jędrzejewska B (2008) Home ranges of red deer in a European old-growth forest. Am Midl Nat 159:75–82

Katona K, Gál-Bélteki A, Terhes A, Bartusz K, Szemethy L (2014) How important is supplementary feeding in the winter diet of red deer? A test in Hungary. Wildl Biol 20:226–234

Kidjo N, Cargnelutti B, Charlton BD, Wilson C, Reby D (2008) Vocal behaviour in the endangered Corsican deer: description and phylogenetic implications. Bioacoustics 18:159–181

Kröning F, Vorreyer F (1957) Untersuchungen über Vermehrungsquoten und Körpergewichte beim weiblichen Rotwild. Z Jagdwiss 3:145–153

Kropil R, Smolko P, Garaj P (2015) Home range and migration patterns of male red deer *Cervus elaphus* in Western Carpathians. Eur J Wildl Res 61:63–72

Kruuk LEB, Clutton-Brock TH, Rose KE, Guinness FE (1999) Early determinants of lifetime reproductive success differ between the sex in red deer. Proc Royal Soc B Biol Sci 266:1655–1661

Kuijper DPJ (2011) Lack of natural control mechanisms increases wildlife-forestry conflict in managed temperate European forest systems. Eur J For Res 130: 895–909

Kurtenbach K, Sewell HS, Ogden NH, Randolph SE, Nuttal PA (1998) Serum complement sensitivity as a key factor in Lyme disease ecology. Infect Immunol 66: 1248–1251

Landete-Castillejos T, García A, Gómez JA, Molina A, Gallego L (2003) Subspecies and body size allometry affect milk production and composition, and calf growth in red deer: comparison of *Cervus elaphus hispanicus* and *Cervus elaphus scoticus*. Physiological and Biochemical Zoology 76:594–602

Langvatn R (1986) Size and weight relationships in Norwegian red deer (*Cervus elaphus atlanticus*). In: Linn S (ed) CIC Symposium Das Rotwild, Graz, pp 244–267

Langvatn R, Loison A (1999) Consequences of harvesting on age structure, sex ratio and population dynamics of red deer *Cervus elaphus* in central Norway. Wildl Biol 5:213–223

Langvatn R, Mysterud A, Stenseth NC, Yoccoz NG (2004) Timing and synchrony of ovulation in red deer constrained by short northern summers. The American Naturalist 163:763–772

Langvatn R, Albon SD, Burkey T, Clutton-Brock TH (1996) Climate, plant phenology and variation in age of first reproduction in a temperate herbivore. J Anim Ecol 65:653–670

Latham J, Staines BW, Gorman ML (1997) Correlations of red (*Cervus elaphus*) and roe (*Capreolus capreolus*) deer densities in Scottish forests with environmental variables. J Zool 242:681–704

Lawson RE, Putman RJ, Fielding AH (2000) Individual signatures in scent gland secretions of Eurasian deer. J Zool 251:399–410

Lawson RE, Putman RJ, Fielding AH (2001) Chemical communication in Eurasian deer (Cervidae): do individual odours also code for attributes? J Zool 253: 91–99

Lemaître J-F, Gaillard J-M, Pemberton JM, Clutton-Brock TH, Nussey DH (2014) Early life expenditure in sexual competition is associated with increased reproductive senescence in male red deer. Proc R Soc B 281:20140792

Letková V, Lazar P, Soroka J, Goldová M, Curlík J (2008) Epizootiology of game cervid cysticercosis. Natura Croatica 17:311–318

Lincoln GA (1971) The seasonal reproductive changes in the red deer stag (*Cervus elaphus*). J Zool 163:105–123

Linnell JDC, Cretois B, Nilsen EB, Rolandsen C, Solberg EJ, Veiberg V, Kaczensky P, Van Moorter B, Panzacchi M, Rauset GR, Kalternborn B (2020) The challenges and opportunities of coexisting with wild ungulates in the human-dominated landscapes of Europe's Anthropocene. Biol Conserv 244:108500, 1–12

Lister AM (1986) New results on deer from Swanscombe, and the stratigraphic significance of deer in the Middle and Upper Pleistocene of Europe. J Archaeol Sci 13: 319–338

Lister AM (1995) Sea levels and the evolution of island endemics: the dwarf red deer of Jersey. Geol Soc Lond, Spec Publ 96:151–172

Loe LE, Mysterud A, Langvatn R, Stenseth NC (2003) Decelerating and sex-dependent tooth wear in Norwegian red der. Oecologia 135:346–253

Loison A, Langvatn R, Solberg EJ (1999) Body mass and winter mortality in red deer calves: disentangling sex and climate effects. Ecography 22:20–30

Lorenzini R, Garofalo L (2015) Insights into the evolutionary history of *Cervus* (Cervidae, tribe Cervini) based on Bayesian analysis of mitochondrial marker sequences, with first indications for a new species. J Zool Syst Evol Res 53:340–349

Lovari S, Cuccus P, Murgia A, Murgia C, Soi F, Plantamura G (2007) Space use, habitat selection and browsing effects of red deer in Sardinia. Ital J Zool 74: 179–189

Lovari S, Ferretti F, Corazza M, Minder I, Troiani N, Ferrari C, Saddi A (2014) Unexpected consequences of reintroductions: competition between reintroduced red deer and Apennine chamois. Anim Conserv 17: 359–370

Lovari S, Lorenzini R, Masseti M, Pereladova O, Carden RF, Brook SM, Mattioli S (2019) *Cervus elaphus*, red deer The IUCN Red List of Threatened Species

Ludt C, Schroeder W, Rottmann O, Kuehn R (2004) Mitochondrial DNA phylogeography of red deer (*Cervus elaphus*). Mol Phyl Evol 31:1064–1183

Mager F (1941) Wildbahn und Jagd Altpreussens. Neudamm, Berlin

Malcicka M (2015) Life history and biology of *Fascioloides magna* (Trematoda) and its native and exotic hosts. Ecol Evol 5:1381–1397

Malo AF, Roldán ERS, Garde J, Soler AJ, Gomendio M (2005) Antlers honestly advertise sperm production and quality. Proc R Soc B-Biol Sci 272(1559):149–157

Martín J, Carranza J, López P, Alarcos S, Pérez-González J (2014) A new sexual signal in rutting male red deer: age related chemical scent constituents in the belly black spot. Mamm Biol 79:362–368

Martin J, Tolon V, Morellet N, Santin-Janin H, Licoppe A et al. (2018) Common drivers of seasonal movements on the migration-residency behavior continuum in a large herbivore. Sci Rep 8:7631, 1–13

Mattioli L, Apollonio M, Mazzarone V, Centofanti E (1995) Wolf food habits and wild ungulate availability in the Foreste Casentinesi National Park, Italy. Acta Theriol 40:387–402

Mattioli S (1993) Antler conformation in red deer of the Mesola Wood, northern Italy. Acta Theriol 38:443–450

Mattioli S (1996) Morfometria dei palchi. In: Mazzarone V, Mattioli S (eds) Indagine sulla popolazione di cervo dell'Acquerino: relazione finale 1993–1995. Regione Toscana, Firenze, pp 13–20. (In Italian)

Mattioli S, De Marinis AM (2009) Guida al rilevamento biometrico degli Ungulati. Documenti Tecnici 28: 1–216. ISPRA (in Italian)

Mattioli S, Ferretti F (2014) Morphometric characterization of Mesola red deer *Cervus elaphus italicus* (Mammalia: Cervidae). Ital J Zool 81:144–154

Mattioli S, Fico R, Lorenzini R, Nobili G (2003) Mesola red deer: physical characteristics, population dynamics and conservation perspectives. Hystrix 14:87–94

Mattioli S, Ferretti F, Nicoloso S, Corlatti L (2021) Spatial variations in antler investment of Apennine red deer. Ecol Evol 11:7850–7864

Mayle BA, Peace AJ, Gill RMA (1999) How many deer? A guide to estimating deer population size. Forestry Commission Field Book 18, Forestry Commission, Edinburgh

Mazzarone V, Siemoni N, Pedone P, Lovari C, Mattioli L (1991) A method of red deer (*Cervus elaphus* L.) census during the roaring period in a forested area of the Northern Apennines. In: Csányi S, Ernhaft J (eds) Transactions of the 20th I.U.G.B. Congress, Gödöllo, pp 140–145

McComb K (1987) Roaring by red deer stags advances the date of oestrus in hinds. Nature 330:648–649

McComb K (1991) Female choice for high roaring rate in red deer, *Cervus elaphus*. Anim Behav 41:79–88

Meiri M, Lister AM, Higham TFG, Stewart JR, Straus LG, Obermaier H, Morales MRG, Marin-Arroyo AB, Barnes I (2013) Late-glacial recolonization and phylogeography of European red deer (*Cervus elaphus* L.). Mol Ecol 22:4711–4722

Meiri M, Kosintsev P, Conroy K, Meiri S, Barnes I, Lister A (2018) Subspecies dynamics in space and time: a study of the red deer complex using ancient and modern DNA and morphology. J Biogeogr 45:367–380

Meisingset EL, Loe LE, Brekkum O, Bisch R, Rivrud IM, Lande US, Zimmermann B, Veiberg V, Mysterud A (2018) Spatial mismatch between management units and movement ecology of a partially migratory ungulate. J Appl Ecol 55:745–753

Milner JM, Bonenfant C, Mysterud A, Gaillard J-M, Csányi S, Stenseth NC (2006) Temporal and spatial development of red deer harvesting in Europe: biological and cultural factors. J Appl Ecol 43:721–734

Milner JM, van Beest FM, Schmidt K, Brook RK, Storaas T (2014) To feed or not to feed? Evidence of the intended and unintended effects of feeding wild ungulates. J Wildl Manag 78:1322–1334

Mitchell B (1967) Growth layers in dental cement for determining the age of red deer (*Cervus elaphus* L.). J Anim Ecol 36:279–293

Mitchell B, McCowan D, Nicholson IA (1976) Annual cycles of body weight and condition in Scottish red deer, *Cervus elaphus*. J Zool 180:107–127

Mitchell B, Staines B, Welch D (1977) Ecology of red deer: a research review relevant to their management in Scotland. Institute of Terrestrial Ecology, Cambridge

Morellet N, Gaillard J-M, Hewison AJM, Ballon P, Boscardin Y, Duncan P, Klein F, Maillard D (2007) Indicators of ecological change: new tools for managing populations of large herbivores. J Appl Ecol 44:634–643

Morellet N, Klein F, Solberg E, Andersen R (2011) The census and management of populations of ungulates in Europe. In: Putman R, Apollonio M, Andersen R (eds) Ungulate management in Europe: problems and practices. Cambridge University Press, Cambridge, pp 106–143

Moyes K, Nussey DH, Clements MN, Guinness FE, Morris A, Morris S, Pemberton JM, Kruuk LEB, Clutton-Brock TH (2011) Advancing breeding phenology in response to environmental change in a wild red deer population. Glob Chang Biol 17:2455–2469

Muir PD, Sykes AR, Barrell GK (1987) Growth and mineralization of antlers in red deer (*Cervus elaphus*). N Z J Agric Res 30:305–315

Murgia A, Mandas L, Mondoloni S, Riga F (2017) Current status, distribution and recovery plan of *Cervus elaphus corsicanus*. In: Bro E, Guillemain M (eds) 33rd IUGB congress, Montpellier, pp 243–244

Mysterud A (2010) Still walking on the wild side ? Management actions as steps toward 'semidomestication' of hunted ungulates. J Appl Ecol 47:920–925

Mysterud A, Loe LE, Zimmermann B, Bischof R, Veiberg V, Meisingset E (2001a) Partial migration in expanding red deer populations at northern latitudes- a role for density dependence. Oikos 120:1817–1825

Mysterud A, Yoccoz NG, Stenseth NC, Langvatn R (2001b) Effects of age, sex and density on body weight of Norwegian red deer: evidence of density-dependent senescence. Proc R Soc Lond B 268:911–919

Mysterud A, Meisingset EL, Langvatn R, Yoccoz NG, Stenseth NC (2005) Climate-dependent allocation of resources to secondary sexual traits in red deer. Oikos 111:245–252

Mysterud A, Bonenfant C, Loe LE, Langvatn R, Yoccoz NG, Stenseth NC (2008) The timing of male reproduction effort relative to female ovulation in a capital breeder. J Anim Ecol 77:469–477

Mysterud A, Loe LE, Zimmermann B, Bischof R, Veiberg V, Meisingset E (2011) Partial migration in expanding red deer populations at northern latitudes – a role for density dependence? Oikos 120:1817–1825

Nettleton PH, Sinclair JA, Herring JA, Inglis DM, Fletcher TJ, Ross HM, Bonniwell MA (1986) Prevalence of herpesvirus infection in British red deer and investigation of further disease outbreaks. Vet Rec 118:267–270

Neumann A (1968) Rotwildpopulation Hohenbucko. Beiträge zur Jagd- und Wildforschung 6:93–101

Niedziałkowska M, Jędrzejewska B, Honnen A-C, Otto T, Sidorovich VE, Perzanowski K, Skog A, Hartl GB, Borowik T, Bunevich AN, Lang J, Zachos FE (2011) Molecular biogeography of red deer *Cervus elaphus* from eastern Europe: insights from mitochondrial DNA sequences. Acta Theriol 56:1–12

Niedziałkowska M, Jędrzejewska B, Wójcik JM, Goodman SJ (2012) Genetic structure of red deer population in northeastern Poland in relation to the history of human interventions. J Wildl Manag 76:1264–1276

Niedziałkowska M, Doan K, Górny M, Sykut M, Stefaniak K et al (2021) Winter temperature and forest cover have shaped red deer distribution in Europe and the Ural Mountains since the Late Pleistocene. J Biogeogr 48: 147–159

Nigsch A, Glawischnig W, Bagó Z, Greber N (2019) *Mycobacterium caprae* infection of Red Deer in western Austria – optimized use of pathology data to infer infection dynamics. Front Vet Sci 21:350

Nussey DH, Coltman DW, Coulson T, Kruuk LEB, Donald A, Morris SJ, Clutton-Brock TH, Pemberton JM (2005) Rapidly declining fine-scale spatial genetic structure in female red deer. Mol Ecol 14:3395–3405

Nussey DH, Pemberton J, Donald A, Kruuk LEB (2006) Genetic consequences of human management in an introduced island population of red deer (*Cervus elaphus*). Heredity 97:56–65

Nussey DH, Metherell B, Moyes K, Donald A, Guinness FE (2007) The relationship between tooth wear, habitat quality and late-life reproduction in a wild red deer population. J Anim Ecol 76:402–412

Nussey DH, Kruuk LEB, Morris A, Clements MN, Pemberton JM, Clutton-Brock TH (2009) Inter- and intrasexual variation in aging patterns across reproduction traits in a wild red deer population. Am Nat 174: 342–357

O'Gara BW (2002) Taxonomy. In: Toweill DE, Thomas JW (eds) North American elk: ecology and management. Smithsonian Institution Press, Washington/London, pp 3–65

Okarma H (1995) The trophic ecology of wolves and their predatory role in ungulate communities of forest ecosystems in Europe. Acta Theriol 40:335–386

Oleaga A, Casais R, González-Quirós P, Prieto M, Gortázar C (2008) Sarcoptic mange in red deer from Spain: improved surveillance or disease emergence? Vet Parasitol 154:103–113

Passilongo D, Reby D, Carranza J, Apollonio M (2013) Roaring high and low: composition and possible functions of the Iberian stag's vocal repertoire. PLoS One 8(5):e63841, 1–9

Pemberton JM, Albon SD, Guinness FE, Clutton-Brock TH, Dover GA (1992) Behavioural estimates of male mating success tested by DNA fingerprinting in a polygynous mammal. Behav Ecol 3:66–75

Pérez-Barbería FJ, Carranza J, Sánchez-Prieto C (2015) Wear fast, die young: more worn teeth and shorter lives in Iberian compared to Scottish red deer. PLoS One 10(8):e0134788

Pérez-Espona S, Hall RJ, Pérez-Barbería FJ, Glass BC, Ward JF, Pemberton JM (2013) The impact of past introductions on an iconic and economically important species, the red deer of Scotland. J Hered 104:14–22

Pérez-González J, Carranza J, Torres-Porras J, Fernández-García J (2010) Low heterozygosity at microsatellite markers in Iberian red deer with small antlers. J Hered 101:553–561

Prokešová J (2004) Red deer in floodplain forest: the browse specialist? Folia Zool 53:293–302

Pucek Z, Bobek B, Łabudzki L, Miłkowski L, Morow K, Tomek A (1975) Estimates of density and number of ungulates. Pol Ecol Stud 1:121–135

Putman RJ, Staines BW (2004) Supplementary winter feeding of wild red deer *Cervus elaphus* in Europe and North America: justifications, feeding practice and effectiveness. Mammal Rev 34:285–306

Putman RJ, Langbein J, Hewison AJM, Sharma SK (1996) Relative roles of density-dependent and density-independent factors in the population dynamics of British deer. Mammal Rev 26:81–101

Putman RJ, Nelli L, Matthiopoulos J (2019) Changes in bodyweight and productivity in resource-restricted populations of red deer (*Cervus elaphus*) in response to deliberate reductions in density. Eur J Wildl Res 65:13

Pyziel AM, Laskowski Z, Demiaszkiewicz AW, Hoglund J (2017) Interrelationships of *Dictyocaulus* spp. in wild ruminants with morphological description of *Dictyocaulus cervi* n. sp. (Nematoda: Trichostrongyloidea) from red deer, *Cervus elaphus*. J Parasitol 103:506–518

Queirós J, Acevedo P, Santos JPV, Barasona J, Beltran-Beck B, Gonzalez-Barrio D, Armenteros JA, Diez-Delgado I, Boadella M, de Mera IF, Ruiz-Fons JF, Vicente J, de la Fuente J, Gortazar C, Searle JB, Alves PC (2019) Red deer in Iberia: molecular ecological studies in a southern refugium and inferences on European postglacial colonization history. PLoS One 14(1):e0210282

Ratcliffe PR (1984) Population dynamics of red deer (*Cervus elaphus* L) in Scottish commercial forests. Proc Roy Soc Edinb 82B:291–302

Ratcliffe PR (1987) Distribution and current status of sika deer, *Cervus nippon*, in Great Britain. Mammal Rev 17: 39–58

Reby D, McComb K (2003) Anatomical constraints generate honesty: acoustic cues to age and weight in the roars of red deer stags. Anim Behav 65:519–530

Rehbein S, Lutz W, Visser M, Winter R (2002) Investigation of the parasite fauna of game animals of Northrhine-Westphalia. 3. Endoparasites of red deer. Z Jagdwiss 48:69–93

Reiner G, Tramberend K, Nietfeld F, Volmer K, Wurmser C, Fries R, Willems H (2020) A genome-wide scan study identifies a single nucleotide substitution in the tyrosinase gene associated with white coat colour in a red deer (*Cervus elaphus*) population. BMC Genet 21:14

Reiner G, Klein C, Lang M, Willelms H (2021) Human-driven genetic differentiation in a managed red deer population. Eur J Wildl Res 67:29

Richard E, Gaillard J-M, Saïd S, Hamann J-L, Klein F (2010) High red deer density depresses body mass of roe deer fawns. Oecologia 163:91–97

Rivals F, Lister AM (2016) Dietary flexibility and niche partitioning of large herbivores through the Pleistocene of Britain. Quat Sci Rev 146:116–133

Rivrud IM, Bischof R, Meisingset EL, Zimmermann B, Loe LE, Mysterud A (2016) Leave before it's too late: anthropogenic and environmental triggers of autumn migration in a hunted ungulate population. Ecol 97:1058–1068

Rivrud IM, Meisingset EL, Loe LE, Mysterud A (2019) Future suitability of the habitat in a migratory ungulate under climate change. Proc R Soc B 286:20190442

Robinson S, Milner-Gulland E, Grachev Y, Zuther S, Orynbaev M, Lushchekina A, Morgan E, Beauvais W, Singh N, Khomenko S, Cammack R, Kock R (2019) Opportunistic bacteria and mass mortality in ungulates: lessons from an extreme event. Ecosphere 10:e02671

Røskaft E (1978) Kraniometriske målinger hos hjort *Cervus elaphus* L. – en vurdering av vekst- og størreiesrelasjoner. (Craniometry of red deer: an assessment of growth and size relationships). Thesis, University of Trondheim (in Norwegian)

Rossi S, Balenghien T, Viarouge C, Faure E, Zanella G, Sailleau C, Mathieu B, Delécolle JC, Ninio C, Garros C, Gardès L, Tholoniat C, Ariston A, Gauthier D, Mondoloni S, Barboiron A, Pellerin M, Gibert P, Novella C, Barbier S, Guillaumat E, Zientara S, Vitour D, Bréard E (2019) Red deer (*Cervus elaphus*) did not play the role of maintenance host for bluetongue virus in France: the burden of proof by long-term wildlife monitoring and *Culicoides* snapshots. Viruses 11:903

Saarinen J, Eronen J, Fortelius M, Seppä H, Lister AM (2016) Patterns of diet and body mass of large ungulates from the Pleistocene of Western Europe, and their relation to vegetation. Palaeontol Electron 19.3.32A:1–58

Sadleir RMFS (1987) Reproduction of female Cervids. In: Wemmer CM (ed) Biology and management of the Cervidae. Smithsonian Institution Press, Washington, DC, pp 123–144

Scagliarini A, Vaccari F, Turrini F, Bianchi A, Cordioli P, Lavazza A (2011) Parapoxvirus infections of red deer, Italy. Emerg Infect Dis 17:684–687

Schmidt KT, Hoi H (2002) Supplemental feeding reduces natural selection in juvenile red deer. Ecography 25: 265–272

Skog A, Zachos FE, Rueness EK, Feulner PGD, Mysterud A, Langvatn R, Lorenzini R, Hmwe SS, Lehoczky I, Hartl GB, Stenseth NC, Jakobsen KS (2009) Phylogeography of red deer (*Cervus elaphus*) in Europe. J Biogeogr 36:66–77

Skonhoft A, Veiberg V, Gauteplass A, Olaussen JO, Meisingsets EL (2013) Balancing income and cost in red deer management. J Environ Manag 115:179–188

Slate J, Kruuk LEB, Marshall TC, Pemberton JM, Clutton-Brock TH (2000) Inbreeding depression influences lifetime breeding success in a wild population of red deer (*Cervus elaphus*). Proc R Soc Lond B 267:1657–1662

Sommer RS, Zachos FE, Street M, Jöris O, Skog A, Benecke N (2008) Late Quaternary distribution dynamics and phylogeography of the red deer (*Cervus elaphus*) in Europe. Quat Sci Rev 27:714–733

Staines BW, Crisp J (1978) Observations on food quality in Scottish red deer (*Cervus elaphus*) as determined by chemical analysis of the rumen contents. J Zool 185: 253–259

Staines BW, Crisp JM, Parish T (1982) Differences in the quality of food eaten by red deer (*Cervus elaphus*) stags and hinds in winter. J Appl Ecol 19:65–77

Staines BW, Langbein J, Burkitt TD (2008) Red deer *Cervus elaphus*. In: Harris S, Yalden DW (eds) Mammals of the British Isles: handbook, 4th edn. The Mammal Society, Southampton, pp 573–587

Stanton DWG, Mulville JA, Bruford MW (2016) Colonization of the Scottish islands via long-distance Neolithic transport of red deer (*Cervus elaphus*). Proc Royal Soc B 283:201600952016

Suttie JM, Mitchell B (1983) Jaw length and hind foot length as measure of skeletal development of red deer (*Cervus elaphus*). J Zool 200:431–434

Szaniawski A (1966) Osteologische Untersuchungen über den Rothirsch in Polen. Acta Theriol 10:195–267

Szunyoghy J (1963) A magyarországi szarvas. Das ungarische Rotwild. Hungarian Natural History Museum, Budapest

Tomek A (2002) Właściwości i struktura populacji jelenia (*Cervus elaphus* L.) w lasach Krynickich (Karpaty). Zeszyty Naukowe AR w Krakowie, Rozprawy 278: 1–100. (in Polish)

Torres-Porras J, Carranza J, Pérez-González J, Mateos C, Alarcos S (2014) The tragedy of the commons: unsustainable population structure of Iberian red deer in hunting estates. Eur J Wildl Res 60:351–357

Tsahar E, Izhaki I, Lev-Yadun S, Bar-Oz G. (2009) Distribution and extinction of ungulates during the Holocene of the southern Levant. PLoS One 4(4):e5316

Tucak Z (1997) Morphometrische Eigenschaften der Rothirsche (*Cervus elaphus* L.) aus dem Donaugebiet in Baranja. Z Jagdwiss 43:141–153

Tucak Z, Banaj B, Šubaric D (1999) Beitrag zur Morphometrie des Rotwildes (*Cervus elaphus* Linné 1758) aus dem Donaugebiet Bačka. Z Jagdwiss 45: 27–133

Turbill C, Ruf T, Mang T, Arnold W (2011) Regulation of heart rate and rumen temperature in red deer: effects of season and food intake. J Exp Biol 214:963–970

Uerpmann H-P (1987) The ancient distribution of Ungulate Mammals in the Middle East Fauna and Archeological sites in Southwest Asia and Northeast Africa. Reichert V., Wiesbaden

Valcárcel F, González J, Tercero Jaime JM, Olmeda AS (2016) Long term study of ixodid ticks feeding on red deer (*Cervus elaphus*) in a meso-Mediterranean climate. Exp Appl Acarol 69:61–72

Valentinčić SI (1958) Beitrag zur Kenntnis der Reproduktionserscheinungen beim Rotwild. Z Jagdwiss 4:105–130

Vannini C, Fattorini N, Mattioli S, Nicoloso S, Ferretti F (2021) Land cover and weather jointly predict biometric indicators of phenotypic quality in a large herbivore. Ecol Indic 128:107818

Veiberg V, Loe LE, Mysterud A, Langvatn R, Stenseth NC (2004) Social rank, feeding and winter weight loss in red deer: any evidence of interference competition? Oecologia 138:135–142

Veiberg V, Nilsen EB, Rolandsen CM, Heim M, Andersen R, Holmstrøm F, Meisingset EL, Solberg EJ (2020) The accuracy and precision of age determination by dental cementum annuli in four northern cervids. Eur J Wildl Res 66:91

Verheyden-Tixier H, Renaud P-C, Morellet N, Jamot J, Besle J-M, Dumond B (2008) Selection for nutrients by red deer hinds feeding on a mixed forest edge. Oecologia 156:715–726

Vicente J, Fierro Y, Martinez M, Gortazar C (2004) Long-term epidemiology, effect on body condition and interspecific interactions of concomitant infection by nasopharyngeal bot fly larvae (*Cephenemyia auribarbis* and *Pharyngomyia picta*, Oestridae) in a population of Iberian red deer (*Cervus elaphus hispanicus*). Parasitology 129:349–361

Vigne J-D (1992) Zooarchaeological and biogeographical history of the mammals of Corsica and Sardinia since the last ice age. Mamm Rev 22:87–96

Vikøren T, Våge J, Madslien KI, Røed KH, Rolandsen CM et al. (2019) First detection of Chronic Wasting Disease in a wild red deer (*Cervus elaphus*) in Europe. J Wild Dis 55:970–972

Vislobokova IA (2008) The major stages in the evolution of artiodactyl communities from the Pliocene to early Middle Pleistocene of northern Eurasia: part 1. Paleontol J 42(3):297–312

Volodin I, Volodina E, Frey R, Carranza J, Torres-Porras J. (2013) Spectographic analysis points to source-filter coupling in rutting roars of Iberian red deer. Acta Ethol 16:57–63

Volodin IA, Nahlik AJ, Tari T, Frey R, Volodina EV (2019) Rutting roars in native Pannonian red deer of Southern Hungary and the evidence of acoustic divergence of male sexual vocalization between Eastern and Western European red deer (*Cervus elaphus*). Mamm Biol 94: 54–65

von Raesfeld F, Reulecke K (1988) Das Rotwild, 9th edn. P. Parey Verlag, Hamburg/Berlin

Wagenknecht E, ed. (1984) Altersbestimmung des erlegten Wildes. Neumann-Neudamm, Melsungen

Wenham G, Adam CL, Moir CE (1986) A radiographic study of skeletal growth and development in fetal red deer. Br Vet J 142:336–349

Wierzbowska I (1999) Zmienność wymiarów czaszek populacji jelenia szlachetnego (*Cervus elaphus* L) w Karpatach (Size variation in skulls of red deer populations of Polish Carpathians). PhD thesis, University of Kraków (in Polish)

Wölfel H (1984) Jugendentwicklung, Mutter-Kind-Bindung und Feindvermeidung beim Rothirsch III. Z Jagdwiss 30:3–14

Wotschikowsky U (2010) Ungulates and their management in Germany. In: Apollonio M, Andersen R, Putman R (eds) European ungulates and their management in the 21st century. Cambridge University Press, Cambridge, UK, pp 201–222

Zachos FE, Hartl GB (2011) Phylogeography, population genetics and conservation of the European red deer *Cervus elaphus*. Mammal Rev 41:138–150

Zachos FE, Althoff C, v. Steynitz Y, Eckert I, Hartl GB. (2007) Genetic analysis of an isolated red deer (*Cervus elaphus*) population showing signs of inbreeding depression. Eur J Wildl Res 53:61–67

Zachos FE, Mattioli S, Ferretti F, Lorenzini R (2014) The unique Mesola red deer of Italy: taxonomic recognition (*Cervus elaphus italicus* nova ssp., Cervidae) would endorse conservation. Ital J Zool 81:136–143

Zachos FE, Frantz AC, Kuehn R, Bertouille S, Colyn M, Niedziałkowska M, Pérez-González J, Skog A, Sprĕm N, Flamand M-C (2016) Genetic structure and effective population sizes in European red deer (*Cervus elaphus*) at a continental scale: insights from microsatellite DNA. J Hered 107:318–326

Zaffaroni E, Manfredi MT, Citterio C, Sala M, Piccolo G, Lanfranchi P (2000) Host specificity of abomasal nematodes in free ranging alpine ruminants. Vet Parasitol 90:221–230

Zanella G, Durand B, Hars J, Garin-Bastuji B, Duvauchelle A, Ferme M, Karoul C, Boschiroli MI (2008) *Mycobacterium bovis* in wildlife in France. J Wildl Dis 44:99–108

Zanella G, Bar-Hen A, Boschiroli ML, Hars J, Moutou F, Garin-Bastuji B, Durand B (2012) Modelling transmission of bovine tuberculosis in red deer and wild boar in Normandy, France. Zoonoses Public Health 59(Suppl 2):170–178

Zanet S, Trisciuoglio A, Bottero E, de Mera IG, Gortazar C, Carpignano MG, Ferroglio E (2014) Piroplasmosis in wildlife: *Babesia* and *Theileria* affecting free-ranging ungulates and carnivores in the Italian Alps. Parasit Vectors 7:70

Zweifel-Schielly B, Leuenberger Y, Kreuzer M, Suter W (2012) A herbivore's food landscape: seasonal dynamics and nutritional implications if diet selection by a red deer population in contrasting Alpine habitats. J Zool 286:68–80

Sika *Cervus nippon* Temminck, 1838

4

Rory Putman, Jarmila Krojerová-Prokešová, and Miroslava Barančeková

Contents

R. Putman (✉)
Institute of Biodiversity, Animal Health and Comparative Medicine, University of Glasgow, Glasgow, UK
e-mail: rory.putman@glasgow.ac.uk

J. Krojerová-Prokešová
Institute of Vertebrate Biology, Czech Academy of Sciences, Brno, Czech Republic
e-mail: krojerova@ivb.cz

M. Barančeková
Department of Infectious Diseases, Veterinary University Brno, Palackeho trida, Brno, Czech Republic
e-mail: barancekovam@vfu.cz

© Springer Nature Switzerland AG 2022

L. Corlatti, F. E. Zachos (eds.), *Terrestrial Cetartiodactyla*, Handbook of the Mammals of Europe,
https://doi.org/10.1007/978-3-030-24475-0_20

Common Names

English	Sika
German	Sika(-hirsch)
French	Cerf sika
Spanish	Sika
Italian	Cervo sika
Russian	Пятнистый олень

While commonly referred to in English as sika deer, the addition of the word "deer" is tautologous, since the word "sika" in Japanese itself means deer.

Taxonomy, Systematics and Paleontology

Sika deer (Fig. 1) originated in the mainland of Northeastern Asia (the Southern Ussuri, Korea, Manchuria, and Northeastern China) and subsequently spread to Southeastern China, Vietnam, Taiwan, and, about 0.43 mya, also to the Japanese Archipelago (Cook et al. 1999; Kawamura 2009; Nagata 2009; Wilson 2000). A small number of fossil records (primarily antlers) of sika *C. nippon* have been dated to the Pleistocene, while numerous fossils of this species have been dated to the Holocene (Kuwabara 2004; Takahashi et al. 2003; Takakuwa 2004, 2006; Kawamura 2009). Based on the dating of the fossil records it can be concluded that *C. nippon* inhabited China since the early Middle Pleistocene and Japan since the Middle Pleistocene (Kawamura 2009).

Based on morphological characters, sika deer in the world were originally classified into 13 subspecies (Ellerman and Morrison-Scott 1951; Whitehead 1993). Six of these subspecies are found in Japan (Ohtaishi 1986): *Cervus nippon nippon* (Kyushu Island, Shikoku Island, and Goto Island), *C. n. yesoensis* Heude, 1884 (Hokkaido Island), *C. n. centralis* Kishida, 1936 (Honshu and Tsushima Island), *C. n. mageshimae* Kuroda and Okada, 1950 (Mageshima and Tanegashima Island), *C. n. yakushimae* Kuroda and Okada, 1950 (Yakushima and Kuchinoerabu Island), and *C. n. keramae* Kuroda, 1924 (Ryukyu Island).

According to Geist (1999) and Groves and Grubb (2011) three different subspecies were

Fig. 1 Sika deer (*Cervus nippon*). (**a**) Female, (**b**) male (photographs © Andrea Dal Pian)

recognized in the Northeastern mainland Asia: *C. n. hortulorum* Swinhoe, 1864, *C. n. mantchuricus* Swinhoe, 1864, and *C. n. dybowskii* Taczanowski, 1876. However, the problem with nomenclature of these subspecies is that only *C. n. dybowskii* was described based on the wild animal from Ussuri region. The other two subspecies were described based on captive animals: *C. n. hortulorum* based on animals from Imperial Summer Palace Beijing (delivered from Altai deer farms, where animals from Far East Russia had been imported) and *C. n. mantchuricus* on individuals from Harbour city of Jingkou. Currently, it is widely accepted that *C. n. hortulorum* and *C. n. dybowskii* are the same and that the correct Latin name for the subspecies is *C. n. hortulorum*.

There remains some dispute as to whether *C. n. hortulorum* and *C. n. mantchuricus* are also the same or not; in the study of Ba et al. (2015; see further below) genetic differentiation between their haplotypes was only 1%, less than that revealed between two different cyt *b* haplotypes detected in the native range of *C. n. hortulorum* in Primorsky Krai (Krojerová-Prokešová et al. 2013), so it is probable that *C. n. hortulorum* and *C. n. mantchuricus* also represent the same subspecies. There are similar discrepancies in English names used for this/these subspecies – three names are used: Manchurian sika, Dybowski's sika, and Ussuri sika. In this handbook chapter we consider all northeastern mainland sika as Dybowski's sika (*C. n. hortulorum*) as these are the names mostly used for mainland Russian forms of sika.

Besides *C. n. hortulorum,* two other subspecies are described in Southern and Western China: *C. n. sichuanicus* Guo, Chen, and Wang, 1978 and *C. n. kopschi* Swinhoe, 1873 (Guo and Zheng 2000). Further, distinct subspecies were present in Vietnam (*C. n. pseudaxis* Gervais, 1876) and on the island of Taiwan (Formosa; *C. n. taiouanus* Blyth, 1860). While the population in Taiwan have subsequently been reintroduced (McCullough 2009a, b), in Vietnam only captive populations survive, with a rather low, but not dangerously low, genetic diversity (Thévenon et al. 2004).

Overall, however the taxonomy is somewhat confused. There has clearly been much mixing of stocks, particularly on the mainland, and more recently in transfer of animals between the various islands of Japan and there is thus in effect a continuum of intermediate forms. Some authorities have also suggested, based on its morphological features and its karyology, that many of the mainland forms may themselves represent ancient hybrids between sika and red deer or sika and Manchurian wapiti (*Cervus canadensis xanthopygus*) (Bartoš and Žirovnický 1981; Goodman et al. 2001; Herzog 1987, 1995; Lowe and Gardiner 1975). This theory was also supported by the fact that that these taxa can naturally hybridize in the wild as well as in captive and introduced populations (Bartoš 2009; Bartoš and Žirovnický 1981; Geist 1999; Mirolyubov and Ryaschenko 1948; Prisyazhnyuk 2005; Sokolov 1959). However, recent phylogenetic analyses rejected this hypothesis and mainland forms of sika are considered to be a clear sika subspecies (Cook et al. 1999; Krojerová-Prokešová et al. 2013; Nagata et al. 1999; Pitra et al. 2004; Randi et al. 2001).

Subspecific classification based on morphology is not completely supported by genetic studies. Mitochondrial DNA (mtDNA), especially the most variable D-loop region, was used as a powerful genetic marker for analyzing the population structure of sika deer in Japan (Cook et al. 1999; Nagata et al. 1995, 1999; Tamate and Tsuchiya 1995; Tamate et al. 1998; Yamada et al. 2006; Yoshio et al. 2008; Yuasa et al. 2007). The maternally inherited markers clearly split Japanese sika populations into two distinct mitochondrial lineages, the southern and the northern group (Nagata et al. 1999), which does not reflect the morphological classification. The present boundary between these two lineages does not correspond to one of the sea channels separating the Japanese islands, but is located in the western part of the Honshu and Shikoku Islands (Nagata et al. 1999; Yamada et al. 2006), where there is no obvious geographical barrier to dispersal. This separation of Japanese sika into these two lineages has subsequently been confirmed also by studies of nuclear DNA, although the border between the two lineages seems to be moved southward due to male-biased dispersal (Krojerová-Prokešová et al. 2017). Furthermore, analyses showed that the Southern Japanese lineage is more closely related to the mainland sika subspecies, than to the Northern Japanese lineage (Nagata 2009) and thus the divergence of these lineages probably occurred before the colonization of Japan from China (Kawamura 2009).

There have been relatively few genetic studies offering reevaluation of the status of putative subspecies on the mainland of Asia. Some kind of distinction between South China sika *C. n. kopschi,* Vietnamese sika *C. n. pseudaxis*, and Dybowski's sika *C. n. hortulorum* was confirmed by studies dedicated to particular subspecies (Barančeková et al. 2012; Wilson 2000; Wu et al. 2004), while Ba et al. (2015) is the only study simultaneously to analyze specimens of all known subspecies together, suggesting two main lineages across mainland Asia (a northern mainland group formed by *C. n. hortulorum/ mantchuricus* and a southern mainland/Taiwan group consisting of *C. n. sichuanicus, C. n. kopschi, C. n. pseudaxis,* and *C. n. taiouanus*) and supported closer relationship of Southern Japanese lineage to mainland Asian subspecies previously detected by Nagata (2009).

In a rather more radical review of the taxonomy, Groves (2006) and subsequently Groves and Grubb (2011) have suggested that sika should be divided into four distinct species: *C. nippon* of Southern Japan, *C. yesoensis* of Central and Northern Japan (styled *C. aplodontus* by Groves and Grubb 2011), *C. taiouanus* of Taiwan, and *C. hortulorum* of the mainland range. A recognition of two distinct genetic types among mainland forms and two in the Japanese islands conforms to a degree with Groves (2006) and Groves and Grubb's (2011) proposal of four distinct "species." However, this specific status is not widely accepted and in these pages we consider all taxa as subspecies of *C. nippon*, pending further information.

Current Distribution

Sika in their native range within Asia have greatly suffered due to the habitat changes (Yuasa et al. 2007) and due to uncontrolled hunting, which led to the extinction of local wild populations in Vietnam, Taiwan, and South Korea (McCullough 2009a, b). The historical range of Dybowski's sika was primarily in Primorsky Krai extending to the borders of China (Aramilev 2009). According to scarce historical records, the population numbers of Dybowski's sika in Far East Russia were substantially reduced at the end of the nineteenth and the beginning of the twentieth century, primarily by overhunting and the capture of animals for deer farms (Makovkin 1999). In the 1940s, the population was estimated to be at about 300 individuals (Bromley 1956), 160 of which were located in the present Lazovsky Reserve (Voloshina and Myslenkov 2009). The population numbers started to increase in the 1970s and by the end of the 1980s the abundance of sika was two to three times higher within the whole area of the Primorsky Krai than when at its lowest ebb. Reoccupation of the original native range consisted of both wild and farm-escaped sika deer, but it is considered that all should be part of the original genetic stock (Aramilev 2009; Krojerová-Prokešová et al. 2013). At present, the Russian population is the most numerous one, estimated at >20,000 individuals and appears to be relatively stable (Aramilev 2009). The size of the current population in China is uncertain and probably consists of sporadic isolated individuals that are likely dispersers across Russian and North Korean borders (McCullough 2009a). Wild sika deer are extinct in South Korea (Won and Smith 1999) and the current status of North Korean population is likewise unknown (McCullough 2009a).

The South China sika deer (*C. n. kopschi*) and Sichuan sika deer (*C. n. sichuanicus*) are the only remaining subspecies occurring in significant numbers in the wild of China. Currently, the Sichuan sika deer is distributed in three unconnected areas located in Northwest Sichuan province and Southwest Gansu province. Sichuan is occupied by the largest population, consisting of about 850 individuals (Guo 2000; Guo and Zheng 2000; He et al. 2014). The South China sika deer population of around 300 individuals currently exists in three isolated populations (Jiangxi, Anhui, and Zhejiang province).

Even though the Japanese island forms of sika are both widespread and abundant in their native Japan, deforestation and uncontrolled hunting in the second half of the nineteenth century together with severe winters in 1879 and 1903 led to the extinction of many local populations (Nagata et al. 1998). Sika markedly increased in numbers and extended its geographic distribution in Japan over the last few decades; the range of sika deer expanded by 1.7 times from 1978–2003 (Saitoh et al. 2015).

Sika have been reintroduced in places within their natural range where populations have become locally extinct (especially island populations: e.g., Taiwan). In addition, they have become introduced into adjacent territory within their natural biogeographic range (e.g., more widely in Russia and in the Ukraine). In addition, over the last two centuries, sika have been introduced into many other parts of the world. Outside Asia, free-ranging populations are established in Australasia (New Zealand); North America (Kentucky, Maryland, North Carolina, Texas, and Virginia); and Europe (Austria, Czech Republic, Denmark, France, Germany, Ireland, Poland, and the UK).

Altogether, sika were introduced to 35 different European countries at the end of the nineteenth century and the beginning of the twentieth century (Bartoš 2009). Most of these introductions date from the last years of the nineteenth century (1890s) through to 1930s, although some have continued until more recently. The most authoritative reviews of distribution, history of introductions, and current status can be found in Banwell (2009), Bartoš (2009), Feldhamer and Demerais (2009), McCullough et al. (2009), Swanson and Putman (2009), and Apollonio et al. (2010), which document dates and history of introductions/reintroductions. See also Eick et al. (1995) for Continental Europe and Ratcliffe (1987) for the UK.

The geographical origin and subspecies status of most of the introduced individuals can no longer be ascertained as the archives of the main supplier, the Hagenbeck Company from Hamburg, were destroyed (Bartoš 2009). Recent phylogenetic studies using mtDNA markers have however confirmed that the majority of established feral populations is of Japanese sika origin (both northern and southern genetic groups); fewer successful introductions would appear to have been made of sika from mainland Asia and the majority of these (usually from Far Eastern Russia) have remained in enclosures rather then becoming established in the wild (Apollonio et al. 2010). For example: of sika introduced to the Czech Republic, recent mtDNA analyses have revealed that introduced individuals belonged to at least four different sika subspecies: *C. n. yesoensis* and *C. n. nippon*, both from Japan, and *C. n. hortulorum* and *C. n. kopschi* from mainland Asia (Barančeková et al. 2012; Krojerová-Prokešová et al. 2013). Although over the years, a number of individuals escaped, forming two free-living populations, which are believed to be essentially of Japanese sika (Vavruněk and Wolf 1977); mainland populations of sika have been maintained until now entirely within deer parks and enclosures (Krojerová-Prokešová et al. 2013).

A significant complication arises from the fact that many introductions were not made directly from Asia but from captive populations where sika had been held in common with other cervids (captive). The analysis of mtDNA in Czech enclosures uncovered that almost one-third of the individuals bred in captivity contained mtDNA haplotypes of other sika subspecies, although these individuals were morphologically determined as Dybowski's sika. These haplotypes clustered with haplotypes of subspecies from Northern Japan, Southern Japan, and Southeast China. This definitively confirms that there has been hybridization between various sika subspecies within the enclosed populations either in the Czech Republic or possibly even among animals imported from other central European collections (Krojerová-Prokešová et al. 2013). In many cases sika are known to have hybridized with red deer or other species before release and, in consequence, many introduced animals may have been of hybrid status when released to form feral populations. Due to this and to subsequent continuing hybridization among wild populations (see section "Genetics": Hybridization) the sika of many free-ranging European populations are known to comprise a mix of genetic stocks and in some cases to be distinct genetic hybrids between sika and red deer. This will be developed further below, but it should be noted that for this account the term "sika" is used to refer to animals of supposed sika phenotype and such animals are not necessarily genetically of a single genetic type, nor necessarily purebred sika.

As well as those populations established in the wild or feral state, sika are widely maintained in deer parks, farms, and other captive collections.

Current distribution of free-ranging populations in Europe is summarized in in Fig. 2 and Table 1.

Clearly, occurrence in most countries is at relatively low density; within Europe the most significant populations are present in the UK and Ireland (estimated at *c.* 50,000), Czech Republic (>10,000), and European Russia (estimated at around 2400), although it is probable that many of these individuals are of hybrid status (see below). Historic populations in Estonia and Lithuania are believed extinct.

Description

Sika deer are medium-sized deer, intermediate between red and roe deer, but even within subspecies are extremely variable (presumably dependent on founder source). Manchurian and Formosan strains are perhaps more uniform in size, but Japanese sika in particular show marked variation in size, with some populations extremely small and stocky (females as little as 50 cm at the shoulder, males 70–80 cm) and others more nearly approaching fallow deer in size (100–120 cm at the shoulder). There is also a pronounced variation in body mass (see below section "Measurements").

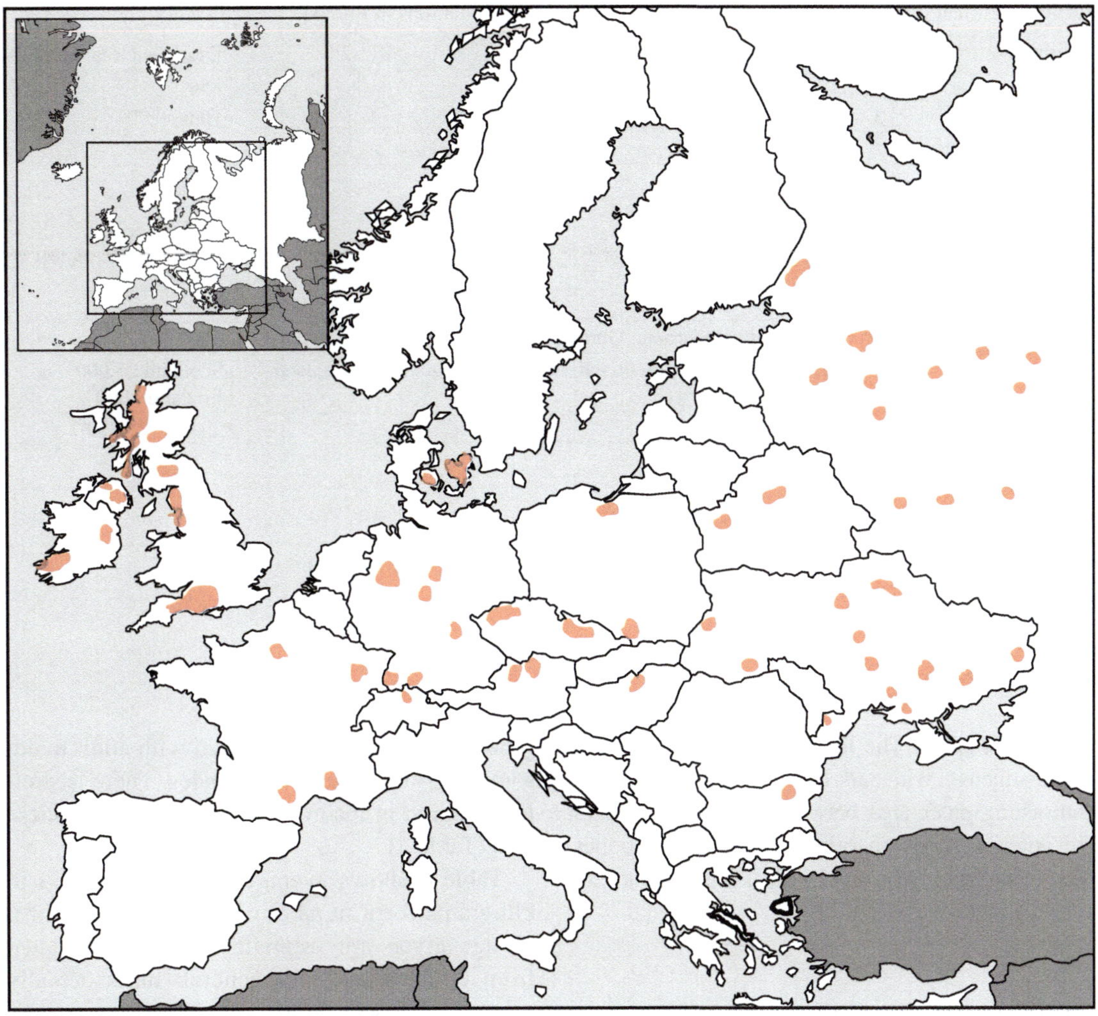

Map template: © Getty Images/iStockphoto

Fig. 2 European distribution range of *Cervus nippon* in Europe. Please see table caption for distribution data source. (Map template: © Copyright Getty Images/iStockphoto)

All races and subspecies show pronounced sexual dimorphism with adult males some 30–40% larger than females on average.

Pelage

In all sika, the summer coat is chestnut or fawn, marked with distinct white spots; in the mainland Manchurian race this base color is a very distinct deep chestnut red. Formosan sika retain their spotted pelage throughout the year; in other races the distinct spotting is lost in winter coat which is grey to almost black. The winter coat is notably thick and dense and mature stags develop a pronounced cape or mane. Newborn calves are various shades of brown ranging from dark chocolate to nearly yellow, marked with white spots; this calf coat is partially retained until the first winter moult in October–November.

Most characteristic is a white caudal patch outlined in black, which may be erected or flared in alarm. The tail, which is white with a black stripe of variable thickness, extends only partway across

Table 1 Current distribution and abundance of free-living sika populations in Europe

Country	Number of populations	Estimated total number
Austria	2	850[a]
Czech Republic	3	10,000[b, c]
European Russia		2400[a]
Denmark	Scattered	500[a]
France	4–6	200[d]
Germany	7	1500[a] – 2500[e]
Hungary	1	100[a, f]
Poland	2	220[a]
Switzerland	Immigrant from Hochrbein, Germany	250[g]
UK	Widespread in Northern Scotland; localized population in England	Scotland 25,000[h] England 2000[h]
Ireland		20,000–25,000[h]

Sources:
[a]Bartoš 2009
[b]Hunting statistics of the Czech Republic 2015 (www.uhul.cz)
[c]Krojerová-Prokešová et al. 2017
[d]Maillard et al. 2010
[e]Wotschikowsky 2010
[f]Csányi and Lehoczki 2010
[g]Imesche-Bebie et al. 2010
[h]Putman 2010

this caudal patch. The facial appearance of sika is also distinctive, with dark lines above the eyes and a contrasting paler area between them emphasizing the anterior raised margins of the frontal bones. Metatarsal glands on hock are usually white and very distinct.

Moult

Moult starts in May and complete by July. Winter coat develops in September and coat change completes by early November. Stags develop thick manes during the rut which persist throughout winter. The first, calf moult occurs 2–3 months from birth, but is inconspicuous and calf coat may be partially retained till first winter coat develops in October–November.

Measurements

There is marked variation between populations in body size with recorded differences in shoulder height for sika as: females: 50–100 cm; males: 70–120. All races and subspecies show pronounced sexual dimorphism with adult males some 30–40% larger than females. There is similar variation in recorded body mass for Japanese sika (Table 2).

Table 2 shows average whole-body mass in kilograms for a number of British populations. Values given are estimates only, recalculated from cull statistics. In general, mass actually recorded derives from culled animals and is taken as carcase mass, measured after the removal of the viscera, or dressed carcase mass, after removal of viscera, head, and feet. There is, however, a fairly precise relationship between these different body masses, with carcase mass approximately 87% of whole carcase mass for males and 90% for females and with dressed carcase mass approximately 76% of actual live body mass for both sexes (Zejda and Horáková 1988). For this table, mean body mass has been recalculated from carcase mass recorded for Dorset and Killarney sika, and from data for dressed carcase mass from New Forest and Scottish animals.

Such variation in mass is not atypical; body mass of Killarney (Birkett, original data) and Dorset animals (Putman and Clifton-Bligh 1997) are

Table 2 Mean body mass, in kilograms, of sika deer from a number of different British populations

	Males		Females	
	Yearlings	Mature	Yearlings	Mature
[Age in months]	16–20	> 20	16–20	> 20
Lulworth, Dorset	42	63	35	40
New Forest, Hampshire	33	50	–	31
Scottish populations, average	–	40	–	31
Killarney, Ireland	39	57	34	44

very similar to those reported from populations at Mohnesee (West Germany; Konig and Eick 1989), or from Northern Moravia in the Czech Republic (Zejda and Horáková 1988), while body mass of animals from the New Forest, or from Scottish forests (Ratcliffe 1991), is very similar to published body mass for sika in Maryland, USA (Feldhamer et al. 1985).

In the larger mainland Dybowski's sika deer (*C. n. hortulorum*), Danilkin (1999) summarizes results from a number of studies, reporting that height at the withers ranges from 96–120 cm for males and 85–106 (extreme 115) cm in females. Males can weigh up to around 170 kg but more commonly are between 86 and 138 kg and females weigh up to 80–90 kg (63–98 kg; Danilkin 1999).

Comparisons of body mass of animals of known age suggest that hinds may not reach actual full body size until 2 or 3 years of age. Stags continue to gain mass for a further 2 years, maturing fully at 5 years (Kaji et al. 1988; Konig and Eick 1989).

Skull and Teeth

The skull is notably shorter than red deer, with short pointed rostrum. For Dybowski sika, within their native range, Danilkin (1999) reports that the maximum length in males is 318–330 (limit 349) mm, in females it is 287–302 (limit 310) mm; the maximum width is 143 (155) mm and 121 (129) mm.

Dental formula: deciduous teeth 0.1.3.0/3.1.3.0; permanent dentition 0.1.3.3/3.1.3.3. Eruption sequence identical to red deer, and as with that species used to estimate age.

Antlers

Sika stags have relatively simple antlers, which characteristically develop up to four points only on each side; the bay tine (bez tine) is absent. Brow tines present at an acute angle to the main beam (cf. red deer where angle between brow tine and main beam is >90°; characteristic).

Stag calves begin to develop pedicles at 6 or 7 months. These produce simple single-spiked antlers in the second year. These unbranched spikes are replaced in the third year by antlers bearing brow, trez, and top tines; the full head of eight points is developed in the following season. Antlers are cast each year around April/May and the new growth remains in velvet through May to August. Antlers are cleaned during the latter part of August and early September, and with a tendency for older stags to finish the development slightly earlier than younger males. It is considered a diagnostic characteristic of Dybowski's sika that the velvet covering the antlers during antler growth is a distinctive red coloration.

Scent Glands

Pronounced metatarsal gland (hock gland) exuding a waxy material impregnates surrounding hair. Also very obvious preorbital gland appears particularly active in rutting males, when it secretes a milky fluid which may appear clearly on cheeks. No interdigital gland, but whole tail is believed to be glandular. Secretions of metatarsal (and possibly preorbital) individually distinctive; metatarsal secretion may also contain coded information of sex and age (Lawson et al. 2000, 2001).

Physiology

Physiology of European sika (reproductive cycle, control of antler growth, ruminant physiology of an intermediate feeder, etc.) is similar to that of red deer. Specific reviews of endocrine physiology and nutritional physiology of sika are offered by Yamauchi and Matsuura (2009) and Masuko and Souma (2009).

Genetics

Chromosomes

$2n = 64–68$ in both pure sika and red x sika hybrids; number variable due to Robertsonian fusions (Ohmura et al. 1983; Zima and Král 1984). The modal number in Northern Japanese sika (*C. n. yesoensis*) from Hokkaido is 68, with a pair of metacentric chromosomes, which is usually considered to be the basic chromosome number in the Cervidae (Slate et al. 2002).

Phylogeography

Phylogenetic studies support diversification of sika within their native range to four different lineages: two present in mainland Asia (Ba et al. 2015) and two in Japan (Cook et al. 1999; Goodman et al. 2001; Krojerová-Prokešová et al. 2017; Nagata et al. 1995, 1999; Tamate and Tsuchiya 1995; Tamate et al. 1998; Yamada et al. 2006; Yoshio et al. 2008; Yuasa et al. 2007). The diversification of sika to a significantly greater number of morphological subspecies (Ellerman and Morrison-Scott 1951; Whitehead 1993) is uncertain. For further details see section on (Taxonomy). The situation is really complicated due to possible admixture among subspecies and further taxonomic revision is required.

It has been established that all four known sika lineages are also present in introduced sika populations in Europe (Barančeková et al. 2012; Biedrzycka et al. 2012; Krojerová-Prokešová et al. 2013; Pitra et al. 2005) although the majority of populations are of Japanese genetic types.

Genetic Diversity

Genetic variability of native as well as introduced sika populations is relatively low (Table 3) in comparison to, e.g., red deer (Feulner et al. 2004; Hajji et al. 2007; Hmwe et al. 2006; Krojerová-Prokešová et al. 2015; Kuehn et al. 2003; Nielsen et al. 2008; Sanchéz-Fernandéz et al. 2008). Generally, lower values of genetic diversity of sika in their native range would appear to be primarily due to their evolutionary history, in relation to the colonization of different Japanese islands during the Pleistocene, as well as secondarily due to human-mediated demographic changes at the turn of nineteenth and twentieth century (Lu et al. 2006; Nagata et al. 1998; Wu et al. 2004). Deforestation and uncontrolled hunting in the second half of the nineteenth century led to the extinction of many local populations and possibly put others through bottleneck events.

Genetic variability of Japanese sika differs within Japanese islands and is the lowest in Southern Japan (Nagasaki) and in Hokkaido Island (Table 3; see also Kaji 1995; Krojerová-Prokešová et al. 2017; Nagata et al. 1998; Tamate et al. 2000). The highest genetic diversity was detected in the population occupying Honshu Island maybe due to contact between the two known maternal lineages of Japanese sika (Goodman et al. 2001; Krojerová-Prokešová et al. 2017; Nagata et al. 1998; Tamate et al. 2000; Yoshio et al. 2009).

Introduced European sika populations generally have lower genetic diversity than the native populations, but some populations, e.g., population in Wicklow (Ireland), in the Czech Republic, or in Kintyre (Scotland) have been found to have a higher genetic variability, possibly due to the crossbreeding with native red deer population (below) or the fact that source populations for such introductions have subsequently lost some of their original genetic diversity in bottleneck events (Table 3).

Table 3 Genetic diversity of nuclear DNA of several native and introduced populations of Japanese as well as mainland Asia subspecies

Location	Population	N	N_A	He	Ho	N_m	Source
Native populations							
China	Captive	113	5.93	0.685	0.568	14	Shen-Jin et al. 2014
China (Tiebu NR)	Wild	149	6.56	0.562	0.756	9	He et al. 2014
Japan	Wild	122	8.36	0.53	0.36	22	Krojerová-Prokešová et al. 2017
Japan (Hokkaido)	Wild	93–108	3.67	0.22	0.21	–	Nagata et al. 1998
	Wild	68	2.19	0.28–0.32	–	9	Goodman et al. 2001
Japan (Honshu)	Wild	39	3.67	0.28	0.23	3	Nagata et al. 1998
	Wild	135	4	0.516	0.524	9	Tamate et al. 2000
	Wild	122	3.38	0.42–0.6	–	9	Goodman et al. 2001
	Wild	177	–	0.58	0.59	17	Okada et al. 2005
	Wild	9	6.22	0.53	0.52	9	Yoshio et al. 2009
	Wild	274	6.65	0.64	0.51	10	Konishi et al. 2017
Japan (Kyushu)	Wild	95	2.55	0.19–0.45	–	9	Goodman et al. 2001
Japan (Shikoku)	Wild	23	2.33	0.39	–	9	Goodman et al. 2001
Russia (Primorski Krai)	Wild	109	10	0.74	0.6	13	Krojerová-Prokešová et al. 2013
Vietnam	Captive	144	4.07	0.59	0.57	9	Thévenon et al. 2004
Introduced populations							
Czech Republic	Feral, captive	221	7.27	0.467	0.373	22	Krojerová-Prokešová et al. 2017
Czech Republic	Captive	84	10.32	0.710	0.617	13	Krojerová-Prokešová et al. 2013
England	Feral	10	2	0.21	–	9	Goodman et al. 2001
Ireland	Feral	45	–	0.21–0.56	0.4	9	McDevitt et al. 2009
Scotland	Feral	49	3.19	0.13–0.38	–	9	Goodman et al. 2001
Scotland	Feral	213	3.05	0.15	0.14	22	Senn et al. 2010

Abbreviations: *N* number of samples, N_A average number of allele/locus, H_E expected heterozygosity, H_o observed heterozygosity, *Nm* number of microsatellite loci used in the study

Hybridization

Where distribution of sika overlaps with that of congeneric red deer, both species can hybridize to produce fertile offspring thus potentially threatening the genetic integrity of native red deer. Hybridization between red and sika has been documented in both the UK and the Republic of Ireland (Diaz et al. 2006; Goodman et al. 1999; Harrington 1973, 1982; Lowe and Gardiner 1975; McDevitt et al. 2009; Pemberton et al. 2006; Pérez-Espona et al. 2009, 2013; Ratcliffe et al. 1992; Senn and Pemberton 2009; Senn et al. 2010a, b; Smith et al. 2014), in the Czech Republic (Bartoš 2009; Bartoš et al. 2010; Bartoš and Žirovnický 1981; Krojerová-Prokešová et al. 2013, 2014, 2017; Zima et al. 1990), and in Poland (Biedrzycka et al. 2012), and there is concern about the potential for hybridization elsewhere (e.g., Germany: Wotschikowsky 2010; Austria: Weisz et al. 2002).

The genetic evidence suggests that the most common mode of hybridization is that of sika stags mating with red hinds. This is because early-generation hybrids (as detected by microsatellites) mostly have red deer mitochondrial (maternal) DNA (Goodman et al. 1999; Pemberton et al. 2006; Senn and Pemberton 2009; Smith et al. 2014). In fact, where substantial populations of both red and sika occur in the wild, both species seem to breed true and hybridization appears to be rather uncommon, with an estimate of perhaps one in 500 to one in 1000 matings estimated in an area of species overlap in Argyll, Scotland, by Goodman et al. (1999). However, where populations are expanding their distributional range into areas where red deer are already present, hybridization may be more frequent. At the leading edge of a wave of expansion young sika stags are encountered at a considerable distance away from the main population center and stags may typically become established in an area from 10–15 years before the first hinds are noted (Ratcliffe 1987; Staines 1998). These individuals usually join red deer groups (Bartoš 2009; Bennetsen 1977). It is interesting that red deer males usually ignore the presence of sika stag and only chase away other red stags (Bartoš and Žirovnický 1982). Moreover, some studies report aggressive behavior of sika males toward young red stags during the rut (Matuszewski and Sumiński 1984).

The best working hypothesis for the circumstances under which hybridization occurs in areas of distributional overlap therefore is that dispersing sika stags on the edge of expanding sika range find themselves in areas that contain red deer hinds but no sika hinds. Once a first cross has been established, further hybridization or backcrossing to either parental type is rapid, resulting in rapid gene introgression. In areas of distributional overlap, an appreciable proportion of red and, particularly, sika deer have a small number of alleles thought to be characteristic of the other taxon (typically at one or two loci of a panel of 10–20 loci: Goodman et al. 1999; Pemberton et al. 2006). It is notable that in such admixed populations, proportion of sika incorporating red deer genetic material is characteristically higher than the proportion of red deer displaying introgression of sika genes (e.g., Swanson and Putman 2009).

The situation is further complicated by the fact that many introductions of sika to the wild were not direct imports from Japan or mainland Asia, but derived as escapes or deliberate releases from deer parks where the animals had commonly been kept in mixed collections with other species of *Cervus* and where hybridization may have been facilitated by the confined conditions so that individuals subsequently released were already of hybrid status (Bartoš 2009; Powerscourt 1884). Certainly many of the releases in Britain and elsewhere in continental Europe were secondary introductions from one of the earliest introduced populations at Powerscourt Park at Enniskerry in Ireland, where these animals had been maintained in close proximity to populations of red deer and other cervid species within the same park. Pérez-Espona et al. (2009), however, have suggested that the sika introduced around the British Isles, even those deriving from Powerscourt, seem unlikely to have been extensively hybridized prior to release, arguing that if the sika had extensive red deer admixture, they would be unlikely to cluster in genetic studies of nuclear DNA, within Japanese sika, and would appear more distant from all true sika, possibly clustering toward the red deer samples included in the same analysis. This is not the case (Goodman et al. 2001). Secondly, in a search for diagnostic markers distinguishing between red and sika (Goodman et al. 1999; Slate et al. 1998), it was relatively easy to find microsatellite markers with large or fixed allele frequency differences between Scottish red and sika deer. Similarly, even though allele frequencies in introduced sika populations in the Czech Republic were clearly differentiated from those detected in native populations (Krojerová-Prokešová et al. 2013, 2017), Czech sika still cluster with native sika if red deer samples are included in the analysis (Krojerová-Prokešová et al., unpublished data).

The most detailed analysis of hybridization between red and sika deer in wild populations has been carried out in Argyll, Scotland (Abernethy 1994; Goodman et al. 1999; Pemberton et al. 2006; Ratcliffe et al. 1992; Smith et al. 2018; Swanson 1999). The pattern is generally one of a moving wave of hybridization related to the expansion of sika into resident red deer areas showing locally high proportions of the population carrying small amounts of hybrid DNA (Pemberton et al. 2006, Smith et al. 2018). In most cases, once the initial wave had passed, although there may be some introgression of genetic material characteristic of the other taxon, there has been little disruption of the overall appearance of each species, which still phenotypically resemble on one or other parent.

The same is noted for hybridization between Dybowski's sika (*C. n. hortulorum*) and Manchurian wapiti (*C. canadensis xanthopygus*) which occasionally hybridize in the area of their natural contact along the Ussuri River on the Russia-China border (Flerov 1952; Heptner et al. 1961; Sokolov 1959). According to an unpublished study by S. Goodman (reported in Aramilev 2009), in the rare cases when this hybridization occurs, the hybrid individuals retain the phenotype of one or other parent, indicating that natural selection may be acting against hybrids with intermediate phenotypes. Recent genetic research conducted in this area during 2007–2013 also revealed only small nuclear gene introgression (Krojerová-Prokešová et al., unpublished data) and no introgression of mtDNA between both species (Krojerová--Prokešová et al. 2013).

In other areas, however, hybridization is followed by a breakdown of assortative mating resulting in development of a "hybrid swarm" with up to 50% of animals in the population showing hybrid characteristics – e.g., in Wicklow, Ireland (McDevitt et al. 2009), in West Loch Awe, Scotland (Goodman et al. 1999; Senn and Pemberton 2009), in Kadyny forest, Poland (Biedrzycka et al. 2012), and in Doupovské hory Mts, the Czech Republic (Krojerová-Prokešová et al. 2014). This last area is one which expanding sika populations reached at the end of 1980s, at the same time as numbers of red deer also started to increase. Thus the contact occurred at the wave of expansion of both species. After 30 years of co-occurrence population numbers of sika have exceeded the numbers of red deer and approximately 10% of the population is formed by hybrids (Krojerová-Prokešová et al. 2014).

The fact that in some areas red deer and sika appear to coexist without recent and/or past hybridization, e.g., in Killarney, Ireland (McDevitt et al. 2009; Smith et al. 2014); in other areas early hybridization events tend to be masked by repeated backcrossing into one or both parental species (Diaz et al. 2006; Senn and Pemberton 2009) and yet in other areas again there appears to be a breakdown in assortative mating leading to this development of hybrid swarms, is curious and has been described as a "mottled hybrid zone," in which the occurrence of hybridization is determined by demographic and environmental stochastic forces (Senn and Pemberton 2009). Swanson and Putman (2009) suggest for the British Isles that the outcome depends in large part on relative population sizes and genetic provenance of the local populations of both red and sika (and the extent to which they might already be of hybrid origins). Further, also extensive body size differentiation between red deer and sika might have played a role in the number of successful crossbred events (Biedrzycka et al. 2012; Krojerová-Prokešová et al., unpublished data; McDevitt et al. 2009).

Life History

Growth

There is no published information on sex-specific growth patterns in European populations.

Reproduction

Sika are seasonal breeders with the rut beginning toward the end of September or early October, depending on location and latitude. However, the sika rut is often more protracted than that of other species and whistling stags may be heard from the end of August through until mid-December or, exceptionally, as late as mid-February. Large variation in timing of rutting activity of sika deer has been reported across Europe and also within very small areas (Bennetsen 1977, Bartoš 2009). Variation in timing of rutting activity results in long calving period. Calves are born from early May to late June after a gestation of 220 days; however, since sika have a less well-synchronized breeding cycle than many other native European species, it is not uncommon to find newborn calves in August or September or even, although infrequently, as late as October. Normally a single calf is born, but infrequent cases of twin births have been reported (Clinton et al. 1992; Davidson 1990).

Most hinds breed successfully for the first time as yearlings, and thereafter breed each adult year with high pregnancy rates maintained till aged 10 years or older. In three English populations (New Forest, Lulworth [East Dorset], and Bovington) conception rates among yearling females were around 80% and those among adult hinds 80% (New Forest) and 90% (Dorset populations), respectively (RP own data; Putman and Clifton-Bligh 1997). In six sika populations studied across Scotland, fertility rates were equally high with yearling conception rates 80% or above, and pregnancy rates among adults mostly between 85% and 100% (Chadwick et al. 1996). Pregnant calves are not uncommonly recorded among animals culled throughout most Scottish and many English populations (Chadwick et al. 1996; Chapman and Horwood 1968) – although whether or not these precocious breeders would have been able to maintain the pregnancy to its full term or successfully rear the resultant calf is perhaps less certain.

Reproductive rates are clearly extremely high and no evidence for any density-dependent reduction in fecundity among British sika was detected (e.g., Chadwick et al. 1996; Putman et al. 1996; Putman and Clifton-Bligh 1997) with recorded densities up to 35 ind./km^2. From a long-term study of an unculled sika population on the Muckross peninsula of Southern Ireland however over 12 years, Raymond (2008) found a reduction in the number of adult females observed pregnant in night counts, and in the number of culled females found pregnant; with increasing population density, density of mature sika accounted for 52% of observed variation in pregnancy rates in culled animals (Raymond 2008). Population density also affected the proportion of yearling females becoming pregnant. In this population, sika numbers increased from 62 (32 km^2) to 106 (54 km^2) between 1992 and 2000 and this was followed by a sharp rise to 181 (92 km^2) in 2003.

Recruitment rates to the adult population are not however as high as high fecundity rates reported might suggest.

Survival

Studies in their native habitat in Japan and in Ireland (O'Donoghue 1991; Raymond 2008) suggest there is a high early juvenile mortality and only 40–50% of calves born may survive to the beginning of their first winter. Density of foxes and badgers were found by Raymond (2008) to be significant predictors of calf mortality suggesting predation as the major factor in juvenile survival.

Further information on survival rates can be found for Japanese populations in Ohtaishi (1978) and Ueno et al. (2018).

Habitat and Diet

Habitat

Characteristically sika are woodland animals, primarily associated with coniferous or mixed woodland. However, they are intermediate feeders (sensu Hofmann 1985) taking grasses and dwarf shrubs as well as browse; both in their native range and where introduced they may feed out into clearings within woodland, or commonly onto heathland or grassland areas beyond the forest edge. A good review of habitat relations in their native range is offered by Takatsuki (2009a, b).

Sika are opportunistic species however and where introduced, while usually still associated with forested habitats, they may secondarily adapt to reedbeds or saltmarsh vegetation (e.g., Diaz et al. 2005; Feldhamer et al. 1978). Studies in the UK suggest that British sika seem primarily associated in their distribution with acid soils, with the majority of populations established in areas of coniferous plantations and adjacent heath. The deer do appear to be dependent on some degree of woodland cover however and seem less able to adapt to completely treeless conditions than red deer in upland Scotland (Staines 1998).

In their most "typical" British habitat of acid coniferous woodland, sika show a very predictable pattern of habitat use, lying up in dense thickets during the day and moving out to feed at night in the more open communities within the forest itself or on open ground beyond. This regular pattern is maintained throughout the year; indeed, the overall pattern of use of available habitats changes little between the seasons (Mann and Putman 1989a). Within the New Forest and associated range in Southern Hampshire, sika occupy a more varied environment of acid grasslands, heathland, and extensive areas of broadleaved as well as coniferous woodland. While the general pattern of habitat use remains much the same, the animals make greater use of these deciduous woodlands for feeding and make far less extensive use of open fields and heaths (Mann and Putman 1989a). Use of cover increases in heavily disturbed areas especially during daylight (Putman and Mann 1990).

In Killarney, Southern Ireland, sika preferentially selected areas of open canopy high forest and successional habitats (regenerating broadleaved woodlands or coniferous plantations of <15 years (Burkitt 2009). Sika significantly avoided closed canopy forest and open habitats such as grasslands and areas of open heath and there was neither selection for, nor avoidance of woodland edge or other ecotone habitats. Sika selected open canopy high forest in spring, summer, autumn, and winter and successional habitats in summer, autumn, and winter. They avoided closed canopy forest in all seasons and more open habitats in summer, autumn, and winter.

Diet

Sika are intermediate feeders (sensu Hofmann 1985) taking grasses and dwarf shrubs as well as browse. There are numerous studies of diet in their native Asia, most recently summarized by Takatsuki (2009a, b). Outside Asia most studies of diet have been carried out for sika of coniferous forest and heathland habitat in Great Britain. All show a high intake of grasses and heather in all seasons, although the actual proportion of grasses to heather differs between studies (64:20 Burkitt 2009; 70:20 Mann 1983; 30:50 Mann 1983; Mann and Putman 1989b; 60:20 Quirke 1991). A variety of other dietary components contributes to the remainder of the diet (pine needles, bark, and gorse) but rarely contribute significantly. Where broadleaved browse is available, however, it may comprise a significantly higher proportion of the diet (up to 23% of the total food intake, Mann and Putman 1989b) although of course this tends to be more significant during the spring and summer, while the proportions of coniferous browse become greater in autumn and winter (Burkitt 2009; Mann and Putman 1989b). Few comparable data are available for, e.g., reedbed populations or others (but see Diaz et al. 2005).

In the Czech Republic studies of diet have mostly been restricted to the winter period (Heroldová 1990; Obrtel et al. 1985a, b), except for Kamler et al. (2007) who studied annual diet of sika. All these studies were done in areas where free-living sika deer do not coexist with red deer. Based on all available data, sika deer diet in the Czech Republic consists mainly of grasses (24%) and forbs (10%), with high amount of grains (20%) and coniferous browse (20%) throughout the year. This indicates high dependence of sika deer on field crops and on artificial feeding over winter.

Behavior

Social Organization

Outside the breeding season, adult males and females are strongly segregated; in most populations stags and hinds occupy distinct geographical ranges for much of the year. Although commonly considered a "herding" species, sika are actually one of the less social of the deer species (Putman and Mann 1990).

From the end of winter through until September, the majority of animals are generally encountered alone, or in the case of hinds, a single animal accompanied by a calf and perhaps a yearling. Thus in studies in Southern England 95% of females seen through spring and summer were encountered as solitary hinds or single females and followers and most males were also encountered individually (Mann 1983; Putman and Mann 1990). The rut in September causes an increase in aggregation and increases the number of groups encountered containing adults of both sex; these larger aggregations persist through until March or April, when females drift away from the groups to calve. Even during this winter period, however, sika are rarely observed in groups of more than five or six.

Groups appear very temporary associations formed as animals coincide in favored feeding areas. Observations on marked sika hinds in Wareham Forest in Dorset showed little consistency of group composition or individual association from one day to the next (Horwood and Masters 1970, 1981); these social "groups" are thus extremely fluid in composition. Group size appears very responsive to habitat; mean group size in denser habitats, such as coniferous woodland or closed oakwoods, tends to consist of one or two individuals, with larger groupings generally being encountered in more open habitats (Table 4; Putman and Mann 1990). Group size even varies in relation to density of different growth stages within commercial coniferous plantations (Chadwick et al. 1996).

Home Range

Individual deer has relatively small home range areas. In the Killarney National Park in Ireland (a semi-agricultural landscape with open mixed woods of 150–250 ha bordering on larger areas of mixed conifer/broadleaved plantations) mature hinds were found to use ranges of only 18–22 ha; adult stags had somewhat larger ranges (between 45 and 55 ha) and young stags ranged more widely still, within ranges of between 60 and 70 ha (Burkitt in Putman 2000). Based on a sample of animals with movements recorded for seven years or more Raymond (2008) reports somewhat smaller ranges still with ranges estimated for stags at between 13 and 41 hectares and those for adult hinds from 4–9 hectares. Radio-tracking studies in two areas of coniferous forestry in Northern Scotland suggest range areas of similar extent (C. Maclean, unpublished data in Putman 2000). On the Arne peninsula of Southern England (an area of coastal heathland and saltmarsh), Uzal et al. (2013), using kernel estimation methods, report significantly larger ranges with 90% ranges of mature hinds from 35.51–159.85 ha (median 87.67 ha; $n = 21$); in a similar area further inland (Hartland Moor) 90% ranges extended from 58–52 to 137.49 ha (median 107.76; $n = 10$) with an overall median for both sites extending to 89.98 ha ($n = 31$). Much larger home ranges are reported from GPS studies in the Doupovské hory Mts in the Czech Republic. Here the average size of male home range was 819 ± 827 ha and varies from 114 up to 3018 ha based on kernel estimation method (Dvořák et al. 2014).

Activity

Sika appear to be active throughout the 24-hour period where undisturbed, with peaks of activity at dawn and dusk; however, populations become more markedly crepuscular or nocturnal in areas of high disturbance (Putman and Mann 1990).

Table 4 Typical group sizes recorded in a range of different habitat types in England. (From Mann 1983; Putman and Mann 1990)

Habitat	Winter	Spring	Summer	Autumn
Deciduous woodlands	2–3	2	1–2	2–3
Plantation conifer	1–2	2–3	1–2	1–2
Prethicket conifer	2	2	1–2	1–2
Polestage conifer	2	1–2	1–2	1–2
Rides and clearings	2	2–3	1–2	1–2
Heathland	2–6	2–4	2	2–3
Agricultural fields	2–7	2–7	2–5	2–7

Senses and Communication

Both sexes utter a brief, high-pitched squeal when alarmed, with the call directed toward the object of suspicion. At the same time the alerted animal often moves gradually toward the source of alarm to investigate more closely; a deeper alarm "bark" precedes immediate flight. Young calves offer a high-pitched wavering bleat, not unlike the mewing of a buzzard, when calling for their mothers, and hinds themselves use a deeper, stronger bleat when themselves searching for their calves. Young stags also make a submissive bleat when threatened by the approach of a dominant male, while mature stags may make a subdued "mee-mee-mee" call as they approach and follow hinds at the time of the rut (Smith 1996). Male vocalizations are at their peak in this time, and in sharp contrast to the roaring or belching calls of red stags or fallow bucks, rutting male sika produce a characteristic high-pitched whistle, audible up to a kilometre away. Whistles are repeated and tend to consist of three to five in any one bout lasting perhaps from seven to ten seconds overall. Stags may commonly also make a deeper, moaning call, during the rut (more like that of fallow deer or red), although this call does not carry over such distances.

Sika also communicate by scent. The preorbital gland of males is particularly active during the rut, when it may visibly secrete a white milky fluid which trickles down the cheeks, but the gland also secretes a simple series of volatiles at other times of year. Analysis of the odor profile from this secretion and that from the metatarsal gland shows that the odors given by different individuals are individually distinct (and thus individually identifiable) and that the secretions from the metatarsal gland also contain coded information about sex and probably age of the signaler (Lawson 1996; Lawson et al. 2000, 2001).

Reproductive Behavior

Early records of rutting behavior of sika deer in Britain suggested that stags mark and defend mating territories in woodland within the hind range (Horwood and Masters 1970, 1981). These territories are marked by fraying and bole-scoring of perimeter trees and thrashed ground vegetation such as heather bushes. More recent evidence makes it clear that mating strategy within sika is extremely flexible, with stags adopting a number of different strategies depending on circumstance (Putman 1993; Putman and Mann 1990; Thirgood et al. 1998). In various different populations now studied males have been found to defend rutting territories, as described, but in other cases to collect and defend a harem, as do red deer, or simply patrol areas of superior food quality within the female range and cover oestrus hinds when encountered (Putman and Mann 1990; Thirgood et al. 1998; see also Endo 2009).

The development of a simple breeding "lek" has also been reported in certain central European populations (Bartoš et al. 1992). It seems probable that, as in other deer species, stags adopt differing strategies depending on the male's own age and dominance status, the density and distribution of females, and the degree of competition experienced from other breeding males (Endo 2009; Langbein and Thirgood 1989; Thirgood et al. 1998).

Endo (2009) and Minami et al. (2009) report that almost all copulations are followed by a period in which the male guards the female from other stags but multiple matings by females with more than one individual male are common.

Parasites and Diseases

Rehbein (2010) provides a comprehensive review of both internal and external parasites affecting sika within their native range across the former Soviet Union and in Japan.

Within Europe, sika may carry a number of external parasites but in all studies the primary ectoparasites are ticks, *Ixodes ricinus*, and deer keds *Liptotena cervi* (Adams and Dannatt 1989; Rehbein 2010; Sleeman 1983). From Ireland, Sleeman (1983) also reports infestations of sucking and biting lice (*Solenopotes burmeisteri* and *Damalinia meyeri sikae*).

Systematic reviews for species of helminths associated with European populations of sika have been presented for, e.g., Poland (Drozdz 1966), Great Britain (Dunn 1967; Hawkins 1988), the Czech Republic (Kotrlá and Kotrlý 1977, 1980), Germany, and Austria (Rehbein 2010). Hawkins compiled a list of some 50 different gastrointestinal helminths recorded from sika overall across their European range, although not all are recorded in all localities and animals rarely carry a heavy individual burden.

In the Czech Republic, Kotrlá and Kotrlý (1977) recovered 16 species of gastrointestinal helminths while in Rehbein's survey of sika culled in six different localities in Germany and Austria, 19 species were recovered (18 different species of gastrointestinal nematodes and one cestode, *Moniezia benedeni*). In all cases, the abomasum was the most heavily parasitized part of the gastrointestinal tract, followed by the small intestine and the large intestine. No parasites were recovered from the rumen. Overall, however, in both studies, both prevalence and individual burdens of different parasitic species were not high, in comparison to the helminth burdens recorded in other cervid species sampled in the same localities. Gastrointestinal burdens of calves were found to be significantly higher than those of deer >1 year and, among adult individuals, males tended to have higher parasitic burdens than females (Rehbein 2010).

In all localities, within Great Britain, Czech Republic, and Austria/Germany, in addition to gastrointestinal burdens, *Cysticercus* spp. were reported from the mesenteries of the gut. Some individuals were suffering from lungworm (*Dictyocaulus* spp) and some were affected by the liver flukes (*Dicrocoelium* spp. and *Fasciola hepatica*) but such infections were generally at low prevalence (e.g., *F. hepatica* reported from 2/139 [1.4%] carcases examined in the New Forest of England; Adams and Dannatt 1989; 2.3% of carcases examined in Austria and Germany by Rehbein 2010). (The species of trematodes recorded differ between studies. Kotrlá and Kotrlý [1977] report *Cysticercus* (*Taenia*), *Dicrocoelium dendriticum*, and *Dictyocaulus viviparus*, while Rehbein [2010] records *Taenia hydatigena* (*Cysticercus tenuicollis*), *Dicrocoelium chinense*, and *Dictyocaulus eckertii*, but it is probable that reports in fact refer to the same species. There is some argument as to whether *Dictyocaulus eckertii* is a different species from *D. viviparus*, and *Cysticerus tenuicollis* and *C. taenia* are also homonyms. While *Dicrocoelium chinensis* and *D. dendriticum* are distinct species based on mitochondrial and nuclear ribosomal DNA sequences [Liu et al. 2014] it is unlikely that they would be accurately distinguished from morphological examination.)

All authors report that the diversity of species found in any individual animal and levels of infestation (actual worm burdens) are significantly lower than those recovered from native herbivores in the same areas. It would seem that most parasites attack sika only opportunistically and are perhaps in general not well adapted to this novel host, despite the fact that they may be regular parasites of the congeneric *Cervus elaphus*. Only *Spiculopteragia* (syn. *Ostertagia*) *asymmetrica* has been widely reported in all European studies at high frequency (Drozdz 1966; Dunn 1967; Hawkins 1988; Kotrlá and Kotrlý 1977, 1980; Rehbein 2010).

Kotrlá and Kotrlý (1977) recorded *Ashworthius sidemi* as the otherwise most abundant helminth parasite in Czech populations: a species introduced with sika from their native range; Rehbein (2010) recovered specimens of *Spiculopteragia houdemeri* and *S.* (syn *Rinadia*) *andreevae* in Germany/Austria, again species recorded for the first time outside their natural range of East Asia. Kotrlá and Kotrlý (1977) reported infestation by the trematode *Fascioloides magna*, a species from North America introduced into Europe alongside introductions of North American wapiti (*Cervus canadensis*) and white-tailed deer (*Odocoileus virginianus*) and highly pathogenic in the red deer and in roe deer (Novobilsky et al. 2006).

Rehbein (2010) reports oocysts of eight morphotypes of the protozoan parasite *Eimeria* spp. in rectal feces and isolated cysts of *Sarcocystis* spp. in the cardiac and/or diaphragmatic myocytes of 32.9% of the sika deer sampled. Sarcocysts were significantly more frequently observed in the muscle samples from sika >1 year (41.4%) than in those from sika deer calves (19.6%). Both bovine and avian tuberculosis (TB) have been recorded from sika in Europe; 2 cases of bovine TB were identified in Knapdale, Argyll (Scotland), between 1990 and 2000 and 5/240 sika culled in the south of England between 1971 and 1996 (Delahay et al. 2002). In Ireland, 6 positives were found among 153 carcases taken in Co. Tyrone [Northern Ireland] in 1996/7 season but 0/28 from C. Fermanagh (again, Delahay et al. 2002); 5 individuals of sika shot in 1984 in Co Wicklow were tested positive (3.8% of sample; Dodd 1984).

Sika also seem particularly prone to a form of white muscle disease – a wasting dystrophy of the muscles, appearing in its most acute form as a post-traumatic myopathy following physical stress such as chases by dogs or live capture operations (Adams and Dannatt 1989; author's own data). This can result in mortality rates over a period from a few hours to as much as 48 h following the incident of stress.

Population Ecology

Throughout most of their European range sika populations are largely controlled by human intervention and mortality imposed through the cull is probably the major source of loss among adults. Maximum recorded lifespan in the wild is 15–16 years, while animals may live upto 26 years in captivity (Ohtaishi 1978). Annual mortality rates for male and female sika in an unculled population in Southern Ireland were respectively 19.75% and 18.05% (Raymond 2008). Studies in Japan and work in Killarney National Park, Ireland (O'Donoghue 1991; Raymond 2008), suggest there is a high juvenile mortality, with only between 40% and 50% of calves surviving to the December following their birth.

Reproductive rates are extremely high and there is no clear evidence for any density-dependent reduction in fecundity (Chadwick et al. 1996; Putman and Clifton-Bligh 1997; Putman et al. 1996) even at densities as high as 35 ind./km^2, although, as reported above, Raymond (2008) detected a reduction in adult pregnancy rates and in the proportion of yearling females found pregnant in an unculled population in Southern Ireland as densities rose from 32 deer per km^2 to 54 per km^2 and then to 92 per km^2. As noted, however, calf mortality may be as high as 50%; density of foxes and badgers were found by Raymond (2008) to be significant predictors of calf mortality suggesting predation as the major factor in juvenile survival. Predation by foxes is not only strongly seasonal but also varies markedly from year to year, dependent on the density of fox populations, the density of roe or sika populations, and the relative availability of alternative prey.

Competition with Other Ungulates

In most parts of their introduced range, sika deer are sympatric with red deer. They are commonly also sympatric with roe deer and with fallow deer. In all these cases there is potential for feeding competition, and with red-sika populations there is also the possibility of mate competition.

Anecdotal evidence has been presented to suggest that there may be some impact on native deer species as a result of competitive interactions. The potential interaction between red and sika has been indicated by correlative results by several studies, suggesting suppression of productivity, or geographical displacement of red deer by sika (Abernethy 1994; Dzieciolowski 1979; Feldhamer and Armstrong 1993; McKelvey 1959), but competition was not unequivocally demonstrated in these cases.

In long-term studies of the ecology of populations of both sika and red deer in different management blocks in Killarney, Ireland, Burkitt (2009) reported that when sika density was high, calculated overlap in habitat use with sympatric red deer was lower than in sites where sika density was low, suggesting some degree of niche shift at high densities. Both species continued to use their preferred habitats but in significantly different proportion. Proportionally, sika made most use of successional habitats (regenerating woodland or coniferous plantations <15 years) and open high forest, while red deer used more open habitats and actively clearly avoided successional habitats, suggesting that increased numbers of sika were leading to competitive interactions between red and sika causing habitat shifts for both species but most markedly in red deer. Higher levels of resource overlap were found in blocks where the combined density of both species was relatively low in comparison with other sites (Burkitt 2009). The effects of sympatry, even in sites where the density of one or other species was higher than the other, did not alter diet selection of either red or sika deer.

While shifts in habitat use in relation to increasing density of both species were more marked in red deer, there would nonetheless appear to have been an actual competitive effect in terms of suppressed productivity. Raymond (2008) reports for the same populations that body condition of adult male sika, yearling male, sand yearling females (as reflected in kidney fat index: KFI) was affected more significantly by density of red deer in the same area than by density of sika, with density of red deer explaining respectively 13.5% (adult males), 52% (yearling males), and 34% (yearling females) of variance in KFI. As already noted, both pregnancy rate and annual post-winter recruitment of juveniles also showed significant negative correlation with density of red deer in a given block (Raymond 2008). The magnitude of the effect of sika on population demography of sympatric red deer is unfortunately not reported.

Makovkin (1999) has suggested a potential for competitive displacement of wapiti by an expansion of sika deer in Primorsky Krai, Far East Russia, but data are largely anecdotal. In addition, a number of authorities have suggested that both roe and red deer decline in forests inhabited by sika, and some data show lower than expected densities of roe in sika-inhabited forest (e.g., Chadwick et al. 1996; Danilkin 1996).

Analyses of overlap in resource utilization with red, roe, and fallow deer (as well as with free-ranging cattle and horses) in the New Forest of Southern England in relation to both habitat use and diet (Putman 1986, 1996) have shown considerable potential for competition with fallow, although no evidence of direct influence on population numbers (Putman and Sharma 1987), perhaps because population numbers of all species are strongly controlled by (human) management.

Conservation Status

Listed by IUCN (2017) in its native range as of Least Concern because of its abundant, increasing population in Japan, and stable population in Russia, the subpopulation in Taiwan is evidently holding steady. However, the large subpopulation in Japan masks all population trends from elsewhere within its range, resulting in this overall listing as of Least Concern. All other free-ranging subpopulations within Asia are very seriously threatened. We note however that within Europe, the species is an introduced nonnative; at present there is no suggestion that European populations represent a significant "reserve" population in case of need for reintroductions within the native range, nor necessarily a valuable repository of lost genetic material, particularly in the light of widespread hybridization between stocks. Purebred populations of Japanese sika are reported in Britain from Dawyck, Peebles-shire, and Morar, Scotland (Ratcliffe 1987), the New Forest in Hampshire (Diaz et al. 2006), Killarney in Ireland (McDevitt et al. 2009), and Bouzovsko, the Czech Republic (Krojerová-Prokešová et al., unpublished data).

Because of the known potential for damage to commercial forestry and natural habitats, as well as fears that continuing hybridization may threaten the genetic integrity of native red deer stocks (Pérez-Espona et al. 2009), movement and release of an animal of genus *Cervus* is now prohibited in Scotland (Section 14 of the Wildlife & Natural Environment Act, Scotland, 2012).

Management

As noted, throughout their European range all sika populations derive from introductions largely through the latter part of the nineteenth century and early part of twentieth century, although some introductions continued to a later date. Almost all introductions outside their native range have been deliberate, although subsequent release to the wild may be accidental (through escape from captivity) or deliberate. Once established in the wild patterns of dispersal and range expansion seem to be of three main types. Some introduced populations appear to remain contained and localized with minimal range expansion (e.g., Morar or Dawyck in Scotland); others appear to be constrained by barriers such as railways and canals, building up high population densities in areas of release and only later showing irruptive expansion of range (e.g., Kintyre peninsula in Scotland; New Forest in Southern England, and Southwestern Bohemia in the Czech Republic). In other cases (in continuous areas of good habitat), they may show a steady expansion in range, estimated in mainland Scotland at between 3 and 5 km per year (Putman 2000).

Rates of expansion appear to be related at least in part to continuity of suitable habitat but may also be related to degree of hybridization with red deer *C. elaphus*.

Impact on Forestry, Agriculture, and Conservation Habitats

Where established at high density, sika may have considerable impact. In areas of local high density, sika may have significant impacts on ground vegetation (Takatsuki and Ito 2009) and where they have been introduced they may have substantial impacts on open heathland and/or wetland areas (reedbeds, saltmarshes) causing significant change in vegetation structure and species composition as well as significant erosion/exposure of soil (Diaz et al. 2005; Uzal et al. 2013; Uzal Fernandéz 2010). In the Killarney National Park of Southern Ireland, sika were found to cause significant browsing impacts on both wet and dry heaths (*Calluna vulgaris* and *Erica* spp.) as well as to *Vaccinium myrtillus* and broadleaved trees such as *Betula pubescens*, *Ilex aquifolium*, and *Fagus sylvatica*, with degree of damage significantly related to sika density (although we may note as above that population densities in this area are particularly high).

Sika will graze on agricultural fields that border woodlands in which they are resident; local populations have been recorded as causing damage to arable crops in England as well as in the Czech Republic but, because populations are only locally distributed, damage to crops is likewise only likely to be of very local significance (e.g., Kamler et al. 2007; Packer et al. 1999; Putman and Moore 1998). By contrast, sika may cause considerable damage to commercial forestry (Abernethy 1998; Chadwick et al. 1996; Lowe 1994; Ratcliffe 1989). Damage may be caused through browsing of both lateral and leading shoots, much as by red deer in similar contexts, and also by bark-stripping in hard winters. The economic significance of such damage may be locally very considerable. An assessment of bark-stripping damage by sika in Craggan and Loch Coire forests, Sutherland, showed that the percentage of damaged trees varied between $10.0 \pm 9.3\%$ and $75.5 \pm 6.9\%$, dependent on forest block. Lodgepole pine (mean 51.8% damaged) appeared more vulnerable than Sitka spruce (10% damaged) or Scots pine (0.0% damaged) (Swanson and Putman 2009).

Mature trees may also suffer additional damage in some areas through "bole-scoring" when sika stags gouge deep vertical grooves into the bole of particular trees during defence, marking, and advertisement of mating territories in the rut. Such bole-scoring damage appears to be a peculiarity of sika (Carter 1984; Larner 1977).

Most European populations are controlled by man and subject to similar legislation in each of the countries where they are established, as applies to other species of ungulates. In many countries they are regarded as regular hunting quarry and sales of stalking and venison may contribute significantly to the overall income received from deer population management (Csányi et al. 2014). Hunting seasons in different European countries are summarized in Apollonio et al. (2010).

Future Challenges for Research and Management

This is a species capable of irruptive growth. Evidence from the rapid recolonization, under protection, of former range in mainland Russia and in Japan, coupled with documented rates of expansion when introduced to free range in Great Britain, Ireland, and the Czech Republic – as well as in New Zealand – establishes sika as a species with a very high invasive potential. Where they become established in the wild state it is clear that sika can reach extremely high density in suitable habitat and at such densities may have serious impact on commercial forestry as well as (more localized) impacts on agricultural crops and native vegetation. There is growing evidence of a significant potential for competition for native species of ungulates from sika and the potential risk of hybridization with native stocks of red deer when expanding sika populations colonize areas already occupied by this latter species.

At present sika are established as free-ranging populations only in nine countries in Western Europe (outside Russia) but the species is held by many zoological gardens and other captive collections of exotic species. It is of utmost importance that high levels of biosecurity are observed over such captive collections and that any escapes are quickly contained. As already noted, movement and release of an animal of genus *Cervus* is now prohibited in Scotland. For the most part, European populations of sika do not represent an important genetic resource in support of the conservation of the species in its native range.

Acknowledgments The authors are grateful to Leonid Baskin and Alexey Danilkin for providing data on measurements of Dybowski's sika from Russia and to others who have provided unpublished data.

References

Abernethy K (1994) The introduction of sika deer, *Cervus nippon nippon*, to Scotland. PhD thesis, University of Edinburgh, Edinburgh

Abernethy K (1998) Sika deer in Scotland. Deer Commission for Scotland, The Stationery Office

Adams JC, Dannatt N (1989) The culling and processing of wild deer. Arun District Council/Forestry Commission, UK

Apollonio M, Andersen R, Putman R (2010) European ungulates and their management in the 21st century. Cambridge University Press, Cambridge, UK

Aramilev VV (2009) Sika deer in Russia. In: McCullough DR, Takatsuki S, Kaji K (eds) Sika deer: biology and management of native and introduced populations. International Springer/Springer, Tokyo, pp 475–499

Ba H, Yang F, Xing, X, Li C (2015) Classification and phylogeny of sika deer (*Cervus nippon*) subspecies based on the mitochondrial control region DNA sequence using an extended sample set, Mitochondrial DNA 26(3):373–379

Banwell DB (2009) The sika in New Zealand. In: McCullough DR, Takatsuki S, Kaji K (eds) Sika deer: biology and management of native and introduced populations. International Springer/Springer Japan, Tokyo, pp 643–656

Barančeková M, Krojerová-Prokešová J, Voloshina IV, Myslenkov AI, Kawata Y, Oshida T, Lamka J, Koubek P (2012) The origin and genetic variability of the Czech sika deer population. Ecol Res 27:991–1003

Bartoš L (2009) Sika deer in Continental Europe. In: McCullough DR, Takatsuki S, Kaji K (eds) Sika deer: biology and management of native and introduced populations. International Springer/Springer Japan, Tokyo, pp 573–594

Bartoš L, Žirovnický J (1981) Hybridisation between red and sika deer. II. Phenotype analysis. Zool Anz Jena 207:271–287

Bartoš L, Žirovnický J (1982) Hybridisation between red and sika deer. III. Interspecific behaviour. Zool Anz Jena 208:30–36

Bartoš L, Zeeb U, Mikes J (1992) Lekking behaviour in sika deer. In: Maruyama N, Bobek B, Ono Y, Regelin W, Bartos L, Ratcliffe PR (eds) International symposium on wildlife conservation – present trends and perspectives for the 21st century. Japan Wildlife Research Center, Tokyo, pp 205–208

Bartoš L, Kotrba R, Pintíř J (2010) Ungulates and their management in the Czech Republic. In: Apollonio M, Andersen R, Putman R (eds) European ungulates and their management in the 21st century. Cambridge University Press, Cambridge, UK, pp 243–261

Bennetsen E (1977) Some facts about sika deer in Denmark. Deer 4:153–157

Biedrzycka A, Solarz W, Okarma H (2012) Hybridization between native and introduced species of deer in Eastern Europe. J Mammal 93(5):1331–1341

Bromley GF (1956) Ecology of wild spotted deer in the Maritime Territory. In: Yurgenson PB (ed) Studies of mammals in government preserves. USSR Ministry of Agriculture, Moscow, pp 152–224

Burkitt TD (2009) A comparison of ecology between sympatric native red deer (*Cervus elaphus* Linnaeus 1758) and introduced Japanese sika deer (*Cervus nippon nippon* Temminck 1836) populations in southwest Ireland. PhD thesis, Manchester Metropolitan University

Carter NA (1984) Bole scoring by sika deer (*Cervus nippon*) in England. Deer 6:77–78

Chadwick AH, Ratcliffe PR, Abernethy K (1996) Sika deer in Scotland: density, population size, habitat use and fertility – some comparisons with red deer. Scott For 50:8–16

Chapman DI, Horwood MT (1968) Pregnancy in a sika deer calf (*Cervus nippon*). Res Vet Sci 29:105–107

Clinton T, Hayden TJ, Lynch JM, Murphy P (1992) A case of twin foetuses in a sika hind (*Cervus nippon*) from County Wicklow, Ireland. Deer 8:437–438

Cook CE, Wang Y, Sensabaugh G (1999) A mitochondrial control region and cytochrome b phylogeny of sika deer (*Cervus nippon*) and report of tandem repeats in the control region. Mol Phylogenet Evol 12:47–56

Csányi S, Lehoczki R (2010) Ungulates and their management in Hungary. In: Apollonio M, Andersen R, Putman R (eds) European Ungulates and their management in the 21st century. Cambridge University Press, Cambridge, UK, pp 291–318

Csányi S, Carranza J, Pokorny B, Putman R, Ryan M (2014) Valuing ungulates in Europe. In: Putman R, Apollonio M (eds) Behaviour and management of European Ungulates. Whittles Publishing, Dunbeath, pp 13–45

Danilkin A (1996) Behavioural ecology of Siberian and European roe deer. London: Chapman and Hall 277 pp

Danilkin AA (1999) Deer (Cervidae). GEOS. 552 pp. (In Russian)

Davidson MM (1990) Sika deer. In: King CM (ed) The handbook of New Zealand mammals. Oxford University Press in association with the Mammal Society, pp 468–477

Delahay RJ, de Leeuw ANS, Barlow AM, Clifton-Hadley RS, Cheeseman CL (2002) The status of *Mycobacterium bovis* infection in UK wild mammals: a review. Vet J 164:90–105

Diaz A, Pinn EH, Hannaford J (2005) Ecological impacts of sika deer on Poole Harbour saltmarshes. In: Humphreys J, May V (eds) The ecology of Poole Harbour. Elsevier

Diaz A, Hughes S, Putman R, Mogg R, Bond JM (2006) A genetic study of sika (*Cervus nippon*) in the New Forest and in the Purbeck region, southern England: is there evidence of recent or past hybridization with red deer (*Cervus elaphus*)? J Zool (Lond) 270:227–235

Dodd K (1984) Tuberculosis in free-living deer. Vet Rec 115:592–593

Drozdz J (1966) Studies on helminths and helminthiasis in Cervidae. II. The helminth fauna in Cervidae in Poland. Acta Parasitol Pol 14:1–13

Dunn AM (1967) Endoparasites of deer. Deer 1:85–90

Dvořák S, Barták V, Macháček Z, Matějů J (2014) Home range size and spatio-temporal dynamics of male sika deer (*Cervus nippon*; Cervidae, Artiodactyla) in an introduced population. Folia Zool 63(2):103–115

Dzieciolowski R (1979) Structure and spatial organisation of deer populations. Acta Theriol 24:3–21

Eick E, Konig R, Willett J (1995) Sika, *Cervus nippon* Temminck, vol 1. International Sika Society, Mohnese

Ellerman JR, Morrison-Scott TCS (1951) Checklist of Palearctic and Indian mammals. British Museum (Natural History), London

Endo A (2009) Variation in mating behaviour of sika deer: mating behavior of sika deer on Nozaki Island. In: McCullough DR, Takatsuki S, Kaji K (eds) Sika deer: biology and management of native and introduced populations. International Springer/Springer Japan, Tokyo, pp 285–296

Feldhamer GA, Armstrong WE (1993) Interspecific competition between four exotic species and native artiodactyls in the United States. Trans N Am Wildl Nat Resour Conf 58:468–478

Feldhamer GA, Demerais S (2009) Free-ranging and confined sika deer in North America: current status, biology and management. In: McCullough DR, Takatsuki S, Kaji K (eds) Sika deer: biology and management of native and introduced populations. International Springer/Springer Japan, Tokyo, pp 615–642

Feldhamer GA, Chapman JA, Miller RL (1978) Sika deer and white-tailed deer on Maryland's eastern shore. Wildl Soc Bull 6:155–157

Feldhamer GA, Stauffer JB, Chapman JA (1985) Body morphology and weight relationships of sika deer (*Cervus nippon*) in Maryland. Z Säugetierkd 50: 88–106

Feulner PGD, Bielfeldt W, Zachos FE, Bradvarovic J, Eckert I, Hartl GB (2004) Mitochondrial DNA and microsatellite analyses of the genetic status of the presumed subspecies *Cervus elaphus montanus* (Carpathian red deer). Heredity 93:299–306

Flerov KK (1952) Fauna SSSR. Tom 1 (2) Kabargi i oleni. Izdatelstvo Akademii Nauk SSSR, Moskva (Academy of Science of the USSR Publishers), Leningrad, USSR. (In Russian)

Geist V (1999) Deer of the world: their evolution, behaviour, and ecology. Swan Hill Press, Shrewsbury

Goodman SJ, Barton NH, Swanson G, Abernethy K, Pemberton JM (1999) Introgression through rare hybridization: a genetic study of a hybrid zone between red and sika deer (genus *Cervus*) in Argyll, Scotland. Genetics 152:355–371

Goodman S, Tamate H, Wilson R, Nagata J, Tatsuzawa S, Swanson G, Pemberton J, McCullough D (2001) Bottlenecks, drift and differentiation: the population genetic structure and demographic history of sika deer (*Cervus nippon*) in the Japanese archipelago. Mol Ecol 10:1357–1370

Groves C (2006) The genus *Cervus* in eastern Eurasia. Eur J Wildl Res 52:14–22

Groves C, Grubb P (2011) Ungulate taxonomy. The John Hopkins University Press, Baltimore

Guo YS (2000) Distribution, numbers and habitat of Sichuan sika deer (*Cervus nippon sichuanicus*). Acta Theriol Sin 20(2):1–87. (in Chinese with English summary)

Guo YS, Zheng H-Z (2000) On the geological distribution, taxonomic status of species and evolutionary history of sika deer in China. Acta Theriol Sin 20:168–179. (in Chinese with English summary)

Hajji GM, Zachos FE, Charfi-Cheikrouha F, Hartl GB (2007) Conservation genetics of the imperilled Barbary red deer in Tunisia. Anim Conserv 10(2):229–235

Harrington R (1973) Hybridisation among deer and its implications for conservation. Irish For J 30:64–78

Harrington R (1982) The hybridisation of red deer (*Cervus elaphus* L. 1758) and Japanese sika deer (*C. nippon* Temminck, 1838). Int Congress Game Biol 14: 559–571

Hawkins D (1988) The parasitic interrelationship of deer and sheep on the Knebworth Park Estate, near Stevenage. Deer 7:296–300

He Y, Wang ZH, Wang XM (2014) Genetic diversity and population structure of a Sichuan sika deer

(*Cervus sichuanicus*) population in Tiebu Nature Reserve based on microsatellite variation. Dongwuxue Yanjiu 35(6):528–536
Heptner VG, Nasimovitch AA, Banikov AG (1961) Mlekopitayushtchie Sovetskogo soyuza. Tom pervyy. Parnokopytnye i neparnokopytnye. Gosudarstvennoe izdatel'stvo 'Vysshaya shkola', Moskva. (In Russian)
Heroldová M (1990) The diet of sika deer (*Cervus nippon*) outside the growing season, with a snow cover. Folia Zool 39:197–206
Herzog S (1987) Mechanisms of karyotype evolution in *Cervus nippon* Temminck. Caryologia 40:347–353
Herzog S (1995) Hybridization and the species concept: implications for wildlife management strategies. In: Eick E, König R, Willett J (eds) Sika, *Cervus nippon* Temminck, 1838, vol II. International Sika Society, Möhnesee, p 3.4:15–20
Hmwe SS, Zachos FE, Eckert I, Lorenzini R, Fico R, Hartl GB (2006) Conservation genetics of the endangered red deer from Sardinia and Mesola with further remarks on the phylogeography of *Cervus elaphus corsicanus*. Biol J Linn Soc 88:691–701
Hofmann RR (1985) Digestive physiology of the deer – their morphophysiological specialisation and adaptation. In: Fennessy PF, Drew KR (eds) Biology of deer production, vol 22. Royal Society of New Zealand Bulletin, pp 393–407
Horwood MT, Masters EH (1970; 1981) Sika deer. British Deer Society, Fordingbridge
Imesch-Bebié N, Gander H, Schnidrig-Petrig R (2010) Ungulates and their management in Switzerland. In: Apollonio M, Andersen R, Putman R (eds) European ungulates and their management in the 21st century. Cambridge University Press, Cambridge, UK, pp 357–391
IUCN (2017) The IUCN red list of threatened species. Version 2017-2 (www.iucnredlist.org)
Kaji K (1995) Deer eruptions – a case study in Hokkaido, Japan. Honyurui Kagaku (Mamm Sci) 35:35–43. (in Japanese)
Kaji K, Koizumi T, Ohtaishi N (1988) Effects of resource limitation on the physical and reproductive condition of sika deer on Nakanoshima Island, Hokkaido. Acta Theriol 33:187–208
Kamler J, Homolka M, Barančeková M, Krojerová J, Dvořák J (2007) [Feeding ecology of sika deer.] In: Sika deer in Western Bohemia, Žlutice 7.-8. 9. 2007: 18–21. [in Czech]
Kawamura Y (2009) Fossil record of sika deer in Japan. In: McCullough DR, Takatsuki S, Kaji K (eds) Sika deer: biology and management of native and introduced populations. International Springer/Springer Japan, Tokyo, pp 11–25
Konig R, Eick E (1989) Variations due to growth and seasonal changes in the body weights of sika deer (*Cervus nippon*) of the Mohnesee population in West Germany. In: Symposium on wildlife in Japan, April 1989. Japan Hunters' Association
Konishi S, Hata S, Matsuda S, Arai K, Mizoguchi Y (2017) Evaluation of the genetic structure of sika deer (*Cevus nippon*) in Japan's Kanto and Tanzawa mountain areas, based on microsatellite markers. Anim Sci J 88. [epub ahead of print]
Kotrlá B, Kotrlý A (1977) Helminths of wild ruminants introduced into Czechoslovakia. Folia Parasitol 24:35–40
Kotrlá B, Kotrlý A (1980) Zur Verbreitung von Helminthen durch eingeführtes Wild. Angew Parasitol 21:79–82
Krojerová-Prokešová J, Barančeková M, Voloshina I, Myslenkov A, Lamka J, Koubek P (2013) Dybowski's sika deer (*Cervus nippon hortulorum*): genetic divergence between natural Primorian and introduced Czech populations. J Hered 104(3):312–326
Krojerová-Prokešová J, Barančeková M, Koubek P (2014) [Application of molecular-genetics methods to reveal the deer origin and the interspecific crossbreeding.] Workshop: Sika deer – overview of knowledge in the Czech Republic. Dubina, Karlovy Vary, 16.5.2014. Czech Forest Society, Prague, pp 9–14. (in Czech)
Krojerová-Prokešová J, Barančeková M, Koubek P (2015) Admixture of Eastern and Western European Red Deer lineages as a result of postglacial recolonization of the Czech Republic (Central Europe). J Hered 106(4): 375–385
Krojerová-Prokešová J, Barančeková M, Kawata Y, Oshida T, Igota H, Koubek P (2017) Genetic differentiation between introduced Central European sika and source populations in Japan: effects of isolation and demographic events. Biol Invasions 19:2125–2141
Kuehn R, Schroeder W, Pirchner F, Rottmann O (2003) Genetic diversity, gene flow and drift in Bavarian red deer populations (*Cervus elaphus*). Conserv Genet 4: 157–166
Kuwabara T (2004) Relative sea-level changes and marine-terrace deposits in Kamikita Plain, northern end of Honshu, Japan. J Geol Soc Jpn 110:93–102. (In Japanese with English abstract)
Langbein J, Thirgood SJ (1989) Variation in mating systems of fallow deer in relation to ecology. Ethology 83: 195–214
Larner JB (1977) Sika deer damage to mature woodlands of southwestern Ireland. In: Proceedings of the 13th international congress of game biology, pp 192–202
Lawson RE (1996) The influences of evolution, habitat and social organisation upon chemical signalling in deer. PhD thesis, University of Southampton
Lawson RE, Putman RJ, Fielding AH (2000) Individual signatures in scent gland secretions of Eurasian deer. J Zool (Lond) 251:399–410
Lawson RE, Putman RJ, Fielding AH (2001) Chemical signalling in Eurasian deer: do individual odours also code for attributes? J Zool (Lond) 253:91–99
Liu GH, Yan HB, Otranto D, Wang XY, Zhao GH, Jia WZ, Zhu XQ (2014) *Dicrocoelium chinensis* and *Dicrocoelium dendriticum* (Trematoda: Digenea) are distinct lancet fluke species based on mitochondrial and nuclear ribosomal DNA sequences. Mol Phylogenet Evol 79:325–331
Lowe R (1994) Deer management: developing the requirements for the establishment of diverse coniferous and broadleaf forests. Unpublished report, Coilte, Bray, Co. Wicklow

Lowe VPW, Gardiner AS (1975) Hybridisation between red deer (*Cervus elaphus*) and sika deer (*Cervus nippon*) with particular reference to stocks in NW. England. J Zool (Lond) 177:553–566

Lu XP, Wei FW, Li M, Yang G, Liu H (2006) Genetic diversity among Chinese sika deer (*Cervus nippon*) populations and relationships between Chinese and Japanese sika deer. Chin Sci Bull 51(4):433–440

Maillard D, Gaillard J-M, Hewison M, Ballon P, Duncan P, Loison A, Toïgo C, Baubet E, Bonenfant C, Garel M, Saint-Andrieux C (2010) Ungulates and their Management in France. In: Apollonio M, Andersen R, Putman R (eds) European ungulates and their management in the 21st century. Cambridge University Press, Cambridge, UK, pp 441–474

Makovkin LI (1999) The sika deer of Lazovsky reserve and surrounding areas of the Russian Far East. Almanac Russki Ostrov, Russia, Vladivostok

Mann JCE (1983) The social organisation and ecology of the Japanese sika deer (*Cervus nippon*) in southern England. PhD thesis, University of Southampton

Mann JCE, Putman RJ (1989a) Habitat use and activity patterns of British sika deer (*Cervus nippon* Temminck) in contrasting environments. Acta Theriol 34:83–96

Mann JCE, Putman RJ (1989b) Diet of British sika deer (*Cervus nippon* Temminck) in contrasting environments. Acta Theriol 34:97–110

Masuko T, Matsuura Y (2009) Nutritional physiology of wild and domesticated Japanese sika deer. In: McCullough DR, Takatsuki S, Kaji K (eds) Sika deer: biology and management of native and introduced populations. International Springer/Springer Japan, Tokyo, pp 61–82

Masuko T, Souma K (2009) Nutritional physiology and wild and domesticated Japanese sika deer. In: McCullough DR, Takatsuki S, Kaji K, editors. Sika deer: biology and management of native and introduced populations. Tokyo: International Springer/Springer Japan 61–82

Matuszewski G, Sumiński P (1984) Sika deer in Poland Deer 6:74–75

McCullough DR (2009a) Sika deer in Korea and Vietnam. In: McCullough DR, Takatsuki S, Kaji K (eds) Sika deer: biology and management of native and introduced populations. International Springer/Springer Japan, Tokyo, pp 541–548

McCullough DR (2009b) Sika deer in Taiwan. In: McCullough DR, Takatsuki S, Kaji K (eds) Sika deer: biology and management of native and introduced populations. International Springer/Springer Japan, Tokyo, pp 549–560

McCullough DR, Takatsuki S, Kaji K (eds) (2009) Sika deer: biology and management of native and introduced populations. International Springer/Springer Japan, Tokyo

McDevitt A, Edwards CJ, O'Toole P, O'Sullivan P, O'Reilly C, Carden RF (2009) Genetic structure of, and hybridisation between, red (*Cervus elaphus*) and sika (*Cervus nippon*) deer in Ireland. Mamm Biol 74:263–273

McKelvey PJ (1959) Animal damage in North Island protection forests. NZ Sci Rev 17:28–34

Minami M, Ohnishi N, Okada A, Takatsuki S (2009) Reproductive ecology of sika deer on Kinkazan Island, Northern Japan: reproductive success of males and multi-mating of females. In: McCullough DR, Takatsuki S, Kaji K (eds) Sika deer: biology and management of native and introduced populations. International Springer/Springer Japan, Tokyo, pp 397–318

Mirolyubov II, Ryaschenko LP (1948) Sika deer. Vladivostok, USSR. (In Russian)

Nagata J (2009) Two genetically distinct lineages of the Japanese sika deer based on mitochondrial control regions. In: McCullough DR, Takatsuki S, Kaji K (eds) Sika deer: biology and management of native and introduced populations. International Springer/Springer Japan, Tokyo, pp 27–41

Nagata J, Masuda R, Yoshida MC (1995) Nucleotide sequences of the cytochrome b and 12S rRNA in the Japanese sika deer *Cervus nippon*. J Mammal Soc Jpn 20:1–8

Nagata J, Masuda R, Kaji K, Ochiai K, Asada M, Yoshida MC (1998) Microsatellite DNA variations of the sika deer, *Cervus nippon*, in Hokkaido and Chiba. Mammal Study 23:95–101

Nagata J, Masuda R, Tamate HB, Hamasaki S, Ochiai K, Asada M, Tatsuzawa S, Suda K, Tado H, Yoshida MC (1999) Two genetically distinct lineages of the sika deer *Cervus nippon*, in Japanese Islands: comparison of mitochondrial D-loop region sequences. Mol Phylogenet Evol 13:511–519

Nielsen EK, Olesen CR, Pertoldi C, Gravlund P, Barker JSF, Mucci N, Randi E, Loeschcke V (2008) Genetic structure of the Danish red deer (*Cervus elaphus*). Biol J Linn Soc 95:688–701

Novobilsky A, Horackova E, Hirtova L, Modry D, Koudela B (2006) The giant liver fluke *Fascioloides magna* (Bassi 1875) in cervids in the Czech Republic and potential of its spreading to Germany. Parasitol Res 100:549–553

O'Donoghue A (1991) Growth, reproduction and survival in a feral population of Japanese sika deer (*Cervus nippon*). Unpublished PhD thesis, University College, Dublin

Obrtel R, Holišová V, Kožená I (1985a) The winter diet of sika deer (*Cervus nippon*) in the Bouzovsko area. Folia Zool 34:1–22

Obrtel R, Holišová V, Kožená I (1985b) The winter diet of sika deer (*Cervus nippon*) in the Manětínsko area. Folia Zool 34:291–302

Ohmura Y, Fukumoto F, Ohtaki K (1983) Chromosomal polymorphism in Japanese sika, *Cervus* (*Sika*) *nippon*. Jpn J Vet Sci 45:23–30

Ohtaishi N (1978) Ecological and physiological longevity in mammals – from the age structures of Japanese deer. J Mamm Soc Jpn 3:130–134

Ohtaishi N (1986) Preliminary memorandum of classification, distribution and geographic variation on sika deer. Mamm Sci 53:13–17. (in Japanese with English summary)

Okada A, Tamate HB, Minami M, Ohnishi N, Takatsuki S (2005) Use of microsatellite markers to assess the spatial genetic structure of a population of sika deer *Cervus nippon* on Kinkazan Island, Japan. Acta Theriol 50: 227–240

Packer JJ, Doney J, Mayle BA, Palmer CF, Cope M (1999) Field and desk studies to assess tolerable damage levels for different habitats and species of deer. Report to the UK Ministry of Agriculture, Fisheries and Foods, on project VC 0315, London, UK

Pemberton J, Swanson G, Barton N, Livingstone S, Senn H (2006) Hybridisation between red and sika deer in Scotland. Deer 13:22–26

Pérez-Espona S, Pemberton JM, Putman RJ (2009) Red and sika deer in the British Isles, current management issues and management policy. Mamm Biol 74: 247–262

Pérez-Espona S, Hall RJ, Pérez-Barbería FJ, Glass BC, Ward JF, Pemberton JM (2013) The impact of past introductions on an iconic and economically important species, the red deer of Scotland. J Hered 104:14–22

Pitra C, Fickel J, Meijaard E, Groves C (2004) Evolution and phylogeny of old world deer. Mol Phylogenet Evol 33:880–895

Pitra C, Rehbein S, Lutz W (2005) Tracing the genetic roots of the sika deer *Cervus nippon* naturalized in Germany and Austria. Eur J Wildl Res 51:237–241

Powerscourt V (1884) On the acclimatization of the Japanese deer at Powerscourt. Proc Zool Soc London 52: 207–209

Prisyazhnyuk VE (2005) Interspecific hybridisation of the sika deer and diagnostics of hybrid populations. In: Spicin VV (ed) Ungulates in zoos and breeding centers: inter-department scientific and methodic articles. Moscow Zoo, pp 85–93

Putman RJ (1986) Competition and coexistence in a multispecies grazing system. Acta Theriol 31:271–291

Putman RJ (1993) Flexibility of social organisation and reproductive strategy in deer. Deer 9:23–28

Putman RJ (1996) Competition and resource partitioning in temperate ungulate assemblies. Chapman and Hall, London, UK

Putman RJ (2000) Sika deer. British Deer Society/Mammal Society, London, UK

Putman RJ (2010) Ungulates and their management in Great Britain and Ireland. In: Apollonio M, Andersen R, Putman R (eds) European ungulates and their management in the 21st century. Cambridge University Press, Cambridge, UK, pp 129–164

Putman RJ, Clifton-Bligh JR (1997) Age-related bodyweight, density and fecundity in a south Dorset sika population (*Cervus nippon*), 1985–93. J Nat Hist 31:649–660

Putman RJ, Mann JCE (1990) Social organisation and behaviour of British sika deer in contrasting environments. Deer 8:90–94

Putman RJ, Moore NP (1998) Impact of deer in lowland Britain on agriculture, forestry and conservation habitats. Mammal Rev 28:141–164

Putman RJ, Sharma SK (1987) Long term changes in New Forest deer populations and correlated environmental change. Symp Zool Soc Lond 58:167–179

Putman RJ, Langbein J, Hewison AJM, Sharma SK (1996) Relative roles of density-dependent and density-independent factors in population dynamics of British deer. Mammal Rev 26:81–101

Quirke K (1991) The diet of red deer, sika deer and Scottish blackface sheep in Killarney National Park, Co. Kerry. MSc thesis, National University of Ireland

Randi E, Mucci N, Carlo-Hergueta F, Bonnet A, Douzery EJP (2001) A mitochondrial DNA control region phylogeny of the Cervinae: speciation in *Cervus* and implications for conservation. Anim Conserv 4:1–11

Ratcliffe PR (1987) Distribution and current status of sika deer (*Cervus nippon*) in Great Britain. Mammal Rev 17:37–58

Ratcliffe PR (1989) The control of red and sika deer populations in commercial forests. In: Putman RJ (ed) Mammals as pests. Chapman and Hall, London, UK, pp 98–115

Ratcliffe PR (1991) Sika deer. In: Corbet GB, Harris S (eds) The handbook of British mammals. Blackwell Scientific, Oxford

Ratcliffe PR, Peace AJ, Hewison M, Hunt E, Chadwick AH (1992) The origins and characterization of Japanese sika deer populations in Great Britain. In: Maruyama N, Bobek B, Ono Y, Regelin W, Bartos L, Ratcliffe PR (eds) International symposium on wildlife conservation: present trends and perspectives for the 21st century. Japan Wildlife Research Center, Tokyo, pp 185–190

Raymond N (2008) The ecology and performance of Japanese sika deer (*Cervus nippon nippon* Temminck, 1838) and their impact on vegetation in Killarney National Park. PhD thesis Manchester Metropolitan University

Rehbein S (2010) Die Endoparasiten des Sikawildes in Deutschland und Österreich. Doctoral dissertation, Paris Lodron-Universität Salzburg

Saitoh T, Kaji K, Izawa M, Yamada F (2015) Conservation and management of terrestrial mammals in Japan: its organizational system and practices. Therya 6(1): 139–153

Sanchéz-Fernandéz B, Soriguer R, Rico C (2008) Cross-species tests of 45 microsatellite loci isolated from different species of ungulates in the Iberian red deer (*Cervus elaphus hispanicus*) to generate a multiplex panel. Mol Ecol Resour 8:1378–1381

Senn HV, Pemberton JM (2009) Variable extent of hybridization between invasive sika (*Cervus nippon*) and native red deer (*C. elaphus*) in a small geographical area. Mol Ecol 18:862–876

Senn HV, Barton NH, Goodman SJ, Swanson GM, Abernethy KA, Pemberton JM (2010a) Investigating temporal changes in hybridization and introgression in a predominantly bimodal hybridizing population of invasive sika (*Cervus nippon*) and native red deer (*C. elaphus*) on the Kintyre Peninsula, Scotland. Mol Ecol 19:910–924

Senn HV, Swanson GM, Goodman SJ, Barton NH, Pemberton JM (2010b) Phenotypic correlates of hybridisation between red and sika deer (genus *Cervus*). J Anim Ecol 79:414–425

Shen-Jin L, Yan Y, Xue-Bin W (2014) Genetic diversity analysis by microsatellite markers in four captive populations of the sika deer (*Cervus nippon*). Biochem Syst Ecol 57:95–101

Slate J, Coltman DW, Goodman SJ, Maclean I, Pemberton JM, Williams JL (1998) Microsatellite loci are highly conserved in red deer (*Cervus elaphus*), sika deer (*Cervus nippon*), and Soay sheep (*Ovis aries*). Anim Genet 29:307–315

Slate J, Van Stijn TC, Anderson RM, McEwan KM, Maqbool NJ, Mathias HC, Bixley MJ, Stevens DR, Molenaar AJ, Beever JE, Galloway SM, Tate ML (2002) A deer (subfamily Cervinae) genetic linkage map and the evolution of ruminant genomes. Genetics 160:1587–1597

Sleeman DP (1983) Parasites of deer in Ireland. J Life Sci Roy Dublin Soc 4:203–209

Smith S (1996) New Forest sika deer. II Vocal Behaviour. Deer 9:643–645

Smith SL, Carden RF, Coad B, Burkitt T, Pemberton JM (2014) A survey of the hybridisation status of *Cervus* deer species on the island of Ireland. Conserv Genet 15: 823–835

Smith SL, Senn HV, Pérez-Espona S, Wyman MT, Heap E, Pemberton JM (2018) Introgression of exotic *Cervus* (*nippon* and *canadensis*) into red deer (*Cervus elaphus*) populations in Scotland and the English Lake District. Ecol Evol 2018:1–13

Sokolov II (1959) Fauna SSSR. Mlekopitayushtchie. Kopytnye zveri (Otryad Perissodactyla I Artiodactyla). Izdatel'stvo Akademii nauk SSSR, Moskva, Leningrad. (In Russian)

Staines BW (1998) Sika deer: their status, distribution and ranging behaviour. In: Staines BW, Palmer SCF, Wyllie I, Gill R, Mayle B (eds) Desk and limited field studies to analyse the major factors influencing regional deer populations and ranging behaviour. Final report to UK Ministry of Agriculture, Fisheries and Food on contract VC 0314, London

Swanson GM (1999) The genetic and phenotypic consequences of translocations of deer (genus *Cervus*) in Scotland. PhD thesis, University of Edinburgh

Swanson GM, Putman RJ (2009) Sika deer in the British Isles. In: McCullough DR, Takatsuki S, Kaji K (eds) Sika deer: biology and management of native and introduced populations. International Springer/ Springer Japan, Tokyo, pp 595–614

Takahashi K, Matsuoka H, Taru H, Yasui K, Hasegawa Y (2003) Vertebrate fossils from the excavation site of the Naumann's elephant at Sahama. Shizuoka Chigaku 87: 15–21. (In Japanese)

Takakuwa Y (2004) Sika deer fossils from the Middle Pleistocene of Nishikigaoka, Shimoda, Kamikita-gun, Aomori Prefecture. Report on sika deer fossils from Nishikigaoka, Shimodacho. Board of Education of Shimoda-cho, Aomori Prefecture, Japan. (In Japanese)

Takakuwa Y (2006) Cervid fossils from the Kiyokawa Formation of Shimosa Group, Sodegaura, Chiba Prefecture, Japan. Quat Res (Daiyonki-Kenkyu) 45: 197–206. (In Japanese with English abstract)

Takatsuki S (2009a) North-South variations in sika deer ecology as a forest-dwelling cervid. In: McCullough DR, Takatsuki S, Kaji K (eds) Sika deer: biology and management of native and introduced populations. International Springer/Springer Japan, Tokyo, pp 217–230

Takatsuki S (2009b) Geographical variations in food habits of sika deer: the northern grazer vs. the southern browser. In: McCullough DR, Takatsuki S, Kaji K (eds) Sika deer: biology and management of native and introduced populations. International Springer/ Springer Japan, Tokyo, pp 231–238

Takatsuki S, Ito TY (2009) Plants and plant communities on Kinkazan Island, north Japan in relation to sika deer herbivory. In: McCullough DR, Takatsuki S, Kaji K, editors. Sika deer: biology and management of native and introduced populations. Tokyo: International Springer/Springer Japan 125–144

Tamate HB, Tsuchiya T (1995) Mitochondrial DNA polymorphism in subspecies of the Japanese Sika deer, *Cervus nippon*. J Hered 86:211–215

Tamate HB, Tatsuzawa S, Suda K, Izawa M, Doi T, Sunagawa K, Miyahira F, Tado H (1998) Mitochondrial DNA variations in local populations of the Japanese sika deer, *Cervus nippon*. J Mammal 78:1396–1403

Tamate HB, Okada A, Minami M, Ohnishi N, Higuchi H, Takatsuki S (2000) Genetic variations revealed by microsatellite markers in a small population of the sika deer (*Cervus nippon*) on Kinkazan Island, Northern Japan. Zool Sci 17:47–53

Thévenon S, Thuy LT, Ly LV, Maudet F, Bonnet A, Jarne P, Maillard J-C (2004) Microsatellite analysis of genetic diversity of the Vietnamese sika deer (*Cervus nippon pseudaxis*). J Hered 95:11–18

Thirgood SJ, Langbein J, Putman RJ (1998) Intraspecific variation in ungulate mating strategies: the case of the flexible fallow deer. Adv Study Behav 28:333–361

Ueno M, Iijima H, Takeshita K, Takahashi H, Yoshida T et al (2018) Robustness of adult femal survival maintains a high-density sika deer (*Cervus nippon*) population following the initial irruption. Wildl Res 45:143–154

Uzal FA (2010) The interaction of Sika deer (*Cervus nippon* Temminck 1838) with lowland heath mosaics. PhD thesis, Bournemouth University

Uzal A, Walls S, Stillman RA, Diaz A (2013) Sika deer distribution and habitat selection: the influence of the availability and distribution of food, cover, and threats. Eur J Wildl Res 59(4):563–572

Vavruněk J, Wolf R (1977) Breeding of red deer in West-Bohemian region. Textb Sci For Inst VŠZ Prague 20: 97–115. (In Czech)

Voloshina IV, Myslenkov AI (2009) Sika deer distribution changes at the northern extent of their range in the Sikhote-Alin Mountains of the Russian Far East. In: McCullough DR, Takatsuki S, Kaji K (eds) Sika deer: biology and management of native and introduced

populations. International Springer/Springer Japan, Tokyo, pp 501–520
Weisz I, Arnold W, Suchentrunk F (2002) Population genetics of red deer (*Cervus elaphus*) in Lower Austria – is there any hybridization between sika deer (*C. nippon*) and red deer? Mamm Biol 67(Suppl):41–42
Whitehead GK (1993) Encyclopedia of deer. Swan Hill Press, Shrewsbury
Wilson RL (2000) An investigation into the phylogeography of sika deer (*Cervus nippon*) using microsatellite markers. MSc thesis, University of Edinburgh
Won C, Smith KG (1999) History and current status of mammals of the Korean peninsula. Mammal Rev 29:3–33
Wotschikowsky U (2010) Ungulates and their management in Germany. In: Apollonio M, Andersen R, Putman R (eds) European ungulates and their management in the 21st century. Cambridge University Press, Cambridge, UK, pp 201–222
Wu H, Wan QH, Fang SG (2004) Two genetically distinct units of the Chinese sika deer (*Cervus nippon*): analyses of mitochondrial DNA variation. Biol Conserv 119:183–190
Yamada M, Hosoi E, Tamate HB, Nagata L, Tatsuzawa S, Tado H, Ozawa S (2006) Distribution of two distinct lineages of sika deer (*Cervus nippon*) on Shikoku Island revealed by mitochondrial DNA analysis. Mammal Study 31:23–28
Yamauchi K, Matsuura Y (2009) Endocrinology of sika deer. In: McCullough DR, Takatsuki S, Kaji K (eds) Sika deer: biology and management of native and introduced populations. International Springer/Springer Japan, Tokyo, pp 83–100
Yoshio M, Asada M, Ochiai K, Goka K, Murase K, Miyashita T, Tatsuta H (2008) Spatially heterogeneous distribution of mtDNA haplotypes in a sika deer (*Cervus nippon*) population on the Boso Peninsula, central Japan. Mammal Study 33:59–69
Yoshio M, Asada M, Ochiai K, Goka K, Miyashita T, Tatsuta H (2009) Evidence for cryptic genetic discontinuity in a recently expanded sika deer population on the Boso Peninsula, Central Japan. Zool Sci 26:48–53
Yuasa T, Nagata J, Hamasaki S, Tsuruga H, Furubayashi K (2007) The impact of habitat fragmentation on genetic structure of Japanese sika deer (*Cervus nippon*) in southern Kantoh, revealed by mitochondrial D-loop sequences. Ecol Res 22:97–106
Zejda J, Horakova M (1988) Three kinds of weights in sika deer (*Cervus nippon*). Folia Zool Brno 37:289–299
Zima J, Král B (1984) Karyotypes of European mammals, Part III. Acta Sci Nat Brno 18(9):1–51
Zima J, Kožená I, Hubálek Z (1990) Non-metrical cranial divergence between *Cervus elaphus*, *C. nippon nippon* and *C. nippon hortulorum*. Acta Sci Nat Brno 24:1–41

Common Fallow Deer *Dama dama* (Linnaeus, 1758)

5

Anna Maria De Marinis, Roberta Chirichella, and Marco Apollonio

Contents

A. M. De Marinis
Italian Institute for Environmental Protection and Research (ISPRA), Ozzano dell'Emilia (BO), Italy
e-mail: annamaria.demarinis@isprambiente.it

R. Chirichella · M. Apollonio (✉)
Department of Veterinary Medicine, University of Sassari, Sassari, Italy
e-mail: rchirichella@uniss.it; marcoapo@uniss.it

© Springer Nature Switzerland AG 2022

L. Corlatti, F. E. Zachos (eds.), *Terrestrial Cetartiodactyla*, Handbook of the Mammals of Europe, https://doi.org/10.1007/978-3-030-24475-0_21

Common Names

English	Common fallow deer
German	Damhirsch
French	Daim
Spanish	Gamo
Italian	Daino
Russian	лань

Taxonomy and Systematics

The International Commission on Zoological Nomenclature with the Opinion 581 (China and Melville 1960) validated *Dama* Frisch, 1775 as the generic name for fallow deer. However, several authors (e.g., Corbet 1978) have continued to consider it congeneric with *Cervus*. Molecular analyses support the monophyly of the genus *Dama* (e.g., Gilbert et al. 2006). Different divergent times were estimated for this genus in relation to the selected calibration point, for example 5.1 mya, just after the start of the Pliocene, or 3 ± 0.4 mya, during the Pliocene, according, respectively, to Pitra et al. (2004) and Gilbert et al. (2006).

The genus *Dama* includes the common fallow deer and the Mesopotamian fallow deer. These deer can be distinguished on the basis of body size (Mesopotamian is larger) and antler shape, and by some minor differences in coat and tail color, rhinarium, and skull morphology. The common and Mesopotamian fallow deer *Dama dama*

Fig. 1 Common fallow deer. A male (left) during the rutting season and a group of females with fawns (right) (photographs by A. M. De Marinis)

(Linnaeus, 1758) (Fig. 1) and *Dama mesopotamica* (Brooke, 1875) have been regarded as sister taxa on the basis of molecular and morphological analyses (e.g., Heckeberg 2020, Pitra et al. 2004). On the contrary, Masseti et al. (2008), estimating a divergence time of 423,021 years (95% CI = 430,000–116,000 years) between haplotypes of these two taxa, supported the subspecific classification of *D. d. dama* and *D. d. mesopotamica*. At present, the taxonomic debate is still open.

Paleontology

The ancestry of the modern fallow deer (Fig. 1) has been tentatively traced back to several Pliocene/Early Pleistocene forms whose relationships to *Dama* are still unclear (see Heckeberg 2020).

The common fallow deer belong to the group of Asian biota that immigrated to the European regions. At the beginning of the Late Pleistocene, fallow deer lived in continental Europe and commonly occur in faunal assemblages from the last interglacial (*c.* 130–115 kya) of mid-latitude Europe (Masseti and Vernesi 2014). During the last glacial period, common fallow deer retreated into the southern areas of its former distribution, and survived in southern Anatolia and perhaps in southern Italy, Sicily, and the southern Balkan Peninsula (Uerpmann 1987; Stuart 1991; Masseti 1996; Masseti and Vernesi 2014). There is no clear archaeological evidence of the early colonization of the Iberian Peninsula. Davis and MacKinnon (2009) suggest that the species was absent in this peninsula before Roman times and that the Romans were responsible for its introduction. Osteological remains from the island of Mallorca, Spain, recovered in Iron Age (Talayot culture) settlements, could instead suggest a pre-Roman introduction chronology for the species, but these remains "came from insecure contexts more likely associated with Roman activity" (Valenzuela et al. 2016). Paleontological evidence of deer presence in glacial refugia is very fragmentary. As far as is presently known, several artistic Epipaleolithic representations of the common fallow deer from southern Italy and Sicily constitute the latest chronological evidence for the persistence of this cervid in Italy during the Late Glacial Maximum (Masseti and Rustioni 1988; Masseti 1996). Subsequently, the fallow deer did not return as far north as its previous range, unlike many other temperate species (Masseti and Vernesi 2014). This lack of expansion from southern refugia is atypical, but not unique (see Bilton et al. 1998).

Post-glacial expansion of the genus *Dama* was limited almost exclusively to the northern Mediterranean and the Near East (Chapman and Chapman 1997; Masseti and Vernesi 2014), where, despite the contiguity of their native geographical distribution, the common and Mesopotamian fallow deer were not sympatric. Fallow deer of the

Taurus range belong to the common form (Harrison 1968), while the original distribution of the Mesopotamian deer ranged from southeastern Turkey and the Levant to eastern Persia (Uerpmann 1987).

Archaeozoological evidence suggests that translocations of fallow deer began early in the Pre-Neolithic Cyprus and in the Neolithic of continental Greece and several Aegean islands (Masseti and Vernesi 2014). Translocations increased during the Bronze Age when this species also began to spread into the western Mediterranean (Masseti and Vernesi 2014). The artificial occurrence of the genus *Dama* in northern Africa, as reported by Chapman and Chapman (1980) and Kitagawa (2008), is questionable.

The Mesopotamian fallow deer qualify as Endangered (EN) in the IUCN Red List under criterion D, due to very small and restricted population numbers in Iran and in Israel. However, the population trend seems to be increasing (www.iucnredlist.org). Reintroduction projects have been undertaken in different Iranian locations and in Israel (Fernández-García 2012; Saltz et al. 2011) and today are still ongoing.

Current Distribution

The common fallow deer are the most widespread deer species in the world (Chapman and Chapman 1997). From as early as the Pre-Neolithic period, the species was exported around the Mediterranean, later to central and northern Europe and, more recently, to the rest of the world. Free-living herds have been established in North and South America, South Africa, New Zealand, and Australia. Fallow deer were introduced on small and medium-sized islands in different parts of the world. Many fallow deer are also maintained in captivity for exhibition, commercial production of meat and antler velvet, or hunting. The distribution of the common fallow deer is, therefore, a direct consequence of human activity, to the point that it can provide information on past patterns of human migration and trade (Masseti 1996, 2002; Sykes et al. 2011).

The common fallow deer are also one of the most widespread introduced mammals in Europe (as it has been established in over 10 countries; Genovesi et al. 2009). At present it is difficult to define in detail the distribution of the species because of many introductions and re-introductions over the course of centuries. Distributional data which has been published in some European countries are poor, especially for free-ranging populations, often referring to different spatial and temporal scales and sometimes conflicting. Hence the distribution map of the species reported in Fig. 2 can be affected by the heterogeneity of the data sources.

The only geographical area where common fallow deer have persisted as a native form is southern Anatolia (Masseti 2002; Masseti et al. 2008). Here its historical range extended from the Marmara region through the coastal mountains to the southeast till the first half of the twentieth century (Borovali 1986; Danford and Alston 1880). This range has drastically decreased during the second half of the last century, when the species was reported only from Antalya and Adana provinces (see Durmuş 2019). Illegal hunting and environmental changes caused by human activities and increasing urbanization have led to the progressive disappearance of the fallow deer from Anatolia. Today a small number of deer have survived in a single site in Düzlerçamı Wildlife Development Area (Antalya province).

Description

Size and Morphology

The common fallow deer is a medium-sized deer. Adult males can reach more than 100 kg while adult females more than 50 kg. As the mass ratio of males to females is between 1.7 and 2.4 (McElligott et al. 2001), this deer is probably the most sexually dimorphic cervid (see Table 1).

Body mass varies throughout the year in relation to reproductive activity. Male body mass changes dramatically during the rutting period, decreasing by 15–30% (Berlioz et al. 2017). In females, body mass changes during gestation and lactation are related to higher energy requirements which in female ungulates are *c.* 40% during late gestation and 150% during peak lactation, compared to maintenance (Loudon 1985). Body mass

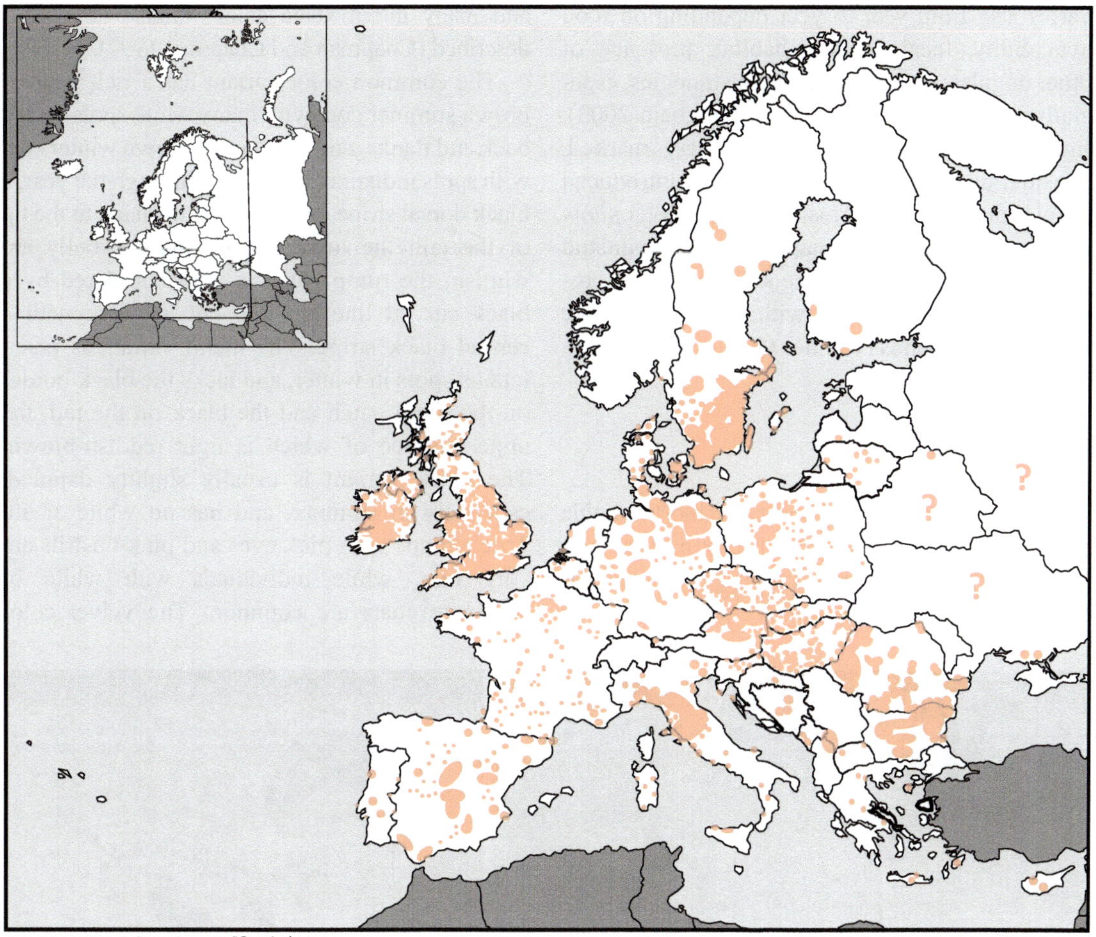

Map template: © Getty Images/iStockphoto

Fig. 2 Distribution map of the common fallow deer in Europe based on the IUCN Red List of Threatened Species. Version 2021–2 modified according to Chapman NG and Chapman DI 1980, Chapman DI and Chapman NG 1997, Masseti 1999, 2012, Apollonio 1999, Long 2003, Apollonio et al. 2010, Ilie and Enescu 2018, and data available from https://species.biodiversityireland.ie/profile.php?taxonId=119294&taxonDesignationGroupId=25, https://www.bds.org.uk/index.php/research/deer-distribution-survey; https://ias.biodiversity.be/species/show/25, https://globalspecies.org/ntaxa/901536#cite_1, https://www.gbif.org/species/5220136 ? = Country with data deficiency (Map template: © Copyright Getty Images/iStockphoto)

Table 1 Body and skull measurements of adult males and females of the common fallow deer

	Females				Males			
Measurement	*n*	Mean	Min	Max	*n*	Mean	Min	Max
HBL (cm)	9	141	114	167	11	159	138	177
SH (cm)	8	81	67	95	11	94	87	101
CBL (mm)	46	247	231	267	47	263	241	283
ZW (mm)	54	109	102	118	69	127	115	140
ML (mm)	49	191	175	203	68	200	185	216

Data from Ueckermann and Hansen (1994) for body measurements of free-ranging deer (8–10 years old) from Germany, and Feldhamer et al. (1988) for skull measurements of free-ranging and park deer (2 years old and over) from England and Scotland

n sample size, *mean* mean value, *min* minimum value, *max* maximum value, *HBL* head-body length, *SH* shoulder height, *CBL* condylobasal length, *ZW* zygomatic width, *ML* mandible length varies also from year to year depending on food availability, local density, habitat, presence of other ungulates, and management strategies, especially in park deer (Putman and Langbein 2003). Interpopulation variation can be very marked, considering that the species was introduced throughout the world. Insular populations show a reduction in size in comparison to their mainland source population; body size shifts are more pronounced for populations with longer residence times on the islands (van der Geer et al. 2018).

Pelage

The common fallow deer have the most variable pelage coloration of any deer species because of a long history of semi-domestication. Four main color variants (common, black, white, and menil; Fig. 3) and many intermediate color variants have been described (Chapman and Chapman 1997).

The common color variant has a rich yellow-brown summer coat with many white spots on the back and flanks, and a dull gray-brown winter coat with spots indistinct or absent. Through the year, a black dorsal stripe extends from the nape to the tip of the tail; the lower flanks and the belly are whitish; the rump patch is white bordered by a black curved line and the tail is white with a central black stripe. The menil variant is paler, retains spots in winter, and lacks the black border on the rump patch and the black on the tail, the upper surface of which is light reddish-brown. The black variant is usually slightly dappled, especially in summer, and has no white at all. True albinos with pink eyes and pink nostrils are very rare, while individuals with white or off-white coats are common. The velvet color

Fig. 3 Color variants of pelage in the common fallow deer: common (upper left), black (upper right), white (bottom left), menil (bottom right, file n. 505984862, www.istock.com, all other photographs by A. M. De Marinis)

tends to match the pelage. Coat colors are alike in both sexes. The fawn coat is similar to the summer coat of the adults except for fawns of white fallow deer which are sandy colored without any spotting at birth and become white in a period of one to several years (Chapman and Chapman 1997). Selection for or against color variants is often practiced in parks (Chapman and Chapman 1997). All the color variants can interbreed, and the offspring are sometimes different from either parent. Spots are dominant over no spots, and white coat color is controlled by a recessive allele. However, the genetics of coat color in fallow deer remains a wide and complex field of research.

A rare variation concerning the hair length was observed in Mortimer Forest (Shropshire, England) where both sexes have unusually long and curly hairs that are the expression of a dominant, autosomal gene. No similar polymorphism in hair length has been reported in any other species of deer (Chapman and Putman 1991).

Males have a prominent brush of hairs from the penis sheath visible from *c.* 3 months while females have a tuft of long hairs (*c.* 12 cm) below the vulva (Chapman and Putman 1991).

The dorsal guard hairs are shielded. A dichotomous key based on macroscopic and microscopic features of the dorsal guard hairs (De Marinis and Asprea 2006) allows the identification of the species and the age class (young prior to the first molt and adults). These data can be used in feeding ecology studies of large predators.

Molt takes place twice a year in spring (April–June) and in autumn (late September–November). The molt period varies in relation to body condition, sex, age, and population, while the sequence of molt changes in relation to the individual (Chapman and Chapman 1997). The fawn coat is retained until the first winter coat develops. Small birds as finches can collect tufts of hair as nesting material (Chapman and Chapman 1997).

The summer coat of Mesopotamian fallow deer has spots blending together to form a continuous line along each side of the brown dorsal stripe, and a white rump patch not completely bordered by a dark curved line with a white tail. It is difficult to find substantial differences in the coats of the common and Mesopotamian fallow deer because of the extremely wide range of the pelage colors of the common fallow deer (Chapman and Chapman 1997).

Skull

The fallow deer skull, as that of other cervids, shows numerous adaptations to support an herbivorous diet, such as a long rostrum providing solid anchorage for masseter muscles, a thin and long mandible, and a dentition able to grind herbaceous as well as woody materials. This species has a slender skull whose width is usually less than one half of the length (see Shackleton 1999). Nasal bones are considerably longer than the upper teethrow with anterior ends showing deep notches and posterior ends reaching and even surpassing the line of the forward edges of the orbita; size and shape of nasal bones are species-specific in cervids. The premaxillary bones articulate with the nasals, as observed only in some deer species. Rostral fenestrae are large and distinctive as in cervids generally. Orbits protrude to either side of the skull and are larger than in *Cervus*. Zygomatic arches protrude less than the orbits, as in other cervids. The braincase is large, rounded, and convex with developed occipital crests. The palate is somewhat wider than the length of the row of the upper molars and equal in females and males. Posterior nares are not divided by the vomer. Pterygoid processes are weakly bent forward and the hooks of the pterygoid bone lie behind the posterior processes of the superior jaw. The auditory bullae are small, inflated, and rounded with short auditory tubes. The foramen magnum has stout occipital condyles. The mandible is long and slender with a high and recurved coronoid process, a small condyle well above the tooth rows, and a large and rounded angular process not projecting behind the condyle. Most of the cranial measurements of adult males exceed those of females by 4–14% (Feldhamer et al. 1988), as reported in Table 1.

Skulls of the common and Mesopotamian fallow deer show some minor differences: the Mesopotamian fallow deer have different sutures between zygomatic and maxilla bones, nasal bones with blunt and much broader anterior ends and a more strongly convex lateral profile (Harrison 1968).

Antlers

As in all cervids, antlers develop from permanent outgrowths of frontal bones called pedicles. When fully grown, fallow deer antlers show a broad flattened "palm" and have only two frontal tines, the well-developed brow tine, just above the coronet, and the small trez tine, just below the palm; the upper and rear edges of the palm end in a series of points or spellers, directed posteriorly, with the longest at the bottom of the palm (Chapman and Chapman 1997). Antlers of the Mesopotamian fallow deer differ distinctly from those of the common fallow deer, being flattened in their basal part, immediately above the short brow tine, and not at all or only slightly palmated in their distal part (Harrison 1968).

Antler size, mass, and complexity are very variable, depending on age, body condition, habitat quality, and genotype and seem to be negatively associated to the degree of teeth wear (De Marinis 2015). The maximum average antler size occurs before 10 years of age (Chapman and Chapman 1997).

The growing antlers are covered with velvet, a specialized skin transformed from pedicle integument, most likely due to a mix of chemical and mechanical induction (e.g., Li 2013). Velvet is thickened in comparison to pedicle epidermis, contains hair follicles without arrector pili muscles, and is connected to extremely large sebaceous glands, but lacks sweat glands. Velvet is richly supplied with arteries and veins and, hence, is the main nutritional source for antler formation.

The antler cycle is closely associated with the photoperiodic regulation of the reproductive activity. Antlers grow during the late spring and early summer, they mineralize before the rut and are cast thereafter. A cascade of events involving several hormones such as melatonin, prolactin, luteinizing hormone, and testosterone mediates the primary effect of the photoperiod. Histogenesis and ossification of antlers were described by Kierdorf H. et al. (1995) and Kierdorf U. et al. (2003), and the mechanism of casting by Goss et al. (1992). In Mediterranean habitats, that are the native environments of the species, velvet shedding occurs in mid-August while antler casting occurs between mid-March and mid-April, with older bucks losing antlers 2 or 3 weeks earlier than younger ones (Ciuti and Apollonio 2011).

Pedicles usually start to develop from 5 to 7 months. First antlers vary from small knobs to spikes and are clean of velvet by about 15 months of age. Some second and third sets of antlers have porous tips because the velvet is shed before mineralization of the antler is complete. Antler growth is faster during the velvet period in territorial males (about 10 g/day per antler) than in non-territorial males (3.6–5.2 g/day per antler), likely because the first adopt better foraging strategies (Ciuti and Apollonio 2011). When the antlers are growing, the diet of males should be richer both in species diversity and nutritive value with increased protein content due to a higher energy requirement (Robbins 2013). Males can be grouped in age classes in relation to antler development. In Mediterranean habitats males show spike antlers when younger than 23 months, branched antlers between 29 and 34 months of age, and palmate antlers when older than 41 months (De Marinis 2015).

Antlers, as other secondary sexual characters, should reflect the condition of their possessors; in particular the symmetrical development of antlers should signal male quality. Fluctuating asymmetry (random difference in size observed between bilateral structures, FA) should reflect the individual capacity to cope with stressful situations (Palmer and Strobeck 1986). Therefore, FA could be used as a measure of ecological stress and give useful information on individual quality. It is this visual asymmetry that would be assessed by potential mates and rivals. Antlers are hypothesized to demonstrate decreasing levels of FA with increasing trait size and decreasing level of FA with increasing age. The studies of FA patterns in fallow deer reported conflicting results: some studies have found a relationship between asymmetry and measures of individual quality while others have not (e.g., Ciuti and Apollonio 2011; Putman et al. 2000). However, the studies of FA pattern in fallow deer, carried out mainly in enclosed populations, could probably be biased by the small sample size and the use of linear measurements to quantify asymmetry of complex structures like antlers. On the contrary, directional asymmetry, DA, which occurs when the character on one side is larger than on the other, is not considered a signal of developmental instability and hence of individual quality (Palmer 1996). In fallow deer DA toward the right side has been reported in several antler measurements (Alvarez 1995; Pélabon and Joly 2000).

Abnormal antlers in size and shape occur due to genetic or physiologic causes or injuries to pedicles, growing antlers, contralateral hind limb, or same-side front limb (Chapman and Chapman 1997). Double-head antlers have been observed in the common fallow deer as in other deer species; this malformation is the result of a new antler growth without the previous casting of the old antler; the double-head's second antler generation develops as a periosteal exostosis of the distal pedicle bone (Kierdorf U. et al. 1994). Antlerless or perruque heads (i.e., with antlers that continue to grow year after year without casting) are rare (Chapman and Chapman 1997).

Dentition

Fallow deer dentition is 0.0.3.0/3.1.3.0 for deciduous teeth and 0.0.3.3/3.1.3.3 for permanent teeth. Incisors and incisiform canines are spatulate; the first permanent incisor is twice as wide as others. There is a thick pad of fibrous connective tissue instead of upper incisors. Several dental eruption sequences were recorded (see De Marinis 2015, Bowen et al. 2016) according to the different criteria adopted to describe the eruption pattern. Teeth wear rate changes in relation to sex, as shown by the variation of I_1 crown height with the age class (De Marinis 2015). Male tooth row length is not significantly different from that of female (Chapman and Chapman 1970) contrary to what one might expect considering that the fallow deer is a strongly dimorphic species. Males consume a higher amount of food compared to females, even of lesser quality, hence the male wear rate is higher. The relatively early depletion of male teeth could be associated to a sex-specific strategy of dentine depletion, as already observed in red deer (Carranza et al. 2008).

Morphological characters for the discrimination between red and fallow deer teeth from archaeological and paleontological remains were described on permanent dentition by Lister (1996).

Dental anomalies have been found very occasionally, except in the population of Richmond Park (England), where the following incidence values were recorded for congenital dental anomalies: absence of one or two lower incisiform teeth in 19% of fawns and 18% of adults and presence of one or two upper canines in 25% of fawns (Chapman and Chapman 1997).

Age Determination

A protocol for estimating the age till 8 years was developed recording the epiphyseal sutural states in known-age deer of an enclosed population (Carden and Hayden 2006). Skeletal development in terms of epiphyseal fusion can be delayed by malnutrition, diseases, and trauma.

The chronology of mandibular tooth eruption varies in relation to the population due to environmental and genetic factors as well as material and methods adopted to study the eruption process. The following data identify the eruption times in months of each mandibular tooth according to several authors (see De Marinis 2015): 4–7 (M_1), 7–12 (I_1), 13–20 (M_2, I_2), 15–20 (I_3, C), 18–28 (M_3, P_4, P_3), 20–30 (P_2).

Scoring schemes were developed from known-age specimens coming from park deer population. A first scheme, based on radiographs of developing permanent molariform teeth, allows age estimation up to 3 years (Brown and Chapman 1990). A second scheme, based on eruption and wear stage of the fourth premolar and molars, should allow age estimation up to 10 years, but there is considerable overlap among wear stages at the upper end of the age spectrum (Bowen et al. 2016). Photographic reference systems and dichotomous keys were developed separately for males and females from known-age specimens coming from a Mediterranean free-ranging population. Taking into account that teeth wear rate depend on sex and habitat, these systems and keys can be used to age estimation of males and females living only in Mediterranean habitats (De Marinis 2015).

Glands

Fallow deer have several scent glands (areas of specialized skin rich in secretory cells) which play an important role in communication among members of the species. Males have preputial glands associated with the penis sheath. They consist of modified sebaceous glands. In fallow deer these glands are associated to sweat glands, unlike in other deer (Odend'hal et al. 1996). The preputial glands become active during the rut, and their secretion is responsible of the strong rutty smell of the urine during this period. Urine scattering, attained by a series of vigorous side-to-side penile movements, is used by fallow deer bucks for habitat and self-marking. Urine odor provides specific information about reproductive state, physical condition, and energy level that is important for both sexes to assess the social and physical status of bucks (Apollonio and Di Vittorio 2004). The pungent smell is believed to help synchronize female estrus (Chapman et al. 1981). Morphological changes in the preputial gland of immature fallow deer start at about 9–12 months of age (hypertrophy of the epidermis of the transition zone, together with some keratosis) and then recommences at about 15 months (hyperkeratosis). Eversion of the transition zone, which is a rutting characteristic of the adult, occurs for the first time when the deer are about 17 months old (Chapman et al. 1981). Accessory glands of reproduction were described by Chapman and Chapman (1979) in male fallow deer; these glands undergo an annual cycle of growth and regression, reaching their maximum development at the time of the rut.

Both sexes have suborbital, rear interdigital, and metatarsal glands. The suborbital glands are situated just below the corner of each eye and consist of pockets of skin, which fit into depressions of the skull (lacrimal fossae) next to the eye sockets. Lawson et al. (2000) highlighted the total lack of detectable volatiles in suborbital gland secretions between November and March, unlike what has been observed in other cervids. Considering that the presence of particular volatiles has been related to the possibility to carry information about individual identity, suborbital gland secretions of fallow deer do not seem to be individually characteristic outside the reproductive period. Rear interdigital glands are situated in pockets of skin between the two cleaves of the hoof. They start their activity within 2–3 weeks of birth and remain active throughout life. The glucidic content of their secretions does not change in relation to sex and age, suggesting that the interdigital glands may play only an accessory role in the production of odoriferous signals in fallow deer (Parillo and Diverio 2009). The metatarsal glands are located on the hock. Changes in their activity are documented in rut and post-rut periods only in adult males (Apollonio and Vailati 1996). Fallow deer have more complex gas chromatographic profiles of volatiles in metatarsal secretions than in suborbital secretions (Lawson et al. 2000). Only metatarsal gland secretions contain coded information about the individual identity of the signaler, and this information remains consistent over time, as shown in individuals sampled in successive years (Lawson et al. 2000). The scent profile of metatarsal gland secretions of cervids seems to change as a function of habitat type and sociality. Species habitually associated with open habitats and living in social groups, like fallow deer, show more complex scent profiles in metatarsal than in suborbital secretions (Lawson et al. 2000).

Physiology

Fallow deer can live in many different habitats showing great adaptability; for this reason, this species does not have any particular physiological adaptation to a specific environment. Body temperature depends on health, season, and activity patterns. The average tympanic temperature is 38.6 ± 0.7 °C, range 37.4–40.8 °C and the average rectal temperature is 40.1 ± 0.8 °C, range 37.5–42.0 °C (Drew 1998). Tympanic temperature seems to be superior to rectal temperature as an indicator of normothermia in fallow deer and may be a better method to assess the individual health status during capture and handling (Drew 1998). Heart and respiratory rates are 24–52 beats/min and 6–28 breaths/min, respectively (Galka et al. 1999).

Different values of blood parameters were recorded in relation to sex and age, as well as genetic, environmental, nutritional, and physiological factors (e.g., Poljičak-Milas et al. 2009). Management practices, capture methods, and blood sampling techniques also result in differences in blood parameters values (e.g., Vengušt et al. 2006): for example, variations in leucocyte count and serum proteins were recorded in relation to transport stress (English and Lepherd 1981). The relative proportions of fetal and adult hemoglobins were recorded in fawns captured during their first week of life in a free-living population of fallow deer in Coto Doñana National Park, southwestern Spain (Schreiber et al. 1992). The percentage of adult hemoglobin at birth seems to be correlated with the survival rate of the fawns. The exact age when the transition to adult hemoglobin is completed remains unresolved. The relative amount of persisting fetal hemoglobin is an indicator of the age of fawns. Blood coagulation parameters were reported by Siroka et al. (2011). A significant difference was found in the average activated partial thromboplastin time between males and females. It may be hypothesized that the differences in clotting times represent a physiological advantage for females to deal with delivery-associated bleeding but a physiological disadvantage for males to deal with injuries (e.g., due to fighting in the rut period).

The male and female reproductive cycle depends on seasonal variation in day length through the hypothalamus and pituitary control. The importance of light in determining the breeding season is confirmed by the fact that in the southern hemisphere fallow deer mate 6 months later than in the northern hemisphere. The male reproductive cycle is characterized by a variation in testes mass. The first stages of spermatogenesis occur at about 7 months of age, while the appearance of spermatozoa is only at about 14–16 months when the testes and epididymes are ten times larger than those of fawns (Chapman and Chapman 1997). Adult deer appear to be in breeding condition over a period of about 6 months with a peak in spermatogenic activity in October/November and are fertile as late as February or March (Chapman and Chapman 1997). There is no apparent age effect on the seasonality of testes and epididymes size fluctuations (Gosch and Fischer 1989). These cyclical changes are positively associated with changes in body mass (Chapman and Chapman 1997). Velvet shedding and antler casting occurred at about 80% and 25%, respectively, of maximal testis volume (Gosch and Fischer 1989). The male sexual activity is dependent not only on spermatogenesis but also on seasonal changes in hormone secretion. Testosterone induces rutting activity, but it is of short duration compared with the length of spermatogenic activity. Thus, high testosterone levels are not essential for fertilization to occur.

The ovaries of fawns are inactive and small and increase in size gradually until the females reach puberty that is at about 16 months of age. The length of the estrus cycle is 20–27 days. The highest percentage of fertilization occurs generally during October in the northern hemisphere, although females are receptive between September and January. Females are polyestrous and, in the absence of pregnancy, may cycle up to six times (Asher 2007). Most of females conceive during their first cycle. The absence of mature, high ranking, older bucks, or even the larger representation of younger bucks can prolong the fertile period of females (Komers et al. 1999). The reproductive state of females is assessed by males through the behavior known as *flehmen*. Multiple successive silent ovulations and short-lived (8–10 days) *corpora lutea* leading up to the start of the breeding season have been observed.

The transient nature of the preliminary *corpora lutea* may serve to promote within-herd synchrony of the first estrus of the season (Asher 2007).

The mammary glands undergo an annual cycle of growth and regression. The mass of these glands at the time of parturition is about 400 to 800 g (Chapman and Chapman 1997). Lactogenesis is strongly influenced by prevailing photoperiod during the last few months of pregnancy (Asher 2007). The energy requirements of lactation exceed those of pregnancy (Loudon 1985). The chemical composition of milk (Malacarne et al. 2015, Wang et al. 2017) changes over the lactation period, as found in other ruminants, and may also change among populations living in different habitats and having different diets.

Genetics

Chromosomes

The diploid number of chromosomes in *Dama dama* is $2n = 68$, and the number of chromosomal arms is FN = 70 (Rubini et al. 1990 and reference therein).

The karyotype of the fallow deer was described by studying various captive or introduced populations (Arslan and Zima 2014 and reference therein). It includes 32 acrocentric autosomal pairs of diminishing size and one medium-sized metacentric autosomal pair. The sex chromosomes are acrocentric (X) and metacentric (Y); X is the largest chromosome and Y is one of the smallest. There are significant amounts of heterochromatin at all autosomal centromeres as well as at X-chromosome, while the metacentric autosome pair shows only slight C-bands and the whole Y chromosome is relatively heterochromatic. The C-banding pattern is apparently rather uniform in the Cervidae family (Arslan and Zima 2014 and reference therein).

The results of the karyotype comparison between red and fallow deer are surprising because their karyotypes differ much less than those of red and sika deer, two easily interbreeding species (e.g., Herzog 1990). This means that the reproductive barrier between red and fallow deer may not primarily be caused by karyological incompatibility but by other physiological (e.g., immunological) or behavioral factors.

The localization of the nucleolus organizer regions (NOR) at the telomeric position in the two largest acrocentric autosomes was described by Mayr et al. (1987). A similar number and distribution of the NORs were reported in other deer species.

Fontana and Rubini (1990) discussed the chromosomal evolution of the Cervidae, pointing out that Robertsonian fusions took part in the karyotypic differentiation of Cervinae.

Phylogeny and Phylogeography

Baker et al. (2017) studied the relative influence of natural as well as anthropogenic processes in shaping the evolution of genetic diversity among European populations using microsatellite and mitochondrial DNA loci. The results confirmed very low levels of diversity within regional populations and revealed a high degree of differentiation among them. These data are consistent with populations being founded by low numbers of individuals and affected by strong genetic drift. The northern and central populations could descend from animals captured in Anatolia and translocated in Europe, while the southern populations could descend from animals forced into Italian, Iberian, and Anatolian peninsulas during the last glacial period. However, the origin of the Iberian lineage and the identification of refugia remain open questions. Marchesini et al. (2021) criticized the methodological approach adopted by Baker et al. (2017) with particular reference to sampling scheme, mitochondrial DNA, and microsatellite dataset, and contested that the genetic "signature reflecting postglacial refugia can still be detected" in the extant populations. By means of a reanalysis of the data, they concluded that the genetic expectations of the glacial refugia hypothesis cannot be found in current fallow deer populations which derive from extensive human-mediated translocations. Baker et al. (2021) revisited the study published in 2017, taking into account the comments by Marchesini et al. (2021). Some further analyses reinforced the original conclusions of the earlier paper: the low levels of diversity within populations and the strong structure among populations due to a combination of natural and anthropogenic processes, and the conservation implications of these data.

Genetic Diversity and Population Structure

Despite the wide distribution of the species and the availability of the samples, fallow deer genetic studies are very few. Initially, enzyme electrophoretic analyses revealed a very low level of genetic variation in British, German, and Italian populations (Hartl et al. 1986; Pemberton and Smith 1985; Randi and Apollonio 1988).

Subsequently, DNA variation analyses confirmed very low levels of diversity within populations and showed a high degree of differentiation among them (Ludwig et al. 2012; Masseti et al. 1997, 2008).

In particular, the analysis performed using the RAPD fingerprinting technique by Masseti et al. (1997) on two historical stocks that still survive in Central Italy showed that a fairly high percentage of the variability is due to differences between populations (75.54%). The Castelporziano stock revealed a higher level of genetic variability in contrast to San Rossore stock. The higher rate of within population genetic variability of the Castelporziano stock may be the result of multiple introductions even in very recent historical times. San Rossore stock generated populations with a low level of genetic variation (e.g., Maremma Regional Park, Central Italy; Scandura et al. 1998).

Moreover, Masseti et al. (2008) analyzing mitochondrial DNA sequences revealed that the Rhodian fallow deer population, founded by humans in Neolithic times, possesses a set of mitochondrial lineages, never found in any other population. Two highly distinct groups of haplotypes can be identified with a divergence time of at least 31 kya. The presence of these haplotypes could be due to translocation from Anatolia to Rhodes of two groups, already well differentiated in the mainland. Rhodian fallow deer are distinct even from the extant deer surviving in Düzlerçamı. The persistence of these haplotypes is particularly significant because human-mediated processes (e.g., domestication) usually result in genetic depletion and erosion of an ancestral genetic pool. Masseti et al. (2008) hypothesized that humans unknowingly preserved a remarkable portion of the original genetic diversity of the source population in the Rhodian fallow deer.

Hybridization

In captivity, the common fallow deer hybridized with the Mesopotamian fallow deer and produced fertile offspring (Zuckerman 1952; Gray 1972).

Hybridization between two different genera of deer is rare. In captivity, the hybridization between the common fallow deer and *Axis porcinus* has been recorded, but it is unknown if the offspring was fertile or how long it survived.

Life History

Growth

As in other cervids, maternal condition plays a determining role in prenatal growth (Pélabon 1997). Fetal body mass is linearly and negatively related to maternal kidney-fat-index and body mass during pregnancy (Ács and Lanszki 2017). Prenatal growth is faster for males (Birgersson and Ekvall 1997). Thus, male fawns appear to have higher body mass at birth than females (4.7 kg vs. 4.3 kg, Putman and Langbein 2003). However, the literature data on body mass by sex are quite different. Significant birth mass differences were recorded between fawns born to yearlings, around 3.5 kg, and those born to adult females, around 4.5 kg (Putman and Langbein 2003). Males grow faster than females, but male fawns of primiparous mothers have a slower growth rate than male fawns of multiparous mothers, whereas no such difference exists for female fawns (Birgersson and Ekvall 1997). According to Braza et al. (2000), male-biased maternal investment exists in fallow deer and it is affected by the environmental conditions such as the earliness of autumn rainfall, and the level of extra-investment made by mother in male fawns during the previous breeding season.

Average birth mass was revealed greater in the middle of the fawning period in a free-ranging population (Ács and Lanszki 2017), but it turned out greater at the beginning of this period in a fenced population (Birgersson and Ekvall 1997). According to Putman and Langbein (2003), the mass of fawns that were born in the first week of June were significantly greater at the beginning of winter than those that were born later. Moreover, pre-winter body mass was constantly greater and more strongly correlated to subadult body mass in males than in females and was positively related to maternal body mass (Birgersson and Ekvall 1997). Fallow deer seem to quadruple their birth mass by 3 to 4 months. Females attain the maximum body mass when about 3 years old while males continue to increase in mass, not reaching their maximum until at least 6 years old; the greatest body length (from head to rump) is not reached until 7–8 years in both sexes, and the shoulder height shows a similar growth pattern (Chapman and Chapman 1997). Mandibles are fully grown by about 2.5–3 years in females and 5 years in males (Langbein 1991).

Reproduction

Puberty occurs in males between about 7 and 14–16 months. Although yearlings are physiologically capable of breeding, males generally do not breed until they are about 4 years old, except in heavily hunted populations (McElligott et al. 2001). The complex social structure of the herds appears to preclude matings until a male reaches social maturity when it can actively compete for reproductive opportunities. Fallow bucks usually get most of the matings between 5 and 8 years of age (Farrell et al. 2011), when males are considered prime-aged (McElligott et al. 2003). Unlike males, females reproduce for the first time when they reach puberty (at around 16 months of age). To date, reproduction in females has been reported until 23 years of age (Farrell et al. 2011).

The gestation period is 225–237 days long; there is no embryonic diapause (Harrison and Hyett 1954). The development goes slowly until the embryo establishes a placental circulation (between 3 and 4 weeks of age). The common fallow deer give birth to a single fawn, twins occur in $<1\%$ of births, and, even if a doe carries twins, there is no guarantee that both will survive to term (Chapman and Chapman 1997). The neonatal sex ratio is 1:1 (Asher and Adam 1985). The sex ratio of adults depends on several factors, including the management regime.

In deer park populations yearling does have a fecundity consistently lower than that of adults, mate and give birth later, and produce fawns lighter at birth (Putman and Langbein 2003). The mean winter live body mass at which 50% of yearling does give birth to fawns was estimated at 32 kg (Putman and Langbein 2003). The average fawning rate varies a lot in yearlings (from 12 to more than 90%) and less in adult females ($>80\%$). The pattern of rearing success reflects that of fawning success with yearlings achieving lower rates than adults (Putman and Langbein 2003). The rearing ability of old females declines with age due to a loss in grazing efficiency causing insufficient milk production.

The effects of age and body condition on female reproductive success have been studied in several fallow deer park populations (Putman and Langbein 2003). In general, the reproductive success is consistently lower in parks that have extensive public access and where are present mixed herds with red deer. A higher level of postnatal mortality could probably explain this low reproductive success. Disturbed does may leave neonates immediately after parturition before they have a chance to establish the mother-infant bond.

Survival

Prenatal mortality is around 10% (Chapman and Chapman 1997). The highest mortality is recorded among neonates. Kjellander et al. (2012) found a death rate of 23.6% without significant differences between sexes in a free-ranging population of southern Sweden. The survival probability of adult females is relatively high until old age. Survival probabilities of males resulted highly variable with age, and have been described in a park deer population using a model incorporating four phases: yearling, pre-reproductive, prime-age, and senescent (McElligott et al. 2002). Yearlings had a survival rate of 70–75%. Pre-reproductive males (2–3 years old) had the highest survival rate (85%) compared to all other males. Prime-aged males (4–9 years old) had a slightly lower rate (between 70 and 80%) due to the intense intrasexual competition, and this rate remained stable up to 9 years of age. Considering reproducing and non-reproducing males separately, there were differences in the survival probabilities, with reproducers consistently surviving better than non-reproducers. The higher survival probabilities of reproducing males could indicate their generally higher phenotypic quality. Survival probabilities declined sharply after 9 years of age. The decrease also in reproduction probabilities, body condition, and antler size indicates that senescence had begun (McElligott et al. 2002).

The lifespan recorded in the fenced populations of Castelporziano estate (Rome, Italy) and Phoenix Park (Dublin, Ireland) is respectively 11.5–13 years for males and 19–23 years for females (see De Marinis 2015). The lifespan in captivity is over 20 years.

Natural mortality is mainly determined by climatic conditions. Severe drought and heavy snowfall can induce abrupt variations in population density; high rainfall can affect the fawn survival in summer when they are more easily chilled and risk hypothermia (Putman et al. 1996). Overwinter losses represent a considerable proportion of the annual mortality among wild populations (Putman and Langbein 2003). Length of the growing season of the previous year, days of snow, late winter temperature, and mean body mass in autumn are significantly correlated with winter losses in park deer populations (Putman and Langbein 2003). A threshold dressed body mass in autumn of 24 kg for adult females and yearling males is suggested as critical for overwinter survival (Putman and Langbein 2003). Mortality due to carnivore predation is low. The vigilance behavior and herding help to protect deer from predation. The contribution of fallow deer (mainly fawns and youngs) is less than 5% to the diet of Iberian lynx *Lynx pardinus* (Coto Doñana National Park, southwestern Spain, Delibes 1980) and less than 3% to that of wolf *Canis lupus* (the Apennines, central Italy, Mattioli et al. 2011; central and western Poland, Nowak et al. 2011). Foxes *Vulpes vulpes* may sometimes predate neonates; eagle *Aquila* sp. and golden jackal *Canis aureus* should be mentioned as other potential predators of fawns. During an experimental evaluation of predator avoidance in Sweden, fallow deer completely avoided areas with bear scent (Sahlen et al. 2016). Hunting, poaching, and to a lesser extent road accidents are other causes of mortality. Few individuals may die for injuries from conspecific during the rut or other accidents (e.g., entanglement of antlers in fences); chases by free-ranging and feral dogs may cause injuries and deaths (Langbein and Putman 1992), especially in those countries where these potential predators can reach high numbers.

Habitat and Diet

Habitat Use

Fallow deer live in a variety of climates ranging from cool-humid to hot-dry areas. However, most populations are found in a hot-humid climate. The most suitable habitats are plains and slightly rugged or hilly areas with a combination of vegetation types. Fallow deer prefer deciduous or mixed mature woodland with established understory, interspersed with grassy and brushy open areas, but can live in a wide variety of environments from Mediterranean forests to conifer plantations (Feldhamer et al. 1988). Fallow deer do not commonly live in high-altitude areas or regions with long snowy winters. Populations are currently expanding into areas composed of a mosaic of housing, gardens, orchards, parks, farmland, and pastures. Deer may be attracted to human environments because of the lack of predators and hunting threats as well as the availability of edge habitats where find food supply and tree cover. In such habitats, human activity may greatly affect the distribution, dynamics, and behavior of deer (Duarte et al. 2016). Habitat use changes seasonally with food availability (deer can feed mainly in open areas during spring and summer and in woods during autumn and winter, Thirgood 1995), differs between daytime and nighttime (woodland are used more by day while open areas more at night, Borkowski and Pudelko 2007), and varies strongly according to sex and age classes (Apollonio et al. 2005).

Sexual segregation is widespread among sexually dimorphic ruminants. Several hypotheses have been formulated to explain this phenomenon. Recent studies have provided evidence that only more hypotheses simultaneously may explain sexual segregation because they account for factors operating at different spatial and temporal scales. Ciuti and Apollonio (2008) analyzed a long-term dataset using a multitemporal (among years and between night and day) and multispatial (among and within habitats) approach to investigate sexual segregation in a fenced population of fallow deer. The predation risk hypothesis (females select safer areas to reduce predation risk even at the expense of nutrient intake) and the indirect competition hypothesis (males are forced into suboptimal foraging habitats through indirect competition by females) can explain segregation on large spatial and temporal scales. The forage selection hypothesis (differences in foraging efficiency and hence food selection between sexes) provides a general explanation of small scale sex differences in habitat use. The predation risk hypothesis and the indirect competition hypothesis give a valid interpretation of the sexual segregation on small temporal and large spatial scales and large temporal and small spatial scales, respectively. The influence of human activities could be a strong drive to sexual segregation, forcing the more vulnerable or sensitive sex into suboptimal habitats. This, in turn, may lead to a not natural browsing pressure on environments, causing habitat deterioration (Apollonio et al. 2005). Contrasting results were found in parks subjected to high public visitor pressure, where human disturbance increased levels of vigilance mainly in females, without changing patterns of habitat use and overall daily time budgets (Langbein and Putman 1992).

Home Range

Fallow deer do not establish territories except males during the rut. Home range size can vary a lot among populations and within the same population among seasons in relation to the availability and distribution of food and shelter, climatic factors, density of animals, and degree of disturbance. Home ranges can overlap extensively. In general, males have home ranges larger than females due to higher energy demands. Female home ranges are usually smaller during summer (e.g., Ciuti et al. 2003) because their movement patterns are restricted by the presence of fawns (Ciuti et al. 2006). As in other cervids, mothers seldom move further than 400 m away from their fawns in the first 2–3 months following parturition (Kjellander et al. 2012). Males have usually larger home ranges during the rutting period (e.g., Davini et al. 2004) due to the movements between breeding and non-breeding grounds. As in other deer, inter-seasonal home range fidelity proved to be remarkably high (Ciuti et al. 2003; Davini et al. 2004).

In Mediterranean habitats, that are the native environments of the species, the mean size of annual home range resulted larger (some hundreds of hectares, e.g., Ciuti et al. 2003, Davini et al. 2004) than those reported in northern Europe (some tens of hectares, e.g., Chapman and Chapman 1997, Putman 1996), mainly because these Mediterranean environments are characterized by a fragmented distribution of trophic resources.

Spatial Movements

The common fallow deer seem to be one of the least dispersive deer species. Large-scale one-directional movements that could be classed as dispersal are unusual. Most dispersal movements tend to be of juvenile males leaving family groups, before the birth of the next fawn. For this reason, the proportion of males among juveniles killed in road accidents is very high. This species is tolerant of very high local densities and hence colonization of new areas is gradual.

This cervid is not noted for any migratory behavior. Erratic movement may occur up to 15 km or more and on these occasions, fallow deer can cross large rivers. Bucks are often responsible of wide movements during the rut and may visit more than one lek or rutting stands. Females move from their home ranges to lek and can travel also more than 8 km, as recorded in a Mediterranean habitat (Imperio et al. 2020). It seems to be a distance threshold over which females are no longer able to compensate travel costs; this distance should change with population, depending in a complex way on topography, vegetation type, physical barriers (roads, channels), presence of predators, hunting, and human disturbance.

Diet

Anatomical considerations based on the ratio of small to large intestines suggest that the common fallow deer is an unselective bulk-feeder (sensu Hofmann 1985). Field studies confirm that this deer is both browser and grazer. Fallow deer show morphological adaptations in the mouth, teeth, and salivary glands for processing and digesting the chemical compounds of the plant cell wall (Pérez-Barbería 2020). Herbaceous monocots contribute to more than 60% of forage intake from March to September, and more than 20% even in winter (e.g., Caldwell et al. 1983). The browsing of young broadleaved trees gives a significant contribution to the diet in autumn and winter. Acorns, chestnuts, beech mast, and other fruits are characteristic food through autumn and early winter. Increasing amounts of dwarf brushes, sedges, rushes, brambles, forbs, mosses, and conifer browse are mainly consumed in winter, if graze material is limited (e.g., Caldwell et al. 1983). The barking of trees occurs in winter; bark is stripped using incisiform teeth and dental pad (Chapman and Chapman 1997). Fallow deer appear to drink rarely, taking water from dew and vegetation.

The quality of the diet and the nutrient composition (protein, digestible fiber, and lignin) depend on food resources, change with the season, year, and habitat, and vary across age and gender classes. Bucks have a higher quality diet in autumn and winter whereas does have a higher quality diet in spring and summer (Putman et al. 1993). Sex-related differences in the diet can be explained because of different metabolic demands (Barboza and Bowyer 2000). In the rutting season, a low diet similarity was found between non-reproductive males and females despite the sex spatial overlap (Azorit et al. 2012).

Food choice is based on constant sampling and evaluation of the variety of food available as recorded in cafeteria (multiple-choice) experiments (Bergvall and Leimar 2005). Deer seem to be more selective when food types are limited in number, occur relatively close to each other, and are characterized by distinct tastes or other cues used for food selection. Therefore, fallow deer should be more or less selective based on the spatial distribution of food resources (Bergvall et al. 2007).

Behavior

Foraging Behavior

Fallow deer tend to lie up by day in cover, where they drowse and ruminate, and to move by night in open areas where they graze. The movements from forestry to foraging areas are usually along well-marked tracks. Peak feeding periods are generally at dusk and dawn. This activity pattern is influenced by many factors, such as cyclical changes in the environment, food availability, foraging efficiency, predation risk, mating activity, intraspecific and interspecific interactions, and last but not least human disturbance. In little disturbed areas fallow deer graze and ruminate in open areas at all hours, with a pattern of alternation between feeding and rumination roughly every 3–4 h (Caravaggi et al. 2018).

Foraging decisions are conditioned by the tradeoff between the benefits of gaining more food and the costs of reduced vigilance or increased toxin ingestion. This trade-off reflects personality dimensions such as boldness (Bergvall et al. 2011). Bold females will raise heavier fawns but these females, and their fawns, will be more prone to predation. On the opposite, shy females and their offspring will be less prone to predation, even though the fawns will be lighter and at greater risk of starvation. Different personality types coexist in a population. Selection pressures on boldness should depend on environmental variation and predator density (Bergvall et al. 2011).

For group-living animals as fallow deer, foraging decisions are probably taken at the hierarchical level: at the higher level the group chooses where to forage, at the lower level the individual chooses what to eat (Stutz et al. 2018), becoming less selective due to the competition for food especially in large groups (Bergvall et al. 2006). Group foraging may prioritize other critical needs such as the management of predation risk rather than foraging efficiency (Stutz et al. 2018). The protection against predators and foraging interference decreases from the center to the periphery of the group. The conflicting requirements among group members can determine the instability of large groups (Focardi and Pecchioli 2005).

Social Behavior

The common fallow deer present a very flexible social organization strongly influenced by environmental context and closely linked to the annual cycle.

Adult females spend most of their time in family groups mainly composed of five or fewer individuals with one or two adult females with their current and sometimes previous year's offspring (e.g., Thirgood 1996). Males may remain in the family group up to 20 months of age. These groups are very stable in time and space and are generally led by a dominant doe (Apollonio et al. 1998). Female groups tend to be two to four times larger in more open habitats than in woodland and larger in winter/spring than in summer/autumn (e.g., Thirgood 1996). When living in heterogeneous environments, group sizes frequently change as animals move between closed and more open habitats, thus few permanent social associations develop. In more homogeneous environments, changes in group size are less apparent and more permanent groupings may persist, developing dominance hierarchies. The groups can reach the size of several hundred individuals where food resources are abundant, but such large aggregations are mostly transient. In general size of herds and social dynamics depend on habitat type/quality, deer density, food resources, degree of disturbance, time of year, and weather conditions. Males are solitary or form unstable groups with unrelated individuals ("bachelor groups"). Groups of adult males and females are separated for much of the year in most populations. The degree of social and spatial segregation shows great variability in relation to habitat, population density, and sex ratio (e.g., Thirgood 1996, Apollonio et al. 2005, Focardi and Pecchioli 2005). For example, where male density is high, bucks tend to remain in single-sexed groups; where it is lower, they tend to form mixed-sex groups; these groups have a high probability of splitting up into single-sex groups (Villerette et al. 2006).

Social interactions such as mutual grooming are relatively rare, except between does and their fawns. Play behavior, such as sudden chases and jumps, is common among young fawns in summer, and sometimes adult females join in.

Inter-male mounting is restricted to the first 3 months of antler regrowth (Holečková et al. 2000). It is typical for young animals and significantly decreases with age, suggesting that the animal needs the experience to cope with changing hormonal states or that the behavior is a form of learning. There is no evidence of bonding between individuals involved in mounting and no apparent dependence of mounting activity on testosterone levels. Inter-male mounting cannot be clearly linked to the roles of a dominant actor and a subordinate recipient. Female-female and female-male mounts are very rarely observed (Holečková et al. 2000).

Mating Behavior

The common fallow deer is a polygynous breeding species. The mating season occurs in October in the northern hemisphere and in April in the southern hemisphere (Chapman and Chapman 1997). The rut period changes with location and latitude as the photoperiod is the key stimulus controlling the deer reproductive processes by neurohormonal changes. The rut lasts approximately 135 days. The time of the mating season is influenced by several factors such as the sex ratio, population structure and density, spatial distribution of animals, and weather conditions (Sándor et al. 2014).

At the beginning of the mating season males start to scrape the ground, mark trees, thrash bushes, and groan. The buck vocalization may serve for threatening males and attracting females. McElligott et al. (1999) identified a strong correlation between the time invested in vocal display and mating success. Bucks use acoustic cues such as call rate and overall call structure to gain information on the dominance rank, motivation, and condition (fatigue) of competitors (Pitcher et al. 2015). Therefore, groans are the product of sexual selection driving the buck vocalizations to transmit multiple data on the caller quality (Pitcher et al. 2015). Since bucks have individually distinctive vocalizations, it was proposed that they may call repeatedly to familiarize females with their vocal characteristics (McElligott et al. 1999). However, the familiarization hypothesis has yet to be tested in fallow deer. Groaning is not fixed over time but is modified dynamically according to male age and rank. Therefore, groans are individually distinctive only within each breeding season (Briefer et al. 2010). However, changes in the structure of groans and groaning rate occur also during the same breeding season and are associated with the declining body condition of males and variation in the availability of mating opportunities (Vannoni and McElligott 2009). The bucks who achieved most of the matings are those who had initiated vocal activity early in the season (also more than 3 weeks before any matings occur) and who had remained vocal on most days. Dominance relationships are established during this pre-rut period, largely by non-contact agonistic interactions depending on body size and body mass. These relationships are in turn tested and modified by fights during the rut to produce the dominance ranks that influence mating success (McElligott et al. 1998, Moore et al. 1995). In summary, the number of matings gained by males is highly variable and is related to factors such as age, body size and mass, vocal display, social dominance rank, fighting success, and spatial strategies together with the experience from previous mating seasons, and that gained as subadult males (see McElligott et al. 2003). Mating success peaks between the ages of 6 and 7 years, when males usually have the highest numbers of matings (McElligott and Hayden 2000, McElligott et al. 2002). Successful males do not appear to suffer a decrease in survival probabilities or a reduction in fecundity, despite the high investment required to gain matings (McElligott et al. 2002).

Rutting activities are linked to pronounced hypophagia among adult males, starting weeks before the mating period and resulting in an important mass loss. Hypophagia and scent urination occur at the same time. The physiological processes inducing preputial gland activity and scent-urination may be the same as for appetite suppression (Apollonio and Di Vittorio 2004). These processes are possibly linked to hormonal changes recorded during the breeding season. Prime-aged males (5–8 years old) lose on average 26% of their body mass. This mass loss is positively correlated with initial mass and not with mating success or activities associated with mating success (McElligott et al. 2003). Foraging strategies adopted before the mating season affect mass loss. All other age and sex classes are unaffected in their feeding behavior by the rutting season. Females and young males even show a marked increase in grazing in response to a higher forage quality from summer to autumn.

Mating System

Langbein and Thirgood (1989) distinguished three main mating strategies: single male territorial strategy (classic rutting stands and temporary stands), multi-male territorial strategy (multiple stands and leks), and non-territorial strategy (dominance within mixed-sex herds, harem holding, and rowing males). Lekking is by far the most demanding mating strategy for males but is also the most successful strategy in terms of matings achieved within the population (e.g., Ciuti et al. 2011). A lek is a communal male display area that females visit for the sole purpose of mating and leave soon after mating. Many different models have been proposed to explain how leks initially form (Beehler and Foster 1988; Bradbury and Gibson 1983; Höglund and Alatalo 2014; Stillman et al. 1993). Leks are usually located in the same area year after year, and multiple factors are thought to be responsible for the persistence of a lek in a specific location, the combination of which depends on local ecological conditions (Apollonio et al. 2014). Therefore, habitat manipulations can affect the use of these traditional mating sites (Apollonio et al. 1998).

The mating system of a population can be single or mixed. The density of bucks and breeding females, as well as habitat structure (environmental heterogeneity, resource distribution, and tree cover) exert the greatest influence on the dominant mating strategy within a population or during a year. A greater variety of concurrent mating strategies was observed in the wild compared to park populations (Thirgood et al. 1999). The mating system adopted by a male appears to be determined primarily by the male's age and dominance status, the density and degree of aggregation of females, and the degree of competition experienced from other breeding males (Langbein and Thirgood 1989).

Individual male reproductive success is highly skewed in all systems (e.g., each year three males accounted for 60–90% of copulations in a lek, Apollonio et al. 1989) as confirmed by paternity assessment based on genetic data (Say et al. 2003). The asymmetry in copulatory success may be associated with male dominance rank and/or female choice. High skew in male reproductive success may contribute to the low genetic diversity recorded in fallow deer, mainly in enclosed populations. Assortative mating can help explain the persistence of genetic variation for male traits associated with reproductive success (Farrell et al. 2011). A small but consistent proportion of females (mainly yearlings) do not mate with the most successful males each year. Yearling females mate later in the rut than older ones and hence they are more likely to mate with younger and low ranking males. The timing of matings by young males coincides with the peak of yearling female matings, while low ranking males attempt to mate at a later stage when highly successful competitors show fatigue. Alternatively, yearling females may not be able to pay the potential costs associated with mating with high quality males or could be less experienced at discriminating between males of differing quality or at avoiding mating with young and low ranking males. Therefore, there is a strong behavioral difference between young and adult females in mate choice (Imperio et al. 2020).

Different reproductive strategies have been reported in female fallow deer. During 10 years of study in a large urban park in Dublin, Ireland, polyandrous and monandrous multiple-mating females as well as single-mating females were observed in a non-lekking population (Briefer et al. 2013). Each year, on average, 12% of all females were polyandrous females. This small but consistent proportion suggests that monandry and polyandry are alternative female strategies in fallow deer. Polyandry was not related to female age, stage of the rut, dominance rank of mates, or number of matings achieved by males during the rut and did not increase offspring viability and quality. The hypothesis of fertility insurance, where females remating if fertilization from the first mating is uncertain due to the possible sperm depletion, could explain polyandry in fallow deer. This strategy is more costly in terms of increased predation risk, energy and time spent searching for mates, selecting mates, and copulating and involves a complex trade-off between fertility success, preferences for high-quality males, and potential costs (Briefer et al. 2013).

Aggression

Studies on the aggressive behavior of territorial males were carried out primarily in enclosed parks where live lekking or non-lekking populations. Before the start of the rut, males become more and more intolerant to the presence of rivals resulting in a marked increase in aggression.

Many agonistic interactions are non-contact displays (such as groaning, parallel walk, antler displaying, or thrashing against vegetation) to avoid combat situations that may result in physical injuries (Bartoš et al. 2007). Nevertheless, direct fights are not uncommon between animals of comparable size and can last several minutes (Festa-Bianchet et al. 1990). There is no evidence that body mass and antler length are important factors in either the duration or intensity of fighting (Jennings et al. 2004). Fights are longer when at least one of the males had previously experienced a victory (Jennings et al. 2004). Lateral palm presentation serves to de-escalate fighting between mature bucks (Jennings et al. 2002). Damage likelihood is associated with dominance status and not with fight duration/number or mating success (Jennings et al. 2017). Chapman and Chapman (1997) described a thickening of the integument of the forehead of fallow bucks (dermal shield) during the rut that gives resilience against knocks, preventing severe damages to the skull. However, death and serious injuries can occur, especially toward later stages of the rut when the most actively rutting bucks lose their strength (e.g., Festa-Bianchet et al. 1990, Moore et al. 1995). As the number of estrus females increases during the rut, there is an increase in fighting (e.g., Jennings et al. 2013). The resulting hierarchy of dominance is linear (see Jennings et al. 2010), and mating success is highly skewed in favor of high ranking males (Moore et al. 1995).

Although most fights tend to involve only pairs of males, approximately 10% of contests are disrupted by the intervention of a third-party male, as recorded in a large enclosed city park in Ireland (Jennings et al. 2009). High ranking males engage in these triadic interactions against lower ranked individuals to prevent the advancing of subordinates in the hierarchy and at the same time to achieve a limited increase in dominance rank (Jennings et al. 2018). Suffering third-party interventions during fighting may reduce the probability of mating during the day on which the interventions occur. Therefore, intervention and suffering intervention might represent an interrelated strategy for which males that suffered an intervention are also highly likely to engage in third-party interventions (Jennings et al. 2017). This kind of behavior is highly variable between and within individuals over days and between years. However, further studies are needed to provide a more complete understanding of the evolution of third-party intervention behavior in fallow deer.

Parental Care

Fawns are born from mid-May to July with a peak in June (Chapman and Chapman 1997). Occasionally fawns can be born as late as November, but they are lighter and have higher mortality rates compared to early-born fawns (Ács and Lanszki 2017). As maternal age increases, fawns are born earlier in the breeding season. After parturition, the doe eats the afterbirth to prevent the attraction of predators.

The fawn is active within few minutes of birth, and suckling occurs within an hour or so. The doe establishes a mother-infant bond immediately after the birth, cleaning and drying the newborn by licking it. After parturition, the females do not return to the herd for 2 to 10 days. The type of mother-infant relationship can be categorized as "hider" type where the mother and the infant stay separated and out of contact for long periods, during which the infant stays hidden in the vegetation, near where it was born. Although the change of hiding place is typically initiated by the female, the exact position of the bed-site is chosen by the fawn. Selected bed-sites show significantly lower visibility and a higher amount of canopy cover than random sites. Fawns seem to select these two bed-site variables independently, perhaps for thermoregulatory reasons (see Kjellander et al. 2012). The newborns generally rest most of the time; the dappled coat provides excellent camouflage. If disturbed, the fawn "freezes"; after a period varying from an individual to the other, the "freezing" behavior changes in "fleeing" behavior. The doe remains separate from the fawn for much of the day but within hearing distance of a fawn distress call. At least for the few days after the birth, the mother recognizes the fawn apparently by smell, later it seems that the recognition is by the mother as well as by the fawn. The doe nurses the fawn on average once every 4 h until it is at least 4 months of age. Fawns to primiparous mothers suckled relatively longer, with no corresponding increase in mass, compared to multiparous mothers (Birgersson and Ekvall 1994). However, during the first 2 weeks, it seems that the transfer of milk is poorly reflected by suckling time since the inexperienced mothers do not terminate the lactation when the udder is empty. Grasses may be eaten a few days after the birth, but rumination does not begin until 2 to 3 weeks of age (Chapman and Chapman 1997). After about a month from birth, youngs and their mothers form small groups. The fawns may be left in the care of few adults while the other does feed. The mother begins weaning the fawn when it is around 20 days old, but weaning continues until the fawn is about 7 months old and can be prolonged up to January–February, even if sporadically. Mothers begin decreasing maternal care by decreasing the duration of tactile contacts (Chapman and Chapman 1997).

Allosuckling (i.e., suckling from a female other than the mother) was observed in captive and wild populations. The onset of allosuckling behavior seems to be related to the increasing ability of the fawns to move around after the first 2 weeks of life. Birth date affects allosuckling, with late-born fawns performing fewer and shorter allosuckling bouts. No difference was found between sexes (Pélabon et al. 1998). Allosuckling was positively correlated with the age of the females and negatively correlated with the rate of their aggressive behavior and it was common between related members in a stable social unit (Ekvall 1998). Allosuckling could be a result of kin selection and/or reciprocal altruism (Ekvall 1998) or could be a mean for offspring to gain important immunological benefits by acquiring antibodies from various lactating females (Roulin and Heeb 1999).

Senses and Communication

The common fallow deer have an acute sense of smell as would be expected in a species well endowed with scent glands and a good sense of hearing. The large ears can be swiveled together or separately through circa 180 degrees, picking up sound from any direction without turning the head.

The eyes, set on the side of the head, give a wide field of vision but poor depth perception. Discrimination of distant stationary objects is very poor, but any movement is quickly detected. Dissection of the eye reveals that deer do not appear to possess a UV filter, suggesting that they may be sensitive to ultraviolet light. Two classes of cone pigments together with a high concentration of rod pigments are present in the retina, with sensitivity in the blue-green part of the electromagnetic spectrum (Jacobs et al. 1994). Fallow deer possess a *tapetum lucidum*, common to all nocturnal mammals. These cells reflect light that would otherwise be lost when it passes through the retina back into the eye, increasing the amount of available light. This combination of adaptations allows a good dichromatic vision both by day and by night. Fallow deer can discriminate colors in the blue-green part of the spectrum (Birgersson et al. 2001). This dichromatic color vision may allow distinguishing between different plant species or different parts of plants that might be of variable nutritional (or toxic) value.

Communication and recognition make use of whole-body odor as well as vocal and visual cues. Tactile signals such as touching, licking, pawning, nipping, and butting occur mainly between mothers and their fawn, but also between adults.

Fallow deer use different types of vocalizations in mother-offspring interactions, during the rut and in a situation of danger: bleating is a short medium-high pitched sound produced by females close to parturition or with their young; peeping is a high pitched sound produced by fawns when are in distress or to establish contact with their mothers; wailing is an intense distress sound produced by a fawn older than 2 days; barking is a short, loud, and explosive alarm call used by females, sometimes repeated several times; mewing is a higher pitched sound given by any deer during submission postures; and groaning is a low belch like the sound produced by rutting males [rarely by subadults ($2<\text{years}<4$) and yearlings] composed by a series of single events up to more than 3000 vocalizations per hour, lasting 0.35/0.5 s and ranging from about 100 to 8000 Hz (Stachowicz et al. 2014).

Visual communication is essentially based on body posture as well as on the position and movement of the tail and the appearance of the rump patch (Alvarez et al. 1976). Alarm signals are one of the most common visual cues and pronking is a characteristic alarm gait of fallow deer as of other ungulates. Pronking is also a signal of the animal's fitness which should deter any pursuit by a predator as well as a play that may prepare young animals for adult life.

Parasites and Diseases

Current knowledge of the common fallow deer diseases is likely to be biased in favor of those pathogens and host populations viewed as important by interest groups as for other deer species. Moreover, the European countries hardly have systematic national disease surveillance for wild ungulates except for epidemic diseases (Putman et al. 2011). Data available are not suitable for the assessment of the possible long-term effects of pathogen infection on deer population dynamics (East et al. 2011).

Ectoparasites

The most common ectoparasites reported for this deer species in Europe (Chapman and Chapman 1997; Pavlásek and Minár 2014; Sleeman 1983; Szczurek and Kadulski 2004) include: lice (*Solenopotes burmeisteri*, *Bovicola tibialis*, and *Damalinia* sp.), keds (*Lipoptena cervi*), warble flies *Hypoderma diana*, flies (*Cephenemyia multispinosa* and *C. auribarbis*), ticks (*Ixodes ricinus*, *Haemaphysalis punctata*, *H. concinna*, and *Dermacentor reticulatus*), fleas (*Paraceras melis*), and mites (*Sarcoptes scabiei*, *Chorioptes texanus*, and *Demodex kutzeri*).

The percentage of deer infested varies with locality and times of the year. For example, the percentage is high in late winter, spring, and early summer for *Damalinia tibialis*, and from August to December for *L. cervi* (Chapman and Chapman 1997). The extent of the infestation is species-dependent. For example, in Poland, the highest percentages were reported for *L. cervi* (76%) and *I. ricinus* (29%), while the lowest for *C. texanus* (<1%); despite the percentage value, the infestations were symptomless (Szczurek and Kadulski 2004).

Fallow deer are among the most widespread ungulates in the world and so are their parasites. Mertins et al. (2011) described the impact on native species of deer of *Bovicola tibialis*, a chewing louse brought to North America with fallow deer. This louse causes a pathologic hair loss syndrome that is most evident in winter and spring. The most severe cases may progress to morbidity and mortality, mainly in young deer.

In the last decades, an increasing incidence of tick-borne zoonoses has been reported in Europe. This trend regards in particular pathogens (viruses, bacteria and protozoa) transmitted by *Ixodes* ticks. Like other deer species, the common fallow deer are becoming more and more important as a tick maintenance host. This deer can reach high density in potential tick habitats such as deciduous and mixed forests or green periurban and urban areas (Jaenson et al. 2018). However, at high density, fallow deer can destroy the ground vegetation to such an extent to reduce tick abundance (van Wieren and Hofmeester 2016). Hence the relationship between deer density and tick abundance is complex, due to a large number of climatic and environmental factors acting simultaneously, and still unclear (Gray et al. 2021).

Lyme borreliosis and tick-borne encephalitis are the two tick-borne diseases with the largest impact on human health (Hönig et al. 2019). Lyme borreliosis is caused by spirochetes of the *Borrelia burgdorferi* sensu lato complex, while encephalitis is caused by the tick-borne encephalitis virus (*Flavivirus*, Flaviviridae) (Hönig et al. 2019). Deer are not reservoirs for *B. burgdorferi* s.l. and encephalitis virus (Gray et al. 2021; Michelitsch et al. 2019). The presence of the zoonotic bacterium *Anaplasma phagocytophilum*, the causative agent of granulocytic anaplasmosis in both humans and animals, has frequently been reported in fallow deer from different regions of Europe with infection rates ranging between 1.5 and 95.4% (see Kazimírová et al. 2018). This variation is mainly due to the locality, examined tissue, and the sensitivity of the detection method used. The fallow deer may act as reservoir host for *A. phagocytophilum* (Kazimírová et al. 2018). Piroplasms are tick-borne protozoa, infecting red and white blood cells of their hosts. The common fallow deer have been identified as susceptible to *Theileria* and *Babesia* infections (e.g., Kazimírová et al. 2018, Yabsley and Shock 2013).

Endoparasites

Chapman and Chapman (1997), Kowal et al. (2012), Rehbein et al. (2014), and Vengušt and Bidovec (2003) provide a review of internal parasites (mainly nematodes, cestodes, and trematodes) affecting fallow deer. These parasites have different prevalence and intensity of invasion in relation to a different susceptibility of fallow deer to infection. The presence of eggs, larvae, and oocysts of gastrointestinal and pulmonary endoparasites was investigated in fecal samples of fawns and yearlings of farmed fallow deer by Rehbein and Bienoschek (1995). Outbreaks of parasitic diseases are rare (Vengušt and Bidovec 2003).

Ashworthius sidemi, an abomasum nematode of Asiatic cervids introduced in many European countries via sika deer, was observed in fallow deer from Poland, Sweden, the Czech Republic, and Russia (Kowal et al. 2012; Kuznetsov 2021; Magdalek et al. 2017). This parasite can cause chronic abomasitis and infection can have health consequences, especially in young animals (Kowal et al. 2012). This nematode may be able to infect cattle and sheep feeding together with deer on the same meadows or pastures (Kornacka et al. 2020).

Even not showing any significant clinical sign if not a moderate decline of physical conditions, fallow deer may be affected by the following trematodes: the common liver flukes *Fasciola hepatica*, a native European endoparasite and the giant liver fluke, *Fascioloides magna*, an alien endoparasite of American origin introduced in Europe in the nineteenth century, probably together with North American elk (e.g., Nagy et al. 2018). The giant liver fluke has established in three permanent natural foci: northern Italy, floodplain forests along the river Danube, and the Czech Republic with southwestern Poland (see Králová-Hromadová et al. 2016). However, the European distribution of *F. magna* is permanently growing and its emergence is expected in new areas (Nagy et al. 2018). This alien species has already replaced the endemic one in some areas (Houszka et al. 2016). Fallow deer is a definitive host of the liver flukes (Malcicka 2015). Fasciolosis is an emerging/re-emerging zoonosis in many countries as a consequence of environmental changes as well as anthropogenic modifications (Mas-Coma et al. 2005). The strong dependence of liver fluke's incidence on weather factors (such as high rainfall) indicates that climate change may have a remarkable influence on the future evolution of this disease (Iglódyová et al. 2017).

Neosporosis caused by *Neospora caninum* is among the main causes of abortion in cattle nowadays. However, many aspects of the life cycle of this protozoan parasite are unknown, and the role of wildlife as reservoirs in this cycle is still not completely explained. The seroprevalence of *N. caninum* in wild and farmed fallow deer varies greatly: from 1% (Bartova et al. 2007) to 13% (Moskwa et al. 2018). Seronegative results were obtained in Spain (Almería et al. 2007) and in Belgium (De Craeye et al. 2011). A fatal case of meningoencephalomyelitis caused by *N. caninum* was diagnosed in a 3 weeks fallow deer in a zoo in Switzerland (Soldati et al. 2004). A study conducted at a fallow deer farm (Cabaj et al. 2017) revealed that the mean vertical transmission ratio of *N. caninum* in naturally infected deer seems to be comparable to those observed in cattle and showed a significant reduction in birth rate. Bucks naturally infected by *N. caninum* cannot transmit the infection to females by natural fertilization.

Toxoplasmosis and sarcosporidiosis (sarcocystosis) are other protozoal diseases causing abortion. The seroprevalence of *Toxoplasma gondii* varies from 10% (Moskwa et al. 2018) to 24% (Bartova et al. 2007). Seronegative results were obtained in the Czech Republic and Belgium (De Craeye et al. 2011; Hejlícek et al. 1999). Coinfections with *N. caninum* seem to be rare (Moskwa et al. 2018). Compared to sheep and goats, fallow deer appear to be at less risk of infection by *T. gondii* and *N. caninum* through the ingestion of food contaminated with oocysts excreted with feces because of the different feeding habits (Moskwa et al. 2018).

The prevalence of *Sarcocystis* infection in fallow deer is very high (>90%, e.g., Poli et al. 1988) and comparable to the prevalence reported in other cervids in Europe (see Prakas et al. 2016). No significant difference between fawns and adults was recorded (Poli et al. 1988). The clinical disease is rare. A detailed microscopic investigation of the ultrastructure of the cyst wall morphology of *Sarcocystis* sp. was carried out by Poli et al. (1988).

Viral Diseases

Foot and mouth disease FMD affects even-toed ungulates, both domesticated and wild. The virus responsible for FMD is a non-enveloped RNA *Aphthovirus*, belonging to the Picornaviridae family. The common fallow deer are susceptible to natural and experimental infection with FMD virus. The disease is mild or inapparent and is characterized by the formation of vesicles and ulcers in the mouth, nose, teats, and feet. Fallow deer are unlikely to play a significant role in the maintenance and transmission of FMD in the long term in the absence of FMD infection in domestic ungulate populations (Weaver et al. 2013). To date, the European countries are recognized as free from FMD without vaccination (https://www.oie.int/en/disease/foot-and-mouth-disease/).

Rabies is a fatal viral infection that causes encephalitis in warm-blooded animals. In deer rabies occurred infrequently. Only one case is reported for fallow deer: an outbreak affecting more than 20 fallow deer in a farm of Mongolia, caused by rabid dog bites (Zhu et al. 2015).

Bacterial Diseases

According to EFSA and ECDC report (2021), in European Union the first and second most reported zoonoses in humans are campylobacteriosis and salmonellosis, respectively. No *Campylobacter* or *Salmonella* species were isolated from wild or farmed fallow deer (Carbonero et al. 2014). Shiga toxin-producing *Escherichia coli* (STEC) infection, yersiniosis, and listeriosis are the other most reported zoonoses in humans (EFSA and ECDC report 2021). *Yersinia* (Syczyło et al. 2018) and *Listeria* (Pomilio 2016) species and Shiga toxin-producing *Escherichia coli* (STEC) strains (Sánchez et al. 2008) were isolated from fallow deer.

Tuberculosis (TB) is a worldwide disease caused by members of the *Mycobacterium tuberculosis* complex. It may cause zoonotic disease in humans; to date, it is a rare zoonosis in the European Union (EFSA and ECDC 2021). Bovine tuberculosis (bTB) has complex epidemiology, which may involve multiple hosts. For example, in Doñana National Park (DNP) in southern Spain, wild boar *Sus scrofa*, red and fallow deer, domestic cattle, and to a lesser extent carnivores as Iberian lynx contribute to the maintaining of *M. bovis* circulation (Gortázar et al. 2008). Infection prevalence in DNP was 13.0% in fallow deer, lower than that recorded in red deer. Differences in grazing patterns and feeding areas as well as use and aggregation around watering areas, especially during summertime, could determine a different exposition to mycobacteria (Gortázar et al. 2008). Deer may be long-lived reservoirs of infection having the potential to initiate new outbreaks well outside currently infected areas or to reinitiate infection after bTB has been eliminated through selective culling of all positively tested animals (Gortázar et al. 2008). bTB may contribute to the periodic die-offs that occur among deer during severe droughts (Gortázar et al. 2008). The importance of wildlife in the epidemiology of *M. bovis* infection in domestic animals may continue to grow in the near future (Gortázar et al. 2010).

Mycobacterium avium subspecies *paratuberculosis* is the etiological agent of paratuberculosis. It is one of the most serious diseases affecting dairy cattle worldwide. The frequency of detection in fallow deer reported in different areas of Europe (see Álvarez et al. 2005) is low, suggesting that free-ranging deer populations are rarely infected by the pathogen, unlike farmed animals. However, deer might contribute to the persistence and spread of the pathogen, particularly where higher densities of animals occur (such as at drinking points and in dens and shelters).

Brucellosis is an infectious disease caused by bacteria of the genus *Brucella*, being considered one of the most important zoonosis worldwide. The fallow deer like other deer have not been identified as a significant brucellosis reservoir for livestock in Spain (Muñoz et al. 2010).

Prion Diseases

Chronic wasting disease (CWD) is the most contagious prion disease affecting wild and farmed cervids; it is a neurodegenerative disease that is always fatal once the clinical signs appear. CWD is enzootic in some areas of North America. The first cases of CWD in Europe were documented in a wild moose and a wild reindeer in Norway (Ricci et al. 2017). In addition to its increased geographic distribution, the known host range of CWD is also expanding (Ricci et al. 2017). The fallow deer seem to have reduced susceptibility to CWD. The absence of clinical disease in individuals exposed to a contaminated environment and infected mule deer for 7 years suggests that a relatively strong species barrier against CWD infection may exist in fallow deer (Rhyan et al. 2011).

Population Ecology

Population Dynamics

Size fluctuations of a population are a result of temporal variation in survival, fecundity, and dispersal rates. This variation can be associated with density-dependent and -independent processes, which primarily affect recruitment rates and levels of mortality within a population (Putman et al. 1996). For example, analyzing the performance of park populations of fallow deer in Great Britain, the first response to density-resource limitation seems to be an increased age at first breeding and a reduction in the proportion of yearlings ovulating and subsequently maintaining pregnancy. Neonate mortality is directly related to weather conditions in the first week of life (Putman et al. 1996).

The importance of each density-dependent and -independent process may vary with latitude and habitat.

In northern/central Europe, resources are scarce in winter because vegetation stops growing, and snow cover renders food inaccessible. The scarcity of resources, combined with high thermoregulation costs, high costs of movement in deep snow, shorter length of growing season, and high susceptibility to predation, constitutes limiting factors for survival. Therefore, the combination of these density-independent factors should represent a crucial determinant of fluctuations of population size. To date, the population dynamics of fallow deer has received poor attention in northern/central Europe.

Some insights in Mediterranean areas come from a fenced estate of central Italy, where a combination of density-dependent and climatic factors plays a relevant role in regulating fallow deer population (Imperio et al. 2012). The intraspecific density-dependent factors are stronger than interspecific ones. The fallow deer appear not to be much influenced by the other herbivores occurring in the same area, probably due to its trophic position (between intermediate and roughage eaters, Hofmann 1989). In Mediterranean environments, the only important climatic factor is presumably associated with spring and summer precipitation. For example, very wet conditions in May–June appear to play a negative role in fawn survival, probably due to the insurgence of thermoregulatory problems (Van Moorter et al. 2009). On the contrary, water availability during the spring of the previous year has a positive effect on fallow deer population size. It improves body conditions in young females and their fecundity, presumably through the increase of primary productivity (Imperio et al. 2012).

Studying fallow deer population dynamics for nearly two decades in two enclosed game reserves of southern Spain allowed a better assessment of the effects of density and hunting pressure in the same climatic regime (López-Montoya et al. 2017). Density-dependent mechanisms have a stronger effect than hunting and climatic factors whereas the populations are close to their carrying capacity. Density-dependent mechanisms compensate for harvest extractions. Hunting causes a reduction in population that in turn relieves intraspecific forage competition, increases performance and reproduction, and leads to subsequent rapid population growth. The weather effect is more important thc morc abundant the deer are.

Effects of Climate Change

There is a growing concern about the potential negative effects of climate change on the behavior and ecology of deer populations (Root et al. 2003; Parmesan 2006, for reviews).

The Mediterranean Basin is facing the effects of climate change more than ever. During the twenty-first century temperature will warm 20% more than the global average and warming will be particularly large in summer, approximately 50% larger than the global average (Lionello and Scarascia 2018). Fallow deer populations with densities far from their carrying capacity and less competition for resources could take advantage of these climatic changes to reach autumn in better body condition. The worst effects of climatic constraints have been detected when populations are closer to their carrying capacity. The importance of weather factors could become much greater if both densities of fallow deer and aridity increase. Therefore, population densities must be well controlled to modulate the negative effects of climatic constraints on fallow deer condition (López-Montoya et al. 2017).

Furthermore, climatic changes could increase the ecological overlap between deer species, emphasizing the negative effects of interspecific competition, for example between fallow and roe deer (Ferretti and Fattorini 2021).

Competition with Other Ungulates

Several studies suggested a potential interspecific competition among the common fallow deer and native/introduced deer species based on behavioral interferences and/or overlapping in habitat and resource use (e.g., Ferretti et al. 2012, Putman 1996). A few studies analyzed changes in deer population density to look for evidence of actual competition among deer. Inverse numerical trends were recorded for fallow and roe deer populations in the New Forest (southern England) over more than 30 years. These changes in deer population density appear to be, for the most part, independent of one another and hence cannot be considered an uncontroversial effect of competition (Putman 1996). On the contrary, the decline of the endemic subspecies of roe deer *Capreolus capreolus italicus* observed in a Mediterranean area can be mainly explained by the increased numbers of fallow deer (Focardi et al. 2006). This species could have changed the vegetative cover in the long term and fragmented the distribution of suitable habitats for roe deer, contributing to the variations in home range size and phenotypic performance of the endemic subspecies. Only when the fallow deer density has increased above some threshold, is it possible to detect the effects of this density increase on roe deer population. These results provide evidence of the real existence of interspecific competition between fallow and roe deer. Evidence was also collected in the State Nature Reserve of Bosco della Mesola, northern Italy, where occurs the only native red deer population of peninsular Italy, recently recognized as a distinct subspecies (*Cervus elaphus italicus*) by Zachos et al. (2014). Modest body size, simplified antlers, and a low reproductive performance characterize Mesola red deer. The common fallow deer was reintroduced in the same area after the Second World War and in the following decades outnumbered the red deer, decreasing the availability of habitat and food resources. In the last decades, actions were conducted to reduce the population size of fallow deer by capture and culling with positive consequences on the Mesola red deer population (Mattioli et al. 2003). However, additional efforts are required to eradicate fallow deer (Lovari and Nobili 2010).

Conservation Status

The common fallow deer are not globally threatened and is categorized as Least Concern (LC) in the IUCN Red List. However, the last autochthonous population in Anatolia and the Rodhian population, the oldest still surviving on a Mediterranean island, are of the utmost importance.

To conserve the Anatolian population, in the 1960s the Turkish government began a breeding program at Düzlerçamı. In 1964 an area of 1750 ha, in which seven individuals were living, was designated as Fallow Deer Conservation Area (see Arslangündoğdu et al. 2010). In the following decades, the conservation area was gradually expanded until when the Düzlerçamı Wildlife Development Area was designated in 2005 (Arslangündoğdu et al. 2010) with a surface of more than 28,000 ha. The fallow deer number of the last autochthonous stock, reported by Arslangündoğdu et al. (2010) from 1966 to 2009, shows strong fluctuations throughout the years. A combination of several factors (increasing human pressure due to urbanization and recreational activities, poaching, poisoning by local people, as well as the presence of shepherds and stray dogs) may have had a negative impact on the fallow deer number (Arslangündoğdu et al. 2010). As far as we know, the latest study carried out in the breeding station using camera traps (Ünala and Çulhacıa 2018), identified 80 adult individuals on the base of spot distribution and antler structure, and estimated a population density of 20.1 deer per km^2. Attempts to reintroduce fallow deer in the former ranges of the species (Gökova and Adaköy, Muğla province; Ayvalık, Balıkesir province and Pos-Çatalan, Adana province), translocating animals from the Düzlerçamı Breeding Station in the 1980s and 1990s, have not been successful (Masseti 1999, 2002). Reintroduction projects, carried out in the second decade of twenty-first century, have instead successfully translocated fallow deer in the National Park of Dilek Peninsula and Büyük Menderes Delta, Aydın province, and in the Köyceğiz Dalyan Special Environmental Protection Area, Muğla Province (Durmuş 2019). Reintroductions result to be an important tool for the conservation of the Asia Minor population.

The number of fallow deer on the island of Rhodes is reputed that today do not exceed a few hundred. Spotlight counts, carried out in the northcentral part of the island, gave a mean value of Kilometric Abundance index of 1.4 deer/km (De Marinis and Masseti 2021). The Rodhian fallow deer appear to be seriously threatened by poaching and by the continual habitat reduction.

Management

Introductions

Movements of fallow deer by humans, including introductions/translocations and release/escape of animals from fenced areas, were a common practice in past centuries and determined the establishment of the populations currently living in Europe and in other countries. The Romans introduced the common fallow deer to most of Europe, from the Iberian Peninsula to Britain (Baker et al. 2017 and references therein). Medieval introductions gave rise to many of the extant populations today. There is historical evidence confirming that, during the 1400s, common fallow deer were still captured in Anatolia and taken to western Europe to embellish palaces and game parks (Masseti 1996). Still today, humans contribute to change the distribution of this species all over the world.

Impact on Human Activities and Collisions

An analysis of methods and systems for assessment, prevention, and control of deer damage in different European countries is reported in Apollonio et al. (2010). This analysis reveals that an ecosystem management approach, instead of a single species approach, become more and more necessary to manage conflicts among relevant stakeholder groups in reaching decisions about the use and conservation of biological resources (Kviberg and Craig 2006).

Fallow deer-vehicle collisions are not distributed randomly in space and time and show daily and seasonal peaks (Langbein et al. 2011; Valente et al. 2020). Deer cross the roads during daily movements to or from foraging areas within their home range, especially at dusk and dawn when they are more active. These movements are rather regular and distributed throughout the year. In addition, there are peaks in road-crossing coinciding with larger-scale dispersal movements of juveniles in late spring and early summer and mating movements of adults in autumn.

The common fallow deer have a rather patchy distribution and therefore might be less vulnerable to traffic than other ungulates. Moreover, major roads or railways seem to be recognized as barriers to movement; the boundaries of home ranges appear to coincide with these major routes. Vehicle collisions involving fallow deer represent only a low percentage of all collisions involving ungulates reported per country. Traffic mortality rates are low in relation to the fallow deer population size. In the UK, the total mortality due to vehicle collisions is estimated to lie between 7 and 13% as a proportion of the spring population size of fallow deer (Langbein et al. 2011). In Finland collision and traffic mortality rates are 3.2% and 2.1% of the winter population, respectively (Niemi et al. 2015). The behavioral responses of fallow deer when crossing roads to some mitigation measures are reported in Langbein et al. (2011). Fallow deer seem to use underpasses more frequently than overpasses and begin using the underpasses a few months after their construction. The ratio between the size of the aperture and the total length of an underpass appears critical: fallow deer seem to avoid the underpasses with a ratio of less than 1.5. Wildlife warning reflectors and auditory deterrents seem ineffective at modifying fallow deer behavior due to habituation. However, the potential effectiveness of these deterrents would be expected to be quite different in different contexts.

Farming

The common fallow deer have become one of the most commonly farmed deer in Europe to the point that its farming significantly contributes to the meat industry (Ward et al. 2014). The ratio of lean meat produced per kg of live mass is high and the meat is highly nutritious, and with a low cholesterol content (e.g., Daszkiewicz et al. 2015).

Velvet antler is regarded as a by-product of fallow deer farming, used in traditional Chinese medicine, and in some countries as a dietary supplement.

Hunting

The legislation regarding hunting (organization, methods, seasons and management objectives), the effectiveness and problems of the current strategies, and the monitoring techniques in different European countries are summarized in Apollonio et al. (2010). However, a brief comparative review of some of these issues based on available data is reported in Table 2 for most of the European countries.

The meat quality of hunted animals is quite different in relation to different hunting methods like dog-driven hunting and harvest culling, which imply different animal stresses before shooting (Cifuni et al. 2014). The value of the carcasses is also influenced by the age of the deer shot. Carcasses of young fallow deer (18–30 months), compared to older animals, are characterized, for example, by a higher dressing proportion, a higher percentage of the most valuable commercial cuts (the saddle, haunch, and shoulder), a high meat yield with the lowest percentage of bones and a lower percentage of skin and head (Żochowska-Kujawska et al. 2019).

Fallow deer are also a popular trophy hunting species because of their "exotic" antlers.

Future Challenges for Research and Management

The common fallow deer are a species capable of rapid population growth, although its spatial dispersal rate is low. This species can reach a high density in suitable habitats where it can have an impact on forestry, crops, and native vegetation. Fallow deer are potential competitors for native ungulates. The attendance of urban areas can result in increasing the likelihood of vehicle collisions and disease transmission to humans, livestock, and pets. This can lead to increased human-deer

Table 2 Hunting bags, use of artificial feeding, monitoring types and techniques, hunting methods, and seasons in different European countries

Country	Number of culled animals[a]	Artificial feeding	Monitoring	Hunting methods	Hunting plans	Hunting period
Austria	2	Obligatory	Estimation based on hunting records	Hunting from high seats; stalking; rarely driving (hunting dogs allowed in most provinces)	Minimum and maximum number to be culled	Buck: 01–08/31–12; doe-fawn: 01–08/15–01
Belgium	1	Sporadic	Estimated spring population (no rules exist concerning methods)	All hunting methods	Escaped individuals	01–10/31–12
Czech R.	4	Obligatory	Visual survey	Hunting from high seats; stalking, usually with dogs	Shooting plans	16–08/31–12
Croatia	2	Obligatory	Counts at feeding sites; vantage point counts	Stalking; tracking; hunting from high seats or from the ground	Shooting plans	Buck: 16–09/14–01; doe-fawn: 01–10/31–12
Denmark	4	Sporadic	Estimation based on hunting records and hunters questionnaires	Driving; stalking	Suggested hunting quotas	Buck: 01–09/31–01; doe-fawn: 01–10/31–01
Estonia	0					
Finland	1	Common	Counts at feeding sites; estimation based on hunting records	Hunting at feeding sites, agricultural fields or flushing by beaters	No shooting plans; number of hunting licenses based on estimated numbers	25–09/31–01
France	3	Usually not used for this species	Total or partial drives	Hunting drives with dogs; stalking; hunting from high seats	Minimum and maximum number to be culled, in some departments no shooting plans according to	01–06/28–02

(continued)

Table 2 (continued)

Country	Number of culled animals[a]	Artificial feeding	Monitoring	Hunting methods	Hunting plans	Hunting period
					the attempt to eradicate this species	
Germany	5	Obligatory (part); common (part)	Estimation based on hunting records and on amount of damage to forest regrowth and forest stand	All hunting methods, even night hunting	Shooting plans	Buck and doe: 10–08/31–12; yearling female: 10–06/31–12; yearling male: 10–06/31–01; fawn: 10–08/31–01
Greece	0					
Hungary	5	Common	No standardized methods (usually vantage point counts; counts at feeding sites; roadside counts)	Stalking on foot or by horse cart; hunting from high seats; females and fawns also in drives without dogs	Shooting plans	Prime-aged male: 01–10/31–12; male yearling and low quality buck: 01–10/31–01; doe, female yearling, fawn: 01–10/28–02
Ireland	3	Sporadic	No standardized methods (drive censuses; vantage point counts; pellet surveys)	Hunting from high seats; stalking	Shooting plans prepared by some landowners on voluntary basis	Northern Ireland = male: 01–08/30–04, female: 01–11/31–03; Ireland = male: 01–09/31–12, female: 01–11/28–02
Italy	3	Usually not used for this species	No standardized methods (usually vantage points counts, drive censuses, spotlight counts in 45% of provinces)	Stalking	Shooting plans	Two months, usually 01–10/30–11, but also 60 days before and after this period, exceptions in Eastern Alps and Emilia Romagna region (longer periods)
Latvia	0					
Lithuania	0					
Luxembourg	2	Not available	Not available	Not available	Not available	15–08/15–12
Macedonia	1	Not available	No censuses performed, only fenced animals	Not available	No shooting plans; no harvest data exist	16–09/31–01
Netherland	2	Forbidden	Drive counts; vantage point counts	Hunting from fixed places	3 types: (1) no management; (2) culled only for public health, public safety,	01–09/15–02

(continued)

Table 2 (continued)

Country	Number of culled animals[a]	Artificial feeding	Monitoring	Hunting methods	Hunting plans	Hunting period
					safety of air traffic, to prevent damage to crops, cattle, forests; (3) shooting plans	
Norway	0					
Poland	4	Obligatory	No standardized methods	Stalking; hunting at high seats or blinds; silent drives	Shooting plans	Male: 01–10/31–01; female and fawn: 01–10/15–01
Portugal	2	Sporadic	Estimation based on hunting records; censuses in few areas (usually open hill counts, drive counts, pellet surveys)	Stalking; sit and wait hunting; spear hunting; drive hunting (only from October to February)	Shooting plans	01–06/31–05 (but usually 01–09/28–02)
Romania	2	Common	Counts of footprints in the snow and other complementary methods (vantage points counts; counts at feeding sites)	Stalking; hunting from high seats	Shooting plans	Buck: 10–10/01–12; cull buck: 01–09/15–12; doe and fawn: 01–09/15–02
Serbia	1	Not available	Estimation, methodology not very scientific	Not available	Shooting plans	Male: 01–02/15–09; female and fawn: 01–02/30–09
Slovakia	4	Obligatory	Transects; vantage point counts; counts at feeding sites	Stalking; hunting from high seats	Shooting plans	01–09/31–12
Slovenia	2	Common	Estimation based on hunting records	Hunting from high seats	Shooting plans	Bucks: 16–08/31–12; does and fawns: 01–09/31–12; yearling: 01–07/31–12
Spain	5	Sporadic	Linear transects; counts at feeding sites	Stalking; Montería (Spanish system)	Shooting plans, except for Montería system based on hunting records	01–10/28–02

(continued)

Table 2 (continued)

Country	Number of culled animals[a]	Artificial feeding	Monitoring	Hunting methods	Hunting plans	Hunting period
Sweden	5	Common	No censuses performed	Drive hunting; stalking even with dogs; hunting from high seats (small hunting grounds)	Minimal management plans	Changes from year to year and between different areas of the country
Switzerland	0					
UK	5	Sporadic	No standardized methods (drive censuses; vantage point counts; pellet surveys)	Hunting from high seats; stalking	Shooting plans prepared by some landowners on voluntary basis	England, Wales = male: 01–08/30–04, female: 01–11/31–03; Scotland = male: 01–08/30–04, female: 21–10/15–02

The hunting bags refer to the last 5 years. Data from Apollonio et al. (2010) and Putman et al. (2011)
Note:
[a]Number of culled animals: 0 = 0; 1 = 1–100; 2 = 101–1,000; 3 = 1,001–5,000; 4 = 5,001–10,000; 5 > 10,001

conflicts. Further introductions, translocations, and escapes from fenced areas should be avoided.

This species needs a broader consideration that allows developing a countrywide standardized system to monitor population densities, damages to ecosystem and human activities, vehicle collisions, and disease impacts.

Actions to contrast factors threatening the conservation of the native population in Turkey should be considered a priority. Other introductions should be carried out to establish new stable nuclei in Turkish territory, avoiding interbreeding of different genetic lineages.

Further interdisciplinary researches, based on ecological, genetic, morphometric, historical, archaeological, and archaeozoological data, are needed to increase the knowledge of the origin, history, genetic variability, and status of the most ancient populations present in some southern European areas such as Castelporziano and San Rossore (central Italy) and in the Aegean island of Rhodes to ensure their conservation.

Acknowledgments We express our appreciation to the editors and the reviewers for suggestions and improvement of this chapter. We are indebted to M. Masseti for the substantial revision and improvement of some parts of the chapter. We are grateful to M. A. De Marco, M. Scandura, and A. Varcasia for insightful comments on an earlier draft and L. Sorbini for the continuous and great assistance in the bibliographic research.

References

Ács K, Lanszki J (2017) Pre-, postnatal growth and maternal condition in a free ranging fallow deer population. Folia Zool 66(1):72–78

Almería S, Vidal D, Ferrer D et al (2007) Seroprevalence of *Neospora caninum* in non-carnivorous wildlife from Spain. Vet Parasitol 143(1):21–28

Alvarez F (1995) Functional directional asymmetry in fallow deer (*Dama dama*) antlers. J Zool 236(4): 563–569

Alvarez F, Braza F, Norzagaray A (1976) The use of the rump patch in the fallow deer (*D. dama*). Behaviour 56: 298–308

Álvarez J, de Juan L, Briones B et al (2005) *Mycobacterium avium* subspecies paratuberculosis in fallow deer and wild boar in Spain. Vet Rec 156:212–213

Apollonio M (1999) *Dama dama* (Linnaeus, 1758). In: Mitchell-Jones AJ, Amori G, Bogdanowicz W et al (eds) The atlas of European mammals. T & AD Poyser Ltd. & Academic Press, London, pp 386–387

Apollonio M, Di Vittorio I (2004) Feeding and reproductive behaviour in fallow bucks (*Dama dama*). Naturwissenschaften 91(12):579–584

Apollonio M, Vailati G (1996) Functional morphology of metatarsal glands in fallow deer (*Cervus dama*). Z Saugetierkd 61:321–326

Apollonio M, Festa-Bianchet M, Mari F (1989) Correlates of copulatory success in a fallow deer lek. Behav Ecol Sociobiol 25(2):89–97

Apollonio M, Focardi S, Toso S et al (1998) Habitat selection and group formation pattern of fallow deer *Dama dama* in a submediterranean environment. Ecography 21(3):225–234

Apollonio M, Ciuti S, Luccarini S (2005) Long-term influence of human presence on spatial sexual segregation in fallow deer (*Dama dama*). J Mammal 86:937–946

Apollonio M, Andersen R, Putman R (eds) (2010) European ungulates and their management in the 21st century. Cambridge University Press, Cambridge

Apollonio M, De Cena F, Bongi P et al (2014) Female preference and predation risk models can explain the maintenance of a fallow deer (*Dama dama*) lek and its 'handy' location. PLoS One 9(3):e89852

Arslan A, Zima J (2014) The C-Banding and Ag-NOR distribution patterns in the fallow deer *Dama dama* (Mammalia: Cervidae) from Turkey. Cytologia 79(2): 181–185

Arslangündoğdu Z, Kasparek M, Sarıbaşak H et al (2010) Development of the population of the European fallow deer, *Dama dama dama* (Linnaeus, 1758), in Turkey (Mammalia: Cervidae). Zool Middle East 49:3–12

Asher GW (2007) Reproductive cycles in female cervids. In: Youngquist RS, Threlfall WR (eds) Current therapy in large animal theriogenology, 2nd edn. Saunders Elsevier, United States, pp 921–931

Asher GW, Adam JL (1985) Reproduction of farmed red and fallow deer in northern New Zealand. In: Fennessy PF, Drew KR (eds) Biology of deer production, Bulletin 22. The Royal Society of New Zealand, pp 217–224

Azorit C, Tellado S, Oya A et al (2012) Seasonal and specific diet variations in sympatric red and fallow deer of southern Spain: a preliminary approach to feeding behaviour. Anim Prod Sci 52(8):720–727

Baker KH, Gray HWI, Ramovs V et al (2017) Strong population structure in a species manipulated by humans since the Neolithic: the European fallow deer (*Dama dama dama*). Heredity 119:16–26

Baker KH, Gray HW, Pekşen ÇA et al (2021) Conservation genetics of the European fallow deer: a reply to Marchesini et al. Mamm Biol 101(3):313–319

Barboza PS, Bowyer RT (2000) Sexual segregation in dimorphic deer: a new gastrocentric hypothesis. J Mammal 81:473–489

Bartoš L, Fričová B, Bartošová-Víchová J et al (2007) Estimation of the probability of fighting in fallow deer (*Dama dama*) during the rut. Aggress Behav 33(1):7–13

Bartova E, Sedlak K, Pavlik I et al (2007) Prevalence of *Neospora caninum* and *Toxoplasma gondii* antibodies in wild ruminants from the countryside or captivity in the Czech Republic. J Parasitol 93(5):1216–1218

Beehler BM, Foster MS (1988) Hotshots, hotspots, and female preference in the organization of lek mating systems. Am Nat 131:203–219

Bergvall UA, Leimar O (2005) Plant secondary compounds and the frequency of food types affect food choice by mammalian herbivores. Ecology 86(9):2450–2460

Bergvall UA, Rautio P, Kesti K et al (2006) Associational effects of plant defences in relation to within- and between-patch food choice by a mammalian herbivore: neighbour contrast susceptibility and defence. Oecologia 147:253–260

Bergvall UA, Rautio P, Luotola T et al (2007) A test of simultaneous and successive negative contrast in fallow deer foraging behaviour. Anim Behav 74(3):395–402

Bergvall UA, Schäpers A, Kjellander P et al (2011) Personality and foraging decisions in fallow deer, *Dama dama*. Anim Behav 81(1):101–112

Berlioz E, Azorit C, Blondel C et al (2017) Deer in an arid habitat: dental microwear textures track feeding adaptability. Hystrix 28(2):222–230

Bilton DT, Mirol PM, Mascheretti S et al (1998) Mediterranean Europe as an area of endemism for small mammals rather than a source for northwards postglacial colonization. Proc R Soc B 265:1219–1226

Birgersson B, Ekvall K (1994) Suckling time and fawn growth in fallow deer (*Dama dama*). J Zool 232(4): 641–650

Birgersson B, Ekvall K (1997) Early growth in male and female fallow deer fawns. Behav Ecol 8:493–499

Birgersson J, Alm U, Forkman B (2001) Colour vision in fallow deer: a behavioural study. Anim Behav 61(2): 367–371

Borkowski J, Pudelko M (2007) Forest habitat use and home-range size in radio-collared fallow deer. Ann Zool Fenn 44:107–114

Borovali Ö (1986) Turkey. In: The SCI record book of trophy animals. Safari Club International, Tucson, pp 528–530

Bowen F, Carden RF, Daujat J et al (2016) *Dama* dentition: a new tooth eruption and wear method for assessing the age of fallow deer (*Dama dama*). Int J Osteoarchaeol 26(6):1089–1098

Bradbury JW, Gibson RM (1983) Leks and mate choice. In: Bateson P (ed) Mate choice. Cambridge University Press, Cambridge, pp 109–138

Braza F, San José C, Aragón S (2000) Variation of male-biased maternal investment in fallow deer (*Dama dama*). J Zool 250:237–241

Briefer E, Vannoni E, McElligott AG (2010) Quality prevails over identity in the sexually selected vocalisations of an ageing mammal. BMC Biol 8(1):1–15

Briefer EF, Farrell ME, Hayden TJ et al (2013) Fallow deer polyandry is related to fertilization insurance. Behav Ecol Sociobiol 67(4):657–665

Brooke V (1875) On a new species of deer from Mesopotamia. Proc Soc Zool London 1875:261–266

Brown W, Chapman NG (1990) The dentition of fallow deer (*Dama dama*): a scoring scheme to assess age from wear of the permanent molariform teeth. J Zool 221(4):659–682

Cabaj W, Bień J, Bogdaszewski M et al (2017) Potential impact of *Neospora caninum* infection on farm productivity of fallow deer (*Dama dama*). Small Rumin Res 156:78–81

Caldwell JF, Chapman DI, Chapman NG (1983) Observations on the autumn and winter diet of fallow deer (*Dama dama*). J Zool 201:559–563

Caravaggi A, Gatta M, Vallely M et al (2018) Seasonal and predator-prey effects on circadian activity of free-ranging mammals revealed by camera traps. PeerJ 6: e5827

Carbonero A, Paniagua J, Torralbo A et al (2014) *Campylobacter* infection in wild artiodactyl species from southern Spain: occurrence, risk factors and antimicrobial susceptibility. Comp Immunol Microbiol Infect Dis 37(2):115–121

Carden RF, Hayden TJ (2006) Epiphyseal fusion in the postcranial skeleton as an indicator of age at death of european fallow deer (*Dama dama dama*, Linnaeus, 1758). In: Ruscillo D (ed) Recent advances in ageing and sexing animal bones. Oxbow Books, Oxford, pp 227–236

Carranza J, Mateos C, Alarcos S et al (2008) Sex-specific strategies of dentine depletion in red deer. Biol J Linn Soc 93(3):487–497

Chapman DI, Chapman NG (1970) Development of the teeth and mandibles of fallow deer. Acta Theriol 15: 111–131

Chapman NG, Chapman DI (1979) Seasonal changes in the male accessory glands of reproduction in adult fallow deer (*Dama dama*). J Zool 189(3):259–273

Chapman NG, Chapman DI (1980) The distribution of fallow deer: a worldwide review. Mammal Rev 10: 61–138

Chapman DI, Chapman NG (1997) Fallow deer: their history, distribution and biology, 2nd edn. Coch-y-bonddu Books, Machynlleth

Chapman NG, Putman R (1991) Fallow deer. In: Corbet GB, Harris S (eds) The handbook of British mammals, 3rd edn. Blackwell Scientific Publications, Oxford

Chapman DI, Chapman NG, Kennaugh JH (1981) Development of the preputial gland of immature fallow deer (*Dama dama* Linnaeus) with particular reference to puberty. Z Saugetierkd 46(5):322–330

China WE, Melville RV (1960) Opinion 581. Determination of the generic names for the fallow deer of Europe and the Virginia deer of America (Class Mammalia). Bull Zool Nomencl 17:267–275

Cifuni GF, Amici A, Contò M et al (2014) Effects of the hunting method on meat quality from fallow deer and wild boar and preliminary studies for predicting lipidoxidation using visible reflectance spectra. Eur J Wildl Res 60(3):519–526

Ciuti S, Apollonio M (2008) Ecological sexual segregation in fallow deer (*Dama dama*): a multispatial and multitemporal approach. Behav Ecol Sociobiol 62: 1747–1759

Ciuti S, Apollonio M (2011) Do antlers honestly advertise the phenotypic quality of fallow buck (*Dama dama*) in a lekking population? Ethology 117(2):133–144

Ciuti S, Davini S, Luccarini S et al (2003) Variation in home range size of female fallow deer (*Dama dama* L.) inhabiting a sub-Mediterranean habitat. Rev Ecol-Terre Vie 58:1–15

Ciuti S, Bongi P, Vassale S et al (2006) Influence of fawning on the spatial behaviour and habitat selection of female fallow deer (*Dama dama*) during late pregnancy and early lactation. J Zool 268:97–107

Ciuti S, De Cena F, Bongi P et al (2011) Benefits of a risky life for fallow deer bucks (*Dama dama*) aspiring to patrol a lek territory. Behaviour 148(4): 435–460

Corbet GB (1978) The Mammals of the Palaearctic Region: a taxonomic review. British Museum (National History) and Cornell University Press, London/Ithaca (NY)

Danford CG, Alston ER (1880) On the mammals of Asia Minor. Part II. Proc Zool Soc London 1880:50–64

Daszkiewicz T, Hnatyk N, Dąbrowski D et al (2015) A comparison of the quality of the *Longissimus lumborum* muscle from wild and farm-raised fallow deer (*Dama dama* L). Small Rumin Res 129:77–83

Davini S, Ciuti S, Luccarini S et al (2004) Home range patterns of male fallow deer *Dama dama* in a sub-Mediterranean habitat. Acta Theriol 49(3):393–404

Davis S, MacKinnon M (2009) Did the Romans bring fallow deer to Portugal? Environ Archaeol 14(1): 15–26

De Craeye S, Speybroeck N, Ajzenberg D et al (2011) *Toxoplasma gondii* and *Neospora caninum* in wildlife: common parasites in Belgian foxes and Cervidae? Vet Parasitol 178:64–69

De Marinis AM (2015) Valutazione dell'età nei Cervidi tramite esame della dentatura. Guida pratica all'identificazione delle classi di età del Daino. Manuali e linee guida 90.3/2013, ISPRA

De Marinis AM, Asprea A (2006) Hair identification key of wild and domestic ungulates from southern Europe. Wildl Biol 12:305–320

De Marinis AM, Masseti M (2021) The spatio-temporal co-occurrence of free-ranging common fallow deer and domestic caprines on the island of Rhodes, Greece. Mammalia 85(3):227–230

Delibes M (1980) Feeding ecology of the Spanish lynx in the Coto Doñana. Acta Theriol 25:309–324

Drew ML (1998) Comparison of tympanic membrane and rectal temperatures of anesthetized fallow deer (*Dama dama*). J Zoo Wildl Med 9(3):338–340

Duarte J, Farfán MA, Fa JE et al (2016) Deer populations inhabiting urban areas in the south of Spain: habitat and conflicts. Eur J Wildl Res 61(3)

Durmuş M (2019) Determining space use and demography of a reintroduced fallow deer (*Dama dama*) population using GPS telemetry in Dilek Peninsula National Park, Turkey. Dissertation, Middle East Technical University

East ML, Bassano B, Ytrehus B (2011) The role of pathogens in the population dynamics of European ungulates. In: Putman R, Apollonio M, Andersen R (eds) Ungulate management in Europe. Problems and practices. Cambridge University Press, New York, pp 319–348

EFSA and ECDC (European Food Safety Authority and European Centre for Disease Prevention and Control) (2021) The European Union One Health 2019 Zoonoses Report. EFSA J 19(2)

Ekvall K (1998) Effects of social organisation, age and aggressive behaviour on allosuckling in wild fallow deer. Anim Behav 56:695–703

English AW, Lepherd EE (1981) The haematology and serum biochemistry of wild fallow deer (*Dama dama*) in New South Wales. J Wildl Dis 17(2):289–295

Farrell ME, Briefer E, Hayden T, McElligott AG (2011) Assortative mating in fallow deer reduces the strength of sexual selection. PLoS One 6(9). https://doi.org/10.1371/annotation/091a063b-f39b-4968-82c4-97db860ecf43

Feldhamer GA, Farris-Renner KC, Barker CM (1988) *Dama dama*. The American Society of Mammalogists. Mamm Species 317:1–8

Fernández-García JL (2012) The endangered *Dama dama mesopotamica* Brooke, 1875: genetic variability, allelic loss and hybridization signals. Contrib Zool 81:223–233

Ferretti F, Fattorini N (2021) Competitor densities, habitat, and weather: effects on interspecific interactions between wild deer species. Integr Zool 16(5):670–684

Ferretti F, Sforzi A, Lovari S (2012) Avoidance of fallow deer by roe deer may not be habitat-dependent. Hystrix 23(2):28–34

Festa-Bianchet M, Apollonio M, Mari F et al (1990) Aggression among lekking male fallow deer (*Dama dama*): territory effects and relationship with copulatory success. Ethology 85:36–246

Focardi S, Pecchioli E (2005) Social cohesion and foraging decrease with group size in fallow deer (*Dama dama*). Behav Ecol Sociobiol 59:84–91

Focardi S, Aragno P, Montanaro P et al (2006) Interspecific competition from fallow deer *Dama dama* reduces habitat quality for the Italian roe deer *Capreolus capreolus italicus*. Ecography 29:407–417

Fontana F, Rubini M (1990) Chromosomal evolution in Cervidae. Biosystems 24(2):157–174

Galka ME, Aguilar JM, Quevedo MA et al (1999) Alpha-2 agonist dissociative anesthetic combinations in fallow deer (*Cervus dama*). J Zoo Wildl Med 30:451–453

Genovesi P, Bacher S, Kobelt M et al (2009) Alien mammals of Europe. In: DAISIE (ed) Handbook of alien species in Europe. Invading nature. Springer series in invasion ecology, vol 3. Springer, Dordrecht, pp 119–128

Gilbert C, Ropiquet A, Hassanin A (2006) Mitochondrial and nuclear phylogenies of Cervidae (Mammalia, Ruminantia): systematics, morphology, and biogeography. Mol Phylogenet Evol 40(1):101–117

Gortázar C, Torres MJ, Vicente J et al (2008) Bovine tuberculosis in Doñana Biosphere Reserve: the role of wild ungulates as disease reservoirs in the last Iberian lynx strongholds. PLoS One 3(7):e2776

Gortázar C, Ferroglio E, Lutton CE et al (2010) Disease-related conflicts in mammal conservation. Wildl Res 37(8):668–675

Gosch B, Fischer K (1989) Seasonal changes of testis volume and sperm quality in adult fallow deer (*Dama dama*) and their relationship to the antler cycle. J Reprod Fertil 85:7–17

Goss RJ, Van Praagh A, Brewer P (1992) The mechanism of antler casting in fallow deer. J Exp Zool 264(4):429–436

Gray AP (1972) Mammalian hybrids: a check-list with bibliography. Commonwealth Agricultural Bureau, Slough

Gray J, Kahl O, Zintl A (2021) What do we still need to know about *Ixodes ricinus*? Ticks Tick-borne Dis 12:101682

Harrison DL (1968) The mammals of Arabia. Vol. II. Carnivora – Artiodactyla – Hyracoidea. Ernest Benn Ltd., London

Harrison RJ, Hyett AR (1954) The development and growth of the placentomes in the fallow deer (*Dama dama* L.). J Anat 88(3):338–355

Hartl GB, Schleger A, Slowak M (1986) Genetic variability in fallow deer, *Dama dama* L. Anim Genet 17:335–341

Heckeberg NS (2020) The systematics of the Cervidae: a total evidence approach. PeerJ 8:e8114

Hejlícek K, Literák I, Vostalová E et al (1999) *Toxoplasma gondii* antibodies in pregnant women in the České Budějovice District (in Czech). J Microbiol Immunol Infect 48:102–105

Herzog S (1990) Constitutive heterochromatin in the fallow deer (*Cervus dama*). Z Säugetierkd 55(1):63–65

Hofmann RR (1985) Digestive physiology of the deer – their morphophysiological specialisation and adaptation. J R Soc N Z 22:393–407

Hofmann RR (1989) Evolutionary steps of ecophysiological adaptation and diversification of ruminants: a comparative view of their digestive system. Oecologia 78: 443–457

Höglund J, Alatalo RV (2014) Leks. Princeton University Press, Princeton

Holečková J, Bartoš L, Tománek M (2000) Inter-male mounting in fallow deer, *Dama dama* – its seasonal pattern and social meaning. Folia Zool 49:175–181

Hönig V, Švec P, Marek L et al (2019) Model of risk of exposure to Lyme borreliosis and tick-borne encephalitis virus-infected ticks in the border area of the Czech Republic (South Bohemia) and Germany (lower Bavaria and upper Palatinate). Int J Environ Res Public Health 16(7):1173

Houszka M, Piekarska J, Podkowik M et al (2016) Morphology and molecular study of *Fascioloides magna* – a growing threat to cervids (Cervidae) in Poland. J Vet Res 60(4):435–439

Iglódyová A, Lazar P, Čurlík J et al (2017) Observations on autochtonous liver flukes in wild ruminants in Slovakia. Helminthologia 54(4):307–313

Ilie A, Enescu CM (2018) Hunting of red deer and fallow deer in Romania. Res J Agric Sci 50(4):184–188

Imperio S, Focardi S, Santini G et al (2012) Population dynamics in a guild of four Mediterranean ungulates: density-dependence, environmental effects and inter-specific interactions. Oikos 121(10): 1613–1626

Imperio S, Lombardi S, De Marinis A et al (2020) Female mating tactics in lekking fallow deer (*Dama dama*): experience explains inter-individual variability more than costs. Sci Rep 10(1):1–12

Jacobs GH, Deegan JF II, Neitz J et al (1994) Electrophysiological measurements of spectral mechanisms in the retinas of two cervids: white-tailed deer (*Odocoileus virginianus*) and fallow deer (*Dama dama*). J Comp Physiol A 174(5):551–557

Jaenson TG, Petersson EH, Jaenson DG et al (2018) The importance of wildlife in the ecology and epidemiology of the TBE virus in Sweden: incidence of human TBE correlates with abundance of deer and hares. Parasites Vectors 11(1):477

Jennings DJ, Gammell MP, Carlin CM et al (2002) Does lateral presentation of the palmate antlers during fights by fallow deer (*Dama dama* L.) signify dominance or submission? Ethology 108:389–401

Jennings DJ, Gammell MP, Carlin CM et al (2004) Effect of body weight, antler length, resource value and experience on fight duration and intensity in fallow deer. Anim Behav 68:213–221

Jennings DJ, Carlin CM, Gammell MP (2009) A winner effect supports third-party intervention behaviour during fallow deer, *Dama dama*, fights. Anim Behav 77(2):343–348

Jennings DJ, Carlin CM, Hayden TJ et al (2010) Investment in fighting in relation to body condition, age and dominance rank in the male fallow deer (*Dama dama*). Anim Behav 79:1293–1300

Jennings DJ, Hayden TJ, Gammell MP (2013) Personality and predictability in fallow deer fighting behaviour: the relationship with mating success. Anim Behav 86(5): 1041–1047

Jennings DJ, Boys RJ, Gammell MP (2017) Investigating variation in third-party intervention behavior during a fallow deer (*Dama dama*) rut. Behav Ecol 28(1): 288–293

Jennings DJ, Boys RJ, Gammell MP (2018) Suffering third-party intervention during fighting is associated with reduced mating success in the fallow deer. Anim Behav 139:1–8

Kazimírová M, Hamšíková Z, Špitalská E et al (2018) Diverse tick-borne microorganisms identified in free-living ungulates in Slovakia. Parasites Vectors 11(1):495

Kierdorf U, Kierdorf H, Schultz M (1994) The macroscopic and microscopic structure of double-head antlers and pedicle bone of Cervidae (Mammalia, Artiodactyla). Ann Anat 176(3):251–257

Kierdorf H, Kierdorf U, Szuwart T (1995) A light microscopic study of primary antler development in fallow deer (*Dama dama*). Ann Anat 177:525–532

Kierdorf U, Stoffels E, Stoffels D et al (2003) Histological studies of bone formation during pedicle restoration and early antler regeneration in roe deer and fallow deer. Anat Rec 273A(2):741–751

Kitagawa C (2008) On the presence of deer in ancient Egypt: analysis of the osteological record. J Egypt Archaeol 94(1):209–222

Kjellander P, Svartholm I, Bergvall UA et al (2012) Habitat use, bed-site selection and mortality rate in neonate fallow deer *Dama dama*. Wildl Biol 18(3):280–291

Komers PE, Birgersson B, Ekvall K (1999) Timing of estrus in fallow deer is adjusted to the age of available mates. Am Nat 153:431–436

Kornacka A, Cybulska A, Bień-Kalinowska J et al (2020) *Ashworthius sidemi* in cattle and wild ruminants in Poland–the current state of play. Ann Parasitol 66(4): 517–520

Kowal J, Nosal P, Bonczar Z (2012) Parasites of captive fallow deer (*Dama dama* L.) from southern Poland with special emphasis on *Ashworthius sidemi*. Ann Parasitol 58(1):23–26

Králová-Hromadová I, Juhásová L, Bazsalovicsová E (2016) The giant liver fluke, *Fascioloides magna*: past, present and future research. Springer, Heidelberg

Kuznetsov D (2021) The first detection of abomasal nematode *Ashworthius sidemi* in fallow deer (*Dama dama*) in Russia. Acta Parasitol:1–4

Kviberg K, Craig J (2006) Woodhill forest: a candidate for ecosystem management. J R Soc N Z 36(3):129–141

Langbein J (1991) Effects of density and age on body condition, reproductive performance, behaviour and survival of fallow deer. Dissertation, University Southampton

Langbein J, Putman RJ (1992) Behavioural responses of park red and fallow deer to disturbance and effects on population performance. Anim Welf 1(1):19–38

Langbein J, Thirgood SJ (1989) Variation in mating systems of fallow deer in relation to ecology. Ethology 83: 195–214

Langbein J, Putman R, Pokorny B (2011) Traffic collisions involving deer and other ungulates in Europe and available measures for mitigation. In: Putman R, Apollonio M, Andersen R (eds) Ungulate management in Europe: problems and practices. Cambridge University Press, Cambridge, pp 215–259

Lawson RE, Putman RJ, Fielding AH (2000) Individual signatures in scent gland secretions of Eurasian deer. J Zool 251:399–410

Li C (2013) Histogenetic aspects of deer antler development. Front Biosci 1(5):479–489

Lionello P, Scarascia L (2018) The relation between climate change in the Mediterranean region and global warming. Reg Environ Chang 18(5):1481–1493

Lister AM (1996) The morphological distinction between bones and teeth of fallow deer (*Dama dama*) and red deer (*Cervus elaphus*). Int J Osteoarchaeol 6:119–143

Long JL (2003) Introduced mammals of the world: their history, distribution and abundance. CABI Publishing, Wallingford

López-Montoya AJ, Moro J, Azorit C (2017) Drivers of population growth variations for two Mediterranean sympatric Deer. Ecol Res 32(5):703–712

Loudon ASI (1985) Lactation and neonatal survival of animals. Symp Zool Soc London 54:183–207

Lovari S, Nobili G (2010) Programma nazionale di conservazione del cervo della Mesola. Quad. Cons. Natura 36, Ministero dell'Ambiente e della Tutela del Territorio e del Mare, Ministero delle Politiche Agricole Alimentari e Forestali - Corpo Forestale dello Stato, I.S.P.R.A.

Ludwig A, Vernesi C, Lieckfeldt D et al (2012) Origin and patterns of genetic diversity of German fallow deer as inferred from mitochondrial DNA. Eur J Wildl Res 58(2):495–501

Magdalek J, Kyrianova IA, Vadlejch J. (2017) *Ashworthius sidemi* in wild cervids in the Czech Republic. 9th workshop on biodiversity, Jevany, pp 66–71

Malacarne M, Franceschi P, Formaggioni P et al (2015) Minerals content and distribution in milk from red deer (*Cervus elaphus*), fallow deer (*Dama dama*) and roe deer (*Capreolus capreolus*). Small Rumin Res 130: 208–215

Malcicka M (2015) Life history and biology of *Fascioloides magna* (Trematoda) and its native and exotic hosts. Ecol Evol 5(7):1381–1397

Marchesini A, Vernesi C, Gandolfi A et al (2021) No genetic signature of glacial refugia in current European fallow deer (*Dama dama dama* L., 1758) populations: a comment on Baker et al. (2017). Mamm Biol 101:305–312

Mas-Coma S, Bargues MD, Valero MA (2005) Fascioliasis and other plant-borne trematode zoonoses. Int J Parasitol 35:1255–1278

Masseti M (1996) The postglacial diffusion of the genus *Dama* Frisch, 1775, in the Mediterranean region. Ricerche Biologia Selvaggina 25:7–29

Masseti M (1999) The fallow deer, *Dama dama* L., 1758, in the Aegean region. Contributions to the Zoogeography and Ecology of the Eastern Mediterranean Region 1(Supplement):17–30

Masseti M (2002) Island of deer. Natural history of the fallow deer of Rhodes and of the terrestrial and fresh water vertebrates of the Dodecanese, 2nd edn. Municipality of Rhodes, Greece

Masseti M (2012) Atlas of terrestrial mammals of the Ionian and Aegean islands. De Gruyter, Berlin

Masseti M, Rustioni M (1988) Considerazioni preliminari sulla diffusione di *Dama dama* (Linnaeus, 1758) durante le epoche tardiglaciale e postglaciale nell'Italia mediterranea. Studi per l'Ecologia del Quaternario 10: 93–119

Masseti M, Vernesi C (2014) Historic zoology of the European fallow deer, *Dama dama dama*: evidence from biogeography, archaeology and genetics. In: Baker K, Carden R, Madgwick R (eds) Deer and people. Windgather Press, Oxford, pp 13–22

Masseti M, Pecchioli E, Romei A et al (1997) RAPD (Random Amplified Polymorphic DNA) fingerprinting analysis of some Italian populations of fallow deer *Dama dama*. Ital J Zool 64(3):235–238

Masseti M, Pecchioli E, Vernesi C (2008) Phylogeography of the last surviving populations of Rhodian and Anatolian fallow deer (*Dama dama dama* L., 1758). Biol J Linn Soc Lond 93(4):835–844

Mattioli S, Fico R, Lorenzini R et al (2003) Mesola red deer: physical characteristics, population dynamics and conservation perspectives. Hystrix 14:87–94

Mattioli L, Capitani C, Gazzola A et al (2011) Prey selection and dietary response by wolves in a high-density multi-species ungulate community. Eur J Wildl Res 57(4):909–922

Mayr B, Krutzler J, Auer H et al (1987) NORs, heterochromatin, and R-bands in three species of Cervidae. J Hered 78:108–110

McElligott AG, Hayden TJ (2000) Lifetime mating success, sexual selection and life history of fallow bucks (*Dama dama*). Behav Ecol Sociobiol 48:203–221

McElligott AG, Mattiangeli V, Mattiello S et al (1998) Fighting tactics of fallow bucks (*Dama dama*, Cervidae): reducing the risks of serious conflict. Ethology 104:789–803

McElligott AG, O'Neill KP, Hayden TJ (1999) Cumulative long-term investment in vocalization and mating success of fallow bucks, *Dama dama*. Anim Behav 57(5): 1159–1167

McElligott AG, Gammell MP, Harty HC et al (2001) Sexual size dimorphism in fallow deer (*Dama dama*): do larger, heavier males gain greater mating success? Behav Ecol Sociobiol 49:266–272

McElligott AG, Altwegg R, Hayden TJ (2002) Age-specific survival and reproductive probabilities: evidence for senescence in male fallow deer (*Dama dama*). Proc R Soc Lond B 269:1129–1137

McElligott AG, Naulty F, Clarke WV et al (2003) The somatic cost of reproduction: what determines reproductive effort in prime-aged fallow bucks? Evol Ecol Res 5(8):1239–1250

Mertins JW, Mortenson JA, Bernatowicz JA et al (2011) *Bovicola tibialis* (Phthiraptera: Trichodectidae): occurrence of an exotic chewing louse on cervids in North America. J Med Entomol 48(1):1–12

Michelitsch A, Wernike K, Klaus C et al (2019) Exploring the reservoir hosts of tick-borne encephalitis virus. Viruses 11:669

Moore NP, Kelly PF, Cahill JP et al (1995) Mating strategies and mating success of fallow bucks in a non-lekking population. Behav Ecol Sociobiol 36:91–100

Moskwa B, Kornacka A, Cybulska A et al (2018) Seroprevalence of *Toxoplasma gondii* and *Neospora caninum* infection in sheep, goats, and fallow deer farmed on the same area. J Anim Sci 96(6): 2468–2473

Muñoz PM, Boadella M, Arnal M et al (2010) Spatial distribution and risk factors of Brucellosis in Iberian wild ungulates. BMC Infect Dis 10(1):46

Nagy E, Jócsák I, Csivincsik Á et al (2018) Establishment of *Fascioloides magna* in a new region of Hungary: case report. Parasitol Res 117(11):3683–3687

Niemi M, Matala J, Melin M et al (2015) Traffic mortality of four ungulate species in southern Finland. Nat Conserv 11:13–28

Nowak S, Mysłajek RW, Kłosińska A et al (2011) Diet and prey selection of wolves (*Canis lupus*) recolonising Western and Central Poland. Mamm Biol 76(6): 709–715

Odend'hal S, Miller KV, Demarais S (1996) Preputial glands in artiodactyla. J Mammal 77(2):417–421

Palmer AR (1996) Waltzing with asymmetry. Bioscience 46:518–532

Palmer AR, Strobeck C (1986) Fluctuating asymmetry: measurement, analysis, pattern. Ann Rev Ecol Syst 17:391–421

Parillo F, Diverio S (2009) Glycocomposition of the apocrine interdigital gland secretions in the fallow deer (*Dama dama*). Res Vet Sci 86(2):194–199

Parmesan C (2006) Ecological and evolutionary responses to recent climate change. Annu Rev Ecol Evol Syst 37: 637–669

Pavlásek I, Minár J (2014) New host and the extend of the host range of warble fly *Hypoderma diana* (Diptera, Hypodermatidae). Acta Mus Sil Sci Nat 63(1):61–64

Pélabon C (1997) Is weight at birth a good predictor of weight in winter for fallow deer? J Mammal 78:48–54

Pélabon C, Joly P (2000) What, if anything, does visual asymmetry in fallow deer antlers reveal? Anim Behav 59:193–199

Pélabon C, Yoccoz NG, Ropert-Coudert Y et al (1998) Suckling and allosuckling in captive fallow deer (*Dama dama*, Cervidae). Ethology 104(1):75–86

Pemberton JM, Smith RH (1985) Lack of biochemical polymorphism in British fallow deer. Heredity 55:199–207

Pérez-Barbería FJ (2020) The ruminant: life history and digestive physiology of a symbiotic animal. In: García-Yuste S (ed) Sustainable and environmentally friendly dairy farms. SpringerBriefs in Applied Sciences and Technology. Springer, Cham, pp 19–45

Pitcher BJ, Briefer EF, McElligott AG (2015) Intrasexual selection drives sensitivity to pitch, formants and duration in the competitive calls of fallow bucks. BMC Evol Biol 15:149

Pitra C, Fickel J, Meijaard E et al (2004) Evolution and phylogeny of old world deer. Mol Phylogenet Evol 33: 880–895

Poli A, Mancianti F, Marconcini A et al (1988) Prevalence, ultrastructure of the cyst wall and infectivity for the dog and cat of *Sarcocystis* sp. from fallow deer (*Cervus dama*). J Wildl Dis 24(1):97–104

Poljičak-Milas N, Marenjak TS, Slavica A et al (2009) Comparative hematological and biochemical values in pregnant and non-pregnant red, *Cervus elaphus*, and fallow deer, *Dama dama*, females. Folia Zool 58(1):36–44

Pomilio F (2016) APHAEA/EWDA diagnosis card: listeriosis. www.ewda.org

Prakas P, Butkauskas D, Rudaitytė E et al (2016) Morphological and molecular characterization of *Sarcocystis taeniata* and *Sarcocystis pilosa* n. sp. from the sika deer (*Cervus nippon*) in Lithuania. Parasitol Res 115(8): 3021–3032

Putman RJ (1996) Competition and resource partitioning in temperate ungulate assemblies. Chapman and Hall, London

Putman R, Langbein J (2003) The deer manager's companion: a guide to the management of deer in the wild and in parks. Swann Hill Press, Shrewsbury

Putman RJ, Culpin S, Thirgood SJ (1993) Dietary differences between male and female fallow deer in sympatry and in allopatry. J Zool 229(2):267–275

Putman RJ, Langbein J, Hewison AJM et al (1996) Relative roles of density-dependent and density-independent factors in the population dynamics of British deer. Mammal Rev 26:81–101

Putman RJ, Sullivan MS, Langbein J (2000) Fluctuating asymmetry in antlers of fallow deer (*Dama dama*): the relative roles of environmental stress and sexual selection. Biol J Linn Soc 70(1):27–36

Putman R, Apollonio M, Andersen R (eds) (2011) Ungulate management in Europe: problems and practices. Cambridge University Press, Cambridge

Randi E, Apollonio M (1988) Low biochemical variability in European fallow deer (*Dama dama* L.): natural bottlenecks and the effects of domestication. Heredity 61(3):405–410

Rehbein S, Bienoschek S (1995) Dynamics of endoparasites in farmed fallow deer (*Dama dama*) from birth to puberty. Appl Parasitol 36(3):212–219

Rehbein S, Visser M, Jekel I et al (2014) Endoparasites of the fallow deer (*Dama dama*) of the Antheringer Au in Salzburg, Austria. Wien Klin Wochenschr 126(1): 37–41

Rhyan JC, Miller MW, Spraker TR et al (2011) Failure of fallow deer (*Dama dama*) to develop chronic wasting disease when exposed to a contaminated environment and infected mule deer (*Odocoileus hemionus*). J Wildl Dis 47:739–744

Ricci A, Allende A, Bolton D et al (2017) Chronic wasting disease (CWD) in cervids. EFSA J 15(1):4667

Robbins CT (2013) Wildlife feeding and nutrition. Academie Press Knc

Root TL, Price JT, Hall KR et al (2003) Fingerprints of global warming on wild animals and plants. Nature 421(6918):57–60

Roulin A, Heeb P (1999) The immunological function of allosuckling. Ecol Lett 2(5):319–324

Rubini M, Negri E, Fontana F (1990) Standard karyotype and chromosomal evolution of the fallow deer (*Dama dama* L.). Cytobios 64(258–259):155–161

Sahlén E, Noell S, DePerno CS et al (2016) Phantoms of the forest: legacy risk effects of a regionally extinct large carnivore. Ecol Evol 6(3):791–799

Saltz D, Bar David S, Zidon R et al (2011) Reintroducing the Persian fallow deer *Dama mesopotamica* in Israel – a chronology. Anim Prod Sci 51(4):251–258

Sánchez S, García-Sánchez A, Martínez R et al (2008) Detection and characterisation of Shiga toxin-producing *Escherichia coli* other than *Escherichia coli* O157: H7 in wild ruminants. Vet J 180(3):384–388

Sándor G, László R, Náhlik A (2014) Determination of time conception of fallow deer in a Hungarian free range habitat. Folia Zool 63:122–126

Say L, Naulty F, Hayden TJ (2003) Genetic and behavioural estimates of reproductive skew in male fallow deer. Mol Ecol 12(10):2793–2800

Scandura M, Tiedemann R, Apollonio M et al (1998) Genetic variation in an isolated Italian population of fallow deer *Dama dama* as revealed by RAPD-PCR. Acta Theriol 5:163–169

Schreiber A, Braza F, San José C et al (1992) Relationships of foetal haemoglobin level to age of newborn fallow deer (*Dama dama*). Can J Zool 70(12):2364–2368

Shackleton DM (1999) Hoofed mammals of British Columbia. UBC Press, Vancouver

Siroka Z, Krocilova B, Pikula J et al (2011) Blood coagulation parameters in fallow deer (*Dama dama*). Vet Med (Praha) 56(3):119–122

Sleeman DP (1983) Parasites of deer in Ireland. Master's thesis. University College Dublin

Soldati S, Kiupel M, Wise A et al (2004) Meningoencephalomyelitis caused by *Neospora caninum* in a juvenile fallow deer (*Dama dama*). J Vet Med A 51(6): 280–283

Stachowicz JBE, Vannoni E, Pitcher BJ et al (2014) Acoustic divergence in the rut vocalizations of Persian and European fallow deer. J Zool 292:1–9

Stillman RA, Clutton-Brock TH, Sutherland WJ (1993) Black-holes, mate retention, and the evolution of ungulate leks. Behav Ecol 4:1–6

Stuart AJ (1991) Mammalian extinctions in the Late Pleistocene of Northern Eurasia and North America. Biol Rev 66:453–562

Stutz RS, Bergvall UA, Leimar O et al (2018) Cohesiveness reduces foraging efficiency in a social herbivore. Anim Behav 135:57–68

Syczyło K, Platt-Samoraj A, Bancerz-Kisiel A et al (2018) The prevalence of *Yersinia enterocolitica* in game animals in Poland. PLoS One 13(3):e0195136

Sykes NJ, Baker KH, Carden RF et al (2011) New evidence for the establishment and management of the European fallow deer (*Dama dama dama*) in Roman Britain. J Archaeol Sci 38:156–165

Szczurek B, Kadulski S (2004) Ectoparasites on fallow deer, *Dama dama* (L.) in Pomerania, Poland. Acta Parasitol 49(1):80–86

Thirgood SJ (1995) The effects of sex, season and habitat availability on patterns of habitat use by fallow deer. J Zool 235:645–659

Thirgood SJ (1996) Ecological factors influencing sexual segregation and group size in fallow deer (*Dama dama*). J Zool 239:783–797

Thirgood S, Langbein J, Putman RJ (1999) Intraspecific variation in ungulate mating strategies: the case of the flexible fallow deer. Adv Study Behav 28:333–359

Ueckermann E, Hansen P (1994) Das Damwild. Verlag Paul Parey, Hamburg

Uerpmann HP (1987) The ancient distribution of ungulate mammals in the Middle East: fauna and archaeological sites in Southwest Asia and Northeast Africa. Beihefte zum Tübinger Atlas des Vorderen Orients, Reihe A, Naturwissenschaften 27:1–173

Ünala Y, Çulhacıa H (2018) Investigation of fallow deer (*Cervus dama* L.) population densities by camera trap method in Antalya Düzlerçamı Eşenadası Breeding Station. Turk J For 19(1):57–62

Valente AM, Acevedo P, Figueiredo AM et al (2020) Overabundant wild ungulate populations in Europe: management with consideration of socio-ecological consequences. Mammal Rev 50(4):353–366

Valenzuela A, Baker K, Carden RF et al (2016) Both introduced and extinct: the fallow deer of Roman Mallorca. J Archaeol Sci Rep 9:168–177

van der Geer AAE, Lomolino MV, Lyras G (2018) "On being the right size" – do aliens follow the rules? J Biogeogr 45:515–529

Van Moorter B, Gaillard JM, McLoughlin PD et al (2009) Maternal and individual effects in selection of bed sites and their consequences for fawn survival at different spatial scales. Oecologia 159(3): 669–678

van Wieren SE, Hofmeester TR (2016) The role of large herbivores in *Ixodes ricinus* and *Borrelia burgdorferi* s.l. dynamics. In: Braks MAH, van Wieren SE, Takken W et al (eds) Ecology and prevention of Lyme borreliosis. Wageningen Academic Publishers, Wageningen

Vannoni E, McElligott AG (2009) Fallow bucks get hoarse: vocal fatigue as a possible signal to conspecifics. Anim Behav 78:3–10

Vengušt G, Bidovec A (2003) Parasites of fallow deer (*Dama dama*) in Slovenia. Helminthologia 40:161–164

Vengušt G, Žele D, Kobal S et al (2006) Haematological and biochemical values of farmed fallow deer (*Dama dama*) after using different methods of capture. Vet Arh 76:S189–S197

Villerette N, Helder R, Angibault JM et al (2006) Sexual segregation in fallow deer: are mixed-sex groups especially unstable because of asynchrony between the sexes? C R Biol 329(7):551–558

Wang Y, Bekhit AEDA, Morton JD et al (2017) Nutritional value of deer milk. In: Watson RR, Collier RJ, Preedy VR (eds) Nutrients in dairy and their implications on health and disease. Academic, pp 363–375

Ward JF, Asher GW, Archer JA et al (2014) Genetic effects on first antler growth in relation to live-weight of red deer farmed in New Zealand. Livest Sci 167: 92–99

Weaver GV, Domenech J, Thiermann AR et al (2013) Foot and mouth disease: a look from the wild side. J Wildl Dis 49(4):759–785

Yabsley MJ, Shock BC (2013) Natural history of zoonotic *Babesia*: role of wildlife reservoirs. Int J Parasitol Parasites Wildl 2:18–31

Zachos FE, Mattioli S, Ferretti F et al (2014) The unique Mesola red deer of Italy: taxonomic recognition (*Cervus elaphus italicus* nova ssp., Cervidae) would endorse conservation. Ital J Zool 81:136–143

Zhu H, Chen X, Shao X et al (2015) Characterization of a virulent dog-originated rabies virus affecting more than twenty fallow deer (*Dama dama*) in Inner Mongolia, China. Infect Genet Evol 31:127–134

Żochowska-Kujawska J, Kotowicz M, Sobczak M et al (2019) Age-related changes in the carcass composition and meat quality of fallow deer (*Dama dama* L.). Meat Sci 147:37–43

Zuckerman S (1952) The breeding season of mammals in captivity. Proc R Soc B 122:827–950

Axis Deer *Axis axis* (Erxleben, 1777)

6

Nikica Šprem and Frank E. Zachos

Contents

N. Šprem (✉)
Department of Fisheries, Apiculture, Wildlife Management and Special Zoology, Faculty of Agriculture, University of Zagreb, Zagreb, Croatia
e-mail: nsprem@agr.hr; nikica.sprem@gmail.com

F. E. Zachos
Natural History Museum Vienna, Mammal Collection, Vienna, Austria

Department of Evolutionary Biology, University of Vienna, Vienna, Austria

Department of Genetics, University of the Free State, Bloemfontein, South Africa
e-mail: frank.zachos@nhm-wien.ac.at

© Springer Nature Switzerland AG 2022

L. Corlatti, F. E. Zachos (eds.), *Terrestrial Cetartiodactyla*, Handbook of the Mammals of Europe,
https://doi.org/10.1007/978-3-030-24475-0_22

Common Names

English	Axis deer, chital, spotted deer
German	Axishirsch
French	Cerf axis, chital
Spanish	Ciervo axis, Ciervo moteado, chital
Italian	Cervo pomellato, chital
Russian	Аксис

Taxonomy, Systematics and Paleontology

The earliest known fossils of *Axis* are dated to the Late Miocene of China. The axis deer is usually considered monotypic (Fig. 1). The monophyly of *Axis* (besides the chital, there are three other acknowledged species: *A. calamianensis*, *A.*

Fig. 1 (**a**) Axis deer from Dugi Otok Island; bucks with and without antlers; (**b**) axis deer hind from Dugi Otok Island (photographs by Igor Ilić – Serval)

kuhlii, and *A. porcinus*) has been corroborated by recent molecular and morphological phylogenetic studies. There seems to be a sister group relationship between *Axis* on the one hand and the barasingha *Rucervus duvaucelii* and the extinct Schomburgk's deer *Rucervus schomburgki* on the other, to the exclusion of Eld's deer which is usually also classified in this genus as *Rucervus eldii* (Gilbert et al. 2006; Hassanin et al. 2012; Heckeberg et al. 2016; Mennecart et al. 2017; but see Price et al. 2005 for a somewhat different phylogeny).

Current Distribution

Axis deer are naturally distributed in India, Sri Lanka, Bangladesh, southern Nepal, and southern Bhutan but have been introduced to many regions all over the world (Duckworth et al. 2015). Axis deer occur in Armenia, the Andaman Islands, Pakistan, Papua New Guinea, Australia, the United States (California, Texas, Florida, Hawaii), and South America (Brazil, Uruguay, Argentina). In Europe, the only successful introductions were to a few Adriatic Islands in Croatia resulting in three free-ranging populations on the Brijuni Islands, on Rab, and on Dugi Otok (Kusak and Krapinec 2010; Šprem 2015) (Fig. 2). Introductions to Ukraine and Moldova as well as to the British Isles are sometimes mentioned, but there is no reliable information available about this or any free-living populations resulting from these alleged introductions.

The first introduction of axis deer to Croatia occurred on the Brijuni Islands in 1911 (Kusak and Krapinec 2010). The genetic origin of the introduced individuals is unknown, but they are believed to have been imported by the German wild animal merchant Carl Hagenbeck (Bojović 1987; Šprem et al. 2008). Several attempts to introduce this species to continental parts of Croatia (Macelj, Posavlje, Božjakovina, Sljeme, Zelendvor, Lukovo, and Bribir Forest above the town of Crikvenica) failed due to the poor adaptation of this species to the continental climate (Krapinec 2001).

The Brijuni National Park (743 ha) is located in the northern Adriatic Sea (44°54′35″N, 13°46′12″E), off the western coast of the Istrian Peninsula. Most of the deer population (*c.* 95%) lives on Veliki Brijun Island (literally "large Brijun"); the remaining 5% live on Mali Brijun Island ("small Brijun"). Some axis deer swim from one island to the other. Veliki Brijun, the largest island of the Brijuni archipelago, is also partly cultivated in a landscape of grasslands, and it contains exceptionally valuable architectural heritage and protected vegetation communities of the Western Istrian climatic type. The deer introduced at the turn of the twentieth century have established a permanent population in a new environment (Kolić 1990). However, the spring of 1960 was an extreme season for the axis deer, with some 550 individuals dying due to a sudden cold spell accompanied by gale force northern winds (*bura*), snow, and rain (Fig. 3). Some 100 axis deer survived, and fallow deer *Dama dama* became the dominant ungulate species up to the present day (Bojović 1987). The axis deer population in Brijuni National Park is stable, with about 150 individuals in 2017 (E. Kolić 2018, pers. comm.). Several cases have been reported of axis deer swimming from Brijuni Islands to the mainland (*c.* 3 km), but establishment of new populations was unsuccessful (Frković 2004).

The Rab Island population in the northeastern Adriatic Sea (44°47′24″N, 14°40′10″E) was founded with eight axis deer captured on the Brijuni Islands (two males and six females) and released in 1974. This population is maintained for hunting purposes (Španjol and Skorup 2007). The current distribution extends over 846 ha on the Kalifront Peninsula (Centore et al. 2018). This area is a forest ecosystem with scrublands and woodlands of Euro-Mediterranean vegetation (Ugarković and Ugarković 2013). The population size is increasing slightly and during the 2015/2016 season, it was estimated at 78 axis deer (15 juveniles, 14 yearlings, and 49 adults) (Tomljanović 2016).

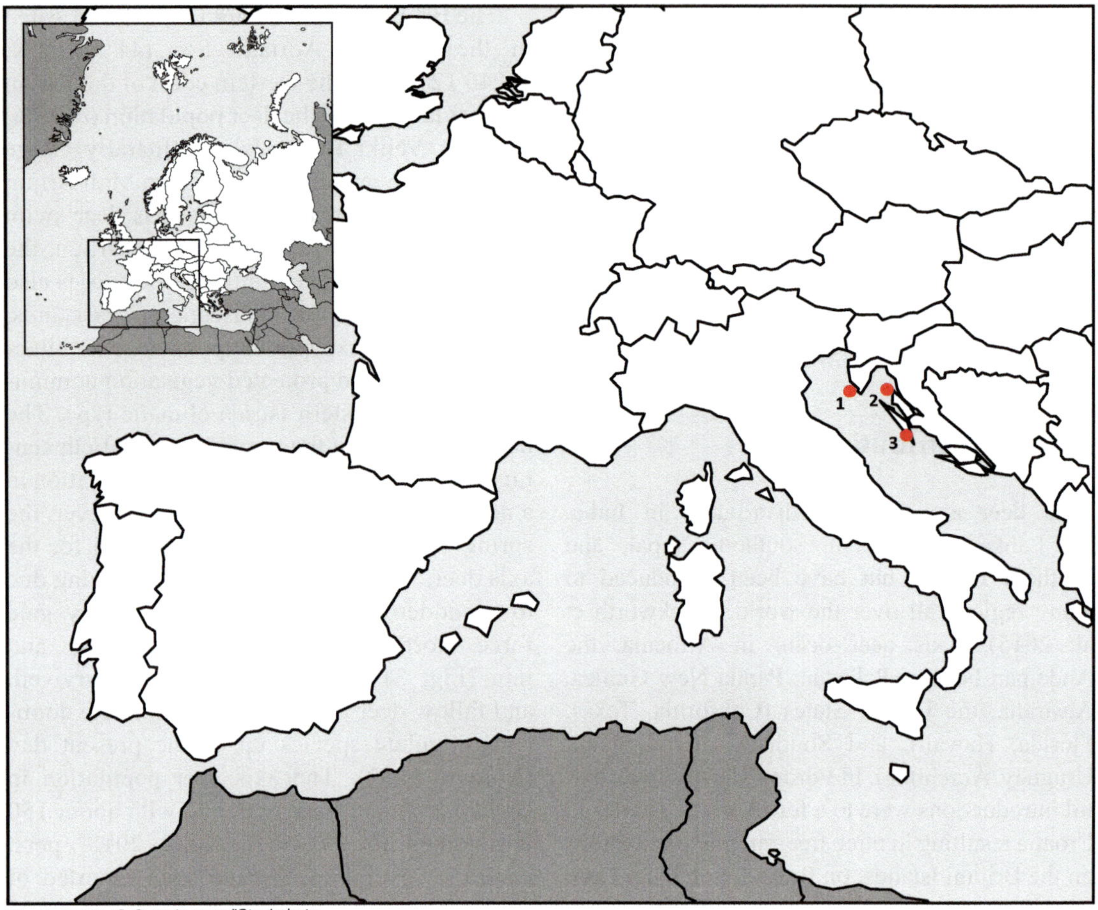

Fig. 2 Distribution range of axis deer; (1) Brijuni Islands, (2) Rab Island, (3) Dugi Otok Island (Map template: © Copyright Getty Images/iStockphoto)

The origin of the axis deer population on the island of Dugi Otok in the central Adriatic Sea (43°57′56″N, 15°06′24″E) goes back to 15 December 2012, when 13 individuals from the Brijuni Islands escaped from a fenced area (Šprem 2015). The total area of the island is 113.3 km^2, but the deer are currently only present in the southern parts. The population is increasing slightly and has been estimated at about 60 individuals (J. Tomljanović 2018, pers. comm.).

In 1953, further introduction attempts occurred on several Adriatic Islands (Cres, Plavnik, Mljet, Jakljan and Lokrum), but numbers gradually declined over the years, and the last individuals were observed in the early 1990s (Frković 2004, 2014).

Description

Axis deer are medium-sized deer with a head-body length of about 140–155 cm, a shoulder height of 70–95 cm, a tail length of 25–30 cm, and a body mass between 45 and 85 kg (though it may reach up to 100 kg in Croatia: Janicki et al. 2007). Males are 40–60% heavier than females (Mattioli 2011). The coat is reddish-brown with small white spots present in all seasons and a white patch on the throat. Males show a dark chevron over the face; does have a paler

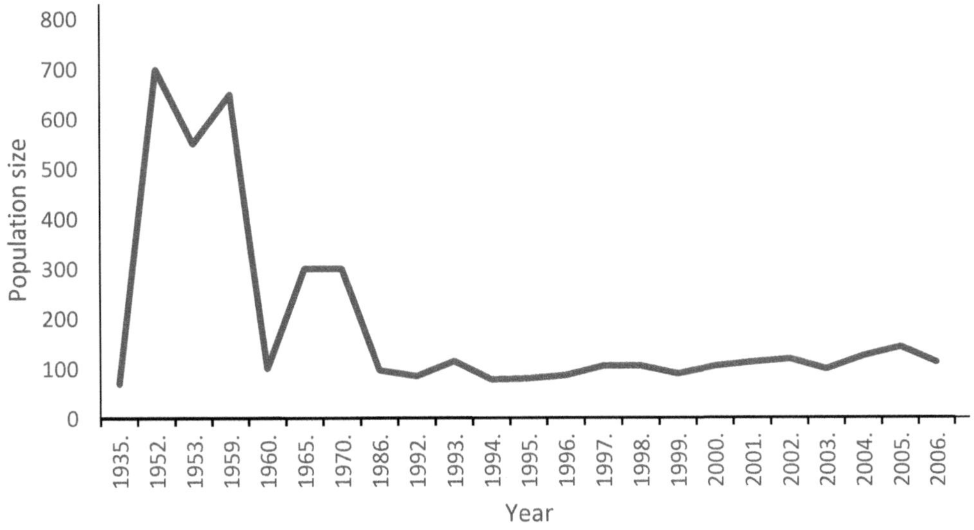

Fig. 3 Temporal change in the number of axis deer on the Brijuni Islands between 1935 and 2006 (cf. Šprem et al. 2008)

color on the face and on the neck. Axis deer have well-developed antorbital, metatarsal, and rear interdigital glands; the antorbital glands have a peculiar hair tuft. They have 32 teeth with the upper canines usually missing. Antlers are three-tined and lyre-shaped, with a long brow tine and a trez tine. The latter is inward-oriented (Mattioli 2011). Antlers are up to more than 90 cm in length, usually with six tines in three characteristic shapes: pear-shaped (most common), parallel, or rounded with a larger volume (Šprem et al. 2009).

Genetics

The karyotype of the axis deer is $2n = 66$ (fundamental number $NF = 70$) (Khongcharoensuk et al. 2017). Genetic diversity and structure have rarely been studied in axis deer, and as yet there are no genetic analyses on the Croatian axis deer. One of the few (and unfortunately somewhat confusing) molecular studies is the mtDNA analysis of 25 Pakistani axis deer by Abbas et al. (2016). Hybridization with sika deer has been verified in a deer farm in the USA. A female's hybrid origin was confirmed by means of gel electrophoresis of blood proteins as well as through karyotypic analysis (Asher et al. 1999). Gray (1954) lists hybridization between axis deer and hog deer as well as alleged hybridizations between axis and red deer and between axis and white-tailed deer (*Odocoileus virginianus*).

Life History

In axis deer, there is no strictly defined mating season, and mating can occur year-round primarily due to the tropical climate of its native habitat. Does are polyestrous with a cycle of 18–19 days. The antler cycle of the axis deer is not strictly linked to seasons but with mating season, and it is thus possible to meet deer in different antler phases (Loudon and Curlewis 1988). Similarly, mating season and birth season periods are not defined (Šprem et al. 2008). Regarding reproductive activity in Brijuni National Park, both young and mature bucks are equally fertile when the antlers have lost their velvet, as also suggested by Chapple et al. (1993). The youngest bucks present during the mating season are 4-year-olds, and the youngest ones that are actively involved in rutting activity are 5-year-olds. Buck battles during rutting are uncommon, and no serious injuries have been recorded. Rutting activities recorded during the night suggest a behavioral shift due to daytime disturbance by tourists. Gestation lasts for 7–8 months, and 85% of the does calve in the warm seasons, i.e., between the beginning of March and the end of September (Šprem et al. 2008). Kirkwood et al. (1987) report different results for captive axis deer born in an enclosure at the Zoological Society of London, with a similar number of births occurring in winter and summer, but with higher mortality of fawns born in winter. After a gestation of 231–235 days, does normally give birth to twins (but the number can vary between one and three fawns) (Frković 2004). Body mass at birth is about 3–4 kg, and it is common for does to have two litters within 15 months (Janicki et al. 2007).

Habitat and Diet

Due to the influence of the Mediterranean climate, the most food-limited seasons are summer and early autumn when the lack of precipitation causes the islands to simply "dry up." Most of the axis deer's daily activities are spent in search of food and feeding. In Brijuni National Park, in addition to grazing grounds, they also seek out leaves of laurel *Laurus nobilis*, ash *Fraxinus ornus*, and holm oak *Quercus ilex* as well as acorns. They occasionally feed on leaves of myrtle *Myrtus communis*, scrape moss off rocks, and take cedar seeds *Cedrus* spp. Unlike European mouflon (*Ovis gmelini musimon*) and fallow deer, axis deer on the Brijuni Islands regularly consume supplementary feed (hay, corn), regardless of the quality of the grassy areas, and come in for supplementary feed much earlier than the other two ungulate species (Šprem et al. 2008). On the contrary, on the island of Rab, axis deer did not consume the provided supplementary food such as hay (Krapinec 2002a).

Based on a study on the island of Rab, axis deer change their preference for plant species over the seasons but mainly feed on shoots from stumps (Krapinec 2002b). During winter, they prefer young shoots of butcher's broom *Ruscus aculeatus*, Spanish broom *Spartium junceum*, wild madder *Rubia peregrina*, and green briar *Smilax aspera*. During spring, the most preferred species was the tree heath *Erica arborea*, while in summer and autumn the most preferred plant species were laurustinus *Viburnum tinus* and black nightshade *Solanum nigrum*.

Behavior

In Brijuni National Park, axis deer live in herds of 10–30 individuals (Šprem et al. 2008). A similar herd size was observed in Whipsnade Park in southern England (Loudon and Curlewis 1988). Herds usually comprise does with fawns, while bucks gather in smaller groups after casting the antlers. Sometimes a young buck (2–3 years of age) can join the female herds. Females within herds are closely related, while relatedness in male groups is low (Šprem et al. 2008). On the island of Rab, axis deer were observed in smaller groups (up to five head) than on the Brijuni Islands due to lower population density (Krapinec 2002a). Herd-living behavior in axis deer is also characteristic in other areas (Choudhury 1966; Miura 1981; Khan and Vohra 1992). After cleaning the antlers, bucks approach the female herd in pursuit of does in estrus (Shankar Raman 1998; Govindhaswamy et al. 2007). Axis deer are very shy and mostly active in the early morning, in the evening, and at night (Krapinec 2002a; Šprem et al. 2008). During the day, they usually rest hidden in the shadows of forests and are only rarely found in open areas. The activity of axis deer reduces in winter and during cold days. With low temperatures and strong northerly wind, they often remain near the feeding stations (Krapinec 2001). On Brijuni Island, axis deer are tolerant of the sympatric European mouflon and fallow deer, and no agonistic behavior has been observed so far. However, on the island of Rab, European mouflon show aggression toward axis deer (Krapinec 2002a). It has been observed that tourism may affect axis deer in their daily activity, and at the peak time of tourist presence, the animals avoid open areas and remain in the forest (Šprem et al. 2008).

Axis deer show a distinct bipedal territorial marking behavior called "preaching" (Schaller 1967; Geist 1998, p. 67f.) in which the bucks rise on their hind legs and mark overhanging branches with their preorbital glands as "signposts." During preaching the trampling of the hind legs creates a distinct spot in the soil.

Parasites and Diseases

There are as yet no studies of parasites and diseases in the Croatian axis deer. Limited data from another allochthonous stock of axis deer in Texas showed that they had low parasite load and that all parasite species (nematodes and arthropods) found in axis deer were known from the native sympatric white-tailed deer (Richardson and Demarais 1992). Based on endogenous fat deposits, body condition in the Texan axis deer (and two other co-occurring introduced species, sika *Cervus nippon* and fallow deer) seems to have been physically better than in white-tailed deer (Richardson and Demarais 1992). Meshram et al. (2008) reported helminthic infections from 200 axis deer in their native range in India, and Basso et al. (2014) report on perinatal mortality in captive axis deer due to infection with the protozoan *Neospora caninum*.

Population Ecology

A camera trap survey was performed on the island of Rab to determine the movement activity pattern of axis deer and European mouflon and the temporal overlap between these two ungulate species. The results indicated similar activity patterns and a high overlap coefficient (Centore et al. 2018). Axis deer showed a bimodal activity pattern with increasing and decreasing movement activity during the day, which is a typical behavioral pattern of cervids (Náhlik et al. 2009). The two activity peaks in axis deer occurred in the morning (between 7:00 and 9:00 am) and in the evening (between 5:00 and 11:00 pm), while activity at midday was avoided throughout the year, but with seasonal variations. During spring and summer, the activity of axis deer increased and was fairly evenly distributed over the day, while in autumn and winter two marked peaks of main activity were found (Centore et al. 2018).

Conservation Status

Axis deer are classified as Least Concern in the IUCN Red List, and they are not listed by CITES. In the Croatian hunting legislation, axis deer are listed as game animals with a regulated hunting season (Official Gazette 2018).

Management

A certain number of axis deer (up to 20) are removed each year from Brijuni National Park for population control. Two main methods are used to reduce population numbers in the park: hunting and capture of live animals and their subsequent sale or transport to other habitats. Both methods are conducted by specialized park staff in accordance with the game breeding program and the game protection program. Hunting is done with hunting rifles, and the venison is offered on the menus of tourist resorts within the park (Grignolio et al. 2014). For the capture of live animals, specially adapted stalls are used. Depending on the requirements of the buyer in terms of sex and age of the live animals, selection and categorization are done, and animals are prepared for further transport. In addition to the regular removal of animals, there is often the need for culling in the case of problematic animals. Problems primarily arise due to the large number of tourists and constant contact with humans, when the animals change their behavior and become habituated to people. Such "tame" animals approach and disturb people during their regular visits to the park, on beaches, in restaurants, and on the golf course. There have been cases where animals approached open restaurants and ate the salad from the tables. Brijuni National Park is a unique example of a strictly protected area where game management (i.e., hunting, feeding) is permitted in Croatia.

Contrary to Brijuni National Park, the aim of game management on the islands of Rab and Dugi Otok is to completely eradicate the axis deer populations by hunting in accordance with the Croatian Hunting Act (Official Gazette 2018). Bucks may only be hunted when they have clean antlers, does only when they are not close to giving birth late in the gestation period, while fawns are always protected and may never be shot (Official Gazette 2017).

With regard to forest regeneration, axis deer pose less of a risk than European mouflon, but they still damage both terminal and lateral shoots. Increase in shoot diameter was directly proportional to shoot length and to the length of the bitten part of the shoot (Krapinec et al. 2000).

Future Challenges for Research and Management

Future genetic studies should shed light on the genetic structure of Croatian axis deer and potential signatures of the small number of founding animals ("founder effect"). Comparison with genetic data from animals in the native range (which is also still largely missing) is required to identify the geographic origin of the introduced deer. Further research should also be conducted on the ecological impact of the non-native axis deer on the island ecosystems.

Acknowledgments The authors would like to express their gratitude to Damir Ugarković, Stefano Mattioli, and Luca Corlatti for suggestions and improvement of this chapter, and to Igor Ilić–Serval for providing the photographs.

References

Abbas G, Nadeem A, Babar ME, Hussain T, Tahir MS, Shehzad W, Aslam N, Tayyab M, Javed M (2016) Sequence diversity and phylogenetic analysis in Pakistani spotted deer (*Axis axis*). Pak J Agric Sci 53:991–998

Asher GW, Gallagher DS, Tate ML, Tedford C (1999) Hybridization between sika deer (*Cervus nippon*) and axis deer (*Axis axis*). J Hered 90:236–240

Basso W, Moré G, Quiroga MA, Balducchi D, Schares G, Venturini MC (2014) *Neospora caninum* is a cause of perinatal mortality in axis deer (*Axis axis*). Vet Parasitol 199:255–258

Bojović D (1987) Naseljavanje jelena axisa. Enciklopedija lova, Beograd, pp 375–376. (in Serbian)

Centore L, Ugarković D, Scaravelli D, Safner T, Pandurić K, Šprem N (2018) Locomotor activity patterns of two recently introduced non-native ungulate species in a Mediterranean habitat. Folia Zool 67

Chapple RS, English AW, Mulley RC (1993) Characteristics of the oestrous cycle and duration of gestation in chital (*Axis axis*) hinds. J Reprod Fertil 98:23–26

Choudhury KCR (1966) Behaviour of chital (*Axis axis* Erxleben). J Bombay Nat Hist Soc 63:747

Duckworth JW, Kumar NS, Anwarul Islam M et al (2015) *Axis axis*. The IUCN Red List of Threatened Species 2015:e.T41783A22158006. Retrieved 03 May 2018

Frković A (2004) Axis deer (*Axis axis* H. Smith). In: Mustapić Z (ed) Lovstvo. Hrvatski lovački savez, Zagreb, pp 61–62. (in Croatian)

Frković A (2014) Introduction of the axis deer (*Axis axis* Erxleben 1977) to the Croatian littoral in 1953. Sumar List 9–10:489–498. (in Croatian with English abstract)

Geist V (1998) Deer of the world. Their evolution, behavior, and ecology. Stackpole Books, Mechanicsburg

Gilbert C, Ropiquet A, Hassanin A (2006) Mitochondrial and nuclear phylogenies of Cervidae (Mammalia, Ruminantia): systematics, morphology, and biogeography. Mol Phylogenet Evol 40:101–117

Govindhaswamy U, Sadanand DS, Reddy S, Shivaji S (2007) Seasonal variations in semen characteristics, semen cryopreservation, estrus synchronization, and successful artificial insemination in the spotted deer (*Axis axis*). Therigenology 67:1371–1378

Gray AP (1954) Mammalian hybrids. A check-list with bibliography. Commonwealth Agricultural Bureaux, Farnham Royal

Grignolio S, Heurich M, Šprem N, Apollonio M (2014) The management of ungulates in protected areas. In: Putman R, Apollonio M (eds) Behaviour and management of European ungulates. Dunbeath, Whittles Publishing, pp 178–191

Hassanin A, Delsuc F, Ropiquet A, Hammer C, Jansen van Vuuren B, Matthee C, Ruiz-Garcia M, Catzeflis F, Areskoug V, Thanh Nguyen T, Couloux A (2012) Pattern and timing of diversification of Cetartiodactyla (Mammalia, Laurasiatheria), as revealed by a comprehensive analysis of mitochondrial genomes. C R Biol 335:32–50

Heckeberg NS, Erpenbeck D, Wörheide G, Rössner GE (2016) Systematic relationships of five newly sequenced cervid species. PeerJ 4:e2307

Janicki Z, Slavica A, Konjević D, Severin K (2007) Game zoology. Faculty of Veterinarian Medicine, University of Zagreb, Zagreb, 216 p. (in Croatian)

Khan JA, Vohra U (1992) Group size and group composition of chital (*Axis axis*) in Gir, Gujarat, India. Mammalia 56:662–665

Khongcharoensuk H, Tanomtong A, Patawang I, Supanuam P, Sornnok S, Pinthing K (2017) Karyotype and ideogram of the axis deer (*Axis axis*, Cervidae) by conventional staining, GTG-, high-resolution GTG-, and Ag-NOR-bandind techniques. Cytologia 82:91–98

Kirkwood JK, Gaskin CD, Markham J (1987) Perinatal mortality and season of birth in captive wild ungulates. Vet Rec 120:386–390

Kolić E (1990) National Park "Brioni" fauna with reference to the study of fallow deer. Graduate thesis, Faculty of Veterinary Medicine, University of Zagreb. (in Croatian)

Krapinec K (2001) Feeding structure of mouflon (*Ovis ammon* L.) and axis deer (*Axis axis* Erx.) on the island of Rab. Master thesis, Faculty of Forestry, University of Zagreb, 107 p. (in Croatian)

Krapinec K (2002a) The results of mouflon (*Ovis ammon musimon* Pall.) and axis deer (*Axis axis* Erx.) interaction with cultivated grasslands and the Jerusalem artichoke (*Helianthus tuberosus* L.) plantation in the Kalifront Hunting ground on the island of Rab. Glas Šum Pokuse 39:1–41

Krapinec K (2002b) Plant preference by mouflon (*Ovis ammon musimon* Pal.) and axis deer (*Axis axis* Erx.) in the forest community of holm oak and manna ash (*Fraxino orni – Quercetum ilicis* H-ić/1956/1958). Glas Šum Pokuse 39:67–102

Krapinec K, Vukelić J, Grubešić M (2000) A contribution to insights into green olive tree (*Phillyrea latifolia* L.) browsing by wild ruminants on the Island of Rab. Sumar List 5–6:285–292. (in Croatian with English abstract)

Kusak J, Krapinec K (2010) Ungulate and their management in Croatia. In: Apollonio M, Andersen R, Putman R (eds) European ungulates and their management in the 21st century. Cambridge University Press, Cambridge, pp 527–539

Loudon ASI, Curlewis JD (1988) Cycles of antler and testicular growth in an a seasonal tropical deer (*Axis axis*). J Reprod Fertil 83:729–738

Mattioli S (2011) Family Cervidae (deer). In: Wilson DE, Mittermeier RA (eds) Handbook of the mammals of the world. Vol. 2. Hoofed mammals. Lynx Edicions, Barcelona, pp 350–443

Mennecart B, DeMiguel D, Bibi F, Rössner GE, Métais G, Neenan JM, Wang S, Schulz G, Müller B, Costeur L (2017) Bony labyrinth morphology clarifies the origin and evolution of deer. Sci Rep 7:13176

Meshram MD, Shirale SY, Khillare KP (2008) Incidence of helminthic infection in axis deer. Vet World 1:10

Miura S (1981) Social behavior of the axis deer during the dry season in Guindy sanctuary, Madras. J Bombay Nat Hist Soc 78:125–138

Náhlik A, Sándor G, Tari T, Király G (2009) Space use and activity patterns of red deer in a highly forested and in a patchy forest agricultural habitat. Acta Silv Lign Hung 5:109–118

Official Gazette, 44/17 (2017) Ordinance on hunting season

Official Gazette, 99/18 (2018) Hunting Act

Price SA, Bininda-Emonds ORP, Gittleman JL (2005) A complete phylogeny of the whales, dolphins and even-toed hoofed mammals (Cetartiodactyla). Biol Rev 80:445–473

Richardson ML, Demarais S (1992) Parasites and condition of coexisting populations of white-tailed and exotic deer in South-Central Texas. J Wildl Dis 28:485–489

Schaller GB (1967) The deer and the tiger: a study of wildlife in India. Chicago University Press, Chicago

Shankar Raman TR (1998) Antler cycles and breeding seasonality of the chital (*Axis axis* Erxleben) in southern India. J Bombay Nat Hist Soc 95:377–391

Španjol Ž, Skorup V (2007) Game management of the island of Rab. Hunting association "Kunić" Rab, 354 p. (in Croatian)

Šprem N (2015) Game management plan for state hunting ground no. XIII/4 – "Dugi Otok" for the period from 01 April 2015 to 31 March 2020. Faculty of Agriculture, Zagreb, 55p. (in Croatian)

Šprem N, Blažina D, Florijančić T, Treer T, Prekalj G (2008) The axis deer (*Axis axis*) in Brijuni National Park. J Cent Eur Agric 9:317–322. (in Croatian with English abstract)

Šprem N, Blažina D, Florijančić T, Treer T, Prekalj G (2009) Antler characteristics of the axis deer (*Axis axis*) in the Brijuni National Park. In: Book of abstracts, 44rd Croatian & 4rd international symposium on agriculture, 16–20 February, pp 209–210

Tomljanović K (2016) Game management plan for state hunting ground no. VIII/6 "Kalifront" for the period from 01 April 2016 to 31 March 2026. Faculty of Forestry, Zagreb, 193p. (in Croatian)

Ugarković NK, Ugarković D (2013) Chemical and fatty acid composition of male mouflon (*Ovis ammon musimon* Pal.) meat. Eur J Wildl Res 59:469–475

European Roe Deer *Capreolus capreolus* (Linnaeus, 1758)

7

Rita Lorenzini, Mark Hewison, Jean-Michel Gaillard, Luisa Garofalo, Luca Rossi, Nicolas Morellet, Hélène Verheyden, Sandro Lovari, Adrian M. Lister, and Stefano Mattioli

Contents

R. Lorenzini (✉)
Istituto Zooprofilattico Sperimentale del Lazio e della Toscana M. Aleandri, Centro di Referenza Nazionale per la Medicina Forense Veterinaria, Grosseto, Italy
e-mail: rita.lorenzini@izslt.it

M. Hewison · N. Morellet · H. Verheyden
CEFS, Université de Toulouse, INRA, Castanet Tolosan, France
e-mail: Mark.Hewison@inra.fr; Nicolas.Morellet@inra.fr; Helene.Verheyden@inra.fr

J.-M. Gaillard
CNRS, Laboratoire de Biométrie et Biologie Evolutive, UMR5558, Université de Lyon, Université Lyon 1, Villeurbanne, France
e-mail: jean-michel.gaillard@univ-lyon1.fr

L. Garofalo
Istituto Zooprofilattico Sperimentale del Lazio e della Toscana M. Aleandri, Rome, Italy
e-mail: luisa.garofalo@izslt.it

L. Rossi
Department of Veterinary Sciences, University of Turin, Turin, Italy
e-mail: luca.rossi@unito.it

S. Lovari · S. Mattioli
Research Unit of Behavioural Ecology, Ethology and Wildlife Management, Department of Life Sciences, University of Siena, Siena, Italy
e-mail: sandro.lovari@gmail.com; steven.mattioli@libero.it

A. M. Lister
Natural History Museum, London, UK
e-mail: a.lister@nhm.ac.uk

© Springer Nature Switzerland AG 2022

L. Corlatti, F. E. Zachos (eds.), *Terrestrial Cetartiodactyla*, Handbook of the Mammals of Europe, https://doi.org/10.1007/978-3-030-24475-0_25

Common Names

English	European roe deer
German	Europäisches Reh
French	Chevreuil européen
Spanish	Corzo europeo
Italian	Capriolo europeo
Russian	Европейская косуля

Taxonomy, Systematics and Paleontology

In the family Cervidae, the roe deer *Capreolus* spp. is placed within the clade of Old World tribes of Odocoileinae (Alceini, Capreolini, and Hydropotini), which may have originated from the early Late Miocene, about 10 mya, up to the late Middle

Miocene, some 14 mya. In line with fossil evidence, molecular data suggest that the lineage leading to the ancestor of *Capreolus* (genus *Procapreolus*) diverged from its sister taxon *Hydropotes inermis* (Chinese water deer) around 11–10 mya, in the Late Miocene (Randi et al. 1998a).

All known fossil roe deer are distributed, like the modern species, in temperate latitudes of Eurasia (Lister et al. 1998). Early species of roe deer are placed within the genus *Procapreolus.* Nine or so fossil species of *Procapreolus* have been named, six of them present in Europe, ranging from around 10–2.5 mya (Late Miocene to Early Pleistocene), and distributed from China to Britain and as far south as the Caucasus (Valli 2010; Croitor 2018). In body mass, they were broadly similar to modern roe (*c.* 25–30 kg), though some species were a little larger (estimated up to 50 kg). The antlers of *Procapreolus* show the general capreoline form with three tines high above the base. Their form varied between species but in general differed from those of modern *Capreolus* in being slenderer (up to 40 cm long), often somewhat laterally flattened, a little more divergent, and lacking the strong pearling, especially near the base, that characterizes modern roe. Those fossil species for which the cranium is preserved show, in contrast to modern roe, strongly emergent upper canines (up to 50 mm long) and enlarged fossae housing the preorbital glands. The metapodials (distal limb elements) were relatively somewhat shorter than in modern roe.

The latest species of *Procapreolus*, *P. cusanus*, is known only from France, Italy, and maybe England and lived between around 3.0–2.5 mya. It appears to have overlapped in time with the earliest *Capreolus*, *C. constantini,* known from the Late Pliocene of the Trans-Baikal (Vislobokova et al. 1995). It seems very likely that *Capreolus* arose from within *Procapreolus* (rendering the latter genus paraphyletic), but the precise relationships are unclear. A nother species, *C. cusanoides*, has been described from deposits around a million years old in central Germany; this form had reverted to weak antler pearling and flattened beam reminiscent of *Procapreolus* (Kahlke 2001).

Fossils referred to the modern species, *C. capreolus,* are known from many sites in Europe starting around 700 kya (Lister et al. 2010). In the sequence of glacial and interglacial stages since that time, roe deer extended their range into northern Europe during forested interglacials and were generally absent there in the colder, mostly open-habitat cold stages. Studies of dental wear indicate a consistent tree/shrub browsing diet with a minimal herb or grass component (Rivals and Lister 2016; Saarinen et al. 2016). The pattern of distributional change has been mapped in most detail for the last glacial/interglacial cycle, based on fossils from almost 3000 sites across Europe (Sommer et al. 2009). In the early part of the last glacial period, between about 60–21 kya, finds are mostly concentrated in the Mediterranean peninsulae and up to central latitudes of Europe, none extending beyond 50 N. From the maximum of the last glaciation at 21 kya, until around 14.5 kya, the range was compressed further south to around 45 N except for isolated records from the Carpathian Mountains. With subsequent warming and northward spread of forests, roe deer expanded from these refugia and by around 14 kya had reached the latitude of Britain. The range contracted southward again during the Younger Dryas cold interval (12.8–11.7 kya) and then rapidly expanded in the early postglacial, reaching northern England, southern Sweden, and the Baltic states by around 11 kya.

Both traditional systematics and recent molecular-based phylogenies recognize two living species, the European roe deer *C. capreolus* and the larger Siberian roe deer *C. pygargus* Pallas 1771. Since their divergence some 2–3 mya (Randi et al. 1998b), they have probably lived in allopatry for most of their evolutionary history, due to climatic barriers during glaciations. In prehistorical times, however, because of alternating contractions and expansions of their overlapping areas, they must have come into contact more than once. Although possible in captivity, interbreeding between European and Siberian roe deer has never been documented in the wild, so that the

Fig. 1 Male and female roe deer (photograph by Andrea Dal Pian)

formation of successful hybrid populations in the suture zones might have been prevented by high levels of reproductive isolation developed during and after speciation.

Traditionally, the European roe deer (Fig. 1) is distributed in the western Palearctic, from Europe to the Caucasus and Near East, while the Siberian roe deer extends from the Pacific coast to eastern Europe. Their current contact zone appears to lie in far eastern Europe, in a narrow range between the rivers Volga and lower Don in Russia. Recent DNA-based studies (Lorenzini et al. 2014; Matosiuk et al. 2014; Olano-Marin et al. 2014) have documented the substantial presence of Siberian mitochondrial lineages in the natural area of *C. capreolus*, for example, the Baltic coast (Poland, Lithuania), Hungary, Ukraine, much farther west than previously thought for the westernmost limit of the *C. pygargus* distribution (Fig. 2). A plausible explanation is the occurrence of Pleistocene/Holocene migration waves from the east that might have naturally expanded the Siberian roe deer westward, into the alleged range of *C. capreolus*, due to environment-dependent factors, with null or minimal impact of anthropogenic changes (Lorenzini et al. 2014). There remain uncertainties as to whether there had been introgression through hybridization during population expansion(s), or whether the two species simply coexist(ed) in the same areas.

In *C. capreolus,* the diploid number of chromosomes ($2n$) is constantly 70 (68 acrocentric autosomes, the Y chromosome, and a submetacentric X chromosome), while *C. pygargus* has $2n = 71$–84, due to the presence of additional heterochromatic B-chromosomes that clinally increase (1 to14) from west to east in the distribution range, varying in number in different tissues of the same individual (mosaicism) and in different individuals from the same population. This feature, together with different body size, could act as a strong reproductive barrier between the two species.

The nominate subspecies *C. c. capreolus* occurs in most of the species range. However, the existence of different southern forms of roe deer in Italy and Iberia was observed due to some peculiar

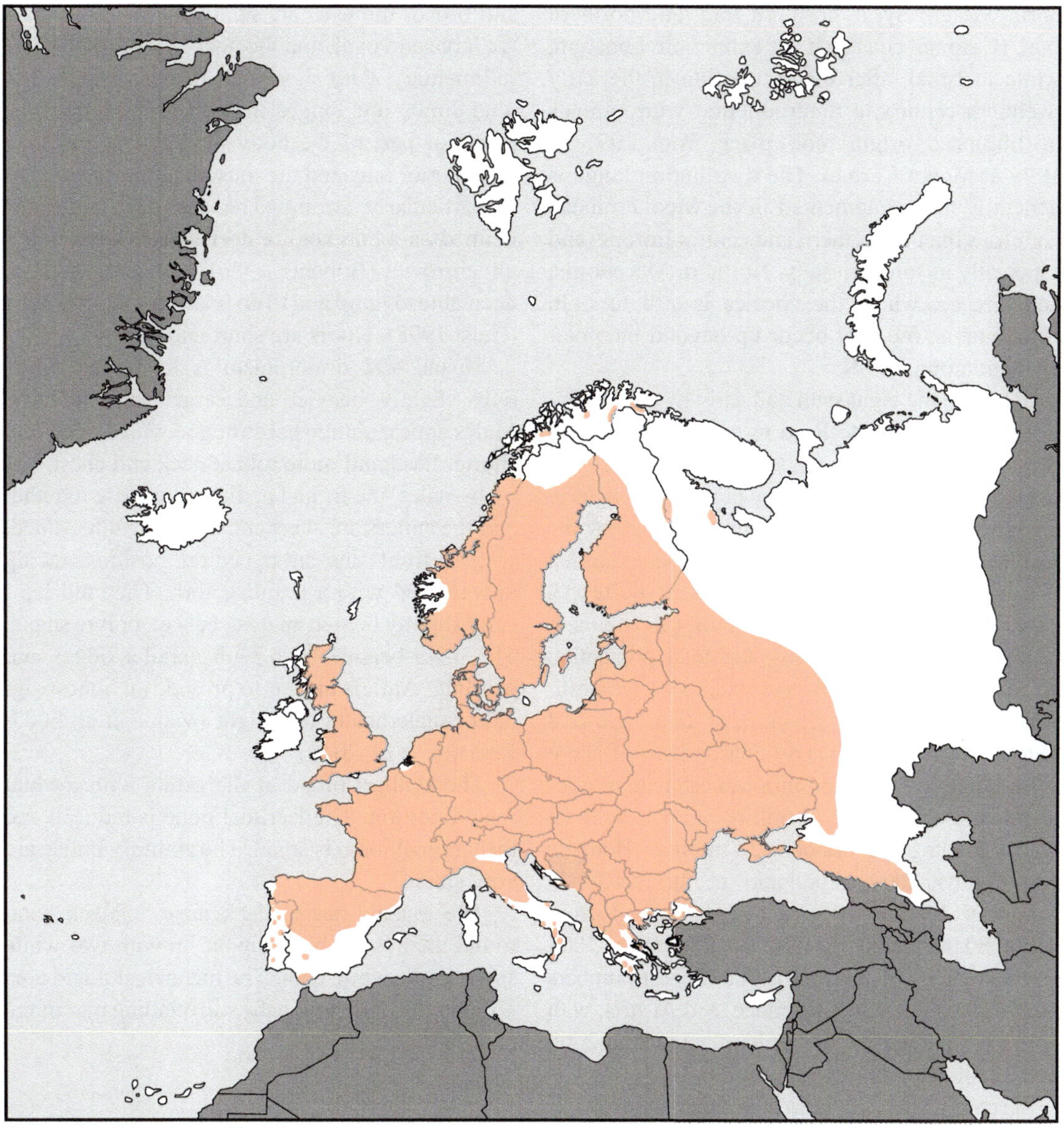

Fig. 2 Current distribution range of *Capreolus capreolus*. (Distributiosn is based on the IUCN Red List of Threatened Species. Version 2019. Map template: © Copyright Getty Images/iStockphoto)

morphological (e.g., skull shape and color pattern of the coat) and behavioral (e.g. food preferences) traits, consequences of adaptation to the diet in Mediterranean environments. Their subspecific status, later confirmed by molecular studies, was suggested as *C. capreolus italicus* Festa 1925 and *C. c. garganta* Meunier 1983, respectively (see the Genetics heading for details).

Current Distribution

In Europe, the roe deer occurs in most of the continent, excluding Iceland, Ireland, and the Mediterranean islands, from Great Britain and the Iberian Peninsula to western Russia (Fig. 2). It also occurs in the Caucasus, northern Turkey, north-western Syria, northern Iraq, and northern Iran (Caspian coast). It is extinct in Lebanon, while in Israel, after the extirpation in the early twentieth century, a reintroduction with animals of European origin took place from 1997 to 1999 at Mount Carmel. The distribution range is generally more fragmented in the Mediterranean countries than in northern and central Europe, and especially in southern Italy, Northern Macedonia, and Greece, where the species is still rare. In Scandinavia, roe deer occur up beyond the polar circle at around 70 N.

Between the eighteenth and early twentieth centuries, the roe deer declined in range and numbers due to over-exploitation and habitat loss. Strict regulations in hunting and forestry practice, changes in land use due to the abandonment of rural areas, the scarcity of natural predators, and reintroductions led to a general recovery of roe deer in most of Europe, especially from the second half of the twentieth century. In the early 1980s, the total population size in the continent (Russia excluded) was estimated at around 6 million head (Burbaite and Csányi 2009), with a yearly total harvest of about 1.7 million. In 2005 the European estimate of population grew to around 10 million animals, with an annual hunting bag of about 2.7 million (Burbaite and Csányi 2009; Apollonio et al. 2010). In European Russia, the roe deer increased from about 80,000 head in 1990 to 98,000 in 2007 (Burbaite and Csányi 2009). The highest numbers are recorded in Germany, France, and Austria, with an overall harvest of 1.87 million animals in 2005 and 2 million in 2015.

According to a review at a continental scale, population densities in 72 localities of 19 countries from the UK and Spain to Belarus and Russia range between 0.1 and 54 individuals/km^2, with an average of 14.5 roe deer/km^2 (Melis et al. 2009).

Description

General Appearance

The European roe deer is a small-sized deer, with a slender build, a relatively short trunk, a rounded rump, and long legs. The head is relatively short and broad, the eyes are relatively large, the ears are large and oval, and the neck is long. The tail is rudimentary, very short, and inconspicuous. The hind limbs are longer than the fore limbs, the posterior part of the body is slightly taller than the anterior one, and the musculation of haunches is particularly strong. The roe deer has been defined as a "duiker," a deer able to dive in the undergrowth (Bubenik 1986), and a "saltatorial" deer, able to jump and to do relatively short sprints (Geist 1998). Hoofs are short and narrow.

Sexual size dimorphism is low, with males only slightly heavier and larger than females. Males appear somewhat stronger, with a relatively shorter head and more robust neck and chest.

In males, the frontal pedicles are close together and the antlers are short and normally three-tined, with a front tine in a central position facing upward and with a terminal fork. They are typically slightly bent in an oval, basket, or lyre shape. The main beam is rich with parallel ridges and pearling. Antlers appear to provide an honest signal of male health and vigor (Vanpé et al. 2007; Lemaître et al. 2018).

The skull is broad at the orbits with a blunt rostral portion. The lacrimal bone is reduced and the lacrimal fossa is small. The auditory bullae are very small.

The muzzle has a black nose, a black band round the nose, a black upper lip with two white spots, and a white chin. The metatarsal gland area is distinctly darker than the surrounding part of the hind foot.

Measurements

Mean whole body mass of adult males ranges from around 18 to 32 kg (pre- and postrut figures pooled, Table 1). The heaviest recorded bucks can reach 40–50 kg prerut mass in June–early July and 34–43 kg postrut mass in late August–early September or winter. This suggests an average loss of mass due to the rut of 7.5–11% (Mattioli and Spada 2009; Apollonio et al. 2020). Mean body mass of adult females normally ranges from 17 to 30 kg (Table 1). Adult males are generally only 5–10% heavier than females. Eviscerated

Table 1 Mean whole body mass of adult male and female roe deer (in kg). For males, mostly pre- and postrut summer mass pooled; for females, mostly autumn and winter mass

	Males	Females	Source
SW Spain	19.8	17.1	Aragón et al. 1995
Bavarian Forest N. P. (Germany)	20.4	20.4	Wotschikowsky 1981
Passiria Valley, E Alps (Italy)	21.2	21.2	Wotschikowsky and Schwab 1994
S Bavaria, lowland (Germany)	22.2	21.7	Lang, unpubl.
Hackel, Saxony (Germany)	23.3	21.6	Stubbe 1990
Dourdan, N France	23.4	21.4	Vincent et al. 1995
Great Britain (United Kingdom)	23.9	22.3	Staines and Ratcliffe 1991
Central Spain	23.9	22.9	Horcajada-Sánchez and Barja 2016
Aurignac, SW France	24.0	22.4	Hewison et al. 2009
Slovakia	24.0	22.6	Danilkin 1999
Massa Carrara, Central Italy	24.6	21.7	Orlandi et al. 2006
Dorset (United Kingdom)	24.6	23.1	Gill 1994
Chizé, W France	23.8	22.8	Douhard et al. 2017
Pistoia, C Italy	25.4	22.9	Visani 2009
N Spain	25.5	23.2	Costa 1992
Grosseto, Central Italy	25.8	22.4	Bruno, unpubl.
S England (United Kingdom)	26.0	24.0	Prior 1968
Trois Fontaines, E France	25.6	24.8	Douhard et al. 2017
Arezzo, Central Italy	26.2	25.3	Mattioli et al. 1995
Bulgaria	26.3	23.9	Danilkin 1999
W Alps, Piedmont (Italy)	26.6	25.1	Meneguz and Tizzani, unpubl.
Serbia	27.3	25.5	Milošević-Zlatanović 2001
Bologna, N Italy	27.8	25.1	Mattioli and Spada 2009
Estonia	28.4	26.7	Danilkin 1999
Bogesund (Sweden)	28.8	27.0	Cederlund and Liberg 1995
Lithuania, woods	29.2	26.1	Bluzma 1976
Storfosna (Norway)	29.4	28.8	Andersen et al. 1995
Grimsö (Sweden)	29.9	27.9	Cederlund and Liberg 1995
Sweden	30.7	27.1	von Essen 1983
Lithuania, fields	31.0	27.7	Petelis and Brazaitis 2003

body mass with head (without thoracic and abdominal organs) represents around 79–80% of live mass in adult bucks and does (Mattioli 2003). Mean head-trunk length ranges from 105 to 120 cm, height at wither around 68 to 80 cm, and hind foot length from 35 to 40 cm. Mean length of the mandible ranges from around 145 to 167 mm, the mean condylo-basal length of the skull from 177 to 196 mm in adult males and 175 to 191 mm in females (cf. Danilkin and Hewison 1996).

Re-analyzing data from Aragon et al. (1998), and taking into account 21 craniometric measures of eight different European populations, skulls are on average 2.5% larger in adult males than in females (0.4–4%, according to the population). In the northern Apennines, mandibles are on average 1.1% longer in adult males than in females (Mattioli 2003).

Mean antler length of adult males ranges from 16 to 22 cm (Table 2), with the longest recorded antlers reaching 30–34 cm (Csányi, pers. comm.). Antler investment is relatively low, with a mean production of around 100–220 g of bony tissue for males weighing after the rut on average 23–28 kg (260–440 g for both antlers and the whole skull). Net antler mass (dry mass of both antlers without skull) represents on average 0.5–0.7% of body mass (almost four times less than an average red deer). Antler mass is correlated with body mass (Stubbe 1990) and linked to habitat quality. Since

Table 2 Mean antler length in adult roe deer from different areas of Europe

	Mean (cm)	n	Source
Jeseniky Mountains (Czechia)	15.6	180	Zejda and Koubek 1988
E Germany	16.4	569	Stubbe 1966
Amsterdam (Netherland)	17.2	305	Pélabon and van Breukelen 1998
Chizé, W France	17.3	171	Lemaître et al. 2018
E Alps, Belluno (Italy)	18.2	799	Ramanzin and Sturaro, unpubl.
Cracow region (Poland)	18.2	401	Wajdzik et al. 2007
Andalusia (Spain)	18.6	137	Caro et al. 2008
Lublin region (Poland)	18.7	5781	Dziedzic 1991
Bogesund (Sweden)	19.0	201	Kjellander, unpubl.
Serbia	19.0	625	Milošević-Zlatanović 2001
C Spain (Spain)	19.4	86	Horcajada-Sánchez and Barja 2016
Cantabrian Range (Spain)	19.4	206	Caro et al. 2008
S Moravia (Czechia)	19.5	642	Zejda and Koubek 1988
Lithuania	19.8	277	Sabalinkiene et al. 2017
Carniola (Slovenia)	19.8	182	Hafner 2004
N Apennine, Massa Carrara (Italy)	19.9	288	Orlandi et al. 2006
N Apennine, Pistoia (Italy)	20.0	624	Visani 2009
Trois Fontaines, E France	20.0	26	Lemaître et al. 2018
Grimsö (Sweden)	20.1	58	Kjellander, unpubl.
W Alps, Piedmont (Italy)	20.4	2909	Meneguz and Tizzani, unpubl.
Rába-Kemeneshát, W Hungary	20.5	10,174	Csányi, unpubl.
N Apennine, Arezzo (Italy)	20.6	127	Mattioli et al. 1995
Baranja, Slavonia (Croatia)	21.3	2086	Degmečić et al. 2010
N Apennine, Bologna (Italy)	21.5	4850	Mattioli and Spada 2009
Nagykunság, C E Hungary	21.5	6748	Csányi, unpubl.

antlers grow in winter, annual antler mass can be strongly affected by the harshness of winter weather. Trophy-sized antlers can reach 800–1000 g of gross mass (Csányi and Sorensen, pers. comm.).

Pelage

The summer coat is typically reddish brown to bright orange, paler underneath, and with a small and undefined yellowish rump patch. The dorsal area of the face is darker and unevenly colored, especially in males. Some individuals can have grayish eye rings, some a white rim above the nose, many a brownish or dark grey forehead and nasal area. The winter coat is mainly grey or greyish brown, with a conspicuous whitish rump patch and generally one or two pale patches on the throat. The rump patch is heart-shaped in females and kidney-shaped in males, flared when the animal is alarmed. Females have a vulvar tuft of hair 5–7 cm long, and males have a distinct preputial tuft of hair 10 cm long.

Newborn fawns have a brown coat flecked by many white spots, which begin to fade at about 6 weeks of age and disappear by 5 months with their first molt. Molting occurs around May and October, starting from the head and neck region.

Glands

Metatarsal glands are present on the outer side of the hind legs. The interdigital glands are present in the front and hind feet, but particularly developed in the hind ones (Stubbe 1990). The skin of the forehead, between the pedicles under the eyes, at the base of the ears and possibly of the chin of males has a diffuse glandular system ("frontal organ," as defined by old literature), with many sebaceous and apocrine glands (Adams and

Johnson 1980). Forehead skin glands and interdigital glands are used by adult bucks during the territorial period when rubbing stems and scraping ground. Preorbital glands are missing, a rare condition in deer.

Antlers

Pedicles begin to develop in male fawns at 3–4 months of age, around September, and individuals in good conditions at 4–5 months of age start to grow small bony buttons 0.5–2 cm, which are cleaned of velvet in December and cast in January. A new antler set is produced in February–March when the fawn is 8–9 months old, cleaned in May and cast in November. Kids of low body mass do not develop buttons in their first autumn and will grow their first set of antlers (mainly buttons or spikes 3–5 cm long) only the following spring (Sempéré 1990). Yearling antlers are typically simple spikes 6–15 cm long, but can be also two- and more rarely three-tined antlers. In adults, velvet shedding occurs around March–April and antler casting around October. New antlers begin to grow 7–12 days after the previous casting, with a rapid and regular growth phase of 60 days (with a speed of 2.6–3.6 mm per day), followed by a period of mineralization of 30–45 days (Sempéré 1990). The culmination of antler development, quite variable and unpredictable, in many cases is attained between 4 and 6 years of age (Stubbe 1990). Improved nutrition can bring forward the peak. A decline in antler size can be seen in senescent males 8 years and older (Vanpé et al. 2007).

Dentition

Milk teeth 0.0.3.0/3.1.3.0; permanent dentition 0.0.3.3/3.1.3.3. The permanent bicuspid third premolar typically erupts at around 13 months of age. Eruption patterns can be influenced by body condition (De Marinis et al. 2018). For animals older than 13 months, age can only be estimated by examining the tooth wear of cheek teeth of the mandible. At about 6–7 years of age, the infundibulum of the first molar tends to disappear. The average estimation error is relatively small (± 1 year) but accuracy decreases with age (Hewison et al. 1999; Høye 2006). Sectioning the incisor or the molar for counting annual layers, a reliable method in many species, in roe deer, when tested on known age mandibles, proved to be not sufficiently accurate.

Physiology

In terms of physiology and hormonal cycles, including reproduction and antler cycle, roe deer are quite different from other cervids. The resting metabolic rate is relatively stable, without distinct seasonal rhythms. In the temperate zone, sexes tend to accumulate low body fat reserves. The seasonal differences in body mass and physical condition are normally of modest amplitude (Hewison et al. 1996). Adult bucks tend to have nearly 7.5–10% mass loss during the rut (Mattioli and Spada 2009; Apollonio et al. 2020) and adult does face a higher metabolism per kg of body mass only in late spring and early summer during lactation, but over-winter decline in body mass and fat levels can be negligible (Mauget et al. 2003). Only in the boreal zone, roe deer accumulate significant fat supplies in September–October, which become depleted in early January and can cover only about 20% of winter energy requirements (Holand et al. 1998).

Normally, cervids of the temperate zone give birth in spring and mate in autumn: In males, testosterone is controlled by decreasing daylength, and antlers grow in spring-early summer and clean their velvet in late summer; females are polyestrous and the ovulation is induced by the shortening of the daylength after the autumn equinox. On the contrary, roe deer give birth in spring but mate in summer. Testosterone is stimulated by the lengthening of photoperiod after the winter solstice in late December and antlers grow in winter and clean in spring (Sempéré et al. 1998). Females are monestrous (with a period of receptivity of just 36 h) and the ovulation is induced by the long photoperiod after the summer solstice. Roe females are the only ungulates with an embryonic diapause. The embryonic reactivation, induced by the short winter photoperiod, occurs in late December–early January. Therefore, long photoperiods have opposite effects on males and females, stimulating the sexual cycle in bucks and inhibiting the sexual and ovarian cycles in does. Testosterone increases from February to July and testicular activity is present from March to late August coinciding with territorial defense and mating. Minimum levels in October induce antler casting.

After fertilization in mid-summer, the blastocyst, a 20- to 30-cell organism less than 1 mm in diameter, enters a 5-month diapause. During the last 6 weeks, it grows to a 100-cell stage and prepares for implantation in the uterine endometrium. The true pregnancy lasts around 150 days.

The release of a specific "pregnancy-associated glycoprotein" (PAG) from the trophoblast cells might trigger the implantation by stimulating the mother's hormonal response (Lambert 2005). Mass-specific resting metabolic rate is higher in newborns and young fawns than in mothers, with a maximum in the first week. Due to very rich milk, fawns grow rapidly (linear in the first 6 weeks) (Mauget et al. 1999). Mass-specific metabolism is higher in primiparous females than in fully grown adults, due to the additional costs for body growth (Mauget et al. 2003).

Genetics

For information on karyotype, see above (Taxonomy, Systematics and Paleontology).

Genetic Variability and Phylogeography

Roe deer are widespread in the Palaearctic and have always been a popular game species, undergoing frequent local extinctions, reintroductions, and translocations due to human activities, basically through hunting and habitat modifications. For these reasons, the genetic structure of roe deer populations in Europe has been extensively studied to address questions concerning the level of intraspecific genetic variability and phylogeography in relation both to natural population dispersal, focusing mainly on the Pleistocene, and to anthropogenic impact in the Holocene.

Below species level, the roe deer harbors an overall high genetic diversity, as revealed by biochemical and molecular results (cf. Hartl et al. 1991; Randi et al. 2004; Lorenzini and Lovari 2006). Molecular information, in particular, told us much about historical dynamics and contemporary demography of populations, their connectivity, as well as the uptake of interactions with the environment, within a so-called landscape genetics framework, that is, the integration of landscape ecology and population genetics. In this context, human impact has played, and still plays, an important role in shaping the genetic structure of contemporary roe deer in much of its current distribution. At small geographic scale, anthropogenic land-use changes (habitat fragmentation) and direct impact on populations (overhunting, translocations) can lead to extinction of local gene pools, hybridization with allochthonous forms, and large genetic distance between populations due to high resistance barriers to functional connectivity (e.g., roads) that reduce gene flow (Coulon et al. 2006a). Landscape genetic tools have been recently used to map the fine-scale distribution in Italy of *C. c italicus* populations and locate their admixture zone with reintroduced European *C. c. capreolus* (Mucci et al. 2012). Furthermore, there is evidence for an impact of anthropogenic barriers, like road and motorway density, on genetic divergence, dispersal and persistence of spatially structured roe deer metapopulations in many areas of the distribution (Breyne et al. 2014; Burkart et al. 2016).

Recently, molecular phylogeographic investigations revealed that the European roe deer harbors high genetic variation and a complex population structure across the entire range. Data from different mitochondrial and nuclear DNA-based studies (Randi et al. 2004; Lorenzini and Lovari 2006; Lorenzini et al. 2014) concordantly suggest the existence of four major genetic lineages of roe deer in Europe (Fig. 3). One central lineage of mitochondrial haplotypes is widely distributed across the whole European continent (from northern Iberia to Scandinavia, down to

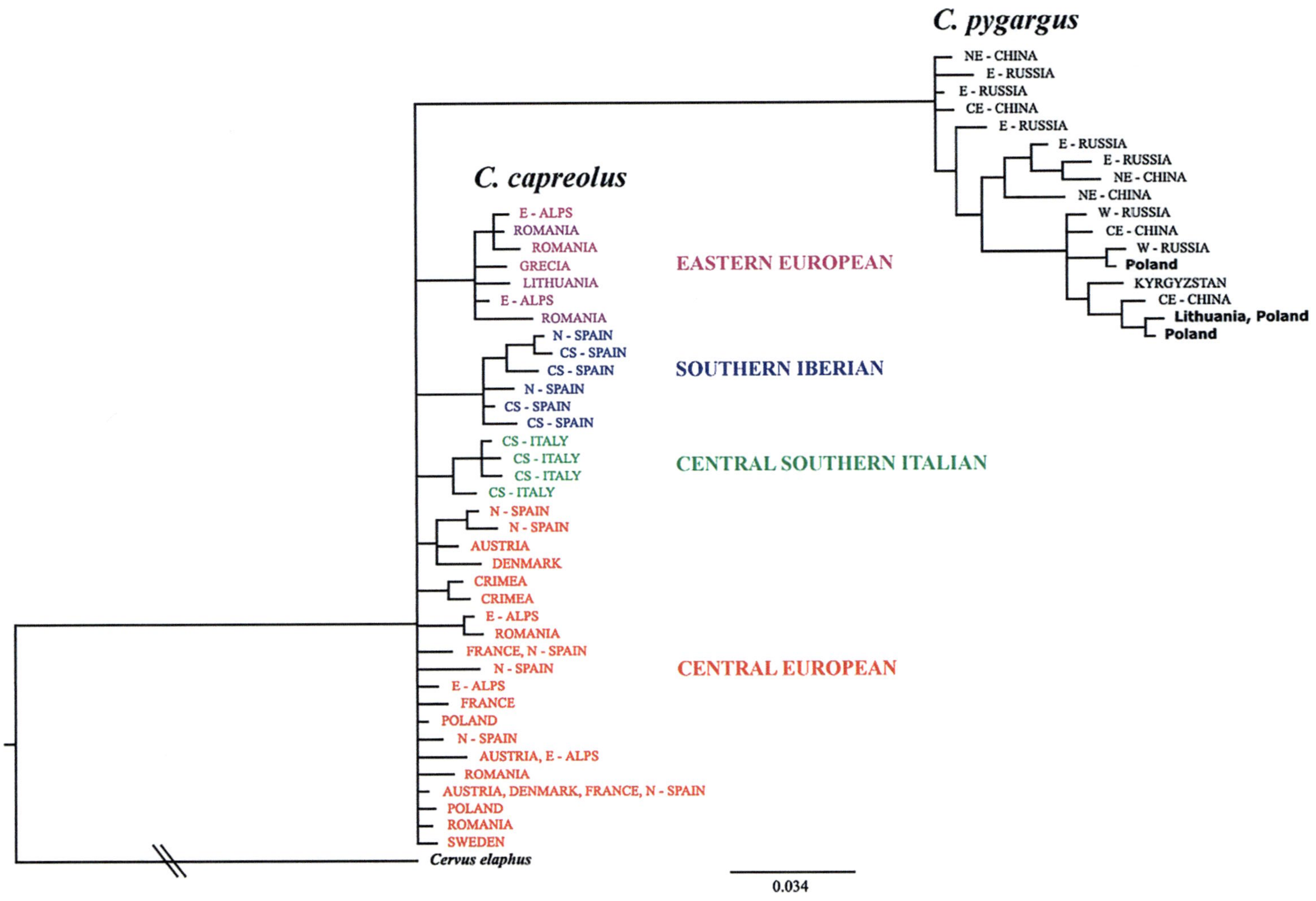

Fig. 3 Phylogeny of *Capreolus* based on mitochondrial DNA haplotypes. Siberian haplotypes found in Europe are indicated in bold within the *C. pygargus* clade. E-Alps: Eastern Alps, N-Spain: Northern Spain; CS-Spain: Central Southern Spain; CS-Italy: Central Southern Italy; NE China: North Eastern China; E-Russia: Eastern Russia; CE-China: Central Eastern China; W-Russia: Western Russia. (Redrawn from source data in Lorenzini et al. (2014))

Germany and central Alps), Ukraine, and Crimea. An Eastern lineage is found in the Balkans up to Lithuania, through the eastern Italian Alps to eastern Europe (Romania, Bulgaria, Slovakia, Poland). A third Southern Iberian (or Western, cf. Randi et al. 2004) lineage comprises the roe deer populations from central-southern Spain and Portugal, that is, the south-western limit of the species' distribution, where the subspecies *C. c. garganta* was proposed (see below). A fourth lineage, or alternatively a subclade within the Central main lineage (cf. Randi et al. 2004), includes roe deer from central southern Italy, supporting the existence of the endemic subspecies *C. c. italicus* (see below).

According to genetic data, the observed phylogeographic patterns have been influenced most strongly by demographic processes due to the climatic dynamics of the Pleistocene. From estimates of coalescent times, the extant mitochondrial lineages predate the Last Glacial Maximum (23–18 kya), and the distribution of this preexisting genetic variability would result from population migrations during the Pleistocene. In this scenario, human interference only minimally affected the natural distribution of variation between roe deer populations. Anthropogenic factors might have played a role only in the last centuries, limited to the local level (Mucci et al. 2012).

The evolutionary history of roe deer, as suggested by molecular evidence, is very similar to what the fossils say. During the glacial stages of the Pleistocene, roe was absent from the northern regions and high altitudes of Europe. In such cold episodes, both fossil remains and genetic data indicate the presence of roe deer in the Mediterranean region (Italy, Iberia, and the Balkans) and several eastern European areas (e.g., the Carpathians), where it existed as refugial populations, along with other temperate species (Lorenzini and Lovari 2006; Sommer et al. 2009). In the interglacials, taking advantage of ice melting and climatic warming, roe moved west- and northward mainly from the eastern refugia, accompanying the spread of woodland, and eventually recolonized central and northern Europe (e.g., Britain and Scandinavia). Movements of populations to the north and their retreat to the south probably occurred several times during the glaciation/interglacial transitions, and this must have considerably affected the genetic architecture of the roe deer in Europe.

According to the traditional models of Quaternary radiations of populations, the Italian and Iberian peninsulas, as well as the Balkans, have been the main glacial refugia for temperate species in the Mediterranean area. Here, the ice sheets of glacial periods probably covered the highest peaks of the mountainous chains, leaving the low altitudes and the southern regions free from ice. The Pyrenees in Iberia, and the central-western Alps in Italy, partly limited the dispersal of populations from southern areas, preventing their massive advance to the north. Thus, the southern regions functioned as an effective refuge, but the roe populations, rather than spreading northward later, remained as geographically independent isolates, failing to act as a main source for postglacial recolonizations, and continued to diverge genetically. Consequently, recolonization routes to northern Europe were probably due to range expansions from one or more refugia in central and eastern Europe (e.g., the Carpathian region), rather than proceeding from the Mediterranean areas. In this scenario, only the Balkans, with mountain chains extending in a north–south direction, appear as a suitable corridor for northward postglacial expansions (but see Sommer et al. 2009).

In the last century, based on morphological observations, *C. c. italicus* and *C. c. garganta* were suggested as different subspecies for the European roe deer of the Mediterranean area (Festa 1925; Meunier 1983). Later, genetic analyses and phylogeographic inferences based on mtDNA diversity (Lorenzini et al. 2002, 2003) revealed high differentiation for the central-southern populations of Iberia and Italy, thus supporting the previous morphological observations. Nuclear data at microsatellite loci further confirmed their genetic uniqueness, explained as a result of prolonged isolation and genetic drift (Randi et al. 2004; Lorenzini and Lovari 2006; Royo et al. 2007). Estimates of divergence time revealed an ancient genetic lineage for the southern Iberian populations, that diverged on average some 92–46 kya, while *C. c. italicus* differentiated more recently, splitting about 17–8 kya from the central European clade (Lorenzini et al. 2014).

Traditionally, the Italian subspecies was confined to the historical populations of the Castelporziano Estate, near Rome, and the Gargano and Pollino National Parks. Recent molecular findings revealed the presence of *italicus* mtDNA haplotypes also in Tuscany and Emilia Romagna, much further north than formerly thought (Lorenzini et al. 2002; Gentile et al. 2009; Mucci et al. 2012; Biosa et al. 2015). Reintroductions and natural expansions from southern populations were responsible for the presence of the subspecies in the northern Apennines, where no roe deer had previously survived. Over the last few hundreds years, however, massive restockings in much of the *italicus* range (e.g., Calabria) for hunting purposes have led to wide zones of sympatry with nonnative roe deer from central and eastern Europe. Currently, they highly threaten the genetic integrity and survival of the Italian subspecies, with great detriment to local biodiversity. The risk of gene flow with allochthonous populations also exists for the *garganta* subspecies in Iberia, where translocations of roe from northern Spain and central Europe are still under debate due to strong hunting pressure.

These two endemic subspecies can be considered full-fledged Evolutionarily Significant Units (ESUs, sensu Moritz 1994) and their genetic structure should be constantly monitored. Furthermore, anthropogenic admixing of roe deer from different gene pools, through further reintroductions and translocations, should be avoided both in Italy and Iberia.

Life History

Roe deer are medium-sized ungulates with a wide geographic range. Originally a forest-dwelling species, roe deer have markedly expanded their distribution in terms of habitat and are currently occurring in most of terrestrial ecosystems, from sea coasts to high mountain ranges. The life history traits and tactic of roe deer facilitates this pronounced ecological plasticity. Roe deer life history is characterized by a high level of iteroparity, a limited intensity of sexual selection, a high female allocation to each reproductive attempt, and low body reserves. Iteroparity describes a tactic of lifetime allocation to reproduction by an individual or a group of individuals (e.g., populations or species) that consists to spread the total lifetime reproductive output throughout several reproductive attempts, leading the reproductive lifetime to match the total lifetime in the most iteroparous species. This is opposed to the allocation tactic of semelparity in which individuals produce the total lifetime reproductive output within a single reproductive event, often after a long maturation period (e.g., bamboos). Iteroparity thus measures the number of reproductive bouts during the lifetime of an organism. The overwhelming prevalence of iteroparity among living organisms (cf. Cole's paradox) is nowadays interpreted by most evolutionary ecologists as an adaptive response to the unpredictability of environmental conditions.

Growth

Roe deer fawns are born between late April and mid-June and births are highly synchronized, with 80% of births typically occurring within about 3 weeks. Median birth date is earliest in lowland forests (mid-May) and latest in the most northern and high elevation areas (early June). Birth dates in a given population are normally distributed. Roe deer fawns are precocial at birth and weigh on average 1.6 kg, with a huge individual variation both within and among populations (e.g., from 0.7 to 2.2 kg at Trois Fontaines, France). They are hiders during most of their first month of life and grow fast (i.e., 100–250 g/day). While they are fully weaned in October, the strict lactation period (sensu Pontier et al. 1989) ends in late July–early August. Roe deer reach about two-thirds of their full mass at the onset of their first winter and their final mass at 3 or 4 years of age. Body mass decreases from 6 to 7 years of age onwards due to senescence (Douhard et al. 2017). Sexual size dimorphism is limited as adult males are only 5–10% heavier than adult females. Birth timing, birth mass, and early growth do not differ between the sexes.

Reproduction

Roe deer females are unique among ungulates in displaying delayed implantation and unique among ungulates in temperate areas in mating during summer. From their third spring onwards, roe deer males establish a territory that they patrol and defend from March to August or September. A very few, typically large, males start defending a territory during their second spring. Rut occurs between mid-July and mid-August, during which most females older than 1 year mate. Females are monestrous and are very active during the rut. Around half of them perform breeding excursions and most mate repeatedly, often with several males, so that sperm competition is likely to be high and multipaternity occurs within litters. A few days after mating, embryonic development stops (at the blastocyst stage) and embryos enter diapause until the winter solstice, when the first embryo implantations take place. Females that go on to give birth in spring will have successfully implanted before mid-January. Implantation failure can be substantial and is more pronounced in old females (i.e., older than 7 years of age; Chirichella et al. 2019), but resorption of implanted embryos is rare. The pregnancy rate is very high, at >0.95 in most populations for does between 3 and 10 years of age. Two-year-old females display similarly high pregnancy rates in populations with abundant resources, but primiparity is delayed by 1 year for up to one third of females in populations with limited resources (i.e., high population density or poor habitat productivity). Pregnancy rates decrease sharply with increasing age from 10 years onwards. The longest-lived females (i.e., 13 years and older) often stop reproducing 1 to 4 years before dying. Most roe deer females that give birth in spring produce between 1 and 3, exceptionally 4–5, newborn fawns (Flajšman et al. 2018a). Litter size increases from 2 to 5–6 years of age and then decreases with increasing age. There is a clear latitudinal cline in litter size, with roe deer females mostly producing 1 or 2 fawns in South Europe, most often 2 around latitude 45°, and 2 or 3 fawns in Scandinavia (Flajšman et al. 2018b). Female reproductive performance is closely related to body mass. Under most environmental conditions, females lighter than 18–20 kg are not pregnant, and litter size increases with female mass. Less is known about male reproductive output. Male roe deer cannot monopolize reproduction of a large number of females, which limits their annual reproductive success. Further studies are required to assess variation in male reproductive success in relation to age, body mass, and environmental conditions.

Survival

Roe deer display the age-specific survival pattern typical of vertebrate species, which includes three distinct stages: a first stage (juvenile) with low and variable survival, an intermediate stage (prime-age) with high and constant survival, and a third stage (senescent) with decreasing survival with increasing age.

The critical stage of roe deer survival is the first summer of life. Depending on the maternal home range (HR) quality, on the abundance of fawn predators such as red fox or lynx, and on climatic conditions during the spring, the yearly summer survival of fawns varies tremendously, typically between 0.20 and 0.80 (Pettorelli et al. 2005). At the individual level, the earlier a fawn is born, the faster is its early growth, and the heavier and longer-lived its mother is, the higher its chance to survive until the onset of the winter (Plard et al. 2014a). Fawns that survive to the onset of the winter have high subsequent survival prospects. Winter fawn survival is typically high, at around 0.85, but in populations with strongly limited resources or in years with particularly severe winters, high fawn mortality occurs. Yearling survival (i.e., survival between 1 and 2 years of age) is high in roe deer except in populations that are intensively hunted or in the presence of large predators (Nilsen et al. 2009). From birth to 2 years of age, survivorship is remarkably similar in male and female roe deer. From 2 years of age to around 8 years of age, yearly survival is high and quite constant in female roe deer, typically being around 0.90 and decreasing to less than 0.80 only in exceptional situations of intense predation, hunting, or exceptionally harsh conditions (e.g., disease outbreaks, climatic events). Due to senescence, yearly survival of females older than 8 years decreases with increasing age, so that females only exceptionally reach 16 years of age or more in the wild. Prime-age survival of males is lower (on average about 0.80–0.85) and more variable among years than that of females. Throughout the adulthood stage (i.e., from 2 years onwards), females outlive males and the between-sex difference in yearly survival consistently increases with age (Gaillard et al. 2004). This means that males reach 12 years or more only exceptionally in the wild. These age-specific survival patterns lead to a balanced sex ratio at recruitment, but an increasingly female-skewed adult sex ratio with increasing age. On average, the adult sex ratio is about 1 male/1.20 females.

Life History Strategy

Roe deer have a life history strategy centered on high female adult survival, high reproductive output, and highly variable early survival. As females do not store body reserves to meet the increased energy expenditure caused by late gestation and early lactation (i.e., income breeder tactic, Andersen et al. 2000), the abundance of high-quality resources in spring in the maternal HR is key to yearly reproductive success (Pettorelli et al. 2005). As roe deer are highly sedentary, a positive autocorrelation of reproductive success occurs throughout a female's lifetime, leading to marked differences in individual fitness (Gimenez and Gaillard 2018). As a general rule, being born early and growing fast generates high survival prospects during early life (Plard et al. 2015), and living in a high-quality HR is key to adult roe deer performance (McLoughlin et al. 2007).

Habitat and Diet

Movement and Space Use

The roe deer was initially described as highly sedentary, with strong spatial fidelity throughout the year (Hewison et al. 1998). However, movement tactics vary over its geographical range and partial migration is common (Cagnacci et al. 2011). Migration propensity increases with the presence of snow, or the presence of snow in interaction with terrain roughness and forest cover (Cagnacci et al. 2011; Peters et al. 2017). Behavioral plasticity in movement tactics is pronounced, with most individuals adopting a multirange tactic, occupying a number of subseasonal functional HRs, particularly when the environment is spatially heterogeneous and/or temporally unpredictable (Couriot et al. 2018). The multirange tactic is widespread, allowing roe deer to track spatio-temporal variation in resource distribution and, thereby, to cope with changes in their local environment (Couriot et al. 2018).

Variations in HR size of roe deer have been widely studied and have provided contrasting conclusions. For example, some authors found a positive link between annual reproductive success and HR size (Saïd et al. 2005), whereas others did not (Bongi et al. 2008). The use of different tracking methodologies, time schedules, HR metrics, or sample size could explain inconsistencies among studies (Borger et al. 2006a). In general, there is no difference in HR size between the sexes, although males may have larger ranges in summer (Kjellander et al. 2004). Range size decreases with the availability of woodland edge (Saïd and Servanty 2005) and increases with increasing landscape openness (Lovari et al. 2017) so that roe deer may occupy a minimum surface area of woodland (Cargnelutti et al. 2002). Home ranges are consistently smaller at high density compared to low density (Kjellander et al. 2004). From a comprehensive analysis of variation in HR size among and within populations across their geographical range, we have now broadly consistent findings across the distributional range of roe deer (Morellet et al. 2013). Among populations, HR size decreases with increasing forage abundance, but increases with increasing seasonality, temperature variation, latitude, and snow cover. Within populations, roe deer HR size is generally greatest in winter and smallest in spring and decreases with increasing forage abundance (Kjellander et al. 2004; Morellet et al. 2013). In a study in Italy, HR size (kernel estimates at 90%) varied from 29.9 to

63.8 ha for biweekly and yearly estimates, respectively, (Borger et al. 2006b). Across populations, HR size varied from 38.6 to 77.4 ha at a weekly scale and from 51.4 to 136.0 ha at a monthly scale (Morellet et al. 2013).

Dispersal

Natal dispersal is a presaturation process in roe deer, highly variable among populations (Wahlström and Liberg 1995a), with no density-dependence (Gaillard et al. 2008) and equally prevalent in both sexes (Debeffe et al. 2012), although females may disperse slightly farther (Coulon et al. 2006b). Dispersal is condition-dependent (Debeffe et al. 2012, 2014b; Hewison et al. 2021), more frequent in more open landscapes (Debeffe et al. 2012), and occurs during spring at around 11 months of age, although some may disperse in the second year (Wahlström and Liberg 1995b). Dispersal may be facilitated by the availability of wooded habitat during transience (Coulon et al. 2004). Rut excursions occur in around half of adult females (Lovari et al. 2008) and may be considered as a form of breeding dispersal (Debeffe et al. 2014a), but otherwise adults of both sexes are highly sedentary over their lifetime (Linnell and Andersen 1998).

Habitat Selection

The roe deer is classically considered a species adapted to wooded habitat (Hewison et al. 1998), with woodland providing important resources like access to cover (Mysterud and Ostbye 1999) and forage, particularly at edges with open areas which provide good feeding resources for browsers (Saïd and Servanty 2005). It is present from the Mediterranean maquis scrub of Andalusia to the boreal forest of Scandinavia. However, roe deer show pronounced behavioral plasticity in habitat selection, as habitat composition, availability, and distribution of resources markedly affect habitat selection (Morellet et al. 2011; Dupke et al. 2017). For example, roe deer strongly select woodlands substituting this habitat by hedgerows when the availability of woodlands decreases, providing roe deer with similar resources (Morellet et al. 2011). This behavioral plasticity has led some authors to define two "ecotypes": forest roe deer and field roe deer. The latter lives year-round in open fields and forms large groups during winter (Bresiński 1982). However, there is no real support for genetic differentiation between these two ecotypes (Kamieniarz et al. 2017). Roe deer often face a trade-off between food and safety in situations where food availability and predation risk peak in the same habitat type (Benhaiem et al. 2008). In particular, roe deer decrease the use of risky habitats during periods of high human disturbance (Padie et al. 2015; Martin et al. 2018). In some parts of its distribution, roe deer are sympatric with large predators such as the lynx, an efficient stalk-and-ambush predator, which can constrain roe deer to spend more time in open areas further away from forest edges in order to reduce the risk of encountering lynx (Gehr et al. 2018, but see Ratikainen et al. 2007). Under additive risk of predation by large carnivores and human hunting, multiple predators may create areas of contrasting risk, as well as increasing the level of risk in a given landscape (Lone et al. 2014). However, there is substantial inter-individual variability in how individuals manage risky situations which impose constraints on how they are able to exploit high-risk habitats, suggesting the existence of a risk management syndrome (Bonnot et al. 2015). Roe deer tend to avoid snow >30–40 cm deep, and snow >50 cm deep is a limiting factor (Baskin and Danell 2003). In the Alps, they reach 2700 m asl in summer and autumn (Büntgen et al. 2017).

Diet Selection

Roe deer are medium-sized herbivores with a moose-type digestive tract, feeding accordingly as a typical browser (Clauss et al. 2010). It selects food with high concentrations of soluble sugars and low fiber content (Tixier et al. 1997; Lechner-Doll et al. 2008), with some tannins (Verheyden-Tixier and Duncan 2000), presumably for their antiparasite properties and benefits for protein assimilation. However, food intake is highly plastic and mainly determined by the habitat in which they live (Cornelis et al. 1999). The roe deer is apparently granivorous or frugivorous when seeds and fruit are sufficiently abundant, but switches to a browser diet when these preferred foods are rare (Tixier and Duncan 1996). For instance, cultivated seeds and acorns are widely consumed when available (Duncan et al. 1998). In agricultural systems, individuals with access to open habitats (Morellet et al. 2011) feed on cultivated plants and have a higher diet quality than the individuals restricted to more wooded habitats (Abbas et al. 2011). Although a typical browser, roe deer may also eat a certain amount of grasses during winter, particularly in open landscapes (Abbas et al. 2013). This dietary flexibility is related to plasticity in gut morphology, with an increase in the size of the distal fermentation chamber compensating for the lower quality of food in winter (Holand 1992) or in very poor habitats (Serrano Ferron et al. 2012). Diet composition varies little with respect to sex, age, or reproductive status (Tixier et al. 1997), but varies markedly among seasons in relation to availability: In woodland habitat, evergreens (bramble, ivy, etc.) dominate in winter, whereas herbs and tree leaves and buds (hawthorn, hornbeam, etc.) are preferred in summer (Cransac et al. 2001).

Roe deer live in sympatry with other wild ungulates, red deer (*Cervus elaphus*), fallow deer (*Dama dama*), wild boar (*Sus scrofa*), moose (*Alces alces*), and mountain ungulates (northern chamois *Rupicapra rupicapra*, southern chamois *R. pyrenaica* and mouflon *Ovis gmelini musimon*). Even though there is substantial diet overlap, the ability of each species to select for different food resources and different bite size may limit potential competition.

Behavior

Foraging Behavior and Activity

Roe deer are small-sized selective feeders with proportionally small rumens (Duncan et al. 1998). Hence, they have relatively short and frequent rumination bouts, approximately every 3 h, but this is highly variable among seasons in relation to forage availability and quality (Cederlund 1989). Activity peaks at dawn and dusk (Pagon et al. 2013), but is also generally higher during night than day, particularly in human-dominated landscapes (Benhaiem et al. 2008; Bonnot et al. 2015). Roe spend around 10–20% of their active time in vigilance (San José et al. 1996) and readily take flight (Bonnot et al. 2017).

Roe deer forage between ground level and around 120 cm, eating between 1.5 and 4 kg of green vegetation per day, depending on season (Duncan et al. 1998). They often move while foraging for dispersed, high quality items, selecting the largest possible bite sizes so that intake rates are limited by food processing (Duncan et al. 1998). Fawns can begin ingesting vegetation at only a few days of age and quickly develop dietary preferences (Tixier et al. 1998).

Social Behavior

Roe deer are typically weakly gregarious and are generally observed alone or in small family groups (Dzięciołowski 1979). However, group size is variable, particularly during winter, and increases with landscape openness (Hewison et al. 2001), cover distribution (San José et al. 1997), and density (Vincent et al. 1995). In intensive agricultural landscapes, they may form groups of up to several hundred animals in winter (Bresiński 1982; Gerard et al. 1995), but associations are often weak and temporary, driven by fusion-fission of family units (Pays et al. 2007).

Roe deer males are highly seasonally territorial, from March to the end of August or the beginning September, maintaining exclusive non-overlapping mating territories (Bramley 1970) which they mark (Johansson and Liberg 1996) and actively defend against rival intruders by display and occasional combat (Hoem et al. 2007). Territory limits often coincide with physical features of the landscape such as paths, roads, habitat edges, or streams (Bramley 1970). Roe deer females are not territorial, but are particularly solitary just prior to and following giving birth (Danilkin and Hewison 1996) with little range overlap, even at high density (Maublanc et al. 2012).

Communication

Roe deer communicate vocally, particularly mother and fawn, but adult males also use barking as a territorial signal, which may have an individual signature (Reby et al. 1999). Barking also serves as an alarm call to a perceived threat, occasionally accompanied by stamping (Danilkin and Hewison 1996). From late winter onwards, adult males mark their territory with olfactory and visual signals, rubbing and fraying young trees and bushes with their antlers and forehead, and scraping the ground to produce bare patches of earth (Johansson et al. 1995). Both these behaviors likely leave scent marks, which may also carry an individual signature (Lawson et al. 2000). Marking occurs throughout the territory and is most intense at the beginning of the territorial period and just prior to the rut (Johansson et al. 1995). Olfactory signals in urine, feces, and from the metatarsal and interdigital glands also likely communicate individual information, but are little studied to date.

Aggression

Male-male aggression is frequent during the territorial season, particularly towards nonterritorial intruding males, but is less common between territorial neighbors once boundaries are established (Danilkin and Hewison 1996). Conflict is often resolved by display and pursuit, while fighting is relatively rare. There appears to be site-specific dominance such that a male generally wins any contest within his territory limits (Hewison et al. 1998; Hoem et al. 2007). In addition, adult males are aggressive towards yearling males at the onset of spring (Strandgaard 1972), particularly those with large antlers (Wahlström 1994), which may drive natal dispersal. Two to three weeks prior to giving birth, females become socially intolerant and may be aggressive towards their young of the previous year. They may also react aggressively towards other unrelated does in the weeks following birth (Roviani 2014). Infanticide has been observed by adult males which may attack fawns with their antlers and/or trample them (Hewison et al. unpubl. data).

Mating Behavior

The mating system is site-defense polygyny where males attempt to maximize the size of their territory so as to overlap with the HRs of as many females as possible (Vanpé et al. 2009a). However, because variation in the number of female HRs within a male territory is generally low (typically maximum of 5, Strandgaard 1972), the level of polygyny in this mating system is also low (Vanpé et al. 2008). This may be increased by female rutting excursions (Richard et al. 2008) which appear to be motivated by active female mate searching (Lovari et al. 2008), perhaps for inbreeding avoidance (Debeffe et al. 2014a). Indeed, multiple paternity within a single litter occurs, but infrequently (Vanpé et al. 2009b).

The rut occurs in mid-summer (second half of July and first half of August, Liberg et al. 1998). Males can become territorial for the first time during their 4th summer (Bramley 1970), but the best-quality individuals may do so a year earlier, so that their first fawns are born when they are 3 years old (Vanpé et al. 2009c). Young nonterritorial males may adopt satellite or peripheral mating tactics (Hewison et al. 1998). Mating involves frequent pursuit of a female in estrus, occasionally including aggressive behavior, forming circles, or a figure of eight (Danilkin and Hewison 1996). Repeated copulation occurs during heat and may involve a second male.

Parental Care

While males provide no parental care, females suckle young frequently (approx. 6–8 times per day) over the first two months of life with 30–80 ml of milk per bout (Espmark 1969). Suckling frequency then decreases, although it can continue sporadically until autumn. Roe deer fawns are hiders for the first 2–3 months of their life, remaining concealed in the vegetation while the mother feeds at a distance, and freezing when initially approached (Linnell et al. 1998). Females may actively defend their young against predation attempts (Jarnemo 2004). Fawns mostly remain associated with their mothers through the winter until emancipation the following spring, although if they remain philopatric, they may re-form a loose family association with their mother as yearlings (Linnell et al. 1998).

Individual Heterogeneity

A number of personality traits have recently been documented in roe deer. Both behavioral and physiological traits are consistent for a given individual (Debeffe et al. 2015; Monestier et al. 2016) and describe among-individual variation in risk management (Bonnot et al. 2015), dispersal propensity (Debeffe et al. 2014c) and neophobia (Monestier et al. 2017). Individuals vary widely in their degree of vigilance and tolerance of threat (Bonnot et al. 2015, 2017).

Parasites and Diseases

Ectoparasites

Insects. Roe deer are often subclinically or mildly clinically affected by strictly host-specific chewing lice (*Damalinia meyeri*) and by *Lipoptena cervi,* a widespread blood-sucking ked of Old World deer. Subcutaneous infestation by larvae of the oestrid fly, *Hypoderma diana* (and to a lesser extent the red deer warble fly, *H. actaeon*), may result in painful nodular skin lesions on the back and hips. Subclinical to deadly (by respiratory failure) infestation may be caused by the pharyngeal bot fly, *Cephenemyia stimulator*, whose developing larvae remarkably grow in size within the throat pouches during springtime (Pewsner et al. 2017).

Arachnida. Roe deer are common maintenance hosts of the "wood tick," *Ixodes ricinus*, and some related tick-borne pathogens, namely, the protozoan *Babesia capreoli* and the bacterium *Anaplasma phagocytophylum,* both deemed emerging zoonotic agents (Overzier et al. 2013). Regionally, infestation by hard ticks of the genera *Dermacentor*, *Haemaphysalis*, and *Rhipicephalus* has also been recorded. Severe crusted mange by *Sarcoptes scabiei* has been observed in outbreak areas in which northern and southern chamois (*Rupicapra rupicapra* and *R. pyrenaica*) were the main affected reservoir hosts. Localized to generalized alopecia associated to heavy infestation by the follicular mite, *Demodex* spp., has been sporadically reported.

Endoparasites

The digestive tract of roe deer is home to nine species of protozoa (*Giardia duodenalis*, *Cryptosporidium parvum*, plus seven host-specific taxa of the genus *Eimeria*) and numbers of helminths, including the generalist tapeworms, *Moniezia expansa* and *M. benedeni,* and approximately two dozen roundworms. Among the latter, the cervid-specific *Ostertagia leptospicularis* and *Spiculopteragia spiculoptera*, with their respective minor morphs *O. kolchida* and *S. mathevossiani*, are the dominant abomasal species while *Nematodirus europaeus* is prevalent in the small intestine (Rossi et al. 1997). Most protozoa and helminths of the gastro-intestinal tract are only mildly pathogenic, with the exception of the large-sized blood-sucking *Haemonchus contortus,* a generalist species also harbored by sheep, goats, and other wild ruminant hosts, which may cause severe anemia and death (Aguirre et al. 1999).

Roe deer in contact with infected livestock may infrequently harbor the liver flukes, *Fasciola hepatica* and *Dicrocoelium dendriticum*. Severe to deadly infestation by the great American liver fluke, *Fascioloides magna*, accidentally introduced to Europe with imported elk and white-tailed deer, is reported for roe in contact with resilient red and fallow deer in countries of the Danube basin, where this "alien" fluke is currently actively spreading (Malcicka 2015). Larvae of *Echinococcus granulosus,* a zoonotic tapeworm with a canid-ruminant life cycle, have been occasionally found in the liver or lungs of roe deer sharing range with infested moose or livestock. The mosquito-transmitted nematode *Setaria tundra* is common in the abdominal cavity of roe deer throughout Europe. Spill-over to semi-domesticated reindeer (*Rangifer tarandus*) with associated mortality episodes have occurred in Scandinavia (Laaksonen et al. 2007).

The respiratory system of roe deer is home to the host-specific nematodes *Dictyocaulus capreolus* and the snail-transmitted *Varestrongylus capreoli*, located in the large and small bronchi and in the lung parenchyma, respectively. While often subclinical, in case of severe and often mixed infections, they may be cause of mass loss, respiratory distress, and gross pulmonary lesions (Simpson and Blake 2018).

In the heart and muscles, five species of Protozoa (*Toxoplasma gondii* and four nonpathogenic species of *Sarcocystis*) and the cystic larval stage of the canid-transmitted tapeworm *Taenia ovis krabbei* are commonly found.

Infectious Diseases

While antibodies to a range of livestock and wildlife pathogens have been detected on occasion of several serosurveys throughout Europe (Boadella et al. 2010), no wide-scale outbreaks by transmissible infectious diseases are known. An unresolved etiological diagnostic dilemma is the so-called "spring diarrhea," occurring in several countries on recurring local scale. Among other suspected agents, enteropathogenic strains of the bacterium *Escherichia coli* (EPEC) have been implicated. In these small foci, mortality may reach 10% (Guberti et al. 2004; Pewsner et al. 2017).

No significant role has been attributed to roe deer in the medium to long-term maintenance of major livestock diseases in Europe such as foot-and-mouth-disease, tuberculosis, brucellosis, and bluetongue disease. Roe deer is rather deemed an accidental spill-over host (hence an occasional "victim") of these feared transmissible infections. At the time being, there is no evidence that roe may contribute to the spread of emerging chronic wasting disease (CWD) in Europe.

Zoonoses

No major risks to human health derive from handling roe deer or dressing their carcass, although the zoonotic agent *T. gondii* has been reported similarly as in other ungulates (see above). Accordingly, pregnant women and immune-compromised patients should prudently refrain from consumption of unfrozen raw or rare venison. Deer keds (*Lipoptena cervi*) have been occasionally reported as the agent of a localized pruritic dermatitis in humans.

Population Ecology

Population Growth and Demographic Metrics

The roe deer life cycle is typical of long-lived iteroparous vertebrates and is characterized by a strong age structure. In productive populations, we can expect female roe deer to produce on average 1 fawn at 2 years of age, 1.3 fawns at 5 years, and 0.5 offspring at 12. Assuming a mean annual survival of 0.85 for yearlings, 0.90 for prime-aged deer, and 0.70 for animals older than 10 years of age, the asymptotic rate of increase of a roe deer population would be 1.282. On average, a 2-year-old female can expect to successfully raise around three fawns during its lifetime. This corresponds to a rapidly growing population, typical of roe deer in the late twentieth century which steadily increased between the 1980s and the early years of the twenty-first century, making the roe deer a success story (Andersen et al. 1998). The maximum rate of increase so far reported in a roe deer population is 1.44 on the Storfosna island (Norway). Some populations facing both hunting and predation by large carnivores in Scandinavia display markedly negative growth rates and cannot persist without immigration (Nilsen et al. 2009). The stable age structure of a productive roe deer population includes about 30% fawns, 20% yearlings, more than 45% prime-aged adults, and less than 5% females of 10 years of age and older. The generation time, which corresponds to the mean age of mothers and is a reliable metric of the pace of life (Gaillard et al. 2005), is slightly more than 4 years. The generation time of a roe deer population provides a reliable indicator of demographic performance (Nilsen et al. 2009) because it increases from fast-growing populations (less than 4 years) to declining populations (more than 6 years).

Relative Contribution of Demographic Rates

As observed in most long-lived iteroparous species, the growth rate of roe deer populations is more sensitive to a given change in prime-age adult female survival than to the same change in any other demographic parameter. Using the same parameters as above, which represent an average performance in Europe, the demographic impact of changing annual adult survival of females between 2 and 10 years of age by 1% is about twice the impact of changing recruitment of all females by 1%. However, in populations of roe deer, by far the most variable parameter over time is early survival (Gaillard et al. 2013). Summer fawn survival is thus the driver of roe deer population dynamics in most cases, except in situations when adult survival decreases (i.e., in the presence of large predators or high hunting pressure). Temporal variation in early survival generates pronounced cohort effects in recruitment. In particular, high population density and occurrence of drought in spring depress recruitment. In addition to this numerical effect, there is also a long-lasting effect of being born in a good or poor year (Garratt et al. 2015).

Factors Influencing Population Dynamics

Both density-dependence and environmental conditions strongly influence roe deer population dynamics. These factors generally target recruitment. At high density, roe deer fawns have lower survival, grow less, and have thereby a lower body mass, which jeopardizes survival and decreases the reproductive output of primiparous females. Spring and summer droughts have very similar effects on roe deer population dynamics. Populations subjected to lynx predation grow less than populations that face no large carnivores (Andrén and Liberg 2015), but the exact demographic mechanisms of this decline remain unknown.

Interspecific Interactions

When resources are scarce, a potential for competition may occur between roe and red deer. In a French forest, high density of red deer was suggested to negatively affect body mass of roe deer fawns (Richard et al. 2010). In Scottish conifer plantations, roe and red deer densities were inversely related (Latham et al. 1997). Partial overlap in resource use (food, habitat) and inverse numerical trends between roe and fallow deer have been recorded, with an increase in fallow deer and a decrease of roe deer abundance (Ferretti and Fattorini 2020). High densities of the former may reduce habitat quality for the latter, resulting in roe achieving smaller body size and larger HRs (Focardi et al. 2006). A significant behavioral interference from fallow to roe deer, including also aggressive interactions, has been reported as a mechanism of competition (Ferretti et al. 2011). Roe deer may outcompete mountain hare (*Lepus timidus*) (Hulbert and Andersen 2001).

Future Trends and Climate Change

The rapid increase of hunting bags observed in the 1980s and 1990s in most European countries has slowed since the beginning of the new century (Apollonio et al. 2010). The inability of roe deer females to track the earlier onset of spring caused by global warming might provide an unanticipated limitation of roe deer performance in forest habitat (Plard et al. 2014b). Indeed, contrary to a large range of animal and plant species, birth timing of roe deer females at a given location has not changed over the last decades, maybe because photoperiod rather than food resources determines the phenology of roe deer reproduction. Under the current climate change, one might thus expect roe deer to increasingly leave forest habitats to colonize more widely open areas that offer a greater temporal distribution of food resources.

Conservation Status

There are no less than 15,000,000 roe deer, distributed all over Europe and the Near East. The species is listed in the Bern Convention (Appendix III) and it has been classified in the IUCN Least Concern risk category since 2008, but before that was listed as Lower Risk/Least Concern from 1996 (Lovari et al. 2016). Therefore, it cannot be considered as a species at risk of extinction. However, the subspecies *C. c. italicus* (central and southern Italy) and *C. c. garganta* (southern Iberia) show relatively narrow ranges and are of conservation concern. *C. c. italicus* numbers less than 10,000 mature individuals and faces serious threats such as cross-breeding with introduced *C. c. capreolus* (southern Tuscany and Pollino National Park, Calabria, and Basilicata), poaching and predation by free-ranging dogs (Gargano National Park, Puglia; Castelporziano Presidential Estate, Latium), as well as some competition with fallow deer (Castelporziano Presidential Estate, Latium; Maremma Regional Park, southern Tuscany) (Lorenzini et al. 2002; Focardi et al. 2006; Ferretti et al. 2011). The small remaining populations in Greece and Northern Macedonia need serious monitoring and deserve conservation actions, as they are currently declining due to poaching and habitat degradation (Papaioannou 2010; Stojanov et al. 2010). Overall, the main threat throughout all Europe is the mixing of different gene pools as a result of translocations (Lovari et al. 2016). Presently, the success of conservation measures looks rather grim for all subspecies at risk.

Management

Probably humans began to impact on roe deer in the Neolithic when they started to clear-cut and fragment forests, favoring the expansion of a species able to benefit from the ecotone between woods and meadows. Small body size, territorial and predictable behavior, and high reproductive potential made roe deer one of the most preferred quarries. During the Middle Ages and Renaissance, roe deer were considered "small game" and therefore were not hunted exclusively by kings and aristocracy. So the species thrived, with humans probably supporting its expansion. However, the diffusion of efficient weapons, lack of strict regulations on hunting, and deforestation were responsible for a dramatic decline and local extinctions up to the nineteenth century.

Restrictive hunting rules (shorter open seasons, harvesting quotas), scarcity of large predators, recovery of bushy vegetation, and woods in areas abandoned by agriculture and pastoral farming, as well as reintroductions, permitted a rapid increase of roe deer in most of Europe. In recent decades, roe deer have further expanded, colonizing highly cultivated areas of lowland with scarce vegetation cover and, more recently, the periphery of urban areas with dense road networks. Species recovery was difficult only in some Mediterranean areas (Andalusia, southern Italy, Northern Macedonia, Greece) where feral/stray dogs and poachers are common.

Roe deer have a relatively low impact on cultivated fields (Kałuziński 1982) and forests (Gill 1992), mainly concentrating on orchards, vineyards, horticultural crops, and conifer plantations. In many parts of Europe, the main objective is to maintain roe deer densities at economically tolerable levels. Compared to other ungulates, roe deer are easy to manage and resilient to errors. The limited HR size facilitates small-scale management (i.e., areas of 500–1500 ha), while general planning is better done on a larger scale.

A significant difference exists between the approach of game managers and the suggestions of researchers. Game managers often rely on guess estimates, sometimes not based on direct observations, and find by trial and error the most appropriated culling quotas that do not impair the population. The purpose is to maintain density levels sufficiently high to exploit in terms of venison and trophies, but low enough to prevent risks to agriculture and forestry. By contrast, researchers advocate a more technical approach, based on counts to provide estimates of abundance, fertility, mortality, and dispersal.

Roe deer are however not easy to count: Especially in the presence of natural predators and high hunting pressure, they can become quite secretive. The most common count methods are 1) direct observation by vantage point at twilight in early spring on pastures and 2) the drive counts on sample areas in late winter or early spring (Meriggi et al. 2008). The first method requires a limited number of surveyors (1–2 / km^2) and is typical of open landscapes with relatively small woodland patches and large fields and meadows. The main risk of this method is the under-estimation of population size, due to a variable proportion of roe deer that do not come into the open. The drive count is typical of forests, most suitable for densities higher than 5–7 animals per km^2 (Borkowski et al. 2011). It requires more personnel (at least 2–3 persons per ha, in dense woods) and a final total coverage of at least 10% of the district extension. Both count methods are often criticized by researchers, who prefer techniques like capture-mark-resighting, the kilometric index, pellet group counts, and thermal imagery (Morellet et al. 2010). Recently, distance sampling (line transects statistically treated through detection functions, cf. Focardi et al. 2002) and camera trapping (Rowcliffe et al. 2008; Marcon et al. 2019) have met the interest of many researchers (cf. Focardi et al. 2002). Most of these count methods are useful in scientific projects, but they are difficult to adopt for game managers and hunters, due to their complexity and high costs.

Given the difficulties of counting roe deer, the use of ecological indicators such as measures of body size, that is, mandible length (Ellenberg 1974), hind foot length (Zannèse et al. 2006), body mass (Maillard et al. 1989), antler mass and antler length, has been suggested instead of population size estimates (Morellet et al. 2007).

Reproduction-based indices, like ovulation or fawn recruitment rates, reveal useful insights into the relationship between roe deer and habitat and into health conditions of the populations. Yearling fertility rates are particularly sensitive to density changes. Unfortunately, all these body size and demographic indicators are still rarely used in routine management practice. However, an effort to know the actual population size should be made, including alternating simple methods and more sophisticated techniques to adjust the estimate.

Simple biometric measurements (body mass, foot length) should be mandatory on all culled animals, while mandible length could be measured for representative samples.

The so-called "selective hunting," that is, the single-hunter hunting with rifle by sex and age classes according to a predetermined shooting plan, should try to maintain the natural structure of the population. Furthermore, strong manipulations of sex ratio may affect genetic variability.

Typically, three age classes are recognized for both sexes: fawns, yearling, and adults.

The possibility to distinguish in the field young adult, fully mature, and senescent bucks is illusory, given the unpredictability of antler development, once used as the discriminating criterion. It is similarly unrealistic to think that removing certain individuals (with abnormal or below-average antlers) by hunting can improve the overall quality of the roe population, which actually is strictly related to density and the quality of the habitat. Hunting seasons should close for males during the peak of rutting and for females and fawns during the period of juvenile dependency (Apollonio et al. 2011).

Drive hunts and hunting with dogs should be prohibited or at least discouraged, due to the difficulty to select sex and age classes of a fleeing animal, the high distress on the deer, and the potential impact on no-target species.

At a small scale, it is easily possible to enhance the habitat quality by planting highly palatable species (Fladenhofer 2013). At a larger scale, only a substantial change in silvicultural systems toward a more nature-oriented forestry can secure higher food supplies.

Future Challenges for Research and Management

The roe deer is a species with pronounced ecological and behavioral plasticity and high dispersal potential and is capable of reaching high densities in different habitats. In spite of being one of the most important game animals in Europe, this species cannot be considered at risk of extinction.

However, predetermined shooting plans based on estimates of abundance, fertility, mortality, and dispersal are desirable to maintain a balanced sex and age structure and sustainable densities of hunted populations.

Noteworthy, local populations, corresponding to the subspecies *C. c. italicus* (central and southern Italy) and *C. c. garganta* (southern Iberia), show relatively narrow ranges and are of conservation concern, with possible extinction of their gene pools, due to cross-breeding with the expanding introduced *C. c. capreolus*. Appropriate management/conservation actions (e.g., translocations of the threatened gene pools to areas where overlapping with the European roe deer is unlikely to occur), assisted by specific research projects, could be of great help to prevent the genetic disappearance of these unique taxa. On the other hand, further genetic studies are needed to show in detail the genetic structure of populations dwelling in the contact zones of European roe deer with the Siberian roe deer, currently lying in a narrow range between the rivers Volga and lower Don in Russia (and possibly elsewhere in north-eastern Europe), to gain more insights into the issue of hybridization between *C. capreolus* and *C. pygargus* and the possible existence of successful hybrid populations in those suture areas.

Acknowledgments We are grateful to Csányi S. (Szent István University, Godollo, Hungary), Lang J. (Justus-Liebig-Universität, Giessen, Germany), Lemaître J.-F. (University of Lyon 1, France), Kjellander P. (University of Agricultural Sciences, Uppsala, Sweden), Meneguz P.G. (University of Turin, Italy), Ramanzin M. (University of Padua, Italy), Sorensen C. (CIC, Budapest, Hungary), Tizzani P. (World Organization for Animal Health, Paris, France), Zlatanovic-Milošević S. (University of Kragujevac, Serbia) for sharing with us their unpublished morphometric data. Robin M.A. Gill (Forestry Commission, Farnham, United Kingdom) kindly reviewed some sections of the chapter.

References

Abbas F, Morellet N, Hewison AJM, Merlet J, Cargnelutti B, Lourtet B, Angibault JM, Daufresne T, Aulagnier S, Verheyden H (2011) Landscape fragmentation generates spatial variation of diet composition and quality in a generalist herbivore. Oecologia 167:401–411

Abbas F, Picot D, Merlet J, Cargnelutti B, Lourtet B, Angibault JM, Daufresne T, Aulagnier S, Verheyden H (2013) A typical browser, the roe deer, may consume substantial quantities of grasses in open landscapes. Eur J Wildl Res 59:69–75

Adams MG, Johnson E (1980) Seasonal changes in the skin glands of roe deer (*Caprelus capreolus*). J Zool 191:509–520

Aguirre AA, Brojer C, Mörner T (1999) Descriptive epidemiology of roe deer mortality in Sweden. J Wildl Dis 35(4):753–762

Andersen R, Linnell JDC, Aanes R (1995) Radyr I kulturlandskapet. NINA Fagrapport 10:1–80

Andersen R, Gaillard JM, Liberg O, San JC (1998) Variation in life-history parameters. In: Andersen R, Duncan P, Linnell JDC (eds) The European roe deer: the biology of success. Scandinavian University Press, Oslo, pp 284–307

Andersen R, Gaillard JM, Linnell JD, Duncan P (2000) Factors affecting maternal care in an income breeder, the European roe deer. J Anim Ecol 69(4):672–682

Andrén H, Liberg O (2015) Large impact of Eurasian lynx predation on roe deer population dynamics. PLoS One 10(3):e0120570

Apollonio M, Andersen R, Putman R (eds) (2010) European ungulates and their management in the 21st century. Cambridge University Press

Apollonio M, Putman R, Grignolio S, Bartos L (2011) Hunting seasons in relation to biological breeding seasons and the implications for the control or regulation of ungulate populations. In: Putman R, Apollonio M, Andersen R (eds) Ungulate management in Europe: problems and practices. Cambridge University Press, Cambridge, UK, pp 80–105

Apollonio M, Merli E, Chirichella R, Pokorny B, Alagic A, Flajsman SPA (2020) Capital-income breeding in male ungulates: causes and consequences of strategy differences among species. Front Ecol Evol 8:521767

Aragón S, Braza F, San JC (1995) Caracteristicas morfologicas de los corzos (*Capreolus capreolus*) de las Sierras de Cádiz-Málaga. Doñana Acta Vertebrata 22: 51–64

Aragón S, Braza F, San Josè C, Fandos P (1998) Variation in skull morphology of roe deer (*Capreolus capreolus*) in western and central Europe. J Mammal 79:131–139

Baskin L, Danell K (2003) Ecology of ungulates: a handbook of species in eastern Europe and northern and central Asia. Springer, Hamburg/Berlin

Benhaiem S, Delon M, Lourtet B, Cargnelutti B, Aulagnier S, Hewison AJM, Morellet N, Verheyden H (2008) Hunting increases vigilance levels in roe deer and modifies feeding site selection. Anim Behav 76:611–618

Biosa D, Scandura M, Tagliavini J, Luccarini S, Mattioli L, Apollonio M (2015) Patterns of genetic admixture between roe deer of different origin in central Italy. J Mammal 96(4):827–838

Bluzma A (1976) The roe deer in Lithuania. Musu gamta 7: 24–25. (in Lithuanian)

Boadella M, Carta T, Oleaga A, Pajares G, Muñoz M, Gortazar C (2010) Serosurvey for selected pathogens in Iberian roe deer. BMC Vet Res 6:51
Bongi P, Ciuti S, Grignolio S, Del Frate M, Simi S, Gandelli D, Apollonio M (2008) Anti-predator behaviour, space use and habitat selection in female roe deer during the fawning season in a wolf area. J Zool 276: 242–251
Bonnot N, Verheyden H, Blanchard P, Cote J, Debeffe L, Cargnelutti B, Klein F, Hewison AJM, Morellet N (2015) Interindividual variability in habitat use: evidence for a risk management syndrome in roe deer? Behav Ecol 26:105–114
Bonnot N, Hewison AJM, Morellet N, Gaillard JM, Debeffe D, Couriot O, Cargnelutti B, Chaval Y, Lourtet B, Kjellander P, Vanpé C (2017) Stick or twist: roe deer adjust their flight behaviour to the perceived trade-off between risk and reward. Anim Behav 124:35–46
Borger L, Franconi N, De Michele G, Gantz A, Meschi F, Manica A, Lovari S, Coulson T (2006a) Effects of sampling regime on the mean and variance of home range size estimates. J Anim Ecol 75:1493–1405
Borger L, Franconi N, Ferretti F, Meschi F, De Michele G, Gantz A, Coulson T (2006b) An integrated approach to identify spatiotemporal and individual-level determinants of animal home range size. Am Nat 168:471–485
Borkowski J, Palmer SCF, Borowski Z (2011) Drive counts as a method of estimating ungulate density in forests: mission impossible? Acta Theriol 56:239–253
Bramley PS (1970) Territoriality and reproductive behaviour of roe deer. J Reprod Fertil Suppl 11:43–70
Bresiński W (1982) Grouping tendencies in roe deer under agrocenosis conditions. Acta Theriol 27:427–447
Breyne P, Mergeay J, Casaer J (2014) Roe deer structure in a highly fragmented landscape. J Wildl Res 60: 909–917
Bubenik AB (1986) Ernährung, Verhalten und Umwelt des Schalenwildes. BLV, München
Büntgen U, Grenter L, Bollmann K, Jenny H, Liebhold A, Galvn JD, Stenseth NC, Andrew C, Mysterud A (2017) Elevational range shifts in four mountain ungulate species from the Swiss Alps. Ecosphere 8(4):e01761
Burbaite L, Csányi S (2009) Roe deer population and harvest changes in Europe. Estonian J Ecol 58:169–180
Burkart S, Gugerli F, Senn J, Kuehn R, Bolliger J (2016) Evaluating the functionality of expert-assessed wildlife corridors with genetic data from roe deer. Basic Appl Ecol 17:52–60
Cagnacci F, Focardi S, Heurich M, Stache A, Hewison AJM, Morellet N, Kjellander P, Linnell JDC, Mysterud A, Neteler M, Delucchi L, Ossi F, Urbano F (2011) Partial migration in roe deer: migratory and resident tactics are end points of a behavioural gradient determined by ecological factors. Oikos 120: 1790–1802
Cargnelutti B, Reby D, Desneux L, Angibault JM, Joachim J, Hewison AJM (2002) Space use by roe deer in a fragmented landscape some preliminary results. Rev Ecol – Terre Vie 57:29–37
Caro JH, Delibes-Mateos M, Cassinello RJ (2008) Estudio morfobiométrico de la cuerna del corzo andaluz. IREC, Ciudad Real
Cederlund G (1989) Activity patterns in moose and roe deer in a north boreal forest. Ecography 12:39–45
Cederlund G, Liberg O (1995) Rådjuret. In: Viltet, ekologin och jakten. Svenska Jägareförbundet, Solna. (in Swedish)
Chirichella R, Pokorny B, Bottero E, Flajšman K, Mattioli L, Apollonio M (2019) Factors affecting implantation failure in roe deer. J Wildl Manag 83(3): 599–609
Clauss M, Hume ID, Hummel J (2010) Evolutionary adaptations of ruminants and their potential relevance for modern production systems. Animal 4:979–992
Cornelis J, Casaer J, Hermy M (1999) Impact of season, habitat and research techniques on diet composition of roe deer (*Capreolus capreolus*): a review. J Zool 248: 195–207
Costa L (1992) Ecologia del corzo en las montañas cantábricas: modelo de gestion. Tesis doctoral, Universidad de Leon (in Spanish)
Coulon A, Cosson JF, Angibault JM, Cargnelutti B, Galan M, Morellet N, Petit E, Aulagnier S, Hewison AJM (2004) Landscape connectivity influences gene flow in a roe deer population inhabiting a fragmented landscape: an individual-based approach. Mol Ecol 13:2841–2850
Coulon A, Guillot G, Cosson JF, Angibault JM, Aulagnier S, Cargnelutti B, Galan M, Hewison AJM (2006a) Genetic structure is influenced by landscape features: empirical evidence from a roe deer population. Mol Ecol 15:1669–1679
Coulon A, Cosson JF, Morellet N, Angibault JM, Cargnelutti B, Galan M, Aulagnier S, Hewison AJM (2006b) Dispersal is not female biased in a resource-defence mating ungulate, the European roe deer. Proc R Soc Lond Ser B 273:341–348
Couriot O, Hewison AJM, Saïd S, Cagnacci F, Chamaillé-Jammes S, Linnell JDC, Mysterud A, Peters W, Urbano F, Heurich M, Kjellander P, Nicoloso S, Berger A, Sustr P, Kroeschel M, Soennichsen L, Sandfort R, Gehr B, Morellet N (2018) Truly sedentary? The multi-range tactic as a response to resource heterogeneity and unpredictability in a large herbivore. Oecologia 187:47–60
Cransac N, Cibien C, Angibault JM, Morellet N, Vincent JP, Hewison AJM (2001) Variations saisonnières du régime alimentaire du chevreuil (*Capreolus capreolus*) selon le sexe en milieu forestier à forte densité (forêt domaniale de Dourdan). Mammalia 65:1–12
Croitor R (2018) Plio-Pleistocene Deer of Western Palaearctic: taxonomy, systematics, phylogeny. Institute of Zoology of the Academy of Sciences of Moldova, Chişinău
Danilkin AA (1999) Oleny (Cervidae). GEOS, Moskva. (in Russian)

Danilkin A, Hewison AJM (1996) Behavioural ecology of Siberian and European roe deer. Chapman and Hall, London
De Marinis AM, Chirichella R, Bottero E, Apollonio M (2018) Ecological factors affecting eruption timing of mandibular teeth in roe deer. Eur J Wildl Res 64:50
Debeffe L, Morellet N, Cargnelutti B, Lourtet B, Bon R, Gaillard JM, Hewison AJM (2012) Condition-dependent natal dispersal in a large herbivore: heavier animals show a greater propensity to disperse and travel further. J Anim Ecol 81:1327–1337
Debeffe L, Focardi S, Bonenfant C, Hewison AJM, Morellet N, Vanpé C, Heurich M, Kjellander P, Linnell JDC, Mysterud A, Pellerin M, Sustr P, Urbano F, Cagnacci F (2014a) A one night stand? Reproductive excursions of female roe deer as a breeding dispersal tactic. Oecologia 176:431–443
Debeffe L, Morellet N, Verheyden H, Hoste H, Gaillard JM, Cargnelutti B, Picot D, Sevila J, Hewison AJM (2014b) Parasite abundance contributes to condition dependent dispersal in a wild population of large herbivore. Oikos 123:1121–1125
Debeffe L, Morellet N, Bonnot N, Gaillard JM, Cargnelutti B, Verheyden H, Vanpé C, Coulon A, Clobert J, Bon R, Hewison AJM (2014c) The link between behavioural type and natal dispersal propensity reveals a dispersal syndrome in a large herbivore. Proc R Soc Lond, Ser B 281(1790):20140873
Debeffe L, Lemaitre JF, Bergvall UA, Hewison AJM, Gaillard JM, Morellet N, Goulard M, Monestier C, David M, Verheyden H, Jaderberg L, Vanpé C, Kjellander P (2015) Short- and long-term repeatability of docility in the roe deer: sex and age matter. Anim Behav 109:53–63
Degmečić D, Florijančić T, Krapinec K, Domić D (2010) Rogovlje srnjaka kao smjernica gospodarenja lokalnom populacijom (Roe deer antlers as a guideline for managing the local population). Šumarski List 134(7–8):335–346
Douhard F, Gaillard JM, Pellerin M, Jacob L, Lemaître JF (2017) The cost of growing large: costs of post-weaning growth on body mass senescence in a wild mammal. Oikos 126:1329–1338
Duncan P, Tixier H, Hofmann RR, Lechner-Doll M (1998) Feeding strategies and the physiology of digestion in roe deer. In: Andersen R, Duncan P, Linnell JDC (eds) The European roe deer: the biology of success. Scandinavian University Press, Oslo, pp 91–116
Dupke C, Bonenfant C, Reineking B, Hable R, Zeppenfeld T, Ewald M, Heurich M (2017) Habitat selection by a large herbivore at multiple spatial and temporal scales is primarily governed by food resources. Ecography 40:1014–1027
Dzięciołowski R (1979) Structure and spatial organization of deer populations. Acta Theriol 24:3–21
Dziedzic R (1991) Ocena wybranych cech fenotypowych samców saren (*Capreolus capreolus* L.) oraz wpływ na nie czynników środowiskowych na przykladzie macroregionu środkowowschodniej polski. (Assessment of selected phenotypic characters of roe deer bucks (*Capreolus capreolus* L.), and the effect of environmental factors on these characters exemplified by the macro-region of central and eastern Poland. PhD thesis, University of Lublin (in Polish)
Ellenberg H (1974) Die Körpergrösse des Rehes als Bioindikator. Verh Ges Oekol Erlangen 1974:141–154
Espmark Y (1969) Mother-young relations and development of behaviour in roe deer (*Capreolus capreolus* L.). Viltrevy 6:461–531
Ferretti F, Fattorini N (2020) Competitor densities, habitat, and weather: effects on interspecific interactions between wild deer species. Integr Zool 16:670
Ferretti F, Sforzi A, Lovari S (2011) Behavioural interference between ungulate species: roe are not on velvet with fallow deer. Behav Ecol Sociobiol 65:875–887
Festa E (1925) Il capriolo dell'Italia centrale. Bol Museo Zool Anat Comparata Univ Torino 37:1–2
Fladenhofer H (2013) Le chevreuil: aménagement des biotopes. Editions du Gerfaut, Paris
Flajšman K, Pokorny B, Chirichella R, Bottero E, Mattioli L, Apollonio M (2018a) I can produce more offspring as you can imagine: first records on exceptionally large litters in roe deer in central/southern Europe. Eur J Wildl Res 63(3):42
Flajšman K, Borowik T, Pokorny B, Jedrejewska B (2018b) Effects of population density and female body mass on litter size in European roe deer at a continental scale. Mamm Res 63:91–98
Focardi S, Isotti R, Raganella P, Iannuzzo D (2002) The use of distance sampling and mark-resight to estimate the local density of wildlife populations. Environmetrics 13:177–186
Focardi S, Aragno P, Montanaro P, Riga F (2006) Inter-specific competition from fallow deer *Dama dama* reduces habitat quality for the Italian roe deer *Capreolus capreolus italicus*. Ecography 29:407–417
Gaillard JM, Viallefont A, Loison A, Festa-Bianchet M (2004) Assessing senescence patterns in populations of large mammals. Anim Biodivers Conserv 27(1): 47–58
Gaillard JM, Yoccoz NG, Lebreton JD, Bonenfant C, Devillard S, Loison A, Pontier D, Allaine D (2005) Generation time: a reliable metric to measure life-history variation among mammalian populations. Am Nat 166(1):119–123
Gaillard JM, Hewison AJM, Kjellander P, Pettorelli N, Bonenfant C, Van Moorter B, Liberg O, Andren H, Van Laere G, Klein F, Angibault JM, Coulon A, Vanpé C (2008) Population density and sex do not influence fine-scale natal dispersal in roe deer. Proc R Soc Lond Ser B 275:2025–2030
Gaillard JM, Hewison AJM, Klein F, Plard F, Douhard M, Davison R, Bonenfant C (2013) How does climate change influence demographic processes of widespread species? Lessons from the comparative analysis of contrasted populations of roe deer. Ecol Lett 16:48–57
Garratt M, Lemaître JF, Douhard M, Bonenfant C, Capron G, Warnant C, Klein F, Brooks RC, Gaillard

JM (2015) High juvenile mortality is associated with sex-specific adult survival and lifespan in wild roe deer. Curr Biol 25(6):759–763

Gehr B, Hofer EJ, Ryser A, Vimercati E, Vogt K, Keller LF (2018) Evidence for nonconsumptive effects from a large predator in an ungulate prey? Behav Ecol 29:724–735

Geist V (1998) Deer of the world: their evolution, behavior, and ecology. Stackpole Books, Mechanicsburg

Gentile G, Vernesi C, Vicario S, Pecchioli E, Caccone A, Bertorelle G, Sbordoni V (2009) Mitochondrial DNA variation in roe deer (*Capreolus capreolus*) from Italy: evidence of admixture in one of the last *C. c. italicus* pure populations from central-southern Italy. Ital J Zool 76(1):16–27

Gerard JF, Le Pendu Y, Maublanc ML, Vincent JP, Poulle ML, Cibien C (1995) Large group formation in European roe deer: an adaptive feature? Rev Ecol – Terre Vie 50:391–401

Gill RMA (1992) A review of damage by mammals in north temperate forests: 1. Deer For 65:145–169

Gill RMA (1994) The population dynamics of roe deer (*Capreolus capreolus* L.) in relation to forest habitat succession. PhD thesis, Open University

Gimenez O, Gaillard JM (2018) Estimating individual fitness in the wild using capture–recapture data. Popul Ecol 60:101–109

Guberti V, Battisti A, De Marco MA, Morabito S, Delogu M, Di Guardo G, Buccella C, Lavazza A, Raganella-Pelliccioni E (2004) *E. coli* epidemic in a Roe deer metapopulation of the Northern Apennines, Italy. J Mt Ecol 7:283–286

Hafner M (2004) Morfološke značilnosti rogovja srnjadi (*Capreolus capreolus* L.) v populaciji v Škofjeloškem hribovju. (Morphological characteristics of antlers in roe deer (*Capreolus capreolus* L.) in the population of the Škofjeloško hribovje area). Gozdarski Vestnik 62: 411–425. (in Slovene)

Hartl G, Reimoser F, Willing R, Köller J (1991) Genetic variability and differentiation in roe deer (*Capreolus capreolus* L.) of Central Europe. Genet Sel Evol 23: 281–299

Hewison AJM, Angibauldt JM, Boutin J, Bideau E, Vincent JP, Sempéré A (1996) Annual variation in body composition of roe deer (*Capreolus capreolus*) in moderate environmental conditions. Can J Zool 74: 245–253

Hewison AJM, Vincent JP, Reby D (1998) Social organisation of European roe deer. In: Andersen R, Duncan P, Linnell JDC (eds) The European roe deer: the biology of success. Scandinavian University Press, Oslo, pp 189–219

Hewison AJM, Vincent JP, Angibault JM, Delorme D, Van Laere G, Gaillard JM (1999) Tests of estimation of age from tooth wear on roe deer of known age: variation within and among populations. Can J Zool 77:58–67

Hewison AJM, Vincent JP, Joachim J, Angibault JM, Cargnelutti B, Cibien C (2001) The effects of woodland fragmentation and human activity on roe deer distribution in agricultural landscapes. Can J Zool 79:679–689

Hewison AJM, Morellet N, Verheyden H, Daufresne T, Angibault JM, Cargnelutti B, Merlet J, Picot D, Rames JL, Joachim J, Lourtet B, Serrano E, Bideau E, Cebe N (2009) Landscape fragmentation influences winter body mass of roe deer. Ecography 32: 1062–1070

Hewison AJM, Gaillard J-M, Morellet N, Cagnacci F, Debeffe L, Cargnelutti B, Gehr B, Kröschel M, Heurish M, Coulon A, Kjellander P, Börger L, Focardi S (2021) Sex differences in condition-dependence of natal dispersal in a large herbivore: dispersal propensity and distance are decoupled. Proc R Soc Lond B 288: 20202947

Hoem S, Melis C, Linnell JDC, Andersen R (2007) Fighting behaviour in territorial male roe deer *Capreolus capreolus*: the effects of antler size and residence. Eur J Wildl Res 53:1–8

Holand Ø (1992) Winter digestive strategy of a concentrate selector in Norway: the European roe deer. Can J Zool 70:1331–1335

Holand Ø, Mysterud A, Wannag A, Linnell JDC (1998) Roe deer in northern environments: physiology and behaviour. In: Andersen R, Duncan P, Linnell JDC (eds) The European roe deer: the biology of success. Scandinavian University Press, Oslo, pp 117–137

Horcajada-Sánchez F, Barja I (2016) Local ecotypes of roe deer populations (*Capreolus capreolus* L.) in relation to morphometric features and fur coloration in the centre of the Iberian Peninsula. Pol J Ecol 64:113–124

Høye TT (2006) Age determination in roe deer: a new approach to tooth wear evaluated on known age individuals. Acta Theriol 51:205–214

Hulbert IA, Andersen R (2001) Food competition between a large ruminant and a small hindgut fermentator: the case of the roe deer and mountain hare. Oecologia 128: 499–508

Jarnemo A (2004) Predation processes: behavioural interactions between red fox and roe deer during the fawning season. J Ethol 22:167–173

Johansson A, Liberg O (1996) Functional aspects of marking behaviour by male roe deer (*Capreolus capreolus*). J Mammal 77:558–567

Johansson A, Liberg O, Wahlström LK (1995) Temporal and physical characteristics of scraping and rubbing in roe deer (*Capreolus capreolus*). J Mammal 76:123–129

Kahlke HD (2001) Neufunde von Cerviden-Resten aus dem Unterpleistozän von Untermassfeld. In: Kahlke RD (ed) Monographien Das Pleistozän von Untermassfeld bei Meiningen (Thüringen). Römisch-Germanisches Zentralmuseum, teil II, Bonn, pp 461–482, tavv. 72–77

Kałuziński J (1982) Composition of the food of red deer living in the fields and the effects of their feeding on plant production. Acta Theriol 27:457–470

Kamieniarz R, Dobek A, Moliński K, Lisowski M, Szwaczkowski T (2017) Does genotype determine habitat preferences? – studies on forest and field roe deer. Folia Biol 65(4):173–179

Kjellander P, Hewison AJM, Liberg O, Angibault JM, Bideau E, Cargnelutti B (2004) Experimental evidence for density-dependence of home-range size in roe deer (*Capreolus capreolus* L.): a comparison of two long-term studies. Oecologia 139:478–485

Laaksonen S, Kuusela J, Nikander S, Nylund M, Oksanen A (2007) Parasitic peritonitis outbreak in reindeer (*Rangifer tarandus tarandus*) in Finland. Vet Rec 160:835–841

Lambert R (2005) A pregnancy-associated glycoprotein (PAG) unique to the roe deer (*Capreolus capreolus*) and its role in the termination of embryonic diapause and maternal recognition of pregnancy. Isr J Zool 51:1–11

Latham J, Staines BW, Gorman ML (1997) Correlations of red (*Cervus elaphus*) and roe (*Capreolus capreolus*) deer in Scottish forests with environmental variables. J Zool 242:681–704

Lawson RE, Putman RJ, Fielding AH (2000) Individual signatures in scent gland secretions of Eurasian deer. J Zool (Lond) 251:399–410

Lechner-Doll M, Lason K, Lang D, Behrend A (2008) Evolutionary aspects of dietary selection and digestion in the European roe deer (*Capreolus capreolus*) with special reference to seasonality. Zoosyst Evol 77:223–227

Lemaître JF, Cheynel L, Douhard F, Bourgoin G, Débias F, Ferté H, Gilot-Fromont E, Pardonnet S, Pellerin M, Rey B, Vanpé C, Hewison AJM, Gaillard JM (2018) The influence of early-life allocation to antlers on male performance during adulthood: evidence from contrasted populations of a large herbivore. J Anim Ecol 87:921–932

Liberg O, Johansson A, Andersen R, Linnell JDC (1998) Mating system, mating tactics and the function of male territoriality in roe deer. In: Andersen R, Duncan P, Linnell JDC (eds) The European roe deer: the biology of success. Scandinavian University Press, Oslo, pp 221–256

Linnell JDC, Andersen R (1998) Territorial fidelity and tenure in roe bucks. Acta Theriol 43:67–75

Linnell JDC, Wahlström LK, Gaillard JM (1998) From birth to independence: birth, growth, neonatal mortality, hiding behaviour and dispersal. In: Andersen R, Duncan P, Linnell JDC (eds) The European roe deer: the biology of success. Scandinavian University Press, Oslo, pp 257–283

Lister AM, Grubb P, Summer SRM (1998) Taxonomy, morphology and evolution of European roe deer. In: Andersen R, Duncan P, Linnell JDC (eds) The European roe deer: the biology of success. Scandinavian University Press, Oslo, pp 23–46

Lister AM, Parfitt SA, Owen FJ, Collinge SA, Breda M (2010) Metric analysis of ungulate mammals in the early Middle Pleistocene of Britain, in relation to taxonomy and biostratigraphy. II. Cervidae, Equidae and Suidae. Quat Int 228:157–179

Lone K, Loe LE, Gobakken T, Linnell JDC, Odden J, Remmen J, Mysterud A (2014) Living and dying in a multi-predator landscape of fear: roe deer are squeezed by contrasting pattern of predation risk imposed by lynx and humans. Oikos 123:641–651

Lorenzini R, Lovari S (2006) Genetic diversity and phylogeography of the European roe deer: the refuge area theory revisited. Biol J Linn Soc 88:85–100

Lorenzini R, Lovari S, Masseti M (2002) The rediscovery of the Italian roe deer: genetic differentiation and management implications. Ital J Zool 69:367–379

Lorenzini R, San José C, Braza C, Aragón S (2003) Genetic differentiation and phylogeography of roe deer in Spain, as suggested by mitochondrial DNA and microsatellite analysis. Ital J Zool 70:89–99

Lorenzini R, Garofalo L, Qin X, Voloshina I, Lovari S (2014) Global phylogeography of the genus *Capreolus* (Artiodactyla: Cervidae), a Palaearctic meso-mammal. Zool J Linnean Soc 170:209–221

Lovari S, Bartolommei P, Meschi F, Pezzo F (2008) Going out to mate: excursion behaviour of female roe deer. Ethology 114:886–896

Lovari S, Herrero J, Masseti M, Ambarli H, Lorenzini R, Giannatos G (2016) Roe deer *Capreolus capreolus*. IUCN Red List, Gland CH (IUCNredlist.org, accessed 8 Nov 2018)

Lovari S, Serrao G, Mori E (2017) Woodland features determining home range size of roe deer. Behav Process 140:115–120

Maillard D, Boisaubert B, Gaillard JM (1989) La masse corporelle: un bioindicateur possible pour le suivi des populations de chevreuils. Gibier Faune Sauvage 6: 57–68

Malcicka M (2015) Life history and biology of *Fascioloides magna* (Trematoda) and its native and exotic hosts. Ecol Evol 5(7):1381–1397

Marcon A, Battocchi D, Apollonio M, Grignolio S (2019) Assessing precision and requirements of three methods to estimate roe deer density. PLoS One 14: e0222349

Martin J, Vourc'h G, Bonnot N, Cargnelutti B, Chaval Y, Lourtet B, Goulard M, Hoch T, Plantard O, Hewison AJM, Morellet N (2018) Temporal shifts in landscape connectivity for an ecosystem engineer, the roe deer, across a multiple-use landscape. Landsc Ecol 33(6): 937–954

Matosiuk M, Borkowska A, Swisłocka M, Mirski P, Borowski Z, Krysiuk K, Danilkin AA, Zvychaynaya EY, Saveljev AP, Ratkiewicz MAW (2014) Unexpected population genetic structure of European roe deer in Poland: an invasion of the mtDNA genome from Siberian roe deer. Mol Ecol 23:2559–2572

Mattioli S (2003) Morphometric characteristics of roe deer from mid-northern Italy. Deer 12(7):395–400

Mattioli S, Spada A (2009) Il capriolo: caratteristiche biometriche della popolazione del territorio bolognese. Il divulgatore 1–63 (in Italian)

Mattioli L, Lovari C, Mazzarone V (1995) Il capriolo nella Provincia di Arezzo: biologia e gestione venatoria. L'Etruria, Cortona. (in Italian)

Maublanc ML, Bideau E, Willemet R, Bardonnet C, Gonzalez G, Desneux L, Cebe N, Gerard JF (2012) Ranging behaviour of roe deer in an experimental high-density population: are females territorial? C R Biol 335:735–743

Mauget C, Mauget R, Sempéré A (1999) Energy expenditure in European roe deer fawns during the suckling period and its relationship with maternal reproductive cost. Can J Zool 77:389–396

Mauget C, Mauget R, Sempéré A (2003) Metabolic cost of first reproduction in young female European roe deer *Capreolus capreolus*. Acta Theriol 48:197–206

McLoughlin PD, Gaillard JM, Boyce MS, Bonenfant C, Messier F, Duncan P, Delorme D, Van Moorter B, Saïd S, Klein F (2007) Lifetime reproductive success and composition of the home range in a large herbivore. Ecology 88(12):3192–3201

Melis C, Jędrzejewska B, Apollonio M, Bartoń KA, Jędrzejewski W, Linnell JDC, Kojola I, Kusak J, Adamic M, Ciuti S, Delehan I, Dykyy I, Krapinec K, Mattioli L, Sagaydak A, Samchuk N, Schmidt K, Shkvyrya M, Sidorovich VE, Zawadzka B, Zhyla S (2009) Predation has a greater impact in less productive environments: variation in roe deer, *Capreolus capreolus*, population density across Europe. Glob Ecol Biogeogr 18:724–734

Meriggi A, Sotti F, Lamberti P, Gilio N (2008) A review of the methods for monitoring roe deer European populations with particular reference to Italy. Hystrix 19:103–120

Meunier K (1983) Das Spanische Reh. In: Hofmann RR (ed) Wildbiologische Informationen für Den Jäger. Jagd+Hege Ausbildungsbuch VI, Berlin, pp 149–153

Milošević-Zlatanović S (2001) Zoogeographical and population differentiation of the roe deer from Yugoslavia. PhD thesis, Beograd (in Serbian)

Monestier C, Gilot-Fromont E, Morellet N, Debeffe L, Cebe N, Merlet J, Picot D, Rames JL, Hewison AJM, Verheyden H (2016) Individual variation in an acute stress response reflects divergent coping strategies in a large herbivore. Behav Process 132:22–28

Monestier C, Morellet N, Verheyden H, Gaillard JM, Bideau E, Denailhac A, Lourtet B, Cebe N, Picot D, Rames JL, Hewison AJM (2017) Neophobia is linked to behavioural and haematological indicators of stress in captive roe deer. Anim Behav 126:135–143

Morellet N, Gaillard JM, Hewison AJM, Ballon P, Boscardini Y, Duncan P, Klein F, Maillard D (2007) Indicators of ecological change: new tools for managing populations of large herbivores. J Appl Ecol 44:634–643

Morellet N, Klein F, Solberg E, Andersen R (2010) The census and management of population of ungulates in Europe. In: Putman R, Apollonio M, Andersen R (eds) Ungulate management in Europe: problems and practices. Cambridge University Press, Cambridge, UK, pp 106–143

Morellet N, Van Moorter B, Cargnelutti B, Angibault JM, Lourtet B, Merlet J, Ladet S, Hewison AJM (2011) Landscape composition influences roe deer habitat selection at both home range and landscape scales. Landsc Ecol 26:999–1010

Morellet N, Bonenfant C, Börger L, Ossi F, Cagnacci F, Heurich M, Kjellander P, Linnell JDC, Nicoloso S, Sustr P, Urbano F, Mysterud A (2013) Seasonality, weather and climate affect home range size in roe deer across a wide latitudinal gradient within Europe. J Anim Ecol 82:1326–1339

Moritz C (1994) Defining evolutionary significant units for conservation. Trends Ecol Evol 9:373–376

Mucci N, Mattucci F, Randi E (2012) Conservation of threatened local gene pools: landscape genetics of the Italian roe deer (*Capreolus c. italicus*) populations. Evol Ecol Res 14(7):897–920

Mysterud A, Ostbye E (1999) Cover as a habitat element for temperate ungulates: effects on habitat selection and demography. Wildl Soc Bull 27:385–394

Nilsen EB, Gaillard JM, Andersen R, Odden J, Delorme D, Van Laere G, Linnell JD (2009) A slow life in hell or a fast life in heaven: demographic analyses of contrasting roe deer populations. J Anim Ecol 78(3):585–594

Olano-Marin J, Plis K, Sönnichsen L, Borowik T, Niedziałkowska M, Jędrzejewska B (2014) Weak population structure in European roe deer (*Capreolus capreolus*) and evidence of introgressive hybridization with Siberian roe deer (*C. pygargus*) in Northeastern Poland. PLoS One 9(10):e109147

Orlandi L, Leonessi L, Visani G (2006) La gestione del capriolo *Capreolus capreolus* L. in Provincia di Massa-Carrara. Provincia di Massa-Carrara (in Italian)

Overzier E, Pfister K, Herb I, Mahling M, Böck jr, Silaghi C. (2013) Detection of tick-borne pathogens in roe deer (*Capreolus capreolus*), in questing ticks (*Ixodes ricinus*), and in ticks infesting roe deer in southern Germany. Ticks Tick-borne Dis 4(4):4320–4328

Padie S, Morellet N, Hewison AJM, Martin JL, Bonnot N, Cargnelutti B, Chamaille-Jammes S (2015) Roe deer at risk: teasing apart habitat selection and landscape constraints in risk exposure at multiple scales. Oikos 124: 1536–1546

Pagon N, Grignolio S, Pipia A, Bongi P, Bertolucci C, Apollonio M (2013) Seasonal variation of activity patterns in roe deer in a temperate forested area. Chronobiol Int 30:772–785

Papaioannou H (2010) Ungulates and their management in Greece. In: Putman R, Apollonio M, Andersen R (eds) Ungulate management in Europe: problems and practices. Cambridge University Press, Cambridge, UK, pp 540–562

Pays O, Benhamou S, Helder R, Gerard JF (2007) The dynamics of group formation in large mammalian herbivores: an analysis in the European roe deer. Anim Behav 74:1429–1441

Pélabon C, van Breukelen L (1998) Asymmetry in antler size in roe deer (*Capreolus capreolus*): an index of individual and population conditions. Oecologia 116:1–8

Petelis K, Brazaitis G (2003) Morphometric data on the field ecotype roe deer in Southwest Lithuania. Acta Zool Lituanica 13:61–64

Peters W, Hebblewhite M, Mysterud A, Spitz D, Focardi S, Urbano F, Morellet N, Heurich M, Kjellander P, Linnell JDC, Cagnacci F (2017) Migration in geographic and ecological space by a large herbivore. Ecol Monogr 87: 297–320

Pettorelli N, Gaillard JM, Yoccoz NG, Duncan P, Maillard D, Delorme D, Van Laere G, Toïgo C (2005) The response of fawn survival to changes in habitat

quality varies according to cohort quality and spatial scale. J Anim Ecol 74(5):972–981
Pewsner M, Origgi FC, Frey J, Ryser-Degiorgis MP (2017) Assessing fifty years of general health surveillance of roe deer in Switzerland: a retrospective analysis of necropsy reports. PLoS One 12(1):e0170338
Plard F, Gaillard JM, Coulson T, Hewison AM, Delorme D, Warnant C, Bonenfant C (2014a) Mismatch between birth date and vegetation phenology slows the demography of roe deer. PLoS Biol 12(4): e1001828
Plard F, Gaillard JM, Coulson T, Hewison AM, Delorme D, Warnant C, Nilsen EB, Bonenfant C (2014b) Long-lived and heavier females give birth earlier in roe deer. Ecography 37(3):241–249
Plard F, Yoccoz NG, Bonenfant C, Klein F, Warnant C, Gaillard JM (2015) Disentangling direct and growth-mediated influences on early survival: a mechanistic approach. J Anim Ecol 84(5):1363–1372
Pontier D, Gaillard JM, Allainé D, Trouvilliez J, Gordon I, Duncan P (1989) Postnatal growth rate and adult body weight in mammals: a new approach. Oecologia 80(3): 390–394
Prior R (1968) The roe deer of Cranbourne chase. Oxford University Press, London
Randi E, Mucci N, Pierpaoli M, Douzery E (1998a) New phylogenetic perspectives on the Cervidae (Artiodactyla) are provided by the mitochondrial cytochrome b gene. Proc R Soc Lond, Ser B 265:793–801
Randi E, Pierpaoli M, Danilkin A (1998b) Mitochondrial DNA polymorphism in populations of Siberian and European roe deer (*Capreolus pygargus* and *C. capreolus*). Heredity 80:429–437
Randi E, Alves PC, Carranza J, Milośević-Zlatanović S, Sfougaris A, Mucci N (2004) Phylogeography of roe deer (*Capreolus capreolus*) populations: the effects of historical genetic subdivisions and recent non-equilibrium dynamics. Mol Ecol 13:3071–3083
Ratikainen I, Panzacchi M, Mysterud A, Odden J, Linnell J, Andersen R (2007) Use of winter habitat by roe deer at a northern latitude where Eurasian lynx are present. J Zool 273:192–199
Reby D, Cargnelutti B, Hewison AJM (1999) Contexts and possible functions of barking in roe deer. Anim Behav 57:1121–1128
Richard E, Morellet N, Cargnelutti B, Angibault JM, Vanpé C, Hewison AJM (2008) Ranging behaviour and excursions of female roe deer during the rut. Behav Process 79:28–35
Richard E, Gaillard J-M, Saïd S, Hamam J-L, Klein F (2010) High red deer density depresses body mass of roe deer fawns. Oecologia 163:91–97
Rivals F, Lister AM (2016) Dietary flexibility and niche partitioning of large herbivores through the Pleistocene of Britain. Quat Sci Rev 146:116–133
Rossi L, Eckel B, Ferroglio E (1997) A survey of the gastro-intestinal nematodes of roe deer (*Capreolus capreolus*) in a mountain habitat. Parassitologia 39(4): 303–312
Roviani D (2014) A lethal interaction between two female roe deer. Ital J Mammal 25:117–118
Rowcliffe JM, Field J, Turvey ST, Carbone C (2008) Estimating animal density using camera traps without the need for individual recognition. J Appl Ecol 45: 1228–1236
Royo LJ, Pajares G, Alvarez I, Fernandez I, Goy-Ache F (2007) Genetic variability and differentiation in Spanish roe deer (*Capreolus capreolus*): a phylogeographic reassessment within the European framework. Mol Phylogenet Evol 42:47–61
Saarinen J, Eronen J, Fortelius M, Seppä H, Lister AM (2016) Patterns of diet and body mass of large ungulates from the Pleistocene of Western Europe, and their relation to vegetation. Palaeontol Electron 19(3):1–58
Sabalinkiene G, Danusevicius D, Manton M, Brazaitis G, Simkevicius K (2017) Differentiation of European roe deer populations and ecotupes in Lithuania based on DNA markers, cranium and antler morphometry. Silva Fennica 51(3):1743
Saïd S, Servanty S (2005) The influence of landscape structure on female roe deer home-range size. Landsc Ecol 20:1003–1012
Saïd S, Gaillard J-M, Duncan P, Guillon N, Guillon N, Servanty S, Pellerin M, Lefeuvre K, Martin C, Van Laere G (2005) Ecological correlates of home-range size in spring-summer for female roe deer (*Capreolus capreolus*) in a decideous woodland. J Zool 267: 301–308
San José C, Lovari S, Ferrari N (1996) Temporal evolution of vigilance in roe deer. Behav Process 38:155–159
San José C, Lovari S, Ferrari N (1997) Grouping in roe deer: an effect of habitat openness or cover distribution? Acta Theriol 42:235–239
Sempéré AJ (1990) The annual antler cycle of the European roe deer (*Capreolus capreolus*) in relation to the reproductive cycle. In: Bubenik GA, Bubenik AB (eds) Horns, pronghorns, and antlers. Springer, New York, pp 396–415
Sempéré AJ, Mauget R, Mauget C (1998) Reproductive physiology of roe deer. In: Andersen R, Duncan P, Linnell JDC (eds) The European roe deer: the biology of success. Scandinavian University Press, Oslo, pp 161–188
Serrano Ferron E, Verheyden H, Hummel J, Cargnelutti B, Lourtet B, Merlet J, Gonzalez-Candela M, Angibault JM, Hewison AJM, Clauss M (2012) Digestive plasticity as a response to woodland fragmentation in roe deer. Ecol Res 27:77–82
Simpson VR, Blake DP (2018) Parasitic pneumonia in roe deer (*Capreolus capreolus*) in Cornwall, Great Britain, caused by *Varestrongylus capreoli* (Protostrongylidae). BMC Vet Res 14(1):198
Sommer RS, Fahlke JM, Schmölcke U, Benecke N, Zachos FE (2009) Quaternary history of the European roe deer *Capreolus capreolus*. Mammal Rev 39:1–16
Staines BW, Ratcliffe PR (1991) Roe deer. In: Corbet GB, Harris S (eds) The hanbook of British mammals. Blackwell Scientific, Oxford, pp 518–525

Stojanov A, Melovski D, Ivanov G, Linnell JDC (2010) Ungulates and their management in Macedonia. In: Apollonio M, Andersen R, Putman R (eds) European ungulates and their management in the 21st century. Cambridge University Press, Cambridge, UK, pp 572–577

Strandgaard H (1972) The roe deer (*Capreolus capreolus*) population at Kalo and the factors regulating its size. Danish Rev Game Biol 7:1–205

Stubbe C (1966) Die Altersbedingte Gehörnentwicklung beim europäischen Rehwild (*Capreolus c. capreolus* L.). Zool Abh 28:293–308

Stubbe C (1990) Rehwild, 4th edn. DL, Berlin

Tixier H, Duncan P (1996) Are European roe deer browsers? A review of variations in the composition of their diets. Rev Ecol – Terre Vie 51:3–17

Tixier H, Duncan P, Scehovic J, Yani A, Gleizes M, Lila M (1997) Food selection by European roe deer (*Capreolus capreolus*): effects of plant chemistry, and consequences for nutritional value of their diets. J Zool 242:229–245

Tixier H, Maizeret C, Duncan P, Bertrand R, Poirel C, Roger M (1998) Development of feeding selectivity in roe deer. Behav Process 43:33–42

Valli AMF (2010) Dispersion of the genus *Procapreolus* and the relationships between *Procapreolus cusanus* and the roe deer (*Capreolus*). Quat Int 212:80–85

Vanpé C, Gaillard JM, Kjellander P, Mysterud A, Magnien P, Delorme D, Van Laere G, Klein F, Liberg O, Hewison AJM (2007) Antler size provides an honest signal of male phenotypic quality in roe deer. Am Nat 169:481–493

Vanpé C, Kjellander P, Galan M, Cosson JF, Aulagnier S, Liberg O, Hewison AJM (2008) Mating system, sexual dimorphism and the opportunity for sexual selection in a territorial ungulate. Behav Ecol 19:309–316

Vanpé C, Morellet N, Kjellander P, Goulard M, Liberg O, Hewison AJM (2009a) Access to mates in a territorial ungulate is determined by the size of a male's territory, but not by its habitat quality. J Anim Ecol 78:42–51

Vanpé C, Kjellander P, Gaillard JM, Cosson JF, Galan M, Hewison AJM (2009b) Multiple paternity occurs with low frequency in the territorial roe deer, *Capreolus capreolus*. Biol J Linn Soc 97:128–139

Vanpé C, Gaillard JM, Morellet N, Kjellander P, Liberg O, Delorme D, Hewison AJM (2009c) Age-specific variation in male breeding success of a territorial ungulate species, the European roe deer. J Mammal 90:661–665

Verheyden-Tixier H, Duncan P (2000) Selection for small amounts of hydrolysable tannins by a concentrate-selecting mammalian herbivore. J Chem Ecol 26: 351–358

Vincent JP, Bideau E, Hewison AJM, Angibault JM (1995) The influence of increasing density on body weight, kid production, home range and winter grouping in roe deer (*Capreolus capreolus*). J Zool 236:371–382

Visani G (2009) Differenze biometriche del capriolo (*Capreolus capreolus*) nei distretti di gestione della Provincia di Pistoia. Master Thesis, University of Florence (in Italian)

Vislobokova I, Dmitrieva E, Kalmykov NP (1995) Artiodactyls from the late Pliocene of Udunga, Western trans-Baikal, Russia. J Vertebr Paleontol 15(1): 146–159

von Essen L (1983) Rådjuret. In: Hamilton H (ed) Svenska hjortdjur Del II. Gothia, Stockholm, pp 567–718. (in Swedish)

Wahlström LK (1994) The significance of male-male aggression for yearling dispersal in roe deer (*Capreolus capreolus*). Behav Ecol Sociobiol 35:409–412

Wahlström LK, Liberg O (1995a) Contrasting dispersal patterns in two Scandinavian roe deer (*Capreolus capreolus*). Wildl Biol 1:159–164

Wahlström LK, Liberg O (1995b) Patterns of dispersal and seasonal migration in roe deer *Capreolus capreolus* populations. J Zool Lond 235:455–467

Wajdzik M, Kubacki T, Kulak D (2007) Diversification of the body weight and quality of the antlers in males of the roe deer (*Capreolus capreolus* L.) in southern Poland exemplified by surroundings of Cracow. Acta Scientiarum Polonorum Silvarum Colendarum Ratio Industria Lignaria 6:99–112

Wotschikowsky U (1981) Rot- und Rehwild (Nationalpark Bayerischer Wald). Nationalpark Bayerischer Wald Wissenschaftliche Schriftenreihe 7:1–111

Wotschikowsky U, Schwab G (1994) Das Rehproject Hahnebaum. Wildbiologische Gesellschaft München e.V

Zannèse A, Baïsse A, Gaillard JM, Hewison AJM, Saint-Hilaire K, Toïgo C, Van Laere G, Morellet N (2006) Hind foot length: a new biological indicator for monitoring roe deer populations at a landscape scale. Wildl Soc Bull 34:351–358

Zejda J, Koubek P (1988) On the geographical variability of roe bucks (*Capreolus capreolus*). Folia Zool 37: 219–229

8 Chinese Water Deer *Hydropotes inermis* Swinhoe, 1870

Arnold S. Cooke

Contents

A. S. Cooke (✉)
Huntingdon, UK
e-mail: arnrosecooke@gmail.com

© Springer Nature Switzerland AG 2022
L. Corlatti, F. E. Zachos (eds.), *Terrestrial Cetartiodactyla*, Handbook of the Mammals of Europe,
https://doi.org/10.1007/978-3-030-24475-0_26

Common Names

The species is commonly referred to in Britain as the Chinese water deer (Fig. 1). Those introduced to England in the latter part of the nineteenth century originally came from China (Fautley 2013; Putman et al. 2021), so they are correctly named. The species is, however, more numerous on the Korean peninsula (Kim et al. 2011; Kim 2017; Chun 2018), and it is referred to below as the water deer.

English	Chinese water deer
German	Chinesisches Wasserreh
French	Hydropote, hydropote chinois
Spanish	Ciervo acuático
Italian	Idropote, capriolo d'acqua
Russian	Водяной олень

Taxonomy, Systematics and Paleontology

Two subspecies have been recognised: *Hydropotes inermis inermis* Swinhoe 1870, the Chinese water deer; and *H. i. argyropus* Heude, 1884, the Korean water deer. However, Schilling and Rössner (2017) reviewed the evidence for two subspecies and found it inconclusive as the available genetic data were insufficient for confirmation.

The water deer is the only cervid to lack antlers. The monospecific subfamily Hydropotinae was considered to be a sister group of the antlered deer, the Cervidae (Groves and Grubb 1987). Later phylogenetic analysis by Randi et al. (1998) indicated that water deer were closely related to roe deer *Capreolus* species, and their antlers were lost secondarily. A broader genetic study by Gilbert et al. (2006) placed the species in the tribe Capreolini in the subfamily Capreolinae. This remains the most accepted interpretation of available information (Schilling and Rössner 2017).

Well-characterized fossils attributed to the species have been found in China but remain scarce (Dong 1993; Schilling and Rössner 2017). Specimens that might be direct predecessors of water deer or intermediate forms between it and other cervids are unknown.

Current Distribution

The first known written record of the species in China dates from about 238 BCE in the Qin Dynasty (Zhang 1996). In historical times, water deer were found in good numbers in wetlands in eastern China (Sheng and Ohtaishi 1993; Zhang 1996). These deer now occur in just a few fragmented locations dispersed across hundreds of kilometers of the country (M. Chen in Fautley 2013; Chen et al. 2016; Putman et al. 2021). Korean water deer remain fairly widespread through the peninsula and are locally abundant, at least in the De-Militarized Zone and in parts of South Korea (Won and Smith 1999; Kim and Cho 2005; Harris and Duckworth 2015; Jung et al. 2016; Kim 2017; Chun 2018; Choi and Lee 2019). In 2019, a water deer was recorded on a camera trap in the Khasansky district of eastern Russia (Darman et al. 2019). The same authors also recorded the movement of water deer from North Korea into northern China.

Fig. 1 A male water deer poised to attack a rival (photograph by and courtesy of M. McKenzie)

Water deer were introduced to captive situations in England at the end of the nineteenth century and a feral population resulted from deer escaping or being released from the middle of the twentieth century (Chapman 1995). In 2016, in the most recent survey organized by the British Deer Society, water deer were found to have significant populations in parts of Norfolk, Suffolk, Cambridgeshire, Bedfordshire, and Buckinghamshire with sporadic records from areas between and on the fringes of these main concentrations (Smith-Jones 2017; Fig. 2). Efforts to determine total numbers in England were reviewed by Cooke (2019), the population being estimated to be in the range 5000–10,000. Further range expansion has been predicted (Acevedo et al. 2010; Croft et al. 2019).

A small feral population was established south of Limoges in France during the second half of the twentieth century after abandonment of a park where they had been kept (Cooke 1999; Haffner and Savouré-Soubelet 2015). This population died out after several decades (G. Dubost, pers. comm.). Several populations remain in France confined within parks (Schilling and Rössner 2017; Putman et al. 2021).

Description

Size and Morphology

Water deer are typically 50–56 cm at the shoulder (Sheng 1992; Cooke and Farrell 2008). Cooke (2019) reviewed mean whole body mass for five samples of deer collected from the wild in China and England, reporting 14.3–15.9 kg for males and 15.2–16.3 kg for females. Mean mass for females was consistently greater than that for males. At Branféré Zoo in Brittany, France,

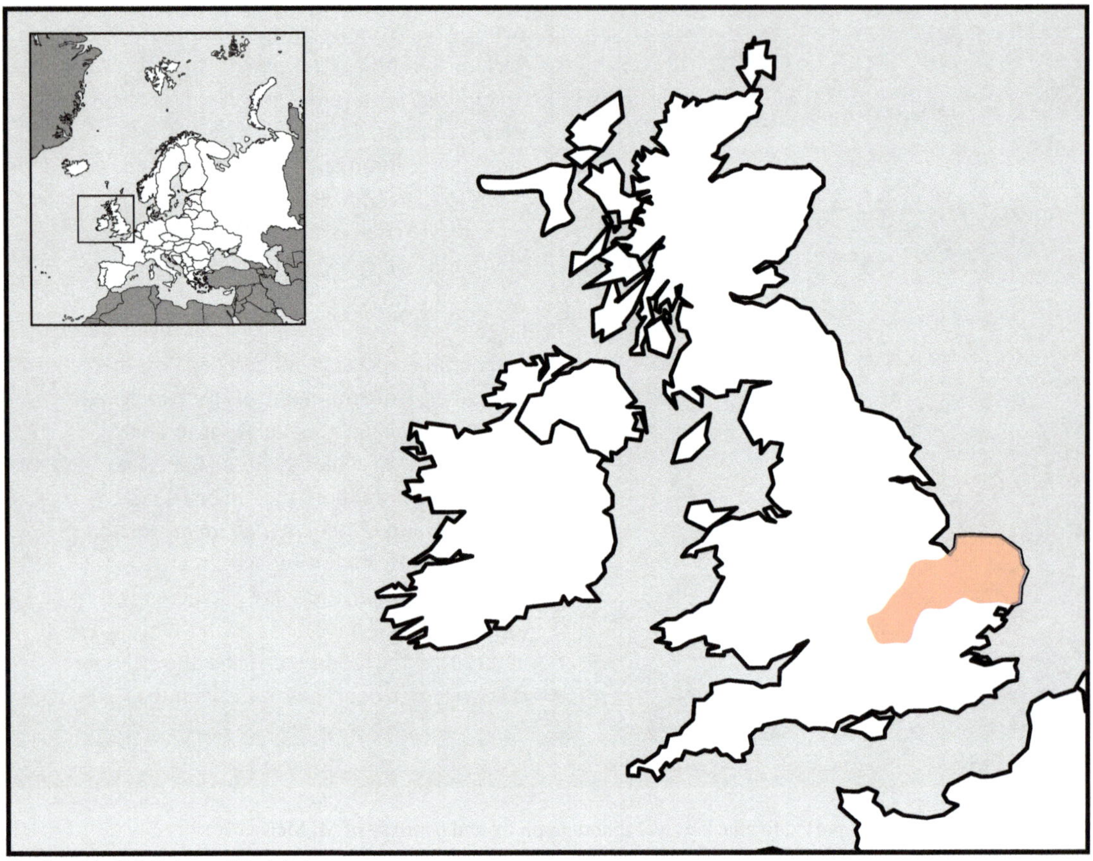

Map template: © Getty Images/iStockphoto

Fig. 2 The main range of water deer based on the British Deer Society's survey in 2016 (Smith-Jones 2017). (Map template: © Copyright Getty Images/iStockphoto)

mean whole body mass was 14.0 kg for males and 15.2 kg for females, and non-gestating females were heavier than males (Dubost et al. 2008). A few deer introduced to the area around Woodwalton Fen in Cambridgeshire in about 1950 (Chapman 1995) developed into a population which has been studied for more than 40 years (Cooke 2019); deer at the Fen have weighed up to 19 kg, but individuals of 20 kg or more occur elsewhere in England.

Water deer have powerful hindquarters and their rump is higher than their shoulders. The back is concave when deer stand upright. The tail is short and brown, being typically 6–7 cm in length. Ears are large, often held erect and are distinctively close together. Eyes and nose can appear as three black buttons when animals are in their thick winter coats.

Mean skull length was found to be 172 mm in China (Sheng 1992) and 167 mm at Woodwalton Fen (Cooke and Farrell 2008). The auditory bullae are unusually large (Allen 1940). The second and fifth metapodia are reduced to distal splinters, the telemetacarpalian condition. This is the only cervid with inguinal glands. The pre-orbital glands are small, and interdigital glands occur on the hind feet and forefeet (Pocock 1923; Ni et al. 1993).

Pelage

In winter, coat color is variable but most are pale brown or a peppery grey-brown (Cooke and Farrell 1998, 2008; Cooke 2019). The obvious spring molt is during April–June, when deer often have an unkempt appearance. Summer coat is usually pale red-brown, with deer molting back into winter coat during September–October. Hair is often shed in tufts when the males fight during the winter rut. It is coarse, hollow, and 40–55 mm in length, being mainly white, but with a dark brown band and a tip of pale brown, buff, or ginger. A band around the top of the nose may be white, grey or black.

Dentition

Permanent dentition is 0.1.3.3/3.1.3.3. Males do not have antlers, but develop long upper canine teeth (tusks), which are obvious in the field (Figs. 1 and 3a). These are loose in the sockets and are held back when the animal is grazing on short vegetation. Maxillary gum tissue is movable and its thickened inferior marginal band loops around the canines (Aitchison 1946). In altercations with rival males, the animal snarls and levator muscles around the lips lift the upper lip, drawing the tusks, which are effectively hinged, forward and inward. Tusks are a male's main weapon when fighting conspecific rivals, although they do not seem to be used in defense if attacked by other species (Cooke and Farrell 2008; Cooke 2019). Females have small canine teeth up to 8 mm in length (Fig. 3b). Tusks erupt around November when a male is 5–6 months of age. By March, tusks are usually approaching about half their final length. In most males, these teeth become close-rooted by or during the second winter. In the reserve at Woodwalton Fen, mean exposed tusk length on mature males was 60 mm with a maximum recorded length of 72 mm. For animals based outside the reserve, the mean length was 48 mm (Cooke 2019), while in China it was 53 mm (Sheng 1992). Broken canines are not unusual, and loss of a tusk is a drawback when fighting.

Age Determination

Dubost et al. (2008) considered that tooth eruption and replacement occurred earlier in captive water deer than in ruminant species of similar or lower body mass; the first molar erupted at 60–65 days, the second at 170 days and the third at 10–12 months. These authors developed a method for ageing water deer based on tooth eruption and wear. Seo et al. (2017) described and illustrated the eruption of posterior teeth in traffic casualties up to 15 months of age. A technique for ageing based on counting cementum layers on the first permanent molar showed promise in a preliminary study (Cooke 2019).

Physiology

Zhang (1996) compared the climate in the species' range in China with that in Bedfordshire and concluded that, although China had hotter and wetter summers, the English climate was evidently suitable. Young deer in Europe have been reported dying from heat stress in poorly-vegetated, artificial habitats in hot summers, whereas those in well-vegetated, semi-natural habitats enjoy good survival in such summers (Chaplin 1977; Dubost et al. 2008; Cooke 2009a, 2019). Thick coats of hollow hair would be expected to offer good protection against cold winter weather, but the winter rut can be disrupted by spells of exceptionally wet or mild weather (Cooke 2009a, 2019). The short tail conforms to Allen's rule that species in cold climes have stumpy tails.

The species offers no resistance to attackers, apart from rival water deer, so that it needs its large, muscular hind legs to permit a rapid escape, although there seems to be no evidence that it can maintain a high speed for very long.

Genetics

Chromosomes

Chromosome number was reported by Zima and Král (1984) as $2n = 70$. The X chromosomes are acrocentric, this species being one of only two telemetacarpalian cervids with this condition (Groves and Grubb 1987).

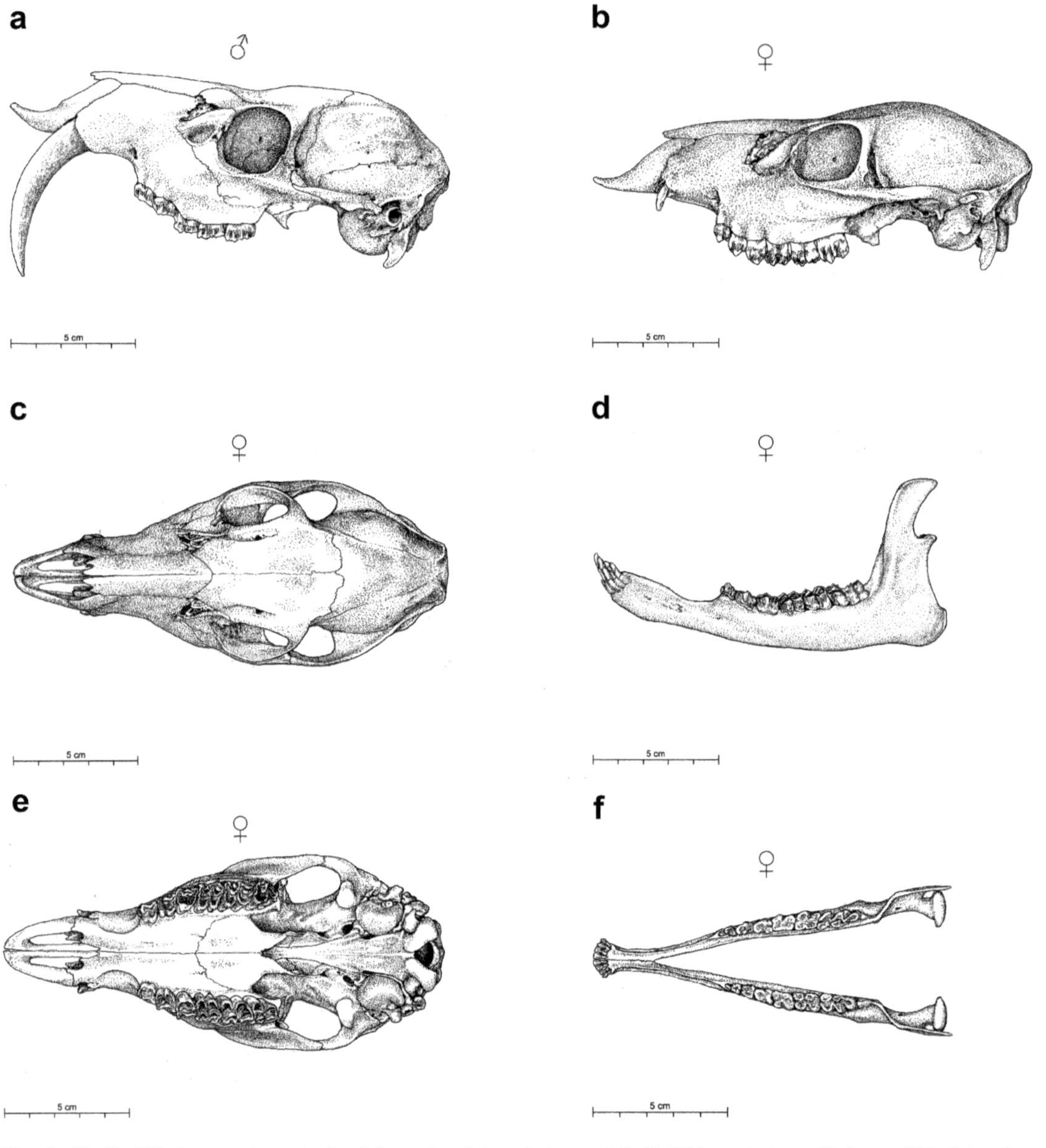

Fig. 3 Skull of *Hydropotes inermis*. (**a**–**c**) Lateral and dorsal views of skull, (**d**) lateral view of left mandible (view from outside), (**e**) ventral view of skull, (**f**) dorsal view of mandible (photographs © Franz Müller)

Genetic Diversity

Designation of the two subspecies was based on morphology, including pelage color (Koh et al. 2009). These last authors reported two sympatric mitochondrial DNA clades, a major clade from China and Korea, and a minor one solely from Korea; a reconsideration of classification was proposed. Recent genetic advances have been reviewed by Schilling and Rössner (2017). Hu et al. (2006), Fautley (2013), and Putman et al. (2021) reported on genetic studies of relevance to conserving the species. In China, populations on the Zhoushan Islands differed genetically from those on the mainland, indicating long-term separation; while genetic diversity was lower in English populations. Introduced populations in England were thought to be descended from an extinct population on the Chinese mainland close to Shanghai.

Life History

Growth

May and June are the main months for giving birth, with most births occurring at the end of May or the beginning of June at Whipsnade Zoo in England, Yancheng Reserve in China, and at Branféré Zoo in France (Zhang 1996; Dubost et al. 2008). Body mass at birth is in the range 0.6–1.1 kg, with an initial gain of about 0.1 kg/day (Chaplin 1977; Cooke and Farrell 1998). Neonate body mass at Branféré Zoo decreased as litter size increased from one to three (Dubost et al. 2008).

The pale spots on the dark brown fawns have disappeared by 2 months of age (Cooke and Farrell 1998; Cooke 2019). Fawns then have a shaggy appearance which lasts for about another 2 months. At the age of about 6 months, they are slightly smaller, slimmer versions of full adults, with a live body mass of 9–12 kg (Chaplin 1977), roughly 60–80% of their eventual size. By the time they are 1 year old, they are largely indistinguishable in the field from older deer.

Reproduction

Young deer reach sexual maturity at 4–7 months in England (Chaplin 1977) and 5–8 months in China (Sheng and Lu 1984), in time to participate in their first rut. Rutting behavior occurs in mid-winter from November until January, generally peaking in December in different populations in Asia and Europe (Cooke 2019).

Water deer produce litters of young. Chaplin (1977) working on populations in the Woburn area of Bedfordshire recorded that, as number of ova per ovulation increased, so implantation rate decreased. He reported as many as seven fetuses, but average number per pregnant female was 2.3. Estimates of length of gestation ranged from 165 to 210 days with an average of about 170 days (Zhang 1996; Dubost et al. 2008; Cooke 2019).

Mean litter size at Poyang Lake and on Zhoushan Island in China was 2.5, and was 2.9 in captive deer (Sheng 1992). Litters of six have been reported by hunters in China, but were exceptionally rare (Sheng and Lu 1984). In England, four was the maximum litter size found by Chaplin (1977), while six litters studied by Zhang (1996) at Whipsnade Zoo were all composed of twins. Childerley (2014) reported that litters of five were not uncommon in a managed population in Bedfordshire. At Branféré Zoo, mean litter size decreased from 2.3 in 2003, to 1.9 in 2004, and to 1.7 in 2005; this decrease was associated with an increase in density in the already high-density, captive population (Dubost et al. 2008).

Survival

Sex ratio at birth has been reported not to deviate from parity (Zhang 1996; Dubost et al. 2008). Sex ratios of adults in wild and captive populations have varied with the widest range being reported for samples from Whipsnade Zoo: females:males 1:0.81 (Middleton 1937) and 1:1.46 (Zhang 1996). The mean reported sex ratio for seven situations was 1:0.97 (Cooke 2019).

In the study at Branféré Zoo, 48% of fawns died during their first month, including 41% during the first week (Dubost et al. 2008). At 3 and 6 months, mortality was 57% and 70%, respectively; after 6 months, mortality rates lessened. Females at Branféré produced on average 0.7 young alive at the age of 1 month (Dubost et al. 2008), whereas, in coastal populations in Jiangsu Province in China, mean production of young per adult female by August was 1.5 (Xu et al. 1996). During the 42-year study at Woodwalton Fen, the average estimate for young deer per female surviving in the population until winter was 0.6 with a maximum of 1.7 (Cooke 2019).

Annual adult mortality at Branféré Zoo was 20% (Dubost et al. 2008), similar to losses experienced at Woodwalton Fen (Cooke 2009a, 2019). At Branféré Zoo, 8 years was the maximum age of a sample of 32 dead yearlings and adults (Dubost et al. 2008); from the rate of tooth wear, deer older than 8 years would be expected to have difficulty grazing. Potential longevity was given as 11 years by Dubost et al. (2011a).

Habitat and Diet

Habitat Selection and Movement

In Yancheng Nature Reserve in China, water deer selected quiet wetland areas with tall vegetation (Zhang et al. 2006). At Dafeng, they left the cover of wetland or grassland at night to feed on arable fields (Zhang and Zhang 2002); and behavior was evidently similar on islands in the Zhoushan Archipelago (Sheng and Lu 1984). In South Korea, Kim et al. (2011) found they occurred at highest densities in lowland areas and came into conflict with farmers over damage to crops. Won and Smith (1999) referred to attraction to forested areas and early successional vegetation. A Habitat Suitability Index was developed and successfully tested by Jung et al. (2016); Korean water deer avoided roads, preferring wetlands and forest areas. Because of its abundance and the extent of knowledge of its habitat requirements in South Korea, the water deer was chosen as a model species to examine the environmental impact of a road construction project (Choi and Lee 2019).

In England, water deer are recorded from a range of inland and coastal habitats including fens, grazing marshes, grassland, arable farmland, and woodland (Cooke and Farrell 1998; Cooke 2019). Their greatest concentrations occur in wetlands in eastern England. Such sites provide cover in the form of woodland, scrub dominated by sallow *Salix* species, beds of common reed *Phragmites australis,* and areas dominated by grasses and sedges. Optimal territories in wetland also have open feeding areas. Dense populations often have access to better feeding on nearby farmland at critical times of year when food is scarce inside the wetland. In counties such as Bedfordshire and Buckinghamshire, many deer live out their lives on predominantly-arable, sparsely-wooded farmland, and aggregations can be seen at certain times of year feeding and resting on open fields (Nau 1992; Childerley 2014; Cooke 2019). In the fens of Cambridgeshire, other populations live in quiet arable areas, making use of cover, such as dry ditches, patches of scrub, and farm crops (Cooke 2010, 2019). Large tracts of woodland do not comprise good habitat, but water deer will readily utilize woodland as part of their home range. In many situations they are "edge" animals moving regularly between cover and feeding areas.

As their name suggests, water deer are good swimmers in both fresh and sea water (Sheng and Lu 1984; Cooke 2019).

At Poyang Lake in China, mean seasonal home ranges in grassland and hills varied between 18 and 46 ha, depending on season and methodology, deer being forced by summer flooding to migrate to the hills (Xiao and Sheng 1990; Sun and Sheng 1990; Sun and Xiao 1995). At Whipsnade Zoo, mean annual home range was 21 ha (Stadler 1991). In contrast, mean seasonal home ranges of deer introduced to a large wildlife sanctuary at Shanghai were 100–300 ha (He et al. 2016). In South Korea, short-term tracking of four individuals revealed home ranges varying from 16 to 643 ha (Kim and Lee 2011).

Rate of dispersal is highly variable. Woodwalton Fen was until recently surrounded by arable farmland; its water deer population was well established by the 1970s, but even by 2014 the area where resident deer could be found was only 10 km by 15 km (Cooke 2019). This can be contrasted with Broadland in eastern England where there are large areas of interconnected wetland. In 1970, the species was still rare there, but it had colonized an area more than 20 km across by 3 years later; and by 2000, deer were more than 30 km away from the center of the Broads in some directions (Cooke 2019).

Feeding

Water deer have been placed between the concentrate selectors and intermediate feeders on the basis of structure of their digestive tract and feeding behavior (Hofmann 1985; Putman 1988). On the Zhoushan Islands off the coast of eastern China, Guo and Zhang (2005) reported that 60%

of plants eaten were forbs, 30% were woody species, 8% were ferns and 2% were grasses. Deer tended to focus on the tender tips of plants with woody stems being ignored. At Whipsnade Zoo, Zhang (1996, 2000a) examined rumen contents and found the diet comprised 66% grasses, 22% forbs and 4% woody species plus 8% unidentified fragments. In Cambridgeshire, rumen contents of eight deer revealed that 34% of drained mass was of browse-species, 62% of graze-species (including 36% of grasses and sedges), and 4% was unidentified (L. Farrell and A. Mitchell-Jones in Cooke 2019). Dietary composition differed in the various studies, indicating that the deer will make use of what is available – although at Whipsnade Zoo, choice appears to have been insufficiently varied (Hofmann et al. 1988).

Behavior

Social Behavior

Chaplin (1977), Stadler (1991), and Zhang (1996, 1998) described in detail maternal behavior toward young fawns. Immediately after the fawn is born, it is groomed, which may impart recognizable scent. A dead fawn elicits no maternal response. Live fawns are initially left in sheltered places and their mother returns four to five times per day to suckle and groom them. Older fawns may be seen accompanying their mothers. There is no significant difference in the attention given to male and female fawns. Amount of suckling decreases from about 3 weeks of age, and they are weaned by 3 months. No durable bond exists between mothers and their young (Dubost et al. 2011b).

There is no mutual grooming by adults, physical contact between individuals being restricted to what is necessary for reproduction and rearing young (Dubost et al. 2011b; Cooke 2019). For much of the year, water deer tend to be solitary. During the midwinter rut, females may shun the company of other females. Published information on group size varies according to circumstance and season. In China, for example, 85% of observations were of solitary animals in early summer, whereas this figure dropped to 55% during the midwinter rut (Sheng 1992). Year-round observations at Woodwalton Fen and Whipsnade Zoo gave figures of 78% and 34% of single animals, respectively (Cooke and Farrell 1981; Stadler 1991), reflecting the artificially-high density at the latter site. At Woodwalton Fen, males tend to be solitary apart from during the rut and when gathering to feed communally on the best grazing grounds in the early months of the year. This latter activity can result in loose associations of 10–20 deer of mixed ages and sexes. For deer resident on farmland, aggregations can also be seen at other times of year.

Mating Behavior

Males defend territories for much of the year, but territories may change in position during that time if the population moves seasonally or because of disturbance or other factors (Cooke 2019). At Poyang Lake in China, water deer returned in early winter to breed on grassland by the lake shore (Xiao and Sheng 1990; Sun and Sheng 1990; Sun and Xiao 1995). Farming activity caused further small-scale movement, with males following the females to good grazing; this was interpreted as female density dictating where the males set up territories. At Whipsnade Zoo, Stadler (1991) described males defending an area during the rut of about 1 ha with resources that were attractive to the females. At Woodwalton Fen, there are usually a number of relatively small, stable territories of up to 15 ha bounded by ditches and containing a mosaic of habitats providing good cover and forage (Cooke 2019). Any buck with such a territory will probably find one or more females based there. There are also areas characterized by substantial open habitat with adjacent patches of sallow that will be occupied by a number of bucks, which emerge to joust at dusk.

As midwinter approaches, males repeatedly test the estrous state of females living and feeding within their territories. When advancing toward a female, the male gives a soft whistle with neck outstretched and head rotating so there is a slapping movement of the ears (Stadler 1991). If the female is not in estrus, she may move away. However, if she is receptive, the pair is likely to stay together for several hours with the male courting and repeatedly mounting the female. Mating has been witnessed as early as mid-November at Branféré Zoo in France (Dubost et al. 2011b) and as late as early January in England (S. Plummer, pers. comm.). Mating lasts less than 20 s (Sun and Dai 1995; Zhang 1996; Dubost et al. 2011b) and is rarely seen in (semi-)natural habitat as it may occur in deep cover and/or at night. The mating system is polygynous (Sun and Dai 1995).

Analysis of fecal androgen levels in males indicated that they peaked when various types of rutting behavior were most frequent (Mauget et al. 2007; Yu et al. 2013). The first authors concluded that the sexually receptive state of females might only last a few hours, but they have earlier peaks in levels of progesterone metabolites without any apparent change in behavior. These observations may explain why males continually assess the females' state of estrus during the rut.

Communication

Males mark territories in a variety of ways. Commonly, there is scraping with urine and dung added; in the rut, pellet group size is reduced to increase the number of places marked (Sun et al. 1994). When defecating, deer adopt an exaggerated posture with the back arched. Males rub their foreheads against vegetation (Feer 1982; Sheng and Lu 1984; Stadler 1991), but the process of marking is not clear. An apparently different method of marking vegetation involves the tusk being hooked around a thin stem and the head being moved up and down (Harris and Duff 1970). It is possible that both methods result in scent being applied from the pre-orbital glands (Sheng 1992; Cooke 2019). The gas chromatographic profiles of scent profiles derived from the pre-orbital glands of 53 water deer were all different indicating that deer can leave their own individual signals (Lawson et al. 2000).

The bark of alarm has been variously described as a growl, scream, or grumble (Cooke and Farrell 1998). Deer are more likely to bark in the summer months (Cooke and Farrell 1981; Stadler 1991) perhaps because barking is associated with maternal care of fawns or because it is an effective method of communicating when vegetation is tall and dense. At Woodwalton Fen, individuals are also more likely to bark in winters when the population is relatively low. A mechanical whickering noise, probably made using the molars, is often heard when a male chases a rival and is occasionally uttered by females (Cooke and Farrell 1981; Stadler 1991). Other sounds include squeaking by males following females, soft squeaking by submissive deer, gentle whistling by mothers to fawns and loud screaming by injured or trapped individuals (Cooke and Farrell 1998).

Activity and Foraging Behavior

Zhang (1996, 2000b) found that deer at Whipsnade Zoo had two main periods of feeding during the day – in the morning and the late afternoon; feeding bouts lasting roughly 20 min were interspersed with periods of rumination. Ma et al. (2013) reported similar findings after studying time budgets and activity rhythms of captive water deer in a park at Shanghai. Recent studies at Woodwalton Fen have demonstrated that much feeding can also occur at night (Cooke 2012a, b; unpublished observations).

Water deer grip palatable vegetation between their incisors and the dental pad, and bite, tear, or pull it off. When feeding in the open on low vegetation, they typically continue to walk slowly. Movement is more limited when tackling a mass of taller vegetation, such as bushes of bramble *Rubus fruticosus* or patches of comfrey *Symphytum officinale*. Water deer have not been reported as standing erect on their hind legs in order to feed on higher vegetation (Cooke 2019).

Aggressive Behavior

Most of the aggression seen in water deer is between males, but females also display aggression and attempt to avoid other deer around the time they are giving birth (Stadler 1991; Dubost et al. 2011b). Chases were most often recorded at Woodwalton Fen during December and from March to May (Cooke and Farrell 1981, 1983, 2001). For much of the year, a male endeavors to maximize his distance from other males. Rivals entering a territory are approached by the resident male with a stiff gait and most are then chased away. If they do not back down, the two rivals may begin parallel walking, which can be a prelude to a fight. Deer attempt to land blows with their front legs and particularly to strike with their tusks (Stadler 1991; Dean and Cooke 2015; Cooke 2019). Fights are rarely seen in semi-natural habitats as they are usually over within a minute and may take place in dense cover. Males can sustain serious injuries when fighting, and death has occasionally been recorded (Stadler 1991). Ear damage is very common and occurs on the left ear more frequently than the right, suggesting that rivals tend to move anticlockwise during a confrontation (Cooke 2013); this suggestion is supported by recent photographic and video evidence (Cooke 2019).

Vigilance and Escape Behavior

Senses of smell, hearing, and, probably especially, sight are important to water deer to avoid danger. Stadler (1991) described deer regularly scanning their immediate environment, while staring behavior helps to evaluate a potential threat. Head bobbing sometimes occurs (Cooke and Farrell 1998; Tian et al. 2012). Deer may stand and bark at a perceived threat or crouch low trying to take advantage of any cover (Cooke 2019). More typically, however, they will run away – not necessarily moving quickly into the nearest cover, but often running straight or in wide arcs to distance themselves from the threat. Sometimes, particularly in open habitats, they bound in an exaggerated fashion flinging their hind legs high into the air. Korean water deer have been found to favor habitat that permits earlier detection of approaching predators (Eom et al. 2018). Similarly, in England, deer in open fields tend to keep away from the edges (Childerley 2014); in addition, being in groups confers extra protection via improved vigilance.

Parasites and Diseases

Ectoparasites reported from England include ticks *Ixodes* species, keds *Lipoptena* species, and unspecified lice, with park populations carrying higher burdens than wild animals living at lower density (Chaplin 1977). Several deer were examined from the mortality incident at Whipsnade Zoo, mentioned above, and were found to contain large numbers of nematodes: cause of death was given as enteritis (Middleton 1937). Survivors were sampled for post-mortem examination, but were found to be in excellent condition. In the study at Branféré Zoo, 25% of dead adults suffered from advanced periodontal diseases (Dubost et al. 2008). At Woodwalton Fen, deformed front legs have been noted on several adults that has made running and even walking difficult (Cooke and Farrell 1998).

Population Ecology

Density

Signs of deer were first seen in Woodwalton Fen in 1962 and, by the winter of 1976/1977, the population was reasonably stable with an estimated density of 40–50 per km^2 (Cooke and Farrell 1981). Density for deer based in the Woodwalton Fen reserve peaked at an unsustainable level of >100 per km^2 in 2011/2012 following destruction of fields of elephant grass *Miscanthus* species on adjacent farmland (Cooke 2019); previously a metapopulation had built up in the area. Densities up to 40 per km^2 have been reported from other English wetlands (Nobbs 2002; Cooke 2019). On farming landscapes in Bedfordshire, published estimates of density have included 0.2–0.5 per km^2 (Nau 1992) and 8–9 per km^2 for a maintained and managed population (Cooke 2009b). However, localized counts on one particular estate in Bedfordshire, where deer are managed for hunting, substantially exceeded 100 per km^2 (S. Plummer, pers. comm.). Densities in populations in China up until the 1990s were 3–90 per km^2 (Wang and Sheng 1990; Sheng 1992; Zhang 1996), but have declined since then (Chen et al. 2009). Across South Korea, landscape densities were 2 per km^2 in upland areas and 7 per km^2 in lowlands (Kim et al. 2011).

Population Dynamics

Zhang (1996) concluded that the water deer has many life-style characteristics that mark it out as being, relative to other cervids, r-selected in the sense discussed by MacArthur and Wilson (1967). Dubost et al. (2011a) collected data on growth, age at maturity, reproduction, and length of life for the water deer for comparison with data for 22 other artiodactyls of similar size and 8 large rodents. They found this species to be the most precocious and prolific ruminant, being comparable to large rodents.

Because of high productivity, a population of water deer can increase very quickly. During 1929 and 1930, 32 were released onto about 60 ha of undeveloped pasture at Whipsnade Zoo, and had increased to about 200 by 1933 (Middleton 1937). At Woodwalton Fen, periods of population increase occurred when the area of suitable habitat increased, and were characterized by high levels of recruitment (Cooke 2009a, 2019). This might have been due to better breeding and/or to reduced dispersal of young deer. During periods of relative stability, mean annual loss and recruitment to the winter population both averaged roughly 20%.

Currently, in England, shooting and road traffic are two major mortality factors for deer aged at least 6 months (Cooke 2009a, 2019). Apart from man, the only predators on adult water deer in England are dogs, including those that are used illegally for coursing purposes (Childerley 2014; Cooke 2019). Significant mortality incidents involving the death of more than 100 deer have been recorded at Whipsnade Zoo due to enteritis (Middleton 1937) and at Yancheng Reserve due to inundation (Zhang 1994; Xu and Lu 1996). Populations in China have been exposed to high levels of hunting; on the Zhoushan Islands during the winter of 1982/1983, 84% of deer were less than 3 years old (Sheng and Lu 1984). Large natural predators are now much rarer in China and Korea, and this is one reason for the recent increase in the deer population in South Korea (Jung et al. 2016; Kim 2017). In South Korea, one third of water deer may die annually from hunting, persecution or in traffic accidents (Kim 2017; Chun 2018). The water deer is the most frequently recorded species of roadkill in the data assembled annually by the South Korea Highway Corporation (Choi and Lee 2019). Water deer are prone to capture myopathy (Cooke and Farrell 1998; Kim and Lee 2011; Smith-Jones 2017).

High mortality of fawns has been found in many situations and has been blamed on a variety of factors including birthing difficulties, exposure, and predation (Chaplin 1977; Zhang 1996; Dubost et al. 2008; Cooke 2019). In England, small fawns are killed by a range of predators, especially the red fox *Vulpes vulpes*, and including avian species, such the carrion crow *Corvus corone*. In China, poaching of fawns for the colostrum in their stomachs to use as a traditional medicine has been a serious problem on the Zhoushan Islands (Zhang and Guo 2000).

At Woodwalton Fen, the main periods of adult mortality during the first 4 years of the study coincided with hard weather, with 64% of dead deer being found during January–March (Cooke and Farrell 1981). Snow cover and flooding forced deer out of the reserve onto the adjacent farmland. There were indications in this study of weather affecting breeding and recruitment (Cooke 2009a, 2019). Thus, both very cold, wet winters, and exceptionally mild winters could disrupt the rut and decrease recruitment the following year. High summer temperatures were associated with good recruitment during the following rut. Conversely, hot summers could lead to death of fawns from heat stress in less well-vegetated situations (Chaplin 1977; Dubost et al. 2008).

Numbers of water deer in woodland in three nature reserves in Cambridgeshire decreased as introduced Reeves' muntjac *Muntiacus reevesi* became established (Cooke 1998, 2006; Cooke and Farrell 2002). The reason for interaction was unclear but camera trap studies in winter revealed that water deer rarely browsed where Reeves' muntjac had previously fed (Cooke 2012b, 2019).

Conservation Status

During the twentieth century, numbers and range were reduced considerably in China by a combination of factors, especially habitat loss, coastal inundation, and hunting for food and traditional medicine (Xu et al. 1998; Harris and Duckworth 2015; Chen et al. 2016). The total population in China appears to have more than halved from at least 10,000 in the early 1990s (Sheng 1992; Sheng and Ohtaishi 1993) to less than 5000 in 2020 (Putman et al. 2021). The species has been reintroduced into the Shanghai area (Chen et al. 2016). Declines also occurred in the past on the Korean peninsula because of hunting and habitat loss (Won and Smith 1999; Kim et al. 2011; Harris and Duckworth 2015). However, the population in South Korea has increased since 1970 and may now be as high as 750,000 due to loss of predators and competitors (National Institute of Biological Resources 2017; Kim 2017; Chun 2018). Nevertheless, the level of persecution plus road deaths in South Korea was said to give cause for concern. Since 1994, the water deer has been categorized globally by the IUCN as Vulnerable (Harris and Duckworth 2015) and now has the same status on the Chinese Red List (Jiang et al. 2015).

Although there is no official recognition in Britain of the global conservation status of water deer, a substantial proportion of its English population resides in wetland reserves protected for other species (Cooke 2019). Water deer in England should be viewed as a valuable conservation resource as the source population in China appears to be extinct (Fautley 2013; Putman et al. 2021). It is conceivable that translocations back to China may take place in the future.

Management

In England, 19 individuals were introduced to Woburn Abbey in Bedfordshire between 1896 and 1913 (Chapman 1995). These bred well and were moved to other collections. Over the years, deer escaped from several sources and there were releases (Chapman 1995). By the early 1960s, small feral populations were established in a number of locations (Whitehead 1964). When the British Deer Society conducted a distribution survey in 2002, the species was well established in parts of eastern England, but isolated populations well away from this main range had failed to prosper (Ward 2005).

In parts of their native range, water deer are considered a pest on farm crops (Harris and Duckworth 2015; Jung et al. 2016; Chun 2018). A risk assessment for the species has been undertaken for the UK Government's Non-Native Species Secretariat (Cooke 2011). In England, observable damage to crops can occur on arable and grass fields, such as those beside dense wetland populations. While this probably does not currently translate into significant economic loss, it is of some concern (Cooke 2009a, 2019). Water deer have not been recorded damaging forestry interests, but they will contribute to road accident costs and to the risk of transmission of diseases to man and livestock. There are no reports of water deer being culled in England to protect conservation interests. Nevertheless, at Woodwalton Fen, they have affected the outcome of conservation management by destroying palatable coppice regrowth at densities estimated to be at least 80 per km^2 (based on the area of the reserve rather than their entire range), and the survival of scarce ground flora could also be at risk in such situations (Cooke 2012b, 2019).

England is one of the few countries outside China and Korea where there are opportunities to hunt water deer, and there is a limited market among trophy hunters for males with long tusks. On a few estates in eastern England, wild populations are managed for shooting (e.g., Childerley 2014) which can result in densities comparable, or even higher, to those in wetland sites. Because the deer are rare and small, the market for their venison is very restricted. In England, a close season for the hunting of water deer during April–October was introduced in 2007 by an amendment to the Deer Act of 1991. In 2010, the species was added to Schedule 9 of the Wildlife and Countryside Act, 1981, which lists animals that may not be released into the wild without an appropriate license.

Future Challenges for Research and Management

The species is viewed by Government and conservation authorities in England with some suspicion regarding its future status and impact. Many observers are independently monitoring its expansion and possible impacts, so it is unlikely to become a serious problem without people noticing. However, it is important that such observations are communicated and collated. Impacts on farmland need to be assessed for their relevance to economic yield and frequency of occurrence now and in the future. Management methods are likely to be the same as are used for other deer species.

Estimation of the national population requires knowledge of distribution through the country and density in utilized habitats. Average or typical densities are poorly researched, especially for a widely-available habitat such as arable land.

Our knowledge of the species would benefit from landscape scale studies of water deer through a range of habitat types and over a period of several years, monitoring movement, colonization and population change, interactions with man and other species, and responses to weather conditions and climate.

Acknowledgments I gratefully acknowledge the considerable contribution of Lynne Farrell to the study at Woodwalton Fen and the support and interest displayed by a succession of staff at the reserve. Mike McKenzie supplied the photograph in Fig. 1, and Marc Baldwin provided help with recent literature. I also thank Norma Chapman, Luca Corlatti, and Frank Zachos for helpful suggestions on earlier drafts of the manuscript.

References

Acevedo P, Ward AI, Real R, Smith JC (2010) Assessing biogeographical relationships of ecologically related species using favorability functions: a case study on British deer. Divers Distrib 16:515–528

Aitchison J (1946) Hinged teeth in mammals: a study of the tusks of muntjacs (*Muntiacus*) and Chinese water deer (*Hydropotes inermis*). Proc Zool Soc London 116:329–338

Allen GM (1940) The mammals of China and Mongolia, Natural history of Central Asia. American Museum of Natural History, New York

Chaplin RE (1977) Deer. Blandford Press, Poole

Chapman NG (1995) Our neglected species. Deer 9:360–362

Chen M, Zhang E, Yang N, Peng Y, Su T, Teng L, Ma F (2009) Distribution and abundance of *Hydropotes inermis* in spring Yancheng coastal wetland, Jiangsu Province, China. Wetland Science 7:1–4. (In Chinese with an English summary)

Chen M, Liu C, He X, Pei E, Yuan X, Zhang E (2016) The efforts to re-establish the Chinese water deer population in Shanghai, China. Anim Prod Sci 56: 941–945

Childerley P (2014) Herd reduction. Sport Rifle 107:90–92

Choi J, Lee S (2019) Application of Habitat Evaluation Procedure with quantifying the eco-corridor in the process of environment impact assessment. Int J Environ Res Public Health 16:1437

Chun K (2018) 171,000 deaths... How did it become a "protected animal"?. Cited 12 July 2018. Available from: www.joins.com/article/22304459. (In Korean)

Cooke AS (1998) Colonisation of Holme Fen National Nature Reserve by Chinese water deer and muntjac, 1976–1997. Deer 10:414–416

Cooke AS (1999) *Hydropotes inermis* Swinhoe, 1870. In: Mitchell-Jones AJ, Amori G, Bogdanowicz W, Kryštufek B, Reijnders PJH, Spitzenberger F, Stubbe M, Thissen JBM, Vohralík V, Zima J (eds) The atlas of European mammals. Poyser, London, pp 398–399

Cooke AS (2006) Monitoring muntjac deer *Muntiacus reevesi* and their impacts in Monks Wood National Nature Reserve. English Nature Research Report 681

Cooke AS (2009a) Chinese water deer *Hydropotes inermis* in Britain. Int Urban Ecol Rev 4:32–43

Cooke AS (2009b) Chinese water deer on farmland. Deer 15(2):14–17

Cooke AS (2010) Chinese water deer on the Huntingdonshire fens. Annu Rep Huntingdon Fauna Flora Soc 62:52–57

Cooke AS (2011) GB non-native organism risk assessment for *Hydropotes inermis*. Cited 15 July 2018. Available from: www.nonnativespecies.org/index.cfm?pageid=143

Cooke AS (2012a) Chinese puzzle. Deer 16(2):10–15

Cooke AS (2012b) The relative contributions of Chinese water deer (*Hydropotes inermis*) and muntjac (*Muntiacus reevesi*) to browsing at Woodwalton Fen National Nature Reserve. Nat Cambs 54:21–26

Cooke AS (2013) A sinister story from the Fen. Deer 16(9): 24–25

Cooke AS (2019) Muntjac and water deer: natural history, environmental impact and management. Pelagic Publishing, Exeter

Cooke AS, Farrell L (1981) The ecology of Chinese water deer (*Hydropotes inermis*) at Woodwalton Fen National Nature Reserve. Nature Conservancy Council, Huntingdon

Cooke AS, Farrell L (1983) Chinese water deer. British Deer Society, Warminster
Cooke AS, Farrell L (1998) Chinese water deer. Mammal Society/British Deer Society, London/Fordingbridge
Cooke AS, Farrell L (2001) Timing of the rut in a wild population of Chinese water deer. Deer 12:22–25
Cooke AS, Farrell L (2002) Colonisation of Woodwalton Fen by muntjac. Deer 12:250–253
Cooke AS, Farrell L (2008) Chinese water deer *Hydropotes inermis*. In: Harris S, Yalden DW (eds) Mammals of the British Isles. Handbook, 4th edn. Mammal Society, Southampton, pp 617–622
Croft S, Ward AI, Aegerter JN, Smith GC (2019) Modeling current and potential distributions of mammal species using presence-only data: a case study on British deer. Ecol Evol 9:8724–8735
Darman YA, Storozhuk VB, Sedash GA (2019) *Hydropotes inermis* (Cervidae), a new species for the Russian fauna registered in the Land of the Snow Leopard National Park (Russia). Nat Conserv Res 4: 127–129. (In Russian with an English summary)
Dean G, Cooke AS (2015) Chinese combat. Deer 17(4): 22–24
Dong W (1993) The fossil records of deer in China. In: Ohtaishi N, Sheng H (eds) Deer of China. Biology and management. Elsevier, Amsterdam, pp 95–102
Dubost G, Charron F, Courcoul A, Rodier A (2008) Population characteristics of a semi-free-ranging polytocous cervid, *Hydropotes inermis*. Mammalia 72: 333–343
Dubost G, Charron F, Courcoul A, Rodier A (2011a) The Chinese water deer, *Hydropotes inermis* – a fast-growing and productive ruminant. Mamm Biol 76: 190–195
Dubost G, Charron F, Courcoul A, Rodier A (2011b) Social organization in the Chinese water deer *Hydropotes inermis*. Acta Theriol 56:189–198
Eom T, Hwang H, Lee J, Rhim S (2018) Ecological factors influencing winter field sign abundance of Korean water deer *Hydropotes inermis argyropus* in a temperate forest in South Korea. Folia Zool 67:173–178
Fautley RG (2013) The ecology and population genetics of introduced deer species. DPhil thesis, Imperial College, London
Feer F (1982) Quelques observations ethologiques sur l'hydropote de Chine *Hydropotes inermis* (Swinhoe, 1870) en captivité. Z Saügetierkd 47:175–185
Gilbert C, Ropiquet A, Hassanin A (2006) Mitochondrial and nuclear phylogenies of Cervidae (Mammalia, Ruminantia): systematics, morphology, and biogeography. Mol Phylogenet Evol 40:101–117
Groves CP, Grubb P (1987) Relationships of living deer. In: Wemmer CM (ed) Biology and management of the Cervidae. Smithsonian Institution Press, Washington, DC, pp 21–59
Guo G, Zhang E (2005) Diet of the Chinese water deer (*Hydropotes inermis)* in Zhoushan Archipelago, China. Acta Theriol Sin 25:122–130
Haffner P, Savouré-Soubelet A (2015) Sur la piste des mammifères sauvages. Dunod, Malakoff
Harris RB, Duckworth JW (2015) *Hydropotes inermis*. IUCN Red List of Threatened Species 2015. Cited 15 May 2018
Harris RA, Duff KR (1970) Wild deer in Britain. David & Charles, Newton Abbot
He X, Chen M, Zhang E (2016) Home range of reintroduced Chinese water deer in Nanhui East Shoal Wildlife Sanctuary of Shanghai, China. Anim Prod Sci 56:988–996
Hofmann RR (1985) Digestive physiology of the deer – their morphophysiological specialisation and adaptation. In: Fennessy PF, Drew KR (eds) Biology of deer production. Royal Society of New Zealand, Wellington, pp 393–407
Hofmann RR, Kock RA, Ludwig J, Axmacher H (1988) Seasonal changes in rumen papillary development and body condition in free ranging Chinese water deer (*Hydropotes inermis*). J Zool 216:103–117
Hu J, Fang S-G, Wan Q-H (2006) Genetic diversity of Chinese water deer (*Hydropotes inermis inermis*): implications for conservation. Biochem Genet 44: 156–167
Jiang Z, Ma Y, Wu Y, Wang Y, Zhou K, Liu S, Feng Z (2015) China's mammal diversity and geographic distribution. Science Press, Beijing
Jung J, Shimizu Y, Omasu K, Kim S, Lee S (2016) Developing and testing a habitat suitability index model for Korean water deer (*Hydropotes inermis argyropus*) and its potential for landscape management decisions in Korea. Anim Cells Syst 20:218–227
Kim B (2017) A harmful wild deer is an internationally endangered species. Cited 6 Mar 2019. Available from: www.pressian.com/news/article/?no=167181. (In Korean)
Kim K, Cho D (2005) Status and ecological resource value of the Republic of Korea's De-militarized Zone. Landsc Ecol Eng 1:3–15
Kim B, Lee S (2011) Home range study of the Korean water deer (*Hydropotes inermis argyropus*) using radio and GPS tracking in South Korea: comparison of daily and seasonal habitat use pattern. J Ecol Field Biol 34: 365–370
Kim B, Oh D, Chun S, Lee S (2011) Distribution, density and habitat use of the Korean water deer (*Hydropotes inermis argyropus*) in Korea. Landsc Ecol Eng 7:291–297
Koh S, Lee B, Wang J, Heo S, Kyung H (2009) Two sympatric phylogroups of the Chinese water deer (*Hydropotes inermis*) identified by mitochondrial DNA control region and cytochrome *b* gene analyses. Biochem Genet 47:11–12
Lawson RE, Putman RJ, Fielding AH (2000) Individual signatures in scent gland secretions of Eurasian deer. J Zool 251:399–410
Ma F, Yu X, Chen M, Liu C, Zhang Z, Ye J (2013) Diurnal time budgets and activity rhythm of captive Chinese water deer (*Hydropotes inermis*) in spring and summer. Acta Theriol Sin 33:1–7
MacArthur RH, Wilson EO (1967) The theory of island biogeography. Princeton University Press, Princeton

Mauget R, Mauget C, Dubost G, Charron F, Courcoul A, Rodier A (2007) Non-invasive assessment of reproductive status in Chinese water deer (*Hydropotes inermis*): correlation with sexual behaviour. Mamm Biol 72:14–26

Middleton AD (1937) Whipsnade ecological survey. Proc Zool Soc London 107:471–481

National Institute of Biological Resources (2017) 2017 Wildlife survey. National Institute of Biological Resources, Incheon. (In Korean)

Nau BS (1992) Chinese water deer in Bedfordshire. Beds Nat 46:17–27

Ni B, Cao H, Sheng H (1993) A morphological study on scent glands of Chinese water deer and forest musk deer. In: Ohtaishi N, Sheng H (eds) Deer of China. Elsevier, Amsterdam, pp 80–84

Nobbs D (2002) Mammals of a Broadland reserve – Wheatfen. In: Keymer I (ed) Norfolk bird and mammal report for 2001. Norfolk & Norwich Naturalists' Society, Norwich, pp 295–300

Pocock RI (1923) On the external characters of *Elaphurus*, *Hydropotes*, *Pudu* and other Cervidae. Proc Zool Soc London 93:181–207

Putman R (1988) The natural history of deer. Comstock, New York

Putman R, Dunn N, Zhang E, Chen M, Miquel C, Savolainen V (2021) Conservation genetics of native and European-introduced Chinese water deer (*Hydropotes inermis*). Zool J Linnean Soc 191:1181–1191

Randi E, Mucci N, Pierpaoli M, Douzery E (1998) New phylogenetic perspectives on the Cervidae (Artiodactyla) are provided by the mitochondrial cytochrome *b* gene. Proc R Soc Lond B 265:793–801

Schilling A-M, Rössner GE (2017) The (sleeping) beauty in the beast – a review on the water deer, *Hydropotes inermis*. Hystrix 28:121–133

Seo H, Kim J, Seomun H, Hwang J, Jeong H, Kim J, Kim H, Cho S (2017) Eruption of posterior teeth in the maxilla and mandible for age determination in water deer. Arch Oral Biol 73:237–242

Sheng H (1992) Chinese water deer *Hydropotes inermis*. In: Sheng H (ed) The deer in China. East China Normal University, Shanghai, pp 96–110. (In Chinese with an English summary)

Sheng H, Lu H (1984) A preliminary study on the river deer population of Zhoushan Island and adjacent islets. Acta Theriol Sin 4:161–166. (In Chinese with an English summary. An English version appeared in 1985 in Contemporary mammalogy in China and Japan, 6–9)

Sheng H, Ohtaishi N (1993) The status of deer in China. In: Ohtaishi N, Sheng H (eds) Deer of China. Elsevier, Amsterdam, pp 1–11

Smith-Jones C (2017) The 2016 BDS deer distribution survey. Deer 18(5):15–18

Stadler SG (1991) Behaviour and social organisation of Chinese water deer (*Hydropotes inermis*) under semi-natural conditions. PhD dissertation, Universitat Bielefeld

Sun L, Dai N (1995) Male and female association and mating system in the Chinese water deer (*Hydropotes inermis*). Mammalia 59:171–178

Sun L, Sheng H (1990) Chinese water deer at the areas of Poyang Lake. J East China Norm Univ (Mamm Ecol Suppl) 9:21–26. (In Chinese with an English summary)

Sun L, Xiao B (1995) The effect of female distribution on male territoriality in Chinese water deer (*Hydropotes inermis*). Z Säugetierkd 60:33–40

Sun L, Xiao B, Dai N (1994) Scent marking behaviour in the male Chinese water deer. Acta Theriol 39:175–184

Tian X, Chen M, Wang H, Pei E, Yuan X, Shen G, Cai F, Xu G (2012) Vigilance pattern of Chinese water deer and the adaptation characteristics using flight initiation distance as a metric. Chin J Zool 47(6):25–30. (In Chinese with an English summary)

Wang H, Sheng H (1990) Population density and habitat selection of Chinese water deer in Zhoushan Islands. J East China Norm Univ (Mamm Ecol Suppl) 9:43–46. (In Chinese with an English summary)

Ward AI (2005) Expanding ranges of wild and feral deer in Great Britain. Mamm Rev 35:165–173

Whitehead GK (1964) The deer of Great Britain and Ireland. Routledge & Kegan Paul, London

Won C, Smith KG (1999) History and current status of mammals of the Korean peninsula. Mamm Rev 29:3–36

Xiao B, Sheng H (1990) Home range and activity patterns of Chinese water deer (*Hydropotes inermis*) in Poyang Lake region. J East China Norm Univ (Mamm Ecol Suppl) 9:27–36. (In Chinese with an English summary)

Xu H, Lu H (1996) A preliminary analysis of population viability for Chinese water deer (*Hydropotes inermis*) lived in Yancheng. Acta Theriol Sin 16:81–88. (In Chinese with an English summary)

Xu H, Lu H, Liu X (1996) The current status and habitat use of Chinese water deer in the coast of Jiangsu Province. Zool Res 17:217–224. (In Chinese with an English summary)

Xu H, Zheng X, Lu H (1998) Impact of human activities and habitat changes on distribution of Chinese water deer along the coast area in northern Jiangsu. Acta Theriol Sin 18:161–167. (In Chinese with an English summary)

Yu X, Chen M, Zhang E, Ye J, Pei E, Yuan X (2013) Correlation between seasonal changes in territory behaviour and fecal testosterone level in captive male Chinese water deer (*Hydropotes inermis*). Chin J Zool 48:43–48. (In Chinese with an English summary)

Zhang E (1994) Chinese water deer in Yancheng Nature Reserve. Unpublished report, University of Cambridge

Zhang E (1996) Behavioural ecology of the Chinese water deer at Whipsnade Wild Animal Park, England. PhD thesis, University of Cambridge

Zhang E (1998) Uniparental female care in the Chinese water deer at Whipsnade Wild Animal Park, England. Acta Theriol Sin 18:173–183

Zhang E (2000a) Ingestive behaviour of the Chinese water deer. Zool Res Sin 21:88–91

Zhang E (2000b) Daytime activity budgets of the Chinese water deer. Mammalia 64:163–172

Zhang E, Guo G (2000) Poaching as a major threat to Chinese water deer in Zhoushan Archipelago, Zhejiang Province, PR China. Deer 11:413–414

Zhang X, Zhang E (2002) Distribution pattern of *Hydropotes inermis* in various habitats in Jiangsu Dafeng Pere David's Deer State Nature Reserve. Sichuan J Zool 21:19–22. (In Chinese with an English summary)

Zhang E, Teng L, Wu Y (2006) Habitat selection of the Chinese water deer (*Hydropotes inermis*) in Yancheng Reserve, Jiangsu Province. Acta Theriol Sin 26:49–53. (In Chinese with an English summary)

Zima J, Král B (1984) Karyotypes of European mammals, part III. Acta Sci Nat Brno 18(9):1–51

Moose *Alces alces* (Linnaeus, 1758) 9

Magdalena Niedziałkowska, Wiebke Neumann, Tomasz Borowik, Marta Kołodziej-Sobocińska, Jonas Malmsten, Jon M. Arnemo, and Göran Ericsson

Contents

M. Niedziałkowska (✉) · T. Borowik · M. Kołodziej-Sobocińska
Mammal Research Institute, Polish Academy of Sciences, Białowieża, Poland
e-mail: mniedz@ibs.bialowieza.pl; tborowik@ibs.bialowieza.pl; mksobocinska@ibs.bialowieza.pl

W. Neumann (✉) · J. Malmsten · G. Ericsson
Department of Wildlife, Fish, and Environmental Studies, Faculty of Forest Sciences, Swedish University of Agricultural Sciences (SLU), Umea, Sweden
e-mail: Wiebke.Neumann@slu.se; jonas.malmsten@slu.se; goran.ericsson@slu.se

J. M. Arnemo
Department of Wildlife, Fish, and Environmental Studies, Faculty of Forest Sciences, Swedish University of Agricultural Sciences (SLU), Umea, Sweden

Department of Forestry and Wildlife, Management, Faculty of Applied Ecology and Agricultural Sciences, Inland Norway University of Applied Sciences, Elverum, Norway
e-mail: jon.arnemo@slu.se

© Springer Nature Switzerland AG 2022

L. Corlatti, F. E. Zachos (eds.), *Terrestrial Cetartiodactyla*, Handbook of the Mammals of Europe,
https://doi.org/10.1007/978-3-030-24475-0_23

Common Names

English	Moose (earlier referred to as Elk in Europe and Moose in North America)
German	Elch
French	Élan
Spanish	Alce
Italian	Alce
Russian	Лось

Taxonomy, Systematics and Paleontology

The moose tribe (Alceini) is most closely related to New World deer and diverged in the Plio-Pleistocene in Europe (Lister 1993; Geist 1998). The earliest form is *Alces* (*Libralces*) *gallicus* which occurred between 2.5 and 1.2 mya in Europe and Western Asia (Lister 1993). It has plesimorphic skull morphology, great length of its antler beams, and smaller body size than modern *A. alces* (Lister 1993). A larger moose species *Alces* (*Cervalces*) *latifrons* (Lister 1993) dated to the early Middle Pleistocene was also found in Europe and Asia. Probably in the middle Pleistocene, it entered North America. In eastern North America, the remains of an endemic form *Alces* (*Cervalces*) *scotti* dated to the Wisconsinan age was found. It is probable that *A. scotti* was an endemic offshoot from *A. latifrons* in North America (Lister 1993 and references therein). The dental and postcranial anatomy, skull structure, and basic antler plan of *A. latifrons* was similar but its size was 50% larger and the length of antler beam was shorter in comparison with *A. gallicus*. The transition from *A. latifrons* to *A. alces* occurred in Eurasia in the late Middle or early Late Pleistocene. The

Fig. 1 Moose (*Alces alces*): (**a**) a female with a calf; (**b**) a male (photographs by: K. Górecki)

contemporary moose species *A. alces* (Fig. 1) is known from remains dated from the early Weichselian (last) cold stage (*c.* 100 kya) up to the present day. In comparison to *A. latifrons*, *A. alces* is smaller, its antler beam has become substantially shortened and the skull architecture is changed (the facial region having deepened and the nasals shortened) (Lister 1993).

The genus *Alces* contains one species, divided into six (according to Whitehead 1993) to eight (Dzięciołowski and Pielowski 1993; Hundertmark 2016) or nine subspecies: *Alces alces alces* (Linnaeus, 1758) in Europe and Asia: *A. a. cameloides* Milne-Edwards 1867, *A. a. pfizenmayeri* Żukowski 1910, *A. a. buturlini* (division according to Hundertmark 2016), and *A. a. caucasicus* Verestsagin 1955 (extinct since the nineteenth century) in Asia, and the other four in North America (Peterson 1955). Hundertmark and Bowyer (2004) provided a distribution map of the eight extant subspecies. This division into subspecies is based on morphology and geographic distribution. Geist (1998) suggested that the species should be separated into two subspecies: *Alces alces alces* Linnaeus, 1758 and *Alces alces americanus* Clinton, 1822. Boeskorov (1999) treated the *alces* and *americanus* subspecies (separated by the Yenisei River) as species; however, according to the analyses of mtDNA there are three forms of moose: the Asian inhabiting Asia, the European occurring in Europe and Western Asia, and the American in North America (Hundertmark et al. 2002). See the Genetics subchapter for more details. Moose in Europe and Western Siberia have $2n = 68$ chromosomes and moose in North America, the Far East and Eastern Siberia have $2n = 70$ chromosomes (Boeskorov 1997).

Current Distribution

The range of the moose extends to both eastern and western hemispheres and is found north of latitude 40° N. In Eurasia the range of the European moose *Alces alces alces* extends from Norway in the northwest, through Sweden, Finland, Baltic States, Belarus, Poland, and Russia (including Western Siberia to the Yenisei River in the east), south to Ukraine (Fig. 2), vagrant in Croatia, Hungary, Romania, and Slovakia (Wilson and Reeder 2005; Jensen et al. 2020). Stable moose population reach their southwestern limit of their geographical range in Poland (Apollonio et al. 2010). Small, isolated populations occur also in southeastern Germany and western Czech Republic (Jensen et al. 2020). The distribution range of Asiatic subspecies of moose extends from western Asia through N Kazakhstan, N China (N Sinkiang), and possibly adjacent parts of Mongolia to Yakutia, Eastern Asia (Hundertmark 2016; Jensen et al. 2020), and Kamchatka.

Description

Moose are very well adapted to cold temperatures, with their anatomical and physiological features in the form of a large body size, long legs, thick fur, and long snout. Next to adaptation to the cold

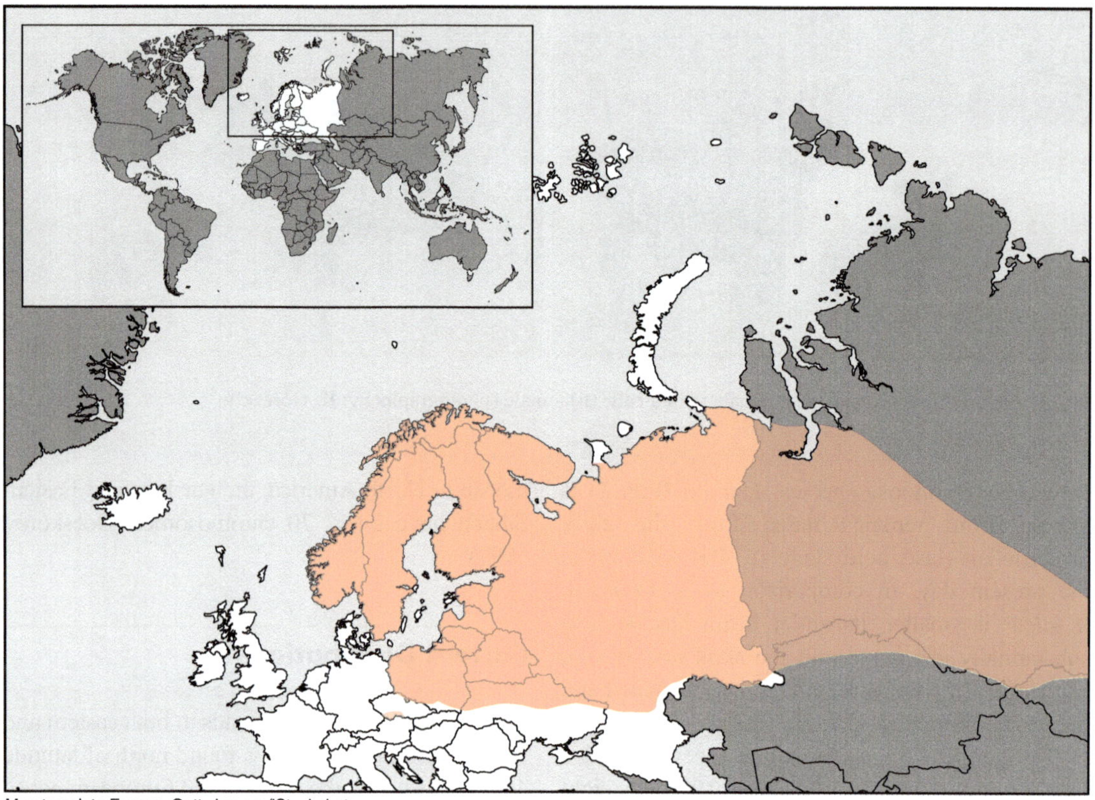

Fig. 2 Continuous range of the European moose (*Alces alces alces*). Distribution is based on the IUCN Red List of Threatened Species. Version 2017–2 and modified according to data published in Niedziałkowska et al. 2014 and Niedziałkowska 2017. Please also consult https://www.iucnredlist.org/about/citationinfo. (Map template: © Copyright Getty Images/iStockphoto)

climate, moose are also well adapted to water conditions. They can swim large distances (several kilometers), both in lakes and in marine systems. Their long legs with splaying hooves give effective support for both swimming and moving on muddy terrain. They can close their nostrils when diving and feeding with heads under water for aquatic forage.

Size and Morphology

Moose are the largest of the deer family. Differences in body size between sexes, however, are negligible at birth. On average, calves weigh between 13 and 15 kg, and as for adult moose, individuals are generally heavier at higher latitude compared to lower latitude (Ericsson et al. 2002, Ericsson, unpublished data). Adult moose weigh between 200 and 550 kg, but can reach a living mass of up to 650 kg and can have a shoulder height up to 2 meters. On average, male moose are about 20% larger than female moose, being both taller and heavier and having a coarser neck and a larger chin beard (Garel et al. 2006). Female moose are fully grown at an age of 3–4 years, whereas male moose grow for a longer period and can become heavier until an age of 8 years (Ericsson and Wallin 2001; Ericsson et al. 2001; Garel et al. 2006; Svensson 2008). Depending on harshness of their environment, morphological characteristics vary among areas. More specifically, moose have larger hooves and longer legs in areas with more severe snow conditions than expected from their size (Lundmark 2008). Long legs may help moose to move easily through snow

(Lundmark 2008). In areas with cold winters, moose are heavier and have shorter ears compared to moose in areas with milder winters (Lundmark 2008). On population scale, vegetation quality and phenology, as well as winter harshness and population density, are important factors that shape geographical gradients in moose body mass (Herfindal et al. 2006a, b). It is important to note, however, that variation in body mass in relation to environmental conditions was largest for moose at the borders of the fundamental niche space (Herfindal et al. 2006b). Next to environmental conditions (i.e., habitat diversity, seasonality, and winter harshness), however, the genetic background influences the variation in moose body mass as well (Herfindal et al. 2014). In moose, sexual size dimorphism is higher in areas with short growing seasons, whereas it is independent of population density (Garel et al. 2006). Yet, in strongly female-biased populations, the strength of female-biased sex ratios affects the sexual size dimorphism with lower male body growth in populations that show a high female to male ratios (Garel et al. 2006). In moose calves, body mass is positively related to the length of the vegetation period and negatively to altitude with a larger altitudinal effect observed for female calves (Ericsson et al. 2002). Following annual variation in weather conditions, the positive relation of calf body mass to length of growing season can generate cohort effects in female moose born at higher altitude (Ericsson et al. 2002). In line with the close link to vegetation length, calving date is an important factor affecting calf body mass next to mother age, thereby generating cohort effects (Solberg et al. 2007).

The moose skull achieves its characteristic elongated shape as animals mature (Svensson 2008). In premature moose, the profile of the skull looks short and juvenile, which grows longer and stronger at beginning of maturity, but still lacks the long and parallel headlines of a mature moose (Svensson 2008). At an age of 3–4 years, the skull of a female moose has attained its typical shape but continues to grow as the individual ages (Svensson 2008). In male moose, the shape of the skull is fully developed by an age of 3–5 years but continues to grow and to become coarser with growing age. In adult moose, skull height (26.1 cm vs. 24.5 cm), width (21.9–22.9 cm vs. 20.7–21.1 cm), and total skull length (58.4 cm vs. 56.8 cm) are on average longer in male moose compared to females, and there are no differences between lower and higher latitudes, except for skull width, with moose at higher latitude having on average wider skulls (Ericsson, unpublished data). Timing of senescence and deterioration of body features vary between the sexes, and are closely related to individuals' tooth wear (Ericsson and Wallin 2001; Svensson 2008). Compared to females, male moose experience a greater rate of tooth wear with age (Ericsson and Wallin 2001). Yet, it is important to note that tooth wear depends on the amount of silicates in the diet and the environment. As a result, moose that e.g., live in sandy environments, or have access to supplemental feed on the ground, have a fast tooth wear.

Pelage

Moose calves are born with a reddish coat, which will transform stepwise into the typical coloring of moose after about 3 months at the beginning of fall (Svensson 2008). The pelage of adult moose can vary in color, from dark brown and almost black to light brown and gray. In extremely rare cases, entirely or partially white moose can occur, which is, however, not indicative of albinism, but is caused by a recessive gene that lead to white fur with specks of brown. The color of the long legs is always light gray, otherwise independent from the individual's hue (Svensson 2008). The bright streak female moose have along the inside of their hind legs during all seasons and which is easily visible from behind as a vulval patch, distinguishes them from male moose year around. Seasonally, moose change their coat only once in late spring/early summer. During the following months, the coat grows to a well-insulated winter pelage. The pelage has two layers – a top layer of long guard hairs and a soft woolly undercoat. The guard hairs are hollow and filled with air, which aids insulation. Generally, thermal radiation is very low in moose (Svensson 2008).

Dentition

Moose have 32 teeth with the following dental formula on each side: Milk teeth 0.1.3.0/3.1.3.0 (Svensson 2008); permanent dentition 0.0.3.3/3.1.3.3 (Bubenik 2007). Calves are born with three incisors, one canine, and three premolars, all being deciduous teeth (Svensson 2008). In summer, the first (permanent) molar adds on. During the deciduous tooth period, the third premolar has three sections, but it is two-crowned as a permanent molar. During normal teeth development, an additional molar grows between December and February, and in the following summer, the yearling will develop its permanent rear molar. During its second fall, most moose have their final permanent dentition (Svensson 2008).

Age Determination

When not handling the animal, the shape of the skull together with body appearance can aid to estimate age class of a moose in the field. Counting cementum annuli requires removal of a tooth, normally the first incisor, and although tooth removal is possible while handling, counting annuli requires lab equipment. The only age determination method for use on live moose is tooth replacement and wear, which is subjective and can be site-specific (Rolandsen et al. 2008; Boertje et al. 2015). In dead moose, section of the first molar to count the annual rings allows for exact age determination (Skuncke 1949; Ericsson et al. 2001).

Antlers

In moose, only males have developed antlers which are replaced regularly. Antlers consist of the same bone tissue as the skeleton, and due to increased blood supply by the blood vessels at the pedicle, the antlers are regrown annually, protected by the velvet (Svensson 2008). Daylight, hormones, and forage control antler growth. While the growth hormone – which dominates between April and September – favors antler growth, testosterone – which increasingly dominates as the rutting period approaches – inhibits growth and promotes calcification (Svensson 2008). Just before the onset of the rutting period, the male removes the velvet skin, which at that point of time has become dry. With the end of the rutting period, levels of testosterone decrease and the growth hormones regain dominance (between December and March), which loosens the antlers' attachment to the pedicle and finally results in the dropping of the antlers (Svensson 2008). Dropping of the antlers has been observed between December and March with older bulls generally losing their antlers first (Svensson 2008). Although being less energy-demanding than gestation and milk production in females, male antler growth is resource-demanding, where phosphorus next to energy are likely the most limiting factors for growth (Moen and Pastor 1998).

Antler size is sensitive to selective harvest with increasing antler growth in young males when released from high harvest pressure, but maybe less affected by the biased sex ratios (Solberg and Saether 1994; Tiilikainen et al. 2010). Geographically, Russian moose antler morphology can vary largely, distinguishing four different groups: European, Siberian, the north of the Far East, the south of the Far East, and suggesting a distinction between the European and the Siberian group at the Ural Mountains (Kolesnikov and Kozlovskii 2014). Within European male moose, three phenotypes of antler exist – non-palmated (also referred to as cervina), intermediate, and palmated – and occurrence of a given type is correlated to age, body size, and antler size characteristics as well as to latitude (e.g., in Finland, Nygren et al. 2007). In younger and very old age groups, the cervina type dominates, whereas male moose at prime age (6.5–10.5 years) carry prevalently the intermediate and palmated types (Fig. 3, Nygren et al. 2007). Older (> 6.5 years) and heavier (> 270 kg carcass mass) bulls produce larger antlers with an average spread of 95 cm to 117 cm, depending on antler type (e.g., Finland, Nygren et al. 2007).

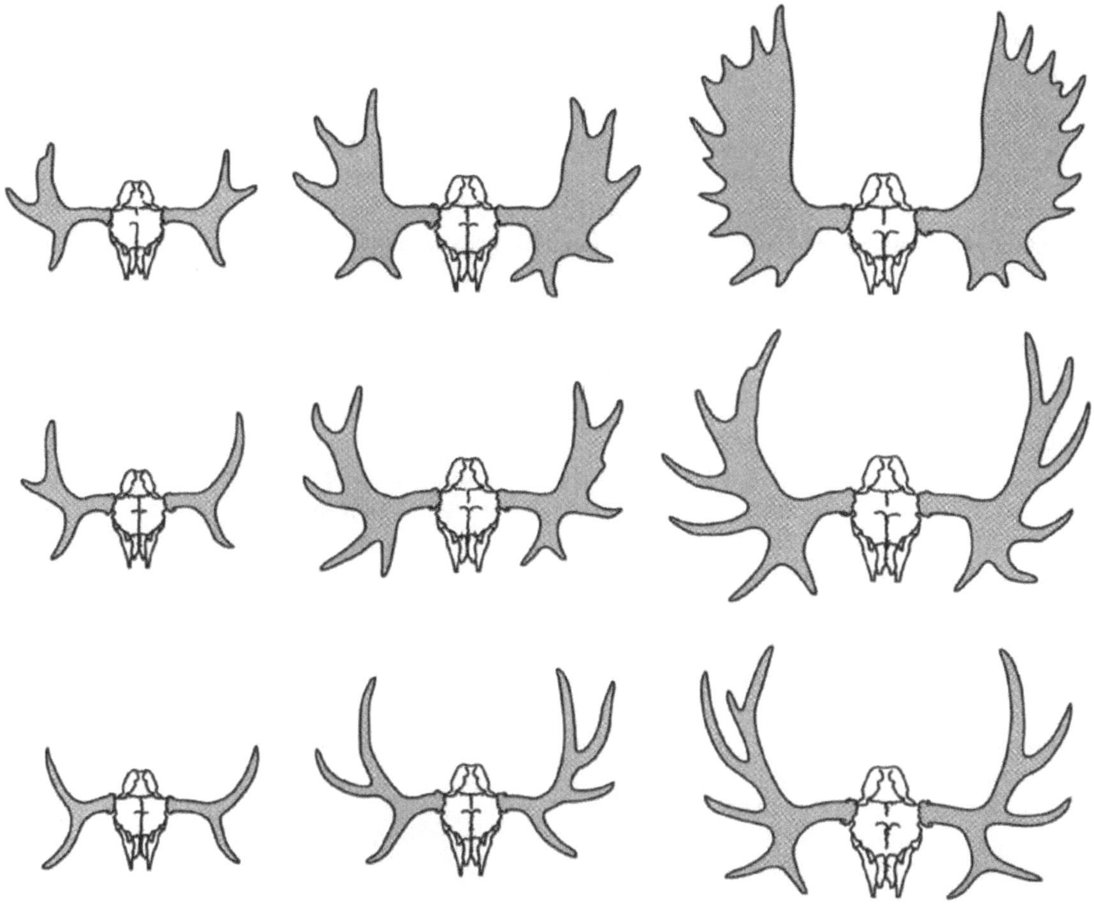

Fig. 3 Illustration from Nygren et al. 2007 (Fig. 1) showing the stylized drawing of three antler types in young (left), middle-aged (middle) and prime (right) bulls. Palmate type (top row), intermediate (center row), and non-palmate (bottom row). (Drawing by Maija Wallén, Nygren et al. 2007). (Reproduced with kind permission from Maija Wallén and Tuire Nygren, 10 Nov 2021)

Physiology

Moose are browsers, and like all other ruminants, they are dependent on microbial fermentation in the rumen to utilize the energy in the forage. They are effective in metabolizing carbohydrates, mostly in the form of cellulose (Solden et al. 2018). In late spring, summer, and early autumn, moose are anabolic and rely on a high protein content in their diet to increase their body condition prior to the winter months. During late autumn, winter, and early spring, they are in a catabolic state, as they rely on a diet rich in fiber.

In Scandinavian female moose, mean daily body temperature ranges between 38.03–38.64 °C with lowest body temperatures recorded in late winter and highest in summer (Græsli et al. 2020). Next to seasonal changes in body temperature, heart rates also vary seasonally with average rates of 40.5 beats per minute (winter) and of 71.9 beats per minute (summer). This results in a decrease of the animals' metabolic rate with 60% from summer to winter (Græsli et al. 2020). Thus, lowered body temperatures and heart rates demonstrate hypometabolism in moose, suggesting a strategy to reduce energy expenditures during a period when resources are limited (Græsli et al. 2020).

Genetics

Chromosomes

Two chromosomal forms exist in moose with moose in Europe and Western Siberia having $2n = 68$ chromosomes and moose in North America, the Far East and Eastern Siberia having $2n = 70$ chromosomes (Boeskorov 1997), although our knowledge of the Asian karyotype is based on a small sample size.

Phylogeny and Phylogeography

There are three mtDNA lineages of moose: Asian, European, and American (Hundertmark et al. 2002). Moose belonging to the European mtDNA genetic lineage inhabit Europe and western Asia and consist of three clades (the eastern the central and the western, Niedziałkowska et al. 2014; Świsłocka et al. 2020). The eastern clade is divided further into four haplogroups (Niedziałkowska 2017). The clades evolved before the Last Glacial Maximum (LGM) and their further evolution took place before, during and after the LGM and during the postglacial expansion (Niedziałkowska et al. 2014; Niedziałkowska 2017; Dussex et al. 2020). The eastern clade is much more complex and its range is much larger than the range of the central and western clades and covers almost the whole of Europe, except Scandinavia (Niedziałkowska 2017, Fig. 4). Also, the effective population size of the eastern clade is much larger than that of the western and central clades. Two of the four haplogroups of the eastern clade are common and occur in the entire mainland part of the continent. The other two have narrow or scattered distributions. The western clade occurs only in Scandinavia and in the south-western range of moose in Europe, and the central clade is endemic and was found almost entirely in eastern Poland (mainly in the Biebrza River Valley, Świsłocka et al. 2008) and western Belarus and their surroundings (Niedziałkowska et al. 2014, Niedziałkowska 2017, Fig. 4).

The highest haplogroup diversity was detected in the contact zones of different clades and haplogroups in western Siberia, the Ural Mountains, northern Finland and in the south-western part of moose range, and the lowest in Scandinavia, where only the western clade was found (Fig. 5, Niedziałkowska 2017). In the Urals, and western and central Siberia, the contact zone of European and Asia-American lineages of moose was identified (Moskvitina et al. 2011; Niedziałkowska 2017; Meiri et al. 2020). In the Ural Mountains individuals belonging to all four haplogroups of the eastern clade were found. In northern Finland, close to the Norwegian and Swedish borders, the contact zone between the western and eastern clades was detected. In the south-western European moose range (in Poland, Belarus, and northern Ukraine) specimens representing the eastern and western and central clades meet in the secondary contact zone (Niedziałkowska et al. 2014; Niedziałkowska 2017). It is also highly probable that the genetic hotspots within the European range of the species represent refugia, where moose survived the recent bottlenecks occurring during the last 3000 years (Niedziałkowska et al. 2014).

Genetic Diversity

Microsatellite DNA studies showed that in most study sites, the hetorozygosity (expected heterozygosity *He*) of European moose was relatively high (range *He* = 0.57–0.75, Haanes et al. 2011, Kangas et al. 2013, Niedziałkowska et al. 2016a) in comparison with other common European ungulate species as red deer *Cervus elaphus* (e.g., Niedziałkowska et al. 2012; Zachos et al. 2016) or roe deer *Capreolus capreolus* (Lorenzini and Lovari 2006; Olano-Marin et al. 2014) and slightly higher than in moose populations in North America (*He* = 0.45–0.64, Hundertmark 2009, Schmidt et al. 2009). Within European populations the highest heterozygosity *He* was detected in Karelia (*He* = 0.75, Kangas et al. 2013) and in Archangelsk, Nizhni Novgorod, Vladimir, and Tver regions of Russia (*He* = 0.73–0.74, Niedziałkowska et al. 2016a) and the lowest in Scandinavia (*He* = 0.53–0.66, Haanes et al. 2011, Niedziałkowska et al. 2016a).

Overall diversity of mitochondrial DNA sequences (haplotype diversity – *Hd*) was greatest

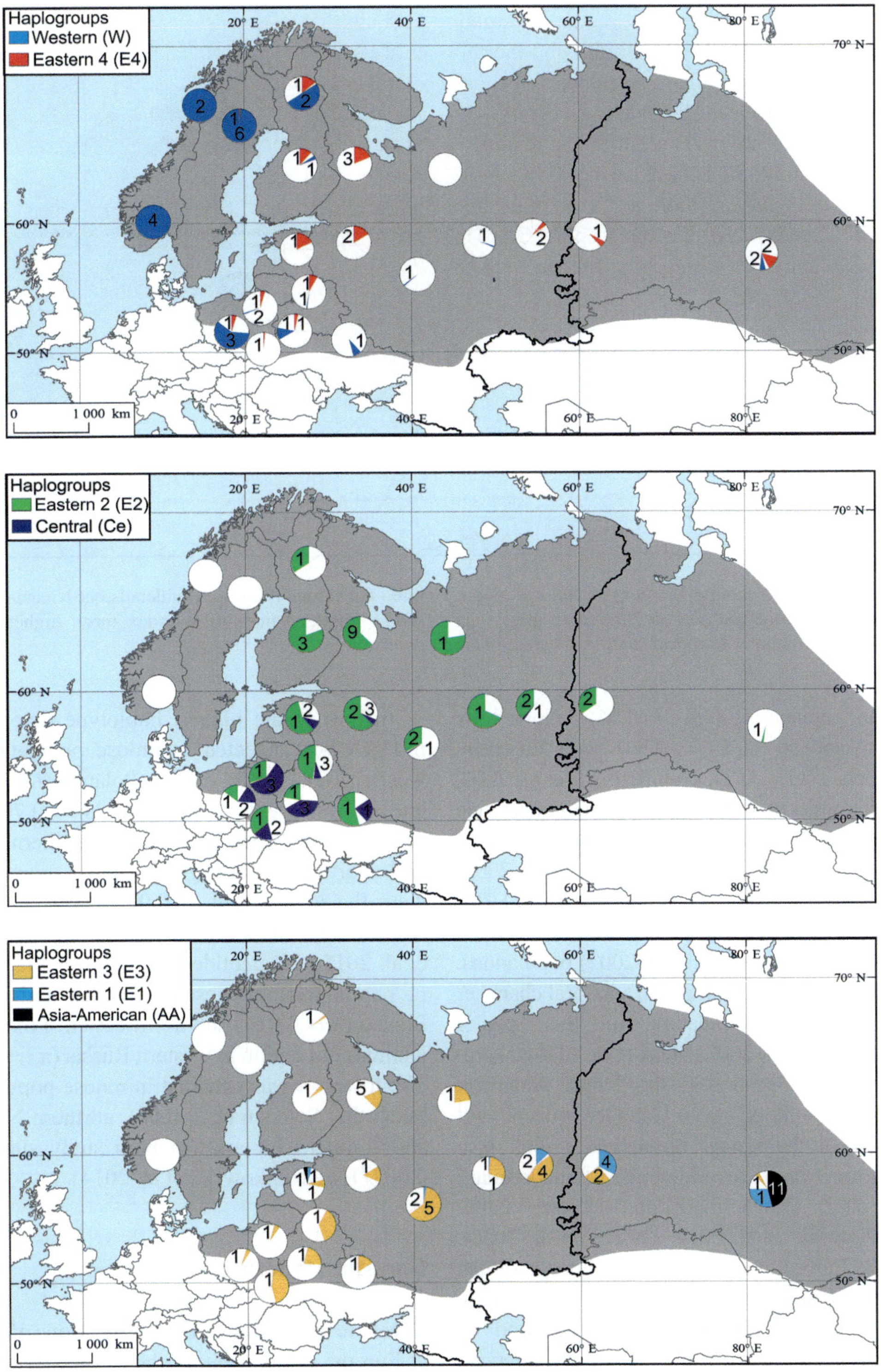

Fig. 4 The proportion of different mtDNA haplogroups (here including also clades and lineages) of moose (*Alces alces*) in 20 local demes in Europe and western Asia. Numbers inside the circles – number of haplotypes of each of the haplogroups, in grey – the range of the continuous European moose lineage, black line – the border of Europe. (Source: Niedziałkowska 2017, modified)

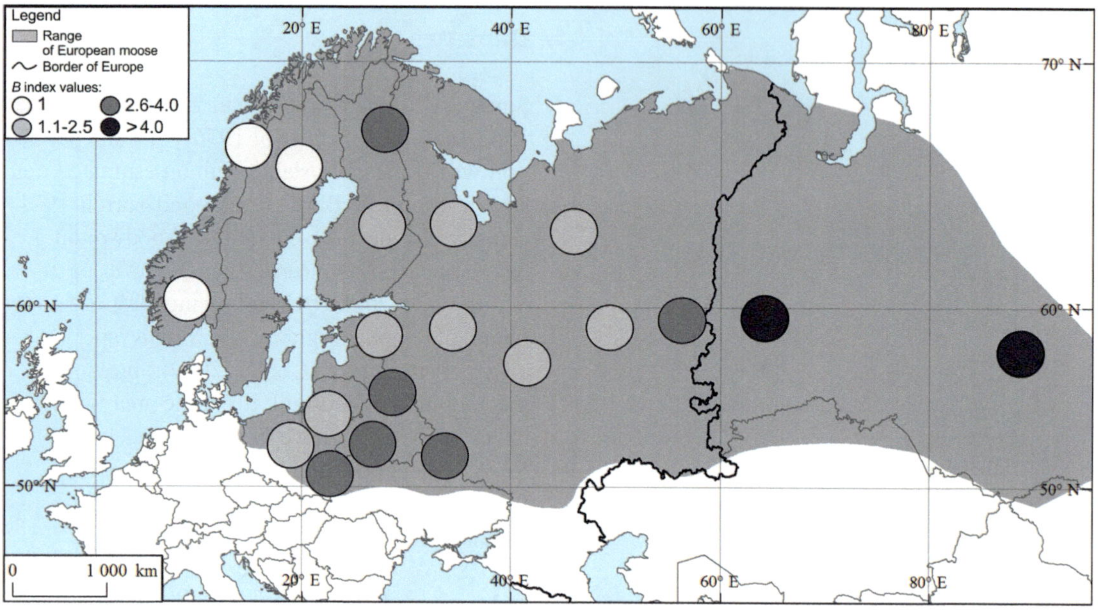

Fig. 5 The genetic diversity (B index) of 20 local moose populations in Europe and western Siberia, in grey – the range of the continuous European moose lineage, black line – the border of Europe (for details, see Niedziałkowska 2017). Higher values of B index mean higher genetic diversity

in Asian populations (*H*d = 0.94) followed by North American (*H*d = 0.86) and European populations (*H*d = 0.74, Hundertmark et al. 2002, Niedziałkowska et al. 2014). The overall nucleotide diversity (π) was also the highest in Asian moose population (π = 0.019, Hundertmark et al. 2002) in comparison with European (π = 0.013, Niedziałkowska et al. 2014) and North American (π = 0.007, Hundertmark et al. 2002) populations, which is probably a result of the ancestral character of the Asiatic population of the species (Hundertmark et al. 2002; Meiri et al. 2020). Moreover, both European and the North American populations suffered from founder effects and bottlenecks in the past as consequences of climatic oscillations and overhunting (Hundertmark et al. 2002; Hundertmark and Bowyer 2004; Niedziałkowska et al. 2014; Dussex et al. 2020). Also, the European moose mtDNA lineage differs from Asian and American mtDNA lineages by a 75-bp length mutation (indel, insertion in European sequences or deletion in Asian-American sequences) within the control region of mtDNA (Hundertmark et al. 2002; Niedziałkowska et al. 2014; Meiri et al. 2020).

In Europe the highest haplotype diversity of mtDNA was detected in moose populations in western Belarus and eastern Poland (*H*d = 0.86) and in south-central Norway (*H*d = 0.70). The haplotype diversity of Swedish, western Finish and northern Norwegian moose populations were the lowest (*H*d = 0.04, 0.009, and 0.29, respectively, Niedziałkowska et al. 2014, Kangas et al. 2015). Nucleotide diversity was the highest in populations of western Belarus and eastern Poland (π = 0.014) and north-eastern Poland and Kaliningrad region in western Russia (π = 0.013). The lowest π was detected in moose populations inhabiting Sweden (π = 0.00), northern Norway, and southern Finland (in both study sites π = 0.003) (Niedziałkowska et al. 2014).

Genetic Structure

There are two major genetic populations of moose in Europe: one inhabiting the Scandinavian Peninsula and the second occurring in the mainland part of the continent. There is a high concordance between genetic structures assessed by analyses of microsatellite and mtDNA (Niedziałkowska et al. 2016a). The time split between these two genetic clusters/mtDNA clades was estimated at approximately 28,000–29,000 years BP (Niedziałkowska et al. 2014, 2016a) indicating the beginning of the Last Glacial Maximum (LGM, Clark et al. 2009) and suggesting different glacial refugia. The main barrier to gene exchange between these two subpopulations is the Baltic Sea and an area near the border between Sweden, Norway, and Finland, where also the contact zone of two mtDNA clades of moose was identified (Niedziałkowska et al. 2016a, b). The admixture rate within the Scandinavian moose subpopulation is lower and gene flow distance is shorter in comparison with gene exchange within the continental moose subpopulation (Niedziałkowska et al. 2016b). However, lower level microsatellite structures were indicated both within the Scandinavian and continental subpopulations. Within the continental subpopulation the genetic structure assessed by microsatellites was also concordant with the distribution of different mtDNA clades. The substructuring of the European population of moose is an effect of several events and processes such as survival in different glacial refugia, postglacial expansion, and mixing of the continental refugial populations, more recent bottlenecks (approximately 3000–1200 years BP) and historical (eighteenth to twentieth centuries) decline in moose numbers (Niedziałkowska et al. 2016a).

Hybridization

There has been alleged hybridization between moose and wapiti *Cervus canadensis* and moose and domestic cow *Bos taurus* but such cases are not well documented (comp. Whitehead 1993) and almost certainly never occurred. Hybridization of European, Asian, and American mtDNA genetic lineages of moose is possible as several hybrids of these lineages were detected in Estonia (according to mtDNA they belong to the Asiatic or American lineage of moose, Niedziałkowska et al. 2014) and according to microsatellite DNA to the European (Niedziałkowska et al. 2016a, Fig. 4). Moreover, two mtDNA genetic lineages: European and Asiatic co-exist in the Urals and western Siberia (Moskvitina et al. 2011; Niedziałkowska 2017; Meiri et al. 2020).

Life History

Growth

At birth, moose calves average 13–15 kg, depending on latitude, and the body mass of a single calf is higher than body mass of twin calves, on average (Ericsson et al. 2002). There is no difference in the body mass of female and male offspring (Schwartz 1997; Ericsson, unpublished data). A young moose, which feeds on milk and forage diet, grows rapidly (self-accelerating phase), while after weaning (*c.* 5 months after birth) its growth rate decreases and continues at a slow rate for several years (Schwartz 1997). In the beginning the growth rate of calves depends predominantly on quality and quantity of the milk consumed. After weaning, due to adverse environmental conditions in the first winter, most of the calves lose mass and body dimensions. The first winter is the time when animal growth switches from the accelerating phase to the self-inhibiting phase. The decline in growth rate, however, is much lower for males than for females. Hence, females complete their body growth earlier in life and have lower body mass than males (4 and 7–9 years, respectively). After reaching reproductive senescence, moose body size decreases (Schwartz 1997).

Reproduction

Moose are short-day breeders, and thus breed seasonally. For most European populations, the rut spans from late August until mid-October (in northern Scandinavia) with the peak occurring in September, depending on latitude (Serafiński 1969; Baskin and Danell 2003; Veeroja et al. 2013; Malmsten et al. 2014a; Neumann and Ericsson 2018). Most females conceive during the first estrus; however, consecutive estruses also occur

(Markgren 1969; Kozhukhov 1989; Malmsten et al. 2014a). Female moose that do not conceive during the first estrus usually ovulate again. As a result, ovulating female moose can also be found considerably later in fall (Malmsten et al. 2014a). Male moose are polygynous and actively fight for access to females (Mysterud et al. 2004). The rutting period corresponds with high testosterone levels, which increase in preceding months causing testicle and antler growth (Schwartz 1997; Malmsten et al. 2015). Low variation in the rut length results in a relatively short calving period occurring from early May to beginning of June (Serafiński 1969; Baskin and Danell 2003). Corresponding to latitudinal difference in timing of estrus and conception, however, moose calves are born earlier at lower latitudes compared to higher ones (Neumann et al. 2020). In moose, the mean length of gestation is 231 $\pm$ 5 (SD) days (Schwartz and Hundertmark 1993). Most often single calf or twins are born (Markgren 1969; Kozhukhov 1989). The birth of triplets occurs but is a rare event that usually comes with a low survival chance of the calves (Neumann et al. 2011a).

In males, puberty is reached by yearlings although body growth is continued. Nonetheless, young bulls are outcompeted by older, prime-aged (7–9 years) males and do not actively breed, unless age ratios are skewed toward younger males by hunting practices (Mysterud et al. 2005). Moose become reproductive at an age of 1.5–2.5 years, depending on body mass (Malmsten et al. 2014a, 2015), which in turn is positively associated to the mass of the testes. In hunted populations, data on male moose of higher age (> 8 years) are limited, but the few examples suggest that senescence starts at 12 years of age and older in male moose (Ericsson et al. 2001; Mysterud et al. 2005). In Sweden, Sand and Cederlund (1996) documented traces of previous reproductive activity in 31.2% of females in the second year of life and 99% in females older than 5 years. The ovulation rate on a population level varies highly across populations and is driven mainly by the greater variability of this parameter in yearlings than in older females. While the probability of ovulation of adult, prime-aged females is usually very high (97% in 4–5-year-olds: Sæther and Haagenrud 1983), the proportion of breeding yearlings can vary considerably (Malmsten et al. 2014a). In Sweden, Markgren (1969) found that in populations from coastal regions 51% of yearlings had reached sexual maturity (passed puberty), while in an inland population it was only 8%. Female moose are reproductive until an age of 15 after which fertility usually decreases (Ericsson et al. 2001). The reproductive onset, ovulation, and twinning rate are positively associated with female body mass, and the effect is higher in yearlings than in adult females (Sand 1996). This relationship varies among populations and habitats exhibiting diverse carrying capacities (Sæther and Haagenrud 1983; Sæther et al. 1996; Sand 1996). However, twinning rate does not seem to be related to ovulation rate in adults (Schwartz 1997). In yearlings, even well-nourished individuals usually produce a single calf but such early reproductive effort can reduce further growth and life time reproductive output (Markgren 1969; Sæther and Heim 1993). Twinning rate of adults varies across European populations (from 1.1 to 1.6 embryos per adult female; Markgren 1969, Baskin and Danell 2003) and increases with female age and habitat quality, reaching a maximum level in prime-aged females (Sæther and Haagenrud 1983).

Sex ratio of calves at birth varies across populations. Although observed in other polygynous ungulates, the sex ratio of moose offspring seems not to be affected by maternal condition or dominance. Nonetheless, Sæther et al. (2004) found that the age composition of males in the population can affect offspring sex ratio. The proportion of male calves increased with the age of their fathers.

Survival

Moose survival is shaped by human-induced and natural factors including hunting, road accidents, predation as well as parasitic and bacterial infections. Annual survival of adult moose ranged from 75% to 94% depending in majority on the site-specific hunting intensity (Van Ballenberghe and Ballard 1997). Nonetheless, in populations with limited hunting harvest natural mortality factors can prevail (Carstensen et al. 2018). The survival of adult moose remarkably differs between sex and age classes (Ericsson and Wallin 2001). For example, in Norway harvest mortality of adult moose increased with age, while in females the highest mortality was indicated among young and old (post-prime) females (Solberg et al. 2000). In calves, stochastic mortality is about 10% during summer (e.g., in system without large predators, Ericsson 1999). Mortality of calves depends on population density and weather conditions, which operate directly and indirectly through plant phenology and maternal body condition (Filonov and Kaletskaya 1988; Stubsjøen et al. 2000; Grøtan et al. 2009; Monteith et al. 2015). In systems where large carnivores occur, mortality of the calves in the first year can be highly influenced by wolf or brown bear predation. Brown bears are effective predators on calves up to a month of age and reduce calf summer survival by 26%, resulting in overall calf mortality rates of about 36% in bear areas (Swenson et al. 2007). Older calves are usually too fast to be killed by a brown bear. In wolf areas, annual calf survival ranges between 56–76% with a summer survival of 87–100%, which suggests that wolves may affect calf survival more strongly later in the year (Sivertsen et al. 2012).

Habitat and Diet

Spatial Movements

In many moose populations, individuals track the changes in food resources and migrate seasonally between winter and summer ranges (partial migration). Moose migration most often occurs in spatiotemporally seasonal heterogeneous environments, while non-migratory behavior is more common in populations inhabiting homogeneous or isolated habitats. At broader geographical scale, latitudinal change in the scale of landscape heterogeneity and forage predictability causes different movement behaviors of moose – from long-distance migrations at higher latitude to short-distance migration or nomadic and resident behavior at lower latitude (Bunnefeld et al. 2011; Singh et al. 2012; van Moorter et al. 2013; Borowik et al. 2020a). However, migration tendency can be affected by other intrinsic and extrinsic factors, e.g., individual age, snow cover, predation risk, human hunting, population density, and road infrastructure (Ball et al. 2001; Seiler et al. 2003; Singh et al. 2012, 2014). There are no differences between the sexes in migratory behavior, but migration distances and thus average home range size can be larger for males (Singh et al. 2012; Allen and Singh 2016). Migration parameters (i.e., migration distance, timing and duration) differ significantly between moose populations. For example, in Sweden, individuals from northern populations migrated up to 122 km in 16 days on average, while in the south it was around 5 km in half a day (Singh et al. 2012). Migration timing is strongly affected by snow cover. In spring, start of migration is related to snow disappearance, while in autumn to snow accumulation (Ball et al. 2001; Singh et al. 2012). Moose from the same population have overlapping home ranges. The size of annual and seasonal home ranges depends on a set of different factors, including: individual sex, age, browse quantity and quality, reproductive status, as well as weather conditions. Home ranges are usually smaller in females, in young individuals, in high quality habitats, and under harsh snow conditions (Cederlund and Sand 1994; Bjørneraas et al. 2012; Allen and Singh 2016). Moreover, due to mobility constraints, females accompanied by offspring tended to have smaller summer home ranges compared to barren females (Van Beest et al. 2011).

Habitat Selection

In Europe, moose occur in a great variety of habitats. Most often moose occupy different types of coniferous and mixed deciduous forests but they can also be found in willow and birch shrubs of the river floodplains and in tundra or subalpine shrub communities (Dzięciołowski and Pielowski 1993; Baskin and Danell 2003; Allen and Singh 2016).

The habitat selection of moose is shaped by a set of interactive factors, i.e., food availability, predation risk, hunting, and weather conditions (ambient temperature, snow cover) (Dussault et al. 2005; Bjørneraas et al. 2012; Van Beest et al. 2012; Mathisen et al. 2018). Moreover, the strength of effects of these factors is different at two spatiotemporal scales. Dussault et al. (2005) found that at the landscape scale moose selected habitats reducing predation risk and providing increased food availability, at the cost of exposure to harsher snow conditions. At the home range level, solitary moose faced a trade-off between snow avoidance and forage availability, whereas females with calves searched for habitats with a lower predation risk. Other surveys found that at the landscape level moose choose habitats providing a high supply of low-quality forage while at the within-home range level moose select for high-quality food (Van Beest et al. 2010) or seek habitats offering cover and low human impact (Herfindal et al. 2009). Moose show also state-dependent habitat selection where reproductive females trade-off cover against forage availability (Bjørneraas et al. 2012). The scale-dependent differences in habitat selection can also express seasonal variation. In Finland, in summer there was only a slight difference between habitat selection at both spatial scales, while in winter, at the landscape level moose preferred habitats providing a high food supply (young successional forests) but within home ranges moose selected areas with diverse habitats (Nikula et al. 2004).

Moose, a concentrate selector (sensu Hofmann 1989), prefer highly productive habitats providing a diverse food supply (Danell et al. 1991). In winter, moose forage is primarily low quality, moose select young forest successional stages (especially pine plantations *Pinus sylvestris*), which are abundant in preferable browse (Bergström and Hjeljord 1987; Bergqvist et al. 2018). In harsh snow conditions, in turn, moose often switch to older tree stands with thinner snow cover and an abundant shrub layer (Peek et al. 1982). In snowless periods, beside young successional forests, moose utilize older pine or other non-pine-dominated forests where they can find important forage plants and where shading facilitates vegetation quality (Hjeljord et al. 1990; Olsson et al. 2011). In many areas, in spring moose move from forest winter ranges to open or semi-open marshlands located around lakes or in river valleys (Gębczyńska and Raczyński 1984). Despite greater exposure to heat stress and insects, moose frequently use wetland habitats, as they offer the high biomass of easily palatable forage which is high in sodium. Moose cope with high ambient temperature and thermoregulatory problems by looking for shelter in higher and denser canopies or cool themselves by wallowing in the water (Van Beest et al. 2012; Melin et al. 2014; Borowik et al. 2020b). Thermoregulatory strategies, however, differ among moose, and those individuals that best adjusted the cost-benefit trade-off of different habitats in relation to variation in thermal conditions generally optimized their body conditions (Van Beest and Milner 2013).

Moose exhibit a distinct sexual size dimorphism that causes different nutritional requirements and habitat selection (Spaeth et al. 2004; Bowyer 2004). Outside the rut period, male and females are spatially segregated. Females select for habitats of high forage quality and low predation risk, whereas males choose habitats offering a high forage supply (Miquelle et al. 1992; Oehlers et al. 2011).

Diet

According to feeding type, moose are selective browsers. Moose have a relatively small ruminoreticular chamber and they have limited ability to digest low-quality food (Hofmann 1989). This physiological constraint, along with high body mass, forces animals to consume large amounts of a high-quality forage that can be easily digested and quickly passed through the digestive system. To meet their nutritional demands, moose consume a variety of plant species. Besides dietary seasonal variations, moose vary their diet also on a daily basis to balance their nutritional intake (e.g., intake of protein versus non-protein macronutrients such as carbohydrates, lipids and fermentable fibers, Felton et al. 2016, 2018). In Eurasian moose, diet consists of around 250 plant species (Dzięciołowski and Pielowski 1993). Baskin and Danell (2003) reported about

83 plant genera eaten by moose in the European part of Russia and Belarus including trees, shrubs, herbs, grasses, ferns, and lichens. Yet, moose consume high quantities of only a few species (Renecker and Schwartz 1997). The most important are woody plants providing browse throughout the year. In European woodlands, these are leaves and twigs of trees and shrubs from these genera: *Salix, Betula, Populus, Pinus, Quercus, Alnus, Padus, Corylus, Sorbus, Acer, Tilia, Rosa, Ribes,* and *Vaccinum* (Dzięciołowski and Pielowski 1993; Spitzer et al. 2020, 2021). Moose diet matches the geographic variation in habitat used, thus its diversity grows southward with increasing habitat fertility.

In the temperate zone, a high forage biomass in the growing season is succeeded by scarcity of food in winter. Hence, moose diet is clearly seasonal and more diverse and nutritious in the growing season than in winter. In the growing season, moose consume young tree shoots and leaves, as well as shrubs, herbs, and grasses. When available, they search for wetland and aquatic vegetation providing moose with a highly palatable and nutritious forage (Tischler 2004). In Scandinavia, berry bushes, e.g., blueberry (*Vaccinium myrtillus*), lingonberry (*Vaccinium vitis-idaea*), and heather (*Calluna vulgaris*), are important food sources in spring and autumn (Mathisen et al. 2010). Leaves from different deciduous trees and different herbs dominate the diet in summer, and twigs of coniferous trees (e.g., pine *Pinus sylvestris*) comprise the diet in winter. In summer, moose diet is dominated by only a few species, i.e., birches (*Betulus* spp.), rowan (*Sorbus aucuparia*), and bilberry (*Vaccinium myrtillus*) among herbaceous forage (Shipley et al. 1998; Wam and Hjeljord 2010). In Norway, moose consumed overall only 15 different browse and 31 herbaceous species (Wam and Hjeljord 2010). From late summer, vegetation senescence (and thus decreasing protein concentration) and lignification cause a deterioration of food quality and a gradual decrease in the diversity of moose diet. In winter, moose rely on browse (up to 99% according to Morow (1976)) and eat predominantly tree and shrub twigs and bark. In Scandinavia, winter diet is dominated by birches, rowan, willow (*Salix* spp.), and pine (*Pinus sylvestris*) (Heikkilä and Mikkonen 1992; Shipley et al. 1998; Wam and Hjeljord 2010). In Poland, at the south-western edge of the species' range, from December to March moose consume predominantly pine, comprising a proportion in moose diets of 92% (Morow 1976; Czernik et al. 2013).

Moose are strongly selective herbivores that can make choices on the basis of both plant morphology and concentrations of plant secondary chemicals (Löyttyniemi 1985; Shipley et al. 1998). However, selection intensity depends on plant abundance and varies between habitats and populations. Many plant species that dominate moose diet are not preferred ones. In winter, low forage availability often forces moose to forage on the most abundant, not necessarily preferred food (Hofmann 1989). For example, although pine constitutes the main share of the winter diet in many locations across Europe, this species is not selected by moose, which prefer trembling aspen (*Populus tremula*), rowan, or buckthorn (*Rhamnus cathartica*) instead (Dzięciołowski and Pielowski 1993). In Scandinavia, in winter moose selected for birches, rowan, willows, and trembling aspen and avoided Norway spruce (*Picea abies*) (Bergström and Hjeljord 1987; Shipley et al. 1998; Wam and Hjeljord 2010). Interestingly, the strength of selectivity for a given species is related to availability of the other preferred species. Wam and Hjeljord (2010) reported that in southern Norway selectivity for birches was negatively associated with availability of more preferred rowan, willows, and trembling aspen.

Moose also utilize human-supplied feeding stations within their home range, in particular during winter, which affects their nutritional balance and habitat utilization (Sahlsten et al. 2010; Mathisen et al. 2014; Felton et al. 2017).

Behavior

Social Organization

Moose are solitary with individuals ranging separately through the landscape, except for female moose and offspring of the year (Månsson et al. 2017). If a female does not reproduce in a given year, she can be accompanied by her offspring from the previous year (Svensson 2008). During certain periods of the year, e.g., in winter when forage availability is limited, moose can accumulate locally by gathering in loose temporary groups formed by animals that occur simultaneously in favored foraging areas. Those moose groups can vary in size from a few to several individuals without any obvious set-up or group structure. Groups can consist of mixed sexes or one sex only (Månsson et al. 2017).

Antipredatory Behavior

In Central Scandinavia, moose are the major prey for wolves (*Canis lupus,* Sand et al. 2012). In contrast, moose contribute considerably less to the diet of wolves in Eastern Europe, where red deer are a sympatric deer species and moose occur in lower densities (Jędrzejewski et al. 2002, 2012). In general, European moose have experienced varying overlap with top predators such as wolves and brown bears *Ursus arctos* over time. Whereas wolves and moose are sympatric in Eastern Europe and Finland without any major temporal disruption, wolves were functionally extinct for more than 100 years in Scandinavia and started to recolonize the peninsula by the end of the nineteenth century (Wabakken et al. 2001). In Norway and Sweden, in particular, hunting has replaced natural predation by wolves and bears during the decades of their absence on the peninsula. Today, Scandinavian moose show no or little behavioral adjustment in their anti-predator behavior toward wolves in spite of the increasing predation risk by wolves (grouping behavior, Månsson et al. 2017; habitat use, Nicholson et al. 2014; wolf predation success, Sand et al. 2006; activity pattern, Eriksen et al. 2011; movement behavior, Wikenros et al. 2017). In contrast, Scandinavian moose reduced the predation risk by brown bears on moose calves within one generation (Berger et al. 2001). Hunting accounts for the major source of mortality in adult moose, also in wolf areas, which seems to constrain the ability of wolves to generate behaviorally mediated trophic cascades, and thus ecosystem effects, through alterations in anti-predator behavior in Scandinavian moose (Gervasi et al. 2013; Nicholson et al. 2014; Sand et al. 2006).

Activity

Based on movement data (meters per hour), moose show distinct diurnal and seasonal activity patterns. Diurnally, moose follow a bimodal activity pattern with higher activity during dusk and dawn (Neumann et al. 2012). Male moose become increasingly more active during autumn, September until mid-October (Neumann and Ericsson 2018). In areas with distinct seasonal changes in timing of sunset and sunrise over the year, diurnal peaks of activity thus can be considerably different over the year. Seasonally, moose are most active in May, September, and October, and show lowest levels of activity during winter (Neumann et al. 2012). Corresponding to their diurnal activity pattern, moose show a temporal shift in habitat use with using more open habitats during the night and habitats with more cover during daytime (Bjørneraas et al. 2011). Moose are heat-sensitive and move less with increasing ambient temperature with a decrease in movement activity starting when ambient temperatures exceed 14 °C (using GPS-collar sensor temperature as a proxy, Ericsson et al. 2015a).

Senses and Communication

Moose have an excellent sense of smell and hearing (Svensson 2008), which are thus their major senses to scan their surrounding environment. When resting, moose select often for minor hills in the landscape that provide better overviews as winds contribute olfactory information from the surroundings. Studies of the visual capability of moose are limited (but see Schiviz et al. 2008). As other deer species, moose can see colors (Schiviz et al. 2008). A higher number of rods than cones, however, suggests better visual capacity for discriminating contrasts rather than colors in moose compared with other deer species (Schiviz et al. 2008).

Reproductive Behavior

During autumn, male moose visit areas that overlap or are separated from areas that they utilize during the summer. Male moose increase their movement activity considerably for about 3 weeks between September and October, whereas female moose do not show such change in movement activity (Neumann and Ericsson 2018), indicating that male moose actively search the area for females to breed with. For calving, female moose select forests with low amounts of vegetation that are below five meters of height and select for forests with dense vegetation following calving (Melin et al. 2016). In predator-free areas, females show site fidelity to their calving areas (e.g., in Norway, Tremblay et al. 2007).

Response to Human Activity and Infrastructures

In contrast to North American moose, female and male moose in Scandinavia are less bold and generally do not show an aggressive behavior, when confronted with humans and human activities. More specifically, moose show a short-term, but distinct behavioral response with moving faster and generally leaving the area when disturbed by humans (recreational activity: Neumann et al. 2010, 2011b; hunting: Ericsson et al. 2015b). Experimental disturbances did not indicate any signs of habituation to disturbances by recreational or hunting activity (Neumann et al. 2011b; Ericsson et al. 2015b). On a temporal scale, moose adjust proximity to human infrastructure such as roads and cabins by remaining at farther distances during daytime (Lykkja et al. 2009; Neumann et al. 2013). Moreover, moose trade off foraging against road proximity with smaller roads having less impact compared to larger roads (Eldegard et al. 2012). Linear infrastructural objects (e.g., roads and power lines) create both barrier and corridor effects in moose moving through the landscape, even though topography and forest cover have larger effects on how moose move (Bartzke et al. 2015). Moose show an avoidance of crossing roads or power lines in forests and are more likely to move along those linear objects when getting closer (Bartzke et al. 2015). In contrast to roads, power lines have no or little impact on moose movement (Neumann et al. 2013; Bartzke et al. 2015). In moose populations with a large portion of migratory individuals, moose road-crossings peak in May and June, and between November and January, corresponding to timing of migration (Neumann et al. 2012; Allen and Singh 2016). On a diurnal base, moose pass roads preferably during twilight when they shift habitat (Neumann et al. 2012; Neumann and Ericsson 2018).

Parasites and Diseases

Ectoparasites

Moose can be a host for many ectoparasite species. The most prevalent are: ticks (Acari: Ixodidae) (Holmes et al. 2018); chorioptic mange mites (Acariformes: Psoroptidae: *Chorioptes*) causing skin disorders (Bochkov et al. 2014; Hestvik et al. 2007; Kadulski 1996), and a haematophagous parasitic fly, the deer ked (*Lipoptena cervi*) (Madslien et al. 2011, 2012; Paakkonen et al. 2012). Ticks, fleas, and mites have been indicated as vectors for numerous viral, bacterial, and protozoan infectious diseases (Puraite et al. 2015, 2016; Tonteri et al. 2016; Vera et al. 2016; Malmsten et al. 2018).

Endoparasites

European moose are hosts for many endoparasite species of different pathogenicity and diverse impact on the health status of the species on both individual and population levels.

The following protozoan parasites were discovered in moose: *Sarcocystis* spp., *Eimeria* spp., *Toxoplasma gondii*, *Neospora caninum*, and *Babesia* spp. (Dahlgren and Gjerde 2008; Moskwa et al. 2014; Puraite et al. 2016; Pyziel and Demiaszkiewicz 2013; Vikøren et al. 2004). Free-ranging cervids are commonly infected by various *Sarcocystis* species (Dahlgren and Gjerde 2010).

In Norwegian moose, several distinct species of *Sarcocystis* were confirmed: *Sarcocystis ovalis*, *S. alces*, *S. scandinavica* (Dahlgren and Gjerde 2008). *Eimeria* in moose are rare and come from: Lithuania where Amastauskene (1974) reported these coccidia for the first time; Belarus, where Soshkin (1997) found *Eimeria alces* in moose; Poland, where the presence of *E. alces* was confirmed and the first report of *E. caturbina* in moose – a parasite typically infecting roe deer (*Capreolus capreolus*) – was documented (Filip and Demiaszkiewicz 2016; Kuligowska et al. 2014; Pyziel and Demiaszkiewicz 2013); and from Norway (Davidson et al. 2015). Another protozoan parasite infecting moose is the zoonotic *Toxoplasma gondii* (Malmsten et al. 2011; Vikøren et al. 2004) – consumption of raw or undercooked meat can be a source of toxoplasmosis in humans. Another protozoan parasite (*Neospora caninum)* was also detected in European moose (Gondim 2006; Moskwa et al. 2014). This parasite is the causative agent of neosporosis and causes high rates of abortion in cattle (Hemphill and Gottstein 2000). In addition, *Babesia*-positive moose (infected by *B. capreoli* and *B. odocoilei*) were found in Norway (Puraite et al. 2016) and in Sweden (*B. venatorum* and *B. divergens*, Malmsten et al. 2018).

Helminth parasites of moose are widespread, prevalent, and exhibit different pathogenicity. Nematodes, particularly species of lungworms and filarioids, are known to cause morbidity and mortality in free-living moose populations. In Norway, Davidson et al. (2015) reported high abomasal gastrointestinal nematode burdens (up to 56,000 parasites in the total abomasal contents). The most prevalent were *Ostertagia antipini* and *Spiculopteragia alcis* – which infected 87% and 80% of studied individuals, respectively (Davidson et al. 2015). Other helminths confirmed in Norwegian moose were: *O. leptospicularis*, *Teladorsagia circumcincta*, *Moniezia* sp., *Trichuris* sp., *Nematodirus* sp., *Strongyloides* sp., *Dictyocaulus* sp. (Davidson et al. 2015). Lungworm infections with *Dictyocaulus* sp., *Varestrongylus alces*, and *Elaphostrongylus alces* were confirmed in Poland (Kowal et al. 2016; Pyziel et al. 2015). *Varestrongylus alces* was also found in Norway (Verocai et al. 2014). In addition, larval forms of cestodes, such as *Cysticercus* sp., *Echinococccus canadensis*, and *Moniezia* sp. were detected in moose (Davidson et al. 2015; Grandi et al. 2018; Haukisalmi 2015; Lavikainen et al. 2006).

In Poland, a pathogenic and invasive blood-sucking nematode of abomasa – *Ashworthius sidemi* was found in moose (Demiaszkiewicz et al. 2013) and in other ruminants, e.g., European bison (*Bison bonasus*) (Demiaszkiewicz et al. 2009; Kołodziej-Sobocińska et al. 2016). This species was probably introduced to Europe with sika deer (*Cervus nippon*) from their native range in Asia (Kotrlá and Kotrlý 1973). Prevalence of *A. sidemi* infections in moose was 20% and intensity reached up to 120 and 200 nematode specimens per animal in Poland and in Russia, respectively (Demiaszkiewicz et al. 2013). This is relatively low when compared with its invasion in the European bison: prevalence 89–100%, infection intensity up to 44,310 nematodes per animal in the wild European bison population from Białowieża Primeval Forest, NE Poland (Kołodziej-Sobocińska et al. 2016). Quite probably, we are observing the beginning of the invasion of this blood-sucking nematode in European moose, which may spread during the next years, as happened in the case of European bison (Kołodziej-Sobocińska et al. 2016).

Apart from nematodes and cestodes, moose can also be infected with trematodes. Giant liver fluke (*Fascioloides magna*) is a trematode of North American origin located in the liver (Pybus et al. 2015). It was found in moose in British Columbia, Canada, in the early 1950s (Cowan 1951). Molecular examination confirmed multiple introductions of *F. magna* to Europe from translocated elk and white-tailed deer (*Odocoileus virginianus*). Its distribution continues to expand (Kralova-Hromadova et al. 2011). The other liver flukes found in moose are *Parafasciolopsis fasciolaemorpha* and *Dicrocoelium lanceatum*. These trematodes were found in eastern Poland and in the Russian Kirov Region (Filip et al. 2016; Maslennikova and Shikhova 2017). *P. fasciolaemorpha* is an obligatory parasite for moose; infection prevalence in Russian moose reaches 70% and the infection intensity has been even up to 48,984 parasite specimens per animal. For the facultative parasite of moose – *D. lanceatum* prevalence was only 2.5% and the infection intensity only up to 102 helminths per animal (Maslennikova and Shikhova 2017).

Infectious Diseases

Vector-borne infectious diseases are an important threat to moose. Vectors for viral diseases of moose can be ticks, e.g., for tick-borne encephalitis virus (TBEV) (Tonteri et al. 2016) or midges, e.g., for Schmallenberg virus (SBV) (Larska et al. 2013). TBEV is a zoonotic arbovirus causing up to 13,000 annual reported human cases of tick-borne encephalitis (TBE) (Süss 2011). Large mammals are secondary hosts for TBEV, and TBE is usually subclinical for them. However, in Sweden, a moose calf with severe behavioral abnormalities was found to be TBEV-positive (Svedmyr et al. 1965). TBEV presence in moose has been recently confirmed in Finland (Tonteri et al. 2016). SBV is a novel *Orthobunyavirus* infecting ruminants that emerged in Germany in 2011 (Hoffmann et al. 2012). For the first time SBV was detected in a moose calf in Poland in Białowieża Primeval Forest (Larska et al. 2013). In the same territory, a high proportion (over 80%) of seropositive European bison was recorded to date (Kesik-Maliszewska et al. 2018). In Sweden, SBV-seropositive moose were detected in 2012 and 2013 (Malmsten et al. 2016), during the first SBV outbreak in domestic animals. Moose may be also infected with foot and mouth disease (FMD) – a highly contagious transboundary and economically devastating acute viral disease of global importance caused by foot-and-mouth disease virus (FMDV) (Frölich 1999; Pal 2018). It is characterized by high morbidity and low mortality (Frölich 1999). Natural infections in Europe were recorded, among others, in UK in five deer species (Gibbs et al. 1975) and in wild boar (*Sus scrofa*) (Croft et al. 2019). Early European case of moose FMD came from the late 1930s from Germany (Magnusson 1939). In Russia, moose is one of the main hosts of FMDV together with roe deer (*Capreolus capreolus*) and saiga antelope (Kruglikov et al. 1985; Kuvshinov 2002). Herpes virus infection leading to malignant catarrhal fever is rare in moose but has been reported in Norway (Vikøren et al. 2006), and in Sweden (Warsame and Steen 1989). Other viral diseases of moose are: malignant catarrhal fever (MCF) diagnosed in Scandinavian and North American moose (Vikøren et al. 2015) and hepatitis E-like virus found in Swedish moose (Lin et al. 2014).

Tick-borne fever (TBV) caused by *Anaplasma phagocytophilum* is a widespread bacterial zoonotic disease of wild and domestic animals (Woldehiwet 2010). *A. phagocytophilum* was detected in moose from Norway (Milner and van Beest 2013; Puraite et al. 2015), Sweden (Malmsten et al. 2014b, 2018), and Poland (Karbowiak et al. 2015). The second vector for TBV could be deer keds (Vichova et al. 2011), as they are for the other zoonotic bacteria – *Bartonella* spp., which were detected in moose in Norway, Finland, and Sweden (Duodu et al. 2013; Vera et al. 2016; Malmsten et al. 2018).

Chronic Wasting disease (CWD) is an inevitably fatal, zoonotic, infectious neurodegenerative prion disease (Saunders et al. 2012). For the first time, CWD was identified in the late 1960s and recognized as a spongiform encephalopathy in 1980 (Williams 2005). Recently, from 2016–2019 almost 100,000 cervid individuals in Norway were tested for CWD and 26 cases were confirmed: six in moose, one in red deer, and 19 in Eurasian tundra reindeer (*Rangifer tarandus tarandus*) (Norwegian Veterinary Institute 2019). In Sweden, 1500 wild cervid (moose, red deer, roe deer) individuals have been tested since 2018, of which four moose cases were confirmed (Swedish National Veterinary Institute 2020).

Zoonoses

Moose, as other wild mammal species, serve as reservoirs for many zoonotic pathogens and may be a reservoir for human-affecting diseases like toxoplasmosis (Malmsten et al. 2011; Vikøren et al. 2004), babesiosis (Puraite et al. 2016; Malmsten et al. 2018), TBE (Süss 2011), TBV (Malmsten 2014b; Woldehiwet 2010), borreliosis (Malmsten et al. 2018), and hepatitis E

(Lin et al. 2014). The food-borne transmission of bovine spongiform encephalopathy (BSE) to humans indicates that the species barrier may not protect humans from animal prion diseases. Although there is no known link between the consumption of CWD-affected meat of wild cervids and human health, hunters are advised against consuming meat from CWD-infected animals (Truong et al. 2018).

Non-infectious Diseases

Non-infectious diseases in moose include selected trace element deficiencies that cause secondary disease in different organs or organ systems. Osteoporosis (bone loss) has been reported in moose from Norway (Ytrehus et al. 1999), and molybdenosis with secondary copper deficiency was reported in Sweden by Frank et al. (2000). In most cases, limited access to forage is suspected to be the cause. Malignant tumors can occur in the ethmoidal bone, located between the brain and the nose, according to Ågren et al. (2014).

Population Ecology

Population Dynamics

Land use such as intensive forestry and agricultural activities shape moose habitats and thus forage ability throughout their Eurasian geographical range. Highest densities occur in Fennoscandia (Jensen et al. 2020). In Fennoscandia (Finland, Norway, and Sweden), a sex- and age-specific harvest over several decades has generated young- and female-biased populations with a very high reproductive potential. Intensive forestry (i.e., high availability of young coniferous forest stands) that produced forage-rich landscapes further increases the reproductive potential of the population (Tiilikainen et al. 2012). Commonly, female moose produce 1–2 calves annually in areas with sufficient suitable habitat (i.e., forest, including young successional forest). In moose, population growth is closely linked to forage availability. Areas with low amounts of suitable habitat and forage (e.g., high amounts of agriculture land compared to forest) decrease female body condition and reduce calf survival (Allen et al. 2017). Specifically, whereas pre-winter conditions influence pregnancy, the females' nutritional condition in winter and over-winter loss in body mass affect reproductive success (Milner et al. 2013). Nutritional stress can lead to delayed first reproduction, pause in reproduction, and lower twinning rates, with a paused reproduction enhancing reproduction the subsequent year (Boertje et al. 2019). Next to forage availability and females' nutritional conditions, recruitment rates correlate also with moose density and adult female-male ratio, with higher rates in areas and years with lower moose densities and more even adult sex ratios (females per males, Tiilikainen et al. 2012). It is important to note that density-dependent factors alone, however, do not have a strong regulatory effect on moose population growth, and without the presence of predators, population fluctuations are likely to surpass carrying capacity (Saether et al. 1996).

In fact, density-dependent factors are less influential on local recruitment rates than climatic conditions (e.g., Norway, Grøtan et al. 2009). On a large scale, moose abundance follows patterns in ambient temperature in addition to habitat quality (e.g., in Russia, Razenkova et al. 2020). Prediction of climate change forecasts higher ambient temperatures, altered precipitation patterns, more frequent extreme weather events, advanced vegetation onset, and longer vegetation periods. Whereas longer vegetation periods likely affect moose positively, estimating the effects of changed precipitation patterns and extreme weather events on moose population dynamics is more difficult. Given moose's heat-sensitivity, warmer environments likely affect moose dynamics negatively, by affecting individual moose both directly (physiologically) and indirectly (altered habitat utilization for thermoregulation affect individuals' body conditions, Van Beest and Milner 2013). Advanced vegetation onsets may have negative effects too. For herbivores like moose that depend on high-quality forage, timing of reproduction is closely linked to peaks in forage availability (i.e., start of vegetation onset,

Neumann et al. 2020). Earlier vegetation onset may thus generate a temporal mismatch as timing of calving and the window of high-quality forage diverge, because females' capability to adjust parturition timing to annual climatic variation seems to be limited (Neumann et al. 2020).

So far, it is difficult to make predictions of the cumulative effects of these altered environmental changes on moose population dynamics in the future. In North America, moose populations at the southern limit of their geographical range show reduced reproductive potential and survival (Lenarz et al. 2009; Ruprecht et al. 2016). Thus, a combination of decreased body condition, lower resilience to environmental stress, increased interspecific interactions, and higher parasite loads affect population dynamics negatively, suggesting a reduction of the species' southern distribution (Weiskopf et al. 2019). For moose in Europe, similar distribution changes may occur, even though recent research documents distribution gains in central Europe (Poland, Germany, Czech Republic, Slovakia, Jensen et al. 2020). Still, most viable populations occur in the northern part of the distribution range (i.e., boreal forest ecoregion, Jensen et al. 2020).

Mortality Factors

In most European countries where moose occur, large predators are functionally extinct, spatially restricted and/or limited in their ecosystem impact, leaving harvest as the major source of mortality in adult moose even in areas where large predators occur (Apollonio et al. 2010; Linnell et al. 2020). Brown bears can kill an adult moose, but only if the moose is in poor condition such as at the end of the winter. In areas without large predators, calf survival during summer is high and may reach up to 90% (Ericsson 1999; Swenson et al. 2007). In contrast to adult moose mortality, large predators can affect calf survival considerably, but the mortality caused by brown bears and wolves usually does not exceed 44% (Swenson et al. 2007; Sivertsen et al. 2012). Wolves are more successful in killing all age categories of moose compared to bears (Sand et al. 2006, 2012; Tallian et al. 2017). Wolf predation on adult and subadult moose (one-year old) can be high in areas with established wolf packs, but still, harvest accounts for the major source of mortality in those age classes (Sand et al. 2006; Gervasi et al. 2013).

In Fennoscandia, moose are harvested on an annual quota system, which is based on an area controlled or leased by collective hunting groups (Apollonio et al. 2010). Hunting pressure is high and the hunting season is long (from September to December/January), and on average one-third of the population is removed annually. In Fennoscandia, harvest usually focuses on the low reproductive part of the population (i.e., males and younger females, Apollonio et al. 2010). As a result, reproductive female moose experience lower mortality risk through harvest compared to other moose categories (Solberg et al. 2000; Ericsson and Wallin 2001). In contrast, male moose experience a high harvest mortality, and few populations have males older than 8 years (Solberg et al. 2000; Ericsson and Wallin 2001). The high turnover and the selective harvest has generated a very skewed age and sex ratio with an overall young moose population and a high proportion of females, and few, but young males (Ericsson 2001; Apollonio et al. 2010).

Interspecific Interactions

Geographically, moose in Europe are sympatric with four other deer species (i.e., roe deer *Capreolus capreolus*, red deer *Cervus elaphus*, fallow deer *Dama dama*, and semi-domesticated reindeer *Rangifer tarandus*). In some areas moose also co-occur with introduced deer species such as white-tailed deer *Odocoileus virginianus* in Finland and sika deer *Cervus nippon* in Poland and Czech Republic (Apollonio et al. 2010). Forage behavior overlaps with all of them, although to varying degrees (Mysterud 2000; Merceron et al. 2014; Spitzer et al. 2020). The largest overlap occurs with roe deer. Yet, considerable overlap also occurs with the mixed feeder, the red deer (Spitzer et al. 2020). In areas with higher density of several ungulate species, competition seems to occur between moose and smaller ungulate species over the browse on *Vaccinium ssp.*, driving moose to increase usage of pine *Pinus sylvestris* instead of *Vaccinium ssp.* (Spitzer et al. 2021). Lowest forage overlap is found with reindeer (Mysterud 2000). In their mountainous summer range, niche overlap of reindeer and moose is negligible, but species may overlap in their use of the mountainous birch forest (Neumann et al., unpublished data).

In spite of their considerable overlap in forage behavior, moose and roe deer differ in the selection of their winter habitats (Torres et al. 2011). Although both species were positively associated with cover of *Vaccinium sp.* in the ground layer, moose and roe deer occurrence differed in habitats connected to human disturbances (Torres et al. 2011). In areas with livestock, moose habitat use can overlap with sheep and cattle with photoperiod and weather conditions affecting the variation in niche overlap (Herfindal et al. 2017).

Conservation Status

IUCN Red list of Threatened Species status: Least Concern (LC)

Population Trend: increasing

The moose is listed in Appendix III of the Bern Convention. It is protected under national legislation in some countries (e.g., in Germany). The major conservation concern is extensive regional and landscape-scale habitat change (Hundertmark 2016).

Management

In Europe, moose occur in heavily human-modified ecosystems, where humans largely define both moose forage availability and mortality. In high density populations moose can cause economically significant damages to forest plantations, especially in pine and middle-age spruce forests (e.g., Apollonio et al. 2010). The species is intensively hunted in northern and eastern Europe for meat, skin, and antlers as trophies and to reduce the damages caused by the species in forests (e.g., in Sweden, Finland, and Russia, Apollonio et al. 2010, Baskin and Danell 2003). In Poland it is a game species but hunting is prohibited since 2001. Except harvest, another significant human-made causes of moose mortality are collisions with cars and trains (e.g., Apollonio et al. 2010).

At the beginning of the twentieth century, moose were at low numbers in all of the countries where reproductive populations exist today (Apollonio et al. 2010). In Fennoscandia, populations recovered to high numbers during the second half of the twentieth century due to changes in hunting regulations that favor harvest of the non/low reproductive part of the population and an increase in forestry activity (Lavsund et al. 2003; Apollonio et al. 2010). In the twentieth century, there were several moose reintroduction attempts in mainland Europe (e.g., in Germany and in Poland) but only one of them in the Kampinos National Park in Poland in the 1950s was successful (Dzięciołowski and Pielowski 1993; Schönfeld 2009). Recently (after the serious decline of the moose numbers in 90ies), population numbers have increased in Poland, Estonia, Latvia, and to a smaller degree in Lithuania. Modest population numbers exist in the Czech Republic and Ukraine (Apollonio et al. 2010; Jensen et al. 2020).

Future Challenges for Research and Management

Like for other cold-adapted species, higher ambient temperature following climate change likely will challenge this large-bodied herbivore physiologically and ecologically. The future geographical distribution of European moose will thus probably decrease with higher ambient temperature reshaping their southern distribution limit and shifting it northwards.

Preliminary results from on-going Swedish research indicate that in some multiple ungulate species systems moose populations do not perform very well (i.e., low reproductive output, lower survival; Ericsson, unpublished data). In those systems, preliminary results suggest that moose can be sensitive to competition with other deer species, probably having worse competition ability compared to smaller deer species.

In areas with populations that have high proportions of migratory moose, browsing damage on pine in concentration areas during winter generates conflicts with forestry, as the majority of moose are not in their winter areas during the hunting season (Singh et al. 2012; Allen and Singh 2016; Ericsson, unpublished data). A major challenge here is to improve the balance of forage in the landscape in relation to ungulate densities (www.naturvardsverket.se). On higher latitudes in particular, the flow of individuals over large areas requires collaboration of many different stakeholders, challenging resource management (Allen and Singh 2016).

References

Ågren EO, Uhlhorn H, Malmsten J, Mörner T, Gavier-Widen D (2014) Continued investigations of ethmoidal Tumours in Swedish moose (*Alces alces*). J Comp Pathol 1(150):112

Allen AM, Singh NJ (2016) Linking movement ecology with wildlife management and conservation. Front Ecol Evol 3:art155

Allen AM, Dorey A, Malmsten J, Edenius L, Ericsson G, Singh NJ (2017) Habitat–performance relationships of a large mammal on a predator-free island dominated by humans. Ecol Evol 7:305–319

Amastauskene T (1974) A new coccidiae species of the elk *Alces alces*. Acta Parasitologica Lithuanica 12:39–42

Apollonio M, Andersen R, Putman R (2010) European ungulates and their management in the 21st century. Cambridge University Press, p 618

Ball JP, Nordengren C, Wallin K (2001) Partial migration by large ungulates: characteristics of seasonal moose ranges in Northern Sweden. Wildl Biol 7:39–47

Bartzke GS, May R, Solberg EJ, Rolandsen CM, Røskraft E (2015) Differential barrier and corridor effects of power lines, roads, and rivers on moose (*Alces alces*) movements. Ecosphere 6:67

Baskin L, Danell K (2003) Ecology of ungulates: a handbook of species in Eastern Europe and Northern and Central Asia. Springer, Berlin/Heidelberg/New York

Berger J, Swenson JE, Persson I-L (2001) Recolonizing carnivores and Naïve prey: conservation lessons from Pleistocene extinctions. Science 291:1036–1039

Bergqvist G, Wallgren M, Jernelid H, Bergstrom R (2018) Forage availability and moose winter browsing in forest landscapes. For Ecol Manag 419:170–178

Bergström R, Hjeljord O (1987) Moose and vegetation interactions in North-Western Europe and Poland. Swedish Wildlife Res Viltrevy Supplement 1(Part 1):213–228

Bjørneraas K, Solberg EJ, Herfindal I, Van Moorter B, Rolandsen CM, Tremblay JP, Skarpe C, Sæther B-E, Eriksen R, Astrup R (2011) Moose *Alces alces* habitat use at multiple temporal scales in a human-altered landscape. Wildl Biol 17:44–54

Bjørneraas K, Herfindal I, Solberg EJ, Sther BE, van Moorter B, Rolandsen CM (2012) Habitat quality influences population distribution, individual space use and functional responses in habitat selection by a large herbivore. Oecologia 168:231–243

Bochkov AV, Klimov PB, Hestvik G, Saveljev AP (2014) Integrated Bayesian species delimitation and morphological diagnostics of chorioptic mange mites (Acariformes: Psoroptidae: *Chorioptes*). Parasitol Res 113(7): 2603–2627

Boertje RD, Ellis MM, Kellie KA (2015) Accuracy of moose age determinations from canine and incisor cementum annuli. Wildl Soc Bull 39:383–389

Boertje RD, Frye GG, Young DD Jr (2019) Lifetime, known-age moose reproduction in a nutritionally stressed population. J Wildl Manag 83:610–626

Boeskorov GG (1997) Chromosomal differences in moose (Mammalia, Artiodactyla, *Alces alces* L.). Genetika 33: 974–978

Boeskorov G (1999) New data on moose (Alces, Artiodactyla) systematics. Säugetierkundliche Mitteilungen 44:3–13

Borowik T, Ratkiewicz M, Maślanko W, Duda N, Kowalczyk R (2020a) The level of habitat patchiness influences movement strategy of moose in eastern Poland. PLoS One 15(3):e0230521

Borowik T, Ratkiewicz M, Maślanko W, Duda N, Kowalczyk R (2020b) Too hot to handle: summer space use shift in a cold-adapted ungulate at the edge of its range. Landsc Ecol 35:1341–1351

Bowyer RT (2004) Sexual segregation in ruminants: definitions, hypotheses, and implications for conservation and management. J Mammal 85:1039–1052

Bubenik AB (2007) Evolution, taxonomy and morphology. In: Franzman AW, Schwartz CC (eds) Ecology and management of the North American Moose by, 2nd edn. University Press of Colorado

Bunnefeld N, Borger L, Van Moorter B, Rolandsen C, Dettki H, Solberg EJ, Ericsson G (2011) A model-driven approach to quantify migration patterns: individual, regional and yearly differences. J Anim Ecol 80: 466–476

Carstensen M, Hildebrand EC, Plattner D, Dexter M, St-Louis V, Jennelle C, Wright RG (2018) Determining cause-specific mortality of adult moose in Northeast Minnesota, February 2013–July 2017. In: Cornicelli L, Carstensen M, Larson MA, Davros N, Davis B (eds) Summaries of wildlife research findings 2016. Minnesota Department of Natural Resources, Saint Paul, pp 236–247

Cederlund G, Sand H (1994) Home-range size in relation to age and sex in moose. J Mammal 75:1005–1012

Clark PU, Dyke AS, Shakun JD, Carlson AE, Clark J, Wohlfarth B, Mitrovica JX, Hostetler SW, McCabe AM (2009) The last glacial maximum. Science 325:710–714

Cowan IM (1951) The diseases and parasites of big game mammals of western Canada. Proc Ann Game Convention 5:37–64

Croft S, Aegerter JN, Massei G, Smith GC (2019) The risk of foot-and-mouth disease becoming endemic in a wildlife host is driven by spatial extent rather than density. PLoS One 14(6):e02118898

Czernik M, Taberlet P, Świsłocka M, Czajkowska M, Duda N, Ratkiewicz M (2013) Fast and efficient DNA-based method for winter diet analysis from stools of three cervids: moose, red deer, and roe deer. Acta Theriol 58:379–386

Dahlgren SS, Gjerde B (2008) *Sarcocystis* in moose (*Alces alces*): molecular identification and phylogeny of six *Sarcocystis* species in moose, and a morphological description molecular description of three new species. Parasitol Res 103(1):93–110

Dahlgren SS, Gjerde B (2010) Molecular characterization of five *Sarcocystis* species in red deer (*Cervus elaphus*), including *Sarcocystis hjorti* n. sp., reveals that these species are not intermediate host specific. Parasitology 137(5):815–840

Danell K, Niemela P, Varvikko T, Vuorisalo T (1991) Moose browsing on scots pine along a gradient of plant productivity. Ecology 72:1624–1633

Davidson RK, Ličina T, Gorini L, Milner JM (2015) Endoparasites in a Norwegian moose (*Alces alces*) population – faunal diversity, abundance and body condition. Int J Parasitol 4(1):29–36

Demiaszkiewicz AW, Lachowicz J, Osińska B (2009) *Ashworthius sidemi* (Nematoda, Trichostrongylidae) in wild ruminants in Białowieża Forest. Pol J Vet Sci 12(3):385–388

Demiaszkiewicz A, Kuligowska I, Lachowicz J, Pyziel A, Moskwa B (2013) The first detection of nematodes *Ashworthius sidemi* in elk *Alces alces* (L.) in Poland and remarks of ashworthiosis foci limitations. Acta Parasitol 58(4):515–518

Duodu S, Madslien K, Hjelm E, Molin Y, Paziewska-Harris A, Harris PD, Colguhoun DJ, Ytrehus B (2013) *Bartonella* infections in deer keds (*Lipoptena cervi*) and moose (*Alces alces*) in Norway. Appl Environ Microbiol 79(1):322–327

Dussault C, Ouellet JP, Courtois R, Huot J, Breton L, Jolicoeur H (2005) Linking moose habitat selection to limiting factors. Ecography 28:619–628

Dussex N, Alberti F, Heino MT, Olsen RA, van der Valk T, Ryman N et al (2020) Moose genomes reveal past glacial demography and the origin of modern lineages. BMC Genomics 21

Dzięciołowski R, Pielowski Z (1993) Łoś [Moose]. Wydawnictwo Anton-5. Sp. Zo.o. Warszawa [In Polish]

Eldegard K, Lyngved JT, Hjeljord O (2012) Coping in a human-dominated landscape: trade-off between foraging and keeping away from roads by moose (*Alces alces*). Eur J Wildl Res 58:969–979

Ericsson G (1999) Demographic and life history consequences of harvest in a Swedish moose population. Dissertation May 27 1999. Department of Animal Ecology. SLU, Umeå

Ericsson G (2001) Reduced cost of reproduction in moose *Alces alces* through human harvest. Alces 37:61–69

Ericsson G, Wallin K (2001) Age-specific moose (*Alces alces*) mortality in a predator-free environment: evidence for senescence in females. Écoscience 8: 157–163

Ericsson G, Wallin K, Ball JP, Broberg M (2001) Age-related reproductive effort and senescence in free-ranging moose, *Alces* alces. Ecology 82: 1613–1620

Ericsson G, Ball JP, Danell K (2002) Body mass of moose calves along an altitudinal gradient. J Wildl Manag 66: 91–97

Ericsson G, Dettki H, Neumann W, Arnemo JM, Singh NJ (2015a) Offset between GPS collar recorded temperature in moose and ambient weather station data. Eur J Wildl Res 61:919

Ericsson G, Neumann W, Dettki H (2015b) Moose antipredator behaviour towards baying dogs in a wolf-free area. Eur J Wildl Res 61:575–582

Eriksen A, Wabakken P, Zimmermann B, Andreassen HP, Arnemo JM, Gundersen H, Milner JM, Liberg O, Linnell J, Pedersen HC, Sand H, Solberg EJ, Storaas T (2011) Activity patterns of predator and prey: a simultaneous study of GPS-collared wolves and moose. Anim Behav 81:423–431

Felton AM, Felton A, Raubenheimer D, Simpson SJ, Krizsan SJ, Hedwall PO, Stolter C (2016) The nutritional balancing act of a large herbivore: an experiment with captive moose (*Alces alces* L). PLoS One 11: e0150870

Felton AM, Felton A, Cromsigt JPGM, Edenius L, Malmsten J, Wam HK (2017) Interactions between ungulates, forests, and supplementary feeding: the role of nutritional balancing in determining outcomes. Mammal Res 62:1–7

Felton AM, Wam HK, Stolter C, Mathisen KM, Wallgren M (2018) The complexity of interacting nutritional drivers behind food selection, a review of northern cervids. Ecosphere 9:e02230

Filip KJ, Demiaszkiewicz AW (2016) A new occurrence of *Eimeria catubrina* (Apicomplexa: Eimeridae) in elk (*Alces alces*) from the Kampinos Forest. Ann Parasitol 62(4):345–347

Filip KJ, Pyziel AM, Demiaszkiewicz AW (2016) A massive invasion of *Parafasciolopsis fasciolaemorpha* in elk (*Alces alces*) in Lublin Province, Poland. Ann Parasitol 62(2):107–110

Filonov KP, Kaletskaya ML (1988) Population studies of moose in the Darwinsky nature reserve. In: Sokolov VE (ed) Populyatsionnye issledovaniya zhivotnykh v zpovednikakh. Nauka, Nauk, pp 63–64. (In Russian)

Frank A, Sell DR, Danielsson R, Fogarty JF, Monnier VM (2000) A syndrome of molybdenosis, copper deficiency, and type 2 diabetes in the moose population of southwest Sweden. Sci Total Environ 249(1–3):123–131

Frölich K (1999) Viral diseases of northern ungulates. Rangifer 20(2–3):83–97

Garel M, Solberg EJ, Saether BE, Herfindal I, Hogda KA (2006) The length of growing season and adult sex ratio affect sexual size dimorphism in moose. Ecology 87: 745–758

Gębczyńska Z, Raczyński J (1984) Habitat preferences and population structure of moose in the Biebrza river valley. Acta Zool Fenn 172:93–94

Geist V (1998) Deer of the world: their evolution, behaviour, and ecology. Stackpole Books, Mechanicsburg

Gervasi V, Sand H, Zimmermann B, Mattisson J, Wabakken P, Linnell JDC (2013) Decomposing risk: landscape structure and wolf behavior generate different predation patterns in two sympatric ungulates. Ecol Appl 23:1722–1734

Gibbs EP, Herniman KA, Lawman MJ, Sellers RF (1975) Foot-and-mouth disease in British deer: transmission of virus to cattle, sheep and deer. Vet Rec 96(26):558–563

Gondim LFP (2006) *Neospora caninum* in wildlife. Trends Parasitol 22(6):247–252

Græsli AR, Thiel A, Fuchs B, Singh NJ, Stenbacka F, Ericsson G, Neumann W, Arnemo JM, Evans AL (2020) Seasonal Hypometabolism in female moose. Front Ecol Evol 8:107

Grandi G, Uhlhorn H, Ågren E, Mörner T, Righi F, Osterman-Lind E, Neimanis A (2018) Gastrointestinal parasitic infections in dead or debilitated moose (*Alces alces*) in Sweden. J Wildl Dis 54:165–169

Grøtan V, Sæther BE, Lillegård M, Solberg EJ, Engen S (2009) Geographical variation in the influence of density dependence and climate on the recruitment of Norwegian moose. Oecologia 161:685–695

Haanes H, Roed KH, Solberg EJ, Herfindal I, Saether BE (2011) Genetic discontinuities in a continuously distributed and highly mobile ungulate, the Norwegian moose. Conserv Genet 12:1131–1143

Haukisalmi V (2015) Checklist of tapeworms (Platyhelminthes, Cestoda) of vertebrates in Finland. ZooKeys 533:1–61

Heikkilä R, Mikkonen T (1992) Effects of density of young scots pine (*Pinus sylvestris*) stand on moose (*Alces alces*) browsing. Acta Forestalia Fennica 231:2–12

Hemphill A, Gottstein B (2000) A European perspective on *Neospora caninum*. Int J Parasitol 30(8):877–924

Herfindal I, Saether B-E, Solberg EJ, Andersen R, Hogda KA (2006a) Population characteristics predict responses in moose body mass to temporal variation in the environment. J Anim Ecol 75:1110–1118

Herfindal I, Solberg EJ, Saether B-E, Hogda KA, Andersen R (2006b) Environmental phenology and geographical gradients in moose body mass. Oecologia 150:213–224

Herfindal I, Tremblay JP, Hansen BB, Solberg EJ, Heim M, Sæther BE (2009) Scale dependency and functional response in moose habitat selection. Ecography 32: 849–859

Herfindal I, Haanes H, Solberg EJ, Roed KH, Hogda K, Saether B-E (2014) Moose body mass variation revisited: disentangling effects of environmental conditions and genetics. Oecologia 174:447–458

Herfindal I, Lande US, Solberg EJ, Rolandsen CM, Roer O, Wam HK (2017) Weather affects temporal niche partitioning between moose and livestock. Wildl Biol

Hestvik G, Zahler-Rinder M, Gavier-Widen D, Lindberg R, Mattsson R, Morrison D, Bornsein S (2007) A previously unidentified *Chorioptes* species infesting outer ear canals of moose (*Alces alces*): characterization of the mite and the pathology of infestation. Acta Vet Scand 49:21

Hjeljord O, Hovik N, Pedersen HB (1990) Choice of feeding sites by moose during summer, the influence of forest structure and plant phenology. Holarct Ecol 13:281–292

Hoffmann B, Scheuch M, Höper D, Jungblut R, Holsteg M, Schirrmeier H, Eschbaumer M, Goller KV, Wernike K, Fischer M, Breithaupt A, Mettenleiter TC, Beer M (2012) Novel Orthobunyavirus in cattle, Europe, 2011. Emerg Infect Dis 18(3):469–472

Hofmann RR (1989) Evolutionary steps of ecophysiological adaptation and diversification of ruminants – a comparative view of their digestive-system. Oecologia 78:443–457

Holmes CJ, Dobrotka CJ, Farrow DW, Rosendale AJ, Benoit JB, Pekins PJ, Yoder JA (2018) Low and high thermal tolerance characteristics for unfed larvae of the winter tick *Dermacentor albipictus* (Acari: Ixodidae) with special reference to moose. Ticks Tick-Borne Dis 9(1):25–30

Hundertmark K (2009) Reduced genetic diversity of two introduced and isolated moose populations in Alaska. Alces 45:137–142

Hundertmark K (2016) *Alces alces*. The IUCN Red List of Threatened Species 2016 Downloaded on 31 August 2018

Hundertmark KJ, Bowyer RT (2004) Genetics, evolution, and phylogeography of moose. Alces 40:103–122

Hundertmark KJ, Shields GF, Udina IG, Bowyer RT, Danilkin AA, Schwartz CC (2002) Mitochondrial phylogeography of moose (*Alces alces*): late Pleistocene divergence and population expansion. Mol Phylogenet Evol 22:375–387

Jędrzejewski W, Schmidt K, Theuerkauf J, Jędrzejewska B, Selva N, Zub K, Szymura L (2002) Kill rates and predation by wolves on ungulate populations in Białowieża Primeval Forest (Poland). Ecology 83:1341–1356

Jędrzejewski W, Niedziałkowska M, Hayward MW, Goszczyński J, Jędrzejewska B, Borowik T, Bartoń K, Nowak S, Harmuszkiewicz J, Juszczyk A, Kalamarz T, Kloch A, Koniuch J, Kotiuk K, Myslajek RW, Nędzynska M, Olczyk A, Teleon M, Wojtulewicz M (2012) Prey choice and diet of wolves related to ungulate communities and wolf subpopulations in Poland. J Mammal 93:1480–1492

Jensen W, Rea R, Penner C, Smith J, Bragina E, Razenkova E, Balciauskas L, Bao H, Bystiansky S, Csányi S, Chovanova Z, Done G, Hackländer K, Heurich M, Jiang G, Kazarez A, Pusenius J, Solberg

EJ, Veeroja R, Widemo F (2020) A review of circumpolar moose populations with emphasis on Eurasian moose distribution and densities. Alces 56:63–78

Kadulski S (1996) Ectoparasites of cervidae in north-East Poland. Acta Parasitol 41(4):204–210

Kangas VM, Kvist L, Laaksonen S, Nygren T, Aspi J (2013) Present genetic structure revealed by microsatellites reflects recent history of the Finnish moose (*Alces alces*). Eur J Wildl Res 59:613–627

Kangas VM, Kvist L, Kholodova M, Nygren T, Danilov P, Panchenko D, Fraimout A, Aspi J (2015) Evidence of post-glacial secondary contact and subsequent anthropogenic influence on the genetic composition of Fennoscandian moose (*Alces alces*). J Biogeogr 42: 2197–2208

Karbowiak G, Víchová B, Werszko J, Demiaszkiewicz AW, Pyziel AM, Sytykiewicz H, Szewczyk T, Peťko B (2015) The infection of reintroduced ruminants – *Bison bonasus* and *Alces alces* – with *Anaplasma phagocytophilum* in northern Poland. Acta Parasitol 60(4):645–648

Kesik-Maliszewska J, Krzysiak MK, Grochowska M, Lechowski L, Chase C, Larska M (2018) Epidemiology of Schmallenberg virus in European bison (*Bison bonasus*) in Poland. J Wildl Dis 54(2):272–282

Kolesnikov VV, Kozlovskii IS (2014) Geographical variation of antler morphology of moose (*Alces alces*) in Russia. Acta Theriol 59:443–448

Kołodziej-Sobocińska M, Demiaszkiewicz AW, Lachowicz J, Borowik T, Kowalczyk R (2016) Influence of management and biological factors on the parasitic invasions in the wild – spread of blood-sucking nematode *Ashworthius sidemi* in European bison (*Bison bonasus*). Int J Parasitol 5(3):286–294

Kotrlá B, Kotrlý A (1973) The first finding of the nematode *Ashworthius sidemi* Schulz, 1933 in *Sika nippon* from Czechoslovakia. Folia Parasitol 20(4):377–378

Kowal J, Kornaś S, Nosal P, Basiaga M, Wajdzik M, Skalska M, Wyrobisz A (2016) Lungworm (Nematoda: Protostrongylidae) infection in wild and domestic ruminants from Malopolska region of Poland. Ann Parasitol 62(1):63–66

Kozhukhov MV (1989) Reproductive potential of *Alces alces* under domestication. Zool Zhurnal 68(5): 150–153. (in Russian)

Kralova-Hromadova I, Bazsalovicsová E, Stefka J, Spakulová M, Vávrová S, Szemes T, Tkach V, Trudgett A, Pybus M (2011) Multiple origins of European populations of the giant liver fluke *Fascioloides magna* (Trematoda: Fasciolidae), a liver parasite of ruminants. Int J Parasitol 41(3–4):373–383

Kruglikov BA, Melnik RI, Nalivaiko VG (1985) Role of wild artiodactyla in the maintenance of foot-and-mouth disease virus. Veterinariya (Moscow) 92:37–38

Kuligowska I, Demiaszkiewicz AW, Kowalczyk R (2014) A new occurrence of *Eimeria alces* (Apicomplexa: Eimeridae) in elk (*Alces alces*) in East Poland. Ann Parasitol 60(4):277–279

Kuvshinov VL (2002) Histopathologuical changes in the central nervous system of moose (*Alces alces*) in the Ivanovo region of Russia. Alces Suppl 2:63–64

Larska M, Krzysiak M, Smreczak M, Polak MP, Mudzinski JF (2013) First detection of Schmallenberg virus in elk (*Alces alces*) indicating infection of wildlife in Białowieża National Park in Poland. Vet J 198(1): 279–281

Lavikainen A, Lehtinen MJ, Laaksonen S, Ågren E, Oksanen A, Meri S (2006) Molecular characterization of *Echinococcus* isolates of cervid origin from Finland and Sweden. Parasitology 133:565–570

Lavsund S, Tuire N, Solberg EJ (2003) Status of moose populations and challenges to moose management in Fennoscandia. Alces 39:109–130

Lenarz MS, Nelson ME, Schrage MW, Edwards AJ (2009) Temperature mediated moose survival in Northeastern Minnesota. J Wildl Manag 73:503–510

Lin J, Norder H, Uhlhorn H, Belak S, Widen F (2014) Novel hepatitis E like virus found in Swedish moose. J Gen Virol 95:557–570

Linnell JDC et al (2020) The challenges and opportunities of coexisting with wild ungulates in the human-dominated landscapes of Europe's Anthropocene. Biol Conserv 244:108500

Lister AM (1993) Evolution of mammoths and moose: the Holarctic perspective. In: Barnosky AD (ed) Quaternary mammals of North America. Cambridge University Press, Cambridge, UK, pp 178–204

Lorenzini R, Lovari S (2006) Genetic diversity and phylogeography of the European roe deer: the refuge area theory revisited. Biol J Linnean Soc 88:85–100

Löyttyniemi K (1985) On repeated browsing of scots pine saplings by moose (*Alces alces*). Silva Fennica 19: 387–391

Lundmark C (2008) Morphological and behavioural adaptations of moose to climate, snow, and forage. Doctoral thesis, Swedish University of Agricultural Sciences 2008:67

Lykkja ON, Solberg EJ, Herfindal I, Wright J, Rolandsen CM, Hanssen MG (2009) The effects of human activity on summer habitat use by moose. Alces 45:109–124

Madslien K, Ytrehus B, Vikøren T, Malmsten J, Isaksen K, Hygen HO, Solberg EJ (2011) Hair-loss epizootic in moose (*Alces alces*) associated with massive deer ked (*Lipoptena cervi*) infestation. J Wildl Dis 47(4): 893–906

Madslien K, Ytrehus B, Viljugrein H, Solberg EJ, Braten KR, Mysterud A (2012) Factors affecting deer ked (*Lipoptena cervi*) prevalence and infestation intensity in moose (*Alces alces*) in Norway. Parasit Vectors 5: 251

Magnusson H (1939) Ein Fall von Maul- und Klauenseuche beim Elch. Deut Tierärztl Woch 47: 509–511

Malmsten J, Jakubek EB, Bjorkman C (2011) Prevalence of antibodies against *Toxoplasma gondii* and *Neospora caninum* in moose (*Alces alces*) and roe deer

(*Capreolus capreolus*) in Sweden. Vet Parasitol 177(3–4):275–280

Malmsten J, Söderquist L, Thulin C-G, Gavier Widén D, Yon L, Hutchings MR, Dalin A-M (2014a) Reproductive characteristics in female moose (*Alces alces*), with emphasis on puberty, timing of oestrus, and mating. Acta Vet Scand 56:23

Malmsten J, Widén DG, Rydevik G, Yon L, Hutchings MR, Thulin C-G, Söderquist L, Dalin A-M (2014b) Temporal and spatial variation in *Anaplasma phagocytophilum* infection in Swedish moose (*Alces alces*). Epidemiol Infect 142:1205–1213

Malmsten J, Söderquist L, Thulin C-G, Dalin A-M (2015) Characteristics of spermatozoa and reproductive organs in relation to age and weight in Swedish moose (*Alces alces*). Anim Reprod Sci 153:76–86

Malmsten A, Malmsten J, Blomqvist G, Näslund K, Vernersson C, Hägglund S, Dalin A-M, Ågren EO, Valarcher J-F (2016) Serological testing of Schmallenberg virus in Swedish wild cervids from 2012 to 2016. BMC Vet Res 13:84

Malmsten J, Dalin A-M, Moutailler S, Devillers E, Gondard M, Felton A (2018) Vector-borne zoonotic pathogens in Eurasian moose (*Alces alces alces*). Vector-Borne and Zoonotic Diseases

Månsson J, Prima MC, Nicholson KL, Wikenros C, Sand H (2017) Group or ungroup – moose behavioural response to recolonization of wolves. Front Zool 14:10

Markgren G (1969) Reproduction of moose in Sweden. Swed Wildl 6:127–299

Maslennikova OV, Shikhova TG (2017) Trematodes of moose on the territory of the Kirov region. Veterinariya 4:37–40

Mathisen KM, Buhtz F, Danell K, Bergstrom R, Skarpe C, Suominen O, Persson IL (2010) Moose density and habitat productivity affects reproduction, growth and species composition in field layer vegetation. J Veg Sci 21:705–716

Mathisen KM, Milner JM, van Beest FM, Skarpe C (2014) Long-term effects of supplementary feeding of moose on browsing impact at a landscape scale. For Ecol Manag 314:104–111

Mathisen KM, Wojcicki A, Borowski Z (2018) Effects of forest roads on oak trees via cervid habitat use and browsing. For Ecol Manag 424:378–386

Meiri M, Lister A, Kosintsev P, Zazula G, Barnes I (2020) Population dynamics and range shifts of moose (*Alces alces*) during the late quaternary. J Biogeogr 47: 2223–2234

Melin M, Matala J, Mehtatalo L, Tiilikainen R, Tikkanen OP, Maltamo M, Pusenius J, Packalen P (2014) Moose (*Alces alces*) reacts to high summer temperatures by utilizing thermal shelters in boreal forests – an analysis based on airborne laser scanning of the canopy structure at moose locations. Glob Chang Biol 20: 1115–1125

Melin M, Matala J, Mehtatalo L, Pusenius J, Packalen P (2016) Ecological dimensions of airborne laser scanning – analyzing the role of forest structure in moose habitat use within a year. Remote Sens Environ 173:238–247

Merceron G, Hofman-Kamińska E, Kowalczyk R (2014) 3D dental microwear texture analysis of feeding habits of sympatric ruminants in the Białowieża Primeval Forest, Poland. For Ecol Manag 328:262–269

Milner JM, van Beest FM (2013) Ecological correlates of a tick-borne disease, *Anaplasma phagocytophilum*, in moose in southern Norway. Eur J Wildl Res 59(3): 399–406

Milner JM, van Beest FM, Solberg EJ, Storaas T (2013) Reproductive success and failure: the role of winter body mass in reproductive allocation in Norwegian moose. Oecologia 172:995–1005

Miquelle DG, Peek JM, Vanballenberghe V (1992) Sexual segregation in Alaskan moose. Wildl Monogr 122: 1–57

Moen R, Pastor J (1998) A model to predict nutritional requirements for antler growth in moose. Alces 34: 59–74

Monteith KL, Klaver RW, Hersey KR, Holland AA, Thomas TP, Kauffman MJ (2015) Effects of climate and plant phenology on recruitment of moose at the southern extent of their range. Oecologia 178: 1137–1148

Morow K (1976) Food habits of moose from Augustów Forest. Acta Theriol 21:101–116

Moskvitina NS, Nemoikina OV, Tyuten'kov OY, Kholodova MV (2011) Retrospective evaluation and modern state of the moose (*Alces alces* L.) population in West Siberia: ecological and molecular-genetic aspects. Contemp Probl Ecol 4:444–448

Moskwa B, Goździk K, Bień J, Kornacka A, Cybulska A, Reiterová K, Cabaj W (2014) Detection of antibodies to *Neospora caninum* in moose (*Alces alces*): the first report in Europe. Folia Parasitol 61(1):34–36

Mysterud A (2000) Diet overlap among ruminants in Fennoscandia. Oecologia 124:130–137

Mysterud A, Langvatn R, Stenseth NC (2004) Patterns of reproductive effort in male ungulates. J Zool 264: 209–215

Mysterud A, Solberg EJ, Yoccoz NG (2005) Ageing and reproductive effort in male moose under variable levels of intrasexual competition. J Anim Ecol 74:742–754

Neumann W, Ericsson G (2018) Influence of hunting on movements of moose near roads. J Wildl Manag 82: 918–928

Neumann W, Ericsson G, Dettki H (2010) Does off-trail backcountry skiing disturb moose? Eur J Wildl Res 56: 513–518

Neumann W, Ericsson G, Dettki H, Bergström R, Hågeryd A, Andersson E, Nordström Å (2011a) Årsrapport GPS-älgarna Referensområde Växjö 2010/ 2011. Tema vilt och skog, p 12. Sveriges lantbruksuniversitet. (In Swedish)

Neumann W, Ericsson G, Dettki H (2011b) The impact of human recreational activities – moose as a case study. Alces 47:17–25

Neumann W, Ericsson G, Dettki H, Bunnefeld N, Keuler N, Helmers D, Radeloff V (2012) Difference in spatiotemporal patterns of wildlife road-crossings and wildlife-vehicle collisions. Biol Conserv 145: 70–78

Neumann W, Ericsson G, Dettki H, Radeloff V (2013) Behavioural response to infrastructure of wildlife adapted to natural disturbances. Landsc Urban Plan 114:9–27

Neumann W, Singh NJ, Stenbacka F, Malmsten J, Wallin K, Ball JP, Ericsson G (2020) Divergence in parturition timing and vegetation onset in a large herbivore – differences along a latitudinal gradient. Biol Lett 16:20200044

Nicholson KL, Milleret C, Månsson J, Sand H (2014) Testing the risk of predation hypothesis: the influence of recolonizing wolves on habitat use by moose. Oecologia 176:69–80

Niedziałkowska M, Jędrzejewska B, Wójcik JM, Goodman SJ (2012). Genetic structure of red deer population in northeastern Poland in relation to the history of human interventions. J Wildlife Manage 76(6):1264–1276

Niedziałkowska M (2017) Phylogeography of European moose (*Alces alces*) based on contemporary mtDNA data and archaeological records. Mammalian Biology – Zeitschrift für Säugetierkunde 84:35–43

Niedziałkowska M, Hundertmark KJ, Jędrzejewska B, Niedziałkowski K, Sidorovich VE, Górny M, Veeroja R, Solberg EJ, Laaksonen S, Sand H, Solovyev VA, Shkvyria M, Tiainen J, Okhlopkov IM, Juškaitis R, Done G, Borodulin VA, Tulandin EA, Jędrzejewski W (2014) Spatial structure in European moose (*Alces alces*): genetic data reveal a complex population history. J Biogeogr 41:2173–2184

Niedziałkowska M, Hundertmark KJ, Jędrzejewska B, Sidorovich VE, Zalewska H, Veeroja R, Solberg E, Laaksonen S, Sand H, Solovyev VA, Sagaydak A, Tiainen J, Juskaitis R, Done G, Borodulin VA, Tulandin EA, Niedziałkowski K (2016a) The contemporary genetic pattern of European moose is shaped by postglacial recolonization, bottlenecks, and the geographical barrier of the Baltic Sea. Biol J Linn Soc 117: 879–894

Niedziałkowska M, Jędrzejewska B, Danyłow J, Niedziałkowski K (2016b) Diverse rates of gene flow and long-distance migration in two moose *Alces alces* subpopulations in Europe. Mammal Res 61:171–178

Nikula A, Heikkinen S, Helle E (2004) Habitat selection of adult moose *Alces alces* at two spatial scales in Central Finland. Wildl Biol 10:1221–1135

Norwegian Veterinary Institute (2019) Skrantesjukestatistikk. http://apps.vetinst.no/skrantesykestatistikk/NO/#omrade. Accessed 26 Nov 2019 [In Norwegian]

Nygren T, Pusenius J, Tiilikainen R, Korpelainen J (2007) Moose antler type polymorphism: age and weight dependent phenotypes and phenotype frequencies in space and time. Ann Zool Fenn 44:445–461

Oehlers SA, Bowyer RT, Huettmann F, Person DK, Kessler WB (2011) Sex and scale: implications for habitat selection by Alaskan moose *Alces alces gigas*. Wildl Biol 17:67–84

Olano-Marin J, Plis K, Sonnichsen L, Borowik T, Niedziałkowska M, Jędrzejewska B (2014) Weak population structure in European roe deer (*Capreolus capreolus*) and evidence of Introgressive hybridization with Siberian Roe Deer (*C. pygargus*) in Northeastern Poland. PLoS One 9:e109147–e109147

Olsson M, Cox JJ, Larkin JL, Widen P, Olovsson A (2011) Space and habitat use of moose in southwestern Sweden. Eur J Wildl Res 57:241–249

Paakkonen T, Mustonen AM, Käkelä R, Laaksonen S, Solismaa M, Aho J, Puukka K, Nieminen P (2012) The effects of an abundant ectoparasite, the deer ked (*Lipoptena cervi*), on the health of moose (*Alces alces*) in Finland. Parasitol Res 111(3):1223–1232

Pal M (2018) Foot and mouth disease: a highly infectious viral zoonosis of global importance. J Appl Microbiol Biochem 2(3):12–13

Peek JM, Scott MD, Nelson LJ, Pierce DJ, Irwin LL (1982) Role of cover in habitat management for big game in northwestern United States. Trans North Am Nat Resour Conf 47:363–373

Peterson RL (1955) North American moose. University of Toronto Press, Toronto

Puraite I, Rosef O, Paulauskas A, Radzijevskaja J (2015) *Anaplasma phagocytophilum* infection in moose (*Alces alces*) in Norway. Microbes Infect 17(11–12):823–828

Puraite I, Rosef O, Radzijevskaja J, Lipatova I, Paulauskas A (2016) The first detection of species of *Babesia starcovici*, 1893 in moose, *Alces alces* (Linnaeus), in Norway. Folia Parasitol 63:009

Pybus MJ, Butterworth EW, Woods JG (2015) An expanding population of the giant liver fluke (*Fascioloides magna*) in elk (*Cervus canadiensis*) and other ungulates in Canada. J Wildl Dis 51(2):431–445

Pyziel AM, Demiaszkiewicz AW (2013) Coccidia (Apicomplexa: Eimeriidae) of elk (*Alces alces*) in Poland. Parasitol Res 112(5):2083–2085

Pyziel AM, Laskowski Z, Hoglund J (2015) Development of a multiplex PCR for identification of *Dictyocaulus* lungworms in domestic and wild ruminants. Parasitol Res 114(10):3923–3926

Razenkova E et al (2020) Vegetation productivity summarized by the dynamic habitat indices explains broad-scale patterns of moose abundance across. Russia Sci Rep 10:836

Renecker LA, Schwartz CC (1997) Food habits and feeding behaviour. In: Franzmann AW, Schwartz CC (eds) Ecology and management of the North American moose. Smithsonian Institution Press, Washington, DC/London, pp 403–440

Rolandsen CM, Solberg EJ, Heim M, Holmstrøm F, Solem MI, Sæther B-E (2008) Accuracy and repeatability of moose (*Alces alces*) age as estimated from dental cement layers. Eur J Wildl Res 54:6–14

Ruprecht J, Hersey K, Hafen K, Monteith K, DeCesare N, Kauffman M, MacNulty D (2016) Reproduction in moose at their southern range limit. J Mammal 97: gyw099

Sæther BE, Haagenrud H (1983) Life history of the moose (*Alces alces*): fecundity rates in relation to age and carcass weight. J Mammal 64:226–232

Sæther BE, Heim M (1993) Ecological correlates of individual variation in age at maturity in female moose (*Alces alces*): the effects of environmental variability. J Anim Ecol 62:482–489

Sæther BE, Hjeljord O, Heim M (1996) Ecological correlates of regional variation in life history of the moose (*Alces alces*). Ecology 77:1493–1500

Sæther BE, Solberg EJ, Heim M, Stacy JE, Jakobsen KS, Olstad R (2004) Offspring sex ratio in moose *Alces alces* in relation to paternal age: an experiment. Wildl Biol 10:51–57

Sahlsten J, Bunnefeld N, Månsson J, Ericsson G, Bergstrom R, Dettki H (2010) Can supplementary feeding be used to redistribute moose *Alces alces*? Wildl Biol 16:85–92

Sand H (1996) Life history patterns in female moose (*Alces alces*): the relationship between age, body size, fecundity and environmental conditions. Oecologia 106: 212–220

Sand H, Cederlund G (1996) Individual and geographical variation in age at maturity in female moose (*Alces alces*). Can J Zool 74:954–964

Sand H, Wikenros C, Wabakken P, Liberg O (2006) Cross-continental differences in patterns of predation: will naïvemoose in Scandinavia ever learn? Proc R Soc B 273:1421–1427

Sand H, Vucetich JA, Zimmermann B, Wabakken P, Wikenros C, Pedersen HC, Peterson RO, Liberg O (2012) Assessing the influence of prey-predator ratio, prey age structure and packs size on wolf kill rates. Oikos 121:1454–1463

Saunders SE, Bartelt-Hunt SL, Bartz JC (2012) Occurrence, transmission, and zoonotic potential of chronic wasting disease. Emerg Infect Dis 18(3): 369–376

Schiviz AN, Ruf T, Kuebber-Heiss A, Schubert C, Ahnelt PK (2008) Retinal cone topography of artiodactyl mammals: influence of body height and habitat. J Comp Neurol 507:1336–1350

Schmidt JI, Hundertmark KJ, Bowyer RT, McCracken KG (2009) Population structure and genetic diversity of moose in Alaska. J Hered 100:170–180

Schönfeld F (2009) Presence of moose (*Alces alces*) in Southeastern Germany. Eur J Wildl Res 55:449–453

Schwartz CC (1997) Reproduction, natality and growth. In: Franzmann AW, Schwartz CC (eds) Ecology and management of the North American moose. Smithsonian Institution Press, Washington, DC/-London, pp 141–172

Schwartz CC, Hundertmark KJ (1993) Reproductive characteristics of Alaskan moose. J Wildl Manag 57: 454–468

Seiler A, Cederlund G, Jernelid H, Grängstedt P, Ringaby E (2003) The barrier effect of highway E4 on migratory moose (*Alces alces*) in the high coast area, Sweden. Habitat fragmentation due to transportation infrastructure. IENE, pp 1–18

Serafiński W (1969) Reproduction and dynamics of moose (*Alces alces* L.) population in Kampinos National Park. Ekologia Polska Seria A 17(37):709–718

Shipley LA, Blomquist S, Danell K (1998) Diet choices made by free-ranging moose in northern Sweden in relation to plant distribution, chemistry, and morphology. Can J Zool 76:1722–1733

Singh NJ, Börger L, Dettki H, Bunnefeld N, Ericsson G (2012) From migration to nomadism: movement variability in a northern ungulate across its latitudinal range. Ecol Appl 22:2007–2020

Singh NJ, Evans AL, Stenbacka F, Ericsson G, Arnemo JM (2014) Neither too hot nor too cold. Thermoregulation in Scandinavian moose. 5th bio-logging science symposium, Strasbourg, 22–26 September 2014

Sivertsen TR, Mysterud A, Gundersen H (2012) Moose (*Alces alces*) calf survival rates in the presence of wolves (*Canis lupus*) in Southeast Norway. Eur J Wildl Res 58:863–868

Skuncke F (1949) Älgen, studier, jakt och vård. P. A. Nordstedts och Söners Förlag, Stockholm. [In Swedish]

Solberg EJ, Saether BE (1994) Male traits as life-history variables – annual variation in body mass and antler size in moose (*Alces alces*). J Mammal 75:1069–1079

Solberg EJ, Loison A, Saether BE, Strand O (2000) Age-specific harvest mortality in a Norwegian moose *Alces alces* population. Wildl Biol 6:41–52

Solberg EJ, Heim M, Grotan V, Saether BE, Garel M (2007) Annual variation in maternal age and calving date generate cohort effects in moose (*Alces alces*). Oecologia 154:259–271

Solden LM, Naas AE, Roux S, Daly RA, Collins WB, Nicora CD, Purvine SO, Hoyt DW, Schückel J, Jørgensen B, Willats W, Spalinger DE, Firkins JL, Lipton MS, Sullivan MB, Pope PB, Wrighton KC (2018) Interspecies cross-feeding orchestrates carbon degradation in the rumen ecosystem. Nat Microbiol 3: 1274–1284

Soshkin DV (1997) Infections of elks (*Alces alces*) with *Eimeria alces* (Eucoccidiida) in the Bryansk region. Parazitologiya (St Petersburg) 31(3):273–274

Spaeth DF, Bowyer RT, Stephenson TR, Barboza PS (2004) Sexual segregation in moose *Alces alces*: an experimental manipulation of foraging behavior. Wildl Biol 10:59–72

Spitzer R (2019) Trophic resource use and partitioning in multispecies ungulate communities. Doctoral thesis. Swedish University of Agricultural Sciences. ISBN 978-91-7760-464-8. https://pub.epsilon.slu.se/16431/

Spitzer R, Coissac E, Felton A, Fohringer C, Juvany L, Landman M et al (2021) Small shrubs with large importance? Smaller deer may increase the moose-forestry conflict through feeding competition over Vaccinium shrubs in the field layer. For Ecol Manag 480: 118768

Spitzer R, Felton A, Landman M, Singh NJ, Widemo F, Cromsigt JPGM (2020) Fifty years of European ungulate dietary studies: a synthesis. Oikos 129:1668–1680

Stubsjøen T, Saether BE, Solberg EJ, Heim M, Rolandsen CM (2000) Moose (*Alces alces*) survival in three populations in northern Norway. Can J Zool 78: 1822–1830

Süss J (2011) Tick-borne encephalitis 2010: epidemiology, risk areas, and virus strains in Europe and Asia – an overview. Ticks Tick-Borne Dis 2(1):2–15

Svedmyr A, Vonzeipe G, Borg K, Hansen HJ (1965) Infections with tick-borne encephalitis virus in Swedish population of elk (*Alces alces*). Acta Pathol Microbiol Scand 65(4):613–620

Svensson R (2008) Der Elch – vom Kalb zum reifen Schaufler. Svenska Jägareförbundets Verlag. ISBN 978-91-88660-30-5 [In German]

Swedish National Veterinary Institute (2020) CWD-provtagning – EU-övervakning. https://www.sva.se/maps/CWD/table_wild.html. Accessed 14 Dec 2020. [In Swedish]

Swenson JE, Dahle B, Busk H, Opseth O, Johansen T, Soderberg A, Wallin K, Cederlund G (2007) Predation on moose calves by European brown bears. J Wildl Manag 71:1993–1997

Świsłocka M, Ratkiewicz M, Borkowska A, Komenda E, Raczyński J (2008) Mitochondrial DNA diversity in the moose, *Alces alces*, from northeastern Poland: evidence for admixture in a bottlenecked relic population in the Biebrza valley. Ann Zool Fenn 45:360–365

Świsłocka M, Matosiuk M, Ratkiewicz M, Borkowska A, Czajkowska M, Mackiewicz P (2020). Phylogeny and diversity of moose (*Alces alces*, Cervidae, Mammalia) revealed by complete mitochondrial genomes. Hystrix, the Italian Journal of Mammalogy 31(1):1–9

Tallian A, Ordiz A, Metz MC, Milleret C, Wikenros C, Smith DW, Stahler DR, Kindberg J, MacNulty DR, Wabakken P, Swenson JE, Sand H (2017) Competition between apex predators? Brown bears decrease wolf kill rate on two continents. Proc R Soc B 284:2016–2368

Tiilikainen R, Nygren T, Pusenius J, Russila V (2010) Variation in growth pattern of male moose *Alces alces* after two contrasted periods of hunting. Ann Zool Fenn 47:159–172

Tiilikainen R, Solberg EJ, Nygrén T, Pusenius J (2012) Spatio-temporal relationship between calf body mass and population productivity in Fennoscandian moose *Alces alces*. Wildl Biol 18:304–317

Tischler KB (2004) Aquatic plant nutritional quality and contribution to moose diet at Isle Royale National Park. MSc Thesis. Michigan Technological University

Tonteri E, Jokelainen P, Matala J, Pusenius J, Vapalahti O (2016) Serological evidence of tick-borne encephalitis virus infection in moose and deer in Finland: sentinels for virus circulation. Parasit Vectors 9:54

Torres RT, Carvalho JC, Panzacchi M, Linnell JDC (2011) Comparative use of forest habitats by roe deer and moose in human-modified landscape in southeastern Norway during winter. Ecol Res 26:781–789

Tremblay J-P, Solberg EJ, Sæther B-E, Heim M (2007) Fidelity to calving areas in moose (*Alces alces*) in the absence of natural predators. Can J Zool 85:902–908

Truong T, Adamowicz W, Boxall PC (2018) Modelling the effect of chronic wasting disease on recreational hunting site choice preferences and choice set formation over time. Environ Resour Econ 70(1):271–295

Van Ballenberghe V, Ballard WB (1997) Population dynamics. In: Franzmann AW, Schwartz CC (eds) Ecology and management of the North American moose. Smithsonian Institution Press, Washington, DC/London, pp 223–245

Van Beest F, Milner JM (2013) Behavioural responses to thermal conditions affect seasonal weight change in a heat-sensitive northern ungulate. PLoS One 8:e65972

Van Beest FM, Mysterud A, Loe LE, Milner JM (2010) Forage quantity, quality and depletion as scale-dependent mechanisms driving habitat selection of a large browsing herbivore. J Anim Ecol 79:910–922

Van Beest F, Rivrud I-M, Loe LE, Milner JM, Mysterud A (2011) What determines variation in home range size across spatiotemporal scales in a large browsing herbivore? J Anim Ecol 80:771–785

Van Beest FM, Van Moorter B, Milner JM (2012) Temperature-mediated habitat use and selection by a heat-sensitive northern ungulate. Anim Behav 84:723–735

Van Moorter B, Bunnefeld N, Panzacchi M, Rolandsen CM, Solberg EJ, Sæther BE (2013) Understanding scales of movement: animals ride waves and ripples of environmental change. J Anim Ecol 82:770–780

Veeroja R, Kirk A, Tilgar V, Tonisson J (2013) Winter climate, age, and population density affect the timing of conception in female moose (*Alces alces*). Acta Theriol 58:349–357

Vera CP, Aaltonen K, Spillmann T, Vapalahti O, Sironen T (2016) Geographic distribution and molecular diversity of *Bartonella* spp. infections in moose (*Alces alces*) in Finland. J Wildl Dis 52(2):209–216

Verocai GG, Hoberg EP, Vikøren T, Handeland K, Ytrehus B, Rezansoff AM, Davidson RK, Gilleard JS, Kutz SJ (2014) Resurrection and redescription of *Varestrongylus alces* (Nematoda: Protostrongylidae), a lungworm of the Eurasian moose (*Alces alces*), with report on associated pathology. Parasit Vectors 7:557

Vichova B, Majlathova V, Novakova M, Majlath I, Curlik J, Bona M, Komjati-Nagyova M, Petko B (2011) PCR detection of re-emerging tick-borne pathogen, *Anaplasma phagocytophilum*, in deer ked (*Lipoptena cervi*) a blood-sucking ectoparasite of cervids. Biologia 66(6):1082–1086

Vikøren T, Tharaldsen J, Fredriksen B, Handeland K (2004) Prevalence of *Toxoplasma gondii* antibodies in wild red deer, roe deer, moose, and reindeer from Norway. Vet Parasitol 120:159–169

Vikøren T, Li H, Lillehaug A, Monceyron Jonassen C, Böckerman I, Handeland K (2006) Malignant catarrhal fever in free-ranging cervids associated with OvHV-2 and CpHV-2 DNA. J Wildl Dis 42(4):797–807

Vikøren T, Klevar S, Li H, Hauge AG (2015) A geographic cluster of malignant catarrhal fever in moose (*Alces alces*) in Norway. J Wildl Dis 51(2):471–474

Wabakken P, Sand H, Liberg O, Bjärvall A (2001) The recovery, distribution, and population dynamics of wolves on the Scandinavian peninsula, 1978–1998. Can J Zool 79:710–725

Wam HK, Hjeljord O (2010) Moose summer and winter diets along a large scale gradient of forage availability in southern Norway. Eur J Wildl Res 56:745–755

Warsame IY, Steen M (1989) Malignant catarrhal fever in wild Swedish moose (*Alces alces* L). Rangifer 9:51–57

Weiskopf SR, Ledee OE, Thompson LM (2019). Climate change effects on deer and moose in the midwest. Journal of Wildlife Management 83(4):769–781

Whitehead GK (1993) The Whitehead encyclopedia of deer. Swan Hill Press, p 309

Wikenros C, Balogh G, Sand H, Nicholson KL, Månsson J (2017) Mobility of moose – comparing the effects of wolf predation risk, reproductive status and seasonality. Ecol Evol 6:8870–8880

Williams ES (2005) Chronic wasting disease. Vet Pathol 42(5):530–549

Wilson DE, Reeder DAM (2005) Mammal species of the world. A taxonomic and geographic reference, 3rd edn. Johns Hopkins University Press. Available from Johns Hopkins University Press, 1-800-537-5487 or (410) 516-6900 http://www.press.jhu.edu

Woldehiwet Z (2010) The natural history of *Anaplasma phagocytophilum*. Vet Parasitol 167 (2–4):108–122

Ytrehus B, Skagemo H, Stuve G, Sivertsen T, Handeland K, Vikøren T (1999) Osteoporosis, bone mineralization, and status of selected trace elements in two populations of moose calves in Norway. J Wildl Dis 35(2):204–211

Zachos FE, Frantz AC, Kuehn R, Bertouille S, Colyn M, Niedziałkowska M, Pérez-González J, Skog A, Sprem N, Flamand MC (2016) Genetic structure and effective population sizes in European red deer (*Cervus elaphus*) at a continental scale: insights from microsatellite DNA. J Hered 107:318–326

Reindeer *Rangifer tarandus* (Linnaeus, 1758)

10

Øystein Holand, Ivan Mizin, and Robert B. Weladji

Contents

Ø. Holand (✉)
Norwegian University of Life Sciences, Aas, Norway
e-mail: oystein.holand@nmbu.no

I. Mizin
Russian Arctic National Park, Arkhangelsk, Russia
e-mail: Ivan_Mizin@mail.ru

R. B. Weladji
Concordia University, Montreal, QC, Canada
e-mail: robert.weladji@concordia.ca

© Springer Nature Switzerland AG 2022

L. Corlatti, F. E. Zachos (eds.), *Terrestrial Cetartiodactyla*, Handbook of the Mammals of Europe,
https://doi.org/10.1007/978-3-030-24475-0_24

Common Names

English	Reindeer (in North America known as Caribou)
German	Rentier
French	Renne
Spanish	Reno
Italian	Renna
Russian	Северный олень

Taxonomy and Systematics

Rangifer tarandus (Fig. 1) is the only species within its genus. An array of *Rangifer* subspecies, and even species, were described in the 1800s and early 1900s without any sound taxonomic platform. Banfield (1961), in his classical work, revised the taxonomy based on morphological (mainly skull measurements) and historical data. He classified *Rangifer* as one species, divided into nine subspecies of which two are extinct. This was followed up by Whitehead (1972) and later by Grubb (2005) and Groves and Grubb (2011) who listed 14 subspecies of which two are extinct.

Three of these subspecies are found in Europe, namely:

- Eurasian tundra reindeer, *Rangifer tarandus tarandus* (Linnaeus, 1758)
- Eurasian forest reindeer *Rangifer tarandus fennicus* Lönnberg, 1909
- Svalbard reindeer, *Rangifer tarandus platyrhynchus* (Vrolik, 1829)

In 2001, the Novaya Zemlya reindeer *Rangifer tarandus pearsoni* Lydekker, 1903 was included in the Red Data Book of the Russian Federation as a separate subspecies (Mizin et al. 2018). See pictures of the four subspecies below.

The current subspecies classification based on morphological characters is outdated and predates the current understanding of *Rangifer* phylogeny and biogeography based on DNA techniques (Gravlund et al. 1998; Flagstad and Røed 2003; Kvie et al. 2016). However, we have to acknowledge the taxonomists' tedious and systematic work through centuries, combining morphological measures with ecological, behavioral, life history, physiological, and geographical information.

Banfield (1961) had applied a functional approach and divided the species into three major ecotypes – the tundra/barren ground, the forest/woodland reindeer, and the high Arctic island type as followed up by Whitehead (1972).

Fig. 1 The four European *Rangifer* subspecies. Upper left: Eurasian tundra reindeer *Rangifer tarandus tarandus* (photograph by O. Natalskaya), upper right: Eurasian forest reindeer *Rangifer tarandus fennicus* (photograph by V. Mamontov), lower left: Svalbard reindeer *Rangifer tarandus platyrhynchus* (photograph by Ø. Holand), and lower right: Novaya Zemlya reindeer *Rangifer tarandus pearsoni* (photograph by I. Mizin)

In Europe, all three ecotypes are represented and comply with the subspecies names: Eurasian tundra reindeer represents the tundra type, the Eurasian forest reindeer complies with the forest type, whereas the Svalbard reindeer and the Novaya Zemlya reindeer are representatives of the high Arctic island type.

This ecotype classification combined with modern DNA techniques enable us to appreciate today's distribution in Europe. Flagstad and Røed (2003) argue that the morphological differences among the types have evolved as adaptive responses to postglacial environmental change. The behavioral and ecological specialization is therefore more recently derived and sets the stage for understanding the adaptive history of *Rangifer* in Europe (Røed et al. 2008).

Rangifer has evolved primarily as a tundra-adapted species. After the last glaciation, as the species was pushed north, some populations stayed behind in higher elevated pockets, and were able to adapt to the growing taiga belt (Matiskainen 1990). Røed et al. (2005) argue that the forest reindeer have adapted to the taiga quite recently after the postglacial forest expansion.

The forest tree line has been dynamic after the last glaciation. Pollen records show that both deciduous and coniferous taxa grew beyond their present northern limits in the Early Holocene (Binney et al. 2017). It is therefore difficult to draw a borderline between the distribution of Eurasian tundra reindeer and the Eurasian forest reindeer. With no clear geographical barrier, there has obviously been introgression and gene flow between the ecotypes. Indeed, their range may partly overlap in space and time, and the genetic structure between the European tundra and forest type is not clear (Røed et al. 2018).

At the northern fringe of the tundra, the animals met the White and Kara Sea. From there, the reindeer were able to colonize Svalbard by trekking over the ice, possibly with Novaya Zemlya and Franz Josef Land as stepping-stones (Syroechkovskii 1995). The strong genetic link between today's population on Svalbard and

Novaya Zemlya and ancient samples from Franz Josef Land (Kvie et al. 2016) supports that the Archipelagos were colonized by reindeer from the Eurasian mainland after the Last Glacial Maximum. Their limited genetic diversity (Kvie et al. 2016) indicates that the populations have been exposed to severe bottlenecks.

Paleontology

Early fossils date to the Early Pleistocene glaciation period (Harington 1999) where a circumpolar periglacial tundra belt evolved (Kahlke 2014). However, other evidence indicates that by the Middle Miocene, *Rangifer* was distinct from other deer in the Cervidae subfamily Odocoileinae (Heckeberg 2020; Chen et al. 2019). This tundra landscape opened niches where a number of cold-adapted mammalian species, including *Rangifer,* were able to thrive (Kahlke 2014). The Pleistocene, spanning from about 2.6 mya to about 12 kya is characterized by marked climatic fluctuations with several overlaying levels of periodicity resulting in glacial and interglacial cycles as well as interstadials and stadials within these cycles (Hayes et al. 1976). In this dynamic environment, *Rangifer* evolved. Geist (1999) referred to *Rangifer* as a typical Ice Age mammal.

The two oldest *Rangifer* fossils date to the Early Pleistocene, with an estimated age of 1.8 and 1.6 million years, respectively, and have been found in the eastern part of Beringia (Harington 1999). This suggests a Beringian origin. The Beringia refugium is known to have played a key role in *Rangifer* evolutionary and biogeographical history (Flagstad and Røed 2003; Weckworth et al. 2012). During the major glacial periods when the sea levels were reduced by up to 120 m (Rohling et al. 2014), Beringia included Eastern Siberia and Alaska joined by a massive land bridge.

The oldest reported European reindeer fossils (but see Croitor 2018) date back to about 620 kya and have been found in Süssenborn, Germany (Kurtén 1968), representing remains from populations living in the periglacial zone south of the fluctuating ice sheet. During the glacials, the tundra steppe stretched throughout Eurasia, south of the ice sheet, all the way to the Iberian Peninsula, and *Rangifer* expanded deep into Eurasia (Kahlke 2014). During the interglacials, *Rangifer* moved north as the ice sheet retreated and probably met the Beringia populations colonizing deglaciated areas from north-east. Although Beringia has been a highly dynamic refugium (Hopkins et al. 1982), it has been the core area of *Rangifer* (Harington 1999).

During the Eemian interglacial (127–117 kya), prior to the last glaciation period (about 115–12 kya), called Weichselian in Europe, the climate was warm (Hofreiter and Stewart 2009). As the cooling started at the border between Eemian/Weichselian, the Euro-Beringian lineage expanded rapidly (Flagstad and Røed 2003; Yannic et al. 2014; Polfus et al. 2017; Taylor et al. 2021) and its maximum distribution during periods of the last glaciation covered most of the Northern Eurasian continent (Fig. 2).

The Weichselian glaciation, which culminated about 18 kya (Patton et al. 2017), contributed to the division into two main genetic lineages (Flagstad and Røed 2003; Yannic et al. 2014). The Euro-Beringian lineage, which is the most diverse, and regarded as ancestral to *Rangifer* in Europe (Flagstad and Røed 2003). Members of the lineage expanded also south in North America as the ice retreated, where they encountered the North American lineage, which had survived south of the extensive North American ice sheet, moving north (Weckworth et al. 2012; Polfus et al. 2017). The woodland caribou seems to originate from this North American lineage (Flagstad and Røed 2003). A third lineage, probably nested within the Euro-Beringia lineage, seems to have a more recent origin in one or several European refugium/refugia expanding north around 15 kya (Flagstad and Røed 2003, Weldenegodguad et al. 2020). Genetic signatures of this lineage are implied to be present in contemporary European reindeer, most pronounced in Fennoscandian reindeer (Flagstad and Røed 2003).

During the Last Glacial Maximum (LGM) (25–20 kya), when the Scandinavian ice sheet extending south to 52° N in Germany (Clark et al.

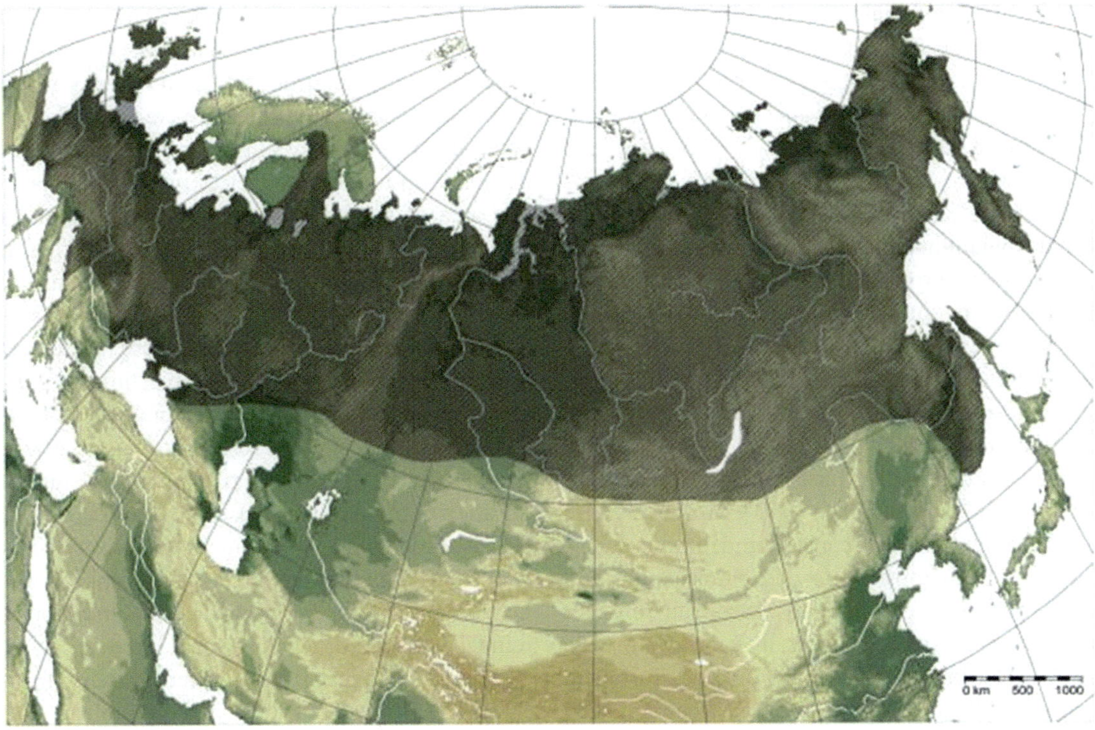

Fig. 2 Last Glacial (Weichselian) maximum Eurasian distribution of *Rangifer* (Kahlke 2014, with permission from Elsevier)

2012), the reindeer and the reindeer hunters thrived in southern Europe (Fontana 2017). The figurative cave paintings in Chauvet-Pont-d'Arc Cave, southern France, are reminders of this blooming culture (Clottes 2003). Also, the Russian plain and stretching up to Ural, hosted people relying heavily on reindeer (Pavlov 2017).

The global warming, although interrupted by cooling events (Clark et al. 2012), following the LGM, extirpated reindeer from southern Europe (Sommer et al. 2014; Costamagno et al. 2016). As the tundra steppe expanded north (Finalyson and Carrion 2007), the cursorial reindeer moved north, and reindeer hunters followed. This expansion north continued into Early Holocene (12–10 kya). The youngest fossils in Denmark date back about 9 kya (Aaris-Sørensen et al. 2007). Reindeer also disappeared from The British Isles (Coard and Chamberlain 1999) and Southern Sweden (Björck et al. 1996) during this period. In the eastern Baltic region, Ukkonen et al. (2006) reported that reindeer vanished around the Pleistocene-Holocene boundary. They argue that this also happened further east at the same latitudes in the European part of Russia as supported by Volokitin and Gribchenko (2017). This rather synchronous disappearance was driven by a continuous warmer climate that affected the environmental conditions for *Rangifer* negatively and pushed the reindeer even farther north and set the stage for today's distribution of *Rangifer*. Lorenzen et al. (2011) estimated that the potential reindeer range decline by 84% between 21 and 6 kya.

As the warming continued the Scandinavian ice sheet withdrew, the reindeer herds spread along the Norwegian ice-free coast where they met members of herds pouring into northern Fennoscandia from the east. Hunting people followed these two main postglacial migration routes (Günther et al. 2018). Physical remains, camp sites (Bang-Andersen 2012), stone-built archery (Selinge 1974), and hunting gear (Pilø et al. 2018), of their hunting activity are well documented.

Current Distribution

Rangifer has a circumpolar distribution inhabiting high arctic, the tundra and taiga region of northern Europe, Siberia and North America, and currently number millions across the Holarctic (Gunn 2016). However, in Europe, the heavy exploitation with organized mass trapping operations during the Medieval time (Indrelid and Hufthammer 2011) paid its toll. The parallel expansion of reindeer pastoralism reduced their number and range (Vorren 1973). Moreover, with the introduction of firearms reindeer was exterminated from most parts of Fennoscandia and large part of Russia. Today European reindeer are found in small pockets in central Norway, eastern Finland, and in the northern European part of Russia, as well as on Svalbard and Novaya Zemlya (Fig. 3).

Indeed, the range of tundra reindeer has contracted during the last centuries. The tundra type was extirpated from Northern Norway in the late 1800s and only remnants were found left in south-central Norway. In Fennoscandia, about 35,000 wild tundra reindeer thrives in mountainous area of southern Norway in 23 populations (Reimers 2018, Fig. 3). Only remnants of tundra dwelling reindeer are found in the Kola Peninsula, the western population considered mountain-tundra reindeer may be classified close to a tundra type (Mizin pers. comm.) with an estimated population size of about 7,500 (Panchenko et al. 2014; Baranova et al. 2016). According to Mizin (pers. comm.), there are no typical wild tundra reindeer in the Arkhangelsk and Komi regions in Russia. Indeed, the taxonomic status of wild reindeer of Arkhangelsk province and Komi Republic, where probably both tundra and forest reindeer are found, remains open.

Forest reindeer was exterminated from Sweden in the late 1700s, and in Finland, the subspecies disappeared in the early 1900s

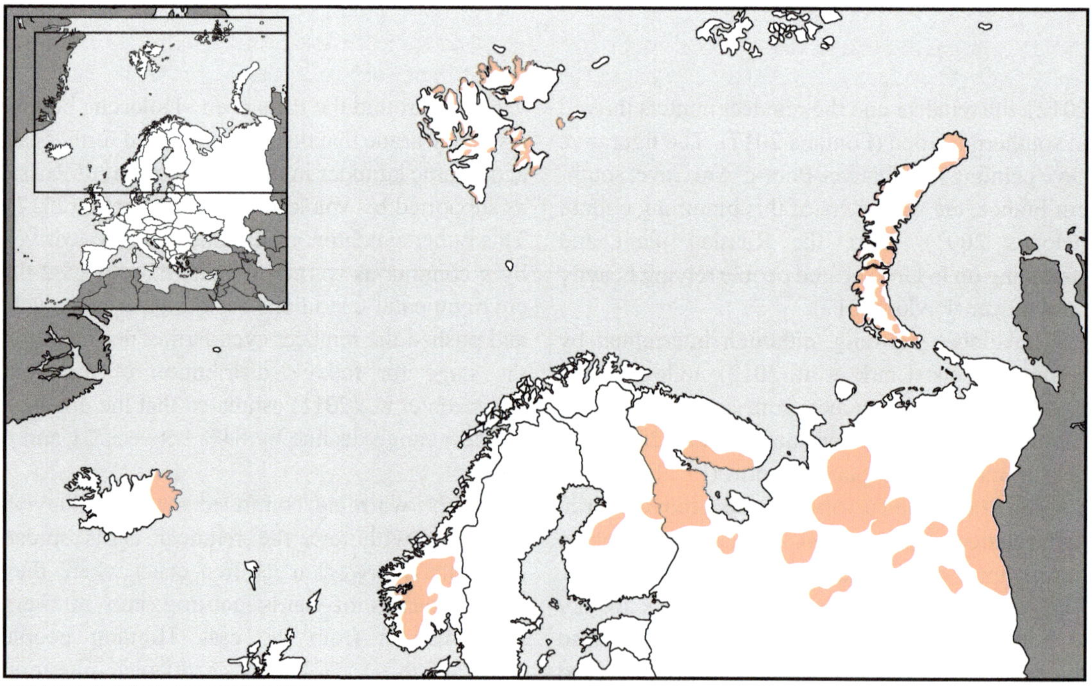

Map template: © Getty Images/iStockphoto

Fig. 3 Main distribution in wild reindeer in Europe. See text for further information. (Map template: © Copyright Getty Images/iStockphoto)

(https://www.suomenpeura.fi/en/wild-forest-reindeer/range.html). They were pushed back and survived in the vast taiga area of western Russia. After the Second World War, forest reindeer were introduced to Finland and the population has gradually increased in number and range. According to the most recent surveys, there are approximately 2300 wild forest reindeer in Finland https://www.luke.fi/en/natural-resources/game-and hunting/finnish-forest-reindeer/. In Russia, the forest reindeer are now separated into two isolated populations: Karelia and East European, estimated to 4,000 and 6,000 head, respectively, based on Mizin et al. (unpublished) and Danilov et al. (2018), those populations themselves being considerably fragmented. However, population estimates are not available for many smaller populations because of insufficient surveys during the last decades (Mizin et al. 2018). Today we find only scattered populations concentrated in the northern areas of Karelia and a smaller population in south east in the Pudozh region extending into Onezhsky district, Arkhangelsk Region. Further, there are scattered population in Arkhangelsk Oblast, mainly living east of the Northern Dvina River. In Nenets Autonomous Okrug, small herds reach the shore of the Chyoshskaya Bay of the Barents Sea. In the Komi Republic, the wild reindeer is widespread, but is not found south of Syktyvkar, and rarely reaches the forest tundra in the north. In the European parts of the Urals, forest and forest-mountain types reindeer (Red Book of the Komi Republic 2019) are common on the slopes of the Nether-Polar Urals and North Urals. On the Kola Peninsula, two reindeer populations are found (Fig. 3), the eastern is classified as forest reindeer, whereas the western is called mountain-tundra reindeer. This is probably a form of native wild reindeer of Kola Peninsula (Mizin pers. comm.).

In European high Arctic, humans took their share. Reindeer inhabiting the Svalbard archipelago were hunted in the 1600s, 1700s, and 1800s by expeditions mainly exploiting the rich marine mammal resources along the coast (Kovacs et al. 2004). In 1925, at the brink of extinction, hunting was banned, and the population has recovered (Kruse 2017) and counts today around 22,000 animals (Le Moullec et al. 2019). They occupy almost all non-glaciated areas of the Archipelago (Fig. 3), with the highest densities in Nordenskiöld Land, Edgeøya, and Barentsøya.

The reindeer on Novaya Zemlya have also been heavily exploited and partly mixed with semi-domestic animals (Mizin et al. 2018). Today the population counts around 5,000 animals (Mizin et al. 2018), found both at the southern and northern island (Fig. 3).

As early as in the late 1700s, domesticated reindeer were brought from Norway to Iceland. Today this feral population numbers around 5,000 winter head (Skarphedinn 2018) situated in East Iceland (Fig. 3).

Description

Rangifer are cold-adapted, medium to large-sized deer. The between and within subspecies variation in morphology, as well as ecology and behavior, reflects the species adaptive potential and plasticity, essential for its wide distribution. *Rangifer* are highly sexual dimorphic, with adult males weighing up to twofold of females (Banfield 1961; Reimers 1982). Both sexes wear antlers. The pelage is thick and furry. The main hooves are big and crescent shaped, well adapted for cratering through snow to reach forage. When spread out they function like snowshoes – floating on top of the snow.

Body Size

All subspecies show pronounced sexual dimorphism with adult males being 30–100% heavier than females (Geist 1999). Additionally, ecotypes and populations vary in body size with recorded shoulder height range of females: 50–100 cm and males: 70–120 cm and body length range of 160–205 cm and 180–215 cm, respectively (Geist 1999). Adult females normally weigh between 60 and 130 kg in fall, whereas for males the range is 120–280 kg. Svalbard reindeer is small and stocky. The subspecies is relatively short-legged, and adult may have a shoulder height of as little as 50 cm in females and 70 cm in males with a body mass in autumn of around 80 and 120 kg, respectively (Reimers 1984), whereas forest males may reach 100–120 cm at the shoulder with a maximum body mass of 250 kg. Tundra reindeer is in between as reflected in variation in height and body mass, normally not exceeding 250 kg in males and 110 kg in females, (Reimers 2018). No measurements of Novaya Zemlya reindeer are available.

Antlers

Reindeer are unique among cervids because females also possess antlers. However, antlerless females are observed in many populations. Its frequency seems to vary over time (reviewed by Reimers 1993; Jacobsen et al. 1998). Antlerless females are rare in populations in good conditions, whereas populations in poor condition seems to have a higher frequency. Lack of antlers in males is very rare. The antlers are the most rapidly growing tissue (Nieminen 1985). In prime males, their growth and regeneration require large quantities of minerals. At the start of the growing season, calcium and phosphorus must be mobilized partly from the skeleton (Moen and Pastor 1998). The increased energy requirements during antler growth in adult males are estimated to 8–16% (Moen and Pastor 1998). Indeed, antler growth is related to nutritional quality of the diet as well as genetic makeup and age, especially in males.

Adult male antlers are characterized with a kink-beamed, a large crown and distinct brow and bez tines (Geist 1999). The structural pattern is among the most complicated among deer species with great variability of the upper zone (Bubenik 1975). The brow tines are well developed usually with one palmate tine dominating (Bubenik 1975). The bez tines are also well developed. The third tine is normally spiked and less advanced. The upper zone is relatively variable and both monopodium and forked appearances are found. The degree of palmation in this terminal zone does vary.

The male tundra ecotype is characterized by long, and cylindrical beams with many tines, whereas the forest type is more massive and show more palmation (Geist 1999). The antlers of male Svalbard and Novaya Zemlya reindeer resemble the tundra form but are smaller and spindly. Energy requirements for antler growth in females are low relative to other energy requirements (Moen and Pastor 1998) because of their much smaller size and simpler antler form, seldom reaching more than 1 kg and a beam length of 50 cm (Skjenneberg and Slagsvold 1979). They reported that semi-domestic males' ossified antlers could weigh up to 10 kg. Geist (1999) suggests that male *Rangifer* have the largest antlers relative to body size among living deer species, pointing towards a highly polygynous mating strategy (Holand and Weladji 2019).

Reindeer males' antlers normally reach their maximum size around 4–7 years, thereafter the relative size is reduced although their main form is kept from year to year (Prichard et al. 1999). They argue that female antlers grow for up to 10 years. However, Melnycky et al. (2013), following individual semi-domestic females over years, found that the growth in females appeared to reach a plateau at age 3 years.

Calves begin to develop pedicles during early post-calving and produce simple single spiked or forked antlers the first fall. Yearling males' antlers resembled fully grown females. Differences in antlers growth, as in body mass, continued to increase between male and female reindeer until at least 5 years of age, and the relative allocation toward antlers tend to increase in males but decrease in females (Melnycky et al. 2013).

The antler growth and casting cycles of females remain out of phase with those of male reindeer (Espmark 1971a; Høymork and Reimers 1999). Adult males cast their antlers after rut normally in early December. Young males and nonpregnant females shed their antlers in March–April (Reimers et al. 2013). Pregnant females normally keep their antlers longer and shed them within a week after parturition (Reimers et al. 2013). The new growth starts in May–June with lactating females being the latest to initiate antler growth. Antlers remain in velvet through the summer until the velvet is shed before the rut in adult males, a little later in young males. The females clean their antlers around rut, whereas the calves keep the velvet on a little longer to the end of the rut.

Pelage

Tundra reindeer have a greyish appearance, characterized by a dark back, dark frontal parts of the legs, a white large rump patch, a small white tail, a white belly and neck with long neck mane, especially pronounced in males, white withers, and dark flank strips (Geist 1999). The stout head is rather dark with a broad muzzle covered by fur and short ears well furred. The two other ecotypes carry basically the same coloration pattern, Svalbard reindeer being paler, while the forest ecotype, is typically darker. The Novaya Zemlya subspecies is lacking the dark side stripe (Mizin et al. 2018). In winter, reindeer get a more whitish appearance as the overcoat is worn (the tips of the hairs break off) and lose some of their pigmentation.

The coat has two layers: a dense woolly undercoat and a fairly long and dense overcoat consisting of hollow, air-filled hairs (Skjenneberg and Slagsvold 1979). While the hollow hairs add buoyancy when reindeer swim, the fur is well adapted to withstand cold and wind, especially the winter pelage of Svalbard reindeer (Cuyler and Øritsland 2002). Although the summer pelage is short and less dense, reindeer can be heat-stressed during warm and sunny summer days. They try to cool down by seeking wind-exposed terrain and snow patches. They also get rid of heat by panting. Newborn calves have a brownish woolly undercoat. The calf's overcoat develops during the summer months and during fall a normal coat is established.

During the molt, which normally starts in May and ends in July, earliest in animals in good condition, the animals look shabby. The thick winter coat is shed as the new shorter darker summer coat come through with males shedding earlier and lactating females later and during the peak of the warble fly season. The later shedding of leg hair may contribute to why females have lower warble infection rates (Cuyler et al. 2012). The summer coat keeps growing and in fall extra overcoat hair is added, and the winter coat is fully developed before the onset of winter (Skjenneberg and Slagsvold 1979).

Skull

Traditionally skull measurements have played an important role in *Rangifer* taxonomy (e.g., Banfield 1961). Nieminen (1980) reported mean basilar length, the length from the median position of the ventral margin of the foramen magnum to the prosthion, in adult females and males, respectively, to be 254 and 267 mm in Svalbard reindeer, 281 and 300 mm in tundra reindeer, and 305 and 363 mm in forest reindeer. This is in line with Banfield (1961) measurements. Based on discriminant analyses of 13 skull measurements, Nieminen (1980) was clearly able to distinguish the three ecotypes, both in females and males. Indeed, the longer and slender skull of forest reindeer compared to the tundra type and the shorter and compact Svalbard reindeer is apparent.

Scent Glands

The antorbital, tarsal, and interdigital, both on the forefoot and hind foot, glands of reindeer were described by Schaffer (1940). Müller-Schwarze et al. (1977) described the caudal gland. In addition to these dermal glands, reindeer possess sudoriferous glands of the apocrine type in the hairy skin (Müller-Schwarze et al. 1979). Their functional importance of intra-herd communication is elaborated in the Behavior section.

Teeth

The calves are normally born with 22 milk teeth (two of the premolars erupt normally after some days); 0.1.3.0/3.1.3.0. The permanent dentition is comprised of 32 (34) teeth; 0.0(1).3.3/3.1.3.3, and is fully developed at age 2 year (the canine teeth in the upper jaw are reduced). The eruption pattern can be used to visually classify: calves, yearlings,

2 years ages, as well as older individuals. Older individual can be aged by cementum layers (Reimers and Nordby 1968). Age related tooth wear has significant consequences for digestive efficiency by affecting rumen particle size (Veiberg et al. 2007). Poor winter foraging condition amplify the wear as seen in Svalbard reindeer (Veiberg et al. 2007). The same has been found on Novaya Zemlya (Sokolov et al. 1996) and under extreme bad winter range condition in Norway (Skogland 1988). They simply wear out their teeth by scraping the vegetation and are susceptible to ingest sand and other mineral particles that accentuate the wear during mastication, affecting old females' body reserves negatively (Kojola et al. 1998).

Hooves

The anatomy of reindeer hooves and lower limbs is complex due to the unusual flexibility of the dew claws (Hull et al. 2021). Their hooves function like snowshoes and when spread out reduce sinking depth and save energy when walking in heavy snow. Floating on top of the snow crust is also important for outcompeting predators. The dew claws contribute to the low hoof loading which is the hoof area relative to body mass and *Rangifer* has an unusually low loading compared to other members of the deer family. The hooves' sharp edge makes them well suited for cratering in snow.

Physiology

Rangifer show a circannual body growth, activity, and bioenergetics as adaptive responses to the highly seasonal environment. The seasonal adjustments harmonized by endogenous signals partly related to the photoperiod, determine the variation in appetite and basal metabolism reflected in their pattern of cyclic growth with maintenance in winter and rapid growth in summer (Klein 1996; Reimers 1983a; Reimers et al. 1983; Skogland 1989). In comparison to summer, the basal metabolic rate in winter is 20–30% lower in non-activity functions (Rusell and Martell 1984).

The different physiological and morphological adaptations (Blix 2016) are most pronounced in the high arctic island representatives where selection pressures have produced phenotypically distinct characteristics and extreme adaptations. Svalbard reindeer is a compact small-bodied subspecies with short legs. Combined with well-developed counter-current heat exchange system in their extremities, elaborate structures in the nasal cavity for restricting evaporative heat and water losses and an extremely well-developed underfur (Blix 2016), they are able to withstand effective temperatures down to −50 °C without having to use extra energy to keep up the core body temperature (Nilssen et al. 1984). Also, the thick subcutaneous fat layer gives extra insulation. In summer, they may exhibit problem to get rid of surplus heat. The fat reserves can amount to 30–40% of their ingesta free body mass (Reimers 1982) and may together with the protein reserves contribute up to 25% of their winter energy requirement (Tyler 1986).

At the other end of the geographical range, Eurasian forest reindeer is found. They are large bodied and long legged, hence well adapted to movement through deep snow and wooded and marshy habitats. Their antlers are big but narrower V-shaped which ease movement in dense forest. Their body reserves in fall are small and not comparable to the high Arctic form.

The tundra form has evolved in open landscape, primarily tundra and alpine habitat. The security in open landscape put pressure on running and endurance. Their traditionally long migration routes enable them to exploit the summer rich lush tundra and alpine areas and to seek refuge in the northern taiga zone in winter. This seasonal cycle has put selection pressure on energy efficiency of locomotion (Fancy and White 1986). Indeed, their relative long legs and low foot loading contribute to their low locomotion expenditures (Fancy and White 1986).

Rangifer is one of the very few ruminants that can utilize lichens effectively as an energy resource, suggesting microbiota adaptations (Salgado-Flores et al. 2016). Their adherence to this energy rich lichen-based winter diet, which is highly digestable but extremely low in protein, has also induced efficient N-cycling mechanisms (Wales et al. 1975).

Genetics

Nes et al. (1965) established the karyotype of *Rangifer t. tarandus* L. followed by Gripenberg et al. (1986) of *Rangifer t. fennicus* Lönnb.; $2n = 70$, $FN = 74$. They found that the autosomes consist of 33 acrocentric pairs and one submetacentric pair. The *Rangifer* karyotype conforms to the basic karyotype of the cervid family (Fontana and Rubini 1990). However, its sex chromosomes are unexpectedly large compared to other cervids (Fraccaro et al. 1968).

Population genetic analyses suggest that the diversity between Eurasian tundra reindeer (*R. t. tarandus*), Eurasian forest reindeer (*R. t. fennicus*), and the Svalbard reindeer (*R. t. platyrhynchus*) and Novaya Zemlya reindeer (*R. t. pearsoni*) reflects both their refugial origins and the colonization of circumpolar regions after the Last Glacial Maximum (Kvie et al. 2016; Weldenegodguad et al. 2020). Svalbard and Novaya Zemlja reindeer, both high arctic ecotypes, clustered separately from the two other subspecies (Weldenegodguad et al. 2020). Their limited genetic diversity (Kvie et al. 2016), especially in the Svalbard reindeer (Weldenegodguad et al. 2020), compared to the two others, makes them more vulnerable to climate change. It indicates recent geographic isolation and genetic bottleneck(s), and supports that the Archipelagos were colonized by reindeer from the Eurasian mainland after the last glacial maximum (Kvie et al. 2016).

Rangifer has adapted to an environment characterized by short daylight and sparse forage resources during the long and cold winter and prolonged daylight and affluent green biomass during the short summer. Weldenegodguad et al. (2020) argue that the changes in their genetic makeup as compared to closely related species are linked to unique phenotypic and functional adaptations to these environments. Comparative genome analyses with other wild and domestic ruminants revealed several expanded gene families related to vitamin D metabolism, retinal development, circadian rhythm, and tolerance to cold temperatures. Indeed, several of these genes show signatures of positive selection (Weldenegodguad et al. 2020).

Life History

Growth

Sexual dimorphism is strong in reindeer with males being larger in both body size, body mass, and antler size (Geist 1999; Høymork and Reimers 1999). For both males and females, body mass grows steadily and reach a plateau, earlier in females than in males (Eloranta and Nieminen 1986; Reimers 1983a). At birth, the calves weigh around 5 kg, the male calves being around half a kg heavier than female calves (Eloranta and Nieminen 1986). The birth mass is positively correlated to the mother's body mass before parturition. In September, the calves reach a body mass of around 40 kg, male calves around 4 kg heavier than female calves, averaging a daily growth rate of around 300 g (Eloranta and Nieminen 1986). Reimers (1983a) reported a growth rate in tundra reindeer calves between 275 and 400 g/day. Indeed, this will vary between subspecies as well as summer range conditions (Reimers 1983a). The females reach fully body size at 3–4 years of age, whereas males continue to gain mass for at least 2 years, maturing fully at 5–6 years (Reimers 1983a). As described in the section "Body Size," mature males may reach a body mass up to twofold mature females. Body mass varies considerably between seasons. Males may lose as much as 30% of their mass during rut (Mysterud et al. 2003) and during extreme winters females may lose up to the same percentage of their autumn body mass (Reimers 1982). Indeed, compensatory summer growth occurs in reindeer, reviewed by Reimers (1997).

Reproduction

Reindeer are strongly polygynous. Dominant males can impregnate several females in a single breeding season, up to 30 under certain circumstances (Røed et al. 2002), while females will give birth to a single calf. Females are normally sexually mature at the age one-and-half year (Reimers 1983b). After the age at first reproduction is reached, female ovulate (and may conceive) nearly every year in the absence of severe food limitation (Skogland 1989). A decrease in reproductive effort (senescence) has been reported after about 10 year of age (Weladji et al. 2010). Variation in age at maturity has been attributed to the difference in body mass within cohort. Generally, females will need to be between 45 and 60 kg to be able to reproduce (Reimers 1983b; Ropstad 2000). The fertility rate is normally above 80% among mature females but can be as low as 10% among 1.5 year old females (see Reimers 1997). Males can breed successfully the first time as yearlings, but their reproductive success is higher between 3.5 and 6.5 years of age. Males with higher social rank control larger mating groups and therefore have higher reproductive success (L'Italien et al. 2012). Selection for antlers in male reindeer is very strong, due to the direct benefit of breeding success via access to females through intrasexual combat (Kiltie 1985). In contrast, antlers offer an indirect benefit to reproduction in females through acquisition and protection of food resources, because larger antlers correspond to increased rank during agonistic intraspecific interactions over craters dug through the snow (Skogland 1989). Indeed, the factors affecting social rank in male and female reindeer include body mass, antler size, and age (Geist 1999; Holand et al. 2004a, b), strongly influence their reproductive fitness.

Reindeer are seasonal breeders with the rut spanning from mid-September to mid-October or even end of October depending on the location as well as the latitude. By taking advantage of the summer forage, both male and female ensure good body condition before entering the breeding season. While prime age females tend to reach estrus earlier, it appears that within an age class, females on better condition (body mass) are at advantage. Approximately 90% of females are successfully copulated in a 10–21-day period, with calving occurring in a highly synchronized manner the following spring (Ropstad et al. 1996; Rowell and Shipka 2009). Being fertilized earlier will allow longer gestation time as calving is highly synchronized, this will often also lead to those females giving birth to calves that are above average body mass (Eloranta and Nieminen 1986). Gestation length vary between 203 and 240 days depending on ambient conditions (Ropstad 2000; Rowell and Shipka 2009). In a semi-domestic reindeer population in Kaamanen, Finland, gestation length mean was 221 days, and it increased with female age and was longer for male than female offspring (Mysterud et al. 2009).

Survival

Indeed, there is a strong correlation between birth mass of calves and their neonatal survival (Eloranta and Nieminen 1986). Moreover, the conditions at birth may provide long-lasting advantages, leading to improved fitness. Reindeer grow fast and mature early, relative to most ungulates (Pedersen 2019). This may secure higher calf survival for *Rangifer*. Nevertheless, as seen in many northern ungulates, mortality is highest during the early life stages during the summer and the first winter (Pedersen 2019; Reimers 2018). Calf mortality during the first 5 months following birth in tundra reindeer varies greatly from almost zero to around 50% (Skogland 1985). A calf's chance of survival appears to be strongly influenced by its mother's body mass, and hence condition, at the time she gives birth (Skogland 1985). The life span of a reindeer may range up to 15–20 years (Syroechkovskii 1995), being shorter for males than for females. Females adopt a conservative strategy to favor own survival and hence secure the potential for future reproduction (Bårdsen et al. 2008). The females' risk-averse life history strategy (Bårdsen et al. 2008) is essential to understand the herd dynamic, which is primarily driven by female body mass and condition (Reimers 1997). The highly variable reproductive success among males implies a bolder strategy which implies combat injuries during rut and depleted body reserves at start of the winter and less resources for winter survival as compared to females. In extreme winter conditions, a winter mortality of around 25% has been reported in Svalbard reindeer population hitting the calf segment hard with an estimated loss of 65% (Reimers 1982). Under such extreme conditions, fetal survival is low (Ropstad 2000) and females may abandon the newborn calves to secure own survival.

Habitat and Diet

Reindeer are medium-sized ruminants classified as intermediate feeders, i.e., flexible in their diet (Hofmann 1989). They are, together with muskoxen (*Ovibos moschatus*), the only truly permanent Arctic ungulates. Reindeer have adapted to these low productive but highly seasonal environments (Bliss et al. 1981) by developing foraging (Blix 2016) and behavioral strategies, including, e.g., seasonal movements (at different scales) in relation to forage quality and quantity. Indeed, their habitats span parts of the northern taiga, tundra, alpine areas, and the high arctic desert. Reindeer are selective foragers, following the phenology progression that is facilitated by movements, aggregation, and dispersal (Skogland 1989). During winter, reindeer feed heavily on lichens (ground lichens like *Cladonia* sp., *Ceteria* sp., as well as aboreal lichens like *Alectoria* sp. and *Bryoria* sp. (Staaland and Nieminen 1993), if available. Ground lichens typically occur as clumps and carpets and can be foraged efficiently when uncovered from snow and ice.

Lichens supply energy to maintain the reindeer over winter but do not prevent undernutrition with a negative balance of nitrogen (Ryg and Jacobsen 1982; Soppela et al. 2008). The exception being Svalbard reindeer, mainly feeding on dried sedges, shrubs, and mosses in winter. Standing biomass of ground lichens tends to decrease at high densities, with resulting negative impact of the animal winter conditions (Skogland 1989; Kumpula et al. 2014). Summer forage contains grass, sedges, herbs, and leaves of willows, while in autumn, they also feed on mushrooms (Staaland and Nieminen 1993). Summer forages are of high quality and allow reindeer to accumulate protein and fat, thereby increasing their body mass. Variation in life history and population parameters of reindeer are related to both winter (Kumpula 2001; Skogland 1985) and summer (Reimers 1997; Reimers et al. 1983) resource availability and accessibility.

Behavior

Social Behavior

Rangifer, mainly the tundra reindeer, is highly gregarious (Geist 1999), forming group of variable sizes being generally smaller during the winter. They have been preyed upon by predators and parasitic insects and evolved an adaptive defense mechanism by clustering. Tundra reindeer herds use specific sites for calving (calving grounds), and most herds move between summer and winter ranges, to find adequate sites to satisfy their nutritional needs, leading to relatively larger groups. Group size also varies by ecotypes. Indeed, Svalbard and Novaya Zemlya reindeer as well as forest reindeer live in small groups have a relatively small home range and remain stationary when conditions are favorable. In their predator-free environment, the high artic populations operate in small groups and spend a high proportion of their time foraging, with short resting/ruminating bouts throughout the bright high arctic summer days (Pedersen 2019; Mizin et al. 2018). Forest reindeer predator defense is based on living in scattered group with only short seasonal migrations within dense taiga habitats (Kojola et al. 2009).

There are advantages but also costs, of living in groups. As group size increases, there are more individuals watching for predators, and each individual can therefore devote more time to foraging without increasing their predation risk. During insect harassment, the animals respond by clumping to reduce the per capita burden (Skogland 1989). The most obvious cost of grouping is competition for food (Skogland 1989). Hence, groups are smaller in the winter when food is sparser and rarer.

Although the hooves are well adapted for cratering snow, cratering for food costs energy (Fancy and White 1985) and is time consuming. Defending craters may therefore pay off, especially for pregnant females and their calves. Their intrasexual rank is influenced by the size of the antlers (Holand et al. 2004a). Indeed, they keep their antlers throughout the winter as opposed to the males and are therefore higher in rank than males in the winter herd.

Energy-Saving Behavior

Rangifer show not only physiological and morphological bioenergetic responses to their highly seasonal environment, but also behavioral plasticity. In winter, they move less to save energy (Skogland 1989; Cuyler and Øritsland 1993). At low effective temperatures, they seek shelter from wind and may clump together to reduce heat losses. During spring migration, strings of animals can be seen along the tundra when the snow often get heavy. Indeed, they may line up to save energy (Skogland 1994). During warm summer days, they may be heat stressed and respond by congregating at snow patches and wind exposed ridges to cool down (Skogland 1989).

Communication and Signals

Reindeer use alarms signs, mother-young follower signals, threat and display (Skogland 1989), typical for herd living species. The scent glands all play important roles in communication. Also, urine is a source of chemical stimuli and the caudal scent is probably incorporated in the female urine. Indeed, males are often seen sniffing the females' anogenital area and the ground where females have urinated during rut to check their estrus status. The caudal gland also plays an important role in mother–infant bonding and often the mother is seen to smell the tail of her calf (Källquist and Mossing 1982). The chemistry of different scent gland secretions has been described by Müller-Schwarze et al. (1977). On the summer range, all female age classes live together in herds. With many calves' present, individual recognition by smell and sound is important at that time. Indeed, in large groups, mothers can recognize their calves by the calves' calls (Espmark 1971b). Raising their tail is also a typical alarm signal in *Rangifer* (Skogland 1989).

Parental Care

Maternal care occurs during gestation, lactation, and post-parturition care of offspring (Holand et al. 2006a). Even postweaning maternal care is documented during the calves' first winter (Holand et al. 2012a), as the mothers protect their calves' position within the herd and by defending and sharing craters (Espmark 1971b). Male reindeer do not provide parental care so male reproductive investment is restricted to the well-defined rutting period.

Mating Behavior

Rangifer is a strongly polygynous species and has a mating system primarily driven by male ability to monopolize access to females in estrous. This is achieved using harem defense, tending and defending individual females, and a mixture between harem and territorial defense depending on the spatial and temporal distribution of receptive females (Espmark 1964; Holand and Weladji 2019; Kojola 1986). Reindeer mating strategy varies considerably between subspecies, demonstrating adaptations to the highly variable environments *Rangifer* inhabits. In the non-migrating high arctic ecotype (Svalbard reindeer) where resources are patchy, female groups are spatially stable and dominant males defend harems at fixed locations (Hætta 2009). Migrating tundra reindeer form large groups during fall migration, which coincide with the breeding season. Dominant males are therefore not able to control these mobile female bands and show a tending strategy following one specific estrous female at a time (Skogland 1989). When resources are more evenly distributed, the mating system is harem-based (Skogland 1989). These mobile harems are unstable due to female movement between groups, creating fission-fusion group dynamics (Body et al. 2015). In addition, instability of male hierarchies (Holand et al. 2012b) may amplify the females' grouping dynamic. The forest ecotype of *Rangifer* forms small groups and dominant males show a combined strategy in defending a territory with a small harem (Kojola 1986); this is close to what is observed in the high arctic ecotype. Males start rutting earlier and end it later than females. The adult males are eagerly fighting for dominance in the early phase of the rut. The fights, with clashing of antlers, probably trigger the females into entering mating mode. In contrast to male mating strategies, females mate choice is less conspicuous and not well understood. Indeed, it is difficult to demonstrate female choice, as male-male competition and male coercion and harassment may override female mate choice or limit its expression. Females may, however, choose their mates on a fine temporal scale. Indeed, during their receptive phase which last only 1–2 days (Ropstad 2000), females court the dominant male to capture his attention (Djaković 2012). Females' major reproductive investment relates to producing and raising their offspring. Accordingly, female *Rangifer* will normally spend more time foraging during the rut to keep up their body condition necessary to overcome harsh winter conditions. Simultaneously, they have to avoid harassment by young males and capture the dominant males' attention to secure early fertilization by good quality males, suggesting there might be some female choice in the process (Body 2014; Holand and Weladji 2019). Indeed, female reindeer are reluctant to mate with young males (Holand et al. 2006b). At the start of the mating season, mature reindeer males drastically reduce feeding and establish dominance hierarchies through aggressive interactions; young males typically do not participate in this capital breeding strategy (Kojola 1991).

Parasites and Diseases

Reindeer health is a holistic and highly dynamic concept, which can be studied at individual and population scales. Both infectious and non-infectious agents may play important roles and interact with other extrinsic and intrinsic factors at the individual level. Indeed, a healthy population may include unhealthy individuals. *Rangifer* health is reflected in their fitness and ability of individuals or populations to cope with disturbances, natural and anthropogenic (Macbeth and Kutz 2019). Also parasites may take their toll. For in-depth information about the theme, we refer to the comprehensive newly published book "Reindeer and Caribou Health and Diseases" edited by Tryland and Kutz (2019).

Much knowledge about reindeer biology, including diseases, comes from experimental, observational, and diagnostical studies of semi-domestic animals. The herding, corralling, and feeding practiced in reindeer husbandry lead to concentration of animals and hence the potential emergence and outbreak of diseases. Indeed, many diseases are connected to semi-domestic herds (Josefsen et al. 2019) and may be spread into wild populations, in areas where they overlap spatially.

Rangifer populations host a variety of infectious agents: viruses, bacteria, helminths, protozoa, and arthropods. They can be classified into parasitic infections and diseases, bacterial infection and diseases, and viral infection and diseases. A prion-related disease (chronic wasting disease) has been diagnosed in one Norwegian population (VKM 2018). In addition, noninfectious diseases and trauma, related to predators, human disturbance, and pollutants, may occur.

Parasitic Infections and Diseases

Parasites, helminths, protozoa, and arthropods, are all important in *Rangifer* biology. Their lifecycles vary, and the transmission may be direct, indirect through vectors, intermediate hosts, and predator-prey interactions. Normally they have a clumped distribution; a few animals being heavily infected. Under such conditions, predators present, stochastic weather events and population fluctuations, these animals will have a higher propensity to die. Thus, there is natural limitation to parasite abundance. However, the parasite load is partly a function of animal density (Bye 1987) and under certain conditions parasites may influence vital rates and hence population dynamics (Albon et al. 2002; Carlsson et al. 2019). The finding of reindeer nematodes in sheep hosted together (Manninen et al. 2014) indicates a risk of nematode cross-infections. This should be investigated further, especially in Norway where overlapping alpine summer grazing ranges between large populations of domestic sheep and small wild reindeer populations are common.

Helminths, roundworms, flatworms, and tapeworms, may infect most organs and their impact varies from subtle to severe. *Ostertagia gruehneri* is the most common nematode in the gastrointestinal tract. The parasite may influence body condition and based on field experimental treatment in Svalbard reindeer, the parasite reduced fecundity although the effect interacted with winter weather suggesting the complexity of describing parasite effects (Albon et al. 2002). *Elaphostrongylus rangiferi*, meningeal worm, is widespread in Eurasian *Rangifer* and may infect other ruminant species as well. The infection is normally symptomless. However, in heavily infected animals, the development of larvae to adult in the central nervous system may cause severe disorder. The temperature-dependent gastropod life stage results in higher clinical disease occurrence after warm summers (Handeland and Slettbakk 1995).

Several arthropods harass *Rangifer*. Ectoparasites, lice, mites, ticks and oestrids, may deplete the animals for resources. Blood-sucking insects, especially mosquitoes and black flies, may drain the animals (Skogland 1989). They are also potential vectors for diseases. Lice, mites, and ticks may cause severe hair breakage and loss during winter, reducing the pelage insulation (Kutz et al. 2019). Blood-sucking deer ked (*Lipoptena cervi*) is spreading north and has been found on forest reindeer (Kynkäänniemi et al. 2014) causing pelage damages.

Two oestrids species, the reindeer warble fly (*Hypoderma tarandi*) and the reindeer throat fly (*Cephenemyia trompe*), parasitize *Rangifer*. In Europe, the flies are found in Fennoscandia and Russia, the exception being Svalbard Archipelago (Halvorsen 2012) and Novaya Zemlya (Mizin et al. 2018). They are also absent from the Icelandic feral population (Sigurdarson and Haugerud 2004).

Heavy invasion of reindeer warble and throat flies, up to several hundred larvae may be found, imposing a heavy physiological load in winter when the larvae are growing rapidly (Halvorsen 2012). The infection load is generally higher on calves and young animals than on adults while bulls tend to be more heavily infected than females (Cuyler et al. 2012). This could partly be due to less antibodies during summer (Åsbakk et al. 2005), which may render them more susceptible to invasion, as well as their lower social status, which may expose them more often to attacks. Mosquitoes and black flies amplify the effect and may render calves and young animals' anemic (Skogland 1989). Both oestrids and blood-sucking insects cause reindeer to aggregate and move more which reduces foraging activity (Folstad et al. 1991; Anderson et al. 1994) and thereby mass gain (Helle and Tarvainen 2010). Indeed, heavy insect summers may induce high die-offs, especially among the youngest age classes, and this may have population consequences (Ballesteros et al. 2012).

Bacterial Infections and Diseases

Bacterial diseases occur in wild reindeer but are not often detected. Here, some of the most common are mentioned.

Anthrax has historically hit the Russian semi-domestic herds hard (Josefsen et al. 2019). The summer 2016 a massive outbreak infected thousands of animals in the Nenets-Yamal Autonomous Okhrug (Josefsen et al. 2019). A major vaccination program and extensive environmental decontamination efforts were initiated. *Bacillus anthracis*, when faced with inhospitable environmental conditions, may form spores that are capable of surviving for decades in soil and water. Thawing permafrost during heat waves, amplified by climate change, may allow the spores to reactivate, spread across the tundra and infect animals (and humans) through ingestion, inhalation, or colonizing open wounds. Biting insects may spread the disease rapidly and within weeks thousands of animals are infected. Indeed, this may hit wild herds with overlapping or adjacent ranges.

Pasteurellosis, caused by *Pasteurella multocida*, outbreaks have been seen in semi-domestic herds in Scandinavia with high mortality and sporadic cases have also been reported in wild tundra reindeer in Norway (Josefsen et al. 2019). Indeed, the disease has the potential to cause epizootics. Brucellosis, caused by *Brucella* spp. in reindeer, is prevalent in Siberia (Zheludkov and Tsirelson 2010), but not reported in wild European populations. Necrobacillosis, caused by *Fusobacterium necrophorum*, attacks primarily the digits and distal feet (Josefsen et al. 2019), and often seen in reindeer husbandry, but occasionally seen in wild reindeer (Handeland et al. 2010).

Viral Infections and Diseases

Parapoxviruses, among them Orf virus is distributed worldwide in small domestic ruminants and may be transferred to reindeer indirectly through sharing pasture and or visiting the same salt blocks. The virus attacks the nose and mouth area and causes contagious ecthyma, diagnosed mainly in semi-domestic herds. Papillomaviruses are rather seldom diagnosed in reindeer but may cause papillomas in the skin. Herpesviruses are rather common and infect the mucosal layer. They have little clinical impact, but mucosal damages may induce secondary bacterial infections (Tryland et al. 2019a). *Rangifer* are susceptible to foot and mouth disease, but has not been reported in wild reindeer. Rabies has been diagnosed in Svalbard reindeer. Arctic fox is the main reservoir (Mørk and Prestrud 2004) and rabies is transmitted to reindeer mainly by bites from infected foxes.

Chronic Wasting Disease

Chronic wasting disease (CWD) is the only prion disease diagnosed in *Rangifer* (Tryland 2019).

In 2016, the first CWD case in free ranging cervids outside North America was detected in a wild (mixed-feral) reindeer population in Southern Norway followed by several more cases. A decision to exterminate the whole population, which border other wild populations as well as a semi-domestic herd, was taken to minimize the risk of spreading. Indeed, CWD in cervids are contagious, spreading between animals via environmental contamination (soil and feces). Mysterud et al. (2020) suggested that the risk of spreading could be amplified by antler chewing. During the winter 2017–2018, the wild reindeer population was culled. Altogether 18 CWD positive reindeer have been diagnosed (out of around 1000 animals). Most wild reindeer shot during the regular hunting seasons 2017–2019 have been tested, in addition to many slaughtered semi-domestic animals, all being negative. The characterizations indicate that the wild reindeer cases are somewhat different from the strains found and investigated in North America (VKM 2018; Pritzkow et al. 2021).

But during the hunt 2020, an 8-year-old *Rangifer* male was diagnosed positive on Hardangervidda, (https://kommunikasjon.ntb.no/pressemelding/bekreftet-funn-av-skrantesjuke-pa-hardangervidda?publisherId=10773547&releaseId=178$91694), hosting the biggest wild populations.

If CWD is established in Norwegian reindeer populations, this will represent a drastic threat to them as well as other cervids. The reindeer industry will be seriously threatened (Maraud and Roturier 2021). The authorities are therefore reconsidering the strategy.

Population Ecology

Population Dynamics

Many studies have assessed changes in reindeer population size and composition (sex and age) and identified the causes of these fluctuations both in wild (e.g., Reimers 1997; Uboni et al. 2016; Albon et al. 2017) and semi-domestic herds (e.g., Tveraa et al. 2014; Weladji 2003). A combination of factors influences reindeer population dynamics, including climatic variability, food limitation, predation, hunting, insect, parasites, disease, competition, and human developments (Klein 1991; Reimers 1997). The combined effects of density dependent and density independent factors have been reported to also affect survival rate, in both juveniles and adults (Skogland 1985; Reimers 1982). Variation in life history traits and population parameters is related to both winter (Kumpula 2001; Skogland 1985) and summer (Reimers 1997; Reimers et al. 1983) resource availability and accessibility. Variable environmental conditions experienced by individuals around birth have the potential to create fitness differences among cohorts (Tyler 2010). As a result, individual reindeer born during favorable conditions may gain consistent selective advantages not shared by individuals born under less favorable conditions. Climatic variations (local and global) affect several life history traits (e.g., body mass, calving date, etc.) and population parameters (e.g., offspring sex ratio) (Weladji 2003). Climate effects on reindeer may be direct (e.g., energetic costs of thermoregulation caused by severe cold, increased energetic costs of moving through deep snow and in accessing forage through snow). Climate also acts indirectly on reindeer life history traits and population parameters (e.g., effect on forage plant biomass and quality, increase level of insect harassment and associated parasitism due to increase temperature during summer). Increasing density significantly reduced body mass and growth rate of reindeer calves, and the effect of climate accentuate this negative effect (Skogland 1984; Weladji and Holand 2003). Because they are density dependent, important driving variables for reindeer population dynamics include animal body size and growth rate, and their relationship to reproduction and mortality (Reimers 1997). Among reindeer, reproductive performance (Reimers 1983b), calving time (Reimers 1983a), calf birth mass (Skogland 1984), and neonatal mortality (Skogland 1984, 1985) are strongly correlated to maternal mass.

Warm summers negatively affect reindeer because of insect harassment (Colman et al. 2003; Hagemoen and Reimers 2002; Weladji et al. 2003b). Indeed, insect harassment reduces the ability of reindeer to feed optimally during this critical period of high forage quality and availability (Hagemoen and Reimers 2002) and could therefore have a negative effect on their body condition and productivity (Reimers 1997; Weladji et al. 2003a).

Intra- and Interspecific Interactions

Among large herbivores, food limitation is reported to be the primary mechanism behind density dependence (Skogland 1985). Indeed, increased density limits per capita food availability; this may cause a reduction in individual's size, one of the most apparent aspects of an organism's life history. Moreover, long-term grazing by populations with stocking rate above carrying capacity may be detrimental for the pasture (i.e., overgrazing), resulting in reduced forage availability and quality; this will also affect herbivore body condition and productivity (e.g., Kumpula 2001). Nevertheless, the effect of reindeer on the vegetation can be either positive or negative, depending on environmental conditions (Bernes et al. 2015). In addition to life history traits (e.g., Forchhammer et al. 2001; Pettorelli et al. 2005), population parameters are also affected by density dependence (e.g., see Post and Stenseth 1999, see also Weladji and Holand 2003).

In Southern Norway, hunting is the dominant top-down mechanism controlling the populations and allowing to avoid food limitation by dampening the effect of harsh winters in population with generally good winter conditions (Reimers 1997). The Svalbard and Novaya Zemyla reindeer live in an area free from predators and oestrid flies, and limited hunting, making the populations being driven mostly by limited food resources and by climatic variability (Aanes et al. 2003; Hansen et al. 2013). Forest reindeer population dynamics in Finland seems limited by predation from wolves *Canis lupus* (Kojola et al. 2009), the other potential predators being brown bear *Ursus arctos* and lynx *Lynx lynx* and also golden eagle *Aquila chrysaetos* (Norberg et al. 2006).

Tundra reindeer overlap to a limited degree with other wild ruminants as they are the only arctic and alpine dwelling large herbivore. Svalbard and Novaya Zemlya reindeer are roaming their Archipelagos without any other large herbivore present. Forest reindeer overlap in time and space with moose and roe deer, both typical browsers (Hofmann 1989). We therefore expect limited interspecific foraging competition.

On Norwegian rangelands, domestic sheep is the most prevalent ruminant during summer, sharing alpine summer ranges with both wild and semi-domestic reindeer. Reindeer and sheep overlap in preferred habitat types and graze on many of the same plant species (Mysterud 2000; Skogland 1984). Several studies have addressed behavioral interactions between the two species (e.g., Moe et al. 1999; Colman et al. 2012). But no real interspecific competition (affecting vital rates) has been documented, although body mass of both reindeer and sheep vary from year to year, and there is a positive relationship between autumn body mass of sympatric reindeer and sheep (Weladji et al. 2003b).

Effects of Climate Change

There is increasing evidence that the globe is currently warming, with changes being more pronounced at northern latitudes (Uboni et al. 2016). Understanding the ecological effects of climatic variability is therefore important. There is recent support for the idea that large-scale atmospheric phenomena, such as the North Atlantic Oscillation (NAO) or Arctic Oscillation, through their effects on vegetation and regional weather conditions, influences several aspects of life histories and population dynamic processes of several mammal species, including reindeer (Aanes et al. 2002; Weladji and Holand 2006). For example, Post and Stenseth (1999) reported effects of the winter NAO index on reindeer calf body mass and adult female fecundity, while Aanes et al. (2002) showed that high values of the Arctic Oscillation index were associated with reduced plant growth and reindeer population growth rate in Svalbard. Reindeer are components of complex ecosystems, which might also be undergoing change in response to climate change. Thus, there will be other ecosystem changes than those known today that are likely to influence reindeer, such as invasion of southern species that may be predators, parasites, or competitors for forage or transmit diseases and parasites. Also, shifts in vegetation distribution or changes in ecosystem processes may influence lichen growth rates and their competition with vascular plants. Effects of global warming on plant productivity and onset of spring is likely to positively affect sub-Arctic reindeer (Tveraa et al. 2013), while its effect through summer heat stress and the related reduction in foraging activity may be negative (Reimers 1997; Weladji et al. 2003a; Weladji and Holand 2006). Although it is shown that local and global climate affect reindeer directly and indirectly, we argue that it is difficult to predict a general pattern of how future climate change will influence reindeer (Weladji and Holand 2006; Albon et al. 2017; Mallory and Boyce 2018). Indeed, the undergoing large changes in climate have already caused unprecedented consequences in the phenology of many plant and animal species, but the magnitude of the phenological responses differ between areas (Elmendorf et al. 2012; Prevéy et al. 2017).

Conservation Status

According to IUCN (Gunn 2016), *Rangifer* is globally classified as Vulnerable. The wild European populations are small and fragmented, and some are threatened. This may be amplified by changing climate (Vors and Boyce 2009; Uboni et al. 2016).

During the last centuries, European wild tundra and forest reindeer have yielded territories for the expanding reindeer pastoralism. Indeed, overlapping spatial and temporal pastoral reindeer herds and wild reindeer are regarded incompatible (Baskin 2005). The last century we have further witnessed a rapid encroachment and exploitation of *Rangifer* ranges, inducing habitat fragmentation and disturbance (Vistnes and Nellemann 2008; Panzacchi et al. 2013; Danilov et al. 2018) detrimental for several populations.

Forest and tundra reindeer in the European part of northern Russia, Murmansk, Karelia, Arkangelsk, and Komi region, have suffered severe to moderate declines the last decades (Danilov et al. 2018; Mizin et al. 2018). However, the population estimates are inconsistent partly due to lack of a unified monitoring system and lack of regular surveys. Their status appears in the Red Book of Russian Federation (2020) as well as in regional Red Books. In the Red Data Book of Murmansk region (2014), both tundra and forest reindeer are classified as Near Threatened. In Karelia republic Red Data Book (2020), forest reindeer is classified as Least Concern, while they are Vulnerable (Mizin, pers. comm.) in Arkhangelsk Red Data Book (2020), and in the Nenets Autonomic District Book (2006, amended in 2019), forest reindeer is classified as Vulnerable. In the Red Data Book of Komi republic (2019), forest reindeer is classified as Rare species. Reindeer of Novaya Zemlya has also declined the recent decades (Mizin et al. 2018). This subspecies is listed in the Arkhangelsk Red Book (2020) as Uncertain.

In Finland, forest reindeer is classified as Near Threatened (https://www.suomenpeura.fi/en/wild-forest- reindeer/conservation-status.html). The eastern Finnish population decreased at the start of the millennium probably due to the return and increase in number of wolves in the area (Kojola et al. 2009). The same happened in the bordering western part of Arkhangelsk oblast. Indeed, such small populations are susceptible to the predator-to-prey ratio and may be "trapped" (Kojola et al. 2009).

Svalbard reindeer live in a predator and insect harassment-free environment but face other challenges. The low genetic variation as compared to the other subspecies (Kvie et al. 2016) makes it vulnerable to climate change. Indeed, an increased frequency of rain on snow events may hit hard causing extreme die offs (Aanes et al. 2002). The total population on the Archipelago has been increasing the last decades and the subspecies is classified as Least Concern (https://www.artsdatabanken.no/Rodliste).

In mountainous area of southern Norway, 35,000 winter head of wild tundra reindeer are found divided into 23 populations (Reimers 2018). Only about 6,000 of these, divided into 4 populations, are regarded not mixed with semi-domesticated reindeer or feral (Røed 2005). They were collectively assessed as Least Concern in 2018 (http://data.artsdatabanken.no/Rodliste). However, their range has contracted the last decades and is confined to rather small and partly fragmented areas (Panzacchi et al. 2013).

The Icelandic feral population, today accounting around 5,000 head, dating back to the late 1750s is thriving (Skarphedinn 2018). Interestingly, in a proposed bill of Nature Conservation in 1999 reindeer was classified as an alien species to Iceland (Skarphedinn 2018).

Management

Rangifer is a key species of the tundra and northern taiga. The formerly continuous reindeer habitat through northern Europe is heavily fragmented being continuously encroached by expanding human activity. Therefore, management and conservation of the reindeer center on conservation of landscapes and their extensive use of grazing lands.

Rangifer central role in the pastoralism livelihood of many indigenous peoples makes it an integrated part of their culture. Today reindeer pastoralism plays an important role for northern people in Eurasia (https://reindeerherding.org/world-reindeer-herders). In Europe, Sami, Komi, and Nenets people are heavily involved in the industry. In Finland, many Finnish people are involved. Locally Norwegian and Swedish farmers are also engaged. In Fennoscandia, we find around 600,000 semi-domesticated reindeer, equally distributed between Norway, Sweden, and Finland. In European part of Russia, semi-domestic reindeer are found in the Murmansk Oblast (~50,000 winter heads), practiced by Sami and Komi and in the Nenets Autonomous Okrug of the Arkhangelsk Oblast (~200,000 heads), practiced by Nenets and Komi and in the Komi Republic (~100,000 heads), where reindeer herders are predominantly Komi (Istomin and Habeck 2016). Indeed, conflicts between wild reindeer conservation and reindeer pastoralism are seen where the ranges overlap in time and/or space. In Europe, such conflicts are seen regionally on the Kola Peninsula (Baranova et al. 2016) and further east, on the border between forest reindeer and domestic herds in Finland and between wild and semi-domestic reindeer in Southern Norway.

Today most European *Rangifer* populations are embraced in management or conservation plans and their conservation status are nationally and regionally assessed. Indeed, *Rangifer* is a high-profile species given high priority in all three European countries hosting them: Russia, Finland, and Norway.

Finland has, in line with EU's Habitat Directive, established conservation areas for protection of forest reindeer. Animals have been reintroduced to former habitats and plans are made for new reintroductions. This may create increased risk of interbreeding between wild forest reindeer and semi-domestic reindeer as documented by Røed et al. (2014). Fences may mitigate further introgressions. Some animals cross seasonally between Finland and Russia and a joint Finnish-Karelian action plan has been developed (www.suomenpeura.fi/media/metsapeura-tiedostot/enpi-dokumentit/wild-forest-reineer-action-plan-en.pdf). A Finnish-Russian Nature Reserve has been established to secure the border population. In Russia, several State Nature Reserves host forest reindeer. In Norway, several of the wild tundra populations roam within National Parks. Most of the Svalbard Archipelago is protected as National Parks and Reserves, making the Svalbard reindeer a well-protected population area-wise. In Russia, the northern part of Novaya Zemlya is part of the Russian Arctic National Park, probably encompassing native Novaya Zemlya reindeer (Mizin et al. 2018).

In 2002, commercial hunting of forest reindeer was banned in Karelia, Archangels, and Komi region. However, poaching is a severe threat to these small populations (Mizin et al. unpublished). In Finland, hunting is strictly regulated and enforced, and only few licenses are issued yearly. In Norway, management systems emphasize tracking population trends and vital rates, and adjusting hunting quota accordingly. For the moment, hunting takes place in all 23 available herds. The total yearly cull has varied between 4,000 and 8,000 animals the last decades (Reimers 2018). Indeed, hunting is the main mortality factor. It is argued that hunting is critical to keep the numbers below carrying capacity in these confined and almost predator-free environments. In Svalbard, limited hunting take place nearby the main settlements, Longyearbyen and Barentsburg. About 150–250 Svalbard reindeer are culled yearly (Reimers 2018).

In Norway, landscape changes, especially transportation infrastructures, hydroelectric power plants, dams and lines, and tourist resorts, may act as barriers for migration and induce further fragmentation of the already small ranges (Panzacchi et al. 2013). Hence, landscape management is important and may include regulation and diversification of land use. Recently an expert group commissioned by the Ministry of Climate and Environment has developed an environmental quality standard for wild reindeer management (Kjørstad et al. 2017). In Svalbard, the situation is somewhat different with large protected areas. Both on Norwegian mainland and on Svalbard, the tourist industry is rapidly expanding and has to be regulated to reduce human disturbances (Gundersen et al. 2021).

In the European North of Russia, the species' habitat is presented by fragments of areas with separate populations different in size, located in the northern taiga and tundra belt. Indeed, the gas and oil industry as well as mining activities are rapidly expanding in northern Russia and pose a real threat to the tundra. Old growth forests, important winter habitats, have dramatically decreased in the recent decades due to large-scale forest cutting operations. The emphasis has been on establishing protected areas (Panchenko et al. 2021). Such areas have to encompass the animals' annual range. Baskin (2005) reported that the migration from the taiga to the tundra has ceased in Arkhangelsk and that the forest-reindeer south of the Nenets reindeer husbandry area (in the Nenets Autonomous Okrug) has almost become extinct. In Komi region, the wild reindeer is almost never found in the forest tundra and tundra, mainly because this territory is intensively used for grazing of domesticated reindeer. On Kola Peninsula, separation of the wild reindeer population into the eastern and western parts occurred more than 60 years ago with the construction of the railway and car road between Murmansk and St Petersburg (Baranova et al. 2016; Mizin et al. unpublished) and hampers their expansion. Therefore, the effectiveness of protected areas, including the Lapland State Nature Reserve, securing key habitats as calving grounds and old growth forest winter habitats, will depend on planning a network and corridors to shelter migration routes.

Future Challenges for Research and Management

Conservation and management must rely on recognizable and functional units with high degree of genetic diversity important for their robustness and adaptive capacity. New genomic tools are surfacing and the reindeer genome has recently been mapped (Li et al. 2017). This will help us to integrate diversity below the species level relevant to its evolutionary legacy. This will also enable us to link genetic differentiation and environmental variation and to study whether genetic divergence among populations have resulted from morphological, life history, and/or behavioral adaptations to different ecological constraints.

However, fragmented populations present a dilemma – whether to keep them separate to conserve local adaptations and minimize the risk of outbreeding – or to manage them as a metapopulation to reduce loss of genetic diversity and minimize inbreeding (Allendorf and Luikart 2009). This can be accentuated in high arctic where the melting sea ice may reduce the connectivity between populations.

Feral and mixed populations also pose challenges. The Hardangervidda herd, accounting for almost one-third of the total Norwegian population (Reimers 2018), is a mixed wild-feral population dating back to the early 1900s when semi-domestic herds were let loose and mixed with the severely depleted wild population. The question remains – is a time span of 100 years enough for treating the Hardangervidda herd as a wild population? Parts of the western European Russian populations may be hybrids (Mizin et al. unpublished), including the populations on Novaya Zemlya and need further investigations. Indeed, introgression between forest reindeer and semi-domestic herds in Finland and Russia, as well as between wild tundra and semi-domestic herds in southern Norway may change the genetic structure and accentuate the "hybrid" management discussion and generate mitigations actions. Actually, in Finland, fences are constructed in an attempt to separate forest and semi-domesticated reindeer (http://www.suomenpeura.fi/en).

In Norway, many herds are monitored. However, lags in management actions, especially population estimates, and the number of hunting licenses issued, may lead to fluctuations (Strand et al. 2012). This is often amplified by weather condition during the hunting season, which influences the harvest heavily. A conservative carrying capacity approach is therefore warranted.

The mechanisms behind the declines in many Russian populations are not well understood, especially the role of predation, migration, nutrition, disease, and parasite interactions (Gunn 2016). Further, the forest reindeer that live in the eastern part of the Arkhangelsk region and Komi republic are not clearly defined and delineated and whether there is gene exchange between them is unknown. Some populations, perhaps mixed with semi-domestic reindeer, migrate north to the tundra in summer and their routes have to be mapped and secured.

Enforcement of the strict hunting regime is essential to reduce poaching. Indeed, sociopolitical factors influencing level of poaching seem to be a main population driver in Russia (Kolpaschikov et al. 2015; Mizin et al. unpublished). In Svalbard reindeer, hunting is not playing an important role in the population dynamic. The same is true for forest reindeer in Finland. Here, the predation pit is a major concern (Kojola et al. 2009).

Domestic reindeer may function as a reservoir of infection for wild reindeer and vice versa (Tryland and Kutz 2019). Impact of infectious agents are often subtle. Nevertheless, they may influence vital rates and hence population dynamics. Monitoring program of reindeer health and disease status are therefore essential. Especially, in a climate change perspective where new parasites and disease vectors are intruding the north (Altizer et al. 2013; Tryland et al. 2019b).

Environmental change due to anthropogenic influence is an increasing threat to many species, especially cold-adapted species (Post et al. 2009). Infrastructure development and new and intensified land use impose landscape changes. Obviously, this will lead to cumulative effects and may influence populations dynamic. But the mechanisms are poorly understood, but progress is being made (Russell et al. 2021). Successful and effective mitigation actions need therefore in-depth knowledge of the drivers involved. Although *Rangifer* show substantial adaptive capacity and phenotypic plasticity, more information is needed to understand how *Rangifer* populations will respond to the ever-changing landscape.

Indeed, reindeer is a key stone species (Olofson et al. 2004) and play an important role in the socio-ecological system of sub-Arctic and Arctic Europe. Their physiological, morphological, and behavioral adaptation render them the dominant large herbivore of the north, hopefully allowing them to thrive and persist across an ever-changing environment. Rapid environmental change may challenge their ability to adapt. Therefore, population monitoring must be given priority. Further, the need for international cooperation, in research and management at all levels, are therefore pressing.

References

Aanes R, Saether BE, Smith FM, Cooper EJ, Wookey PA, Oritsland NA (2002) The Arctic Oscillation predicts effects of climate change in two trophic levels in a high-arctic ecosystem. Ecol Lett 5:445–453

Aanes R, Saether BE, Solberg EJ, Aanes S, Strand O, Oritsland NA (2003) Synchrony in Svalbard reindeer population dynamics. Can J Zool 81:103–110

Aaris-Sørensen K, Muhldorff R, Petersen EB (2007) The Scandinavian reindeer (*Rangifer tarandus* L.) after the last glacial maximum: time, seasonality and human exploitation. J Archaeol Sci 34:914–923

Albon SD, Stien A, Irvine RJ, Langvatn R, Ropstad E, Halvorsen O (2002) The role of parasites in the dynamics of a reindeer population. Proc R Soc B Biol Sci 269: 1625–1632

Albon SD, Irvine RJ, Halvorsen O, Langvatn R, Loe L, Ropstad E, Veiberg V, van der Wal R, Bjørkvoll EM, Duff EI, Hansen BB, Lee AM, Tveraa T, Stien A (2017) Contrasting effects of summer and winter warming on body mass explain population dynamics in a food-limited Arctic herbivore. Glob Chang Biol 23: 1374–1389

Allendorf FW, Luikart GH (2009) Conservation and the genetics of populations. Blackwell Publishing, Malden

Altizer S, Ostfeld RS, Johnson PTJ, Kutz S, Harwell CD (2013) Climate change and infectious diseases: from evidence to a predictive framework. Science 341: 514–519

Anderson JR, Nilssen AC, Folstad I (1994) Mating-behavior and thermoregulation of the reindeer warble fly, *Hypoderma tarandi L* (Diptera, Oestridae). J Insect Behav 7:679706

Åsbakk K, Oksanen A, Nieminen M, Haugerud RE, Nilssen AC (2005) Dynamics of antibodies against hypodermin C in reindeer infested with the reindeer warble fly, *Hypoderma tarandi*. Vet Parasitol 129:323–332

Ballesteros M, Bardsen BJ, Langeland K, Fauchald P, Stien A, Tveraa T (2012) The effect of warble flies on reindeer fitness: a parasite removal experiment. J Zool (Lond) 287:34–40

Banfield AWF (1961) A revision of the reindeer and caribou, genus *Rangifer*. Bulletin of the National Museum of Canada, Biological Services, Ottawa

Bang-Andersen S (2012) Colonizing contrasting landscapes. The pioneer coast settlement and inland utilization in Southern Norway 10,000–9,500 years before present. Oxf J Archaeol 31:103–120

Baranova AI, Panchenko DV, Kholodova MV, Tirronen KF, Danilov PI (2016) Genetic diversity of wild reindeer *Rangifer tarandus* L. from the eastern part of the Kola Peninsula: polymorphism of the mtDNA control region. Biol Bull 43:567–572

Bårdsen BJ, Fauchald P, Tveraa T, Langeland K, Yoccoz NG, Ims RA (2008) Experimental evidence of a risk-sensitive reproductive allocation in a long-lived mammal. Ecology 89:829–837

Baskin LM (2005) Number of wild and domestic reindeer in Russia in the late 20th century. Rangifer 25:51–57

Bergerud AT (1974) Rutting behaviour of newfoundland caribou. In: Geist V, Walther F (eds) The behaviour of ungulates and its relation to management. IUCN, Morges, pp 395–435

Bernes C, Bråthen KA, Forbes BC, Speed JMD, Moen J (2015) What are the impacts of reindeer/caribou (*Rangifer tarandus* L.) on arctic and alpine vegetation? A systematic review. Environ Evid 4:4

Binney H, Edwards M, Macias-Fauria M, Lozhkin A, Anderson P, Kaplan JO, Andreev A, Bezrukova E, Blyakharchuk T, Jankovska V, Khazina I, Krivonogov S, Kremenetski K, Nield J, Novenko E, Ryabogina N, Solovieva N, Willis K, Zernitskaya V (2017) Vegetation of Eurasia from the last glacial maximum to present: key biogeographic patterns. Quat Sci 157:80–97

Björck S, Ekström J, Iregren E, Larsson L, Liljegren R (1996) Reindeer, and palaeoecological and palaeogeographic changes in South Scandinavia during late-glacial and early post-glacial times. Arkæologiske Rapporter, Esbjerg Museum, pp 195–214

Bliss LC, Heal O, Moore J (1981) Tundra ecosystems: a comparative analysis. Cambridge University Press, Cambridge, UK

Blix AS (2016) Arctic animals and their adaptations to life on the edge. Tapir Academic, Trondheim

Body G (2014) Importance of group dynamics on female mating behaviour in reindeer *Rangifer tarandus*. Ph.D. thesis, Concordia University, Canada

Body G, Weladji RB, Holand Ø, Nieminen M (2015) Fission-fusion group dynamics in reindeer reveal an increase of cohesiveness at the beginning of the peak rut. Acta Ethol 18:101–110

Bubenik AB (1975) Taxonomic value of antlers in genus *Rangifer*. Biological papers of the University of Alaska, Special report 1: 41–63

Bye K (1987) Abomasal nematodes from three Norwegian wild reindeer populations. Can J Zool 65:677–680

Carlsson AM, Dobson A, Kutz SJ (2019) The impact of infectious agents on *Rangifer* populations. In: Tryland M, Kutz S (eds) Reindeer and caribou: health and disease. CRC Press, Boca Raton, pp 315–351

Chen L, Qiu Q, Jiang Y, Wang K, Lin Z, Li Z, . . ., Wang W (2019) Large-scale ruminant genome sequencing provides insights into their evolution and distinct traits. Science 364

Clark PU, Shakun JD, Baker PA, Bartlein PJ, Brewer S, Brook E, Carlson AE, Cheng H, Kaufman DS, Liu Z, Marchitto TM, Mix AC, Morrill C, Otto-Bliesner BL, Pahnke K, Russell JM, Whitlock C, Adkins JF, Blois JL, Clark J, Colman SM, Curry WB, Flower BP, He F, Johnson TC, Lynch-Stieglitz J, Markgraf V, McManus J, Mitrovica JX, Moreno PI, Williams JW (2012) Global climate evolution during the last deglaciation. Proc Natl Acad Sci U S A 109:1134–1142

Clottes JAM (2003) Return to Chauvet Cave: excavating the birthplace of art: the first full report. Thames & Hudson, London, UK

Coard R, Chamberlain AT (1999) The nature and timing of faunal change in the British Isles across the Pleistocene/Holocene transition. The Holocene 9:372–376

Colman JE, Pedersen C, Hjermann DO, Holand Ø, Moe SR, Reimers E (2003) Do wild reindeer exhibit grazing compensation during insect harassment? J Wildl Manag 67:11–19

Colman JE, Tsegaye D, Pedersen C, Eidesen R, Arntsen H, Holand Ø, Mann A, Reimers E, Moe SR (2012) Behavioral interference between sympatric reindeer and domesticated sheep in Norway. Rangel Ecol Manag 65:299–308

Costamagno S, Barshay-Szmidt C, Kuntz D, Laroulandie V, Pétillon J-M, Boudadi-Maligne M, Langlais M, Mallye J-B, Chevallier A (2016) Reexamining the timing of reindeer disappearance in southwestern France in the larger context of late glacial faunal turnover. Quat Int 414:34–61

Croitor R (2018) Plio-Pleistocene deer of western Palearctic: taxonomy, systematics, phylogeny. Institute of Zoology of the Academy of Sciences of Moldova, Elan Poligraf, pp 1–140

Cuyler LC, Øritsland NA (1993) Metabolic strategies for winter survival by Svalbard reindeer. Can Zool 71:1787–1792

Cuyler LC, Øritsland NA (2002) Effect of wind on Svalbard reindeer fur insulation. Rangifer 22:93–99

Cuyler LC, White RR, Lewis K, Soulliere C, Gunn A, Russell DE, Colin D (2012) Are warbles and bots related to reproductive status in West Greenland caribou? Rangifer 32:243–257

Danilov PI, Panchenko DV, Tirronen KF, Fyodorov FV, Belkin VV (2018) Changes in mammal Fauna of the Northern Palearctic and living range dynamics of the component species. Biol Bull 45:265–277

Djaković N (2012) Female mating strategy in reinder (*Rangifer tarandus*): the importance of male age structure and availability in the social organisation, behaviour and mate choice. Ph.D. thesis, Norwegian University of Life Sciences, Norway

Elmendorf SC, Henry GHR, Hollister RD, Björk RG, Bjorkman AD, Callaghan TV, Collier LS, Cooper EJ, Cornelissen JHC, Day A, Fosaa AM, Gould WA, Grétarsdóttir J, Harte J, Hermanutz L, Hik DS, Hofgaard A, Jarrad F, Jónsdóttir IS, Keuper F, Klanderud K, Klein JA, Koh S, Kudo G, Lang SI, Loewen V, May JL, Mercado J, Michelsen A, Molau U, Myers-Smith IH, Oberbauer SF, Pieper S, Post E, Rixen C, Robinson CH, Schmidt NM, Shaver GR, Stenström A, Tolvanen A, Totland Ø, Troxler T, Wahren C-H, Webber PJ, Welker JM, Wookey PA (2012) Global assessment of experimental climate warming on tundra vegetation: heterogeneity over space and time. Ecol Lett 15:164–175

Eloranta E, Nieminen M (1986) Calving of the experimental reindeer herd in Kaaamanen during 1970–85. Rangifer 1:115–121

Espmark Y (1964) Rutting behaviour in reindeer (*Rangifer tarandus* L.). Anim Behav 12:159–160

Espmark Y (1971a) Antler shedding in relation to parturition in female reindeer. J Wildl Manag 35:175–177

Espmark Y (1971b) Mother-young relationship and ontogeny of behaviour in reindeer (*Rangifer tarandus*). Z Tierpsychol 29:42–81

Fancy SG, White RG (1985) Energy expenditures by caribou while cratering in snow. J Wildl Manag 49: 987–993

Fancy SG, White RG (1986) Predicting energy expenditures for activities of caribou from heart rates. Rangifer 6:123–130

Finalyson C, Carrion JS (2007) Rapid ecological turnover and its impact on Neanderthal and other human populations. Trends Ecol Evol 22:213–222

Flagstad O, Røed KH (2003) Refugial origins of reindeer (*Rangifer tarandus* L.) inferred from mitochondrial DNA sequences. Evolution 57:658–670

Folstad I, Nilssen A, Halvorsen O, Andersen J (1991) Parasite avoidance: the cause of post-calving migrations in *Rangifer*? Can J Zool 69:2423–2429

Fontana L (2017) The four seasons of reindeer: non-migrating reindeer in the Dordogne region (France) between 30 and 18 k? Data from the Middle and Upper Magdalenian at La Madeleine and methods of seasonality determination. J Archaeol Sci Rep 12:346–362

Fontana F, Rubini M (1990) Chromosomal evolution in Cervidae. Biosystems 24:157–174

Forchhammer MC, Clutton-Brock TH, Lindström J, Albon SD (2001) Climate and population density induce long-term cohort variation in a northern ungulate. J Anim Ecol 70:721–729

Fraccaro M, Gustavsson I, Hultén M, Lindsten J, Tiepolo L (1968) Chronology of DNA replication in the sex chromosomes of the reindeer (*Rangifer tarandus* L.). Cytogenet Genome Res 7:196–211

Geist V (1999) Deer of the world. Swan Hill Press, Shresbury

Gravlund P, Meldgaard M, Pääbo S, Arctander P (1998) Polyphyletic origin of the small-bodied, high arctic subspecies of tundra reindeer (*Rangifer tarandus*). Mol Phylogenet Evol 10:151–159

Gripenberg U, Wessman M, Nieminen M (1986) The chromosomes of reindeer (*Rangifer tarandus*). Rangifer 1:109

Groves C, Grubb P (2011) Ungulate taxonomy. Johns Hopkins University Press, Baltimore

Grubb P (2005) Order Artiodactyla. In: Wilson DE, DeeAnn MR (eds) Mammal species of the world: a taxonomic and geographic reference, 3rd edn. Johns Hopkins University Press, Baltimore, pp 637–722

Gundersen V, Myrvold KM, Rauset GR, Selvaag SK, Strand O (2021) Spatiotemporal tourism pattern in a large reindeer (*Rangifer tarandus tarandus*) range as an important factor in disturbance research and management. J Sustain Tour 29:21–39

Gunn A (2016) *Rangifer tarandus*. The *IUCN* Red List of Threatened Species::e.T29742A22167140

Günther T, Malmström H, Svensson EM, Omrak A, Sánchez-Quinto F, Kılınç GM, Krzewińska M, Eriksson G, Fraser M, Edlund H, Munters AR, Coutinho A, Simões LG, Vicente M, Sjölander A, Sellevold BJ, Jørgensen R, Claes P, Shriver MD, Valdiosera C, Netea G, Apel J, Lidén K, Skar B, Storå J, Götherström A, Jakobsson M (2018) Population genomics of Mesolithic Scandinavia: investigating early postglacial migration routes and high-latitude adaptation. PLoS Biol 16:e2003703

Hætta MJ (2009) The mating strategy of female Svalbard reindeer (*Rangifer tarandus*). MSc thesis, University of Tromsø, Norway

Hagemoen RIM, Reimers E (2002) Reindeer summer activity pattern in relation to weather and insect harassment. J Anim Ecol 71:883–892

Halvorsen O (2012) Reindeer parasites, weather and warming of the Arctic. Polar Biol 35:1749–1752

Handeland K, Slettbakk T (1995) Epidemiological aspects of cerebrospinal elaphostrongylosis in small ruminants in northern Norway. Zentralbl Veterinarmed B 42: 110–117

Handeland K, Boye M, Bergsjø B, Isaksen BH, Agerholm JS (2010) Digital necrobacillosis in Norwegian wild tundra reindeer (*Rangifer tarandus tarandus*). J Comp Pathol 143:29–38

Hansen BB, Aanes R, Herfindal I, Kohler J, Saether B-E (2013) Climate, icing, and wild arctic reindeer: past relationships and future prospects. Ecology 92: 1917–1923

Harington CR (1999) Caribou. Beringian Res Notes 12:1–4

Hayes JD, Imbrie J, Shacklton NJ (1976) Variation in the Earth's orbit: pacemaker of the ice ages. Science 194: 1121–1132

Heckeberg NS (2020) The systematics of the Cervidae: a total evidence approach. PeerJ 8

Helle T, Tarvainen L (2010) Effects of insect harassment on weight gain and survival in reindeer calves. Rangifer 4: 24–27

Hofmann RR (1989) Evolutionary steps of ecophysiological adaptations and diversification of ruminants: a comparative view of their digestive systems. Oecologia 78:443–457

Hofreiter M, Stewart J (2009) Ecological change, range fluctuations and population dynamics during the Pleistocene. Curr Biol 19:584–594

Holand Ø (2003) Reindrift – samisk næring i brytning mellom tradisjon og produksjon. Gan Forlag, Oslo. (in Norwegian)

Holand Ø, Weladji RB (2019) *Rangifer* mating strategies. In: Tryland M, Kutz S (eds) Reindeer and caribou: health and disease. CRC Press, Boca Raton, pp 79–89

Holand Ø, Røed KH, Gjøstein H, Kumpula J, Nieminen M (2003) Effects of age, density and sex ratio on reproductive effort in male reindeer (*Rangifer tarandus*). J Zool (Lond) 261:341–344

Holand Ø, Gjøstein H, Losvar A, Kumpula J, Smith ME, Røed KH, Nieminen M, Weladji RB (2004a) Social rank in female reindeer (*Rangifer tarandus*): effects of body mass, antler size and age. J Zool (Lond) 263: 365–372

Holand Ø, Weladji RB, Gjøstein H, Kumpula J, Smith ME, Nieminen M, Røed KH (2004b) Reproductive effort in relation to maternal social rank in reindeer (*Rangifer tarandus*). Behav Ecol Sociobiol 57:69–76

Holand Ø, Weladji RB, Røed KH, Gjøstein H, Kumpula J, Gaillard J-M, Smith ME, Nieminen M (2006a) Male age structure influences females' mass change during rut in a polygynous ungulate: the reindeer (*Rangifer tarandus*). Behav Ecol Sociobiol 59: 682–688

Holand Ø, Askim KR, Røed KH, Weladji RB, Gjøstein H, Nieminen M (2006b) No evidence of inbreeding avoidance in a polygynous ungulate: the reindeer (*Rangifer tarandus*). Biol Lett 3:36–39

Holand Ø, Weladji RB, Mysterud A, Røed K, Reimers E, Nieminen M (2012a) Induced orphaning reveals post-weaning maternal care in reindeer. Eur J Wildl Res 58: 589–596

Holand Ø, L'Italien L, Weladji RB, Djaković N, Røed K, Hovland AL, Nieminen M (2012b) Shit happens – a glimpse into males' mating tactics in a polygynous ungulate – the reindeer. Rangifer 32:65–71

Hopkins DM, Matthews JV Jr, Schweger CE, Young SB (1982) Paleoecology of Beringia. Academic, New York

Høymork A, Reimers E (1999) Antler development in reindeer in relation to age and sex. Rangifer 22:75–82

Hull E, Semeniuk M, Puolakka HL, Kynkäänniemi SM, Niinimäki S (2021) Tendons and ligaments of the *Rangifer tarandus* metapodial and hoof. Polar Biol 44:1–14

Indrelid S, Hufthammer AK (2011) Medieval mass trapping of reindeer at the Hardangervidda mountain plateau, South Norway. Quat Int 238:44–54

Istomin KV, Habeck JO (2016) Permafrost and indigenous land use in the northern Urals: Komi and Nenets reindeer husbandry. Polar Sci 10:278–287
Jacobsen BW, Colman JE, Reimers E (1998) The frequency of antlerless females among Svalbard reindeer. Rangifer 18:81–84
Josefsen TD, Mørk T, Nymo IH (2019) Bacterial infections and diseases. In: Tryland M, Kutz S (eds) Reindeer and caribou: health and disease. CRC Press, Boca Raton, pp 237–272
Kahlke R-D (2014) The origin of Eurasian Mammoth Faunas (*Mammuthus – Coelodonta* Faunal Complex). Quat Sci Rev 96:32–49
Källquist L, Mossing T (1982) Olfactory recognition between mother and calf in reindeer (*Rangifer tarandus* L.). Appl Anim Ethol 8:561–565
Kiltie RA (1985) Evolution and function of horns and hornlike organs in female ungulates. Biol J Linn Soc 24:299–320
Kjørstad M, Bøthun SW, Gundersen V, Holand Ø, Madslien K, Mysterud A, Myren IN, Punsvik T, Røed KH, Strand, Tveraa T, Tømmervik H, Ytrehus B, Veiberg V (2017) Environmental quality standard for wild reindeer – suggestions from an expert group. Trondheim, NINA Report 1400:1–193. (In Norwegian, English summary)
Klein DR (1991) Caribou in the changing North. Appl Anim Behav Sci 29:279–291
Klein DR (1996) Arctic ungulates at the northern edge of terrestrial life. Rangifer 16:51–56
Kojola I (1986) Rutting behaviour in an enclosured group of wild forest reindeer (*Rangifer tarandus fennicus* Lönnb.). Rangifer 6:173–179
Kojola I (1991) Influence of age on the reproductive effort of male reindeer. J Mammal 72:208–210
Kojola I, Helle T, Huhta E (1998) Foraging conditions, tooth wear and herbivore body reserves: a study of female reindeer. Oecologia 117:26–30
Kojola I, Tuomivaara J, Heikkinen S, Heikura K, Kilpeläinen K, Keränen J, Paasivaara A, Ruusila V (2009) European wild forest reindeer and wolves: endangered prey and predators. Ann Zool Fenn 46:416–422
Kolpaschikov L, Makhailov V, Russell DE (2015) The role of harvest, predators, and socio-political environment in the dynamics of the Taimyr wild reindeer herd with some lessons for North America. Ecol Soc 20:9
Kovacs KM, Gjertz I, Lydersen C (2004) Marine mammals of Svalbard. Norwegian Polar Institute, Tromsø
Kruse F (2017) Catching up: the state and potential of historical data from Svalbard in European Arctic. Polar Rec 53:520–533
Kumpula J (2001) Productivity of the semi-domestic reindeer (*Rangifer t. tarandus* L.) stock and carrying capacity in Finland during 1960s–1990s. PhD thesis, University of Oulu, Finland
Kumpula J, Kurkilahti M, Helle T, Colpaert A (2014) Both reindeer management and several other land use factors explain the reduction in ground lichens (*Cladonia* spp.) in pastures grazed by semi- domesticated reindeer in Finland. Reg Environ Chang 14:541–559
Kurtén B (1968) Pleistocene mammals of Europe. Weidenfeld & Nicolson, London
Kutz SJ, Laaksonen S, Åsbakk K, Nilssen AC (2019) Parasitic infection and diseases. In: Tryland M, Kutz S (eds) Reindeer and caribou: health and disease. CRC Press, Boca Raton, pp 177–235
Kvie KS, Heggenes J, Anderson DG, Kholodova MV, Sipko T, Mizin I, Røed KH (2016) Colonizing the high Arctic: mitochondrial DNA reveals common origin of Eurasian Archipelagic reindeer (*Rangifer tarandus*). PLoS One 11:1–15
Kynkäänniemi S-M, Kettu M, Kortet R, Härkönen L, Kaitala A, Paakkonen T, Mustonen A-M, Härkönen S, Ylönen H, Laaksonen S (2014) Acute impacts of the deer ked (*Lipoptena cervi*) infestation on reindeer (*Rangifer tarandus tarandus*) behaviour. Parasitol Res 113:1489–1497
Le Moullec M, Pedersen ÅØ, Stien A, Rosvold J, Hansen BB (2019) A century of conservation: the ongoing recovery of Svalbard reindeer. J Wildl Manag 83: 1676–1686
Li Z, Lin Z, Ba H, Chen L, Yang Y, Wang K, Qiu Q, Wang W, Li G (2017) Draft genome of the reindeer (*Rangifer tarandus*). Gigascience 6:1–5
L'Italien L, Weladji RB, Holand Ø, Røed KH, Nieminen M, Côté SD (2012) Mating group size and stability in reindeer *Rangifer tarandus*: the effects of male characteristics, sex ratio and male age structure. Ethology 118:783–792
Lönnberg E (1909) Några fynd av subfossila vertebrater. Arkiv för Zoologi 6:1–28
Lorenzen ED, Nogués-Bravo D, Orlando L, Weinstock J, Binladen J, Marske KA, Ugan A, Borregaard MK, Gilbert TMP, Nielsen R, Ho SYW, Goebel T, Graf KE, Byers D, Stenderup JT, Rasmussen M, Campos PF, Leonard JA, Koepfli K-P, Froese D, Zazula G, Stafford TW Jr, Aaris-Sørensen K, Batra P, Haywood AM, Singarayer JS, Valdes PJ, Boeskorov G, Burns JA, Davydov SP, Haile J, Jenkins DL, Kosintsev P, Kuznetsova T, Lai X, Martin LD, McDonald HG, Mol D, Meldgaard M, Munch K, Stephan E, Sablin M, Sommer RS, Sipko T, Scott E, Suchard MA, Tikhonov A, Willerslev R, Wayne RK, Cooper A, Hofreiter M, Sher A, Shapiro B, Rahbek C, Eske W (2011) Species-specific responses of Late Quaternary megafauna to climate and humans. Nature 479:359–365
Lydekker R (1903) Mostly Mammals, Zoological Essays. Hutchinson & Company, London
Macbeth B, Kutz SJ (2019) *Rangifer* health: a holistic perspective. In: Tryland M, Kutz S (eds) Reindeer and caribou: health and disease. CRC Press, Boca Raton, pp 91–105
Mallory CD, Boyce MS (2018) Observed and predicted effects of climate change on Arctic caribou and reindeer. Environ Rev 26:13–25
Manninen SM, Thamsborg SM, Laaksonen S, Oksanen A (2014) The reindeer abomasal nematode (*Ostertagia gruehneri*) is naturally transmitted to sheep when sharing pastures. Parasitol Res 113:4033–4038

Maraud S, Roturier S (2021) Chronic wasting disease (CWD) in Sami Reindeer Herding: the socio-political dimension of an epizootic in an indigenous context. Animals 11:297

Matiskainen H (1990) Mesolithic subsistence in Finland. In: Vermeersch PM, Van Peer P (eds) Contributions to the Mesolithic in Europe: Papers Presented at the Fourth International Symposium, The Mesolithic in Europe, Leuven, 1990. Leuven University Press, Leuven, pp 211–214

Melnycky NA, Weladji RB, Holand Ø, Nieminen M (2013) Scaling of antler size in reindeer (*Rangifer tarandus*): sexual dimorphism and variability in resource allocation. J Mammal 94:1371–1379

Mizin IA, Sipko TP, Davydov AV, Gruzdev AR (2018) The wild reindeer (*Rangifer tarandus*: Cervidae, Mammalia) on the arctic islands of Russia: a review. Nat Conserv Res 3:1–14

Moe S, Holand Ø, Colman J, Reimers E (1999) Reindeer (*Rangifer tarandus*) response to feces and urine from sheep (*Ovis aries*) and reindeer. Rangifer 19:55–60

Moen R, Pastor J (1998) Simulating antler growth and energy, nitrogen, calcium and phosphorus metabolism in caribou. Rangifer 10:85–97

Mørk T, Prestrud P (2004) Arctic rabies – a review. Acta Vet Scand 45(1–2):1–9

Müller-Schwarze D, Quay WB, Brudin A (1977) The caudal gland in reindeer (*Rangifer tarandus* L.): its behavioral role, histology and chemistry. J Chem Ecol 3:591–601

Müller-Schwarze D, Källquist L, Mossing T (1979) Social behavior and chemical communication in reindeer (*Rangifer tarandus* L.). J Chem Ecol 5:483–517

Mysterud A (2000) Diet overlap among ruminants in Fennoscandia. Oecologia 124:130–137

Mysterud A, Holand Ø, Røed, KH, Gjøstein H, Kumpula J, Nieminen M (2003). Effects of age, density and sex ratio on reproductive effort in male reindeer (*Rangifer tarandus*). J Zool (Lond) 261:341–344

Mysterud A, Roed KH, Holand O, Yoccoz NG, Nieminen M (2009) Age-related gestation length adjustment in a large iteroparous mammal at northern latitude. J Anim Ecol 78:1002–1006

Mysterud A, Ytrehus B, Tranulis MA, Rauset GR, Rolandsen CM, Strand O (2020) Antler cannibalism in reindeer. Sci Rep 10:22168

Nes N, Amrud J, Töndevold OB (1965) Kromosomstudier hos rein. (*Rangifer tarandus*). Norsk Veterinærmedisinsk Tidsskrift 17:589–593

Nieminen M (1980) The evolution and taxonomy of the genus Rangifer in northern Europe. In: Reimers E, Gaare E, Skjenneberg S (eds) Trondheim: proceedings of the 2nd reindeer/caribou symposium, Røros, pp 379–391

Nieminen M (1985) Reinens gevir – vekst og betydning – en oversikt (On growth and importance of the reindeer antler). Rangifer 5:26–33

Nilssen KJ, Sundsfjord JA, Blix AS (1984) Regulation of metabolic rate in Svalbard and Norwegian reindeer. Am J Physiol 247:837–841

Norberg H, Kojola I, Aikio P, Nylund M (2006) Predation by golden eagle *Aquila chrysaetos* on semi-domesticated reindeer *Rangifer tarandus* calves in northeastern Finnish Lapland. Wildl Biol 12:393–402

Olofson J, Stark S, Oksanen L (2004) Reindeer influence on ecosystem processes in the tundra. Oikos 105: 386–396

Panchenko DV, Danilov PI, Tirronen KF, Blyudnik LV (2014) The issue of the status of the reindeer in the White Sea region. Vestn Okhotoved 11:171–175

Panchenko D, Paasivaara A, Hyvärinen M, Krasovskij Y (2021) The wild forest reindeer, *Rangifer tarandus fennicus*, in the Metsola Biosphere Reserve, Northwest Russia. Nat Conserv Res 6:116–126

Panzacchi M, Van Moorter B, Jordhøy P, Strand O (2013) Learning from the past to predict the future: using archaeological findings and GPS data to quantify reindeer sensitivity to anthropogenic disturbance in Norway. Landsc Ecol 28:847–859

Patton H, Hubbard A, Andreassen K, Auriac A, Whitehouse PL, Stroeven AP, Shackleton C, Winsborrow MCM, Heyman J, Hall AM (2017) Deglaciation of the Eurasian ice sheet complex. Quat Sci Rev 169:148–172

Pavlov PY (2017) Middle and upper paleolithic sites in the Northeast of the East European plain. In: Kotlyakov VM, Velichko AA, Vasil'ev SA (eds) Human colonization of the Arctic: the interaction between early migration and the paleoenvironment. Academic, London, UK, pp 137–151

Pedersen ÅØ (2019) *Rangifer* biology and adaptations. In: Tryland M, Kutz S (eds) Reindeer and caribou: health and disease. CRC Press, Boca Raton, pp 54–69

Pettorelli N, Weladji RB, Holand Ø, Mysterud A, Breie H, Stenseth NC (2005) The relative role of winter and spring conditions: linking climate and landscape-scale plant phenology to alpine reindeer body mass. Biol Lett 1:24–26

Pilø L, Finstad E, Ramsey CB, Martinsen JRP, Nesje A, Solli B, Wangen V, Callanan M, Barrett JH (2018) The chronology of reindeer hunting on Norway's highest ice patches. R Soc Open Sci 5:171738

Polfus JL, Manseau M, Klütsch CFC, Simmons D, Wilson PJ (2017) Ancient diversification in glacial refugia leads to intraspecific diversity in a Holarctic mammal. J Biogeogr 44:386–396

Post E, Stenseth NC (1999) Climatic variability, plant phenology, and northern ungulates. Ecology 80: 1322–1339

Post E, Forchhammer MC, Bret-Harte MS, Callaghan TV, Christensen TR, Elberling B, Fox AD, Gilg O, Hik DS, Hoye TT, Ims RA, Jeppesen E, Klein DR, Madsen J, McGuire AD, Rysgaard S, Schindler DE, Stirling I, Tamstorf MP, Tyler NJC, van der Wal R, Welker J, Wookey PA, Schmidt NM, Aastrup P (2009) Ecological dynamics across the arctic associated with recent climate change. Science 325:1355–1358

Prevéy J, Vellend M, Rüger N, Hollister RD, Bjorkman AD, Myers-Smith IH, Elmendorf SC, Clark K, Cooper EJ, Elberling B, Fosaa AM, Henry GHR, Høye TT,

Jónsdóttir IS, Klanderud K, Lévesque E, Mauritz M, Molau U, Natali SM, Oberbauer SF, Panchen ZA, Post E, Rumpf SB, Schmidt NM, Schuur EAG, Semenchuk PR, Troxler T, Welker JM, Rixen C (2017) Greater temperature sensitivity of plant phenology at colder sites: implications for convergence across northern latitudes. Glob Chang Biol 23:2660–2671

Prichard AK, Finstad GL, Shain DH (1999) Factors affecting velvet antler weights in free-ranging reindeer in Alaska. Rangifer 19:71–76

Pritzkow S, Gorski D, Ramirez F, Telling GC, Benestad SL, Soto C (2021) North American and Norwegian Chronic wasting disease prions exhibit different potential for interspecies transmission and zoonotic risk. J Infect Dis:jiab385

Red Book of Russian Federation (2020) Ministry of Natural Resources and Ecology of the Russian Federation, Moscow, Order No. 162 dated March 24, 2020

Red Data Book of Karelia Republic (2020) Belgorod, Konstanta. ISBN 978-5-907380-16-5

Red Data Book of Komi Republic (2019) Syctyvkar, Russia. Komi republic print. ISBN 978-5-7934-0814-1

Red Data Book of the Arkhangelsk region (2020) Arkhangelsk, Russia. NarFU. ISBN 978-5-261-01436-2

Red Data Book of the Murmansk region (2014) Kemerovo, Russia. Asia-print. ISBN 978-5-85905-446-6

Red Data Book of the Nenets Autonomous (2006) Department of Natural Resources, Ecology and Agroindustrial Complex of the Nenets Autonomous District, Naryan-Mar, Russia, Order No. 25-dated June 06, 2019

Reimers E (1982) Winter mortality and population trends of reindeer on Svalbard, Norway. Arct Alp Res 14: 295–300

Reimers E (1983a) Growth rate and body size differences in *Rangifer*, a study of causes and effects. Rangifer 3: 3–15

Reimers E (1983b) Reproduction in wild reindeer in Norway. Can J Zool 61:211–21725

Reimers E (1984) Body composition and population regulation of Svalbard reindeer. Rangifer 4:16–21

Reimers E (1993) Antlerless females among reindeer and caribou. Can J Zool 71:1319–1313

Reimers E (1997) *Rangifer* population ecology: a Scandinavian perspective. Rangifer 17:105–118

Reimers E (2018) Våre hjortedyr en bok til undring, glede og kunnskap. Yrkeslitteratur, Oslo

Reimers E, Nordby Ø (1968) Relationship between age and tooth cementum layers in Norwegian reindeer. J Wildl Manag 32:957–961

Reimers E, Klein DR, Sorumgard R (1983) Calving time, growth-rate, and body size of Norwegian reindeer on different ranges. Arct Alp Res 15:107–118

Reimers E, Nieminen M, Tsegaye D (2013) Antler casting in relation to parturition in semi-domesticated female reindeer. Rangifer 33:17–24

Røed KH (2005) Refugial origin and postglacial colonization of Holarctic reindeer and caribou. Rangifer 25:19–30

Røed KH, Holand Ø, Smith ME, Gjostein H, Kumpula J, Nieminen M (2002) Reproductive success in reindeer males in a herd with varying sex ratio. Mol Ecol 11: 1239–1243

Røed KH, Holand Ø, Gjøstein H, Hansen H (2005) Variation in male reproductive success in a wild population of reindeer. J Wildl Manag 69:1163–1170

Røed KH, Flagstad Ø, Nieminen M, Holand Ø, Dwyer MJ, Røv N, Vilà C (2008) Genetic analyses reveal independent domestication origins of Eurasian reindeer. Proc R Soc B 275:1849–1855

Røed KH, Bjørnstad G, Flagstad Ø, Haanes H, Hufthammer AK, Jordhøy P, Rosvold J (2014) Ancient DNA reveals prehistoric habitat fragmentation and recent domestic introgression into native wild reindeer. Conserv Genet 15:1137–1149

Røed KH, Côté S, Yannic G (2018) *Rangifer tarandus*: classification and genetic variation. In: Tryland M, Kutz S (eds) Reindeer and caribou: health and disease. CRC Press, Boca Raton, pp 3–13

Rohling EJ, Foster GL, Grant KM, Marino G, Roberts AP, Tamisiea ME, Williams F (2014) Sea-level and deep-sea-temperature variability over the past 5.3 million years. Nature 508:477–482

Ropstad E (2000) Reproduction in female reindeer. Anim Reprod Sci 60–61:561–570

Ropstad E, Forsberg M, Kindahl H, Dahl E (1996) Aspects of reproduction in female reindeer. Reprod Domest Anim 31:495–499

Rowell JE, Shipka MP (2009) Variation in gestation length among captive reindeer (*Rangifer tarandus tarandus*). Theriogenology 72:190–197

Rusell DE, Martell AM (1984) Winter ecology of caribou (*Rangifer tarandus*). In: Olson R, Hastings R, Geddes F (eds) Northern ecology and resource management. University of Alberta Press, Edmonton, pp 117–144

Russell D, Gunn A, White RG (2021) A decision support tool for assessing cumulative effects on an Arctic migratory tundra caribou population. Ecol Soc 26:4

Ryg M, Jacobsen E (1982) Seasonal changes in growth, feed intake, growth hormone and thyroid- hormones in young male reindeer (*Rangifer tarandus tarandus*). Can J Zool 60:15–23

Salgado-Flores A, Hagen LH, Ishaq SL, Zamanzadeh M, Wright A-DG, Pope PB et al (2016) Rumen and cecum microbiomes in reindeer (*Rangifer tarandus tarandus*) Are Changed in response to a lichen diet and may affect enteric methane emissions. PLoS ONE 11(5):e0155213

Schaffer J (1940) Die Hautdrüsenorgane der Säugetiere, und Bemerkungen über die Proktodäaldrüsen. Urban and Schwarzenberg, Berlin

Selinge KG (1974) Pitfalls, the most common cultural monuments in Jämtland, Sweden. Jämtlands läns museum, Østersund, Fornvårdaren 12:1–39. (In Swedish)

Sigurdarson S, Haugerud RE (2004) "Wild reindeer" in Iceland. In: Ulvevadet B, Klokov K (eds) Family-based reindeer herding and hunting economies, and the status of management of wild reindeer/caribou populations. Centre for Sami Studies, University of Tromsø, Norway, pp 159–162

Skarphedinn GT (2018) Population dynamics and demography of reindeer (*Rangifer tarandus* L.) on the East

Iceland highland plateau 1945–2015. MSc thesis, Agricultural University of Iceland, Iceland

Skjenneberg S, Slagsvold L (1979) Reindeer husbandry and its ecological principles. DU Department of Interior, Bureau of Indian Affairs, Juneau

Skogland T (1984) The effects of food and maternal conditions on fetal growth and size in wild reindeer. Rangifer 4:39–46

Skogland T (1985) The effects of density-dependent resource limitation on the demography of wild reindeer. J Anim Ecol 54:359–374

Skogland T (1988) Tooth wear by food limitation and its life history consequences in wild reindeer. Oikos 51: 238–242

Skogland T (1989) Comparative social organization of wild reindeer in relation to food, mates and predator avoidance. Adv Ethol 29:1–77

Skogland T (1994) Villrein. Fra urinnvåner til miljøbarometer. Teknologisk forlag, Drammen. (In Norwegian)

Sokolov AA, Gruzdev AR, Pronyaev AV (1996) Patterns of cheek tooth wearing in reindeer (*Rangifer tarandus*) from different populations. Zool zhurnal 75:1579–1580

Sommer RS, Kalbe J, Ekstrström J, Benecke N, Liljegren R, Svenning JC (2014) Range dynamics of the reindeer in Europe during the last 25,000 years. J Biogeogr 41:298–306

Soppela P, Saarela S, Heiskari U, Nieminen M (2008) The effects of wintertime undernutrition on plasma leptin and insulin levels in an arctic ruminant, the reindeer. Comp Biochem Physiol B Biochem Mol Biol 149: 613–621

Staaland H, Nieminen M (1993) World reindeer herding: origin, history, distribution, economy. Proceedings VII world conference on animal production. Edmonton Canada, pp 161–203

Strand O, Nilsen EB, Solberg EJ, Linnell JDC (2012) Can management regulate the population size of wild reindeer (*Rangifer tarandus*) through harvest? Can J Zool 90:163–171

Syroechkovskii EE (1995) Wild reindeer. Science Publisher, Washington, DC

Taylor R, Manseau M, Klütsch CFC, Polfus J, Steedman A, Hervieux D, Kelly A, Larter N, Gamberg M, Schwantje H, Wilson P (2021) Population dynamics of caribou shaped by glacial cycles before the Last Glacial Maximum. Mol Ecol 00:1–23

Tryland M (2019) Prions and Chronic wasting disease (CWD). In: Tryland M, Kutz S (eds) Reindeer and caribou: health and disease. CRC Press, Boca Raton, pp 305–314

Tryland M, Kutz S (2019) Reindeer and caribou: health and disease. CRC Press, Boca Raton

Tryland M, das Neves CG, Klein J, Mørk T, Hautaniemi M, Wensman JJ (2019a) Viral infections and diseases. In: Tryland M, Kutz S (eds) Reindeer and caribou: health and disease. CRC Press, Boca Raton, pp 273–303

Tryland M, Ravolainen V, Pedersen AØ (2019b) Climate change potential impacts on pasture resources, health and diseases of reindeer and caribou. In: Tryland M, Kutz S (eds) Reindeer and caribou: health and disease. CRC Press, Boca Raton, pp 493–514

Tveraa T, Stien A, Bardsen BJ, Fauchald P (2013) Population densities, vegetation green-up, and plant productivity: impacts on reproductive success and juvenile body mass in reindeer. PLoS One 8. 0056450

Tveraa T, Stien A, Broseth H, Yoccoz NG (2014) The role of predation and food limitation on claims for compensation, reindeer demography and population dynamics. J Appl Ecol 51:1264 1272

Tyler NJC (1986) The relationship between the fat content of Svalbard reindeer in autumn and their death from starvation in winter. Rangifer 1:311–314

Tyler NJC (2010) Climate, snow, ice, crashes, and declines in populations of reindeer and caribou (*Rangifer tarandus* L.). Ecol Monogr 80:197–219

Uboni A, Horstkotte T, Kaarlejärvi E, Sévêque A, Stammler F, Olofsson J, Forbes BC, Moen J (2016) Long-term trends and role of climate in the population dynamics of Eurasian reindeer. PLoS One 11. 0158359

Ukkonen P, Lõugas L, Zagorska I, Luksevica L, Luksevics E, Daugnora L, Jungner H (2006) History of the reindeer (*Rangifer tarandus*) in the eastern Baltic region and its implications for the origin and immigration routes of the recent northern European wild reindeer populations. Boreas 35:222–230

Veiberg V, Mysterud A, Bjorkvoll E, Langvatn R, Loe LE, Irvine RJ, Bonenfant C, Couweleers F, Stenseth NC (2007) Evidence for a trade-off between early growth and tooth wear in Svalbard reindeer. J Anim Ecol 76: 1139–1148

Vistnes I, Nellemann C (2008) The matter of spatial and temporal scales: a review of reindeer and caribou response to human activity. Polar Biol 31:399–407

VKM (2018) Factors that can contribute to spread of CWD – an update on the situation in Nordfjella, Norway. Report from the Norwegian Scientific Committee for Food and Environment (VKM):16, Oslo

Volokitin AV, Gribchenko YN (2017) North of the East European plain. In: Kotlyakov VM, Velichko AA, Vasil'ev SA (eds) Human colonization of the Arctic: the interaction between early migration and the paleoenvironment. Academic, London, pp 75–104

Vorren Ø (1973) Some trends of the transition from hunting to nomadic economy in Finnmark. In: Berg G (ed) Circumpolar problems: habitat, economy and social relations in the arctic. Pergamon Press, Oxford, UK, pp 185–194

Vors LS, Boyce MS (2009) Global declines of caribou and reindeer. Glob Chang Biol 15:2626–2633

Vrolik WT (1829) Over eene vermoedelijke tweede soort van Rendier (Vol. 2). Muller en comp

Wales R, Milligan L, McEwan, EH (1975) Urea recycling in caribou, cattle and sheep. Biological Papers of the University of Alaska, Special Report 1: 297–307

Weckworth BV, Musiani M, McDevitt AD, Hebblewhite M, Mariani S (2012) Reconstruction of caribou evolutionary history in Western North America and its implications for conservation. Mol Ecol 21: 3610–3624

Weladji RB (2003) Climatic influence on the life history and population dynamics of a Northern ungulate, *Rangifer tarandus*. PhD thesis, Norwegian University of Life Sciences, Norway

Weladji RB, Holand Ø (2003) Global climate change and reindeer: effects of winter weather on the autumn weight and growth of calves. Oecologia 136:317–323

Weladji RB, Holand Ø (2006) Influences of large-scale climatic variability on reindeer population dynamics: implications for reindeer husbandry in Norway. Clim Res 32:119–127

Weladji RB, Holand Ø, Almøy T (2003a) Use of climatic data to assess the effect of insect harassment on the autumn weight of reindeer (*Rangifer tarandus*) calves. J Zool (Lond) 260:79–85

Weladji RB, Steinheim G, Holand Ø, Moe SR, Almoy T, Adnoy T (2003b) Temporal patterns of juvenile body weight variability in sympatric reindeer and sheep. Ann Zool Fenn 40:17–26

Weladji R, Holand Ø, Gaillard JM, Yoccoz NG, Mysterud A, Nieminen M, Stenseth NC (2010) Age-specific changes in different components of reproductive output in female reindeer: terminal allocation or senescence? Oecologia 162:261–271

Weldenegodguad M, Pokharel K, Ming Y, Honkatukia M, Peippo J, Reilas T, Røed KH, Kantanen J (2020) Genome sequence and comparative analysis of reindeer (*Rangifer tarandus*) in northern Eurasia. Sci Rep 10:8980

Whitehead GK (1972) Deer of the world. Viking Press, New York

Yannic G, Pellissier L, Le Corre M, Dussault C, Bernatchez L, Côté SD (2014) Temporally dynamic habitat suitability predicts genetic relatedness among caribou. Proc R Soc B Biol Sci 281: 201440502

Zheludkov MM, Tsirelson LE (2010) Reservoirs of *Brucella* infection in nature. Biol Bull 37:709–715

White-Tailed Deer *Odocoileus virginianus* (Zimmermann, 1780)

11

Jenni Poutanen, Mikael Wikström, and Jon E. Brommer

Contents

J. Poutanen · J. E. Brommer (✉)
Department of Biology, University of Turku, Turku, Finland
e-mail: jenni.poutanen@utu.fi; jon.brommer@utu.fi

M. Wikström
Finnish Wildlife Agency, Helsinki, Finland
e-mail: mikael.wikstrom@riista.fi

© Springer Nature Switzerland AG 2022

L. Corlatti, F. E. Zachos (eds.), *Terrestrial Cetartiodactyla*, Handbook of the Mammals of Europe,
https://doi.org/10.1007/978-3-030-24475-0_27

Common Names

English	White-tailed deer
German	Weißwedelhirsch
French	Cerf de Virginie
Spanish	Ciervo de Virginia
Italian	Cervo della Virginia
Russian	Белохвостый олень

Taxonomy, Systematics and Paleontology

Within the family of Cervidae, the white-tailed deer (Fig. 1) belongs to the subfamily of Capreolinae. Its closest relative is the mule deer *Odocoileus hemionus*. The species is native to the Americas. Within its native range, 38 subspecies are commonly recognized, although more have been described (Heffelfinger 2011). In Europe, the white-tailed deer subspecies introduced in Finland is *O. virginianus borealis*. In the Czech Republic, repeated introductions were made involving several subspecies. The oldest fossils identified as *Odocoileus* date to the middle Pliocene, about four million years ago (Oelrich 1953).

Current Distribution

The white-tailed deer is an American species with a native distribution that spans a remarkably broad latitudinal range, from the equatorial part of South America to the boreal zone of North America. The distribution, biology, and management of the white-tailed deer have been the subject of considerable research across its native range, as summarized in the edited volumes of Halls (1984) and Hewitt (2011).

The species was introduced to a number of European countries, but most introductions failed (Heffelfinger 2011) (Fig. 2).

The most successful introduction of white-tailed deer in Europe was in Finland. Finnish emigrants who had settled in Minnesota (USA) shipped white-tailed deer fawns from Minnesota as a gift to their motherland (Nygren 1984). One male and four female white-tailed deer fawns were released in Finland in 1934 and were kept in an enclosure until 1938. One female died before reproducing in 1937 and the maximal number of

Fig. 1 Male white-tailed deer (photograph © Mikael Wikström)

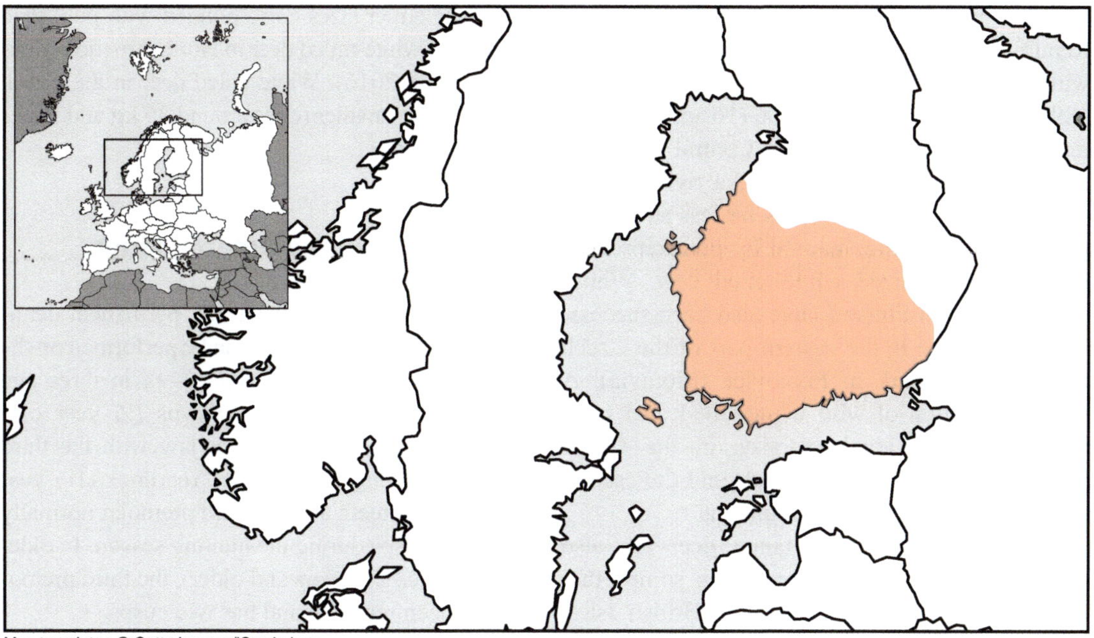

Map template : © Getty Images/iStockphoto

Fig. 2 White-tailed deer distribution in Europe. Distribution data source – see text. (Map template: © Copyright Getty Images/iStockphoto)

individuals in the first introduction therefore was four (one male and three females). A potential re-stocking consisted of four 1-year-old individuals released in 1949. These individuals are, however, thought not to have survived for a long period after their release (Nygren 1984). In any case, after the 1934 introduction, the white-tailed deer rapidly increased in numbers and public records detail their steady increase in numbers from four in 1934 until there were about 1000 individuals in 1961 when harvesting became legal (Kekkonen et al. 2012). The species is slowly expanding its range from the point of introduction, although also a number of translocations within Finland were conducted to increase the white-tailed deer's range. At present, white-tailed deer is found in Finland as a free-ranging population occurring primarily in the south-western part of Finland, with occasional observations outside this core region. The distribution range covered approximately 40% of Finland's surface in 2019. The post-harvest white-tailed deer population was estimated to approximately 111,500 individuals with an average density of approximately 0.5 individuals per square kilometer (Kukko and Pusenius 2019). In the core area, densities of roughly 5–10 individuals per square kilometer occurred (post-harvest population). In terms of habitat, this part of Finland is mainly covered by managed boreal forest interspersed with agriculture and human settlements. Compared with boreal forests in the white-tailed deer native range, the European boreal forests are relatively poor in plant species. The most common tree species are pine (*Pinus sylvestris*), spruce (*Picea abies*), and birch (*Betula pendula* and *B. pubescens*) with usually one tree species dominating a forest stand. A number of dwarf shrubs are common, including (in the "dry forest type," typically dominated by pine) heather (*Calluna vulgaris*), lingonberry (*Vaccinium vitis-idaea*), and ground species like reindeer lichens (*Cladonia* spp.) and red-stemmed feather moss (*Pleurozium schreberi*). In the "moist forest type" (dominated by spruce), the dwarf shrub billberry (*Vaccinium myrtillus*) and feather mosses (e.g., *Hylocomium splendens*) dominate the undergrowth.

White-tailed deer introduction in the Czech Republic was as early as the latter half of 1800s, with main releases in 1853 (7 animals) and 1892–93 (15 animals) and 1906 (16 animals). Primarily, releases were made in Central Bohemia, between the Vltava and Berounka river (Dobris Forest). This introduction was far less successful than the Finnish one, marked by poor reproduction and small sized white-tailed deer. White-tailed deer from Finland have also been successfully introduced in the eastern part of the Czech Republic in a park in Frycovice (Moravia). A small population of wild white-tailed deer of at most 1000 individuals persists in the Czech Republic, in the Brdy Highlands and the eastern foothills of the Krkonoše Mountains.

Introductions of white-tailed deer are also documented to have taken place in some other European countries, including the British Isles, Austria, Bulgaria, and in Serbia and Croatia and western Russia (Heffelfinger 2011). None of these introductions produced populations that are currently persisting; many of the populations introduced early in the nineteenth century did not survive the world wars. The white-tailed deer released in Bulgaria (in 1977) and western Russia (early 1970s) came from Finland.

Description

Size and Morphology

White-tailed deer are medium-sized ungulates with reddish-brown pelage in spring and summer and a more gray-brown winter coat. Fawns have spots until September. Appropriately named, when startled white-tailed deer will raise their long tail in an upright position thereby displaying the tail's white underside. White-tailed deer in Finland are 90–110 cm in height, with a body length of 150–180 cm. Bucks' body mass can be up to 150 kg (typically 70 to 130 kg), and the smaller sized does' body mass is less than 100 kg (typically 40–90 kg). The length of the hoof print is 7–12 cm. In a study considering 451 individuals harvested in a single hunting season in southern Finland, the distribution of age-specific body, skull morphology, and male antler sizes corresponded with published records of white-tailed deer in North America (Kekkonen et al. 2016). White-tailed deer in the Czech Republic are smaller (does around 35 kg and bucks around 55 kg).

Dentition and Age Determination

Deciduous teeth 0.0.3.0/3.1.3.0; permanent dentition 0.0.3.3/3.1.3.3. Aging can be performed on the basis of teeth (Severinghaus 1949), in three age groups (Gee et al. 2002). Fawns (½ year old) have 4–5 teeth in their lower jaw with the third premolar having three cusps. Yearlings (1½ year old) have six teeth and the third premolar normally has three cusps during the hunting season. In older animals (i.e., 2½ years and older), the third premolar has been replaced and has two cusps.

Antlers

Males cast their antlers in winter and regrow them annually from February to March onwards. White-tailed deer antlers have a main beam with single points emerging directly from the main beam typically pointing up (Fig. 1). The two main beams gently curve towards each other. Although antlers increase in size with age, antler growth is highly variable; already yearling males (1½ year old) can have many antler points (up to 8 points) under favorable conditions. Antlers can grow in size until the males are 6½–7½ years old (Häkkinen 1986; Kekkonen et al. 2016). A prime-age deer ($\geq$4½ years old) typically has antlers with an inner width between the antler beams of at least 50 cm, the beam length is at least 54 cm, and the circumference of the antler beam is at least 12 cm. A very low proportion of females (1:10,000) grow small antlers (Kekkonen et al. 2016).

Physiology

The current knowledge of physiology of white-tailed deer is based on North American studies (reviewed e.g., by Ditchkoff 2011). We here focus on reproductive physiology. The endocrine cycle of does follows the usual hormonal cycle of mammals controlled by the hypothalamus and ovaries. The onset of reproduction is under photoperiodic control and hence varies considerably across latitude. Estrous commences with an increase in gonadrotopin releasing hormone (GnRH) which stimulates a sharp peak in luteinizing hormone (LH) which in turn stimulates ovulation. Peak concentrations of LH in serum on the day of ovulation reported vary from 25.4 ng/mL to 60.1 ng/mL (Plotka et al. 1980; Knox et al. 1992). Usual level of LH is about 1 ng/mL (Plotka et al. 1980), although elevated levels some days before ovulation and even prior to estrous have been reported (Knox et al. 1992). Estrogen levels are normally 5–30 pg/mL and rise to peak at the day of estrous. In case the doe does not get pregnant, the corpora lutea decrease and progesterone levels increase until about 1 week prior to the next ovulation. The estrous period lasts typically about 24 h, although estrus may be prolonged for up to 48 h if the doe remains unbred (Knox et al. 1988). The doe can re-enter estrus 25–30 days later (Plotka et al. 1977); this estrous cycle can be repeated up to seven times in a breeding season (Knox et al. 1988). A number of studies testing different substances for hormonal birth control of white-tailed deer have been conducted in the USA, but their use as a means to limit population reproduction of wild populations is limited due to challenges in administrating these drugs (Ditchkoff 2011).

Gestation length is about 200 days (ranging between 190 and 220 days) (Kairikko and Ruola 2004; Verme 1965). Near the end of gestation, mammary tissue responds to a rise in prolactin and starts to develop milk. The first few days of a newborn critically depend on the mother's milk for nutrition, but also for the development of its immune system. The colostrum, which is present in the milk provided to the fawn during its first 24 h post-partum, contains immunoglobulins presenting the mother's antibodies. This transfer of antibodies hence provides the newborn with a first line of passive immunity. Indeed, compromised transfer can lead to immune system deficiencies with consequences for survival of newborn white-tailed deer (Sams et al. 1996; Ditchkoff et al. 2001a). The composition of white-tailed deer milk has been analyzed by Silver (1961) and depends primarily on the age of the fawn, with fat content increasing toward weaning. Malnutrition of females does not alter composition, but lowers the quantity of milk produced.

The rut, growth, and development of antlers in the white-tailed deer are under photoperiodic control and hence show considerable variation across the species' native range. In general, the testosterone levels in bucks start to increase in July–August and reach the highest peak at the end of October stabilizing again on the lowest levels in the middle of December (Verme and Ullrey 1984). Peak levels of testosterone circulating can be 20 ng/mL (Mirarchi et al. 1978) but the peak concentration level also increases as a male ages until the age of five to seven years (Ditchkoff et al. 2001b). In response to testosterone, secondary sexual characteristics such as neck swelling and aggressive behavior increase and testes and epididymides, which are atrophied outside the breeding season, increase in size. Older males develop larger testes and epididymides (Lambiase et al. 1972) and typically produce more spermatozoans (Mirarchi et al. 1977).

Genetics

White-tailed deer karyotype is $2n = 70$. White-tailed deer are an introduced species in Europe. In their native range, white-tailed deer show considerable genetic diversity, both within and between populations. However, at the beginning of the 1900s they were severely overharvested in many parts of North America, and restoration of the North American populations was achieved by restocking with white-tailed deer from a diverse set of source populations. As a consequence, there has been considerable admixture in North American white-tailed deer (e.g., DeYoung et al. 2003).

There occurs no hybridization between white-tailed deer and other ungulates in Europe. According to an individual-based population genetic model, the current genetic diversity of white-tailed deer in Finland is consistent with being founded by the three females and one male which were originally introduced in 1934 (Kekkonen et al. 2012). Despite being originated from only a few individuals and living in isolation without gene flow from other populations, the white-tailed deer population in Finland has retained a high heterozygosity thanks to its rapid population increase (Kekkonen et al. 2012). Nevertheless, the founder effect of the introduction has reduced allelic diversity compared to the North American source population (Kekkonen et al. 2012). There is some evidence of inbreeding depression for body mass in Finnish white-tailed deer (Brommer et al. 2015). In general, however, the Finnish white-tailed deer population is genetically diverse and shows little evidence of being adversely affected by genetic factors.

Life History

Growth

After the first 40 days of gestation, the fetus grows about 3 mm in a day (Armstrong 1950). The gain in fetal body mass is fastest during late gestation. When the fetus is about 150 days old it has reached half of the birth mass (Armstrong 1950, Hamilton et al. 1985). Females increase clearly in size (skull and body mass) until they are 2 years old when growth levels off. Males increase in skull and body size until 4 years of age (Kekkonen et al. 2016), and are in their prime age between 4½ and 8½ years.

Reproduction

In Finland, the rut starts around November. Females give birth to one or two fawns, sometimes triplets (Koivisto 1986) between the end of May and the beginning of July (Lahtinen 1996; Kairikko and Ruola 2004). Fawns weigh about 2.5–3.5 kg at birth (Haugen and Davenport 1950; Kairikko and Ruola 2004). During their first summer fawns gain about 250 g per day and this growth starts to decrease during their first autumn. Fawns reach their maximum mass in their first December after which the mass decreases a bit.

White-tailed deer females are reproductively most productive at 3–7 years old (Verme and Ullrey 1984) after which fecundity starts to decrease. Some very well-developed females can get pregnant already during their first winter but usually this happens at the age of one year. About 10%–15% of 5- to 7-year-old female white-tailed deer give birth to triplets. The mean number of fawns per mother is often between 1.6 and 1.9 (Häkkinen 1986). Depending on the age structure of the population, this reproductive potential translates to an average of 1.1 to 1.4 fawns per adult female per year (Häkkinen 1986). Because of the high reproductive output of white-tailed deer, yearly huntable net productivity of the population can be 50–60%, provided the density and the age and sex structure of the population is suitable.

Survival

The ratio of X:Y sperm is 1:1 in white-tailed deer (DeYoung et al. 2004), but at birth, 50–55% of fawns are males. A slightly higher adult mortality of males equalizes the sex ratio in adults. Thus, sex ratio of unharvested white-tailed deer populations is considered to be approximately one female per one male, possibly slightly female-biased. Clearly, in hunted populations adult sex ratio may deviate from parity. There is, however, a lack of research studying possible biases in sex and age ratio of the white-tailed deer populations in Europe. Kekkonen et al. (2016) found a deficit of older males among white-tailed deer harvested, but because this study only considered one hunting season its relevance for the age structure in the white-tailed deer population remains unclear. Usually, white-tailed deer live up to 13–15 years in the wild but can even reach the age of 20 years.

Habitat and Diet

Space Use

The size of the white-tailed deer home range needs to be sufficiently large to encompass the individual's need for cover, food, and water;

white-tailed deer depend on woody vegetation for cover and food. The size of the white-tailed deer home range furthermore depends on the density of the population: the yearly home range size is smaller in high density populations and larger when the population density is lower. Male white-tailed deer in low density populations (e.g., less than approximately 1 individual per square kilometer) in Finland have yearly home ranges of 50–70 km^2 measured as minimum convex polygons while in higher density populations (e.g., approximately 1.5 to 4 individuals per square kilometer) yearly home ranges of males are 10–40 km^2. Females have slightly smaller home range sizes than males. The annual home range is often from 5 to 50 km^2 depending on the density. In North America, where white-tailed deer density is higher than in Finland, the yearly home range of white-tailed deer varies from 1 to 10 km^2 (an overview of studies is presented by DeYoung and Miller 2011). White-tailed deer often occupy the same areas in consecutive years.

In Finland, fawns mainly disperse from their natal areas during their first March or April (Wikström 2012) although some fawns still move with their mothers at that time (Kairikko and Ruola 2004). Studies on white-tailed deer movement in North America have found that 59–86% of males and 0–29% of females disperse from their natal areas (Nelson and Mech 1984, 1987; Dusek et al. 1989; Nelson 1993; Kilgo et al. 1996). A meta-analysis of dispersal studies conducted in North America found that dispersal distances of male white-tailed deer were mainly determined by forest cover: dispersal was longer in open and fragmented landscapes (Long et al. 2005). A study on Finnish white-tailed deer showed that dispersal distance from the natal area varied considerably, from only a few kilometers up to 44 km (Honzová 2013). Nevertheless, home range, space usage and movement of white-tailed deer in Finland remain poorly studied.

Daily Movement

White-tailed deer are most active around dusk and dawn. Daily movement distances vary with the age and sex of the individuals but are also dependent on, among other things, season and habitat. In North America (Oklahoma), GPS-collared white-tailed deer males moved about 7.4 km/day during the rut and about 6.2 km/day after the rut, whereas females moved 2.6–3.3 km/day depending on parturition (Webb et al. 2010). Daily movement can decrease during the winter depending on the weather and food availability. Deep snow can hamper movement and a decrease in food consumption during the winter time can deteriorate the body condition of deer which may affect movement (Andersson and Koivisto 1980). During the whole year, the daily movement among GPS-collared adult white-tailed deer in Southern Finland varied from approximately 1.5 km/day to 3.5 km/day (Wikström 2012). Nevertheless, the number of daily GPS locations in this study was considerably lower than in the fine-scale movement study conducted by Webb et al. (2010) which may partly explain why a lower daily movement distance was observed in Finland compared to the USA.

Diet

White-tailed deer are an adaptive species inhabiting a large variety of terrestrial habitats from forest and savannas to deserts and coastal marine habitat, and it can feed on various vegetation types. Diet quality is, nevertheless, very important. In particular, white-tailed deer are relatively small ruminants and cannot digest fiber efficiently. In general, in Finland, white-tailed deer diet consists of twigs of dwarf shrubs (e.g., bilberry *Vaccinium myrtillus*, lingonberry *Vaccinium vitis-idaea*, heather *Calluna vulgaris*), as well as lichen and grass. Different cereal crops (wheat, barley, rye) are likely an important food source in autumn. White-tailed deer also feed on trees (seedlings, crown, leaves, buds, and – in winter – branches and bark), for example, juniper *Juniperus communis*, European aspen *Populus tremula*, willows *Salix* ssp., Scots pine *Pinus sylvestris*, and rowan *Sorbus aucuparia* (Andersson and Koivisto 1980). Snow cover and temperature during winter affect the availability of food. In Finland, providing supplementary food to white-tailed deer and other deer is allowed. Supplementary food is typically provided throughout the winter and is considered an important winter food source for the population.

Behavior

Mating Behavior

White-tailed deer do not defend territories nor have harems. Males search mating partners actively inside their home ranges, but GPS-based studies found that male white-tailed deer do sometimes make long-distance excursions outside their typical home range during the rut (DeYoung and Miller 2011). Males prepare for the rutting season already in September when they start leaving scent marks. Gland secretions are an important aspect of olfactory communication in white-tailed deer. Seven regions with glands are identified, and the chemical composition of some of the gland secretions has been studied in some detail (Gassett et al. 1997).

White-tailed deer males show signpost behavior; they make rubs (by rubbing their antlers against trees and bushes) and scrapes (by scraping the soil) combined with scent from urine, saliva, or secretion from a number of glands to present both visual and olfactory cues to conspecifics (DeYoung and Miller 2011). During the rut, adult males usually move more slowly and cautiously and threaten other males. They fight each other by using the antlers to determine a hierarchy among bucks (Hirth 1977). When females are fertile, males tend their partners for up to 3 days minimizing does' copulation with other bucks. Bucks may mate with several does during the rut, and a single pair may mate several times. That is, white-tailed deer have a polygynous mating system. Nevertheless, the demanding combination of multiple-day tending of females that are spatially scattered makes it unlikely that dominant males can monopolize matings, and results in paternity being distributed across males in the population (DeYoung et al. 2006). Genetic paternity assignment in white-tailed deer in the USA showed that older bucks ($\geq$3½ years) were not able to monopolize matings, and that younger, physically immature, bucks managed to sire 30–59% of the offspring (Sorin 2004; DeYoung et al. 2002; DeYoung et al. 2009; Turner et al. 2016). In addition, white-tailed deer does can be promiscuous; in some (captive) populations about 20% of twins have multiple fathers (DeYoung et al. 2002; Sorin 2004), although such polyandry has not been found in other populations (Turner et al. 2016). At present, genetic studies of the mating system of the white-tailed deer in Europe have not been conducted.

Grouping

The sexes are segregated for much of the year. Research conducted in the USA shows that white-tailed deer form small matriarchal social groups, typically staying associated with each other throughout the year, except during fawning season (e.g., Hirth 1977; Aycrigg and Porter 1997). The matriarchal group is composed of relatives, a female and her offspring of one or multiple breeding seasons, although genetic analyses revealed that not all members in the social group are necessarily closely related (Miller et al. 2010). Males are solitary during the breeding season, but form so-called bachelor groups outside the breeding season which consist of multiple unrelated males (Hirth 1977). This male grouping behavior starts when a male is about 1½ years old, after dispersal from its natal range. These bachelor groups start to break up toward the rutting season.

Parental Care

As newborns, fawns stay hidden for 4 to 8 weeks after which they start to move around with their mothers. Siblings are usually hidden at separate places (Halls 1984). Lactation lasts from the fawns' birth to their first September or October but they start gradually eating plant material and ruminating already at 2 weeks old (Plotka et al. 1977; Short 1964).

Parasites and Diseases

One factor behind the successful introduction of the white-tailed deer in Finland is likely that several parasites and diseases are absent. In particular the meningeal worm or brainworm *Parelaphostrongylus tenuis,* which is a common nematode in white-tailed deer in North America, did not become established in Finland (Andersson 1964), either because the introduced deer were not infected or because there was no suitable intermediate host (a gastropod) for the parasite in Finland. Czech white-tailed deer are commonly infected with large liver flukes, *Fasciola hepatica*.

A relatively newly described parasite for Finnish white-tailed deer is a lymphatic dwelling filarioid nematode, *Rumenfilaria andersoni*, but the effect of this nematode on white-tailed deer health is not well known (Laaksonen et al. 2015). This parasite (which is nowadays found in all Finnish cervids) likely was introduced to Finland as part of the introduction of white-tailed deer to the country (Laaksonen et al. 2015). Other parasites found in Finnish white-tailed deer include *Toxoplasma gondii* (Jokelainen et al. 2010) and deer ked *Lipoptena cervi*. Of future concern is chronic wasting disease (CWD), which is a lethal disease caused by a prion protein in which the infected animal loses mass and wastes away even if it eats. CWD is found in cervids in North America, including white-tailed deer. CWD was recorded in 2016 in Norway. The first case of CWD was found in moose *Alces alces* in Finland in 2018 (Finnish Food Authority 2019).

Population Ecology

In Finland and in the Czech Republic, white-tailed deer co-occur with the native roe deer *Capreolus capreolus*. Both white-tailed deer and roe deer are concentrate selectors, and their diet has many plant species in common; these cervids are hence likely competing. However, roe deer occur in Finland only locally and in small numbers, although the roe deer population is increasing in the southwest of the country. Overall, competition with the smaller roe deer is therefore not considered to be a main factor affecting white-tailed deer populations, but more detailed wildlife studies on the competition between these species are needed. Finland also locally has low numbers of fallow deer *Dama dama*, which is another potential competitor. Natural predators for white-tailed deer in Finland are gray wolf *Canis lupus*, Eurasian lynx *Lynx lynx*, and brown bear *Ursus arctos*. Wolf and lynx prey on both adults (lynx only occasionally) and fawns, and bear prey mainly on small fawns. The major cause for white-tailed deer mortality in Finland is hunting as well as traffic collisions.

In the Czech Republic, white-tailed deer are part of a larger community of cervids including sizeable numbers of red deer *Cervus elaphus* and fallow deer, as well as locally abundant numbers of introduced sika deer. Competition with these cervid species likely has been a factor hindering the white-tailed deer establishment in the Czech Republic. At the same time, the white-tailed deer is thought to also receive some benefits from the coexistence with roe deer, red deer, and fallow deer in terms of antipredatory group advantages during foraging (Bartos et al. 2002).

Conservation Status

The white-tailed deer is classified as Least Concern in the IUCN Red List (Gallina and Lopez Arevalo 2016). In Finland and the Czech Republic, the species is considered an alien species and hence has no conservation status.

Management

The white-tailed deer is an alien species in Finland. The National Strategy on Invasive Alien Species by the Finnish Ministry of Agriculture and Forestry (2012) considers white-tailed deer as potentially or locally harmful and in need of monitoring. The strategy states that the spreading of the white-tailed deer from Finland to other countries must be prevented. Deer-vehicle collisions are one of the damages that the white-tailed deer cause in Finland. The vehicle collision rate (i.e., percentage of the population that is involved in a traffic accident) of white-tailed deer is the highest (8%) of all cervid species in Finland (Niemi et al. 2015). In the densest population areas, white-tailed deer also cause damages to agriculture and forestry by eating, for example, vegetable crops and tree seedlings as well as causing damage to people's gardens in populated areas.

To minimize economic damages and traffic accidents the Finnish population size is regulated through intensive hunting. White-tailed deer are the second most important game species in Finland (after moose), and its management is regulated by the Finnish hunting legislation. The white-tailed deer population in Finland is managed via annual licenses. During the hunting season 2018–2019 approximately 53,000 white-tailed deer were harvested in Finland. The meat value of the harvest was calculated to be about 14 million euros, and the recreational value was calculated at approximately 20 million euros, which makes a total calculated value of approximately 34 million euros. In those parts of southern Finland where the white-tailed deer is common, hunting of white-tailed deer is nowadays equally or more economically important than moose hunting.

In the Czech Republic, the white-tailed deer population is small in numbers compared with the abundance of other cervids (red deer, fallow deer, roe deer, sika deer). The white-tailed deer also occurs in very local populations. As a consequence, there is no overarching white-tailed deer management scheme in the Czech Republic, but the species is managed locally.

Future Challenges for Research and Management

White-tailed deer population numbers have rapidly increased in Finland over the last decade. The challenge for management is to develop ways to control the population growth in a sustainable manner. Compared with the body of research on white-tailed deer in its native range, the European populations of white-tailed deer remain understudied. While many aspects of white-tailed deer biology in Europe likely are similar to the species' biology in its native range, it is also clear that white-tailed deer ecology and behavior is very adaptable to its environment. From that perspective, future research challenges for the white-tailed deer include improving our understanding of diet and habitat selection, dispersal, mating system, competition with roe deer and other deer species, and the population's age and sex distribution across its European range.

References

Andersson RC (1964) Neurologic disease in moose infected experimentally with *Pneumostrongylus tenuis* from white-tailed deer. Vet Pathol 1:289–322

Andersson E, Koivisto I (1980) Valkohäntäpeuran talviravinto ja vuorokausirytmi [White-tailed deer's winter food and diurnal rhythm]. Suomen Riista 7:84–92

Armstrong RA (1950) Fetal development of the northern white-tailed deer (*Odocoileus virginianus borealis* Miller). Am Midl Nat 43:650–666

Aycrigg JL, Porter WF (1997) Sociospatial dynamics of white-tailed deer in the Central Adirondack Mountains, New York. J Mammal 78:468–482

Bartos L, Vankova D, Miller KV, Siler J (2002) Interspecific competition between white-tailed, fallow, red, and roe deer. J Wildl Manag 66:522–527

Brommer JE, Kekkonen J, Wikström M (2015) Using heterozygosity-fitness correlations to study inbreeding depression in an isolated population of white-tailed deer founded by few individuals. Ecol Evol 5:357–367

DeYoung RW, Miller KV (2011) White-tailed deer behavior. In: Hewitt DG (ed) Biology and management of white-tailed deer. CRC Press, Boca Raton

DeYoung RW, Demarais S, Gonzales RA, Honeycutt RL, Gee KL (2002). Multiple paternity in white-tailed deer (*Odocoileus virginianus*) revealed by DNA microsatellites. J Mammal 83(3):884–892

DeYoung RW, Demarais S, Honeycutt RL, Rooney AP, Gonzales RA, Gee KL (2003) Genetic consequences of white-tailed deer (*Odocoileus virginianus*) restoration in Mississippi. Mol Ecol 12:3237–3252

DeYoung RW, Muller LI, Demarais S, Guthrie HD, Welch GR, Engelken TJ, Gonzales RA (2004) Do *Odocoileus virginianus* males produce Y-chromosome-biased ejaculates? Implications for adaptive sex ratio theories. J Mammal 85(4):768–773

DeYoung RW, Demarais S, Honeycutt RL, Gee KL, Gonzales RA (2006) Social dominance and male breeding success in white-tailed deer: instability, unpredictability, and individual heterogeneity. Wildl Soc Bull 34: 131–136

DeYoung RW, Demarais S, Gee KL, Honeycutt RL, Hellickson MW, Gonzales RA (2009) Molecular evaluation of the white-tailed deer (*Odocoileus virginianus*) mating system. J Mammal 90:946–953

Ditchkoff SS (2011) Anatomy and physiology. In: Hewitt DG (ed) Biology and management of white-tailed deer. CRC Press, Boca Raton

Ditchkoff SS, Sams MG, Lochmiller RL, Leslie DM (2001a) Utility of tumor necrosis factor-α and interleukin-6 as predictors of neonatal mortality in white-tailed deer. J Mammal 82(1):239–245

Ditchkoff SS, Spicer LJ, Masters RE, Lochmiller RL (2001b) Concentrations of insulin-like growth factor-I in adult male white-tailed deer (*Odocoileus virginianus*): associations with serum testosterone, morphometrics and age during and after the breeding season. Comp Biochem Physiol A Mol Integr Physiol 129(4):887–895

Dusek GL, MacKie RJ, Herriges JD, Compton BB (1989) Population ecology of white-tailed deer along the lower Yellowstone River. Wildl Monogr 104:3–68

Finnish Food Authority (2019) Chronic wasting disease (CWD) and other TSE diseases in cervids. https://www.ruokavirasto.fi/en/farmers/animal-husbandry/animal-health-and-diseases/animal-diseases/wildlife/chronic-wasting-disease-cwd-in-cervids/

Finnish Ministry of Agriculture and Forestry (2012) Finland's national strategy on invasive alien species. https://mmm.fi/documents/1410837/1894125/Finlands_national_strategy_on_invasive_alien_species.pdf/61137cdf-92ad-4ac8-8b6d-0feaeaedfe74

Gallina S, Lopez Arevalo H (2016) *Odocoileus virginianus*. The IUCN Red List of Threatened Species 2016: e.T42394A22162580. https://doi.org/10.2305/IUCN.UK.2016-2.RLTS.T42394A22162580.en. Downloaded on 03 May 2019

Gassett JW, Wiesler DP, Baker AG, Osborn DA, Miller KV, Marchinton RL, Novotny M (1997). Volatile compounds from the forehead region of male white-tailed deer (*Odocoileus virginianus*). J Chem Ecol 23(3):569–578

Gee LK, Holman JH, Causey K, Rossi AN, Armstrong JB (2002) Aging white-tailed deer by tooth replacement and wear: a critical evaluation of a time-honored technique. Wildl Soc Bull 30:387–393

Häkkinen I (1986) Metsästyssuunnitelma. In: Moilanen P, Vikberg P (eds) Valkohäntäpeura – elintavat, metsästys, riistanhoito. Otava, Helsinki

Halls LK (1984) White-tailed deer: ecology and management. Stackpole Books, Harrisburg

Hamilton RJ, Tobin ML, Moore WG (1985) Aging fetal white-tailed deer. In: Proceedings of the annual conference Southeastern Association of Fish and Wildlife Agencies, p 39

Haugen A, Davenport L (1950) Breeding records of white-tail deer in the upper Peninsula of Michigan. J Wildl Manag 3:290–295

Heffelfinger JR (2011) Taxonomy, evolutionary history and distribution. In: Hewitt DG (ed) Biology and management of white-tailed deer. CRC Press, Boca Raton

Hewitt DG (2011) Biology and management of white-tailed deer. CRC Press, Boca Raton

Hirth DH (1977) Social behavior of white-tailed deer in relation to habitat. Wildl Monogr 53:3–55

Honzová M (2013) Analysis of habitat size and migration of roe and white-tailed deer in Finnish lake district, Finland. Diploma thesis, Mendel University in Brno

Jokelainen P, Näreaho A, Knaapi S, Oksanen A, Rikula U, Sukura A (2010) *Toxoplasma gondii* in wild cervids and sheep in Finland: north-south gradient in seroprevalence. Vet Parasitol 171:3–4

Kairikko JK, Ruola J (2004) Valkohäntäpeura. The Hunting Association in Finland

Kekkonen J, Wikström M, Brommer JE (2012) Heterozygosity in an isolated population of a large mammal founded by four individuals is predicted by an individual-based genetic model. PLoS One 7:e43482

Kekkonen J, Wikström M, Ala-Ajos I, Lappalainen V, Brommer JE (2016) Growth and age structure in an introduced and hunted cervid population: white-tailed deer in Finland. Ann Zool Fenn 53:69–80

Kilgo JC, Labisky RF, Fritzen DE (1996) Directional long-distance movements by white-tailed deer *Odocoileus virginianus* in Florida. Wildl Biol 2:289–292

Knox WM, Miller KV, Marchinton RL (1988) Recurrent estrous cycles in white-tailed deer. J Mammal 69: 384–386

Knox WM, Miller KV, Collins DC, Bush PB, Kiser TE, Marchinton RL (1992) Serum and urinary levels of reproductive hormones associated with the estrous cycle in white-tailed deer (*Odocoileus virginianus*). Zoo Biol 11(2):121–131

Koivisto I (1986) Valkohäntäpeuran sopeutuminen Suomeen. In: Moilanen, Wikberg (eds) Valkohäntäpeura, elintavat, metsätys, riistanhoito. Otava, Helsinki

Kukko T, Pusenius J (2019) Valkohäntäpeurakanta talvella 2018–2019. http://wordpress1.luke.fi/riistahavainnot-hirvielaimet/wp-content/uploads/sites/5/2019/03/Valkoh%C3%A4nt%C3%A4peurakanta-talvella-2018_2019.pdf

Laaksonen S, Oksanen A, Hoberg E (2015) A lymphatic dwelling filarioid nematode, *Rumenfilaria andersoni* (Filarioidea; Splendidofilariinae), is an emerging parasite in Finnish cervids. Parasit Vectors 8:228

Lahtinen J (1996) Riistannisäkkäiden tarhaus. Finnish National Agency for Education

Lambiase JT Jr, Amann RP, Lindzey JS (1972) Aspects of reproductive physiology of male white-tailed deer. J Wildl Manag:868–875

Long ES, Diefenbach DR, Rosenberry CS, Wallingford BD, Grund MD (2005) Forest cover influences dispersal distance of white-tailed deer. J Mammal 86: 623–629

Miller BF, DeYoung RW, Campbell TA, Laseter BR, Ford WM (2010) Fine-scale genetic and social structuring in a Central Appalachian White-Tailed Deer herd. J Mammal 91:681–689

Mirarchi RE, Scanlon PF, Kirkpatrick RL (1977) Annual changes in spermatozoan production and associated organs of white-tailed deer. J Wildl Manag:92–99

Mirarchi RE, Howland BE, Scanlon PF, Kirkpatrick RL, Sanford LM (1978) Seasonal variation in plasma LH,

FSH, prolactin, and testosterone concentrations in adult male white-tailed deer. Can J Zool 56(1):121–127
Nelson ME (1993) Natal dispersal and gene flow in white-tailed deer in northeastern Minnesota. J Mammal 74: 316–322
Nelson ME, Mech LD (1984) Home-range formation and dispersal of deer in northeastern Minnesota. J Mammal 65:567–575
Nelson ME, Mech LD (1987) Demes within a northeastern Minnesota deer population. In: Chepko-Sade BD, Halpin ZT (eds) Mammalian dispersal patterns: the effects of social structure on population genetics. The University of Chicago Press, Chicago
Niemi M, Matala J, Melin M, Eronen V, Järvenpää H (2015) Traffic mortality of four ungulate species in southern Finland. Nat Conserv 11:13–28
Nygren K (1984) Finland. In: Halls LK (ed) White-tailed deer: ecology and management. Stackpole Books, Harrisburg
Oelrich TM (1953) Additional mammals from the Rexroad fauna. J Mammal 34:373–378
Plotka ED, Seal US, Schmoller GC, Karns PD, Keenlyne KD (1977) Reproductive steroids in the white-tailed deer (*Odocoileus virginianus borealis*). Biol Reprod 16:340–343
Plotka ED, Seal US, Verme LJ, Ozoga JJ (1980) Reproductive steroids in deer. III. Luteinizing hormone, estradiol and progesterone around estrus. Biol Reprod 22(3):576–581
Sams MG, Lochmiller RL, Qualls CW Jr, Leslie DM Jr, Payton ME (1996) Physiological correlates of neonatal mortality in an overpopulated herd of white-tailed deer. J Mammal 77(1):179–190
Severinghaus CW (1949) Tooth development and wear as criteria of age in white-tailed deer. J Wildl Manag 13: 195–216
Short HL (1964) Postnatal stomach development of white-tailed deer. J Wildl Manag 28:445–458
Silver H (1961) Deer milk compared with substitute milk for fawns. J Wildl Manag 25(1):66–70
Sorin B (2004) Paternity assignment for white-tailed deer (*Odocoileus virginianus*): mating across age classes and multiple paternity. J Mammal 85:356–362
Turner MM, Deperno CS, Booth W, Vargo EL, Conner MC, Lancia RA (2016) The mating system of white-tailed deer under Quality Deer Management. J Wildl Manag 80(5):935–940
Verme LJ (1965) Reproduction studies on penned white-tailed deer. J Wildl Manag 29:74–79
Verme LJ, Ullrey DE (1984) Physiology and nutrition. In: Halls LK (ed) White-tailed deer: ecology and management. Stackpole Books, Harrisburg
Webb SL, Gee KL, Strickland BK, Demarais S, DeYoung RW (2010) Measuring fine-scale white-tailed deer movements and environmental influences using GPS collars. Int J Ecol 2010:Article ID 459610
Wikström M (2012) Home range and site fidelity of white-tailed deer (*Odocoileus virginianus borealis*) in Finland. University of Helsinki

European Bison *Bison bonasus* (Linnaeus, 1758) 12

Rafał Kowalczyk and Glenn Plumb

Contents

R. Kowalczyk (✉)
Mammal Research Institute, Polish Academy of Sciences, Białowieża, Poland
e-mail: rkowal@ibs.bialowieza.pl

G. Plumb
International Union for Conservation of Nature, Species Survival Commission, Bison Specialist Group, Livingston, MT, USA

© Springer Nature Switzerland AG 2022

L. Corlatti, F. E. Zachos (eds.), *Terrestrial Cetartiodactyla*, Handbook of the Mammals of Europe,
https://doi.org/10.1007/978-3-030-24475-0_28

Common Names

English	European bison
German	Wisent
French	Bison d'Europe
Spanish	Bisonte Europeo
Italian	Bisonte Europeo
Russian	Зубр

Taxonomy, Systematics and Paleontology

The European bison (*Bison bonasus,* Linnaeus, 1758) is a large herbivore belonging to the order Cetartiodactyla, suborder Ruminantia, family Bovidae, and genus *Bison* that also includes the American bison (*Bison bison,* Linnaeus, 1758) (Fig. 1). Some authors suggest the European bison species to be within the *Bos* genus due to its close association with other species of wild cattle (Groves and Grubb 2011; Soubrier et al. 2016). Despite a rich fossil record, the taxonomy, evolutionary history, and paleobiogeography of the European bison are still being developed and debated (Massilani et al. 2016; Soubrier et al. 2016).

The genus *Bison* experienced several intervals of expansion, contraction, and local extinction during the last 50,000 years in Europe, culminating in reduced genetic diversity during the Holocene (Gautier et al. 2016; Massilani et al. 2016). The oldest skeletal remains of European bison are dated to >50 kya. The temporal distribution of genotyped individuals reveals that European bison mitochondrial lineages are observed before 50 kya and after 34 kya, when steppe bison (*Bison priscus*) appear to be largely absent from the European continent (Soubrier et al. 2016). Paleontological and archeological findings suggest that the species distribution during the Late Pleistocene and Holocene extended east from France to the Ural and Northern Caucasus, and north from Bulgaria to southern Sweden and portions of the North Sea that were unflooded during the Late Pleistocene (Benecke 2005; Soubrier et al. 2016; Hofman-Kamińska et al. 2019) (Fig. 2).

Traditionally, it has been considered that the European bison developed within one single phylogenetic line (including *Bison priscus*, *B. schoetensacki*, *B. bonasus*); or at least two parallel lines, with one being the line of *B. schoetensacki* (Freudenberg, 1910) called also forest bison, which evolved to the recent

Fig. 1 Male (left) and female with calf (right) of European bison (photograph © Rafał Kowalczyk)

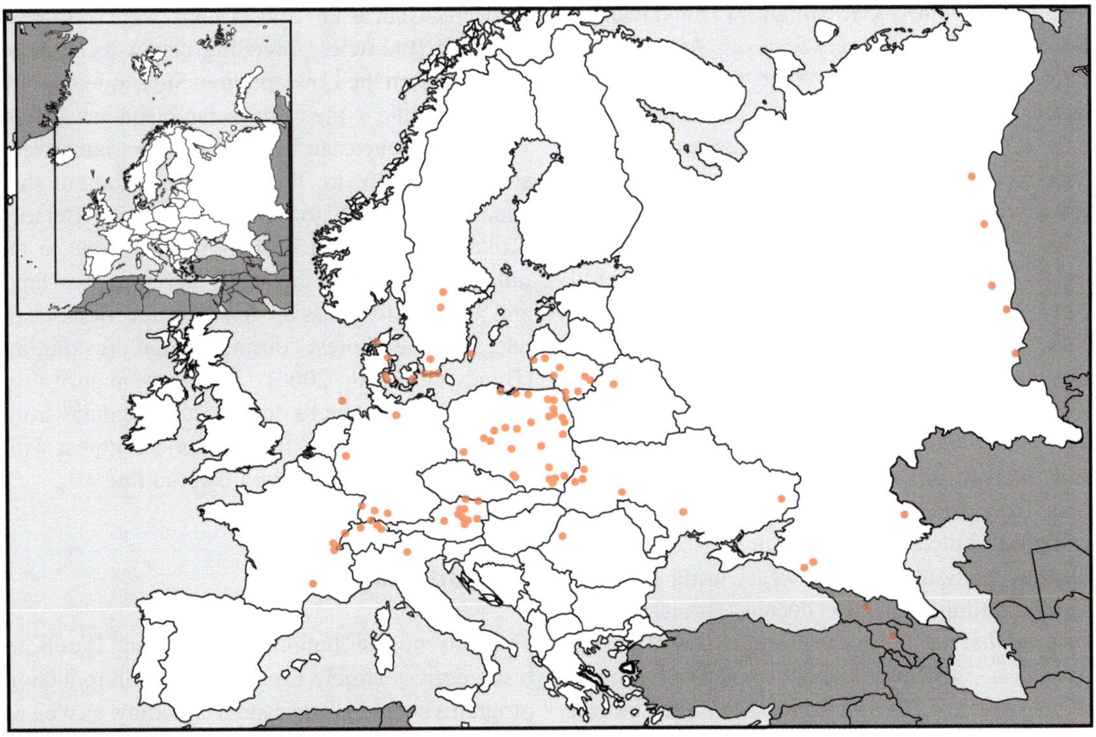

Map template: © Getty Images/iStockphoto

Fig. 2 Historical (Late Pleistocene and Holocene) distribution of European bison in Europe. Dots show locations of historical skeletal remains. (Based on Soubrier et al. (2016) and Hofman-Kamińska et al. (2019)) (Map template: © Copyright Getty Images/iStockphoto)

B. bonasus, and the other being the line of the steppe bison *B. priscus*. However, validity of *B. schoetensacki* is questioned (Drees 2005), and recent genomic analysis place *B. schoetensacki* in one of the *B. bonasus* clades (Soubrier et al. 2016; Massilani et al. 2016; Palacio et al. 2017) and should be renamed accordingly (see details in "Genetics" section). During the Holocene, the species distribution range declined extensively due to meta-population extirpation through an interaction of human persecution and broad environmental change from extensive open landscapes to forested habitats (Hofman-Kamińska et al. 2019). Since the sixteenth century following the medieval period, the European bison persisted in the wild only through royal protection. By the nineteenth century, wild, free-ranging European bison were limited to only the Białowieża Forest of northeast Poland and western Belarus, and Caucasus mountains (Pucek et al. 2004).

Presently, two subspecies are recognized as the Lowland (Białowieża) European bison (*Bison bonasus bonasus*) and the Caucasian (mountain) European bison (*Bison bonasus caucasicus,* Turkin and Satunin, 1904). Some authors earlier recognized the Carpathian (Transylvanian) European bison (*Bison bonasus hungarorum,* Kretzoi, 1946), though that description was based on the morphological identification of only a single bone, and this subspecies has not been recognized in genetic analysis conducted on fossil material (Soubrier et al. 2016; Massilani et al. 2016). Lowland bison were distributed across western and eastern Europe (Węcek et al. 2017; Hofman-Kamińska et al. 2019), while the Caucasian bison occurred only in the northern Caucasus mountains and foothills (Heptner et al. 1966). Compared to the Lowland bison, Caucasian bison morphology included smaller body size, darker coloration, more hairy head and body front, and shorter and more rounded hooves, being attributed to geographical isolation and local adaptation (Flerov 1979). During the nineteenth century, royally decreed translocations ensured that the European bison survived in captivity at several locations across Europe (Krasiński and Krasińska 2017). Yet, by the early twentieth century, the species was increasingly imperiled in the wild, with the Lowland bison becoming extinct in the wild in 1919, and the Caucasian bison extinct in the wild by 1927 (Pucek et al. 2004). As the species was going extinct in the wild, 54 remaining European bison were registered in captivity in the early 1920s (Raczyński 1978; Pucek 1991). A detailed pedigree analysis of these 54 bison indicates that all contemporary European bison are descendants of only 12 founders with individual genotypes (Slatis 1960). Among the bison that survived in captivity there was only one Caucasian male. Captive breeding was then undertaken to create two isolated genetic lines of the European bison, being a Lowland line through seven founders and a Lowland-Caucasian hybrid line with 12 founders (Pucek et al. 2004). Restoration of the species into native habitat started at the Białowieża Forest in 1929 and the history of the many challenges, failures, and successes of this restoration program have been well described (Pucek et al. 2004; Krasińska and Krasiński 2013; Krasińska et al. 2014; Krasiński and Krasińska 2017). In 1950, following sufficient expansion of the Białowieża Forest population, all known individuals of the Lowland-Caucasian hybrid line were removed from the Białowieża breeding center in order to focus only on the Lowland line. Subsequent DNA analysis of the extinct Caucasian bison has shown that some Caucasian bison genetic variants were still detectable in the bison population that occupies the Białowieża Forest in Belarus (Tokarska et al. 2015). This is likely due to an unknown extent of breeding between Lowland and Lowland-Caucasian bison in the Belarusian part of the Forest during initial restoration (Bunevich et al. 2006). The vast majority of European bison herds in Belarus originate from the Belarusian part of the Białowieża Forest with uncertain affiliation to the Lowland line.

Current Distribution

The current distribution of free-living European bison derives largely from multiple reintroduction programs sourced from captive breeding, as well as some recent translocations from wild, free-ranging herds (Fig. 3). In 1952, the European bison was first restored as free-ranging wildlife to the Białowieża Forest in NE Poland, with successive successful reintroductions to locations in Belarus, Lithuania, Poland, Russia, and Ukraine. More recently, free-ranging European bison herds were created in Slovak Republic (2004), Germany (2013), Romania (2014), Bulgaria (2019), and Latvia (2019). In 2019, there were 6244 wild, free-living European bison (EBPB 1987–2020) distributed in 47 herds isolated by geographical distance

Map template: © Getty Images/iStockphoto

Fig. 3 Distribution and ranges of free-living European bison populations (Map template: © Copyright Getty Images/ iStockphoto)

Table 1 Distribution and abundance of 47 free-living populations of European bison (EBPB 1987–2020 and Plumb et al. 2020)

Country	Number of populations	Total population size
Belarus	10	2020
Bulgaria	1	7
Germany	1	26
Latvia	1	5
Lithuania	2	284
Poland	6	2048
Romania	3	107
Russia	16	1381
Slovak Republic	1	48
Ukraine	6	315
Total	**47**	**6244**

or barriers (e.g., border fence) (Fig. 3), with additional new reintroductions being planned. Another 2217 European bison occur at over 200 captive centers and semi-captive herds in nearly 30 countries (EBPB 1987–2020). The Lowland bison line is generally the focus of restoration in Belarus, Lithuania, and northeast Poland, while the Lowland-Caucasian line is the focus in Germany, southern Poland, Romania, Russia, Slovak Republic, and Ukraine (EBPB 1987–2020) (Table 1).

Description

Size and Morphology

The European bison is the largest terrestrial mammal of Europe. A compact body and large head set on a strong neck, combined with a pronounced hump and horns twisted inwards, give the European bison a widely recognized iconic appearance (Fig. 1). The European bison displays a distinct sexual dimorphism, with males 33% bigger than females (Krasińska and Krasiński 2002). Bison cows are characterized by a more delicate construction of the front part of the body, less pronounced hump, narrow head with thinner and more twisted horns, compared to males (Fig. 1). Body mass of mature males is 436–840 kg, and 340–540 kg for mature females (Krasińska and Krasiński 2002). New-born calves weigh an average 26 kg with an observed range of 15–35 kg (Krasińska and Krasiński 2002). Head and body length of adult males vary between 258 and 323 cm, whereas that of females varies between 222 and 292 cm, with shoulder height of mature males reaching 188 cm and 167 cm for mature females (Kowalczyk R., unpublished data on file at zoological collection of the Mammal Research Institute PAS). The tail of both sexes reaches 30–60 cm long and drapes to the heels or below. Black horns are on the sides of the head for both sexes, and become increasingly worn with age for males, and increasingly twisted for females. Maximal distance between horn curves reaches 87 cm with horn length reaching 65 cm (Krasińska and Krasiński 2002, Kowalczyk, unpublished data on file at zoological collection of the Mammal Research Institute PAS). The muzzle is wide and dark gray. At a glance, the European bison closely resembles the American bison, though with smaller and less sloping hindquarters, and some differences in hair coverage and coloration (Krasińska et al. 2014).

Pelage

European bison have a dense, dark brown to golden–brown coat, with some individual variety of coloration shades. The sides of the head and legs are darker in both sexes. In males, the top of the head, chin, neck, shoulders, hump, upper parts of front legs, and prepuce are covered with longer hair. The longest hairs (up to 50 cm) form the tuft at the end of the tail. The rest of the body is coated with short fur; however, the border between the coat covering forequarters and hindquarters is not so distinctly marked as in American bison. New-born calves are red-brown, and after first moulting (3–4 months after birth) their coat coloration change to dark brown. Adult bison begin to moult at the end of winter, usually in March and continues for next 4–5 months (Kiseleva 1974). Both sexes of the European bison intensively rub against trees, broken trunks, and stumps to enhance moulting. The winter coat begins to develop in September and is complete by early November.

Head and Dentition

The skull is wide and massively built, with pronounced protruding bony rims around eye sockets. Males have a greater cavity volume, and the skull is narrower in females. Mean length of the skull is 500 mm (maximum of 542 mm) in males, and 417 mm (maximum of 458 mm) in females. The mean width is 298 mm (maximum of 342 mm) in males, and 240 mm (maximum 277 mm) in females (Szara et al. 2003). Both sexes exhibit two teeth generations: milk teeth 0.0.3.0/3.1.3.0; permanent dentition 0.0.3.3/3.1.3.3, in total 32 teeth. Milk teeth are replaced between 22 and 44 months of age, while molars erupt over a relatively long period between 6 and 43 months of age (Węgrzyn and Serwatka 1984).

Physiology

Physiology of European bison (reproductive cycle, ruminant physiology of an intermediate feeder, etc.) is similar to that of other temperate European ungulates. Body temperature range is 38.1–38.4 °C. Mean heart rate is 106 ± 19/min. Respiratory movements are 10–18/min. Blood pressure (systolic/diastolic) is 132 ± 13/117 ± 7 mmHg (Gill 1999). Main blood parameters and their values are presented in Table 2.

Genomic analysis of genes under selection showed adaptation of European bison to colder climate conditions, which is confirmed by development of thick pelage and lack of historical occurrence in warmer environments of southern Europe (Gautier et al. 2016; Hofman-Kamińska et al. 2019).

Table 2 Hematologic values of European bison in the Białowieża Forest (Anusz et al. 2007)

Parameter	Unit	Mean ± SE	Min–Max
Erythrocytes	T/l	7.4 ± 1.7	2.1–11.1
Hematocrit	l/l	0.3 ± 0.0	0.1–0.5
Hemoglobin	g/l	128.9 ± 25.2	51.00–177.0
Leukocytes	G/l	4.7 ± 1.4	1.3–9.1
Lymphocytes	%	72.1 ± 13.5	34.0–95.0
Monocytes	%	1.3 ± 0.4	1.0–2.0

Genetics

Chromosomes

$2n = 60$ chromosomes, of which 58 are acrocentric autosomes, and two are the sex chromosomes (Melander 1959; Fedyk and Sysa 1971).

Phylogeny and Phylogeography

While some genomic, paleogenomic, morphometric, and paleoecological studies have elucidated large parts of the evolution of bison populations during the Late Pleistocene and Holocene in Eurasia (Soubrier et al. 2016; Massilani et al. 2016), the origin of European bison remains an ongoing subject of scientific debate, with some data interpreted contradictorily (Soubrier et al. 2016; Massilani et al. 2016; Wang et al. 2017; Grange et al. 2018). The first interpretation of ancient mitochondrial genomes and genome-wide nuclear DNA surveys suggests that the European bison is a hybrid between the extinct steppe bison (*B. priscus*) and the aurochs (*Bos primigenius*), the ancestor of modern cattle (Soubrier et al. 2016). The second interpretation of metagenome data proposes that *B. bonasus* mitogenome lineage is more closely related to the *Bos p. taurus* lineage than to the *B. priscus–B. bison* lineages. The lineages maintained parallel evolutionary paths with gene flow during a long period of incomplete speciation. Genetic affiliation between the European bison and cattle mitogenomes results from incomplete lineage sorting (Massilani et al. 2016; Grange et al. 2018). Time for the node separating the *Bos p. taurus–B. bonasus* and the *B. priscus–B. bison* lineages was estimated at 927 (1064–790) kya, and the node separating the *B. p. taurus and B. bonasus* lineages at 768 (886–657) kya (Massilani et al. 2016). It is agreed that the *B. bonasus* mitogenome lineage can be subdivided into two sublineages: Bb1 also named BisonX, which went extinct at the onset of the Holocene, and Bb2, which gave rise to modern-day European bison (Soubrier et al. 2016; Grange et al. 2018). The divergence between BisonX and modern bison lineages occurred at 120 (92–152) kya, likely during the last (Eemian) interglacial.

Genetic Diversity

Contemporary genetic variability of European bison is relatively low, being an effect of population bottleneck due to species extirpation in the wild in the early twentieth century and subsequent restoration from limited number of captive survivors (Pucek et al. 2004; Wójcik et al. 2009; Tokarska et al. 2011). Descendants of only 12 individuals were successfully used in the recovery of the Lowland Caucasian line, while Lowland line derives only from seven founders. Furthermore, the share of genes of individual founders is extremely unequal in the two lines, with domination of a pair of bison (named Planta and Plebejer) constituting over 80% of those in Lowland line, and 50% in the Lowland-Caucasian line (Olech 2003). There is only a single male line in the Lowland bison line, and only three males in the Lowland-Caucasian line (Tokarska et al. 2011). Mean expected heterozygosity, calculated on the basis of microsatellite data, has been estimated as 0.29 (29%) for the Lowland line and 0.35 (35%) for the Lowland-Caucasian line (Tokarska, unpublished). Of more than 52,000 analyzed cattle SNP (single nucleotide polymorphism) loci amplified in European bison, only 900 were found to be polymorphic (Pertoldi et al. 2010). Despite the low genetic variability, deleterious signs of inbreeding depression are rarely observed in European bison. Some slight skull (elongation and narrowing of splanchnocranium) and fore limb shape modifications were found in Lowland-Caucasian line (Kobryńczuk 1985). In general, the lack or limited impact of inbreeding depression on European bison vitality and viability may be due to rapid demographic recovery of the species after the bottleneck, which minimizes the negative impact on the genetic variability by purging the genetic load. It is also possible that genetic depletion took place before the founder event, or because the founders' genomes were, by chance, free from significant genetic load (Tokarska et al. 2011; Tokarska 2013).

Hybridization

European bison may readily hybridize with American bison, which is privately farmed in Europe, to produce fertile offspring. It is unclear to what extent privately farmed European-American hybridized bison could impact the genetic integrity of ongoing and future wild, free-ranging European bison restoration efforts. A hybridized population of European-American bison originating from Askania-Nova, with 95% gene pool coming from *B. bonasus*, was translocated to the wild at the Caucasian Biosphere reserve in 1940, and thrived and increased to approximately 1500 by the 1990s (Sipko et al. 2010). Crossing of European bison and domestic cattle through artificial insemination has proved to be difficult and of very low efficiency (Krasińska and Krasiński 2013). European bison x cattle hybrids are distinguishable from bison on the basis of body mass (>1000 kg) and variable coloration related to the breed of cattle used (Krasińska 1988). First-generation (F_1) male hybrids are infertile, and further crossing only possible by having hybrid females covered by back-crossing (Krasińska 1988). Cross-breeding of bison with domestic cattle has never been observed in the wild, despite close encounter of bison with cattle when moving from forested areas onto adjacent agricultural lands.

Life History

Growth

At birth, the mean body mass of male bison is 28 ± 6 kg (mean ± SD), and 24 ± 4 kg in females, with no significant difference between sexes (Krasińska and Krasiński 2002). Calves double their body mass by 3 months age, and do not differ significantly between sexes during the first two years of age. As observed in the field, the age of bison may be determined by the combination of body size, and horn size and shape. For both sexes, there is strong growth of body mass and markedly inward-curving horns by year 3 of age, with female bison fully grown by 5 years, and males fully grown by 7 years of age (Krasińska and Krasiński 2002). Sex-related differences in both size and shape of horns are pronounced earlier than differences in body structure. Male horns grow continuously until full development by 7–8 years age, with horn tips then frequently becoming worn down and rounded, due to rubbing against trees and aggression with other males. Female horns grow longer and more curved with age, with 20+-year-old females retaining sharp horn tips (Krasińska and Krasiński 2013).

Reproduction

In wild, free-ranging populations, age at primiparity is generally 3 years, with two-day estrus intervals repeating across the annual rut during August to October (Krasiński and Raczyński 1967; Daleszczyk 2002; Krasińska and Krasiński 2013). In captivity, age of primiparity ranges between 2 and 5 years (Jaczewski 1958; Daleszczyk 2011). Average gestation lasts for 264 days (range: 254–270) (Jaczewski 1958; Krasiński and Raczyński 1967; Kiseleva 1974). Parturition usually lasts 1–2 h, with up to 3.5 h observed (Daleszczyk and Krasiński 2001; Zięba 2007). Wild free-ranging European bison males generally begin to exhibit sexual activity at 3 years age (Korochkina 1971), with fully developed spermatogenesis at 4 years age (Czykier et al. 1999). Free-ranging European bison cows almost exclusively deliver a single calf (Krasińska and Krasiński 2013), with only one case of twin calves reported in a wild, free-ranging population in the Vologda region of Russia (Tyapougin and Gusarov 2004; Gusarov 2011). Twin calves are also very rare in captivity (Kelterborn et al. 2009). Nearing parturition, a cow typically leaves the group, and delivers a calf in a relatively secluded place. A newborn calf stands up after 20–45 min and thereafter follows the mother, that is a behavior trait adapted to open habitats, and the cow-calf pair then rejoins a group with a 1–2 days (Daleszczyk and Krasiński 2001). Also like the American bison, the European bison exhibits birth synchrony with a majority of calves (70–80%) born between May and July (Krasiński and

Raczyński 1967; Krasiński 1978); however, births during early and late months of the year have also been observed (Krasiński 1978; Kelterborn et al. 2009), especially in supplementary fed herds (R. Kowalczyk, unpublished).

Sex ratio at birth may differ between populations. In the Białowieża Forest it was found that increased population density and reduced female body mass led to increasing female-biased calf sex ratios, whereas years with oak *Quercus* sp. seed masting (abundant food resources) corresponded to male-biased sex ratios (Hayward et al. 2011), suggesting alignment with the Trivers-Willard hypothesis that offspring sex ratio is responsive to maternal condition (see Rutberg 1986; Hewison and Gaillard 1999). Calves follow their mothers for the first year of their life, and thereafter may remain in a mother's group for several years (Krasińska et al. 1987).

Survival

European bison are generally long-lived, with females reaching 25 years age and males rarely exceeding 20 years (Pucek et al. 2004; Krasińska and Krasiński 2013; Krasińska et al. 2014). Main sources of mortality are diseases and parasitic infections, and injuries due mainly to traffic collisions (Krasińska and Krasiński 2013). Locally, poaching may play important role. Mortality is higher during severe winter (Mysterud et al. 2007). Mysterud et al. (2007) also reported that generally low adult natural mortality in the Polish Białowieża population increases when reduced oak *Quercus* sp. masting combines with increased winter severity. Daleszczyk and Bunevich (2009) reported low but variable age-sex structured mortality rates up to 3 years age between the adjacent Polish and Belarusian Białowieża Forest populations (Krasińska and Krasiński 2013; A.N. Bunevich and Białowieża National Park, unpublished data), with variability in sex structured mortality between the populations then diminished after 4 years of age. Most recently, overall adult annual survival rate of 0.88 ± 0.09 (mean $\pm$ SD) was estimated for the Polish Białowieża population from radio-tracking data, being lower for males (0.85 ± 0.14) compared to females (0.91 ± 0.07) (R. Kowalczyk, unpublished data).

Habitat and Diet

Habitat

Like the American bison, the European bison is well adapted to foraging in open and mixed habitats, with a wide muzzle, hypsodont (i.e., high-crowned) teeth, and functional length of the anterior part of the jaw that facilitate consumption of a large volume of herbaceous primary production (Mendoza and Palmqvist 2008). At the beginning of Holocene (10–12 kya), bison roamed in open habitats across Europe as indicated by stable isotope analysis (Bocherens et al. 2015; Hofman-Kamińska et al. 2019). Isotopic signatures indicate subsequent shifts in habitat use from open landscapes to forests at the Mezolithic to Neolithic transition 7.5–6 kya years ago (Hofman-Kamińska et al. 2019). Kerley et al. (2012) have argued that replacement of open tundra-steppe by forest cover after the last postglacial period and increasing human pressure related mainly to development of Neolithic agriculture, forced bison into forests as a refuge habitat. As described earlier, royal protection of wild European bison following the medieval period focused on forested habitats (Pucek et al. 2004). Following upon this historical pattern, initial European bison restoration efforts focused mainly on forested habitats in Eastern Europe (Pucek et al. 2004; Krasińska and Krasiński 2013; Krasińska et al. 2014). Restoration continues to focus primarily on forested habitats across Europe including mixed and deciduous forests intersected by abandoned agricultural lands/open river valleys/forest glades, coniferous forests with little herbaceous understory, foothills, and higher elevation transition habitats of mountainous areas, as well as southern taiga (Krasińska and Krasiński 2013).

The Refugee Species Hypothesis proposed by Kerley et al. (2012) suggests that the European bison is not an obligate forest specialist, and that continuation of restoration emphases on forested habitats alone risks confining the species to suboptimal or marginal habitats, with important density-dependent consequences for species fitness and long-term recovery. Stable isotope analysis showed strong plasticity and variation in habitat use and preference among modern bison populations in response to the proportion of forest cover (Hofman-Kamińska et al. 2018a). When reintroduced into forested habitats, supplementary feeding provides incentive for forest habitat association by bison (Kuemmerle et al. 2018), yet free-ranging bison exhibit habitat selection preference for open and wet/open habitats and abandoned fields over the proportionally (in relation to available area) less-preferred forest habitats (Kuemmerle et al. 2010, 2018; Zielke et al. 2019). Among mixed-forest habitats, coniferous forests are avoided, while alderwoods are preferred in summer and autumn due to higher humidity and persistence of lush vegetation when compared to other forests (Daleszczyk et al. 2007). In some populations, movement from forests into adjacent open habitats increases in winter (Kowalczyk et al. 2013). Kerley et al. (2012, 2020) suggest that supplemental feeding during winter seriously disrupts the species natural habitat ecology and thus reinforces an unsustainable refugee cycle for the European bison.

Foraging and Diet

The European bison is adapted to uptake large amounts of different plants, including fodder with lower digestibility such as senescent graminoids and fibrous plants (Hofmann 1989; Gautier et al. 2016) with daily fresh matter intake of 23–50 kg (Gębczyńska and Krasińska 1972; Holodova and Belousova 1989). Despite their morphological adaptations to grazing, stomach content analysis, DNA metabarcoding of feces, and teeth microwear analysis indicate that European bison are mixed feeders or browsers (Gębczyńska et al. 1991; Kowalczyk et al. 2011, 2019; Bocherens et al. 2015; Merceron et al. 2015; Hofman-Kamińska et al. 2018b). In the Białowieża Forest, European bison diet consists of 454 vascular plant species, including naturally browsed and delivered with supplementary fodder (Korochkina 1972; Jaroszewicz and Pirożnikow 2008; Kowalczyk et al. 2019). DNA-based analysis of feces showed that during growing season Europe bison diet consists mainly of woody species (30% trees, 30% shrubs), followed by herbaceous forbs (34%), graminoids (4%), and cryptogams (2%) (Kowalczyk et al. 2019), with red raspberry (*Rubus idaeus*), European hornbeam (*Carpinus betulus*), wood avens (*Geum* sp.), stinging nettle (*Urtica dioica*), and meadowsweet (*Filipendula ulmaria*) being among the most highly consumed species (Kowalczyk et al. 2019). In winter bison diet is strongly influenced by access to supplementary feeding. With increasing supplementary feeding, European bison decrease intake of woody materials (65% in non-fed bison utilizing forest habitats to 16% in intensively fed herds) and increased intake of herbaceous forages (32% in non-fed bison utilizing forest habitats to 82% in intensively fed herds) (Kowalczyk et al. 2011). The species of trees mainly browsed by bison are of lower economic importance for forest management, including hornbeam (*Carpinus betulus*), birches (*Betula sp.*), and willows (*Salix sp.*). Thus, the supplementary winter feeding that occurs for many reintroduced European bison populations has a strong influence on foraging ecology (Kerley et al. 2012; Hofman-Kamińska et al. 2018a).

In the Carpathian Mountains, the most important forage species during winter is the bramble (*Rubus fruticosus*) (Pčola et al. 2006; Aleksandrowicz et al. 2009; Mazurek 2010). In natural conditions of Dutch dunes, bison predominantly fed on grasses across all seasons with only 20% of diet consisting of woody forages (Cromsigt et al. 2018). The mixed foraging ecology of the European bison facilitates cascading ecological process such as dispersal of seeds in forest ecosystems (Jaroszewicz et al. 2009). The total number of plant species dispersed by bison is approximately 2–3 times higher than for wild or domestic large ungulates (Jaroszewicz et al. 2013). This is particularly important for plant species with no specific dispersal adaptations. Dung deposition may result in increase in species richness, especially in coniferous forest patches, by introducing new species not recorded previously (Jaroszewicz 2013).

Close to 25% of the species registered on bison dung piles has not been previously registered locally (Jaroszewicz et al. 2009).

Home Range

Home range sizes of European bison females and mixed groups are influenced by the distribution and availability of forage resources, while home range for mature males is more related to reproductive behavior and activity than food-related factors. During the growing season in the Polish Białowieża Forest, home ranges of mixed age and sex groups covered 69 km^2 on average (range: 45–97 km^2) and were similar to those of bulls (70 km^2) (range: 29–152 km^2) (Krasińska et al. 2000). Younger males (4–6 years old) occupy smaller home ranges (44 km^2), while fully mature males ($\geq$6 years) occupy much larger home ranges (84 km^2) due to increased mobility during the breeding season (Krasińska et al. 2000). The Knyszyn Forest of northeast Poland is dominated by coniferous tree stands with reduced understory forages, and thus bison occupy larger home ranges (130 km^2) than in mixed deciduous forests of Białowieża (Kowalczyk 2010, Kerley et al. 2020). In the Bieszczady Mountains of southeast Poland, individual bison home range size varies between 59 and 123 km^2 in winter and 1 and 62 km^2 in summer (Perzanowski and Januszczak 2004). Summer home ranges of bison introduced to the German highlands reached 42.5 km^2 (Schmitz et al. 2015). Within a mixed forest landscape such as the Białowieża Forest that includes also open habitats (forest glades, meadows, river valleys), European bison disproportionately use their home ranges, with core areas overlapping with open habitats, which indicate their natural preference (Kowalczyk 2010).

Migration, Dispersal, and Range Expansion

Consistent with the *Bison* genus (see Plumb et al. 2009), European bison exhibit a complex and dynamic movement ecology underpinned by multiple interactive drivers across multiple spatial and temporal scales, including seasonal variation in forage quality and availability, weather (ambient temperature, snow depth, and ice-crusting), competitive mate selection during the seasonal rut, and density-dependent intraspecific competition (Krasińska et al. 2000; Kowalczyk et al. 2013). This complex movement ecology includes extensive ongoing intra-season local movements, seasonal migrations, and dispersal tied to range expansion. Although all free-ranging bison populations will exhibit dynamic local movement with some preference for open habitats as discussed above, not all bison populations exhibit clearly defined seasonal migration. Dispersal generally occurs by individual male bison (usually in age of 3–7 years) that may disperse over large distances ranging up to 700 km (Krasińska and Krasiński 2013). In expanding populations, dispersal by mixed age-sex groups is observed as the result of density-dependent intraspecific competition (Kowalczyk et al. 2013; Krasińska et al. 2014; Plumb et al. 2014). In complex mixed forest-open landscapes with little elevation variation, bison seasonally move between summer and winter areas, whereas in more simple landscapes of suboptimal forest habitats, bison often attempt to move to open habitats in winter leading to agriculture depredation (Hofman-Kamińska and Kowalczyk 2012). In mountainous area, bison seasonal migrations respond to changes in weather and habitat quality along elevational gradients (Krasińska et al. 2014; Plumb et al. 2014). In the Carpathian Mountains, seasonal migrations by European bison are triggered by significant changes in ambient temperature (decrease in late autumn and increase in early spring) and by the appearance of first snowfalls in autumn, with maximum migration distances of 13–19 km for mixed herds, and 5–23 km for old males (Perzanowski et al. 2012). There is evidence of attempts at range expansion in 70% of free-ranging European bison populations, where this density-dependent process is driven by mixed age-sex groups attempting to utilize open habitats in the vicinity of forests, but where management interdiction otherwise prevents such use of open habitats (Kerley et al. 2012). Daily movement distance of mixed age-sex groups ranges between 1.8 and 9.1 km (Rouys 2003; Schmitz et al. 2015); however, during rut bison males can move several kilometers a day.

Behavior

Group Size

The European bison exhibits a well-developed and dynamic social organization (Krasińska et al. 2014). The primary social unit is a mixed age-sex class group of up to 20 individuals generally, including adult cows, 2–3-year-old subadults, and calves. The size and composition of mixed groups are dynamic and change seasonally, with regular rotation and frequent exchanges of some individuals (Krasińska and Krasiński 2013). Generally, the number of groups decreases as average group size increases during the annual rut and in winter, and the number of groups then increases as average group size decreases during spring-summer. Within any season, groups' size tends to be larger in open habitats than in forests. The mean size of free-ranging mixed groups in the Polish Białowieża Forest was 13 animals (maximum 92, Krasińska and Krasiński 2013), and 21 animals (maximum 120) in the Belarusian Białowieża Forest (Kozlo and Bunevich 2009). In the Bieszczady Mountains of southeast Poland, regardless of the season, mixed groups included either 3–10 individuals (29–64% of observations) or 11–20 individuals (16–32% of observations) (Perzanowski et al. 2015). Mixed groups are led by mature cows who lead the group's local movements in search for optimal foraging conditions. Outside of the breeding season, adult males and females are separated; in most populations bulls and mixed groups occupy distinct ranges for majority of the year. Mature males join mixed groups during the rut, but also in winter, when bison aggregate in supplementary feeding sites at fixed locations in the Białowieża Forest. Sexually mature bulls (4–5 years old) often abandon mixed groups after the rut and remain in bachelor groups consisting of 2–8 individuals. Bachelor groups change their size and composition frequently, with 60% of adult bulls greater than six years old remaining solitary or in pairs (Krasińska and Krasiński 1995).

Activity

European bison exhibit multi-phasic rhythm of activity typical of large ruminants, with foraging bouts of 15–315 min alternating with resting bouts of 15–255 min devoted primarily to rumination (Caboń-Raczyńska et al. 1987). In the growing season, bison spend *c.* 60% of their daily activity on feeding, 30% for resting, and 10% on movements (Caboń-Raczyńska et al. 1987). Observations in the Białowieża Forest identified four feeding bouts coinciding with dawn and dusk, and two periods during mid-day (Caboń-Raczyńska et al. 1987). European bison exhibit limited activity during night, especially between 2300 h and 0200 h (Rouys et al. 2001). Winter supplemental feeding essentially inverts the typical activity budgets of free-ranging European bison populations so that daily time spent resting (60%) is twice as much as time spent feeding (30%) (Caboń-Raczyńska et al. 1983, 1987).

Mating Behavior

The annual rut typically starts in August and ends by mid-October, with some initial increase in time spent by mature males by mid-July in beginning to search for receptive females. A tending-bond mating system is polygynous and based on non-territorial males courting individual estrous females (Krasińska et al. 2014). The highest intensity of rutting activity is observed in August–September. Winter supplementary feeding can influence mature female body condition and alter typical estrus patterns, resulting in disrupted birth synchrony with parturition occurring as late as October–November (Krasiński 1978; Caboń-Raczyńska et al. 1983). During the rut, non-territorial males actively search for receptive estrous females roaming between mixed groups or follow a specific herd. A male bison uses chemical cues in the urine to detect females coming into early heat. The male sniffs the urine and performs a lip curl (flehmen behavior) standing for several seconds with raised head and opened mouth. If a female is in heat, the male will follow and court her. European bison exhibit competitive mate selection in which mature males typically dominate younger, smaller subordinate males, and thus mate more successfully. Mate competition typically includes mature males testing younger rivals with aggressive behaviors, including encircling a rival, hoofing the ground, wallowing, shaking the head, and damaging young trees. If the other male does not retreat or display submissive posture, and rather tries to re-demonstrate aggressive behavior, then the probability for a direct contest increases, though physical fights between males during the rut are observed only rarely in the wild (Krasińska and Krasiński 2013). Aggression is rarely observed among free-ranging cows, subadults of either sex, or calves, though some mild aggression has been observed during unnatural aggregations at winter feeding sites (Krasińska and Krasiński 2013).

Communication and Senses

The call of European bison is best described as grunting, and they never roar like mature American bison males. The call of males is lower and hoarser than that of females. Females grunt usually when communicating with calves and males during the rut (Gill 1999; Krasińska and Krasiński 2013). Gill (1999) notes that the olfactory sense of bison is very well developed and essential in detecting danger and in reproductive communication, including locating other animals or groups by smelling tracks of cows left on the ground. Like many other ungulates, bison have an auxiliary olfactory sense organ (vomeronasal or Jacobson's organ) located between the roof of the mouth and the palate, with the flehmen behavior facilitating the organ function (Gill 1999). The European bison is a near sighted animal, yet is able to distinguish larger objects from a distance of several hundred meters (Gill 1999, Kowalczyk, pers. obs.). They have a well-developed sense of hearing. European bison are able to run quite fast, but only for a short distance.

Parasites and Diseases

Endoparasites

Karbowiak et al. (2014a, b) described 88 species of endoparasites in European bison, with species richness, prevalence, and intensity of infections increasing in multiple populations (Drożdż 1995). Endoparasite species richness and individual animal loads have also been documented to increase when bison are unnaturally aggregated in winter at fixed locations for supplementary feeding (Radwan et al. 2010; Kerley et al. 2012; Karbowiak et al. 2014b; Kołodziej-Sobocińska et al. 2016a, b). Endoparasites often found in bison include the liver fluke *Fasciola hepatica* (prevalence of 44%), the lungworm *Dictyocaulus viviparus* (prevalence of 58%) (Demiaszkiewicz et al. 1999), several species of the round worm *Ostertagia* (*O. ostertagi, O. lyrata, O. leptospicularis, O. kolchida, O. antipini*), and multiple *Nematodirus* species (*N. helvetianus, N. roscidus, N. europaeus*) (Demiaszkiewicz et al. 2012). Endoparasites can sometimes infect up to 100% of individuals (Karbowiak et al. 2014a). One of the most pathogenic parasites in European bison is the blood-sucking nematode *Ashworthius sidemi* (Schulz 1933) that was first found in free-ranging bison in Bieszczady Mountains in Poland in 1997, then in Białowieża Forest in 2000, in Knyszyn Forest in 2009, and in Borki Forest in 2016 (Dróżdż et al. 1998; Demiaszkiewicz et al. 2009, 2018). Another blood-sucking nematode *Haemonchus contortus* was initially described in captive breeding bison in the Białowieża Forest in the 1960s (Dróżdż 1961, 1967). The prevalence and intensity of *A. sidemi* infection in European bison in the Polish Białowieża Forest increased rapidly and reached 100% prevalence in tested animals four years after detection, with maximal median intensity of 8200 nematodes per animal and significant deterioration of blood parameters (Kołodziej-Sobocińska et al. 2016c). Among the factors that influenced infection intensity were the number of years since introduction, herd size, age and sex of bison, suggesting that management practices can also have a strong influence in the spread of a newly detected parasites (Kołodziej-Sobocińska et al. 2016b).

Infectious Diseases

While several diseases are known to occur in European bison, including blue tongue virus, foot-and-mouth disease, respiratory viruses, and bovine tuberculosis (Larska and Krzysiak 2019), they are not as broadly distributed within or across free-ranging populations as the infectious endoparasites described above. The bluetongue virus (BTV) was first detected in a captive group of European bison in 2007 (strain BTV-8) resulting in substantial mortality (Glunz 2008). The virus is transmitted by blood-sucking *Culicoides* spp. midges, and, surprisingly, a different strain (BTV-14) was detected soon thereafter in both cattle and European bison in Northeastern Poland (Orłowska et al. 2016). Clinical signs of BTV include fever, salivation, nasal discharge, edema of the head, congestion and ulceration of the oral mucosa, weakness, depression, and sometimes cyanosis of the tongue (hence the name bluetongue), with morbidity dependent on animal age and the BTV strain involved (Schwartz-Cornil et al. 2008). Although BTV is not common in European bison, it is a mandatory reportable disease that can trigger prescriptive management (OIE 2008). Foot-and-mouth disease (FMD) in European bison in the Polish Białowieża Forest was described in the early twentieth century, but has been absent in the species since 1950s (Krasińska and Krasiński 2013). Following a 2011 outbreak of FMD in cattle in Bulgaria, there is no evidence for the maintenance of FMD in wildlife in Europe (Weaver et al. 2013). Surprisingly high seroprevalence of respiratory viruses such as bovine adenovirus type 3 (BAdV-3), bovine parainfluenza type 3 (PIV-3), and bovine respiratory syncytial virus (BRSV) have been detected in European bison resulting in pathogenesis of the lungs or upper respiratory tract (Salwa et al. 2007; Krzysiak et al. 2018). These respiratory viruses seem to be circulating more freely among free-ranging bison than captive bison, suggesting possible spillover from domestic livestock (Larska and Krzysiak 2019). Likewise, potential spillover of domestic ruminant diseases such as bovine herpesvirus 1 (BHV-1) and the Schmallenberg arbovirus (SBV) are also an emerging concern for European bison (Urban-Chmiel et al. 2017; Kęsik-Maliszewska et al. 2018). Bovine tuberculosis *Mycobacterium bovis* is not a common disease in European bison, and was detected in European bison for the first time in the 1990s at the Bieszczady Mountains (Poland), with some spillover into cattle (Żurawski and Lipiec 1997). This disease was then not detected again in European bison until 2010–2011 through a necropsy of one cow and culling of one bull (Welz et al. 2005; Brewczyński and Welz 2011). It was also detected in one of Polish breeding centers in 2013 (Krajewska et al. 2016). Larska and Krzysiak (2019) make a compelling argument that continuing changes in climate and land use will drive the prevalence of emerging and re-emerging diseases in wild, free-ranging European bison and that comprehensive, effective, and efficient long-term disease monitoring is therefore crucial to the long-term recovery of the European bison.

Population Ecology

Population Dynamics

Many restored European bison populations are now subjected to periodic culling to remove sick or injured animals, and more commonly to reduce population size (Kerley et al. 2012; Krasińska and Krasiński 2013). Annual mortality rarely exceeds 6% in larger populations such as the well-studied Białowieża Forest population (Krasińska and Krasiński 2013) (see details in "Survival" section). As such, established free-ranging European bison populations in Poland are characterized by relatively high annual growth rate of 3–12% (Kowalczyk unpublished data); however, demographic stochasticity underpins highly variable annual growth rates in small populations (EBPB 1987–2020; Krasińska and Krasiński 2013). Long-term data indicates negative density dependence in free-ranging European bison populations (Jędrzejewska et al. 1997; Mysterud et al. 2007; Samojlik et al. 2019). Historical data from Białowieża Forest shows that European bison annual population increase was ~3%, and was negatively correlated to bison density and total biomass per unit area of other wild ungulates

(Jędrzejewska et al. 1997). Contemporary data from the Polish Białowieża Forest showed that recruitment rates were best predicted by population density, with recruitment rates declining as population size increase (Mysterud et al. 2007). Additional factors that influence annual growth rate include spring temperature, Oak forest mast pulses (cyclic increased acorn production by oak tress) in the previous year, and winter severity (Mysterud et al. 2007). Historically, large predators are not thought to have affected European bison density nor population increase rate (Jędrzejewska et al. 1997).

Competition with Other Ungulates

Little information is available on competition of bison with other ungulates. At high ungulate densities observed in the Białowieża Forest in the nineteenth century, the growth rate of the European bison population was negatively correlated to its own density and to the total biomass (per unit area) of other wild ungulates, including (*Alces alces*), red deer (*Cervus elaphus*), roe deer (*Capreolus capreolus*), wild boar (*Sus scrofa*), and fallow deer (*Dama dama*) (Jędrzejewska et al. 1997), but there is little information available on niche separation and dietary overlap between European bison and sympatric wild ungulates. Other species of ungulates generally avoid close encounter with bison, though aggression by bison against wild boars and domestic horses has been reported (Krasińska and Krasiński 2013).

Conservation Status

In 2000, the Red List of the International Union for Conservation of Nature (IUCN) Species Survival Commission (SSC) included the European bison as an Endangered species, and based on extensive restoration activities, the IUCN Red List status was upgraded in 2008 to Vulnerable D1 (Olech 2008). The species Red List status was further upgraded in 2020 to Near-Threatened (Plumb et al. 2020). The European bison is listed as a Protected Fauna Species in Appendix III of the Bern Convention (Council of Europe 1979), and as a Priority Species in Annexes II (including animal and plant species of community interest whose conservation requires the designation of special areas of conservation) and IV (animal and plant species of community interest in need of strict protection) of the European Union Habitats and Species Directive (European Union 2013). The European bison is also included in the European Endangered Species Programme (EEP) for zoos established by the European Association of Zoos and Aquaria (EAZA) in 1996. One of the important tools in conservation management is the European Bison Pedigree Book (EBPB) that has been published annually since 1932 and includes lists of all known European bison individuals in captive and wild-living populations. The species abundance and distribution are increasing both in captivity and in the wild with a total of 8461 bison registered in 2019, including 1738 captive, 479 semi-free-living, and 6244 free-living individuals (EBPB 1987–2020). Despite increase in total abundance, there is also concern about long-term population viability. In 2019, only 2518 mature individuals occurred in eight isolated wild free-living sub-populations greater than minimum viable population (e.g. 150 mature animals), and no sub-population was greater than 500 mature animals (Plumb et al. 2020).

In 2004, the IUCN-SSC-Bison Specialist Group (BSG) published a report entitled "European Bison Conservation Status and Action Plan" (CAP, Pucek et al. 2004). With the overall population growing from ~3000 in 2003 to >8000 in 2019, there is now a clear need to undertake collaborative conservation planning in order to update the 2004 CAP. Key issues to be examined include climate and environmental change, science advances and needs, increased interest in restoration programs involving large mammals, meta-population dynamics, conservation genetics, disease ecology, habitat availability and shifting land use practices, restoration and translocation priorities, human dimensions, in situ and ex situ management, and variable national and European Union legal and policy status. Indeed, an updated CAP will be an important milestone for its potential to empower new initiatives and result in better alignment of multinational conservation strategies and actions. Accordingly, the IUCN-SSC-BSG has formally launched a collaborative multi-stakeholder conservation planning process with the explicit objective to develop an updated CAP that adopts the "One Plan" approach has a very strong scientific basis for actionable consensus developed through transparent multi-stakeholder deliberations (Plumb, pers. comm.).

Management

In general, the European bison has been, and continues to be, intensively managed as a forest specialist (Kerley et al. 2012, 2020). Within many of these forested locations, the species dietary needs and distribution are managed through supplementary winter fodder. Supplemental feeding is a management practice dating to times of royal protection that is aimed at reducing damage to regenerating forest stands and agricultural crops, while also limiting bison dispersal to open habitats (Hayward et al. 2015; Samojlik et al. 2019). Winter supplemental feeding also results in unnatural bison aggregations that subsequently elevates parasitic prevalence, and disrupts natural animal behavior, habitat selection, social organization, and movement ecology (Krasińska et al. 2000; Kołodziej-Sobocińska et al. 2016a; Haidt et al. 2018). Experimental introduction to a coastal dune area in which forest patches intertwine with open grasslands and shrubberies shows that European bison can live in natural environment without provision of additional food (Cromsigt et al. 2018). Many free-ranging populations, including small ones, are controlled by culling to reduce population size, regulate sex-age structure, and remove sick and aggressive individuals (Kerley et al. 2012). In lieu of management culling, some commercial hunts for bison are being organized in Belarus and Poland, and it is currently unclear how bison hunting may affect individual population viability or long-term recovery of the species. Despite active management interventions intended to disrupt dispersal and range expansion, restored bison are beginning to exhibit preference for open grassland and agricultural croplands adjacent to traditional forest habitats (Kowalczyk et al. 2013).

Impact on Agriculture and Forestry

The increasing number of bison and the expected expansion of bison populations out of forest habitats, as well as the potential creation of new free ranging herds, is expected to increase risks of human-bison conflict (Hofman-Kamińska and Kowalczyk 2012). Farm crop depredation by bison in areas neighboring forest habitats is emerging as a key human-bison conflict. Incidences of crop damage increase with decreasing distance from the woodland patches in northeast Poland, with 69% of cases of crop depredation adjacent to the Białowieża Forest, and 80% of cases adjacent to the Knyszyn Forest occurring within 0.5 km from nearest woodland patch (Hofman-Kamińska and Kowalczyk 2012). The majority of crop depredation in Lithuania and Poland occurs during winter (December–March) and focuses on cereals, hay, maize, and rape (Hofman-Kamińska and Kowalczyk 2012; Kibiša et al. 2017). Poland and Lithuania have established crop depredation compensation programs funded by state environmental agencies, with annual compensation levels reaching 300,000 euros in Poland, and 100,000 euros in Lithuania (Hofman-Kamińska and Kowalczyk 2012; Kibiša et al. 2017). Bison damage to tree stands generally has low economic impact, and includes mainly browsing and debarking, though increased debarking has been recorded around feeding sites during winter aggregation (Krasińska and Krasiński 2013). While the amount of woody materials consumed by bison in winter changes with access to supplementary fodder, preferred tree species browsed by bison in the Polish Białowieża Forest are hornbeam *Carpinus betulus*, birch *Betula* sp., and willow *Salix* sp., that are of lower economic importance for forest management (Kowalczyk et al. 2011, 2019).

Future Challenges for Research and Management

The European bison narrowly escaped extinction almost 100 years ago, and through multiple examples of personal perseverance and determined national effort, the species has increased in numbers from 54 animals in captivity to over 8000 animals across the European continent in an array of captive situations and free ranging herds. Yet, evidence is accruing that simply continuing within the scope of recent restoration activities is unlikely to achieve full ecological recovery for the species. Rather, it is more likely that sustainable long-term ecological recovery of the European bison across the historic range will be achieved through innovative collaborations among a broader forward-looking coalition of conservation actors that comprehensively address the full suite of emerging science, threats, and opportunities. As noted by Kerley et al. (2020), the European bison has already gone extinct in the wild once, and it would be a tragedy if we were to place it at risk again through incomplete conservation planning, deficient conservation science, or absence of adaptive management.

Conservation Planning

It is clearly time to undertaking a new collaborative multi-stakeholder conservation planning process to produce an update to the 2004 CAP that includes a long-term conservation action plan with a very strong scientific basis and actionable consensus. The IUCN-SSC-BSG is formally partnering with an array of collaborators to undertake this conservation planning to produce an updated IUCN CAP that will serve as an innovative, efficient, and effective milestone for its potential to empower new initiatives and result in better alignment of multinational conservation strategies and actions.

Science Needs

There is now a critical need to formally organize and inaugurate a European Bison Science Network to enhance levels of collaborative and comparative science. Enhancing the effectiveness and sustainability of restoration activities should include testing of alternative hypotheses about landscape ecology, analyzing habitat availability for optimal restoration designs, and comparative analyses of population viability and potential meta-population management strategies (see Daleszczyk and Bunevich 2009; Hartway et al. 2020). Comparative analyses of the effects of interventionist management such as hunting, culling, and supplemental feeding will be important for improving the efficiency and effectiveness of local conservation management. Also needed are comparative analyses of bison ecology across the historic range (e.g., population ecology, stress physiology, foraging ecology, competition with sympatric wild ungulates, ecological cascades, ethology), comprehensive disease and parasitology monitoring, and innovative social science that addresses the human dimensions of bison recovery.

Adaptive Management

Successfully achieving the full ecological recovery of the European bison conservation will require collaboration of researchers, managers, and policymakers to develop and implement science-based adaptive management (Kerley and Knight 2010). Improved ecological knowledge is needed to devise appropriate management regimes counteracting actual and potential threats to the bison and resulting in the wider naturalization of the species. Adaptive management will be essential to learn from emerging science, as well as restoration and management successes and failures. As an example, recent evidence now recognizes the European bison as a refugee species being managed as a forest specialist despite its evolutionary background as a mixed-feeding species inhabiting open or mosaic habitats (Mendoza and Palmqvist 2008; Kerley et al. 2012, 2020; Bocherens et al. 2015). Adaptive management is the widely recognized approach that would purposefully examine emerging science and alternative hypotheses to critically address the extent to which interventionist management actually contravenes long-term conservation of the species and its evolved ecology. There is a need to institutionalize the capacity to learn and adjust management as needed to restore bison to optimal habitats that secure the needs of the species throughout the year and spacious enough to maintain viable populations (estimated minimum 250 individuals). It is very important to adaptively manage for an effective and efficient balance of interventionist management (e.g., supplementary feeding and culling) with conservation of the full extent of the species naturally evolved ecology. Adaptive management can also serve as a framework to restore large heterogeneous landscapes that include forests, meadows, and open habitats for both existing and future bison herds; to gain and employ improved understanding of the human dimensions of bison recovery and thereby implement effective communication efforts to improve social understanding and acceptance; and to strive for effective genetic conservation and population viability by establishing sufficiently large meta-populations that link isolated populations across regional geographic ranges.

Acknowledgments We thank Tomasz Kamiński from Mammal Research Institute PAS for distribution map preparation

References

Aleksandrowicz K, Perzanowski K, Cătănoiu S, Deju R (2009) Browse and ground flora supply in selected tree stands of Vanatori Neamt Nature Park. Stud Res Vanatori Neamt Nature Park 2:40–46

Anusz K, Kita J, Zaleska M, Salwa A, Malicka E, Bielecki W, Osińska B (2007) Some morphological, biochemical and immunological blood parameters in European bison with and without lesions in the digestive, respiratory, urinary and reproductive tracts. Pol J Vet Sci 10:137–142

Benecke N (2005) The Holocene distribution of European bison: the archeozoological record. Munibe (Antropoliogie-Arkeologie) 57:421–428

Bocherens H, Hofman-Kamińska E, Drucker DG, Schmölcke U, Kowalczyk R (2015) European bison as a refugee species? Evidence from isotopic data on early holocene bison and other large herbivores in Northern Europe. PLoS One 10(2):e0115090

Brewczyński P, Welz M (2011) Threat of tuberculosis of European bison in the Bieszczady. Eur Bison Conserv Newsl 4:63–70

Bunevich AN, Krasińska M, Daleszczyk K (2006) Creation and growth of the free-ranging population of Lowland European bison, *Bison bonasus bonasus* (L.) in the Belarusian part of Białowieża Forest. Parki Narodowe i Rezerwaty Przyrody 25:101–118

Caboń-Raczyńska K, Krasińska M, Krasiński ZA (1983) Behaviour and daily activity rhythm of European bison in winter. Acta Theriol 28:273–299

Caboń-Raczyńska K, Krasińska M, Krasiński ZA, Wójcik JM (1987) Rhythm of daily activity and behaviour of European bison in Białowieża Forest in the period without snow cover. Acta Theriol 32:335–372

Council of Europe (1979) Convention on the Conservation of European Wildlife and Natural Habitats. European Treaty Series-No. 104. Appendix III–protected fauna species. Council of Europe, Strasbourg

Cromsigt JPGM, Kemp YJM, Rodriguez E, Kivit K (2018) Rewilding Europe's large grazer community: how functionally diverse are the diets of European bison, cattle, and horses? Restor Ecol 26:891–899

Czykier E, Sawick B, Krasińska M (1999) Postnatal development of the European bison spermatogenesis. Acta Theriol 44:77–90

Daleszczyk K (2002) Breeding behaviour of European bison. PhD thesis. University of Warmia and Mazury, Olsztyn, Poland. [in Polish]

Daleszczyk K (2011) Some factors influencing reproductive parameters of European bison cows. Eur Bison Conserv Newsl 4:45–54

Daleszczyk K, Bunevich AN (2009) Population viability analysis of European bison populations in Polish and Belarusian parts of Białowieża Forest with and without gene exchange. Biol Conserv 142:3068–3075

Daleszczyk K, Krasiński ZA (2001) Parturition behaviour of European bison living in reserves. Folia Zool 50:75–78

Daleszczyk K, Krasińska M, Krasiński ZA, Bunevich AN (2007) Habitat structure, climatic factors and habitat use by European bison (*Bison bonasus*) in Polish and Belarusian parts of Białowieża Forest, Poland. Can J Zool 85:261–272

Demiaszkiewicz A, Dróżdż J, Lachowicz J (1999) Occurence of lung nematodes in red deer of Białowieża Forest. Med Weter 55:519–520

Demiaszkiewicz AW, Pyziel AM, Lachowicz J, Kuligowska I (2009) New outbreak of ashworthiosis in European bison in Knyszyn Forest. In: Proceedings of International conference "80 years of European bison restoration in the Białowieża Forest", Białowieża, pp 12–13

Demiaszkiewicz AW, Pyziel AM, Kuligowska I, Lachowicz J, Krzysiak MK (2012) Nematodes of the large intestine of the European bison of the Białowieża National Park. Ann Parasitol 58:9–13

Demiaszkiewicz AW, Bielecki W, Rodo A, Pyziel AM, Filip KJ (2018) Parasitofauna of European bison *Bison bonasus* (L.) in Borecka Forest. Med Weter 74:253–256

Drees M (2005) Sexual dimorphism in Pleistocene *Bison priscus* (Mammalia, Bovidae) with a discussion on the position of *Bison schoetensacki*. Senckenberg. Lethaea 85:153–157

Dróżdż J (1961) A study on helminths and helminthiases in bison, *Bison bonasus* (L.) in Poland. Acta Parasitol 9:55–96

Dróżdż J (1967) The state of research on the helminthofauna of European bison. Acta Theriol 12:377–384

Dróżdż J (1995) Polymorphism in the Ostertagiinae Lopez-Neyra, 1947 and comments on the systematics of these nematodes. Syst Parasitol 32:91–99

Dróżdż J, Demiaszkiewicz AW, Lachowicz J (1998) *Ashworthius sidemi* (Nematoda, Trichostrongylidae) a new parasite of the European bison *Bison bonasus* (L.) and the question of independence of A. gagarini. Acta Parasitol 43:75–80

EBPB (European Bison Pedigree Book) (1987–2020) Raczyński J (ed) Białowieża National Park, Białowieża

European Union (2013) Council Directive 2013/17/EU on the conservation of natural habitats and of wild fauna and flora. Off J Eur Union 158:193

Fedyk S, Sysa P (1971) Chromosomes of European bison, domestic cattle and their hybrids. Acta Theriol 16: 465–470

Flerov KK (1979) Systematic and evolution. In: Sokolov VE (ed) European bison. Morphology, systematics, evolution, ecology. Izdatelstvo Nauka, Moscow, pp 1–127. [in Russian]

Gautier M, Moazami-Goudarzi K, Leveziel H, Parinello H, Grohs C, Rialle S, Kowalczyk R, Flori L (2016) Deciphering the wisent demographic and adaptive histories from individual whole-genome sequences. Mol Biol Evol 33:2801–2814

Gębczyńska Z, Krasińska M (1972) Food preferences and requirements of the European bison. Acta Theriol 17:105–117

Gębczyńska Z, Gębczyński M, Martynowicz E (1991) Food eaten by free-living European bison. Acta Theriol 36:307–313

Gill J (1999) Outline of European bison physiology. Severus, Warsaw

Glunz R (2008) Bluetongue disease in European bison—Symptoms, section results, treatment, vaccination. In: Proceedings of 6th Conference 'Bison in the Network Natura 2000', European Bison Friends Society, Cisna, Poland, pp 18–19

Grange T, Brugal J-P, Flori L, Gautier M, Uzunidis A, Geigl E-M (2018) The evolution and population diversity of bison in Pleistocene and Holocene Eurasia: sex matters. Diversity 10:65

Groves C, Grubb P (2011) Ungulate taxonomy. The Johns Hopkins University Press, Baltimore

Gusarov I (2011) Condition of population of European bison in northern region of Russia. Eur Bison Conserv Newsl 4:117–120

Haidt A, Kamiński T, Borowik T, Kowalczyk R (2018) Human and the beast—Flight and aggressive responses of European bison to human disturbance. PLoS One 13(8):e0200635

Hartway C, Hardy A, Jones L, Moynahan B, Traylor-Holzer K, McCann B, Aune K, Plumb G (2020) Long-term viability of Department or Interior bison under current management and potential metapopulation strategies. Natural Resource Report NPS/NRSS/BRD 2020/2097. National Park Service, Fort Collins

Hayward M, Kowalczyk R, Krasiński Z, Dackiewicz J, Cornulier T (2011) Restoration and intensive management have no effect on evolutionary strategies. Endanger Species Res 15:53–61

Hayward MW, Ortmann S, Kowalczyk R (2015) Risk perception by endangered European bison *Bison bonasus* is context (condition) dependent. Landscape Ecology 30:2079–2093

Heptner VG, Nasimovic AA, Bannikov AG (1966) Die Säugetiere der Sowjetunion. Paarhufer und Unpaarhufer. G. Fischer Verl, Jena, pp 1–939

Hewison AJ, Gaillard J-M (1999) Successful sons or advantaged daughters? the Trivers-Willard model and sex-biased maternal investment in ungulates. Trends Ecol Evol 14:229–234

Hofman-Kamińska E, Kowalczyk R (2012) Farm crops depredation by European bison (*Bison bonasus*) in the vicinity of forest habitats in northeastern Poland. Environ Manag 50:530–541

Hofman-Kamińska E, Bocherens H, Borowik T, Drucker DG, Kowalczyk R (2018a) Stable isotope signatures of large herbivore foraging habitats across Europe. PLoS One 13:e0190723

Hofman-Kamińska E, Merceron G, Bocherens H, Makowiecki D, Piličiauskiene G, Ramdarshan A, Berlioz E, Kowalczyk R (2018b) Foraging habitats and niche partitioning of European large herbivores during the Holocene – insights from 3D dental microwear texture analysis. Palaeogeogr Palaeoclimatol Palaeoecol 506:183–195

Hofman-Kamińska E, Bocherens H, Drucker DG, Fyfe RM, Gumiński W, Makowiecki D, Pacher M, Piličiauskienė G, Samojlik T, Woodbridge J, Kowalczyk R (2019) Adapt or die – response of large herbivores to environmental changes in Europe during the Holocene. Glob Chang Biol 25(9):2915–2930

Hofmann RR (1989) Evolutionary steps of ecophysiological adaptation and diversification of ruminants: a comparative view of their digestive system. Oecologia 78:443–457

Holodova MV, Belousova IP (1989) Digestibility of nutrients by European bison (*Bison bonasus*). Zool Z 68: 107–117. [in Russian]

Jaczewski Z (1958) Reproduction of the European bison, *Bison bonasus* (L.) in reserves. Acta Theriol 1:333–376

Jaroszewicz B (2013) Endozoochory by European bison influences the build-up of the soil seed bank in subcontinental coniferous forest. Eur J For Res 132:445–452

Jaroszewicz B, Pirożnikow E (2008) Diversity of plant species eaten and dispersed by the European bison *Bison bonasus* in Białowieża Forest. Eur Bison Conserv Newsl 1:14–29

Jaroszewicz B, Pirożnikow E, Sagehorn R (2009) Endozoochory by European bison (*Bison bonasus*) in Białowieża Primeval Forest across a management gradient. For Ecol Manag 258:11–17

Jaroszewicz B, Pirożnikow E, Sondej I (2013) Endozoochory by the guild of ungulates in Europe's primaeval forest. For Ecol Manag 305:21–28

Jędrzejewska B, Jędrzejewski W, Bunevich AN, Miłkowski L, Krasiński ZA (1997) Factors shaping population densities and increase rates of ungulates in Białowieża Primeval Forest (Poland and Belarus) in the 19th and 20th centuries. Acta Theriol 42:399–451

Karbowiak G, Demiaszkiewicz AW, Pyziel AM, Wita I, Moskwa B, Werszko J, Bień J, Goździk K, Lachowicz J, Cabaj W (2014a) The parasitic fauna of the European bison (*Bison bonasus*) (Linnaeus, 1758) and their impact on the conservation. Part 1 The summarising list of parasites noted. Acta Parasitol 59: 363–371

Karbowiak G, Demiaszkiewicz AW, Pyziel AM, Wita I, Moskwa B, Werszko J, Bień J, Goździk K, Lachowicz J, Cabaj W (2014b) The parasitic fauna of the European bison (*Bison bonasus*) (Linnaeus, 1758) and their impact on the conservation. Part 2 The structure and changes over time. Acta Parasitol 59:372–379

Kelterborn T, Zenter F, Zacharias K (2009) 52 years of European bison breeding on the Wisent-Island in the heart of Mecklenburg-Vorpommern. Eur Bison Conserv Newsl 2:172–181

Kerley GIH, Knight M (2010) Science and evidence based management of mega-herbivores: linking elephants and bison. In: Kowalczyk R, Ławreszuk D, Wójcik JM (eds) European bison conservation in the Białowieża Forest. Threats and prospects of the population development. Mammal Research Institute, Polish Academy of Sciences, Białowieża, pp 179–188

Kerley GIH, Kowalczyk R, Cromsigt JPGM (2012) Conservation implications of the refugee species concept and the European bison: king of the forest or refugee in a marginal habitat? Ecography 35:519–529

Kerley GIH, Cromsigt JPGM, Kowalczyk R (2020) European bison conservation cannot afford to ignore alternative hypotheses: a commentary on Perzanowski et al. (2019). Anim Conserv

Kęsik-Maliszewska J, Krzysiak MK, Grochowska M, Lechowski L, Chase C, Larska M (2018) Epidemiology of Schmallenberg virus in European bison (*Bison bonasus*) in Poland. J Wildl Dis 54:272–282

Kibiša A, Marozas V, Talijūnas D, Papšys R, Sabalinkienė G, Šimkevičius K (2017) Impact of free-ranging European bison to ecosystems in fragmented landscape, Lithuania. Balkan J Wildl Res 4(2):18–25

Kiseleva EG (1974) Reproduction of European bison in Okski Reserve. Berezinskii Zapovednik, Issledovaniya 3:103–138. [In Russian]

Kobryńczuk F (1985) The influence of inbreeding on the shape and size of the skeleton of the European bison. Acta Theriol 30:379–422

Kobryńczuk F, Krasińska M, Szara T (2008) Sexual dimorphism in skulls of the lowland European bison, *Bison bonasus bonasus*. Ann Zool Fenn 45:335–340

Kołodziej-Sobocińska M, Pyziel AM, Demiaszkiewicz AW, Borowik T, Kowalczyk R (2016a) Pattern of parasite egg shedding by European bison (*Bison bonasus*) in the Białowieża Primeval Forest, Poland. Mammal Res 61:179–186

Kołodziej-Sobocińska M, Demiaszkiewicz A, Lachowicz J, Borowik T, Kowalczyk R (2016b) Influence of management and biological factors on the parasitic invasions in the wild-spread of blood-sucking nematode *Ashworthius sidemi* in European bison (*Bison bonasus*). Int J Parasitol Parasites Wildl 5:286–294

Kołodziej-Sobocińska M, Demiaszkiewicz AW, Pyziel AM, Marczuk B, Kowalczyk R (2016c) Does the blood-sucking nematode *Ashworthius sidemi* (Trichostrongylidae) cause deterioration of blood parameters in European bison (*Bison bonasus*)? Eur J Wildl Res 62:781–785

Korochkina LN (1971) Food resources and adaptation of European bison to the conditions of Białowieża Forest. Belovezhskaya Pushcha, Issledovaniya 5:164–176. [in Russian]

Korochkina LN (1972) Monocot plants in the diet of European bison in the Białowieża Forest. Issledovanija 3:204–221

Kowalczyk R (2010) European bison – king of the forest or meadows and river valleys? In: Kowalczyk R, Ławreszuk D, Wójcik JM (eds) European bison conservation in the Białowieża Forest. Threats and prospects of the population development. Mammal Research Institute, Polish Academy of Sciences, Białowieża, pp 123–134

Kowalczyk R, Taberlet P, Coissac E, Valentini A, Miquel C, Kamiński T, Wójcik JM (2011) Influence of management practices on large herbivore diet-Case of European bison in Białowieża Primeval Forest (Poland). For Ecol Manag 261:821–828

Kowalczyk R, Krasińska M, Kamiński T, Górny M, Struś P, Hofman-Kamińska E, Krasiński ZA (2013) Movements of European bison (*Bison bonasus*) beyond the Białowieża Forest (NE Poland): range expansion or partial migrations? Acta Theriol 58:391–401

Kowalczyk R, Wójcik JM, Taberlet P, Kamiński T, Miquel C, Valenti A, Crainec JM, Coissac E (2019) Foraging plasticity allows a large herbivore to persist in a sheltering forest habitat: DNA metabarcoding diet analysis of the European bison. For Ecol Manag 449:117474

Kozlo PG, Bunevich AN (2009) The European bison in Belarus. Belaruskaya Navuka, Minsk. [in Russian]

Krajewska M, Orłowska B, Anusz K, Welz M, Bielecki W, Wojciechowska M, Lipiec M, Szulowski K (2016) Bovine tuberculosis in the bison herd in Smardzewice. Życie Weterynaryjne 91:50–53

Krasińska M (1988) Variability of the skull shape in hybrids between European bison and domestic cattle. Acta Theriol 33:147–186

Krasińska M, Krasiński ZA (1995) Composition, group size, and spatial distribution of European bison bulls in Białowieża Forest. Acta Theriol 40:1–21

Krasińska M, Krasiński ZA (2002) Body mass and measurements of the European bison during postnatal development. Acta Theriol 47:85–106

Krasińska M, Krasiński ZA (2013) European bison: the nature monograph. Springer, Heidelberg/New York/Dordrecht/London

Krasińska M, Caboń-Raczyńska K, Krasiński ZA (1987) Strategy of habitat utilization by European bison in the Białowieża Forest. Acta Theriol 32:147–202

Krasińska M, Krasiński ZA, Bunevich AN (2000) Factors affecting the variability in home range size and distribution in European bison in the Polish and Belarussian parts of the Białowieża Forest. Acta Theriol 45:321–334

Krasińska M, Krasiński Z, Olech W, Perzanowski K (2014) European bison. In: Meletti M, Burton J (eds) Ecology, evolution and behaviour of wild cattle: implications for conservation. Cambridge University Press, Cambridge, pp 115–173

Krasiński ZA (1978) Restitution and evolution of European bison population in Poland. In: Human and science. Wiedza Powszechna, Warsaw, pp 207–229. [in Polish]

Krasiński ZA, Krasińska M (2017) Catching the Lowland European bison *Bison bonasus bonasus* (L.) in Białowieża Forest in 1821–1918. Natl Parks Nat Reserv 36(3):91–98

Krasiński Z, Raczyński IJ (1967) The reproduction biology of European bison living in reserves and freedom. Acta Theriol 12:407–444

Krzysiak MK, Jabłoński A, Iwaniak W, Krajewska M, Kęsik-Maliszewska J, Larska M (2018) Seroprevalence and risk factors for selected respiratory and reproductive tract pathogen exposure in European bison (*Bison bonasus*) in Poland. Vet Microbiol 215:57–65

Kuemmerle T, Perzanowski K, Chaskovskyy O, Ostapowicz K, Halada L, Bashta A-T, Hostert P, Waller DM, Radeloff VC (2010) European bison habitat in the Carpathian Mountains. Biol Conserv 143:908–916

Kuemmerle T, Levers C, Bleyhl B, Olech W, Perzanowski K, Reusch C, Kramer-Schadt S (2018) One size does not fit all: European bison habitat selection across herds and spatial scales. Landsc Ecol 33:1559–1572

Larska M, Krzysiak MK (2019) Infectious disease monitoring of European bison (*Bison bonasus*). In: Ferretti M (ed) Wildlife population monitoring. IntechOpen, Rijeka, pp 248–269

Massilani D, Guimaraes S, Brugal J-P, Bennett EA, Tokarska M, Arbogast R-M, Baryshnikov G, Boeskorov G, Castel J-C, Davydov S, Madelaine S, Putelat O, Spasskaya NN, Uerpmann H-P, Grange T, Geigl E-M (2016) Past climate changes, population dynamics and the origin of Bison in Europe. BMC Biol 14:93

Mazurek A (2010) The assessment of ground flora and litter standing crop in selected habitats of Magurski National Park. BSc thesis, Catholic University of Lublin. [in Polish]

Melander Y (1959) The miotic chromosomes of some cavicorn mammals (*Bos taurus* L., *Bison bonasus* L. and *Ovis aries* L.). Hereditas 45:649–664

Mendoza M, Palmqvist P (2008) Hypsodonty in ungulates: an adaptation for grass consumption or for foraging in open habitat? J Zool 274:134–142

Merceron G, Hofman-Kamińska E, Kowalczyk R (2015) 3D dental microwear texture analysis of feeding habits of sympatric ruminants in the Białowieża Primeval Forest, Poland. For Ecol Manag 328:262–269

Mysterud A, Bartoń K, Jędrzejewska B, Krasiński ZA, Niedziałkowska M, Kamler JF, Yoccoz NG, Stenseth NC (2007) Population ecology and conservation of endangered megafauna: the case of European bison (*Bison bonasus*) in Białowieża Primeval Forest, Poland. Anim Conserv 10:77–87

OIE (2008) Terrestrial Animal Health Code, Seventeenth edition. World Organisation for Animal Health, Paris, France

Olech W (2003) Influence of individual inbred and mother's inbred on calves' survival in European Bison (*Bison bonasus*). Rozprawy Naukowe i Monografie, Wydawnictwo SGGW, Warszawa. [In Polish with English summary]

Olech W (2008) *Bison bonasus*. The IUCN red list of threatened species. IUCN, Gland

Orłowska A, Trębas P, Smreczak M, Marzec A, Żmudziński JF (2016) First detection of bluetongue virus serotype 14 in Poland. Arch Virol 161:1969–1972

Palacio P, Berthonaud V, Guérin C, Lambourdière J, Maksud F, Philippe M, Plaire D, Stafford T, Marsolier-Kergoat M-C, Elalouf J-M (2017) Genome data on the extinct *Bison schoetensacki* establish it as a sister species of the extant European bison (*Bison bonasus*). BMC Evol Biol 17:48

Pčola Š, Adamec M, Pčola Š (2006) Reintroduction of European bison (*Bison bonasus*) in Poloniny National Park. In: Olech W (ed) Prospects of European bison population development. Artisco, Goczałkowice-Zdrój, pp 43–54

Pertoldi C, Wójcik JM, Tokarska M, Kawałko A, Kristensen TN, Loeschcke V, Gregersen VR, Coltman D, Wilson GA, Randi E, Henryon M, Bendixen C (2010) Genome variability in European and American bison detected using the BovineSNP50 BeadChip. Conserv Genet 11:627–634

Perzanowski K, Januszczak M (2004) Preliminary assessment of the home range dynamics of the European bison *Bison bonasus* in Bieszczady Mountains. Parki Narodowe i Rezerwaty Przyrody 23:639–646. [In Polish with English summary]

Perzanowski K, Januszczak M, Wołoszyn-Gałęza A (2012) Seasonal movements of wisents (*Bison bonasus* L., 1758) in the Bieszczady Mountains (SE Poland). Biol Lett 49:11–17

Perzanowski K, Januszczak M, Wołoszyn-Gałęza A (2015) Group stability – a pilot study of a wisent herd of Bieszczady Mountains. Eur Bison Conserv Newsl 8:33–40

Plumb GE, White PJ, Coughenour M, Wallen R (2009) Carrying capacity, migration and dispersal in Yellowstone bison. Biol Conserv 142:2377–2387

Plumb GE, White PJ, Aune K (2014) American bison *Bison bison* (Linnaeus, 1758). In: Melletti M (ed) Wild cattle of the world. Cambridge University Press, Cambridge, pp 83–114

Plumb G, Kowalczyk R, Hernandez-Blanco JA (2020) *Bison bonasus*. The IUCN Red List of Threatened Species 2020: e.T2814A45156279

Pucek Z (1991) History of the European bison and problems of its protection and management. In: Bobek B, Perzanowski K, Regelin W (eds) Global trends in wildlife management. Trans. 18th IUGB Congress Kraków 1987. Świat Press, Kraków-Warszawa, pp 19–39

Pucek Z, Belousova IP, Krasińska M, Krasiński ZA, Olech W (2004) European bison: status survey and conservation action plan. UK IUCN/SSB Bison Specialist Group IUCN, Gland/Cambridge

Raczyński J (1978) European bison. Państwowe Wydawnictwo Rolnicze i Leśne, Warsaw, pp 1–246. [in Polish]

Radwan J, Demiaszkiewicz AW, Kowalczyk R, Lachowicz J, Kawałko A, Wójcik JM, Pyziel AM, Babik W (2010) An evaluation of two potential risk factors, MHC diversity and host density, for infection by an invasive nematode *Ashworthius sidemi* in endangered European bison (*Bison bonasus*). Biol Conserv 143:2049–2053

Rouys S (2003) Winter movements of European bison in the Białowieża Forest, Poland. Mamm Biol 68:122–125

Rouys S, Theuerkauf J, Krasińska M (2001) Accuracy of radio-tracking to estimate activity and distances walked by European bison in the Białowieża Forest, Poland. Acta Theriol 46:319–326

Rutberg AT (1986) Lactation and fetal sex ratios in American bison. Am Nat 127:89–94

Salwa A, Anusz K, Arent Z, Paprocka G, Kita J (2007) Seroprevalence of selected viral and bacterial pathogens in free-ranging European bison from the Białowieża Primeval Forest (Poland). Pol J Vet Sci 10:19–23

Samojlik T, Fedotova A, Borowik T, Kowalczyk R (2019) Historical data on European bison management in Białowieża Primeval Forest can contribute to a better contemporary conservation of the species. Mammal Res 64:543–557

Schmitz P, Caspers S, Warren P, Witte K (2015) First steps into the wild – exploration behavior of European bison after the first reintroduction in Western Europe. PLoS One 10(11):e0143046

Schwartz-Cornil I, Mertens PP, Contreras V, Hemati B, Pascale F, Breard E, Mellor PS, MacLachlan NJ, Zientara S (2008) Bluetongue virus: virology, pathogenesis and immunity. Vet Res 39:46

Sipko T, Trepet S, Gogan PJP, Mizin I (2010) Bringing wisents back to the Caucasus mountains: 70 years of a grand mission. Eur Bison Conserv Newsl 3:33–44

Slatis HM (1960) An analysis of inbreeding in the European bison. Genetics 45:275–287

Soubrier J, Gower G, Chen K, Richards SM, Llamas B, Mitchell KJ, Ho SYW, Kosintsev P, Lee MSY, Baryshnikov G, Bollongino R, Bover P, Burger J, Chivall D, Crégut-Bonnoure E, Decker JE, Doronichev VB, Douka K, Fordham DA, Fontana F, Fritz C, Glimmerveen J, Golovanova LV, Groves C, Guerreschi A, Haak W, Higham T, Hofman-Kamińska E, Immel A, Julien M-A, Krause J, Krotova O, Langbein F, Larson G, Rohrlach A, Scheu A, Schnabel RD, Taylor JF, Tokarska M, Tosello G, van der Plicht J, van Loenen A, Vigne J-D, Wooley O, Orlando L, Kowalczyk R, Shapiro B, Cooper A (2016) Early cave art and ancient DNA record the origin of European bison. Nat Commun 7:13158–13158

Szara T, Kobryńczuk F, Kobryń F, Bartyzel B, Nowicka A (2003) Sex dimorphism of the scapula in the European bison (*Bison bonasus* L.). Veterinarija Ir Zootechnika 23:60–62

Tokarska M (2013) Aspects of the genetics of European bison. In: Krasińska M, Krasiński Z (eds) European bison. The nature monograph, 2nd edn. Springer, Berlin/Heidelberg, pp 35–40

Tokarska M, Pertoldi C, Kowalczyk R, Perzanowski K (2011) Genetic status of the European bison *Bison bonasus* after extinction in the wild and subsequent recovery. Mammal Rev 41:151–162

Tokarska M, Bunevich AN, Demontis D, Sipko T, Perzanowski K, Baryshnikov G, Kowalczyk R, Voitukhovskaya Y, Wójcik JM, Marczuk B, Ruczyńska I, Pertoldi C (2015) Genes of the extinct Caucasian bison still roam the Białowieża Forest and are the source of genetic discrepances between Polish and Belarusian populations of the European bison, *Bison bonasus*. Biol J Linn Soc 114:752–763

Tyapougin YA, Gusarov IV (2004) Creation of population of European bison in condition of northern region of Russia. In: Krasińska M, Daleszczyk K (eds) Proceedings of the conference: European Bison Conservation 30 September–2 October 2004 Białowieża, Poland. Mammals Research Institute PAS, Białowieża, pp 137–139

Urban-Chmiel R, Wernicki A, Stęgierska D, Marczuk J, Rola J, Socha W, Valverde Piedra JL (2017) Detection of BHV-1 and BRSV viruses in European bison in the Białowieża Forest: a preliminary study. J Appl Anim Res 45:170–172

Wang K, Wang L, Lenstra JA, Jian J, Yang Y, Hu Q, Lai D, Qiu Q, Ma T, Du Z, Abbot R, Liu J (2017) The genome sequence of the wisent (*Bison bonasus*). GigaScience 6:1–5

Weaver GV, Domenech J, Thiermann AR, Karesh WB (2013) Foot and mouth disease: a look from the wild side. J Wildl Dis 49:759–785

Węcek K, Hartmann S, Paijmans JLA, Taron U, Xenikoudakis G, Cahill JA, Heintzman PD, Shapiro B, Baryshnikov G, Bunevich AN, Crees JJ, Dobosz R, Manaserian N, Okarma H, Tokarska M, Samuel T, Turvey ST, Wójcik JM, Żyła W, Szymura JM, Hofreiter M, Barlow A (2017) Complex admixture preceded and followed the extinction of wisent in the wild. Mol Biol Evol 34:598–612

Węgrzyn M, Serwatka S (1984) Teeth eruption in the European bison. Acta Theriol 29:11–121

Welz M, Anusz K, Salwa A, Zaleska M, Bielecki W, Osińska B, Kaczor S, Kita J (2005) Bovine tuberculosis in European bison in the Bieszczady region. Med Weter 61:441–444

Wójcik JM, Kawałko A, Tokarska M, Jaarola M, Valenback P, Pertoldi C (2009) Post-bottleneck mtDNA diversity in a free-living population of European bison: implications for conservation. J Zool 277:81–87

Zięba K (2007) Behaviour during parturition and breeding results of lowland European bison (*Bison bonasus bonasus*) in Pszczyna herd. In: Olech W (ed) Proceedings of the conference: the importance of ex-situ breeding for the restoration of European bison. European Bison Friends Society, Gołuchów, pp 37–60

Zielke L, Wrage-Mönnig N, Müller J, Neumann C (2019) Implications of spatial habitat diversity on diet selection of European bison and Przewalski's horses in a rewilding area. Diversity 11:63

Żurawski C, Lipiec M (1997) A case of bovine tuberculosis in European bison. Med Weter 53:90–92

Muskox *Ovibos moschatus* (Zimmermann, 1780)

13

Niels Martin Schmidt and Mikkel Stelvig

Contents

N. M. Schmidt (✉)
Aarhus University, Roskilde, Denmark
e-mail: nms@bios.au.dk

M. Stelvig
Copenhagen Zoo, Frederiksberg, Denmark
e-mail: mst@zoo.dk

© Springer Nature Switzerland AG 2022

L. Corlatti, F. E. Zachos (eds.), *Terrestrial Cetartiodactyla*, Handbook of the Mammals of Europe,
https://doi.org/10.1007/978-3-030-24475-0_29

Common Names

English	Muskox
German	Moschusochse
French	Boeuf musqué
Spanish	Buey almizclado
Italian	Bue muschiato
Russian	овцебык

Taxonomy, Systematics and Paleontology

The muskox (Fig. 1) likely originated in the Pliocene, with the fossil record suggesting radiation in the Pleistocene (Lent 1988). The extinct *Praeovibos* is regarded as the earliest known representative of the muskox group (Kahlke 2014), though there are indications of *Praeovibos* being merely early morphotypes of *O. moschatus* (Campos et al. 2010a). The extinct helmeted muskox *Bootherium bombifrons* is regarded as the closest prehistoric relative of *O. moschatus* (Bover et al. 2018). The muskox belongs to the tribe Ovibovini, subfamily Caprinae, but has no close relatives. The closest relatives are members of the genera *Capricornis* and *Naemorhedus* (Zhou et al. 2019). The muskox is considered monotypic for the genus *Ovibos* (family Bovidae, subfamily Caprinae). Two subspecies have been suggested (*O. m. moschatus* and *O. m. wardi*), but not supported by morphological nor early genetic studies (Lent 1988; Groves 1997a). Inter-breeding between the two suggested subspecies in the wild has been suggested (Groves 1997a) and observed in captive muskoxen. Recent genetic analyses, however, has revealed large genetic distance between muskoxen in the Canadian mainland (*sensu lato O. m. moschatus*) and in Greenland (*sensu lato O. m. wardi*) (Hansen et al. 2018).

Current Distribution

Today, populations of native muskoxen are found in the Canadian mainland, the Canadian Islands, and North and Northeast Greenland, but the species used to roam across the circum-Arctic region, including Alaska and Russia. The fossil record suggests that muskoxen roamed in Scandinavia from about 9000 years ago (Borgen 1979). Following regional loss of the species, reintroductions have been conducted, with muskoxen from Greenland being translocated into Alaska, and from Canada and Alaska into Russia (Thulin et al. 2011; Cuyler et al. 2020). In Scandinavia, muskoxen from Northeast Greenland have been (re)introduced into several European regions (Thulin et al. 2011; Cuyler et al. 2020), including into Svalbard in 1929, into Iceland in 1929, and into Dovrefjell in Southern Norway (Fig. 2) on several occasions, starting in the 1930s (Alendal 1980). The first Icelandic muskoxen quickly died, and the second release of muskoxen, this time from Norway, was also unsuccessful (Alendal 1980). The Svalbard population initially thrived, but went extinct in the early 1980s. Also, the Dovrefjell muskoxen went extinct during the Second World War, but the population was reestablished when more muskoxen from Northeast Greenland were released into the area in the late 1940s and early 1950s. Today the Dovrefjell population is well established. In 1971, a small group of muskoxen left the Dovrefjell population and into Sweden, where they now form a small, separate population in Härjedalen (Fig. 2).

Since their establishment, the Norwegian muskox population increased, particularly in the last 20 years, while the Swedish population after an initial increase has been declining (Fig. 3).

Fig. 1 Muskox group with two adult males to the right, and three adult females and a calf to the left (photograph © Lars Holst Hansen)

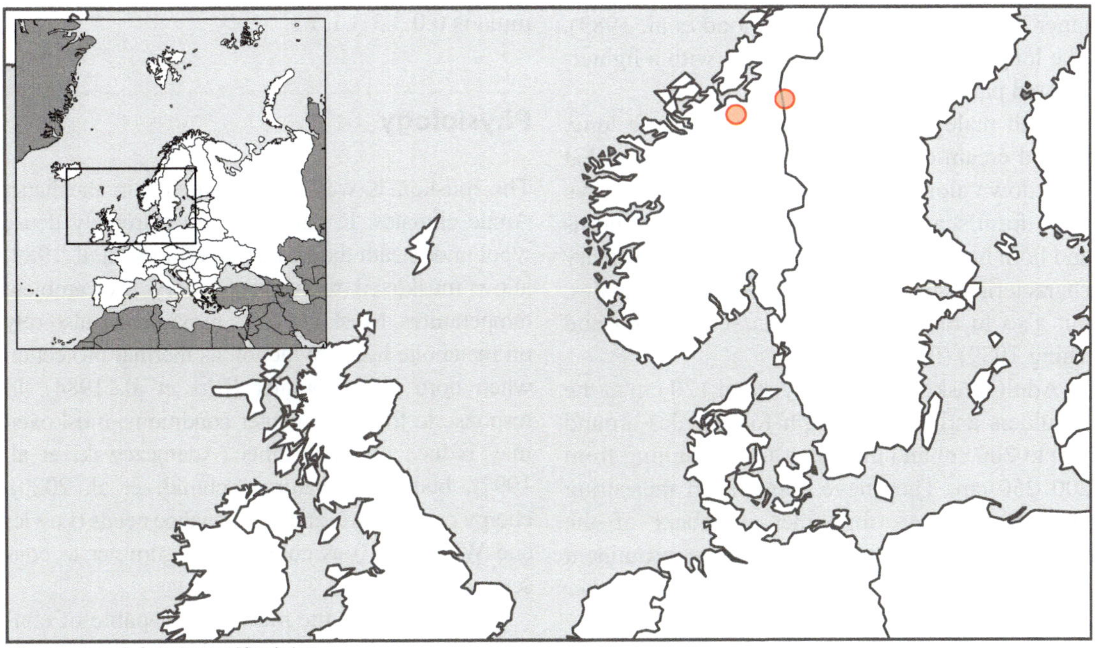

Map template: © Getty Images/iStockphoto

Fig. 2 Location of the two introduced Scandinavian muskox populations, with the Norwegian population to the left and the Swedish to the right. Due to small geographical ranges, dots only indicate overall locations. (Sources: https://miljostatus.miljodirektoratet.no/tema/arter/fremmede-arter/moskus/ and Thulin et al. (2011)) (Map template: © Copyright Getty Images/iStockphoto)

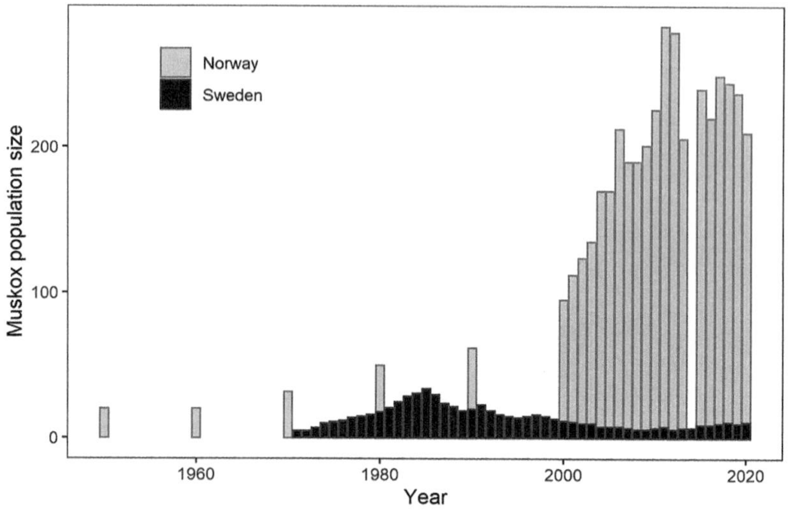

Fig. 3 Sizes of the two Scandinavian muskox populations. (Data from Norway, courtesy to https://miljostatus.miljodirektoratet.no/tema/arter/fremmede-arter/moskus/, and from Sweden to Hielke Chaudron, Myskoxcentrum)

Description

Muskoxen are barrel shaped with short legs. They are covered with fur except for the small area between the nostrils and lips. The fur consists of two types, the outer layer of guard hair and the inner layer of wool or qiviut (Flood et al. 1989). The long guard hair is dark brown with a lighter-colored patch on the back.

Both male and female muskoxen have long, curved cream-colored horns with black tips that drop down along the side of the head, then curve up to form sharp hooks. Males have larger horns and horn bases than females, which is the primary characteristic to recognize males from females, but also to distinguish age classes (Olesen and Thing 1989) (Fig. 1).

Adult muskoxen typically stand 120 cm at the shoulders and have a length for females around 180 to 200 cm and the larger males ranging from 200–250 cm. They have a small tail measuring only 10 cm. As for other members of the Caprinae, the forequarters are more prominent than the hind quarters, making muskoxen rather agile even on rugged slopes.

The average body mass for adult muskox is around 285 kg, with females ranging between 180 and 250 kg and the bigger males with a range of 300–400 kg. During winter, the muskox to a large extent relies on the large fat deposits build up over summer and autumn, and body mass therefore fluctuates seasonally across the year (Adamczewski et al. 1998).

The teeth are typical of bovids, well adapted for handling often rough forage. The dental formula is 0.0.3.3/3.1.3.3.

Physiology

The muskox is well adapted to life in the harsh Arctic climates. In particular, the extremely dense wool underneath the guard hairs (Flood et al. 1989) allow muskoxen to withstand very low ambient temperatures. Muskox calves in particular also rely on metabolic heat production as thermal protection when born in late winter (Blix et al. 1984). In response to the harsh winter conditions, muskoxen may reduce organ weights (Adamczewski et al. 1997), body temperature (Schmidt et al. 2020), energy expenditure, and maintenance needs (Lawler and White 1997) as compared to summer to conserve energy.

As a ruminant, the muskox is capable of consuming large amounts of often low-quality forage, which is then digested during rumination bouts. During the plant growing season muskoxen may however also actively select for nutrient-rich forbs and young shrub leaves. In winter, the muskox consumes forage of very low quality, but this appears not compensated for by larger consumption (Blix et al. 2012), and food intake actually appears reduced in winter (Adamczewski et al. 1994). In winter, the muskox gets water by consuming snow and ice. Warming of this cold ingesta may be energetically expensive, though the heat increment of feeding may compensate this (Crater and Barboza 2007). Being pregnant during the period of scarcity, the muskox must balance maintenance costs and fetal investment to enhance survival (Barboza et al. 2020).

Genetics

Chromosomes

The muskox has 48 chromosomes (12 bi-armed and 34 acrocentric autosomes), an acrocentric X chromosome and a small metacentric Y chromosome (Desaulniers et al. 1989).

Genetic Diversity

Across their entire distributional range, the genetic diversity in extant muskox populations is very low (Thulin et al. 2011; Hansen et al. 2018; Prewer et al. 2019) compared to the Pleistocene (Campos et al. 2010b). The species has undergone several expansions and contractions since then (Campos et al. 2010b). More recently, successive founder effects, as the species expanded from mainland Canada into the Canadian Islands, and from there into Greenland, have resulted in a gradually declining genetic diversity, resulting in an extremely low genetic diversity in muskoxen in Northeast Greenland (Hansen et al. 2018). As the Norwegian muskox population originates from this region, their genetic diversity is even lower (Thulin et al. 2011). Surprisingly, the Swedish population has a slightly higher genetic diversity than its ancestral Norwegian population, albeit still low compared to other populations (Thulin et al. 2011).

Life History

Growth

Muskox calves are born weighing approximately 10–14 kg and grow relatively fast until sexual maturity. Males generally mature at the age of 3–4 years, whereas females reach maturity between 1 and 4 years of age depending on body condition (Jingfors and Klein 1982; Gray 1990; Olesen et al. 1994; Adamczewski et al. 1998). Muskox horns develop quickly within the first 4–5 years. Horns on calves can be seen as small nubs but grows rather fast, and already yearlings have clearly visible horns. Compared to females, male muskoxen have more massive horn bases, and thicker and longer horns (Olesen and Thing 1989).

Reproduction

First calving is usually seen when females are 3 years old (Adamczewski et al. 1997), but in some areas females may calve as 2-year olds (Olesen et al. 1994). After a gestation period of 235 (Rowell et al. 1993) to 250 (Hubert 1977) days, calves are born in late winter/early spring (April–June) where low ambient temperatures may still prevail. Calves are born with only a short layer of guard hairs over the thick layer of qiviut, and therefore rely heavily on their mothers for milk, warmth, and protection (Groves 1997b). Calves are however able to consume grass and leaves already within their first weeks (Gray 1987). Females give birth to a single offspring, but twinning has been observed (Wilkinson 1971). Typically, most of the females in the herd will calve every second year (Thing et al. 1987), but if conditions are favorable muskoxen may give birth every year (Jingfors and Klein 1982).

Survival

In the wild, muskoxen have attained an age of more than 20 years (Buckley et al. 1954; Aastrup 2003). Calves and yearlings in particular are prone to high mortality rates, and there are indications of bull muskoxen to be more prone to winter mortality than females due to increased energy expenditure during the rut (Gunn et al. 1989).

Habitat and Diet

Globally, muskoxen are able to adapt to a variety of habitats, from the margins of sub-Arctic boreal forest to the high Arctic tundra. In its native range, the muskox is a tundra animal, utilizing a variety of tundra habitats (Beumer et al. 2019). In Europe, such tundra-like habitats are found in Alpine environments. In summer, muskoxen select habitats with high plant productivity (Beumer et al. 2019; Tomassini et al. 2019), while in winter they select for habitats with limited snow accumulation such as *Dryas* heaths, allowing them to access their plant forage (Nellemann 1998). Graminoids constitute an important fraction of both muskox summer and winter diets, while in winter shrubs are also important (Klein and Bay 1994; Forchhammer and Boomsma 1995; Larter and Nagy 1997; Kristensen et al. 2011). Muskox grazing activities are known to alter plant communities markedly (Post and Pedersen 2008; Mosbacher et al. 2019), even in areas where muskoxen consume only a limited fraction of the plant forage available (Mosbacher et al. 2016).

Muskoxen have small home ranges and are therefore often regarded as rather sedentary (Gustine et al. 2011; Beumer et al. 2019). Nonetheless, muskoxen may move hundreds of kilometers during the course of the year, but do so within a rather small geographical area (Schmidt et al. 2016). Both seasonal migrations and longer directional movements have however been observed (Reynolds 1998; Aastrup 2003), and the Swedish muskox population was established by a group wandering off from the Dovrefjell population in Norway and into Sweden.

Behavior

Herd Structure

Muskoxen are social, gregarious animals living in herds up to 50 animals (Gray 1987). Herds consist of both males and females, as well as yearlings and calves. Herd structure is rather loose, and herds can frequently change in both size and composition (Gray 1987). Within the herd there is a hierarchy of dominance with the males dominating the females and a single male dominating the other males (Gray 1987). However, in the snow-free season, adult females appear to initiate various behaviors more often than males (Ihl and Bowyer 2011). Males outside a herd can either group together in all-male herds or be seen alone.

Herd size varies throughout the season, being largest in winter (Schmidt et al. 2015). When disturbed, for instance by a predator, muskox gather in a tight group and creates a circle or semicircle defense formation with the front facing the enemy (Fig. 4). Calves and yearlings are placed in the center or behind the adults.

Mating Behavior

Muskoxen have a polygynous mating system, and in late summer the dominant bull starts chasing away other reproductively active males in the herd to form a harem, and thus to monopolize the females in the herd. Male dominance fights occur during the rutting period. The male-male encounters involve roaring, horning, and pawing of ground, lateral display, head-swinging displays, charging, and clashing (Gray 1984). The gland-rubbing behavior where the preorbital gland is rubbed against the foreleg is a very common behavior in these male encounters, but is also seen in other stressful situations where the male is showing agonistic behaviors. The agonistic displays can escalate into a fight where the males attack each other with repeated head-on charges at up to 40 km/h, smashing the horn bases together.

Activity Patterns

Muskox activity patterns consist of different phases cycling between resting/ruminating, foraging, and relocating, and the time allocation between these activities changes with the season (Beumer et al. 2020). During the summer, muskoxen spent most of their time foraging, intercepted by resting bouts. In winter, muskoxen spent approximately half their

Fig. 4 Muskoxen forming a protective semicircle (photograph © Lars Holst Hansen)

time foraging and half their time resting (Beumer et al. 2020). In summer, muskoxen are therefore active round the clock, whereas they in winter exhibit two to three activity peaks.

Parasites and Diseases

A number of endoparasites and diseases may impact muskoxen negatively, and in Canada major mass mortalities have been associated with the bacteria *Erysipelothrix rhusiopathiae* (Kutz et al. 2015). Also the Dovrefjell muskox population has suffered from multiple disease outbreaks, including pneumonia (Ytrehus et al. 2008; Handeland et al. 2014), contagious ecthyma (orf) (Vikøren et al. 2008) and ocular diseases (Handeland et al. 2020). Moreover, the population is subject to a high degree of polyparasitism, which in combination with the warm environment they live in challenges the health of the Dovrefjell population (Davidson et al. 2014).

Population Ecology

In native populations, muskox population dynamics is largely governed by calf recruitment, which in turn is determined by winter conditions, with snow-rich winter resulting in low calf recruitment and vice versa (Schmidt et al. 2015). In some areas, diseases (Kutz et al. 2015) and predation (Arthur and Del Vecchio 2017) may impact population dynamics markedly, whereas the importance of predation appears low in some areas (Thing et al. 1987; Adamczewski et al. 1997), and is negligible in Scandinavia. In the managed Scandinavian populations, human regulation, traffic, and disease are the major mortality factors (Gundersen et al. 2005). Noteworthy is that in comparison with introduced populations in for instance Alaska (Reynolds 1998) and Greenland (Hansen et al. 2018), the predation-free Scandinavian populations have exhibited low rates of increases, also in the periods without culling in Norway. While the causality behind these low Scandinavian growth rates is not known, the Swedish population may be living in generally suboptimal habitats (Thulin et al. 2011) and in Norway the high prevalence of disease and parasites (Davidson et al. 2014) may contribute to the observed low population growth. Concerns have been raised that muskoxen may negatively impact reindeer in Dovrefjell (Bevanger 2005) and elsewhere (Larter and Nagy 1997), through competition for resources such as forage. In Scandinavia, such competition currently appears unlikely, but suitable winter areas are limited in Dovrefjell and competition cannot be ruled out (Bevanger 2005). Muskox cratering in winter have been suggested to facilitate access to forage for other resident herbivores (Schmidt et al. 2018).

Climate change is expected to impact muskoxen in numerous ways, including increased mortality due to extreme weather and changes in exposure to parasite and disease (Cuyler et al. 2020). Compared to the native populations in Canada and Greenland, the Scandinavian muskox populations can be said to already live in future climates, experiencing many of the anticipated (negative) effects of climate change (Davidson et al. 2014).

Conservation Status

The muskox is listed as of least concern by IUCN (Gunn and Forchhammer 2008). A recent inventory of the status and trends of all wild muskox populations in the Northern hemisphere suggested a global number of some 170,000 muskoxen (Cuyler et al. 2020), a number higher than the one reported by IUCN (Gunn and Forchhammer 2008). Both muskox populations in Scandinavia originate from introduced animals and are not regarded as part of the native fauna.

Management

In Scandinavia, the muskox is considered an introduced species. In Norway, the management plan for the Dovrefjell muskox population depicts that the population is to be limited to restricted area within Dovrefjell national park and to number 200 individuals in winter (Rangbru and Seljevoll 2017). The muskox population is therefore monitored and actively regulated through culling to achieve these goals, and to minimize conflicts with humans (Gundersen et al. 2005; Rangbru and Seljevoll 2017). In Sweden, the muskox is protected from hunting but currently no real management plan exists.

Future Challenges for Research and Management

Both Scandinavian populations and in particular the Swedish population are rather small and were founded by few individuals only. The Norwegian population appears generally to thrive, though the health situation needs to be improved (Rangbru and Seljevoll 2017). The Swedish population is declining and shows sign of inbreeding (Laikre et al. 1997). One bull and one cow has been added to the Swedish population from zoos, and one free-ranging cow gave birth in the wild after being mated in a zoo. However, more long-term management actions are needed to ensure a viable population, mainly there through further introductions but potentially also relocation of the population to more suitable areas (Thulin et al. 2011).

The ongoing warming of the planet and the concomitant environmental change is expected to impact muskox populations throughout its distributional range (Cuyler et al. 2020) and thus also the Scandinavian muskox populations.

Acknowledgments We thank Anne Gunn for helpful comments and suggestions to the chapter, and Tord Bretten, Jørn Våge, Bjørnar Ytrehus, and Angelica Åsberg for contributing with knowledge, data, or feedback for this chapter.

References

Aastrup P (2003) Muskox site fidelity and group cohesion in Jameson Land, East Greenland. Polar Biol 27:50–55

Adamczewski JZ, Flood PF, Chaplin RK, Schaefer JA (1994) Seasonal variation in intake and digestion of a high-roughage diet by muskoxen. Can J Anim Sci 74: 305–313

Adamczewski JZ, Flood PF, Gunn A (1997) Seasonal patterns in body composition and reproduction of female muskoxen (*Ovibos moschatus*). J Zool 241: 245–269

Adamczewski JZ, Fargey PJ, Laarveld B, Gunn A, Flood PF (1998) The influence of fatness on the likelihood of early-winter pregnancy in muskoxen (*Ovibos moschatus*). Theriogenology 50:605–614

Alendal E (1980) Overføringer av moskusfe fra Øst-Grønland i perioden 1899–1969. Norsk Polarinstitutt Meddelelser 107:1–42

Arthur SM, Del Vecchio PA (2017) Effects of grizzly bear predation on muskoxen in northeastern Alaska. Ursus 28:81–91

Barboza PS, Shively RD, Gustine DD, Addison JA (2020) Winter is coming: conserving body protein in female reindeer, caribou, and muskoxen. Front Ecol Evol 8:150

Beumer LT, van Beest FM, Stelvig M, Schmidt NM (2019) Spatiotemporal dynamics in habitat suitability of a large Arctic herbivore: environmental heterogeneity is key to sedentary lifestyle. Glob Ecol Conserv 18: e00647

Beumer LT, Pohle J, Schmidt NM, Chimienti M, Desforges JP, Hansen LH, Langrock R, Pedersen SH, Stelvig M, van Beest FM (2020) An application of upscaled optimal foraging theory using hidden Markov modelling: year-round behavioural variation in a large arctic herbivore. Mov Ecol 8:25

Bevanger K (2005) Interaksjoner mellom moskus og villrein – en kundskabsoversikt [in Norwegian]. 1–23. NINA Raport:7

Blix AS, Grav HJ, Markussen KA, White RG (1984) Modes of thermal protection in newborn muskoxen (*Ovibos moschatus*). Acta Physiol Scand 122:443–453

Blix AS, Ness J, Lian H, Callaghan TV (2012) Experimental reduction in food quality is not compensated with increased food intake in high-arctic muskoxen (*Ovibos moschatus*). Br J Nutr 108:1337–1340

Borgen U (1979) Ett fynd av fossil myskoxe i Jämtland och något om myskoxarnas biologi och historia. Fauna Och Flora 74:1–12

Bover P, Llamas B, Thomson VA, Pons J, Cooper A, Mitchell KJ (2018) Molecular resolution to a morphological controversy: the case of North American fossil muskoxen *Bootherium* and *Symbos*. Mol Phylogenet Evol 129:70–76

Buckley JL, Spencer DL, Adams P (1954) Muskox (*Ovibos moschatus*) Longevity. J Mammal 35:456

Campos PF, Sher A, Mead JI, Tikhonov A, Buckley M, Collins M, Willerslev E, Gilbert MT (2010a) Clarification of the taxonomic relationship of the extant and extinct ovibovids, *Ovibos*, *Praeovibos*, *Euceratherium* and *Bootherium*. Quat Sci Rev 29:2123–2130

Campos PF, Willerslev E, Sher A, Orlando L, Axelsson E, Tikhonov A, Aaris-Sørensen K, Greenwood AD, Kahlke RD, Kosintsev P (2010b) Ancient DNA analyses exclude humans as the driving force behind late Pleistocene musk ox (*Ovibos moschatus*) population dynamics. Proc Natl Acad Sci 107:5675–5680

Crater AR, Barboza PS (2007) The rumen in winter: cold shocks in naturally feeding muskoxen (*Ovibos moschatus*). J Mammal 88:625–631

Cuyler C, Rowell J, Adamczewski J, Anderson M, Blake J, Bretten T, Brodeur V, Campbell M, Checkley SL, Cluff HD, Côté SD, Davison T, Dumond M, Ford B, Gruzdev A, Gunn A, Jones P, Kutz S, Leclerc LM, Mallory C, Mavrot F, Mosbacher JB, Okhlopkov IM, Reynolds P, Schmidt NM, Sipko T, Suitor M, Tomaselli M, Ytrehus B (2020) Muskox status, recent variation, and uncertain future. Ambio 49:805–819

Davidson RK, Amundsen H, Lie NO, Luyckx K, Robertson LJ, Verocai GG, Kutz SJ, Br Y (2014) Sentinels in a climatic outpost: endoparasites in the introduced muskox (*Ovibos moschatus wardi*) population of Dovrefjell, Norway. Int J Parasitol 3:154–160

Desaulniers DM, King WA, Rowell JE, Flood PF (1989) The banded chromosomes of the muskox (*Ovibos moschatus*). Can J Zool 67:1155–1158

Flood PF, Stalker MJ, Rowell JE (1989) The hair follicle density and seasonal shedding cycle of the muskox (*Ovibos moschatus*). Can J Zool 67:1143–1147

Forchhammer MC, Boomsma JJ (1995) Foraging strategies and seasonal diet optimization of muskoxen in West Greenland. Oecologia 104:169–180

Gray DR (1984) Dominance fighting in muskoxen of different social status. Biol Pap Univ Alaska Spec Rep 4: 118–122

Gray DR (1987) The Muskoxen of polar bear pass. Fitzhenry & Whiteside

Gray DR (1990) Muskox biology. International studbook for Muskox (*Ovibos moschatus*). Copenhagen Zoo:23–48

Groves P (1997a) Intraspecific variation of mitochondrial DNA of muskoxen based on control region sequences. Can J Zool 75:568–575

Groves P (1997b) Muskox. Alaska Geographic 23:56–86

Gundersen LH, Whist AC, Bretten T, Røtvei I, Klemetsen CE, Linnell JDC, Asbjørnsen E, Andersen R, Lillehaug A, Arnemo JM (2005) Dødsårsaker hos moskus på Dovrefjell 1953–2005. Norsk Veterinærtidsskrift 8:613–619

Gunn A, Forchhammer M (2008) *Ovibos moschatus*. The IUCN Red List of Threatened Species 2008: e. T29684A86066477

Gunn A, Miller FL, Mclean B (1989) Evidence for and possible causes of increased mortality of bull muskoxen during severe winters. Can J Zool 67: 1106–1111

Gustine DD, Barboza PS, Lawler JP, Arthur SM, Shults BS, Persons K, Adams LG (2011) Characteristics of foraging sites and protein status in wintering muskoxen: insights from isotopes of nitrogen. Oikos 120:1546–1556

Handeland K, Tengs T, Kokotovic B, Vikøren T, Ayling RD, Bergsjø B, Sigurðardóttir ÓG, Bretten T (2014) *Mycoplasma ovipneumoniae*-a primary cause of severe pneumonia epizootics in the Norwegian Muskox (*Ovibos moschatus*) population. PLoS One 9:e106116

Handeland K, Madslien K, Bretten T, Røtvei I, Våge J, Tengs T (2020) *Mycoplasma conjunctivae*-Associated Keratoconjunctivitis in Norwegian Muskox (*Ovibos moschatus*). J Wildl Dis 56:489–491

Hansen CCR, Hvilsom C, Schmidt NM, Aastrup P, Van Coeverden de Groot PJ, Siegismund HR, Heller R (2018) The muskox lost a substantial part of its genetic diversity on its long road to Greenland. Curr Biol 28: 4022–4028

Hubert BA (1977) Estimated productivity of muskox on Truelove Lowland. In: Bliss LC (ed) Truelove lowland. A High Arctic Ecosystem. The University of Alberta Press, Devon Island, pp 467–491

Ihl C, Bowyer RT (2011) Leadership in mixed-sex groups of muskoxen during the snow-free season. J Mammal 92:819–827

Jingfors KT, Klein DR (1982) Productivity in recently established muskox populations in Alaska. J Wildl Manag 46:1092–1096

Kahlke RD (2014) The origin of Eurasian mammoth faunas (*Mammuthus-Coelodonta* faunal complex). Quat Sci Rev 96:32–49

Klein DR, Bay C (1994) Resource partitioning by mammalian herbivores in the high Arctic. Oecologia 97: 439–450

Kristensen DK, Kristensen E, Forchhammer MC, Michelsen A, Schmidt NM (2011) Arctic herbivore diet can be inferred from stable carbon and nitrogen isotopes in C-3 plants, faeces and wool. Can J Zool 89: 892–899

Kutz S, Bollinger T, Branigan M, Checkley S, Davison T, Dumond M, Elkin B, Forde T, Hutchins W, Niptanatiak A (2015) *Erysipelothrix rhusiopathiae* associated with recent widespread muskox mortalities in the Canadian Arctic. Can Vet J/La Revue Veterinaire Canadienne 56: 560–563

Laikre L, Ryman N, Lundh NG (1997) Estimated inbreeding in a small, wild muskox *Ovibos moschatus* population and its possible effects on population reproduction. Biol Conserv 79:197–204

Larter NC, Nagy JA (1997) Peary caribou, muskoxen and Banks Island forage: assessing seasonal diet similarities. Rangifer 17:9–16

Lawler JP, White RG (1997) Seasonal changes in metabolic rates in muskoxen following twenty-four hours of starvation. Rangifer 17:135–138

Lent PC (1988) *Ovibos moschatus*. Mamm Species 302: 1–9

Mosbacher JB, Kristensen DK, Michelsen A, Stelvig M, Schmidt NM (2016) Quantifying muskox plant biomass removal and spatial relocation of nitrogen in a High Arctic tundra ecosystem. Arct Antarct Alp Res 48:229–240

Mosbacher JB, Michelsen A, Stelvig M, Hjermstad-Sollerud H, Schmidt NM (2019) Muskoxen modify plant abundance, phenology, and nitrogen dynamics in a High Arctic fen. Ecosystems 22:1095–1107

Nellemann C (1998) Habitat use by muskoxen (*Ovibos moschatus*) in winter in an alpine environment. Can J Zool 76:110–116

Olesen CR, Thing H (1989) Guide to field classification by sex and age of the muskox. Can J Zool 67:1116–1119

Olesen CR, Thing H, Aastrup P (1994) Growth of wild muskoxen under two nutritional regimes in Greenland. Rangifer 14:3–10

Post E, Pedersen C (2008) Opposing plant community responses to warming with and without herbivores. Proc Natl Acad Sci 105:12353–12358

Prewer E, Kutz S, Leclerc LM, Kyle CJ (2019) Already at the bottom? Demographic declines are unlikely further to undermine genetic diversity of a large Arctic ungulate: muskox, *Ovibos moschatus* (Artiodactyla: Bovidae). Biol J Linn Soc 129:459–469

Rangbru B, Seljevoll KV (2017) Forvaltningsplan for moskusbestanden på Dovrefjell. Rapport nr. 4, 1–37. Fylkesmannen i Sør-Trøndelag, Miljøvernavdelingen

Reynolds PE (1998) Dynamics and range expansion of a reestablished muskox population. J Wildl Manag 62: 734–744

Rowell JE, Pierson RA, Flood PF (1993) Endocrine changes and luteal morphology during pregnancy in muskoxen (*Ovibos moschatus*). J Reprod Fertil 99: 7–13

Schmidt NM, Pedersen SH, Mosbacher JB, Hansen LH (2015) Long-term patterns of muskox (*Ovibos moschatus*) demographics in high arctic Greenland. Polar Biol 38:1667–1675

Schmidt NM, van Beest FM, Mosbacher JB, Stelvig M, Hansen LH, Grøndahl C (2016) Ungulate movement in an extreme seasonal environment: year-round movement patterns of high-arctic muskoxen. Wildl Biol 22: 253–267

Schmidt NM, Mosbacher JB, Vesterinen EJ, Roslin T, Michelsen A (2018) Limited dietary overlap amongst resident Arctic herbivores in winter: complementary insights from complementary methods. Oecologia 187:689–699

Schmidt NM, Grøndahl C, Evans AL, Desforges J-P, Blake J, Hansen LH, Beumer LT, Mosbacher JB, Stelvig M, Greunz EM, van Beest FM (2020) On the interplay between hypometabolism and reproduction in a high arctic ungulate. Sci Rep U K 10:1514

Thing H, Klein DR, Jingfors K, Holt S (1987) Ecology of muskoxen in Jameson Land, Northeast Greenland. Holarct Ecol 10:95–103

Thulin CG, Englund L, Ericsson G, Spong G (2011) The impact of founder events and introductions on genetic variation in the muskox *Ovibos moschatus* in Sweden. Acta Theriol 56:305–314

Tomassini O, van Beest FM, Schmidt NM (2019) Density, snow and seasonality lead to variation in muskox habitat selection during summer. Can J Zool 97: 997–1003

Vikøren T, Lillehaug A, Åkerstedt J, Bretten T, Haugum M, Tryland M (2008) A severe outbreak of contagious ecthyma (orf) in a free-ranging musk ox (*Ovibos*

moschatus) population in Norway. Vet Microbiol 127: 10–20
Wilkinson PF (1971) The first verifed occurrence of twinning in the muskox. J Mammal 52:238
Ytrehus B, Bretten T, Bergsjo B, Isaksen K (2008) Fatal pneumonia epizootic in musk ox (*Ovibos moschatus*) in a period of extraordinary weather conditions. EcoHealth 5:213–223
Zhou M, Yu J, Li B, Ouyang B, Yang J (2019) The complete mitochondrial genome of *Budorcas taxicolor tibetana* (Artiodactyla: Bovidae) and comparison with other Caprinae species: insight into the phylogeny of the genus *Budorcas*. Int J Biol Macromol 121:223–232

Further Reading

Gunn A, Adamczewski J (2003) Muskox. In: Feldhamer GA, Thompson BC, Chapman JA (eds) Wild mammals of North America. The Johns Hopkins University Press, Baltimore
Gray DR (1987) The muskoxen of polar bear pass. Fitzhenry & Whiteside
Lent PC (1999) Muskoxen and their hunters. University of Oklahoma Press
Berger J (2018) Extreme conservation – life at the edges of the world. The University of Chicago Press

Northern Chamois *Rupicapra rupicapra* (Linnaeus, 1758) and Southern Chamois *Rupicapra pyrenaica* Bonaparte, 1845

14

Luca Corlatti, Juan Herrero, Francesco Ferretti, Pia Anderwald, Ricardo García-González, Sabine E. Hammer, Carlos Nores, Luca Rossi, and Sandro Lovari

Contents

L. Corlatti (✉)
ERSAF - Stelvio National Park, Office of Scientific Research and Conservation, Bormio (SO), Italy

Chair of Wildlife Ecology and Management, University of Freiburg, Freiburg, Germany
e-mail: luca.corlatti@ersaf.lombardia.it; luca.corlatti@wildlife.uni-freiburg.de

J. Herrero
Technical School, University of Zaragoza, Huesca, Spain
e-mail: herreroj@unizar.es

F. Ferretti · S. Lovari
Research Unit of Behavioural Ecology, Ethology and Wildlife Management, Department of Life Sciences, University of Siena, Siena, Italy
e-mail: ferretti13@unisi.it; sandro.lovari@gmail.com

P. Anderwald
Swiss National Park, Zernez, Switzerland
e-mail: pia.anderwald@nationalpark.ch

R. García-González
Instituto Pirenaico de Ecología (IPE-CSIC), Jaca, Spain
e-mail: ricardogarciagonzalez15@gmail.com

S. E. Hammer
Institute of Immunology, University of Veterinary Medicine, Vienna, Austria
e-mail: sabine.hammer@phylo-dat.net

C. Nores
INDUROT, University of Oviedo, Mieres, Asturias, Spain
e-mail: cnores@uniovi.es

L. Rossi
Department of Veterinary Sciences, University of Turin, Turin, Italy
e-mail: luca.rossi@unito.it

© Springer Nature Switzerland AG 2022

L. Corlatti, F. E. Zachos (eds.), *Terrestrial Cetartiodactyla*, Handbook of the Mammals of Europe,
https://doi.org/10.1007/978-3-030-24475-0_30

Common Names

	Rupicapra rupicapra	*Rupicapra pyrenaica*
English	Northern chamois	Southern chamois
German	Alpengämse	Pyrenäen-Gämse
French	Chamois	Isard
Spanish	Rebeco septentrional	Rebeco meridional
Italian	Camoscio settentrionale	Camoscio meridionale
Russian	Северная серна	Южная серна

Taxonomy and Systematics

Chamois are mountain caprines (Family Bovidae, Subfamily Caprinae) that inhabit most of the medium to high altitude mountain ranges of Southern Europe and the Near East. The taxonomy of the genus has been subject to continuous revisions since the beginning of the twentieth century. Neumann (1899) described *R. rupicapra* on the Alpine arch and *R. ornata* on the Central Apennines. Later on, Lydekker (1913) lumped them into one species, *R. rupicapra*, but Camerano (1914) proposed a third species, *R. pyrenaica*, from the Cantabrian Mountains and the Pyrenees. *R. ornata* and *R. pyrenaica* share a number of striking morphologic traits, which were ignored by Couturier (1938) and Dolan (1963) who supported Lydekker's view. Almost 20 years later, Lovari and Scala (1980) reanalyzed Couturier's biometric data and, with further behavioral (Lovari 1987) and genetic (Nascetti et al. 1985) information, provided support for the existence of two species of chamois, *R. pyrenaica* and *R. rupicapra*, later on confirmed by Hammer et al. (1995) using molecular investigations (Table 1). Eventually, Lovari and Scala (1980) and Masini and Lovari (1988) suggested that a first wave of chamois ancestors may have arrived in Europe from Central Asia during the Middle Pleistocene, some 250–150 kya, reaching its lowest latitudes on the Iberian Peninsula and in Central-Southern Italy (but see details in the section "Paleontology"). Later on, 80–60 kya, a further wave of chamois from Asia Minor or Central-Eastern Europe could have moved westward and southward during a cold period. This picture has held for more than two decades, although alternative scenarios were proposed on the basis of the latest DNA data (cf. section "Phylogeny and Phylogeography" for further details). Currently, the most accepted classification of chamois considers two species, the Southern chamois *R. pyrenaica* and the Northern chamois *R. rupicapra* (Fig. 1) (Wilson and Reeder 2005; Corlatti et al. 2011): *Rupicapra pyrenaica* (with the subspecies *parva* Cabrera, 1911, *pyrenaica* Bonaparte, 1845, and *ornata* Neumann 1899) from South-Western Europe, and *R. rupicapra* (with the subspecies *cartusiana* Couturier 1938, *rupicapra* Linnaeus, 1758, *tatrica* Blahout, 1971, *carpatica* Couturier 1938, *balcanica* Bolkay, 1925, *asiatica* Lydekker, 1908, and *caucasica* Lydekker, 1910) from Central-Eastern Europe. Based on this classification,

Table 1 Main differences between the Northern and the Southern species of chamois (modified from Masini and Lovari 1988)

	Significant differences	
	Rupicapra rupicapra	*Rupicapra pyrenaica*
Craniometrics		
Ethmoidal vacuity	Usually present	Always absent
Distance between horns at base	Larger	Smaller
Skull length	Larger	Smaller
Skull width (parietals)	Larger	Smaller
Angle horn cores/ frontal bones	Smaller	Greater[a]
Genetics	(further details: cf. "Genetics")	
Behavior	Three courtship patterns: fore-foot stamping (cf. herding, Krämer 1969), poke, and kick	Two courtship patterns: bunting and low stretch
Morphology	Winter coat pattern with three small spots	Winter coat pattern with five large and one small spots

[a]Quantitatively assessed for *R. p. ornata* only

Fig. 1 Northern male chamois (left) and Southern female chamois (right) (photographs by Denis Bertanzetti and Javier Ara)

Rupicapra rupicapra and *Rupicapra pyrenaica* are described in individual sections.

Paleontology

From the Late Miocene to the Early Pliocene, the Villafranchian and the Middle Pleistocene, the Rupicaprini dispersed from Asia – where it originated – to Europe (*Rupicapra*) and to North America (*Oreamnos*). The Villafranchian small-sized chamois-antelope *Procamptoceras brivatense* is the closest form to the unknown ancestor of chamois. According to Kurtén (1968), chamois appeared suddenly during the Middle to the Late Pleistocene, in Eastern Europe. In fact, one of the oldest known fossils of chamois in Europe, found in the Arago cave in the Southern French Pyrenees, dates back 440 kya and it was associated to *Rupicapra pyrenaica* (Rivals 2004). Recently, however, it has been suggested that the arrival of *Rupicapra* in Europe likely dates back to nearly 800 kya (Fernandez and Crégut 2007); the existence of chamois in Eastern Europe can thus be considered probable since the end of the Early Pleistocene (Crégut-Bonnoure and Dimitrijevic 2006). Most likely, the two extant and closely related species, the Northern *Rupicapra rupicapra* and the Southern *Rupicapra pyrenaica* chamois, existed already at the beginning of the Würm glaciations, some 80 kya, respectively, from the Iberian Peninsula to the Central-Southern Apennines and from the Alpine arch to the Caucasus (Masini and Lovari 1988). While the Southern chamois may have differentiated in Western Europe from ancestors migrated in the Middle Pleistocene, the Northern species possibly originated in Eastern Europe, moving west after the advent of the dry climatic waves in the East Mediterranean and Pontic regions. However, the latter must have failed to reach Southern Europe, where the former has survived in refuge areas (Masini and Lovari 1988). There are differences in the morphology, biometrics, behavior, and genetics (Table 1, Fig. 2) which separate the Southern and the Northern chamois at a specific level, although some recent genetic information has failed to confirm the above scenario (cf. section "Phylogeny and Phylogeography" for further details).

Current Distribution

Excellently adapted for life in rocky terrains and cold climates, the chamois has colonized most of the main mountain massifs in Europe. The lack of continuity of these habitats and overhunting in post-Neolithic times (Baumann et al. 2005) have highly fragmented its current distribution (Fig. 3).

The Alpine chamois *R. r. rupicapra* is the most abundant subspecies with nearly 500,000 individuals shared among France (*c.* 103,000 ind.), Italy (*c.* 137,000 ind.), Switzerland (*c.* 91,000 ind.), Liechtenstein (*c.* 800 ind.), Germany (*c.* 21,000 ind.), Austria (*c.* 130,000 ind.: R. Reiner 2021 pers. comm.), Slovenia (*c.* 12,000 ind.), and Croatia (*c.* 400 ind.). Introduced to New Zealand in 1907 and 1914 as a gift from the Austrian Emperor, it has spread over much of the mountainous areas of South

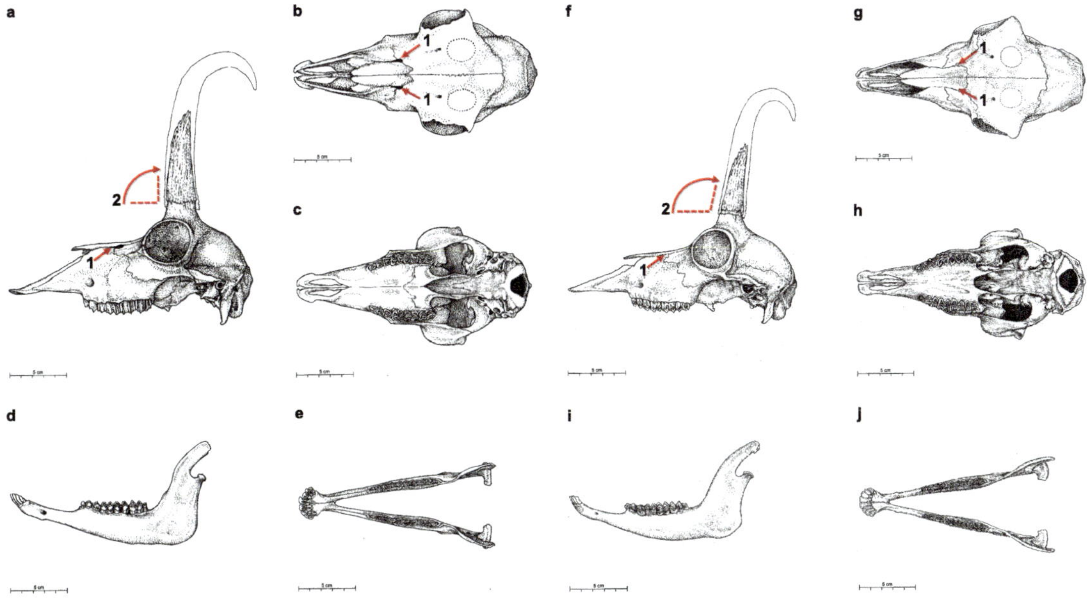

Fig. 2 Left (**a–e**): skull of Northern chamois, *Rupicapra rupicapra*. Right (**f–j**): skull of Southern chamois, *Rupicapra pyrenaica*. Letters indicate the lateral (**a**, **f**), upper (**b**, **g**), and lower (**c**, **h**) view of the upper skull; and the lateral (**d**, **i**) and upper (**e**, **j**) view of the lower jaw. Numbers indicate the ethmoidal vacuity (1), usually present in Northern chamois, always absent in Southern chamois, and the angle horn cores/frontal bones (2), smaller in Northern chamois than in Southern chamois

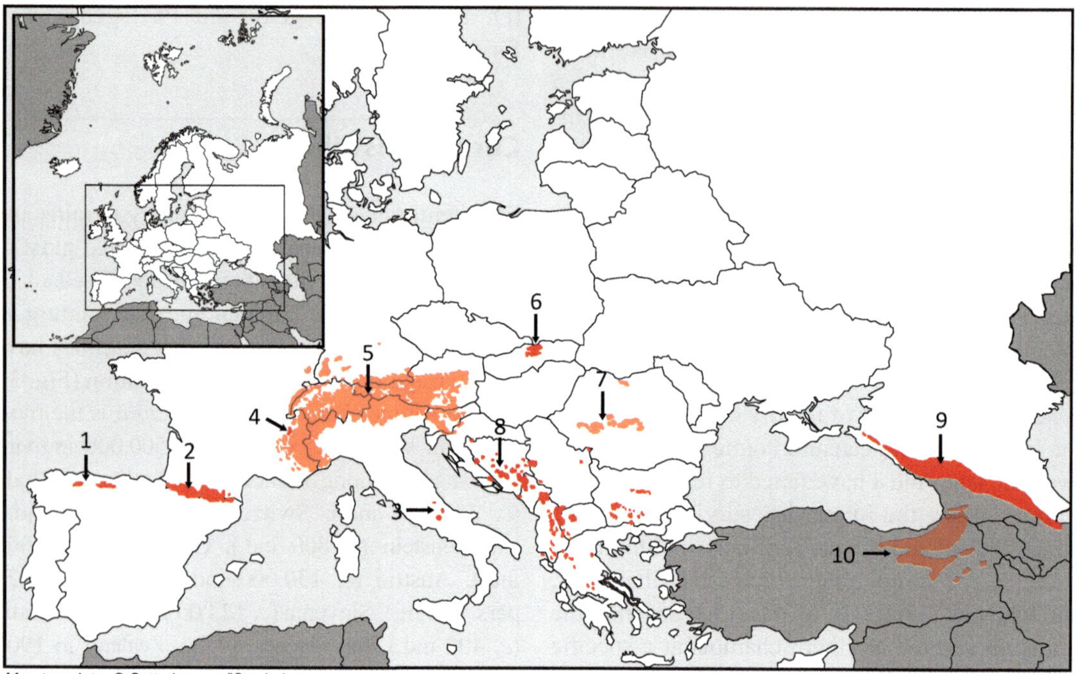

Map template: © Getty Images/iStockphoto

Fig. 3 Distribution range of *Rupicapra* spp. *Rupicapra pyrenaica*: (1) *parva*, (2) *pyrenaica*, (3) *ornata*; *Rupicapra rupicapra*: (4) *cartusiana*, (5) *rupicapra*, (6) *tatrica*, (7) *carpatica*, (8) *balcanica*, (9) *caucasica*, and (10) *asiatica*. Distribution was modified from the IUCN Red List of Threatened Species (Version 2008) based on data from the authors (Map template: © Copyright Getty Images/iStockphoto)

Island where its populations are currently estimated at *c.* 18,000 individuals. Alpine chamois was also introduced to Czech Republic and Slovakia between 1911 and 1924, where currently 350 and 120 individuals are estimated, and to Argentina (Province of Neuquén) for which recent data are lacking, though the population is likely extinct (Guichón et al. 2016). An unsuccessful attempt to introduce Alpine chamois to Norway was conducted in 1863 (Couturier 1938). The Chartreuse chamois *R. r. cartusiana* lives with some 1,500 individuals on the Chartreuse massif in South-Eastern France, possibly including hybrids with Alpine chamois (Iacolina et al. 2019). On the High Tatras and Belianske Tatras, between Poland and Slovakia, live approximately 1,400 individuals of the Tatra chamois *R. r. tatrica* (Ciach and Pęksa 2018). Approximately 9,000 individuals of the Balkan chamois *R. r. balcanica* inhabit the mountainous regions of a wide area including Croatia (*c.* 500 ind.: Šprem and Buzan 2016), Bosnia and Herzegovina (less than 1,000 ind.: cf. Korjenić et al. 2009), Serbia (*c.* 700 ind.: D. Gačić pers. comm.), Montenegro (*c.* 1,400 ind.: Đurović 2018), North Macedonia (*c.* 1,400 ind.: V. Maletić pers. comm.), Kosovo (*c.* 200–300 ind.), Albania (*c.* 450–600 ind.: F. Bego pers. comm.), Bulgaria (*c.* 2,500 ind.: Markov et al. 2016), and Greece (*c.* 1,500 ind.: Papaioannou et al. 2019). The Carpathian chamois *R. r. carpatica* inhabits the Maramureş massif and the Carpathians, both in Romania, with some 8,000 individuals. About 600 Anatolian chamois *R. r. asiatica* inhabit the mountainous areas of North-Eastern Turkey, and another 200 are estimated to live in Georgia. Caucasian chamois *R. r. caucasica* populations were estimated at *c.* 9,000 individuals in the late 2010s on the Caucasus range between Armenia, Georgia, Azerbaijan, and the Russian Federation, and are likely declining at the present time. The Pyrenean chamois *R. p. pyrenaica* occurs on the Pyrenees with *c.* 53,000 individuals shared among Spain, France, and Andorra, while on the Cantabrian mountains (Northern Spain), the populations of Cantabrian chamois *R. p. parva* are estimated at *c.*

16,000 individuals, divided into three subpopulations. The Apennine chamois *R. p. ornata* currently occurs in three main populations along the Apennine chain in Italy, with no less than 2,500 individuals (Antonucci and Di Domenico 2015). Unless otherwise indicated, data are taken from Anderwald et al. (2021), Herrero et al. (2008, 2020), Damm and Franco (2014), and Castelló (2016).

Northern Chamois *Rupicapra rupicapra* (Linnaeus, 1758)

Description

Size and Morphology

The Northern chamois is a medium-small goat-antelope of robust forms (Fig. 1). Morphological differences among subspecies are not obvious and mainly relate to craniometric measures. In all subspecies, sexual dimorphism is weak.

Biometric measures in Northern chamois may show pronounced variations depending on local characteristics. Fully grown individuals may reach a height of 80–85 cm in males and of 75–80 cm in females, and a full body length of 125–135 cm (Couturier 1938). Age- and sex-specific body mass data are abundant for the Alpine subspecies and almost entirely derive from hunted individuals: confounding factors such as shooting date and treatment of carcasses may thus lead to highly variable raw data. Garel et al. (2009) analyzed nearly 5,000 individuals hunted in the French Alps, adjusting eviscerated carcass mass data to October 15: the sex- and age-specific full body mass resulting from their growth model is reported in Table 2 (0.72 is used as a factor to convert eviscerated body mass to full carcass mass, Garel et al. 2009). The table also reports unadjusted live body mass data from Couturier (1938) and converted (by 0.72) dressed mass data from Bassano et al. (2003). Seasonal body mass variations are very pronounced in chamois, and so are variations in (male-biased) sexual mass dimorphism, that may reach a maximum of *c.* 40% in late October but declines to about 4% in spring (Rughetti and Festa-Bianchet 2011).

Both sexes possess short ebony dark horns that curve sharply backward towards the tip (Fig. 4). Horns are visible by the second month of life, and their growth pattern, which largely depends on the geological substrate and related ecological conditions that influence food availability (Chirichella et al. 2013b), is summarized in Table 2 for the Alpine subspecies (Bassano et al. 2003; Corlatti et al. 2015c). The two sexes show limited differences in horn length (Table 2). Horn growth normally halts between November and March,

Fig. 4 Horns of male (left) and female (right) Alpine chamois (photograph by Bogna Sudolska)

Table 2 Sex- and age-dependent values of live body mass and horn length for Alpine chamois

	Males						Females			
Age (in months)	6	18	30	42	54	>66[a]	6	18	30	>42[a]
Body mass (in kg ± SD)[b]										
Garel et al. (2009)	14.2	28.1	34.9	38.1	39.7	41.1	16.1	27.0	29.9	31.0
Bassano et al. (2003)	NA	20.4 ± 4.2	24.9 ± 4.7	33.2 ± 6.1	36.7 ± 5.4	~38.1 ± 6.3	NA	20.4 ± 2.6	22.8 ± 4.2	~27.9 ± 4.6
Couturier (1938)	12–17	21–26	26–33	35–44	35–44	35–44	12–17	18–20	22–25	26–28
Horn length (in cm ± SD)										
Corlatti et al. (2015c)	NA	16.9 ± 1.7	20.9 ± 1.5	22.3 ± 1.6	22.5 ± 1.7	~22.6 ± 1.8	NA	14.3 ± 1.6	18.1 ± 1.5	~20.1 ± 1.6
Bassano et al. (2003)	NA	15.6 ± 1.9	18.2 ± 2.2	21.5 ± 1.8	22.9 ± 2.1	~23.0 ± 1.8	NA	13.7 ± 1.5	15.6 ± 2.6	~19.1 ± 1.7

[a]Asymptotic body mass and horn length, assumed to be reached at 3.5 years in females and at 5.5 years in males (Bassano et al. 2003; Garel et al. 2009)
[b]Body mass estimated in October

creating sharp annuli between consecutive growth segments that allow for accurate age estimation (Schröder and von Elsner-Schack 1985).

Information on biometric measures of other subspecies is poor: available data suggest values similar to those in Table 2, except for greater body mass and horn length (by about 25%) reported for Carpathian chamois (Couturier 1938).

In fully grown individuals, total skull length is 200–220 mm and total width is 100–110 mm (Couturier 1938). Differences in skull length and width, alongside other craniometric features such as the presence of ethmoidal vacuity and the angle between horn cores and frontal bones, have been used to help discriminate between chamois species (Masini and Lovari 1988; cf. Table 1, Fig. 2).

Robust and relatively short limbs, the semi-flexed posture of the main limb segments and the peculiar foot structure – including long and independent fingers connected by a skin fold – make chamois performant on steep, rugged but also snow-covered terrains (Couturier 1958).

Most characteristic in chamois is the presence of supraoccipital oval-shaped glands behind the horns. These scent glands are visible from 16 months of age and are present in both sexes, although they are seasonally much bigger in males than in females (Couturier 1938; Tosi et al. 1990). Their maximum size (3.0–3.5 cm high, 2.5–3.0 cm wide: Couturier 1938) is reached during rut. In males, the secretion of sebum from supraoccipital glands is associated with specific displays of dominance such as "marking" (cf. section "Communication" and Lovari 1985 for the Apennine chamois). Smaller scent glands are also present in the interdigital region.

Pelage

The color of the coat varies seasonally, from reddish brown in summer to dark brown/black in winter. Spring molt starts in March and completes by June, winter coat develops in August/September and is complete by November. Most characteristic in Northern chamois are white patches on the sides of the head, with pronounced black stripes running from the eyes to the muzzle, whose potential functions are unknown, although their "blackness" might signal dominance, at least in females (Corlatti and Sivieri 2020). Other peculiarities include a white rump, black stockings, and a black mane on the backbone which, in the cold season, can grow over 25 cm in males (Couturier 1938). Cases of albinism and melanism (either full or partial) are frequently reported (Couturier 1938).

Dentition

Milk teeth 0.0.3.0/3.1.3.0; permanent dentition 0.0.3.3/3.1.3.3. Sequence of eruption and tooth wear may be used to estimate age but, as in other bovids, enumeration of horn rings is preferable.

Age and Sex Determination

The age of captured individuals can be determined by counting the boundaries between consecutive horn annuli, which form from the interruption of horn growth (Schröder and von Elsner-Schack 1985; Corlatti et al. 2015c). Horn growth stops in November – when the animal is x years and 6 months old – and normally starts again in March – when the animal is x years and 10 months old. Age determination in the field relies on both morphological and behavioral characteristics. Kids (less than 1 year of age) are conspicuously smaller than adults and always stay with their mother, though they may form juvenile groups ("kindergarten"). Yearlings (between 1 and 2 years of age) have a body size intermediate between kids and adults, and horns seldom exceed the ear length. They are mainly found in female groups. Subadults (*c.* 2–3 years old) have horns about 1.5 times longer than the ears, and forelegs appear relatively longer than in adults. Adults (*c.* 4+ years old) are fully grown, and exact age determination in the field becomes impossible. In old individuals (from *c.* 12–14 years of age), the pelage color becomes lighter, and the behavior more solitary (Couturier 1938). Sexual dimorphism in chamois is weak, and sex determination in the field (possible only from the age of 2+) relies on the combination of several behavioral and morphological characteristics. The former mainly include urination posture, behavioral displays during the rut (cf. section "Communication") and social behavior: subadult females tend to stay with nursery groups, possibly including their mother, while males may form small unisexual groups; adult males tend to be solitary, females in groups. The latter mainly include noticeable penile brush in males, and horn features: male horns tend to be thicker and more hooked, while female horns tend to be more parallel towards the base (Fig. 4).

Physiology

Chamois are well adapted to life at high altitude and low temperatures. They cope with low air oxygen by means of voluminous lungs, a powerful heart (300–350 g in an adult male) and a high number of red blood cells (well above ten million / ml) (Catusse et al. 1996). In addition, hemoglobin shows more efficient oxygen-binding properties in chamois than in lower-altitude ruminants (Ascenzi et al. 1993). Winter cold is counteracted with a thick and dark insulating coat composed of three layers, including longer guard hair and much denser down hair than in the summer coat (Catusse et al. 1996). Recently, an energy-saving mechanism – implying diminished heart rate and body temperature in winter – has been revealed (Arnold 2020), and the lower critical temperature of the thermoneutral zone has been identified at *c.* -10 °C (Haymerle 2013).

The digestive tract of chamois is well adapted to seasonal changes in diet quality, typical of the mountain context. Striking variations in the capacity of forestomach and cecum and in the shape and volume of rumen papillae, hence in the surface of rumen adsorptive mucosa, have been described. These variations allow chamois to shift from efficient browsers during the vegetative phase, to almost pure grazers during winter (Hofmann 1984). Body fat reserves accumulated in summer therefore act as an important metabolic fuel in winter (Arnold 2020). A further adaptation to winter harshness and short vegetative period is the rapid growth of kids, mediated by the high content of fat and protein in milk, approximately three- and twofold higher than in domestic goat *Capra hircus*, respectively (Gibert 2017).

Stress physiology also reveals adaptations to cold climates. Unlike in ungulates less adapted to low temperatures (e.g., red deer *Cervus elaphus*, see Huber et al. 2003), in Northern chamois climatic stressors seem to have minor effects on fecal cortisol metabolite levels (Corlatti et al. 2014). Glucocorticoid levels in male Alpine chamois appear to be mainly influenced by age (positive relationship) and social status (greater in dominant males) during the rutting season (Corlatti et al. 2014; Corlatti 2018). In Tatra chamois, fecal cortisol metabolite levels increased in presence of tourist disturbance (Zwijacz-Kozica et al. 2013). In Alpine chamois, the occurrence of environmental stressors in early life (i.e., high population density, harsh winter, and lower forage quality) is reflected in deviations from bilateral symmetry in horns (Chirichella et al. 2020).

Genetics

Chromosomes

$2n = 58$. Cytogenetic studies of somatic cells in the chamois revealed a karyotype with 54 acrocentric chromosomes and with one pair of large metacentrics comprising the autosomes (Gallagher and Womack 1992).

Phylogeny and Phylogeography

Depending on the molecular marker system, phylogenetic studies present contrasting patterns of differentiation pointing towards a complex evolutionary picture for the genus *Rupicapra*. The monophyly or polyphyly of *Rupicapra*, or the number of its species and subspecies, is still under debate (reviewed in Corlatti et al. 2011; Pérez et al. 2017). To date, the most widely accepted taxonomic classification considers two species, *R. pyrenaica* (with three subspecies) from South-Western Europe and *R. rupicapra* (with seven subspecies) from Central-Eastern Europe (e.g., Masini and Lovari 1988; Wilson and Reeder 2005). Cf. section "Taxonomy and Systematics" for further details.

Phylogenies based on mitochondrial DNA (mtDNA) sequences reflect three monophyletic groups (from the Early Pleistocene, 1.9 mya) with a clear geographical signal, suggesting that there could be more than two nominal species (Rodríguez et al. 2009, 2010; Crestanello et al. 2009). Consequently, modern chamois would have occurred in Europe well before the Late to Middle Pleistocene, contrary to available paleontological evidence. The extremely low genetic diversity of multiple autosomal introns (Pérez et al. 2017) and Y-chromosome markers (Pérez et al. 2011) might support the one species classification proposed by Couturier in 1938. In contrast, genetic studies using microsatellites (Pérez et al. 2002) and the melanocortin-1 receptor gene (MC1R) (Pérez et al. 2013) seem to favor the current taxonomy by defining at least two separated clusters corresponding to *R. rupicapra* and *R. pyrenaica*. Recent molecular analyses of complete mitochondrial genomes (Iacolina et al. 2021) confirm the previously reported genus subdivision in three clades (Pérez et al. 2017): two coincident with classical species (*R. pyrenaica* and *R. rupicapra*) and a third group composed by the subspecies *R. p. ornata* and *R. r cartusiana*.

The evolutionary background of the genus *Rupicapra* remains unclear. The use of markers with different modes of evolution may result in conflicting phylogenies, especially when hybridization between divergent lineages is involved. The differences in patterns of variation among Y-chromosome, mtDNA, biparental microsatellites, and other nuclear markers reflect the evolutionary characteristics of the different markers as well as the effects of sex-biased dispersal and species phylogeography. Consequently, more comprehensive marker systems and indicators are needed to widen the view of the diversification processes of the genus *Rupicapra* and to provide evidence for putative taxonomic revisions.

Genetic Diversity

The zoological, historical, and conservation value of geographic populations of chamois is beyond any molecular dispute, regardless of the taxonomic status assigned by scientists and of the degree of genetic diversity (Table 3) (cf. section "Current Distribution" for further details). The Chartreuse population bears mitochondrial haplotypes from the central genetic lineage (*sensu* Rodríguez et al. 2010) but microsatellite alleles from the Eastern clade. Intermediate morphometric features between Iberian and Alpine chamois were previously observed in *R. r. cartusiana* by Lovari and Scala (1980). They attributed them to hybridization events which may have occurred

Table 3 Summary of genetic variability of *Rupicapra rupicapra* subspecies

Species	Subspecies	N	nh	% π	% h	% He	% Ho	A	Source
R. rupicapra	*rupicapra* (W)	20 (18)	9	1.756	90.8	58.1	52.7	4.85	Rodríguez et al. (2010)
		22 (22)	3	3.740	54.0	58.0	59.0	5.09	Crestanello et al. (2009)
		34 (–)	–	–	–	73.8	71.0	7.20	Soglia et al. (2010)
	rupicapra (E)	11 (11)	9	0.709	94.5	55.4	53.2	4.20	Rodríguez et al. (2010)
		28 (28)	4	0.760	75.0	54.0	57.0	5.27	Crestanello et al. (2009)
		52 (–)	–	–	–	68.6	68.0	6.5	Soglia et al. (2010)
	cartusiana	8 (8)	3	0.041	46.4	42.0	43.7	2.90	Rodríguez et al. (2010)
	carpatica	17 (16)	11	0.617	87.5	43.4	35.3	3.35	Rodríguez et al. (2010)
	tatrica	7 (6)	5	0.180	86.0	33.0	42.0	2.18	Crestanello et al. (2009)
		10 (10)	4	0.166	73.3	33.4	32.5	2.45	Rodríguez et al. (2010)
	balcanica	9 (9)	6	1.262	88.9	55.0	38.9	4.00	Rodríguez et al. (2010)
		16 (–)	–	–	–	64.7	58.9	4.87	Šprem and Buzan (2016)
		52 (55)	17	3.46	74.7	53.6	38.8	6.39	Papaioannou et al. (2019)
	asiatica	1 (1)	1	–	–	–	52.6	–	Rodríguez et al. (2010)
	caucasica	10 (6)	4	0.616	80.0	42.5	40.5	3.80	Rodríguez et al. (2010)

N, number of individuals analyzed at microsatellite loci – in parentheses, number of individuals typed for mtDNA; nh, number of mitochondrial haplotypes; π, nucleotide diversity; h, mitochondrial haplotype diversity; He (Ho), expected (observed) heterozygosity at microsatellite loci; A, mean number of alleles; E, Eastern Alps; W, Western Alps during the last Quaternary glaciations in the Western Alps, when *R. rupicapra* and *R. pyrenaica* came into contact. Although neither species is threatened, there is reason for concern about the conservation of several subspecies. Translocations of chamois for hunting purposes, from different geographic populations or subspecies, have increased the risk of losing differentiated gene pools by hybridization, e.g., for *R. r. cartusiana*, *R. r. balcanica*, and *R. r. tatrica* (Lovari 1984a). Effective conservation measures, as with other taxa, should be established while populations of *Rupicapra rupicapra* subspecies are still abundant, rather than when it is too late for them to survive (Corlatti et al. 2011).

Hybridization

Evidence for ancient hybridization events between *Rupicapra rupicapra* and *Rupicapra pyrenaica* following postglacial recolonization has been reported by Rodríguez et al. (2009). Evidence for recent hybridization, on the other hand, has been reported only between different *Rupicapra rupicapra* subspecies (reviewed in Iacolina et al. 2019). The occurrence of hybrids between Balkan and Alpine chamois was confirmed by mtDNA analysis in the Velebit Mountains (Croatia), where both subspecies were introduced for hunting purposes (Šprem and Buzan 2016). Similar patterns may occur in the Rhodope Mountains in Bulgaria, although the signals of introgression (using microsatellites) for the Balkan population are equivocal (Markov et al. 2016) and further screenings are desirable (Valchev et al. 2006). Interbreeding has been suggested to occur also between Balkan and Carpathian chamois in the contact zone of the two subspecies in Djerdap National Park (Serbia) (Damm and Franco 2014). In the Low Tatras (Slovakia), the occurrence of hybrids between the endemic Tatra chamois and Alpine individuals introduced for hunting purposes has been confirmed in several molecular studies (Crestanello et al. 2009; Zemanová et al. 2015). Although lacking evidence, hybridization has been suggested to occur in the Chartreuse Massif in the French Alps, where the genetic identity of the endemic *cartusiana* subspecies has been put at risk by translocations of individuals of the Alpine subspecies for hunting purposes (Roucher 1999). Overall, the consequences of hybridization between chamois subspecies remain unknown and may include disruption of local adaptations, outbreeding depression, disappearance of genetic "types," but also increase of genetic variation, and enhanced adaptability to environmental changes (Zemanová et al. 2015). Nevertheless, future translocations should be avoided or carefully monitored to avoid genetic extinction of taxa due to hybridization.

Life History

Growth

Chamois kids are born from May to late June in Europe (November–December in New Zealand) after a gestation of about 175–185 days. Mass at birth is *c.* 2.4–2.7 kg, and development in the first 6 months is negatively affected by population density (Bauer 1982; Garel et al. 2011b). In Alpine chamois, body mass and horn size increase rapidly in the first 2 years of life, after which there is evidence for partial compensatory growth, so that yearling body mass and horn length are not strongly related to adult characteristics (Rughetti and Festa-Bianchet 2010a). Compensatory horn growth also occurs in other subspecies such as *R. r. balcanica* (Massei et al. 1994), although, interestingly, it appears to be weak in *R. r. rupicapra* × *R. r. balcanica* hybrids, possibly suggesting an effect of heterosis, i.e., the occurrence of hybrid vigor (Kavčić et al. 2018). Body mass and horn size reach asymptotic values at about 3.5 years in females and 5.5 years in males (cf. section "Description"). Sexual dimorphism in body mass is highly seasonal, rapidly declining from 26–40% in October to about 4% in spring (Garel et al. 2011a; Rughetti and Festa-Bianchet 2011). This suggests a unique conservative strategy to accumulate fat resources in summer to be used up during the rut, when males may lose as much as 28% of their body mass (Mason et al. 2011).

Reproduction

Body mass and age at first reproduction are closely associated in chamois (Bauer 1987), and females usually give birth when they approach asymptotic body mass, after the third year of age (Schröder 1971). Sexual maturity, however, may already occur at 18 months, and in New Zealand, the presence of an embryo was confirmed in 7-month-old females (Bauer 1987). Age of primiparity thus depends on ecological characteristics affecting body mass such as density or climatic conditions (Bauer 1987), as well as on individual heterogeneity. Females with rapid horn growth as yearlings, for example, tend to reproduce earlier (Rughetti and Festa-Bianchet 2010b). Pregnancy probability is age-dependent: Valentinčič et al. (1974) report values of 0%, 50%, 90%, and 83% in 1–2, 3, 4–10, and 11+ year old individuals, respectively, for a population in the Eastern Alps. Females give birth to one kid; twins are rare (Couturier 1938). No reliable data exist on twinning probability, although it may presumably vary between 1% and 3% among reproducing females. Breeding success (i.e., kids to adult females ratio in summer) is inversely related to population density, especially in young females, and ranges from *c.* 0.52 to 0.90 (Loison et al. 1996; Pioz et al. 2008). Breeding success is strongly age-dependent: Morin et al. (2016) found a peak of 0.70 for prime-aged females (4–7 years), with reproductive senescence occurring at 8+ years of age. The pattern of senescence in female chamois, however, differs between studied populations. In the Swiss Alps, for example, breeding success declined only in females older than 16 years (Tettamanti et al. 2015). Individual heterogeneity plays a major role also in shaping breeding success, as successful female chamois tend to do consistently better in reproduction in subsequent years than unsuccessful ones. This applies especially to the oldest age classes, resulting in no detectable costs of reproduction (Morin et al. 2016, but see Richard et al. 2017 for Pyrenean chamois). This suggests that some old females may reduce survival costs by skipping reproduction, making the female reproductive strategy increasingly conservative with age (Morin et al. 2016).

Reproductive success in male chamois is largely unknown, owing to the lack of robust DNA-based paternity assignments. Preliminary data suggest that in non-hunted populations mainly males $\geq$6 years successfully sire kids (Corlatti et al. 2015a), but in hunted populations – where older males are rare – they reproduce earlier. Together with a weak sexual dimorphism and high survival probabilities (see below), this suggests that male chamois may enjoy prolonged breeding tenure, possibly until 13–14 years of age (Corlatti et al. 2015a). Reproductive allocation, however, is rather flexible: evidence has been found for either a decline in reproductive effort after prime-age or terminal investment, depending on environmental conditions (Mason et al. 2011). Furthermore, social status also appears important, and territorial males may have greater reproductive success than non-territorial ones (cf. section "Mating Behavior"; Corlatti et al. 2015a). Reproductive benefits of territoriality, however, might be counterbalanced by greater risks of injuries, higher consumptions of fat reserves, and higher cortisol and testosterone levels, which may favor a decrease in parasite resistance, possibly increasing mortality rates (Corlatti et al. 2012a). Data on tactic-specific trade-offs between reproduction and survival, however, are still missing, and the two tactics might as well coexist with similar survival probabilities but different reproductive success (cf. section "Mating Behavior").

Survival

The sex ratio is 1:1 at birth (Henderson and Clarke 1986) and slightly biased towards females in adults (*c.* 1:1.2) (Bubenik and Schwab 1975). This suggests similar mortality patterns between male and female chamois, recently confirmed by studies conducted on the Alpine subspecies (Bocci et al. 2010; Corlatti et al. 2012b). Survival rates are surprisingly high until late age in both sexes in a non-hunted and undisturbed population in the Swiss National Park: 0.90 at 1 year, 0.91 between 2 and 7 years, and 0.92 over 8 years (Corlatti et al. 2012b). For the same population, sex- and age-specific survival probabilities using more detailed age classes are reported in Fig. 5. Nearly unbiased sex-specific survival may be explained by greater accumulation of fat reserves by males before the rut, which may reduce overwinter mortality (Bocci et al. 2010). Survival senescence in chamois may thus be weaker than in other ungulates (Loison et al. 1994), although senescence patterns are not fixed. Female survival senescence begins at *c.* 7 years of age in a population in the hunted population of the Bauges massif, French Alps, whereas in the

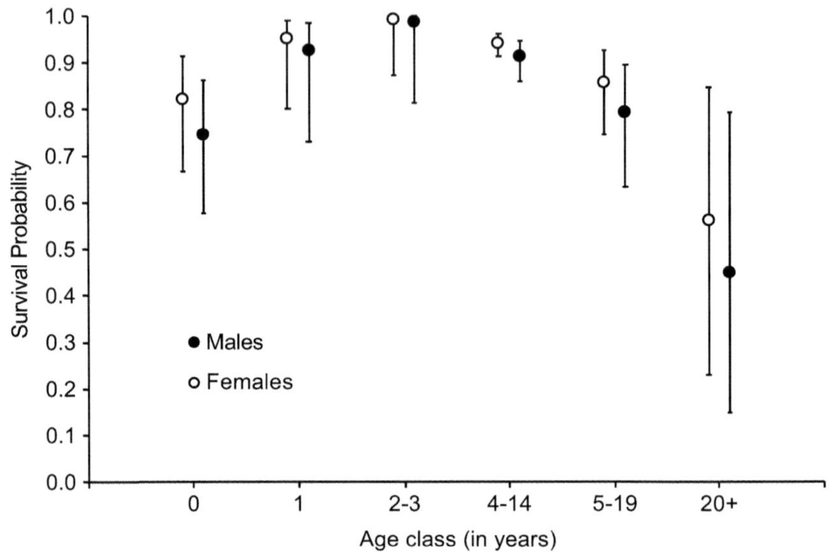

Fig. 5 Sex- and age-specific survival probabilities of 123 Alpine chamois in the Swiss National Park between 1996 and 2008 (data from Corlatti et al. 2012b, refitted using a model with different age classes; sample size for kids was only n = 7). Vertical lines represent 95% confidence interval of the estimate

Swiss National Park, the onset is at *c.* 12 years (Bleu et al. 2015). This suggests that faster life history strategies may be selected for by hunting, which may shift age at first reproduction and possibly alter senescence patterns (Bleu et al. 2015).

In protected populations, there is a negative, though nonsignificant trade-off between early horn growth and natural survival in males and females (Corlatti et al. 2017; Bleu et al. 2014), whereas in hunted populations, this trade-off may be stronger in both sexes because of hunter selectivity, thus depending on the hunting regime adopted (Corlatti et al. 2017). Maximum longevity recorded in the wild (Swiss National Park) is 21 years for males and 22 years for females in the Alpine subspecies (Corlatti et al. 2012b).

Habitat and Diet

Habitat Selection and Movement

Northern chamois mostly occur from the treeline to elevations up to 3,000 m in summer and descend to lower elevations in conifer forests in winter (Ballo 2010; Lovari et al. 2006; Nesti et al. 2010; Anderwald et al. 2015, 2016). However, some populations spend the entire year in forested areas. Stable populations between 100 and 300 m above sea level, for example, are present in Croatia (Šprem, pers. comm.) and in Italy. Based on archeozoological data, Baumann et al. (2005) showed that the native habitat of chamois during the Pleistocene included both alpine grassland and steep forests, so that the recent expansion of the Alpine subspecies' range to forested regions such as the Jura Mountains in Switzerland represents a recolonization of natural habitat. The most likely reasons for the post-Neolithic range contraction into alpine areas was environmental change combined with overexploitation by humans at lower elevations. Accordingly, extant Alpine chamois populations are found in a wide range of habitats: besides alpine and subalpine meadows, pastures, forests, and clearings, the species also occurs on scree slopes, in rocky areas, and in shrublands (Ballo 2010; Nesti et al. 2010; Darmon et al. 2012; Anderwald et al. 2015). An increase in the use of forest habitats in recent years, for example, has been shown in chamois populations in the Tatra mountains (Ciach and Pęksa 2019). However, habitat choice does seem to depend on the availability of steep cliffs or slopes as escape terrain (e.g., von Elsner-Schack 1985a) and absence of human disturbance (e.g., Kati et al. 2020).

In their alpine habitat, chamois show a preference for meadows on relatively steep (20–35°) southern, eastern, and south-eastern slopes, where snow cover is reduced during winter (von Elsner-Schack 1985a; Nesti et al. 2010). Preferred meadows are often dominated by *Sesleria* spp. and sedges (Darmon et al. 2012). Typical altitudinal ranges in the Gran Paradiso National Park (Italian alps) are 1,900–2,600 m in summer and 1,800 –2,200 m in winter, with resident and seasonally migrant males showing altitudinal segregation (Lovari et al. 2006; Nesti et al. 2010): resident males remained at lower elevations throughout the year, while seasonal migrants spent the summer at similar elevations as females (2,100–2,600 m).

Seasonal migrations mainly take the form of short-distance altitudinal movements with the onset being dependent on weather conditions and allow individuals to exploit vegetation phenology and more favorable temperatures (cf. Corlatti et al. 2021). Although home range sizes do not appear to differ between males and females during winter, females show larger summer home ranges than resident males (e.g., a median 95% kernel of 711 ha for females *vs.* 49 ha for resident males in Gran Paradiso (Nesti et al. 2010), or a range of 270–598 ha for females *vs.* 34–98 ha for males in the Swiss National Park (Boschi and Nievergelt 2003)). Earlier studies suggested mean summer ranges of 60 ha and winter ranges of 74 ha in females, and fairly stable home ranges (*c.* 20–40 ha) in males in the Karwendel mountain range of Northern Tyrol, Austrian Alps (Hamr 1984b, 1985). Females are highly philopatric, whereas young males have a greater tendency to disperse, especially before reproduction, which avoids inbreeding (Loison et al. 1999a). Long migrations of up to 20 km were observed in male chamois in Austria and peaks of 50 km in New Zealand (Hamr 1985). Chamois also appear to have good swimming skills: recently, an individual was reported to have crossed the *c.* 3-km wide stretch that separates the island of Rab from the mainland of Croatia (Safner et al. 2019), and other similar cases have been reported elsewhere (Kavčić et al. 2020).

Diet

Alpine chamois are intermediate feeders (Hofmann 1989) with a diet consisting mostly of graminoids throughout the year (e.g., Schröder and Schröder 1984; Bertolino et al. 2009; Redjadj et al. 2014).

Besides graminoids, forbs, dwarf shrubs, and conifer material also contribute to the diet seasonally. While their relative importance varies by region according to availability, their seasonal representation is largely comparable between different studies. Forbs are taken primarily in summer, while the consumption of conifer material is mostly restricted to winter months, and Ericaceae are taken mainly in autumn (Bertolino et al. 2009; La Morgia and Bassano 2009; Anderwald et al. 2015; Andreoli et al. 2016). The diet is very similar between the sexes (Andreoli et al. 2016). As to food habits, female chamois are particularly selective, with individuals feeding on relatively high-quality plants and on <18% of the available biomass (Dupart et al. 2020).

There is considerable overlap between the dietary composition of Alpine chamois and other ungulates such as mouflon *Ovis gmelini musimon* (Bertolino et al. 2009; Redjadj et al. 2014), red deer (Homolka 1996; Homolka and Heroldová 2001; Bertolino et al. 2009; Anderwald et al. 2015), Alpine ibex *Capra ibex* (Anderwald et al. 2015), and domestic sheep *Ovis aries* (La Morgia and Bassano 2009). La Morgia and Bassano (2009) suggested a temporary shift in the summer diet of chamois from forbs to qualitatively inferior Cyperaceae in the Gran Paradiso National Park while sheep grazed in the same range as chamois.

Behavior

Social Behavior

The mother-kid dyad is the only fixed social unit in chamois (Krämer 1969). After birth in May–June, the period of maternal care extends to early October but the bond between mothers and juveniles lasts until about 1 year of age, possibly favoring social cohesion and learning processes through the synchronization of activities (Ruckstuhl and Ingold 1999). Females and offspring form instable groups that often fuse and split up. Group size is thus sensitive to variations in local conditions, correlating positively with population density, visibility, and food resources, and negatively with human disturbance and proximity to refuge areas (von Elsner-Schack 1985b; Chirichella et al. 2015). The largest assemblages are found in snow-free periods (von Elsner-Schack 1985b): in Alpine chamois, female group size between May and October was found to range from 1 to 25 (Chirichella et al. 2015), although much larger groups (>100 individuals) are possible (Couturier 1938). Yearling males often clump in small flocks, while adult males live solitarily or in small groups of 2–3 individuals for most of the year (Krämer 1969). Sexual segregation is thus generally strong, especially in summer, but group composition is flexible and varies over the year. In winter, the presence of mixed herds is largely due to foraging constraints, whereas in autumn, intersexual interactions are favored by the onset of the breeding season.

Mating Behavior

The rut normally occurs between early November and early December in Europe (Krämer 1969). The level of mating monopolization in chamois is unknown, although there is some support for weak polygyny (cf. Corlatti et al. 2015a). Despite the weak sexual size dimorphism, male-male competition during the rut is intense. At least two reproductive tactics have been recognized in sexually mature individuals: territorial males, which actively defend a small area from intruders (Krämer 1969; von Hardenberg et al. 2000; Corlatti et al. 2012a), and non-territorial males, which adopt female-following behavior (Krämer 1969; Corlatti et al. 2012a). Territories are normally below 2,000 m of elevation and they may be occupied in spring already (von Hardenberg et al. 2000). The maintenance of alternative behaviors in chamois is not well understood, and male types might coexist with either equal or unequal lifetime reproductive success (cf. section "Life History"). In the first case, frequency-dependent selection might play a prominent role. Furthermore, environmental stochasticity may also contribute to the long-term maintenance of alternative male types: in years with high snow cover, females would be forced to move to lower elevations, where mainly territorial males are located, while non-territorials may gain advantages in years of low snow cover by following females at higher elevations (Lovari et al. 2006; Corlatti et al. 2020). Coexistence with unequal lifetime reproductive success, on the other hand, may be mediated by condition-dependent selection. As the two male types do not show significant biometric differences (Corlatti et al. 2012a), variation in reproductive success of sexually mature males might be conditional on competitive skills, rather than on horn size or age (Corlatti et al. 2015a). Agility and speed, among other factors, are thought to play a major role in male mating success (Rughetti and Festa-Bianchet 2010a). Greater reproductive benefits may be traded-off against higher physiological costs, primarily associated with energy expenditure and immune-mediated parasite infection during the rut, thus higher mortality (Corlatti et al. 2012a, c). Evidence of such a trade-off, however, is still missing and individual heterogeneity might as well favor coexistence of different male types with similar survival probabilities but unequal lifetime reproductive success.

Communication

During the rut, male chamois show a rich behavioral repertoire, made up of 31 behavioral patterns (Krämer 1969; Corlatti et al. 2013b). Male-male communication is largely based on indirect aggression, the most frequent behaviors being "marking" (the supraoccipital glands are rubbed on a grass stem or a thin twig to deposit scent) and "neck up" (a male stands on stiff legs, moving slowly along a tangential line to the opponent, the head is held high, the mane is fully erect, and the mouth is often open) (Fig. 6), whereas the most frequent courtship pattern is "head up" (the male approaches the female with abrupt steps, lifting up its muzzle) (Corlatti et al. 2013b). Information on female behavioral patterns is lacking in Northern chamois, but data available for the Southern species suggest a marked behavioral sexual dimorphism, with a much poorer repertoire in females than in males, and frequent use of direct aggression (Lovari 1985; Locati and Lovari 1990). Vocalizations include the "rut call," an indirectly aggressive grunt of relatively low pitch, emitted through the nose and the mouth by males during the mating season, a loud whistling snort emitted

Fig. 6 Alpine chamois displaying the "neck up" posture during the rut (photograph by Marco Coraglia)

through the nose by both sexes in alert situations throughout the year, and bleating in kids (rare in adults, only in the presence of a serious threat).

Activity

Chamois are generally more active in spring-summer than in winter (Brivio et al. 2016). Daily activity has long been considered bimodal, with peaks at dawn and dusk (Hamr 1984a), but chamois may display multiple patterns (unimodal, bimodal, and trimodal) over the year (Brivio et al. 2016). Nocturnal movements may occur throughout the year (Carnevali et al. 2016). In the Swiss National Park, nocturnal activity correlated negatively to diurnal activity in summer and positively in winter, possibly to optimize food intake (Grignolio et al. 2018). Nocturnal activity increases during moonlit nights (Carnevali et al. 2016), especially in the mating season (Grignolio et al. 2018). Human disturbance may negatively affect activity rhythms, as shown in Tatra chamois (Pęksa and Ciach 2018). A unimodal pattern of activity with a peak in the early morning hours has been observed in a forested area in Croatia, possibly reflecting a temporal niche shift to avoid predation by wolf *Canis lupus* (Šprem et al. 2015).

Foraging Behavior

On the Tatra mountains, time spent foraging by chamois was affected nonlinearly by time of day, herd size, and human disturbance, and represented on average 46% of their overall time budget. Males spent less time foraging than females (Pęksa and Ciach 2018). Notably, in the Swiss National Park, males showed higher bite rates and lower step rates than females, suggesting higher food selectivity in the latter (Puorger et al. 2018; cf. Ferretti et al. 2014 for the Apennine chamois).

Different patterns of time spent foraging are associated with alternative male tactics. Foraging time in territorial males increases in summer, sharply drops in November (hypophagia: Willisch and Ingold 2007; Corlatti and Bassano 2014) and increases again in winter. Foraging time of non-territorial males shows smaller variations, decreasing gradually from spring to autumn, and increasing in winter (Corlatti et al. 2013a). As dietary quality is similar in territorial and non-territorial males throughout the year (Corlatti et al. 2013a), these foraging patterns might reflect different investments made by alternative male tactics before the rutting season. A greater amount of time spent foraging might be associated with more resources acquired and thus with better body conditions before the rut in territorial males (Corlatti et al. 2013a). Alternatively, a greater amount of time spent foraging in spring and summer may simply reflect greater selectivity in food choice in territorial males, who inhabit lower-food-quality habitats compared to non-territorial ones (Corlatti et al. 2021). Tactic-specific body mass data before the rut would help clarifying the mechanisms underlying different foraging patterns in male chamois.

Parasites and Diseases

Ectoparasites

Insects. Chamois are often subclinically affected by host-specific chewing lice *Damalinia alpina* (syn. *Bovicola alpinus*) and biting lice *Linognathus rupicaprae*, and two species of blood-sucking keds, *Melophagus rupicaprinus* and *Lipoptena cervi*, which are common in Old World Caprinae and Cervidae, respectively. However, hair loss associated with massive infestation by chewing lice is frequently found in individuals starving to death in late winter (Boch and Schneidawind 1988). Subcutaneous infestation by larvae of the Oestrid fly *Hypoderma diana* has been occasionally observed where chamois coexist with infested roe deer *Capreolus capreolus*, the main reservoir of this Dipteran. No bot fly infestation is known in *Rupicapra* spp.

Arachnida. In forested areas, chamois are exposed to infestation by the "wood tick," *Ixodes ricinus*, and other hard ticks, *Dermacentor marginatus* and *Haemaphysalis punctata*. *I. ricinus* is the recognized vector of the protozoan *Babesia capreoli*, responsible for fatal babesiosis in chamois (Hoby et al. 2007). Accordingly, increasing risk is associated with cohabitation with cervids in a global context of upslope shift displayed by *I. ricinus*, likely due to climate changes. Larvae of the seasonally abundant chigger mite *Kepkatrombicula* (syn. *Neotrombicula*) *desaleri* may accumulate in typical bright orange crusts around the mouth and eyes and on the pinnae. The burrowing mite *Sarcoptes scabiei* is the agent of scabies or sarcoptic mange, the most severe and persistent epidemic disease in chamois and sympatric Alpine ibex. Scabies is currently endemic in the Eastern Alps in Austria, Germany, Slovenija, and Italy. Clinical features are pruritus, hair loss, and generalized crusty lesions. Death is due to toxemia and malnutrition, and usually occurs within 3–4 months from exposure by contacts with infected individuals or fomites. Mortality rate is related to the life history of affected populations, namely to the occurrence (or absence) of previous contacts with the disease. In case of a novel epidemic wave of sarcoptic mange in previously unaffected areas, population size may decrease by approximately two-thirds, with maxima well above 80%, while during subsequent epidemic waves, occurring at *c.* 10 / 15-year intervals, mortality rarely exceeds 25% (Rossi et al. 2007).

Endoparasites

The digestive tract of chamois is home to at least 7 species of protozoa (*Giardia duodenalis*, *Cryptosporidium parvum*, plus 5 host-specific taxa of the genus *Eimeria*) and a number of helminths, including 3 generalist tapeworms, *Avitellina centripunctata*, *Moniezia expansa*, and *Moniezia benedeni*, and approximately 50 roundworms (Durand and Gauthier 1996). Among the latter, the Bovid specialists *Teladorsagia circumcincta* and *Marshallagia marshalli*, with their respective minor morphs, are the dominant abomasal species, while *Nematodirus* spp. is prevalent in the small intestine. Most protozoa and helminths of the gastrointestinal tract of chamois are weakly pathogenic, with the exception of the large-sized blood-sucking *Haemonchus contortus*, a generalist species also harbored by domestic sheep, goats, and other wild ruminant hosts, which may cause severe anemia and death. A trade-off between mating effort, immunological defense, and resilience to subclinical gastrointestinal parasitism has recently been highlighted (Corlatti et al. 2012a).

Chamois in contact with infected livestock may unfrequently harbor the liver flukes *Fasciola hepatica* and *Dicrocoelium dendriticum*. Larvae of *Echinococcus granulosus*, a zoonotic tapeworm with a canid-ruminant life cycle, have occasionally been found in the liver or lungs of chamois sharing range with infected sheep and goats, so the definitive host is mostly shepherd dogs. Larvae of *Taenia marginata* (*Cysticercus tenuicollis*), a tapeworm with a similar life cycle as *E. granulosus*, are frequently observed on serous surfaces.

The lung parenchyma is home to slender snail-transmitted nematodes of the genera *Muellerius* and *Neostrongylus*. Typical nodules formed by masses of adult worms, embryonated eggs, and coiled larvae develop on dorsal regions of the caudal lobes. Lungworm infections are most often subclinical. Larger nematodes (*Protostrongylus* spp. and *Dictyocaulus filaria*) may be present in the bronchial tree and the trachea.

In the heart and muscles of chamois, cyst-forming Protozoa (*Toxoplasma gondii* and 2–3 nonpathogenic species of the genus *Sarcocystis*) have been detected.

Larvae of *Taenia multiceps* (*Coenurus cerebralis*), another tapeworm with a canid-ruminant life cycle, may develop and enlarge in the central nervous system, causing obvious nervous signs (e.g., circling) and death.

Infectious Diseases

Two infectious diseases can have major demographic consequences in chamois: infectious kerato-conjunctivitis (IKC) and transmissible pneumonia.

Outbreaks of IKC are due to virulent strains of an atypical bacterium, *Mycoplasma conjunctivae*. Transmission occurs via direct contact and eye-to-eye transport by flies over short distances. Symptoms are partial or complete blindness and the associated difficulty to move on rough terrain. Affected individuals may fall from cliffs or die of starvation in case of bilateral eye perforation. Nevertheless, spontaneous recovery is a frequent outcome. IKC is characterized by high prevalence, relatively low mortality, and a biased distribution of clinical cases (more often females than males, and more adult females than kids) (Giacometti et al. 2002). Population decrease ranges from *c.* 5% to 30%. Spread may occur over large areas at the speed of over 15 km/year (much faster than sarcoptic mange), depending on the virulence of *Mycoplasma* strains and the connectivity between social units of the host. Sporadic cases and small foci with limited spatial spread of IKC are also known in the Alps. Contrary to sarcoptic mange, IKC signs do not persist in chamois groups longer than a few months after an epidemic wave. Spillover of *M. conjunctivae* from cohabiting domestic sheep, goats, and Alpine ibex is likely.

Transmissible pneumonia affecting chamois of all ages, with die-offs characterized by respiratory signs (e.g., cough and air hunger behavior), has been reported. Lesions include pleuritis and extensive fibrino-suppurative, partly necrotizing bronchopneumonia. Their spreading potential is much lower than that of IKC and sarcoptic mange, and persistence lasts no longer than several months to 1 year. Mortality may be over 30%. While a major role is played by bacteria of the family Pasteurellaceae, the etiology of transmissible pneumonias is still insufficiently defined (Posautz et al. 2014). Accordingly, little can be inferred on the cross-transmission of candidate pathogens at the livestock/chamois or other wild caprines/chamois interface.

Outbreaks of relatively limited demographic impact are caused by the pox-like contagious ecthyma virus. The main sign is the development of crusty lesions on the lips associated with inflamed gums, implying limited feeding activity. Kids are particularly sensitive.

While contact with the large majority of ruminant pathogens has been revealed, no significant role is attributed to Northern chamois in the medium to long-term maintenance of major livestock diseases such as foot and mouth disease, tuberculosis, brucellosis, and bluetongue.

Zoonoses

No major risks to human health derive from handling Northern chamois or dressing their carcasses or using their meat or nonmeat products.

Population Ecology

Population Dynamics

In increasing chamois populations, annual growth rate generally varies between 5% and 15% as a function of density and habitat quality, but can occasionally reach 25% (Corti 2002; Corlatti et al. 2019a). On the other hand, the rate of decrease appears much more variable. Apart from hunting by humans, major limiting factors of chamois populations are the synergistic effect of high density and harsh winters – with overall mortalities up to *c.* 30% (Corlatti et al. 2019a) but with maxima of over 80% depending on sex- and age-classes (Rughetti et al. 2011) – and outbreaks of diseases such as sarcoptic mange or keratoconjunctivitis – with overall mortalities up to *c.* 80% and 30%, respectively (cf. section "Parasites and Diseases" for further details). However, Alpine chamois populations can normally recover from such events within a few years through subsequent higher reproductive performance (e.g., Rossi et al. 1995; Loison et al. 1996). Density dependence affects juvenile body mass (Garel et al. 2011b), mortality rate, and juvenile to female ratio (Capurro et al. 1997; Willisch et al. 2013). Chamois appear to be particularly sensitive to adverse weather conditions during late winter and early spring, with decreases in juvenile body mass during years characterized by late springs (Garel et al. 2011b) and decreased population growth following severe snow conditions in late winter, especially when population density is high (Willisch et al. 2013; Corlatti et al. 2019a). Summer conditions can also severely affect chamois population dynamics, with high temperatures hampering the survival of kids in the following winter (Chirichella et al. 2021) and higher precipitation favoring juvenile to female ratio in the following year (Donini et al. 2021). Ultimately, hot and dry summers may negatively affect the nutritional quality of forage available to females and offspring, thereby affecting energy intake and survival (Loison et al. 1999b).

Kids are vulnerable to predation from golden eagles *Aquila chrysaetos*, and all age classes are taken by wolves and lynx *Lynx lynx* where they co-occur. In the Swiss Jura Mountains, lynx killed a maximum of 11% of the chamois spring population (Molinari-Jobin et al. 2002).

Competition with Other Ungulates

Alpine chamois often appear to be the inferior competitor in interactions with other sympatric ungulates. Spatial/habitat displacement has been reported in response to mouflon (Chirichella et al. 2013a), red deer (Anderwald et al. 2015), and domestic sheep (Mason et al. 2014a) in the European Alps.

Mouflon. In the Italian Alps, female chamois increase their group size in the proximity of mouflon and increasing proportion of mouflon rams in the group, either to avoid spatial displacement or in response to it (Chirichella et al. 2015).

Red deer. With increasing red deer density in an area of high productivity in the Swiss National Park, chamois decreased their use of meadows and forested areas in favor of scree slopes, and juveniles showed reduced horn growth (Anderwald et al. 2015). On the other hand, there was no evidence for competition in an adjacent valley of lower productivity, where red deer density was lower and there was strong altitudinal segregation between the two species, suggesting that chamois can find competition refuges outside resource hotspots (Anderwald et al. 2016). A negative effect of increasing deer density on chamois population growth rate was found in the Stelvio National Park, Italy (Corlatti et al. 2019a), mainly acting on the kid-female segment of the population (Donini et al. 2021; Gamelon et al. 2020).

Sheep. In addition to a shift to a poorer summer diet induced by temporary range overlap with domestic sheep (La Morgia and Bassano 2009), Alpine chamois show a displacement response to less suitable habitat by sheep presence that exceeds the apparent altitudinal shift due to climate change (Mason et al. 2014a), although it is unclear whether other associated disturbance factors (e.g., shepherd dogs) were involved.

Alpine ibex. Alpine chamois appear to have a stronger influence on the habitat use of ibex than vice versa (Anderwald et al. 2015), and the effect is seemingly stronger on female ibex than on males (Herfindal et al. 2019). Despite the high overlap in both habitat and dietary niche, however, there is no clear evidence of competitive interactions with respect to population density or body condition between these two native Alpine species.

Effects of Climate Change

Due to differences in vulnerability to winter conditions between the sexes and different age classes

(Willisch et al. 2013), as well as age-specific survival rates between populations (Bleu et al. 2015), the effects of climate change on Alpine chamois populations are difficult to predict. Mason et al. (2014b) report long-term decreases in the body mass of yearlings linked to both increasing population size and increasing spring and summer temperatures. As chamois are sensitive to high temperatures, and no temporal change in resource productivity and phenology was found in their study area, they attribute this observation to either an increase in intraspecific competition or a decrease in the time the animals spend foraging due to higher thermoregulatory demands. Similarly, decreases in body mass have been linked to additive effects of warm springs and summers during the animals' first 2 years of life (Rughetti and Festa-Bianchet 2012), although the presence of forest cover seems to mitigate this temperature-induced decline in body mass (Reiner et al. 2021). Owing to the ecological plasticity of the species, response to climate change is likely site-specific (cf. Loison et al. 1999b; Chirichella et al. 2021; Donini et al. 2021; Reiner et al. 2021).

Conservation Status

The Northern chamois is listed in Appendix III of the Bern Convention and in Annex V of the EU Habitat Directive. The subspecies *balcanica* and *tatrica* are listed in Annexes II and IV of the EU Habitat Directive. As a species, the Northern chamois is abundant and widespread. Despite the declining trend of some subspecies, the bulk of the population, represented by the Alpine subspecies, is relatively secure. Consequently, its status is assessed by the IUCN Red List as Least Concern (LC) (Anderwald et al. 2021). Several subspecies, however, require conservation measures. In particular, *R. r. cartusiana* is listed as Vulnerable (VU) D1 + 2, with a population confined to a single mountain massif (Anderwald et al. 2021). *R. r. tatrica* is currently listed as Endangered (EN) B1 + 2ab, given the restricted range, the small population and the issues related to potential interbreeding with the Alpine subspecies (Anderwald et al. 2021). While robust data on population size and structure of *R. r. asiatica* are needed to properly assess its conservation status, recent information suggests that the subspecies has experienced catastrophic declines in population size in the recent decades and it should be listed as Endangered (EN) C1 + 2a(i) (Anderwald et al. 2021). Owing to suspected ongoing population decline, *R. r. caucasica* is listed as Vulnerable (VU) C1 (Anderwald et al. 2021).

Management

The Northern chamois does not show major conflicts with human-related activities, although negative browsing impacts on silver fir *Abies alba* saplings have been reported (Kupferschmid et al. 2014). The species is hunted throughout its distribution range, with a variety of hunting regimes (Damm and Franco 2014). Harvesting plans can be based on the adoption of simplified age classes (i.e., yearlings and older individuals, as in France and in some hunting districts of the Italian Alps), or on more sophisticated grouping systems (e.g., 1, 2–3, 4–10, and ≥11 years of age for both sexes in Italy; 1–3, 4–9, ≥10 years of age for females, 1–3, 4–7, ≥8 years of age for males in some provinces in Austria). Hunting of kids occurs in many chamois populations over the Alpine arch, e.g., in Austria, Germany, Switzerland, and in France (Damm and Franco 2014). Opportunities and pitfalls of hunting lactating female chamois, in particular, have recently been discussed by Rughetti and Festa-Bianchet (2014) and Corlatti et al. (2019b). In the Alps, hunting generally follows the principle of sustainable hunting: in increasing populations (e.g., below carrying capacity) chamois growth rate normally varies between 5% and 15% (cf. section "Population Ecology"), and harvest rate, on a national level, is between 7% and 15%, with very few exceptions (Damm and Franco 2014). Given the biological and ecological characteristics of the species, hunting quotas above 17% and highly skewed sex-ratio should be avoided (cf. Skonhoft et al. 2002). Locally, there are concerns about relative over-harvesting of mature males (Damm and Franco 2014). Outside of the Alpine range, information on chamois abundance and harvest rate is fragmentary

(Damm and Franco 2014). Besides sustainable hunting, general management recommendations for the Northern chamois include the reduction of poaching and human disturbance (Shackleton 1997), and the protection of the genetic integrity of non-Alpine subspecies – threatened by introductions of Alpine individuals for hunting purposes (reviewed in Iacolina et al. 2019). Robust monitoring of abundance is of paramount importance, especially for subspecies whose populations are small and/or declining. In this respect, population estimates and monitoring of chamois are normally conducted using block counts. Although this method leads to underestimates due to imperfect detectability (Corlatti et al. 2015b), it may provide robust indices of relative abundance (Loison et al. 2006; Reiner et al. 2020) and thus appears appropriate to track long-term changes in population size while reducing monitoring costs. When funding and absence of logistical constraints allow for the use of alternative methods, capture-recapture methods, either based on resightings or on DNA sampling, are desirable, while distance sampling appears problematic owing to the difficulties of meeting basic assumptions in rugged terrains (Corlatti et al. 2015b). The use of statistical methods applied to camera trap data to estimate absolute population size, especially in forested populations, is promising but still needs to be thoroughly explored, as least when marked individuals are not available (cf. Gilbert et al. 2021). In hunted populations, cohort analysis may be used to reconstruct past population size up to 10 years before present, and obtain information on population sex- and age-structure (Reiner et al. 2020).

For *R. r. rupicapra*, concerns mainly derive from overharvesting, sex-biased harvesting towards mature males (Aulagnier et al. 2008), poaching and livestock management favoring disease outbreaks (Rossi et al. 2007, 2019). Competition with other species such as red deer and mouflon may also represent a threat for some populations (Chirichella et al. 2013a; Corlatti et al. 2019a). Recreational activities, if uncontrolled, may negatively affect some populations (Schnidrig-Petrig and Ingold 2001). *R. r. cartusiana* is threatened by competition for resources with domestic livestock, red deer, and mouflon. Hybridization has been anecdotally reported with introduced Alpine chamois (Roucher 1999), though a reliable assessment of the genetic status is lacking. Overharvesting and poaching, forestry, and recreational activities may also represent a threat to its populations (Aulagnier et al. 2008). For *R. r. tatrica*, poaching was a major threat in the past, but after implementation of strict conservation measures, hunting is banned and incidents of poaching are rare (Anderwald et al. 2021). Currently, major threats include tourist disturbance (Zwijacz-Kozica et al. 2013), habitat loss, disease transmission from domestic livestock (Anderwald et al. 2021), and interbreeding with individuals of the Alpine subspecies introduced for hunting purposes (Zemanová et al. 2015). Populations are still relatively small, thus stochastic events may represent a threat to their long-term viability. No major threats are reported for *R. r. carpatica*, although grazing by domestic sheep in summer and poaching may represent a source of disturbance for chamois (Aulagnier et al. 2008). Additional concerns are small, fragmented populations (Anderwald et al. 2021). Major threats for *R. r. balcanica* are habitat loss – at least in some regions of Albania and Greece – owing to land demands from human populations and livestock grazing, poaching (whose extent is unknown) and, locally, genetic introgression with Alpine chamois (e.g., Bulgaria: Markov et al. 2016) and predation by feral dogs (e.g., Greece: Hatzirvassanis 1991). *R. r. asiatica* and *R. r. caucasica* are the least known of the Northern chamois subspecies, and their populations are likely declining. Major threats for *R. r. asiatica* are poaching, trophy hunting, habitat degradation, and isolation due to the construction of new roads to alpine habitats for mass tourism and of hydroelectric power plants, as well as increasing tourism activities (Anderwald et al. 2021). For *R. r. caucasica,* major threats are represented by poaching related to political unrest and, possibly, competition with livestock and with tur *Capra caucasica*, red deer, and roe deer (Anderwald et al. 2021; Aulagnier et al. 2008).

Future Challenges for Research and Management

Most subspecies, populations, and subpopulations of the Northern chamois are not monitored, and this should be a high priority for any conservation or management action. The status of the Anatolian and Caucasian subspecies, in particular, needs to be thoroughly assessed. Priorities for the species as a whole include investigations of the demographic and/or genetic impacts of the main drivers of chamois population dynamics. These include hunting management, climate change, disease outbreaks, and interspecific interactions (i.e., competition and predation). In particular, after the wolf got extinct due to anthropogenic pressure, the consequences of the recolonization this large predator on chamois will require investigations. The performance of different methods for monitoring abundance should be assessed: these may include, for example, camera trap distance sampling, other camera-trap methods based on unmarked individuals (cf. Gilbert et al. 2021), as well as DNA-based (also spatially-explicit) capture recapture models. Reliable DNA-based paternity data are necessary to assess the opportunity for sexual selection in chamois and hence, its level of polygyny; the role of females in mate choice and the mechanisms of maintenance of alternative mating tactics, including tactic-specific survival probabilities, offer exciting opportunities for future research. The fitness consequences of hybridization and gene introgression need to be clarified. The eco-ethology of chamois population in non-Alpine habitats (e.g., montane/subalpine forests) needs investigation.

Southern Chamois *Rupicapra pyrenaica* Bonaparte, 1845

Description

Size and Morphology

As its Northern congeneric, the slightly more slender Southern chamois (Fig. 1) is well adapted to rocky terrains, showing a number of adaptations to life on mountains (cf. *R. rupicapra*). In Pyrenean *R. p. pyrenaica* and Cantabrian *R. p. parva* chamois, adult body mass is *c.* 22–32 kg (Crampe 1997; Pérez-Barbería and Pérez-Fernández 2009) and varies seasonally, especially in males, reaching its peak in autumn, before the rut. In Apennine chamois *R. p. ornata*, mean adult body mass is *c.* 27–30 kg (Lovari and Bruno 2003). In Pyrenean and Cantabrian chamois, head-body length is *c.* 96–117 cm and height at the withers is *c.* 61–84 cm (Pérez-Barbería et al. 2017). Pyrenean and Apennine chamois have a larger cranium than Cantabrian chamois, but the latter shows greater sexual dimorphism (Scala and Lovari 1984; Fernández-López and García-González 1986). In Southern chamois, both sexes have black horns of circular section, similar in height (Scala and Lovari 1984; Cabrera 1914) and less divergent than in Northern chamois (García-González and Barandalla 2002), showing an apical hook and a "V" shape in frontal view, especially in males. Horns are longer in adult Apennine chamois than in the other subspecies (mean length: *c.* 25–28 cm in Apennine chamois; *c.* 18–21 cm in Pyrenean chamois; and *c.* 14–17 cm in Cantabrian chamois) (Scala and Lovari 1984; Fernández-López and García-González 1986; Locati and Lovari 1991). As in the Northern chamois, sexual dimorphism is very weak. Males show a visible penis brush in winter, are generally stockier, and tend to have more pronounced hooks and thicker horns than females. Both sexes, in particular males, rub their horns on shrubs (*Juniperus* spp.) and small trees (*Pinus* spp.), attaching resin on to them, as in Northern chamois. Similar to the Northern chamois, scent glands are present in the supraoccipital and interdigital regions (Couturier 1938). The secretion of sebum from the supraoccipital glands is associated with specific displays of dominance (e.g., "marking," Lovari 1985).

Pelage

The coat undergoes two seasonal molts. The summer coat is reddish with a short fur; the winter coat is dark brown, slightly darker in males, with long fur, five large and whitish spots (one from the throat to the breast, with a dark band at both sides of the neck; two at the sides of the neck, up to the shoulders; two on the rump), and a smaller frontal band of the same color (Neumann 1899; Couturier 1938). *R. p. parva* has a more reddish summer coat and a slightly paler throat compared to the rest of the body (Cabrera 1914). On top of the eyelid and on the border of the dark lateral band, there is a long patch similar to an eyebrow.

Dentition

Milk teeth 0.0.3.0/3.1.3.0; permanent dentition 0.0.3.3/3.1.3.3.

Age and Sex Determination

As in the Northern species, age can be determined by counting annual horn rings (Couturier 1938). In Pyrenean and Cantabrian chamois, at least four classes are clearly identifiable in the field: kids (less than 1 year of age), with visible horns after August; yearlings (between 1 and 2 years of age), with horns shorter than ear length; adult males and adult females. Apennine chamois horns grow particularly long segments during the first 4 years of life, in the warm months, which leads to longer and taller horns compared to other chamois (Lovari 1985). Six age classes have been identified by Lovari (1985), on the basis of the visual estimate of the ratio of horn height over ear length, in a mature population in Abruzzo, Lazio and Molise National Park: kids; yearlings; subadults (2–3 years old); young adults (4–5 years old); mature adults (6–8 years old); and fully mature adults (9 + years old). As in the Northern chamois, sexual dimorphism in Southern chamois is weak, and sex determination in the field (possible only from the age of 2+) is based on the combination of several behavioral and morphological characteristics. The former mainly include urination posture, behavioral displays during the rut and social behavior: subadult females tend to stay with nursery groups, possibly including their mother, while males may form small unisexual groups; adult males tend to be solitary, females in groups. The latter mainly include noticeable penile brush in males, and horn features: male horns tend to be thicker and more hooked, while female horns tend to be more parallel towards the base.

Physiology

Adaptations to high elevations and cold climates similar to those described for the Northern species also occur in the Southern chamois. Published information on body condition is available for *R. p. parva* (Pérez-Barbería et al. 1998). Physical condition varies depending on food availability and population density. The male kidney fat index is minimal in spring (May and June) and at its maximum in autumn, before the rut (September and October). Female kidney fat shows a less marked and slightly anticipated cycle, with a minimum at the end of winter and a maximum in summer (from August to October). Nonpregnant females tend to have better physical conditions than pregnant females (Pérez-Barbería et al. 1998). A simple, low cost, and noninvasive proxy biometric indicator for monitoring chamois body condition is provided by Flores-Saavedra et al. (2018), based on the ordinary least squares residuals from the regression between body mass and hind foot length.

As opposed to the Alpine subspecies, a significant negative relationship between endogenous stress response (fecal cortisol metabolites) and minimum ambient temperature was found in Pyrenean chamois (Dalmau et al. 2007), possibly owing to better adaptations to colder climates in the former than in the latter. In *R. p. ornata*, recent studies have suggested the role of some environmental and endogenous factors in promoting female endogenous stress response, as well as levels of testosterone metabolites, in the warm months (Fattorini et al. 2018a, b). Specifically, levels of cortisol and testosterone metabolites increased with resource depletion, were highest in prime-aged females (4–8 years old), i.e., the most dominant and aggressive individuals, and peaked in early summer, when the lactation effort was greatest. Furthermore, stress response in *R. p. ornata* may also be triggered by the interaction with livestock and red deer, and by disturbance caused by hikers (Formenti et al. 2018). Other hemato-biochemical parameters seem to be influenced by seasonality, with urea and albumin being highest in spring/summer (Fraquelli et al. 2015).

Genetics

Chromosomes

$2n = 58$. Cytogenetic studies of somatic cells in the chamois revealed a karyotype with

54 acrocentric chromosomes and with one pair of large metacentrics comprising the autosomes. The sex chromosomes consist of an acrocentric (X) and a metacentric (Y) chromosome, respectively (Gallagher and Womack 1992).

Phylogeny and Phylogeography

The most widely accepted classification considers two species, *R. pyrenaica* (with three subspecies) from South-Western Europe and *R. rupicapra* (with seven subspecies) from Central-Eastern Europe (e.g., Masini and Lovari 1988; Wilson and Reeder 2005). However, the phylogenetic position of the Apennine chamois, *R. pyrenaica ornata*, remains controversial as there is evidence for three conspicuous clades arguing in favor of the old classifications that distinguished the species *R. p. ornata* from *R. rupicapra* and the two other Iberian chamois, *R. p. pyrenaica* and *R. p. parva* (Pérez et al. 2013). In comparison, genetic distances between pairs of populations were highly correlated with geographical distance indicating a deep divergence between the two recognized species and pointing to isolation by distance as the primary agent for differentiation in chamois (Pérez et al. 2002). In the 1920s and during World War II, *R. p. ornata* was close to extinction and today has grown to about 2,500 individuals (Antonucci and Di Domenico 2015). Whole mitochondrial DNA (mtDNA) sequence data comparisons (Pérez et al. 2014) corroborate the morphometric findings of Lovari and Scala (1980) that *R. r. cartusiana* bears some intermediate morphological phenotypes between *R. pyrenaica* and *R. rupicapra*. Cf. sections "Taxonomy and Systematics" and "Phylogeny and Phylogeography" of *Rupicapra rupicapra* for further details.

Genetic Diversity

Although the overall status of this species has greatly improved since 1990, the subspecies *ornata* harbors an extremely low level of variability compared with the other subspecies (Table 4, cf. section "Current Distribution" for further details). The genetic impoverishment, due to long-lasting isolation at small population sizes, may threaten its long-term survival. Intermediate features between Iberian and Alpine chamois were previously observed in *R. r. cartusiana* based on morphometrics (Lovari and Scala 1980) and to some extent in *R. p. ornata* based on mtDNA sequence comparisons (Pérez et al. 2014). Pérez et al. (2014) attributed them to hybridization events which may have occurred during the last Quaternary glaciations in the Western Alps, when *R. rupicapra* and *R. pyrenaica* came into contact. Although neither species is threatened, there is reason for concern about the conservation of several subspecies, e.g., for *R. p. ornata*. Effective conservation measures, as with other taxa, should be established while populations of *R. pyrenaica* subspecies are still abundant rather than when it is too late for them to survive (Herrero et al. 2020; Corlatti et al. 2011).

Hybridization

As the occurrence of the three *Rupicapra pyrenaica* subspecies is mainly restricted to non-overlapping distribution ranges (Pyrenean chamois *R. p. pyrenaica* occurs on the Pyrenees, Cantabrian chamois *R. p. parva* on the Cantabrians, Apennine

Table 4 Summary of genetic variability of *Rupicapra pyrenaica* subspecies

Species	Subspecies	N	nh	% π	% h	% He	% Ho	A	Source
R. pyrenaica	*parva*	40 (15)	9	0.827	92.4	51.3	47.0	4.45	Rodríguez et al. (2010)
	pyrenaica	41 (26)	13	0.516	91.1	51.7	48.3	5.20	Rodríguez et al. (2010)
		9 (9)	6	1.690	89.0	54.0	57.0	4.00	Crestanello et al. (2009)
	ornata	12 (12)	2	0.010	16.7	3.1	3.3	1.15	Rodríguez et al. (2010)
		11 (–)	2	0.030	33.0	–	–	–	Crestanello et al. (2009)
		58 (–)	–	–	–	44.1	43.8	2.1	Lorenzini (2005)

N, number of individuals analyzed at microsatellite loci – in parentheses, number of individuals typed for mtDNA; nh, number of mitochondrial haplotypes; π, nucleotide diversity; h, mitochondrial haplotype diversity; He (Ho), expected (observed) heterozygosity at microsatellite loci; A, mean number of alleles chamois *R. p. ornata* along the Apennine chain in Italy), inter-subspecies hybridization is unlikely. Evidence for hybridization of *Rupicapra pyrenaica* subspecies with any of the *Rupicapra rupicapra* subspecies, in recent times, has not been reported (reviewed in Iacolina et al. 2019, cf. section "Hybridization" in Northern chamois for further details).

Life History

Growth

Births take place in May–June (Lovari and Bruno 2003; Kourkgy et al. 2016; Garin and Herrero 1997). There is no published information on body mass at birth, early growth, and development of kids, for the Apennines. For the Cantabrian chamois, a mechanism of compensatory horn growth after the second year of age – similar to the Northern chamois – has been reported (Pérez-Barbería et al. 1996).

Reproduction

As in the Northern chamois, females give birth to one kid, mainly in the first fortnight of May; twins are rare, but adoptions have been reported (Pérez-Barbería et al. 1998; Scornavacca et al. 2018), particularly during keratoconjunctivitis outbreaks (Arnal et al. 2013). In stable populations, primiparity normally occurs at 3 years of age, whereas in expanding populations, females can reproduce already at 2 years of age (Loison et al. 2002; Crampe et al. 2004). In a Cantabrian population close to carrying capacity, for example, primiparity occurred at 3 years (Pérez-Barbería et al. 1998). The age at first reproduction seems to be at 3 years in the Apennines, too (Lovari and Bruno 2003). In the Pyrenees, the peak in female breeding success (i.e., kids to females ratio) occurs between 4 and 13 years of age (Crampe et al. 2004). Reproductive senescence in females in a stable Cantabrian population occurred after the age of 11 (Pérez-Barbería et al. 1998). Generally, female breeding success in Cantabrian and Pyrenean chamois is highly variable, ranging between 0.5 and 0.9, with an average of 0.7 (Pérez-Barbería et al. 2017) and no apparent differences between hunted and non-hunted populations (Pérez-Barbería et al. 2017). Climate seems to be one of the main factors affecting breeding success (Crampe 1997; González-Quirós et al. 2009), though density dependence has also been described in Cantabrian populations (Palacios 2009). In the Apennines, female breeding success is usually between 0.6 and 0.8, with peaks of 1 kid/female (Locati and Lovari 1988; Latini et al. 2015). In the Pyrenees, a positive relationship between gestation and survival was found, suggesting that nonbreeding females are in poor condition. Furthermore, early lactation decreased the probability of future reproduction, and reproduction the year before reduced kid survival in spring (Richard et al. 2017).

No information is available about male reproductive success in Southern chamois, although in Apennine chamois, males ≥6 years tend to monopolize mating events until at least 13–14 years of age (Locati and Lovari 1988).

Survival

In the Pyrenees, kid survival is variable and related to winter harshness (Crampe et al. 2002). Kid survival in the first year of life has been relatively low in the Apennines, i.e., less than 50%, on average, over the last 3 decades (Dupré et al. 2001; Latini et al. 2015).

Data on prime-aged (2–7 years) adults in non-hunted populations are available for the Pyrenees and show high overall survival probabilities (i.e., over 0.9, cf. Crampe 1997 and Loison et al. 1999c). Age-specific survival shows a similar trend in both sexes, though it decreases more rapidly in males after the age of 13 years (Gonzalez and Crampe 2001; Loison et al. 1999c). No published information on adult survival is available for the Apennines. The slightly more rapid decline in survival probabilities in males may be the reason for the marginally skewed sex ratio towards females also in non-hunted populations, *c.* 1:1.1–1.2 (Gonzalez and Crampe 2001; Pérez-Barbería et al. 2017; Arnal et al. 2013). Overall, sex-ratios estimated through counts in summer usually return a biased value towards females (Herrero et al. 2002; Latini et al. 2015), partly because of the lower male detectability, as males mainly frequent forested areas at lower altitudes (Lovari and Cosentino 1986) and show more solitary behavior. In the Pyrenees, maximum lifespan is *c.* 16–19 years in males, and 21 years in females (Gonzalez and Crampe 2001; Loison et al. 1999c), while females reach 19 years in the Cantabrian chamois (Pérez-Barbería and Pérez-Fernández 2009). There is no compelling information about survival of Apennine chamois: a maximum lifespan of 15 (males) and 19 (females) years has been suggested (Artese and Damiani 2015), but data come from a reintroduced population and need confirmation. In the French Pyrenees, evidence of a negative relationship between horn length and the probability of reaching 9 years of age was found in Pyrenean chamois males, but not in females (Douhard et al. 2020).

Habitat and Diet

Habitat Selection and Movement

Habitat use of Southern chamois depends on habitat characteristics, sex/age, and seasonality (Lovari and Cosentino 1986; Herrero et al. 1996; García-González and Cuartas 1996). In the Cantabrian Mountains and the Pyrenees, chamois are typical ecotonal species, moving between the forest limit and areas above the timberline, in proximity to escape terrain (i.e., cliffs and steep slopes) (Pérez-Barbería and Pérez-Fernández 2009). In the Apennines, upper meadows, rocky areas, and forests (mainly beech *Fagus sylvatica*) at lower altitudes are the main habitats used (Lovari and Cosentino 1986). The overall altitudinal distribution usually ranges between 1,000 and 2,800 m above sea level, although chamois can live at 800 m in the Pyrenees, in *Pinus nigra* plantations, or at 400 m in the Cantabrian mountains, near the coast.

In spring, as snow melts, chamois progressively move upslope, taking advantage of the tender high-quality grass sprouting between snow patches (García-González et al. 1990). Females give birth in the steepest areas, and when kids are able to follow their mothers, they gather in productive meadows (Lovari and Cosentino 1986; Pérez-Barbería and Nores 1996). In summer, Pyrenean and Cantabrian chamois typically occupy high-altitude areas with meadows and rocks, generally above the location of cattle herds, and in winter, they prefer to inhabit valley bottoms and sunny slopes which are less exposed to the wind. On northerly exposed slopes (e.g., in the French Pyrenees) and in winter, chamois prefer steep and rocky slopes where little snow accumulates ("refuge-station" *sensu* Berducou 1982). They also use these areas after late snowfalls in spring (Pépin et al. 1997). Some populations use forest habitats throughout the year, especially in areas of high population density, in the absence or with little extension of an area above the timberline, or where upper elevation areas are heavily used by livestock or tourists (García-González et al. 1992; Herrero et al. 1996).

With regard to sex differences, in the Eastern Pyrenees, males use bush areas with low presence of forage throughout the year, while females select grassland areas; females display seasonal altitudinal migration to a larger extent than males do, with the highest probability of being present (with offspring) in the highest altitudes in summer (Dalmau et al. 2013).

In the Apennines, mixed groups of females, kids, and yearling/subadult males use the upper meadows from early summer to late autumn/early winter (above 1700 m, Lovari and Cosentino 1986), where females select vegetation patches dominated by highly nutritious legumes (mainly *Trifolium thalii*: Ferrari et al. 1988). In the cold period, these individuals move to wooded, steep areas at lower altitudes, where steep slopes prevent snow from accumulating (Lovari and Cosentino 1986). Adult males select forested habitats all year round except for the mating period, i.e., November, when they join females on the upper meadows (Lovari and Cosentino 1986).

Seasonal migrations between winter and summer headquarters normally take place each year, though a part of the population may remain sedentary (e.g., 60% of marked females in Cauterets Massif, Crampe et al. 2007). Seasonal movements are shorter in females than in males, although the former can move up to *c.* 7 km (García-González et al. 1992). Dispersal (i.e., individuals leaving and never coming back), on the other hand, is rare in females, which tend to be philopatric: in the Orlu Reserve, 95% of females *vs.* 68% of males were philopatric, possibly due to inbreeding-avoidance (Loison et al. 1999a). No published information is available for movements of Apennine chamois, except for newly released individuals (Bocci et al. 2016; see also Rossetti et al. 2015). Female philopatry has also been suggested for Apennine chamois, whereas most young males tend to disperse, showing nomadic habits between 3 and 7–9 years of life (Lovari and Bruno 2003). Individuals released in the Sibillini National Park tended to use seasonal home ranges of 50–200 ha (Rossetti et al. 2015), with distances travelled peaking on the first weeks after the release, and stabilizing after *c.* 2–3 months (Bocci et al. 2016). Southern chamois also appear to have good swimming skills: recently, a Cantabrian chamois was found several kilometers from the coast in the Cantabrian sea (Kavčić et al. 2020).

Diet

Southern chamois are intermediate feeders (*sensu* Hofmann 1989), closer to concentrate selectors than to grazers (Ferrari et al. 1988; García-González and Cuartas 1996). Through various strategies of active selection (space-time movements, species, and plant parts selection) chamois can, in fact, feed on a higher quality diet than available on average (Aldezabal and García-González 2004). For example, the main grassland species in the Pyrenean chamois diet (e.g., *Festuca rubra*) shows particularly high values of protein content (Marinas and García-González 2006). Over the year, chamois tend to eat more herbaceous plants than woody plants (Aldezabal and García-González 2004), even when they live in forest environments (García-González and Cuartas 1996). Females feed more on herbaceous plants (81.6%) than males (65.6%) (Pérez-Barbería et al. 1997). Only when snow covers most of the pastures chamois increase browse consumption (Pérez-Barbería et al. 1997; Villamuelas et al. 2016). In these circumstances, they feed on conifers, heather, and other shrubs that stand out from the snow, and when population density is locally high, chamois can cause damage to forest plants (Pyrenees: Berducou 1972).

The trophic adaptation of the Pyrenean chamois has been investigated in the Eastern Pyrenees, showing that chamois are well-adapted to the variations in the seasonal phenology of plants in alpine habitats (Espunyes et al. 2019a). Furthermore, Pyrenean chamois shows high similarity index between diet and available vegetation, which is expected to indicate high adaptability, thereby suggesting a lower potential for trophic disturbance for chamois in a future scenario of global change (Espunyes et al. 2019b).

Contrary to the Pyrenees, Cantabrian chamois increases the proportion of ligneous plants also in their summer diet, particularly heather *Erica* sp. Pérez-Barbería et al. (1997) attribute this fact to the possible competition with livestock, which is normally more abundant on the summering pastures of the Cantabrian mountains than on the Pyrenees.

In Apennine chamois, females, immatures, and kids select for highly nutritious legumes belonging to the *Festuco-Trifolietum thalii* community, which is rare and extrazonal in the Apennines (Ferrari et al. 1988). This vegetation type supplies chamois with a protein-rich diet in the nursing/weaning period (Ferrari et al. 1988). In particular, female diet is dominated by grass and forbs from spring to autumn, with a growing occurrence of trees/shrubs throughout summer-autumn months (Lovari et al. 2014; Ferretti et al. 2015).

Behavior

Social Behavior

Information on social behavior is mostly available for Apennine chamois. Sexual segregation tends to occur almost year-round, with the exception of the mating period (November), a pattern observed also in the Pyrenean chamois (Crampe et al. 2021). For Apennine chamois, 31 behavior patterns have been described for intraspecific communication, including threats, dominance displays, and submission patterns (Lovari 1985; Locati and Lovari 1990, 1991; Lovari et al. 2020b). In female groups, a dominance hierarchy among individuals is positively correlated with age, horn size and, especially, body mass (Locati and Lovari 1991). Prime-aged females are more dominant than other females (Lovari and Rosto 1985; Locati and Lovari 1991; Fattorini et al. 2018b). The rate of aggressive interactions, frequency of vigilance and cortisol/testosterone levels peak in summer, when more nutritious food is available and when nursing costs are higher, whereas intensity of aggression (i.e., threats, in respect to displays) increases throughout summer-autumn, i.e., in parallel with the seasonal decrease of food availability (Fattorini et al. 2018b). Among males, fully mature individuals (>8 years old) are the most dominant and tend to monopolize mating events (see below). Compared to females, males use indirect forms of aggression relatively more often than direct threats in intra-sexual interactions (Locati and Lovari 1990).

Nursing behavior peaks in summer and decreases throughout autumn (Scornavacca et al. 2016, 2018). Occurrence of several kids suckling from the same female at the same time has been reported (*c.* 60% of suckling events in a food-rich area, Scornavacca et al. 2018). No information is available for the other subspecies.

Mating Behavior

The rut occurs in November (Lovari 1984b; Kourkgy et al. 2016). In this period, fully mature males (>8 years old) of Apennine chamois defend territories with groups of females, keeping the same territory for about 2–3 years (Lovari 1984b). Group defense occurs through direct and indirect forms of aggression, often ending with a chase or a rush, and courtship behavior patterns. Three male behavior patterns of Apennine chamois have not been described for the Northern chamois, and two of the latter have not been reported in the former (Masini and Lovari 1988; see also Corlatti et al. 2011, 2013b). Two of these patterns specific to Apennine chamois are relevant to courtship and may have been a behavioral barrier against hybridization between Northern and Southern chamois, if they were sympatric at the beginning of the Würm glaciation (Lovari 1984a; Masini and Lovari 1988). Information on rutting behavior in Cantabrian chamois is not available, whereas for Pyrenean chamois some preliminary information was reported in Brun and García-González (1989).

Activity

In summer, Pyrenean chamois spend *c.* 50–65% of their diurnal time grazing, less than 10% of their time moving, and about 20–30% resting, if not disturbed (Pépin et al. 1991; Aldezabal et al. 1999). This strategy apparently aims to minimize energy costs and increase the acquisition of energy and nutrients. For Pyrenean chamois, a repertoire of 24 different behaviors (excluding displacements) was found in summer, 54% of which corresponded to grazing, 36% to resting, and the rest to other behaviors (Brun and Martínez-Rica 1983). The distribution of grazing time is multimodal in summer, with several peaks of activity throughout the day (Aldezabal et al. 1999). However, with the shortening daylight in winter, two activity peaks occur, at dawn and dusk (Pépin et al. 1991; Ciofolo and Boissier 1992). A similar pattern was observed for Apennine chamois in the 1980s, with multimodal activity in summer and two peaks of activity in autumn (Bruno and Lovari 1989).

Foraging Behavior

Cantabrian females dedicate more time to foraging than males (53% *vs.* 37%) (Pérez-Barbería et al. 1997). No differences between reproductive and nonreproductive females in the time dedicated to feeding have been detected, leading to poorer physical conditions in reproductive females because of reproductive costs (Pérez-Barbería and Nores 1996). Female groups avoid the presence of livestock when they overlap in the summer pastures, so they are displaced to the highest elevations (Pérez-Barbería and Pérez-Fernández 2009). Female Pyrenean chamois spent more time foraging than males, while males spent more time being vigilant, especially the individuals occupying central position of the groups. Group size does not seem to affect vigilance behavior (Dalmau et al. 2010).

In Apennine chamois, males and females have been shown to select nutritious forb-dominated patches in the warm months, and to spend a comparable amount of time feeding, but males appear less selective than females, showing relatively greater feeding intensity and a lower food searching behavior (Ferretti et al. 2014). This finding suggests that males enhance their energy and mass gain by increasing their food intake in the warm season, to meet the costs of mating (cf. section "Mating Behavior"). Females show a higher bite rate in summer (i.e., when nursing effort peaks, Scornavacca et al. 2016, 2018) than in autumn, when plant senescence probably leads to a lower feeding intensity (Bruno and Lovari 1989; Ferretti et al. 2014; Lovari et al. 2014).

Parasites and Diseases

Ectoparasites and Endoparasites

Most ectoparasites and endoparasites recorded in Northern chamois also occur in the Southern chamois, and none is specific to the latter. Since the 1990s, mange caused by *Sarcoptes scabiei* has affected the Eastern subpopulation of Cantabrian chamois with obvious demographic impacts (González-Quirós et al. 2009). The origin of the ongoing outbreak was attributed to contacts with infested domestic goats (Lavín et al. 2000). Mange has never been diagnosed in the *pyrenaica* and *ornata* subspecies.

Infectious Diseases

Outbreaks of infectious kerato-conjunctivitis (IKC) caused by *Mycoplasma conjunctivae* have been reported in the Cantabrians and the entire Pyrenees chain since the 1950s. In a recent large-scale IKC outbreak, the reduction in population size was approximately 25% (Arnal et al. 2013). *M. conjunctivae* has frequently been isolated from the eyes of healthy domestic sheep and goats in the Pyrenees (Fernández-Aguilar et al. 2013). Another important and novel outbreak disease so far exclusive to the Pyrenees is pestivirosis, caused by a specific variant of the border disease virus. Contagion by this virus from the "traditional" domestic sheep reservoir was dated back to the late 1990s (Luzzago et al. 2016). Symptoms are variable degrees of cachexia, alopecia (often associated with skin hyperpigmentation), and neurological disorders prior to death. The immunosuppressive effect of the virus may also result in secondary infection, e.g., bacterial bronchopneumonia (Marco 2012). A large-scale study was carried out to determine the prevalence of antibodies against pestivirus species in wild ruminants in mainland Spain between 2000 and 2017, with the highest values (30%) found in Southern chamois (Jiménez-Ruiz et al. 2020). The demographic effects of pestivirosis are extremely variable, from a mild impact on reproductive performance to severe die-offs with mortality rates between 40% and 85%, as observed in the Eastern Pyrenees (Marco et al. 2009). Recently, Lambert et al. (2018) showed how stochastic processes play a major role in governing epidemiological patterns of pestivirus spread in both protected and hunted populations of Pyrenean chamois, and suggested that a "do nothing" approach during epidemics could be the best management strategy. Neither IKC nor pestivirosis outbreaks have been reported for Apennine chamois. Small-scale outbreaks of "transmissible pneumonias" of uncertain and still debated etiology have been reported in the Pyrenees. Similar to the Northern chamois, no significant role has been attributed to its Southern relative in the medium to long term maintenance of major livestock diseases in Europe, foot and mouth disease, tuberculosis, brucellosis and bluetongue, among others.

Zoonoses

No major risks to human health derive from handling Southern chamois or dressing their carcasses or using their meat or nonmeat products.

Population Ecology

Population Dynamics

Information on population dynamics is mainly available for the Pyrenean and Cantabrian subspecies. The maximum theoretical growth rate is about 25%, although actual mean rates of increase are much lower, about 9.6% in the Cantabrian Mountains (Nores and González-Quirós 2009) and 10% in the Pyrenees (Arnal et al. 2013). While some populations seem to be stable (Crampe 1997), others show greater variations due to outbreaks of keratoconjunctivitis by *Mycoplasma conjunctivae* (Pañella et al. 2010), whereas other populations have shown increases in numbers and distribution (Herrero et al. 2010; Pérez-Barbería et al. 2009). Growth of most populations is limited by density, or serious outbreaks, such as sarcoptic mange in the Cantabrian Mountains. Population variation, around a new level of stability lower than before disease outbreaks, seems to be influenced by density-dependent contagion (Nores and González-Quirós 2009). All these circumstances make demographic parameters (e.g., sex ratio, fertility, mortality, growth rate) change in time and space. For example, on the Cantabrians mortality rate was 0.26, 0.48, and 0.18 before, during, and after a mange outbreak, respectively. Well after the epizootics, mortality returned around values of 0.29 (Nores and González-Quirós 2009). Furthermore, in endemic areas, *Sarcoptes scabiei* may alter sex-ratio by affecting more males than females. In a Cantabrian population monitored for 16 years, the sex ratio changed from 0.7 males per female before the mange outbreak to 0.3 afterwards (Nores and González-Quirós 2009). In some areas of the Pyrenees, keratoconjunctivitis caused by *Mycoplasma conjunctivae* caused higher mortality in adult females than in males, leading to a relatively balanced sex-ratio (Arnal et al. 2013). For Apennine chamois, the population living in the Abruzzo, Lazio and Molise National Park is showing a recent decline (see section "Management"). Populations newly established in other National/Regional Parks are increasing (Antonucci and Di Domenico 2015).

On the Cantabrians, frequent predators of chamois are wolf, golden eagle, and red fox *Vulpes vulpes*. One fourth of wolf diet can be based on chamois if it is locally abundant (Pérez-Barbería et al. 2017). Occasionally, brown bear *Ursus arctos* can prey on kids (Blanco et al. 2011). Chamois are also consumed by scavengers, among which the most important are Eurasian griffon *Gyps fulvus*, Eurasian black vulture *Aegypius monachus*, Egyptian vulture *Neophron percnopterus*, and bearded vulture *Gypaetus barbatus*. In the Apennines, Southern chamois are potential prey to wolves, brown bears, and golden eagles, but actual predation appears rare (Patalano and Lovari 1993, for wolf; Fico et al. 1984 and Di Domenico et al. 2012, for brown bear; Locati 1990 and Scornavacca and Brunetti 2015, for golden eagle).

Competition with Other Ungulates

There is no indication in the literature of a potential competition with other wild herbivores except for red deer, which can show overlap with chamois in diet and use of grasslands (for the Apennines: Lovari et al. 2014; Ferretti et al. 2015). Resource exploitation by red deer is an important factor affecting winter survival of kids – through a negative influence on food resources and, consequently, diet quality for females – in sectors of Abruzzo, Lazio and Molise National Park (Lovari et al. 2014; Ferretti et al. 2015, 2019). Locally, competition with livestock affects the habitat use of Southern chamois (Pérez-Barbería and Pérez-Fernández 2009).

Effects of Climate Change

Environmental changes are likely to influence important aspects of chamois ecology. Moderate shrubification, i.e., woody plant expansion on the Pyrenees does not seem to affect negatively the feeding efficiency of Pyrenean chamois (Espunyes et al. 2019b). Conversely, the best food resources for Apennine chamois include cold-adapted, nutritious vegetation patches dominated by legumes: this vegetation is linked to grounds with long-lasting snow cover (Ferrari et al. 1988), which is thus potentially threatened by ongoing climatic changes (Lovari et al. 2020a). The depletion of these nutritious food resources during the warm months can negatively affect diet quality, foraging, and nursing behavior of female Apennine chamois, increasing their intra-sexual aggression, vigilance, and endogenous stress response, which can ultimately affect winter survival of chamois kids (Lovari et al. 2014; Ferretti et al. 2015; Scornavacca et al. 2016; Fattorini et al. 2018a). Long-term increases in temperature, vegetation dynamics in secondary meadows (with the spread of unpalatable tall grass), and resource exploitation by red deer are likely determinants of this food depletion (Lovari et al. 2014; Ferretti et al. 2015, 2019; Corazza et al. 2016). In the Pyrenees, winter survival of adult female chamois was positively influenced by high precipitation and low temperature in the previous spring (Loison et al. 1999b). In the Apennines, high temperature and low rainfall during the growing season of vegetation have been suggested as a potential limiting factor to chamois through their negative effects on cold-adapted, nutritious food resources (Ferretti et al. 2019; Lovari et al. 2020a).

Conservation Status

Southern chamois is listed in Appendix III of the Bern Convention and Annex V of the EU Habitat Directive, as part of *R. rupicapra sensu lato*. It is catalogued as Least Concern (LC) in the IUCN Red List (Herrero et al. 2020). The Pyrenean and Cantabrian subspecies are allowed to be hunted in Andorra, France, and Spain, except for the Navarre region (Spain). The Apennine chamois is Vulnerable (VU) D1+2 according to the IUCN Red List, as this subspecies lives in very small populations and has restricted area of occupancy (Herrero et al. 2020). It is also listed in Appendix II of the Bern Convention, Annexes II and IV of EU Habitat Directive, Appendix II of CITES, and under Italian law n. 157/1992 ("especially protected species").

Management

Pyrenean and Cantabrian chamois do not show any conflict with human activities such as livestock keeping, forest management, agricultural damages, car strikes, or hunting. On the Cantabrians, chamois are subject to general hunting laws in the vast majority of their range, and starting in 1982, reintroductions have occurred into areas where chamois had been previously extirpated (González-Quirós 2009). In the Pyrenees, more than 600 chamois have been moved from the central to the western side of the mountain chain for hunting purposes (Berducou et al. 2004) and translocations are continuing to date.

Apennine chamois are protected by International/National laws (see above). Throughout the last few centuries, habitat alteration by humans, hunting, and competition with livestock caused this chamois to disappear from most of its range; it survived only in the Abruzzo, Lazio and Molise National Park (established in 1923) until 1991 (Dupré et al. 2001). Enhanced habitat protection, an increase in protected areas and reintroductions / "conservation introductions" (*sensu* IUCN 2012) helped the total number to increase up to no less than *c.* 2500 individuals in the wild (in 4 national parks and 1 regional park). Numbers are generally increasing at the release sites. However, in the Abruzzo, Lazio and Molise National Park, which holds the "mother" population (Lovari et al. 2014), alarming decreasing trends up to over 50% have been reported from important historical sectors during the last two decades (Lovari et al. 2014; Ferretti et al. 2015). Reduced genetic variability (Nascetti et al. 1985; Lorenzini 2005), competition with an increasing number of red deer (Lovari et al. 2014, 2020a; Ferretti et al. 2015) and livestock, the effects of climatic changes (Lovari et al. 2020a), as well as those of vegetation dynamics in secondary meadows (Lovari et al. 2014; Corazza et al. 2016), on its main food resources are all threats to its conservation.

As for the Northern chamois, population estimates and monitoring in the Southern chamois have traditionally been conducted through block counts (Berducou et al. 1982). Although block counts typically lead to underestimates due to imperfect detectability, which may vary depending on landscape features and seasonal variations in chamois behavior (Herrero et al. 2011), they proved robust indices of relative abundance to track population changes (Loison et al. 2006).

Future Challenges for Research and Management

Priorities for the species as a whole include extending monitoring to all populations, increase the knowledge on demography and the impact of hunting and disease outbreaks. Research and monitoring outside protected areas would be particularly important to evaluate effects of hunting on chamois populations. Climate and global changes are important drivers of environmental conditions in mountain areas occupied by chamois. The decreases in pasture surfaces (shrub encroachment) and their quality reduce the carrying capacity of mountain habitats for chamois and at the same time increase habitat for forest ungulates such as wild boar *Sus scrofa*, red deer, and roe deer. The role of interspecific interactions (predation; competition; interactions with livestock) should therefore be clarified. As for the Northern chamois, the performance of different abundance monitoring methods should be assessed.

Acknowledgments We would like to thank all the agencies and administrations who supported our research and monitoring of chamois subspecies (Alpine chamois: LC, FF, PA, SH, LR, SL; Cantabrian and Pyrenean chamois: JH, RGG, CN; Apennine chamois: FF, SL) and all the people (rangers, students, volunteers) who participated in fieldwork activities throughout these years. We also express our appreciation to Frank Zachos (Natural History Museum Vienna), Marco Festa-Bianchet (University of Sherbrooke), Nikica Šprem (University of Zagreb), and Klaus Hackländer (University of Natural Resources and Life Sciences Vienna) for suggestions and improvement of this chapter.

References

Aldezabal A, García-González R (2004) La alimentación del sarrio en el Pirineo central. In: Herrero J, Escudero E, Fernández de Luco D, García-González R (eds) El sarrio pirenaico *Rupicapra p. pyrenaica*: Biología, Patología y Gestión. Consejo de Protección de la Naturaleza de Aragón, Zaragoza, pp 169–189

Aldezabal A, Garin I, García-González R (1999) Activity rhythms and the influence of some environmental variables on summer ungulate behaviour in Ordesa-Monte Perdido National Park. Pirineos 153:145–157

Anderwald P, Herfindal I, Haller RM, Risch AC, Schütz M, Schweiger AK, Filli F (2015) Influence of migratory ungulate management on competitive interactions with resident species in a protected area. Ecosphere 6:228

Anderwald P, Haller RM, Filli F (2016) Heterogeneity in primary productivity influences competitive interactions between red deer and alpine chamois. PLoS One 11:e0146458

Anderwald P, Ambarli H, Avramov S, Ciach M, Corlatti L, Farkas A, Jovanovic M, Papaioannou H, Peters W, Sarasa M, Šprem N, Weinberg P, Willisch C (2021) *Rupicapra rupicapra* (amended version of 2020 assessment). The IUCN red list of threatened species 2021, e. T39255A195863093

Andreoli E, Roncoroni C, Gusmeroli F, Della Marianna G, Giacometti G, Heroldová M, Barbieri S, Mattiello S (2016) Feeding ecology of alpine chamois living in sympatry with other ruminant species. Wildl Biol 22:78–85

Antonucci A, Di Domenico G (eds) (2015) Chamois international congress proceedings, 17–19 June 2014. Lama dei Peligni, Majella National Park, Italy, 272 p

Arnal MC, Herrero J, de la Fe C, Revilla M, Prada C, Martínez-Durán D, Gómez-Martín Á, Fernández-Arberas O, Amores J, Contreras A, García-Serrano A, Fernández de Luco D (2013) Dynamics of an infectious keratoconjunctivitis outbreak by *Mycoplasma conjunctivae* on Pyrenean chamois *Rupicapra p. pyrenaica*. PLoS One 8:1–12

Arnold W (2020) Review: seasonal differences in the physiology of wild northern ruminants. Animal 14: 124–132

Artese C, Damiani G (2015) Observation of the longevity of the individuals of Apennine chamois that were released for the reintroduction in the Gran Sasso-Laga National Park and the survival in the first year of life of the kids from 1995 to 2008. In: Antonucci A, Di Domenico G (eds) Chamois international congress proceedings, 17–19 June 2014. Lama dei Peligni, Majella National Park, Italy, p. 227

Ascenzi P, Clementi ME, Condò S, Coletta M, Petruzzelli R, Polizio F, Rizzi M, Giunta C, Peracino V, Giardina B (1993) Functional, spectroscopic and structural properties of haemoglobin from chamois (*Rupicapra rupicapra*) and steinbock (*Capra hircus ibex*). Biochem J 296:361–365

Aulagnier S, Giannatos G, Herrero J (2008) *Rupicapra rupicapra*. In: Anonymous 2010. IUCN red list of threatened species. Version 2010.3. http://www.iucnredlist.org. Downloaded 1 Dec 2018

Ballo M (2010) Spatial and vertical distribution of chamois (*Rupicapra rupicapra tatrica*) in the Western Tatra Mountains. Oecologia Montana 19:35–41

Bassano B, Perrone A, Von Hardenberg A (2003) Body weight and horn development in alpine chamois, *Rupicapra rupicapra* (Bovidae, Caprinae). Mammalia 67:65–73

Bauer JJ. 1982. Untersuchungen zur Dynamik von stabilen und kolonisierenden Gemsenpopulationen (*Rupicapra rupicapra* L.) Neuseelands. Inaugural-Dissertation zur

Erlangung der Doktorwürde der Biologischen Fakultät der Albert-Ludwigs-Universität zu Freiburg i. Brsg. 205 p
Bauer JJ (1987) Factors determining the onset of sexual maturity in New Zealand chamois (*Rupicapra rupicapra* L.). Zeitschrift für Säugertierkunde 52:116–125
Baumann M, Babotai C, Schibler J (2005) Native or naturalized? Validating alpine chamois habitat models with archaeozoological data. Ecol Appl 15:1096–1110
Berducou C (1972) L'isard nuit-il à l'environnement? Pirineos 105:119–127
Berducou C (1982) A propos de la biologie hivernale de l'isard: la notion de station-refuge et ses implications. Pirineos 117:79–90
Berducou C, Besson J-P, Gardes Moniteurs du PNPO (1982) Dynamique des populations d'isards du Parc National des Pyrénées Occidentales de 1968–1981. Acta Biol Montana 1:153–175
Berducou C, Novoa C, Saint Hilaire K, Appolinaire J, Menaut P (2004) Etat des populations et modes de gestion de l'isard aux Pyrénées françaises. In: Herrero J, Escudero E, Fernández de Luco D, García-González R (eds) El sarrio pirenaico *Rupicapra p. pyrenaica*: Biología, Patología y Gestión. Consejo de Protección de la Naturaleza de Aragón, Zaragoza, pp 69–82
Bertolino S, Di Montezemolo NC, Bassano B (2009) Food-niche relationships within a guild of alpine ungulates including an introduced species. J Zool 277:63–69
Blanco JC, Ballesteros F, García-Serrano A, Herrero J, Nores C, Palomero G (2011) Behaviour of brown bears killing wild ungulates in the Cantabrian Mountains, Southwestern Europe. Eur J Wildl Res 57:669–673
Bleu J, Loison A, Toïgo C (2014) Is there a trade-off between horn growth and survival in adult female chamois? Biol J Linn Soc 113:516–521
Bleu J, Herfindal I, Loison A, Kwak AMG, Garel M, Toïgo C, Rempfler T, Filli F, Sæther B-E (2015) Age-specific survival and annual variation in survival of female chamois differ between populations. Oecologia 179:1091–1098
Bocci A, Canavese G, Lovari S (2010) Even mortality patterns of the two sexes in a polygynous, near-monomorphic species: is there a flaw? J Zool 280:379–386
Bocci A, Menapace S, Alemanno S, Lovari S (2016) Conservation introduction of the threatened Apennine chamois *Rupicapra pyrenaica ornata*: post release dispersal differs between wild-caught and captive founders. Oryx 50:128–133
Boch J, Schneidawind H (1988) Krankheiten des jagdbaren Wildes. Verlag Paul Parey, Hamburg/Berlin
Boschi C, Nievergelt B (2003) The spatial patterns of Alpine chamois (*Rupicapra rupicapra rupicapra*) and their influence on population dynamics in the Swiss National Park. Mamm Biol 68:16–30
Brivio F, Bertolucci C, Tettamanti F, Filli F, Apollonio M, Grignolio S (2016) The weather dictates the rhythms: Alpine chamois activity is well adapted to ecological conditions. Behav Ecol Sociobiol 70:1291–1304
Brun JM, García-González R (1989) Differential behavioural patterns of Pyrenean chamois during rut. Symposium Chamois, Ljubljana 1988. Conseil International de la Chasse et de la Conservation du Gibier, Paris, pp 109–125
Brun JM, Martínez-Rica JP (1983) Datos sobre conducta del sarrio, *Rupicapra rupicapra* (L.) Artiodactila, Bovidae, en el Pirineo Aragonés. Pirineos 119:29–53
Bruno E, Lovari S (1989) Foraging behaviour of adult female Apennine chamois in relation to seasonal variation in food supply. Acta Theriol 34:513–523
Bubenik AB, Schwab P (1975) Structure des populations de chamois. La simulation et sa signification pour la régulation des effectifs. Bulletin Mensile Office Nationale de la Chasse 3:195–198
Cabrera Á (1914) Fauna Ibérica, Mamíferos. Museo Nacional de Ciencias Naturales, Madrid
Camerano L (1914) Ricerche intorno ai camosci (Parte Ia, IIa, IIIa). Memorie della Regia Acc. di Scienze di Torino (Cl. Scienze Fisiche Matematiche Naturali) 64(4):1–82, 64(12):1–88, 65(14):1–82
Capurro AF, Gatto M, Tosi G (1997) Delayed and inverse density dependence in a chamois population of the Italian Alps. Ecography 20:37–47
Carnevali L, Lovari S, Monaco A, Mori E (2016) Nocturnal activity of a "diurnal" species, the Northern chamois, in a predator-free Alpine area. Behav Process 126: 101–107
Castelló JR (2016) Bovids of the world. Princeton University Press, Princeton/Oxford
Catusse M, Corti R, Cugnasse J-M, Dubray D, Gibert P, Michallet J (1996) La grande faune de montagne. ONC, Ed. Hatier, Paris
Chirichella R, Ciuti S, Apollonio M (2013a) Effects of livestock and non-native mouflon on use of high-elevation pastures by Alpine chamois. Mamm Biol 78:344–350
Chirichella R, Ciuti S, Grignolio S, Rocca M, Apollonio M (2013b) The role of geological substrate for horn growth in ungulates: a case study on Alpine chamois. Evol Ecol 27:145–163
Chirichella R, Mustoni A, Apollonio M (2015) Ecological drivers of group size in female Alpine chamois, *Rupicapra rupicapra*. Mammalia 79:375–383
Chirichella R, Rocca M, Brugnoli A, Mustoni A, Apollonio M (2020) Fluctuating asymmetry in Alpine chamois horns: an indicator of environmental stress. Evol Ecol 34:573–587
Chirichella R, Stephens PA, Mason THE, Apollonio M (2021) Contrasting effects of climate change on Alpine chamois. J Wildl Manag 85:109–120
Ciach M, Pęksa L (2018) Impact of climate on the population dynamics of an alpine ungulate: a long-term study of the Tatra chamois *Rupicapra rupicapra tatrica*. Int J Biometeorol 62:2173–2182
Ciach M, Pęksa Ł (2019) Human-induced environmental changes influence habitat use by an ungulate over the long term. Curr Zool 65:129–137
Ciofolo I, Boissier M (1992) Diurnal fluctuations in activity in the izard. J Ethol 10:1–5
Corazza M, Tardella FM, Ferrari C, Catorci A (2016) Tall grass invasion after grassland abandonment influences the availability of palatable plants for wild herbivores: insight

into the conservation of the Apennine chamois *Rupicapra pyrenaica ornata*. Environ Manag 57:1247–1261

Corlatti L (2018) Fecal cortisol metabolites under anonymized sampling: robust estimates despite significant individual heterogeneity. Ecol Indic 95:775–780

Corlatti L, Bassano B (2014) Contrasting alternative hypotheses to explain rut-induced hypophagia in territorial male chamois. Ethology 120:32–41

Corlatti L, Sivieri S (2020) Face markings in Northern chamois: cues of dominance? Mamm Biol 100:559–565

Corlatti L, Lorenzini R, Lovari S (2011) The conservation of the chamois *Rupicapra* spp. Mammal Rev 41:163–174

Corlatti L, Béthaz S, von Hardenberg A, Bassano B, Palme R, Lovari S (2012a) Hormones, parasites and male mating tactics in Alpine chamois: identifying the mechanisms of life history trade-offs. Anim Behav 84: 1061–1070

Corlatti L, Lebl K, Filli F, Ruf T (2012b) Unbiased sex-specific survival in Alpine chamois. Mamm Biol 77:135–139

Corlatti L, Bassano B, Valencak TG, Lovari S (2013a) Foraging strategies associated with alternative reproductive tactics in a large mammal. J Zool 291: 111–118

Corlatti L, Caroli M, Pietrocini V, Lovari S (2013b) Rutting behaviour of territorial and nonterritorial male chamois: is there a home advantage? Behav Process 92:118–124

Corlatti L, Palme R, Lovari S (2014) Physiological response to etho-ecological stressors in male Alpine chamois: timescale matters! Naturwissenschaften 101: 577–586

Corlatti L, Bassano B, Poláková R, Fattorini L, Pagliarella MC, Lovari S (2015a) Preliminary analysis of reproductive success in a large mammal with alternative mating tactics, the Northern chamois, *Rupicapra rupicapra*. Biol J Linn Soc 116:117–123

Corlatti L, Fattorini L, Nelli L (2015b) The use of block counts, mark-resight and distance sampling to estimate population size of a mountain-dwelling ungulate. Popul Ecol 57:409–419

Corlatti L, Gugiatti A, Imperio S (2015c) Horn growth patterns in Alpine chamois. Zoology 118:213–219

Corlatti L, Storch I, Filli F, Anderwald P (2017) Does selection on horn length of males and females differ in protected and hunted populations of a weakly dimorphic ungulate? Ecol Evol 7:3713–3723

Corlatti L, Bonardi A, Bragalanti N, Pedrotti L (2019a) Long term dynamics of Alpine ungulates suggest interspecific competition. J Zool 309:241–249

Corlatti L, Ferretti F, Lovari S (2019b) Hunting lactating female ungulates deserves caution: the case of the chamois. Ethol Ecol Evol 31:293–299

Corlatti L, Lorenzetti C, Bassano B (2019c) Parasitism and alternative reproductive tactics in Northern chamois. Ecol Evol 9:8749–8758

Corlatti L, Bassano B, Lovari S (2020) Weather stochasticity and alternative reproductive tactics in Northern chamois. Biol J Linn Soc 130:359–364

Corlatti L, Cotza A, Nelli L (2021) Linking alternative reproductive tactics and habitat selection in Northern chamois. Ecol Evol 11:7057–7068

Corti R (2002) Le chamois et l'isard. Paris: Off. Natl. Chasse Faune Sauvage, broch. n. 18. 36 p

Couturier M (1938) Le Chamois. Arthaud, Grenoble

Couturier M (1958) Parallèle anatomique physiologique et écologique entre le pied du bouquetin (*Capra ibex ibex*) et celui du chamois (*Rupicapra rupicapra rupicapra*) en rapport avec l'adaptation à la montagne des deux espèces. Mammalia 22:76–89

Crampe J-P (1997) Caractéristiques bio-démographiques d'une population d'isards (*Rupicapra p. pyrenaica*) non-chasée dans le Parc National des Pyrénées, en vallée de Cauterets. Documents scientifiques du Parc National des Pyrénées 31:1–169

Crampe J-P, Gaillard J-M, Loison A (2002) L'enneigement hivernal: un facteur de variation du recrutement chez l'isard (*Rupicapra pyrenaica pyrenaica*). Can J Zool 80:1306–1312

Crampe J-P, Caens P, Florence E, Gaillard J-M, Loison A (2004) Effets de l'age sur la reproduction des femelles d'isards *Rupicapra pyrenaica* dans un population du Parc National des Pyrénées. In: Herrero J, Escudero E, Fernández de Luco D, García-González R (eds) El sarrio pirenaico *Rupicapra p. pyrenaica*: Biología, Patología y Gestión. Consejo de Protección de la Naturaleza de Aragón, Zaragoza, pp 69–82

Crampe J-P, Bon R, Gérard J-F, Serrano E, Caens P, Florence E, Gonzalez G (2007) Site fidelity, migratory behaviour, and spatial organization of female isards (*Rupicapra pyrenaica*) in the Pyrenees National Park, France. Can J Zool 85:16–25

Crampe J-P, Gerard J-F, Goulard M, Milleret C, Gonzalez G, Bon R (2021) Year-round sexual segregation in the Pyrenean chamois, a nearly monomorphic polygynous herbivore. Behav Process 184:104300

Crégut-Bonnoure E, Dimitrijevic V (2006) *Megalovis balcanicus* sp. nov. and *Soergelia intermedia* sp. nov. (Mammalia, Bovidae, Caprinae) new Ovibovini and the early Pleistocene of Europe. Revue de Paléobiologie de Genève 25:723–773

Crestanello B, Pecchioli E, Vernesi C, Mona S, Martínková N, Janiga M, Hauffe HC, Bertorelle G (2009) The genetic impact of translocations and habitat fragmentation in chamois (*Rupicapra*) spp. J Hered 100:691–708

Dalmau A, Ferret A, Chacon G, Manteca X (2007) Seasonal changes in fecal cortisol metabolites in Pyrenean chamois. J Wildl Manag 71:190–194

Dalmau A, Ferret A, Manteca X (2010) Vigilance behavior of pyrenean chamois *Rupicapra pyrenaica pyrenaica*: effect of sex and position in the herd. Curr Zool 52: 232–237

Dalmau A, Ferret A, Torre JLR, Manteca X (2013) Habitat selection and social behaviour in a Pyrenean chamois population (*Rupicapra pyrenaica pyrenaica*). J Mt Ecol 9:83–102

Damm GR, Franco N (2014) CIC Caprinae Atlas of the World. CIC International council for game and wildlife conservation, Budakeszi, Hungary. Rowland Ward Publications, Johannesburg

Darmon G, Calenge C, Loison A, Jullien J-M, Maillard D, Lopez J-F (2012) Spatial distribution and habitat selection in coexisting species of mountain ungulates. Ecography 35:44–53

Di Domenico G, Tosoni E, Boitani L, Ciucci P (2012) Efficiency of scat-analysis lab procedures for bear dietary studies: the case of the Apennine brown bear. Mamm Biol 77:190–195

Dolan JM (1963) Beitrag zur systematischen Gliederung des Tribus Rupicaprini Simpson 1945. Zeitschrift für Zoologische Systematik und Evolutionsforschung 1: 311–407

Donini V, Pedrotti L, Ferretti F, Corlatti L (2021) Disentangling demographic effects of red deer on chamois population dynamics. Ecol Evol 11:8264–8280

Douhard M, Crampe J-P, Loison A, Bonenfant C (2020) A negative association between horn length and survival in a weakly dimorphic ungulate. Ecol Evol 10: 2793–2802

Dupart A, Garel M, Marchand P, Dubray D, Maillard D, Loison A (2020) Through the taste buds of a large herbivore: foodscape modeling contributes to an understanding of forage selection processes. Oikos 129:170–183

Dupré E, Monaco S, Pedrotti L (2001) Piano d'azione nazionale per il Camoscio appenninico (*Rupicapra pyrenaica ornata*). Quad. Cons. Natura, 10, Min. Ambiente – Ist. Naz. Fauna Selvatica

Durand T, Gauthier D (1996) The helminths of the Chamois *Rupicapra rupicapra*: host-parasite-environment relationships and wildlife health management. Vie et Milieu 46:333–343

Đurović M (2018) Ecology characteristics of Balkan chamois (*Rupicapra rupicapra balcanica* Bolkay, 1925) in Montenegro with a special focus on National Park "Durmitor". Master thesis. University of Sarajevo

Espunyes J, Bartolomé J, Garel M, Gálvez-Cerón A, Fernández Aguilar X, Colom-Cadena A, Calleja JA, Gassó D, Jarque L, Lavín S, Marco I, Serrano E (2019a) Seasonal diet composition of Pyrenean chamois is mainly shaped by primary production waves. PLoS One 14:e0210819

Espunyes J, Lurgi M, Büntgen U, Bartolomé J, Calleja JA, Gálvez-Ceron A, Peñuelas J, Claramunt-López B, Serrano E (2019b) Different effects of alpine woody plant expansion on domestic and wild ungulates. Glob Chang Biol 25:1808–1819

Fattorini N, Brunetti C, Baruzzi C, Macchi E, Pagliarella MC, Pallari N, Lovari S, Ferretti F (2018a) Being "hangry": food depletion, aggression and cascading effects on social behavior. Biol J Linn Soc 125:640–656

Fattorini N, Lovari S, Brunetti C, Baruzzi C, Cotza A, Macchi E, Pagliarella MC, Ferretti F (2018b) Age, seasonality and correlates of aggression in female Apennine chamois. Behav Ecol Sociobiol 72:171

Fernandez P, Crégut E (2007) Les Caprinae (Rupicaprini, Ovibovini, Ovini et Caprini) de la séquence pléistocène de Kozarnika (Bulgarie du Nord): morphométrie, biochronologie et implications phylogéniques. Revue de Paléobiologie, Museum d'Histoire Naturelle de la Ville de Geneve 26:425–503

Fernández-Aguilar X, Cabezón O, Marco I, Mentaberre G, Frey J, Lavín S, López-Olvera JR (2013) *Mycoplasma conjunctivae* in domestic small ruminants from high mountain habitats in Northern Spain. BMC Vet Res 9:253

Fernández-López JM, García-González R (1986) Craniométrie comparée entre le chamois pyrénéen et cantabrique. Mammalia 50:87–87

Ferrari C, Rossi G, Cavani C (1988) Summer food habits and quality of female, kid and subadult Apennine chamois, *Rupicapra pyrenaica ornata* Neumann, 1899 (Artiodactyla, Bovidae). Zeitschrift Säugetierkunde 53:170–177

Ferretti F, Costa A, Corazza M, Pietrocini V, Cesaretti G, Lovari S (2014) Males are faster foragers than females: intersexual differences of foraging behaviour in the Apennine chamois. Behav Ecol Sociobiol 68:1335–1344

Ferretti F, Corazza M, Campana I, Pietrocini V, Brunetti C, Scornavacca D, Lovari S (2015) Competition between wild herbivores: reintroduced red deer and Apennine chamois. Behav Ecol 26:550–559

Ferretti F, Lovari S, Stephens PA (2019) Joint effects of weather and interspecific competition on foraging behaviour and survival of a mountain herbivore. Curr Zool 65:165–175

Fico R, Locati M, Lovari S (1984) A case of brown bear predation on Abruzzo chamois. Säugetierkundliche Mitteilungen 31:185–187

Flores-Saavedra W, Espunyes J, Fernández-Aguilar X, Colom-Cadena A, Velarde R, Mentaberre G, Lavín S, López-Olvera JR, Serrano E (2018) Fat reserve assessment in Pyrenean chamois using body measurements. Mamm Biol 89:79–83

Formenti N, Viganò R, Fraquelli C, Trogu T, Bonfanti M, Lanfranchi P, Palme R, Ferrari N (2018) Increased hormonal stress response of Apennine chamois induced by interspecific interactions and anthropogenic disturbance. Eur J Wildl Res 64:68

Fraquelli C, Viganò R, Formenti N, Palme R, Lanfranchi P (2015) Haemato-biochemical parameters and stress markers in the Apennine chamois (*Rupicapra pyrenaica ornata*). In: Antonucci A, Di Domenico G (eds) Chamois international congress proceedings, 17–19 June 2014. Lama dei Peligni, Majella National Park, pp 51–59

Gallagher DS, Womack JE (1992) Chromosome conservation in the Bovidae. J Hered 83:287–298

Gamelon M, Filli F, Sæther BE, Herfindal I (2020) Multi-event capture-recapture analysis in Alpine chamois reveals contrasting responses to interspecific competition, within and between populations. J Anim Ecol 89: 2279–2289

García-González R, Barandalla I (2002) Sexual dimorphism of Pyrenean chamois (*Rupicapra p. pyrenaica*) based on skull morphometry. Pirineos 157:25–37

García-González R, Cuartas P (1996) Trophic utilization of a montane/subalpine forest by chamois (*Rupicapra pyrenaica*) in the Central Pyrenees. For Ecol Manag 88: 15–23

García-González R, Hidalgo R, Montserrat C (1990) Patterns of time and space use by livestock in the Pyrenean summer ranges: a case study in the Aragon valley. Mt Res Dev 10:241–255

García-González R, Hidalgo R, Ameztoy JM, Herrero J (1992) Census, population structure and habitat use of a Chamois population in Ordesa N.P. living in sympatry with Pyrenean Wild Goat. In: Spitz F, Janeau G, González G, Aulagnier S (eds) Ongulés/Ungulates 91. S.F.E.P.M. - I.R.G.M., Paris/Toulouse, pp 321–325

Garel M, Loison A, Jullien J-M, Dubray D, Maillard D, Gaillard J-M (2009) Sex- specific growth in Alpine chamois. J Mammal 90:954–960

Garel M, Forsyth DM, Loison A, Dubray D, Julien J-M, Tustin KG, Maillard D, Gaillard J-M (2011a) Age-related male reproductive effort in two mountain ungulates of contrasting sexual size dimorphism. Can J Zool 89:929–937

Garel M, Gaillard J-M, Jullien J-M, Dubray D, Maillard D, Loison A (2011b) Population abundance and early spring conditions determine variation in body mass of juvenile chamois. J Mammal 95:1112–1117

Garin I, Herrero J (1997) Distribution, abundance and demographic parameters of the Pyrenean chamois (*Rupicapra pyrenaica pyrenaica*) in Navarre, Western Pyrenees. Mammalia 61:55–63

Giacometti M, Janovsky M, Belloy L, Frey J (2002) Infectious keratoconjunctivitis of ibex, chamois and other Caprinae. Rev Sci Technol 21:335–345

Gibert P (2017) Surveillance sanitaire de la faune sauvage. Ed. Point Veterinaire, Puteaux

Gilbert NA, Clare JDJ, Stenglein JL, Zuckerberg B (2021) Abundance estimation of unmarked animals based on camera-trap data. Conserv Biol 35:88–100

Gonzalez G, Crampe J-P (2001) Mortality patterns in a protected population of isards (*Rupicapra pyrenaica*). Can J Zool 79:2072–2079

González-Quirós P (2009) Capturas y reintroducciones de rebecos en la Cordillera Cantábrica. In: Pérez-Barbería FJ, Palacios B (eds) El Rebeco Cantábrico (*Rupicapra pyrenaica parva*). Conservación y Gestión de sus poblaciones. Ministerio de Medio Ambiente, Medio Rural y Marino. Naturaleza y Parques Nacionales, Madrid, pp 448–473

González-Quirós P, Silva P, Solano S (2009) Dinámica poblacional del rebeco cantábrico en las reservas regionales de caza de Asturias. In: Pérez-Barbería, FJ, Palacios B (eds) El Rebeco Cantábrico (*Rupicapra pyrenaica parva*). Conservación y Gestión de sus poblaciones. Ministerio de Medio Ambiente, Medio Rural y Marino. Naturaleza y Parques Nacionales, Madrid, pp 126–153

Grignolio S, Brivio F, Apollonio M, Frigato E, Tettamanti F, Filli F, Bertolucci C (2018) Is nocturnal activity compensatory in chamois? A study of activity in a cathemeral ungulate. Mamm Biol 93:173–181

Guichón ML, Monteverde M, Piudo L, Sanguinetti J, Di Martino S (2016) Mamíferos introducidos en la Provincia de Neuquén: Estado actual y prioridades de manejo. Mastozoología Neotropical 23:255–265

Hammer S, Nadlinger K, Hartl GB (1995) Mitochondrial DNA differentiation in chamois (genus *Rupicapra*): implications for taxonomy, conservation, and management. Acta Theriol 3:145–155

Hamr J (1984a) Home range size and determinant factors in habitat use and activity of the chamois (*Rupicapra rupicapra* L.) in Northern Tyrol, Austria. Doctoral Thesis, Leopold-Franzens University, Innsbruck

Hamr J (1984b) Home range sizes of males chamois (*Rupicapra rupicapra*) in the Tyrolean Alps, Austria. Acta Zool Fenn 171:293–298

Hamr J (1985) Seasonal home range size and utilisation by female chamois (*Rupicapra rupicapra* L.) in Northern Tyrol. In: Lovari S (ed) The biology and management of mountain ungulates. Croom Helm, London, pp 106–116

Hatzirvassanis V (1991) The status of chamois (*Rupicapra rupicapra balcanica* Bolkay, 1925) in Greece. Biologia Gallo-hellenica 18:3–44

Haymerle A (2013) Physiological adaptation to extreme climates in Alpine chamois. In: Proceedings of the international workshop on Chamois, Rhêmes, 26 November. pp 57–59

Henderson RJ, Clarke CMH (1986) Physical size, condition, and demography of chamois (*Rupicapra rupicapra* L.) in the Avoca River region, Canterbury, New Zealand. N Z J Zool 13:65–73

Herfindal I, Anderwald P, Filli F, Campell Andri S, Rempfler T (2019) Climate, competition and weather conditions drive vertical displacement and habitat use of an alpine ungulate in a highly topographic landscape. Landsc Ecol 34:2523–2539

Herrero J, Garin I, García-Serrano A, García-González R (1996) Habitat use in a *Rupicapra pyrenaica pyrenaica* forest population. For Ecol Manag 88:25–29

Herrero J, Garin I, García-Serrano A, Aldezabal A, García-González (2002) Grouping patterns of a forest dwelling Pyrenean Chamois *Rupicapra p. pyrenaica* population. Mountain ungulates research 2002. In: García-González R, Herrero J, Fernández de Luco D (eds) Mountain ungulate research 2002. A selection of communications from the III World Conference on Mountain Ungulates. Pirineos, vol 57, pp 89–101

Herrero J, Escudero E, Fernández de Luco D, García-González R (2008) El sarrio pirenaico *Rupicapra p. pyrenaica*: biología, patología y gestión. In: Proceedings of the first meeting of the Pyrenean chamois. Consejo de Protección de la Naturaleza de Aragón and Fundación para el Estudio y Defensa de la Naturaleza y la Caza. Serie Investigación, 46, ISBN: 84-89862-47-8. Second edition

Herrero J, Garin I, Prada C, García-Serrano A (2010) Interagency coordination fosters the recovery of the Pyrenean chamois *Rupicapra pyrenaica pyrenaica* at its western limit. Oryx 44:529–532

Herrero J, García-Serrano A, Prada C, Fernández-Arberas O (2011) Using block counts and distance sampling to estimate populations of chamois. Pirineos 166:123–133

Herrero J, Lovari S, Nores C, Toigo C (2020) *Rupicapra pyrenaica*. The IUCN red list of threatened species 2020, e.T19771A171131310

Hoby S, Robert N, Mathis A, Schmid N, Meli ML, Hofmann-Lehmann R, Lutz H, Deplazes P, Ryser-Degiorgis MP (2007) Babesiosis in free-ranging chamois (*Rupicapra r. rupicapra*) from Switzerland. Vet Parasitol 148:341–345

Hofmann RR (1984) Feeding habits of mouflon (*Ovis ammon musimon*) and chamois (*Rupicapra rupicapra*) in relation to the morphophysiological adaptation of their digestive tracts. In: Hoefs M (ed) Proceedings of the fourth Biennial symposium of the Northern Wild Sheep and Goat Council, pp 341–355

Hofmann RR (1989) Evolutionary steps of ecophysiological adaptation and diversification of ruminants: a comparative view of their digestive system. Oecologia 78: 443–457

Homolka M (1996) Foraging strategy of large herbivores in forest habitat. Folia Zool 45:127–136

Homolka M, Heroldová M (2001) Native red deer and introduced chamois: foraging habits and competition in a subalpine meadow–spruce forest area. Folia Zool 50:89–98

Huber S, Palme R, Arnold W (2003) Effects of season, sex, and sample collection on concentrations of fecal cortisol metabolites in red deer (*Cervus elaphus*). Gen Comp Endocrinol 130:48–54

Iacolina L, Corlatti L, Buzan E, Safner T, Šprem N (2019) Hybridisation in European ungulates: an overview of the current status, causes, and consequences. Mammal Rev 49:45–59

Iacolina L, Buzan E, Safner T, Basic N, Geric U, Tesija T, Lazar P, Arnal MC, Chen J, Han J, Sprem N (2021) A mother's story, mitogenome relationships in the Genus *Rupicapra*. Animals 11:1065

Jiménez-Ruiz S, Vicente J, García-Bocanegra I, Cabezón O, Arnal MC, Balseiro A, Ruiz-Fons F, Gómez-Guillamón F, Lázaro S, Escribano F, Acevedo P, Domínguez L, Gortázar C, Fernández de Luco D, Risalde MA (2020) Distribution of *Pestivirus* exposure in wild ruminants in Spain. Transbound Emerg Dis 68:1577–1585

Kati V, Kassara C, Vassilakis D, Papaioannou H (2020) Balkan Chamois (*Rupicapra rupicapra balcanica*) avoids roads, settlements, and hunting grounds: an ecological overview from Timfi Mountain, Greece. Diversity 12:124

Kavčić K, Brivio F., Grignolio S, Ugarković D, Stankić I, Safner T, Apollonio M, Šprem N (2018) Is chamois hybridization in the northern Dinaric Mountains an important factor for horn development? Wildlife Biol, wlb.00461

Kavčić K, Corlatti L, Rodriguez O, Kavčić B, Šprem N (2020) From the mountains to the sea! Unusual swimming behavior in chamois *Rupicapra* spp. Ethol Ecol Evol 32:402–408

Korjenić E, Džano A, Šuta N (2009) Population size of chamois – *Rupicapra rupicapra* (Linné) in Diva Grabovica mountain chain. Veterinaria 58:77–81

Kourkgy C, Garel M, Appolinaire J, Loison A, Toïgo C (2016) Onset of autumn shapes the timing of birth in Pyrenean chamois more than onset of spring. J Anim Ecol 85:581–590

Krämer A (1969) Soziale Organisation und Sozialverhalten einer Gemspopulation (*Rupicapra rupicapra* L.) der Alpen. Z Tierpsychol 26:890–964

Kupferschmid AD, Wasem U, Bugmann H (2014) Light availability and ungulate browsing determine growth, height and mortality of *Abies alba* saplings. For Ecol Manag 318:359–369

Kurtén B (1968) Pleistocene mammals of Europe. Weidenfeld & Nicholson, London

La Morgia V, Bassano B (2009) Feeding habits, forage selection, and diet overlap in Alpine chamois (*Rupicapra rupicapra* L.) and domestic sheep. Ecol Res 24:1043–1050

Lambert S, Ezanno P, Garel M, Gilot-Fromont E (2018) Demographic stochasticity drives epidemiological patterns in wildlife with implications for diseases and population management. Sci Rep 8:16846

Latini R, Gentile L, Asprea A, Pagliaroli D, Di Pirro V, Argenio A, Monaco A (2015) Life is a risky business: status and conservation perspectives of the last ancient population of endemic Apennine chamois. In: Antonucci A, Di Domenico G (eds) Chamois international congress proceedings, 17–19 June 2014, Lama dei Peligni, Majella National Park, pp 103–109

Lavín S, Ruiz-Bascaran M, Marco I, Fondevila MD, Ramis AJ (2000) Experimental infection of chamois (*Rupicapra pyrenaica parva*) with *Sarcoptes scabiei* derived from naturally infected goats. J Vet Med B Infect Dis Vet Public Health 47:693–699

Locati M (1990) Female chamois defends kids from eagle attacks. Mammalia 54:155–156

Locati M, Lovari S (1988) La socialità nel camoscio appenninico *Rupicapra pyrenaica ornata* (Neumann, 1899): confronto tra i sessi e suggerimenti di gestione. Atti I Convegno Nazionale di Biologia della Selvaggina, Bologna, pp 561–567

Locati M, Lovari S (1990) Sexual differences in aggressive behaviour of the Apennine Chamois. Ethology 84:295–306

Locati M, Lovari S (1991) Clues for dominance in female chamois: age, weight, or horn size? Aggress Behav 17: 11–15

Loison A, Gaillard J-M, Houssin H (1994) New insight on survivorship of female chamois (*Rupicapra rupicapra*) from observation of marked animals. Can J Zool 72:591–597

Loison A, Gaillard J-M, Jullien J (1996) Demographic patterns after an epizootic of keratoconjunctivitis in a chamois population. J Wildl Manag 60:517–527

Loison A, Jullien JM, Menaut P (1999a) Subpopulation structure and dispersal in two populations of chamois. J Mammal 80:620–632

Loison A, Jullien JM, Menaut P (1999b) Relationship between chamois and isard survival and variation in global and local climate regimes: contrasting examples from the Alps and Pyrenees. Ecol Bull 47:126–136

Loison A, Festa-Bianchet M, Gaillard J-M, Jorgenson JT, Jullien J-M (1999c) Age-specific survival in five populations of ungulates: evidence of senescence. Ecology 80:2539–2554

Loison A, Toïgo C, Appolinaire J, Michallet J (2002) Demographic processes in colonizing populations of isard (*Rupicapra pyrenaica*) and ibex (*Capra ibex*). J Zool 256:199–205

Loison A, Appolinaire J, Jullien J-M, Dubray D (2006) How reliable are total counts to detect trends in population size of chamois *Rupicapra rupicapra* and *R. pyrenaica*? Wildl Biol 12:77–88

Lorenzini R (2005) A panel of polymorphic microsatellites in the threatened Apennine chamois (*Rupicapra pyrenaica ornata*). Mol Ecol Notes 5:372–374

Lovari S (1984a) Il popolo delle rocce. Rizzoli Editore, Milan

Lovari S (1984b) Herding strategies of male Abruzzo chamois on the rut. Acta Zool Fenn 172:91–92

Lovari S (1985) Behavioural repertoire of the Abruzzo chamois, *Rupicapra pyrenaica ornata* Neumann, 1899 (Artiodactyla: Bovidae). Säugetierkundliche Mitteilungen 32:113–116

Lovari S (1987) Evolutionary aspects of the biology of Chamois, *Rupicapra* spp. (Bovidae, Caprinae). In: Soma H (ed) The biology and management of Capricornis and related mountain Antelopes, 1st edn. Croom Helm, London

Lovari S, Bruno E (2003) *Rupicapra pyrenaica*. In: Boitani L, Lovari S, Vigna-Taglianti A (eds) Fauna d'Italia: 697 Mammalia III, Carnivora-Artiodactyla. Calderini – Il Sole 24 Ore, Bologna, pp 403–412

Lovari S, Cosentino R (1986) Seasonal habitat selection and group size of the Abruzzo chamois (*Rupicapra pyrenaica ornata*). Ital J Zool 53:73–78

Lovari S, Rosto G (1985) Feeding rate and social stress of female chamois foraging in groups. In: Lovari S (ed) The biology and management of mountain ungulates. Croom Helm, London, pp 102–105

Lovari S, Scala C (1980) Revision of *Rupicapra* genus. 1. A statistical re-evaluation of Couturier's data on the morphometry of six chamois subspecies. Ital J Zool 47: 113–124

Lovari S, Sacconi F, Trivellini G (2006) Do alternative strategies of space use occur in male Alpine chamois? Ethol Ecol Evol 18:221–231

Lovari S, Ferretti F, Corazza M, Minder I, Troiani N, Ferrari C, Saddi A (2014) Unexpected consequences of reintroductions: competition between increasing red deer and threatened Apennine chamois. Anim Conserv 17:359–370

Lovari S, Franceschi S, Chiatante G, Fattorini L, Fattorini N, Ferretti F (2020a) Climatic changes and the fate of mountain ungulates. Clim Chang 162:2319–2337

Lovari S, Mori E, Procaccio EL (2020b) On the behavioural biology of the Mainland serow: a comparative study. Animals 10:1669

Luzzago C, Ebranati E, Cabezón O, Fernández-Sirera L, Lavín S, Rosell R, Veo C, Rossi L, Cavallero S, Lanfranchi P, Marco I, Zehender G (2016) Spatial and temporal phylogeny of border disease virus in Pyrenean Chamois (*Rupicapra p. pyrenaica*). PLoS One 11: e0168232

Lydekker R (1913) Catalogue of the ungulate mammals of British Natural History Museum. London

Marco I (2012) Pestivirus of chamois and border disease. In: Gavier-Widen D, Duff JP, Meredith A (eds) Infectious diseases of wild mammals and birds in Europe. Wiley-Blackwell, West Sussex, pp 147–152

Marco I, Rosell R, Cabezón O, Mentaberre G, Casas E, Velarde R, Lavín S (2009) Border disease virus among chamois, Spain. Emerg Infect Dis 15:448–451

Marinas M, García-González R (2006) Preliminary data on nutritional assessment of abundant species in alpine pastures of the Pyrenees. Pirineos 161:85–109

Markov G, Zhelev P, Ben Slimen H, Suchentrunk F (2016) Population genetic data pertinent to the conservation of Bulgarian chamois (*Rupicapra rupicapra balcanica*). Conserv Genet 17:155–164

Masini F, Lovari S (1988) Systematics, phylogenetics relationships, and dispersal of the chamois (*Rupicapra* spp.). Quat Res 30:339–349

Mason THE, Chirichella R, Richards SA, Stephens PA, Willis SG, Apollonio M (2011) Contrasting life histories in neighbouring populations of a large mammal. PLoS One 6:e28002

Mason THE, Stephens PA, Apollonio M, Willis SG (2014a) Predicting potential responses to future climate in an alpine ungulate: interspecific interactions exceed climate effects. Glob Chang Biol 20:3872–3882

Mason THE, Apollonio M, Chirichella R, Willis SG, Stephens PA (2014b) Environmental change and long-term body mass declines in an alpine mammal. Front Zool 11:69

Massei G, Randi E, Genov P (1994) The dynamics of the horn growth in Bulgarian chamois *Rupicapra rupicapra balcanica*. Acta Theriol 39:195–199

Molinari-Jobin A, Molinari P, Breitenmoser-Würsten C, Breitenmoser U (2002) Significance of lynx *Lynx lynx* predation for roe deer *Capreolus capreolus* and chamois *Rupicapra rupicapra* mortality in the Swiss Jura Mountains. Wildl Biol 8:109–115

Morin A, Rughetti M, Rioux-Paquette S, Festa-Bianchet M (2016) Older conservatives: reproduction in female Alpine chamois (*Rupicapra rupicapra*) is increasingly risk-averse with age. Can J Zool 94:311–321

Nascetti G, Lovari S, Lanfranchi P, Berducou C, Mattiucci S, Rossi L, Bullini L (1985) Revision of *Rupicapra* genus. III. Electrophoretic studies demonstrating species distinction of chamois populations of the Alps from those of the Apennines and Pyrenees. In: Lovari S (ed) Biology and management of mountain ungulates. Croom-Helm, London, pp 56–62

Nesti I, Posillico M, Lovari S (2010) Ranging behaviour and habitat selection of Alpine chamois. Ethol Ecol Evol 22:215–231

Neumann O (1899) Die Gemse der Abruzzen. Annali del Museo Civico di Storia Naturale di Genova 2:40–44

Nores C, González-Quirós P (2009) Cambios demográficos proveocados por la sarna. In: Pérez-Barbería FJ, Palacios B (eds) El Rebeco Cantábrico (*Rupicapra pyrenaica parva*). Conservación y Gestión de sus poblaciones. Ministerio de Medio Ambiente, Medio Rural y Marino. Naturaleza y Parques Nacionales, Madrid, pp 338–359

Palacios B (2009) Seguimiento poblacional de rebeco cantábrico en el Parque Nacional de los Picos de Europa. In: Pérez-Barbería FJ, Palacios B (eds) El Rebeco Cantábrico (*Rupicapra pyrenaica parva*). Conservación y Gestión de sus poblaciones. Ministerio de Medio Ambiente. Medio Rural y Marino. Naturaleza y Parques Nacionales, Madrid, pp 210–266

Pañella P, Herrero J, Canut J, García-Serrano A (2010) Long-term monitoring of Pyrenean chamois in a protected area reveals a fluctuating population. Hystrix 21:183–188

Papaioannou H, Fernández M, Pérez T, Domínguez A (2019) Genetic variability and population structure of chamois in Greece (*Rupicapra rupicapra balcanica*). Conserv Genet 20:939–945

Patalano M, Lovari S (1993) Food habits and trophic niche overlap of the wolf (*Canis lupus*, L. 1758) and the red fox *Vulpes vulpes* (L. 1758) in a Mediterranean mountain area. Revue d'Ecologie 48:279–294

Pęksa L, Ciach M (2018) Daytime activity budget of an alpine ungulate (Tatra chamois *Rupicapra rupicapra tatrica*): influence of herd size, sex, weather and human disturbance. Mammal Res 63:443–453

Pépin D, Gonzalez G, Bon R (1991) Le Chamois et L'isard. Rev Ecol (Terre Vie) Suppl 6:111–153

Pépin D, Joachim J, Ferrie E (1997) Variability of spring habitat selection by isards (*Rupicapra pyrenaica*). Can J Zool 75:1955–1965

Pérez T, Albornoz J, Domínguez A (2002) Phylogeography of chamois (*Rupicapra* spp.) inferred from microsatellites. Mol Phylogenet Evol 25:524–534

Pérez T, Hammer SE, Albornoz J, Domínguez A (2011) Y-chromosome phylogeny in the evolutionary net of chamois (genus *Rupicapra*). BMC Evol Biol 11:272

Pérez T, Essler S, Palacios B, Albornoz J, Domínguez A (2013) Evolution of the melanocortin-1 receptor gene (MC1R) in chamois (*Rupicapra* spp.). Mol Phylogenet Evol 67:621–625

Pérez T, González I, Essler SE, Fernández M, Domínguez A (2014) The shared mitochondrial genome of *Rupicapra pyrenaica ornata* and *Rupicapra rupicapra cartusiana*: Old remains of a common past. Molecular Phylogenetics and Evolution 79:375–379

Pérez T, Fernández M, Hammer SE, Domínguez A (2017) Multilocus intron trees reveal extensive male-biased homogenization of ancient populations of chamois (*Rupicapra* spp.) across Europe during late Pleistocene. PLoS One 12:e0170392

Pérez-Barbería FJ, Nores C (1996) Grazing activity of breeding and non-breeding female Cantabrian chamois *Rupicapra pyrenaica parva*. Ethol Ecol Evol 8:353–363

Pérez-Barbería FJ, Pérez-Fernández E (2009) Identificación, biología y ecología del rebeco cantábrico. In: Pérez-Barbería FJ, Palacios B (eds) El Rebeco Cantábrico (*Rupicapra pyrenaica parva*). Conservación y Gestión de sus poblaciones. Ministerio de Medio Ambiente, Medio Rural y Marino. Naturaleza y Parques Nacionales, Madrid, pp 26–69

Pérez-Barbería FJ, Robles R, Nores C (1996) Horn growth pattern in Cantabrian chamois *Rupicapra pyrenaica parva*: influence of sex, location and phaenology. Acta Theriol 41:83–92

Pérez-Barbería FJ, Oliván M, Osoro K, Nores C (1997) Sex, seasonal and spatial differences in the diet of Cantabrian chamois *Rupicapra pyrenaica parva*. Acta Theriol 42:37–46

Pérez-Barbería FJ, Mutuberria G, Nores C (1998) Reproductive parameters, kidney fat index, and grazing activity relationships between the sexes in Cantabrian chamois *Rupicapra pyrenaica parva*. Acta Theriol 43: 311–324

Pérez-Barbería FJ, Palacios B, González-Quirós P, Cano M, Nores C, Díaz A (2009) La evolución de la población del rebeco en la cordillera Cantábrica. In: Pérez-Barbería FJ, Palacios B (eds) El Rebeco Cantábrico (*Rupicapra pyrenaica parva*). Conservación y Gestión de sus poblaciones. Ministerio de Medio Ambiente, Medio Rural y Marino. Naturaleza y Parques Nacionales, Madrid, pp 106–125

Pérez-Barbería FJ, García-González R, Palacios B (2017) Rebeco-*Rupicapra pyrenaica*. In: Salvador A, Barja I (eds) Enciclopedia Virtual de los Vertebrados Españoles. Museo Nacional de Ciencias Naturales, Madrid

Pioz M, Loison A, Gauthier D, Gibert P, Jullien J-M, Artois M, Gilot-Fromont E (2008) Diseases and reproductive success in a wild mammal: example in the alpine chamois. Oecologia 155:691–704

Posautz A, Loncaric I, Kübber-Heiss A, Knoll A, Walzer C (2014) Acute die-off of chamois (*Rupicapra rupicapra*) in the Eastern Austrian alps due to bacterial bronchopneumonia with Pasteurellaceae. J Wildl Dis 50:616–620

Puorger A, Rossi C, Haller RH, Anderwald P (2018) Plastic adaptations of foraging strategies to variation in forage quality in Alpine chamois (*Rupicapra rupicapra*). Can J Zool 96:269–275

Redjadj C, Darmon G, Maillard D, Chevrier T, Bastianelli D, Verheyden H, Loison A, Saïd S (2014) Intra- and interspecific differences in diet quality and composition in a large herbivore community. PLoS One 9:e84756

Reiner R, Zedrosser A, Zeiler H, Hackländer K, Corlatti L. 2020. Deterministic reconstruction as an informative tool for monitoring chamois populations. Wildlife Biol, wlb.00757

Reiner R, Zedrosser A, Zeiler H, Hackländer K, Corlatti L (2021) Forests buffer the climate-induced decline of body mass in a mountain herbivore. Glob Change Biol 27:3741–3752

Richard Q, Toïgo C, Appolinaire J, Loison A, Garel M (2017) From gestation to weaning: combining robust design and multi-event models unveils cost of lactation in a large herbivore. J Anim Ecol 86:1497–1509

Rivals F (2004) Les petits bovidés (Caprini et rupicaprini) pléistocènes dans le bassin méditerranéen et le Caucase: étude paléontologique, biostratigraphique, archéozoologique et paléoécologique. Br Archaeol Rep 1327:1–252

Rodríguez F, Hammer S, Pérez T, Suchentrunk F, Lorenzini R, Michallet J, Martínková N, Albornoz J, Domínguez A (2009) Cytochrome b phylogeography of chamois (*Rupicapra* spp.). Population contractions, expansions and hybridizations governed the diversification of the genus. J Hered 100:47–55

Rodríguez F, Pérez T, Hammer SE, Albornoz J, Domínguez A (2010) Integrating phylogeographic patterns of microsatellite and mtDNA divergence to infer the evolutionary history of chamois (genus *Rupicapra*). BMC Evol Biol 10:222

Rossetti A, Menapace S, Morandi F, Alemanno S, Bocci A, Salvi P, Mari F, Perco F (2015) The Apennine chamois reintroduction in the Monti Sibillini National Park: population monitoring and status at six years from the reintroduction beginning. In: Antonucci A, Di Domenico G (eds) Chamois international congress proceedings. 17–19 June 2014, Lama dei Peligni, Majella National Park, pp 21–31

Rossi L, Meneguz PG, De Martin P, Rodolfi M (1995) The epizootiology of sarcoptic mange in chamois, *Rupicapra rupicapra*, from the Italian eastern Alps. Parassitologia 37:233–240

Rossi L, Fraquelli C, Vesco U, Permunian R, Sommavilla GM, Carmignola G, Da Pozzo M, Meneguz PG (2007) Descriptive epidemiology of a scabies epidemic in chamois in the Dolomite Alps, Italy. Eur J Wildl Res 53:131–141

Rossi L, Tizzani P, Rambozzi L, Moroni B, Meneguz PG (2019) Sanitary emergencies at the wild/domestic caprines interface in Europe. Animals 9:922

Roucher F (1999) The fate of the Chartreuse chamois. CAPRINAE – Newsletter of the IUCN/SSC Caprinae Specialist Group, January. pp 4–5

Ruckstuhl KE, Ingold P (1999) Aspects of mother-kid behavior in alpine chamois, *Rupicapra rupicapra rupicapra*. Zeitschrift für Säugetierkunde 64:76–84

Rughetti M, Festa-Bianchet M (2010a) Compensatory growth limits opportunities for artificial selection in Alpine chamois. J Wildl Manag 74:1024–1029

Rughetti M, Festa-Bianchet M (2010b) Effects of early horn growth on reproduction and hunting mortality in female chamois. J Anim Ecol 80:438–447

Rughetti M, Festa-Bianchet M (2011) Seasonal changes in sexual-size dimorphism in northern chamois. J Zool 284:257–264

Rughetti M, Festa-Bianchet M (2012) Effects of spring-summer temperature on body mass of chamois. J Mammal 93:1301–1307

Rughetti M, Festa-Bianchet M (2014) Effects of selective harvest of non-lactating females on chamois population dynamics. J Appl Ecol 51:1075–1084

Rughetti M, Toïgo C, von Hardenberg A, Rocchia E, Festa-Bianchet M (2011) Effects of an exceptionally snowy winter on chamois survival. Acta Theriol 56:329–333

Safner T, Buzan E, Rezić A, Šprem N (2019) Small scale spatial genetic structure of Alpine chamois (*Rupicapra rupicapra*) in Northern Dinarides. Eur J Wildl Res 65:23

Scala C, Lovari S (1984) Revision of *Rupicapra* Genus. II. A skull comparison of *Rupicapra rupicapta ornata* and *R. rupicapra pyrenaica* chamois. Bollettino di Zoologia 51:285–294

Schnidrig-Petrig R, Ingold P (2001) Effects of paragliding on alpine chamois *Rupicapra rupicapra rupicapra*. Wildl Biol 7:285–294

Schröder W (1971) Untersuchungen zur Ökologie des Gamswildes (*Rupicapra rupicapra* L.) in einem Vorkommen der Alpen – I. teil. Z Jagdwiss 17:113–168

Schröder J, Schröder W (1984) Niche breadth and overlap in red deer *Cervus elaphus*, roe deer *Capreolus capreolus* and chamois *Rupicapra rupicapra*. Ann Zool Fenn 172:85–86

Schröder W, von Elsner-Schack IV (1985) Correct age determination in chamois. In: Lovari S (ed) The biology and management of mountain ungulates. Croom Helm, London, pp 67–70

Scornavacca D, Brunetti C (2015) Cooperative defence of female chamois successfully deters an eagle attack. Mammalia 80:453–456

Scornavacca D, Lovari S, Cotza A, Bernardini S, Brunetti C, Pietrocini V, Ferretti F (2016) Pasture quality affects juvenile survival through reduced maternal care in a mountain-dwelling ungulate. Ethology 122: 807–817

Scornavacca D, Cotza A, Lovari S, Ferretti F (2018) Suckling behaviour and allonursing in the Apennine chamois. Ethol Ecol Evol 30:385–398

Shackleton DM (1997) Wild sheep and goats and their relatives. IUCN, Gland

Skonhoft A, Yoccoz NG, Stenseth NC, Gaillard J-M, Loison A (2002) Management of chamois (*Rupicapra rupicapra*) moving between a protected core area and a hunting area. Ecol Appl 12:1199–1211

Soglia D, Rossi L, Cauvin E, Citterio C, Ferroglio E, Maione S, Meneguz PG, Spalenza V, Rasero R, Sacchi P (2010) Population genetic structure of Alpine chamois (*Rupicapra r. rupicapra*) in the Italian Alps. Eur J Wildl Res 56:845–854

Šprem N, Buzan E (2016) The genetic impact of chamois management in the Dinarides. J Wildl Manag 80:783–793

Šprem N, Zanella D, Ugarković D, Prebanić I, Gančević P, Corlatti L (2015) Unimodal activity pattern in forest-dwelling chamois: typical behaviour or interspecific avoidance? Eur J Wildl Res 61:789–794

Tettamanti F, Grignolio S, Filli F, Apollonio M, Bize P (2015) Senescence in breeding success of female Alpine chamois (*Rupicapra rupicapra*): the role of female quality and age. Oecologia 178:187–195

Tosi G, Vailati G, Arduino S, Pacchetti G (1990) Morphology and functionality of the supraoccipital glands in the alpine chamois *Rupicapra rupicapra rupicapra*. Ital J Zool 57:131–138

Valchev K, Andonov K, Popgeorgiev G, Plachijski D, Avramov S (2006) Action plan for the Chamois in Bulgaria: 2007–2016. BBF-SFA, Sofia

Valentinčič VS, Bavdec S, Kusej M (1974) Gravidität der Gamsgaissen in den Jülischen Alpen. Z Jagdwiss 20: 50–53

Villamuelas M, Fernández N, Albanell E, Gálvez-Cerón A, Bartolomé J, Mentaberre G, López-Olvera JR, Fernández-Aguilar X, Colom-Cadena A, López-Martín JM, Pérez-Barbería J, Garel M, Marco I, Serrano E (2016) The enhanced vegetation index (EVI) as a proxy for diet quality and composition in a mountain ungulate. Ecol Indic 61:658–666

von Elsner-Schack I (1985a) What is good chamois habitat? In: Lovari S (ed) The biology and management of mountain ungulates. Croom Helm, London, pp 71–76

von Elsner-Schack I (1985b) Seasonal changes in the size of chamois groups in the Ammergauer mountains, Bavaria. In: Lovari S (ed) The biology and management of mountain ungulates. Croom Helm, London, pp 148–153

von Hardenberg A, Bassano B, Peracino A, Lovari S (2000) Male Alpine chamois occupy territories at hotspots before the mating season. Ethology 106:617–630

Willisch CS, Ingold P (2007) Feeding or resting? The strategy of rutting male Alpine chamois. Ethology 113:97–104

Willisch CS, Bieri K, Struch M, Franceschina R, Schnidrig-Petrig R, Ingold P (2013) Climate effects on demographic parameters in an unhunted population of Alpine chamois (*Rupicapra rupicapra*). J Mammal 94:173–182

Wilson DE, Reeder DAM (2005) Mammal species of the world: a taxonomic and geographic reference, 3rd edn. Johns Hopkins University Press, Baltimore

Zemanová B, Hájková P, Hájek B, Martínková N, Mikulíček P, Zima J, Bryja J (2015) Extremely low genetic variation in endangered Tatra chamois and evidence for hybridization with an introduced Alpine population. Conserv Genet 16:729–741

Zwijacz-Kozica T, Selva N, Barja I, Silván G, Martínez-Fernández L, Illera JC, Jodłowski M (2013) Concentration of fecal cortisol metabolites in chamois in relation to tourist pressure in Tatra National Park (South Poland). Acta Theriol 58:215–222

Barbary Sheep *Ammotragus lervia* (Pallas, 1777)

15

Nikica Šprem, Pavao Gančević, Toni Safner, Klemen Jerina, and Jorge Cassinello

Contents

N. Šprem (✉) · P. Gančević
Department of Fisheries, Apiculture, Wildlife Management and Special Zoology, Faculty of Agriculture, University of Zagreb, Zagreb, Croatia
e-mail: nsprem@agr.hr; nikica.sprem@gmail.com; pavao.gancevic@gmail.com

T. Safner
Department of Plant Breeding Genetics, Biometrics and Experimentation, Faculty of Agriculture, University of Zagreb, Zagreb, Croatia
e-mail: tsafner@gmail.com

K. Jerina
Department of Forestry, Biotechnical Faculty, University of Ljubljana, Ljubljana, Slovenia
e-mail: klemen.jerina@gmail.com

J. Cassinello
Estación Experimental de Zonas Áridas (EEZA-CSIC), Almería, Spain
e-mail: jorge.cassinello@csic.es

© Springer Nature Switzerland AG 2022

L. Corlatti, F. E. Zachos (eds.), *Terrestrial Cetartiodactyla*, Handbook of the Mammals of Europe,
https://doi.org/10.1007/978-3-030-24475-0_35

Common Names

English	Barbary sheep, aoudad, uaddan
German	Mähnenschaf, Berberschaf
French	Mouflon à manchettes, mouflon de Barbarie, arui
Spanish	Arrui, carnero de berbería, muflón del Atlas
Italian	Ammotrago, muflone africano, pecora crinita, capra berbera
Russian	Гривистый баран

Taxonomy, Systematics and Paleontology

The oldest fossil remains of Barbary sheep were found in Haua Fteah, a large natural cave on the northern coast of Cyrenaica, Libya, in deposits aged between 85 and 2 kya (Gray and Simpson 1980). The species was also noted among the Pleistocene fauna in Magreb (Vaufrey 1955).

Barbary sheep (Fig. 1) shares morphological characteristics with the genera *Ovis* and *Capra* (Cassinello 1998), making its phylogeny based on morphological characters complicated and controversial. Historically, some authors placed Barbary sheep in either *Capra* or *Ovis* (see details in Cassinello 1998), though currently it is recognized as a single species within the genus *Ammotragus* Blyth, 1840. This status is further supported by the mtDNA sequence phylogeny reported by Hassanin et al. (2009), where *Ammotragus* is in a separate clade from both *Ovis* and *Capra* species.

In addition, six Barbary sheep subspecies have been described, based on their distribution and morphological differences in coat color and horns: *Ammotragus lervia lervia* Pallas, 1777; *Ammotragus lervia ornata* I. Geoffroy Saint-Hilaire, 1827; *Ammotragus lervia sahariensis* Rothschild, 1913; *Ammotragus lervia blainei* Rothschild, 1913; *Ammotragus lervia angusi* Rothschild, 1921; *Ammotragus lervia fassini* Lepri, 1930 (Cassinello 2013). *Ammotragus lervia lervia* was the only subspecies imported to Europe in the late 1800s for zoos and was at the origin of all European free-ranging populations (Cassinello 1998).

Fig. 1 (**a**) Male Barbary sheep (photograph by EEZA-CSIC); (**b**) female Barbary sheep (photograph by Jorge Cassinello)

Current Distribution

The Barbary sheep is a caprid (goat-antelope) native to the Atlas Mountains of North Africa, from Mauritania in the West to the Red Sea in the East. Cassinello (2015) provides the most recent account on the worldwide distribution of the Barbary sheep, including native and exotic populations. In its African native domains, information on population size and density is lacking: sighting data are scarce and only rough estimates are available. Six subspecies have traditionally been recognized based on the distribution range, and differentiated by coat and horn morphological differences (Cassinello 2013), although genetically there appear to be fewer units (T.L. Silva, pers. comm.). In Europe, only four countries are inhabited by free ranging Barbary sheep populations: Croatia, Czech Republic, France, Italy, and Spain.

In Croatia, five Barbary sheep from Czech Republic, Germany, and Slovakia (three females and two males) were illegally released (according to the local police) in 2002 (P. Gančević, unpublished data). They were initially kept within an enclosure by an unknown person in Southern Dalmatia (Dinarides mountain range – Mt. Mosor), and subsequently released. The current population size is estimated at around 140 individuals (Lazarus et al. 2019).

The population in Czech Republic was established following the escape of several individuals from the Plzeň Zoo in 1976, and according to Bartoš et al. (2010), this population consists of a few dozen animals. Currently, it is not possible to confirm with certainty whether this population still exists in the wild (S. Cupic, pers. comm.).

In France, one population has been reported. In 1980, some animals escaped from an enclosure into the Montagne Sainte-Victoire. Not much is known about this population, but it is assumed that there are several dozen individuals and that they reproduce successfully (Cugnasse and Rigaux 2021).

In Italy two populations have been reported: the first, in the province of Varese where, in 1993, a group of Barbary sheep originated following the escape of five females and one adult male from captivity (private holding). The authorities started a removal plan of some 25–30 animals that were present at the time, but only three of them were culled; the remaining animals could not be found, probably because they were killed illegally (A. Martinoli, pers. comm.). Some authors consider this population still present (Mori et al. 2017), whereas others consider it eradicated (A. Martinoli, pers. comm.). The second population of a few dozen individuals is present in the Beigua Regional Park in the province of Savona, and currently shows an increasing trend (S. Grignolio, pers. comm.).

Barbary sheep were successfully introduced to public and private lands in Spain in the early 1970s (Damm and Franco 2014). The current distribution of Spanish Barbary sheep populations is difficult to estimate accurately, as there are many private hunting areas where introductions were performed, such as in Andalusia and Castile-La Mancha (Cassinello 1998). Therefore, in this chapter, we address only those populations that are present in natural parks and public reserves. Free ranging individuals are currently distributed in most mountain ranges in the Region of Murcia, North of the Guadalentin River. These populations originated from 36 individuals from the Frankfurt Zoo in Germany and the Ain Sebad Zoo in Casablanca, Morocco, then released in Sierra Espuña Natural Park (Cassinello 1998). The current population size is estimated at around 2000 individuals with an increasing trend (Cassinello 1998). Also, a free ranging population, numerically increasing since the introduction of 16 individuals from Sierra Espuña in 1972 is present on La Palma Island, Canary Islands (Cassinello 1998) (Fig. 2).

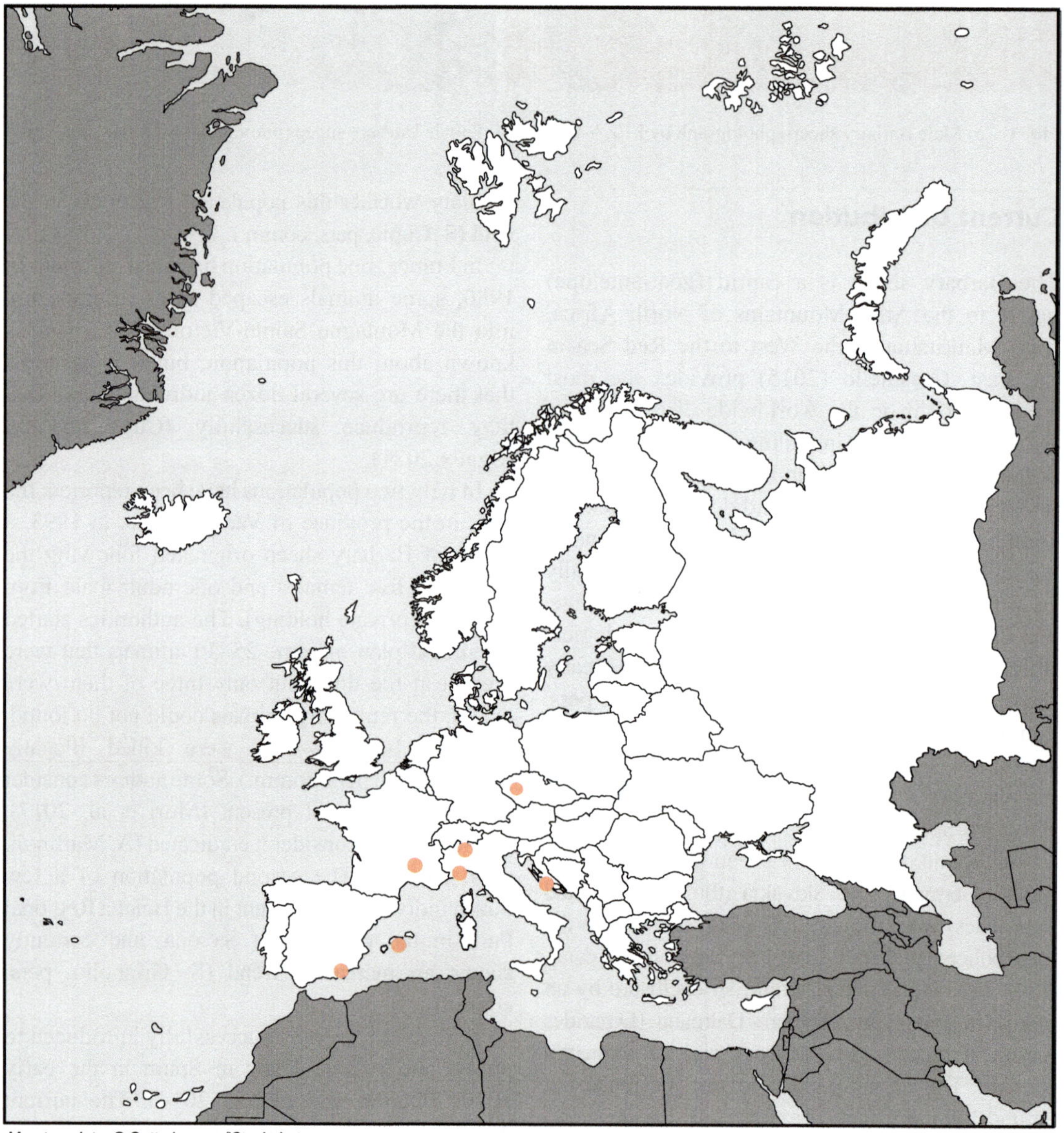

Map template: © Getty Images/iStockphoto

Fig. 2 Distribution of Barbary sheep. Distribution is based on documented records on Barbary sheep presence in Europe (Map template: © Copyright Getty Images/iStockphoto)

Description

Size and morphology

Morphologically, this species shares characteristics with both sheep and goat (Geist 1971; Schaffer and Reed 1972). In general, it resembles a robust goat, with relatively long head, short and stubby legs, and a compact body (Fig. 1). The tail is relatively long (15 cm) and hairy on the dorsal side. However, unlike goats, Barbary sheep is not bearded. Its unique feature is a mane that extends below the throat, along the front of the neck to the chest, where it forks and continues down the forelegs. Sexual dimorphism is particularly strong in adults (Cassinello 1997a), with males that are heavier and larger than females (body mass: 82 kg *vs.* 41 kg; body length: 147 cm vs. 128 cm). Tolić (2005) reported data from an enclosure in Croatia for mature individuals: body mass was 120–150 kg in males and 70 kg in females; head-tail body length was 155–165 cm, with tail length of 20–25 cm; height at the withers was 100–110 cm in males and 90–100 cm in females; horn length was 85 cm in males and 50 cm in females. In free-ranging Barbary sheep in Mt. Mosor in Croatia the full body mass is *c.* 90 kg in males and *c.* 61 kg in females. This Croatian population appears well adapted to the Mediterranean habitat and showed higher (>2–10%) mean values of body and horn measurements and higher growth rates than other populations from Spain and Algeria (Kavčić et al. 2020).

Pelage

The coloration of Barbary sheep may vary depending on the subspecies, from light yellow, brownish grey to dark red, with or without black stripes on the face and back. The winter coat is darker than the summer coat.

Dentition

Permanent dental formula is 0.0.3.3/3.1.3.3.

Sex and age determination

Sex and age can be identified according to facial and horn morphology (horn rings). The horns of males are more massive than those of females, they curve outward, backward, and point inward toward the neck. The distance between horn bases appears to be greater in females than in males, despite the fact that the female skull is narrower and smaller. The circumference of the horn base is greater in males (*c.* 28 cm) than in females (*c.* 19 cm) (Kavčić et al. 2020). Prominence of the withers, or dorsal scapular protrusion, is greater in males than in females. The mane and chap hair are thicker and more abundant in males than in females. The relative abdominal dorsal-ventral depth is greater in females than in males; consequently, males display a well-defined belly tuck-up, even when compared with nulliparous females (Gray and Simpson 1980).

Physiology

The average duration of the estrus cycle is 23 days, with an average luteal phase of 17 days and an inter-luteal phase of 6 days (Abaigar et al. 2012).

The study reporting contents of stabile elements in the tissue (muscle, kidney, and liver) of Barbary sheep from Mt. Mosor showed low levels of toxic elements As, Cd, Hg, and Pb. Also, radionucleotide values for ^{137}Cs and ^{40}K in muscle tissue were found at low level and significantly below the statutory values of 600 Bq kg^{-1} (Lazarus et al. 2019).

Genetics

Chromosomes

Barbary sheep has a karyotype of $2n = 58$ (Gray and Simpson 1980) composed of one pair of biarmed and 27 pairs of acrocentric autosomes, a large acrocentric X chromosome and a minute biarmed Y chromosome (Nadler et al. 1974), as in *Ovis vignei*.

Phylogeny

Protein and DNA sequences were used to clarify the phylogenetic relationship of *Ammotragus* to other Caprini, though the results were dependent on the type of used marker. While serum protein analyses by Schmitt (1963) and immunoglobulin cross-reactivity studies by Curtain and Fudenberg (1973) indicated a close relationship between *Ammotragus* and *Ovis,* the amino acid sequence of various hemoglobin chains examined by Manwell and Baker (1975) showed that *Ammotragus* haemoglobin was more similar to that of the domestic goat, *Capra hircus*. The complete nucleotide mtDNA sequence of Barbary sheep was published by Mereu et al. (2008). Comparison of the sequence with other available sequences of Caprini species showed that the Caprini species share a common ancestor and that the relationship of *Ammotragus* with *Capra* is closer than with *Ovis* (Mereu et al. 2008).

Genetic diversity

There are no compelling estimates of genetic diversity of the European Barbary sheep populations, but since they were all founded by a small number of animals, it is expected to be relatively small. This was supported by Stuhne (2018), who detected only one mtDNA haplotype in 12 samples from the Croatian population. Safner et al. (2018) detected four mtDNA haplotypes in 66 samples from Croatian and Spanish populations. More data on genetic diversity are required to allow for better estimation of genetic status of the European populations.

Hybridization

There is no confirmed evidence of Barbary sheep hybridization in the wild. The only documented record comes from controlled experiments: viable offspring with $2n = 59$ chromosomes were born after natural mating between a male Barbary sheep and female goat (Gray 1972; Bunch et al. 1977, Moore et al. 1981), while crossbreeding attempts between male Barbary sheep and domestic female sheep were unsuccessful (Benirschke 1971; Gray 1972).

Life History

Growth

Sexual maturity occurs at 14 months for males and 9 months for females (Cassinello 1997a). The species is sexually dimorphic, with males already heavier at birth than females when singletons (4.1 kg and 3.8 kg, respectively); no differences emerged between twins, which tended to be lighter than singletons (Cassinello and Gomendio 1996). Body mass at birth is affected by hierarchy and age of the mother, and by inbreeding levels. High-ranked mothers tend to produce heavier calves than low-ranked ones (Cassinello and Gomendio 1996); older mothers give birth to heavier calves than younger mothers, and highly inbred families produce lighter calves than genetically diverse families (Cassinello 1997a).

In a sample of 30 Barbary sheep (20 males and 10 females) from Croatia, males showed faster increase in several morphological metrics (i.e., body length, withers height, chest circumference, anterior cannon circumference, horn base circumference, total body mass, hog-dressed mass, meat mass, and dressing) than females, except in total horn length (Kavčić et al. 2020). In the same study, the authors speculate that environmental covariates, anthropogenic pressure, genetic makeup of individuals, and physiological conditions were the main factors affecting the body and horn growth, although lack of data did not allow to test such hypotheses.

Reproduction

The gestation period is 5.5 months, and most parturitions occur in spring (Ogren 1965; Cassinello and Alados 1996). Newborns are precocious and able to follow their mother within just a few hours from birth. In captivity, the mean interval between parturitions is 10 months, while weaning takes place at 8 months of age (Cassinello and Alados 1996).

Adult females reproduce every year, giving birth to twins (25% of births in captivity), and exceptionally triplets (Arman 1991; Cassinello and Alados 1996), when females are in good conditions (Cassinello and Gomendio 1996). In enclosures with sufficient food and living space, females give birth twice a year and mate immediately after birth, and mature females often have three offspring (M. Olujić, pers. comm.).

Cassinello and Alados (1996) analyzed four components of female reproductive success in captive Barbary sheep: longevity, conception rate (i.e., total number of conceptions divided by the reproductive life span), offspring 1-month survival rate, and age of primiparity. Longevity accounts for 69.9% of the variance of reproductive success, conception rate for 54.2%, offspring 1-month survival rate for 29.8%, and the age of primiparity for 10.4% (see Brown 1988). A detailed study of these components led to the following conclusions: (a) longevity is higher in individuals in better physical condition; (b) conception rate is related to age and social rank; (c) heavier offspring at birth have a higher probability of surviving during their first month of life; and (d) age of primiparity is delayed by high population density, inbreeding coefficients, and birth body mass. On the other hand, high-ranked females are characterized by shorter inter-birth intervals and give birth to a higher proportion of twins.

Sex ratio at birth is influenced by maternal social rank at the time of conception. Cassinello and Gomendio (1996) stated that, for single and twin births, as maternal rank increases, the following sequence is observed: single female/twin females < single male < twin male and female < twin males. From an evolutionary perspective, this indicates that producing twin males is more costly than any other calf combination, followed by both sex twins, single males, and, finally, both single and twin females, which showed no cost differences (see also Williams 1979). Despite this, the species shows lower levels of maternal investment in females than in males (Cassinello 1996), and a single female offspring is more likely to survive than a single male or twins (Cassinello and Gomendio 1996).

Survival

There are no data on mortality and average lifespan in free-ranging populations, but in captivity individuals may live up to 20 years, according to the Studbook records at the EEZA-CSIC (Almería, Spain). For Croatia, Tolić (2005) reported an average lifespan of 12 years, though Barbary sheep may live up to 24 years.

Habitat and Diet

Habitat selection and movement

The native habitat of the Barbary sheep comprises any mountainous landscape present in the North of Africa, from the Atlas Mountains to the Sahara and Sahel regions (see Cassinello 1998). Habitat selection of Barbary sheep exotic populations has been analyzed in the USA (Ogren 1965; Dickinson and Simpson 1980; Johnston 1980) and Spain (Cassinello 2000; Acevedo et al. 2007; Anadón et al. 2018).

Barbary sheep tend to select steep and rocky areas, as well as scrublands and mountain forests, from the sea level to peaks up to 2000 m in Africa. They avoid snowy areas. During the calving season, females seek out shelter and more inaccessible areas at high elevations, while males can be found in a variety of areas, including scrublands and woodlands. In summer, Barbary sheep are more often found in forested areas with milder temperatures, and during the rutting season they select open pastures and brushwood habitats. Barbary sheep prefer regions with low winter rainfall regimes and high average summer temperatures, characteristic of semiarid lands (Cassinello 1998).

Successfully established exotic populations in USA and Spain have shown a formidable capacity to settle, grow, and spread from their release areas (Cassinello 1998). The home range of American Barbary sheep averages between 259 and 3367 ha, with main dispersal movements occurring in summer (Simpson et al. 1978, 1980; Dickinson and Simpson 1980). Preliminary results from a study conducted in Croatia with the use of GPS-GSM collars suggested that Barbary sheep movements are limited to Mt. Mosor and animals are unlikely to expand their range. Generally, tracked animals moved mainly along the mountain ridges of Mt. Mosor and covered the entire range (approx. 25 km) in a relatively short time (2–4 days) several times per year. On the other hand, movements off the main mountain range were short and usually stopped at the foothills, where the vegetation cover is taller, and the terrain becomes less rugged. The monthly home-range size varied between 0.6 and 17.0 km^2 (mean: 5.9 km^2). The cumulative distance covered by Barbary sheep per day was 1.63 km with an annual home-range size of 26.0 km^2 (Gančević et al. 2016a). Pascual-Rico et al. (2018) estimated an average core home range of 9 GPS tracked Barbary sheep from Sierra Espuña to be 1.65 ± 0.67 km^2.

Feeding

With respect to their feeding habits, studies carried out in Tunisia (Ben Mimoun and Nouira 2015) and Spain (Miranda et al. 2012) show a marked preference for feeding on the herbaceous layer (grasses and forbs), although shrubs are also selected year-round. Contrary to previous expectations, the species shows marked grazing behavior, similar to European mouflon (*Ovis gmelini musimon*) (Miranda et al. 2012). In Croatia, Barbary sheep in Mt. Mosor is extremely plastic in the utilization of the available food resources, feeding mainly on grasses in spring, autumn, and winter (up to 80%), while in summer the main diet is composed of deciduous trees (Lazarus et al. 2019). Qualitative dietary information has been provided for the Canary Island populations (Rodríguez Piñero and Rodríguez Luengo 1992).

Finally, it is worth mentioning that in Southeastern Spain, the incidence of Barbary sheep on shrubs is similar to that of the native herbivore ungulates (red deer *Cervus elaphus*, and Iberian wild goat *Capra pyrenaica*), and it does not show a particularly high negative impacts on threatened plants (Velamazán et al. 2017).

Behavior

Social behavior

Barbary sheep is a social species (Solbert 1980) characterized by a hierarchical system where all mature individuals hold social ranks according to their size and age (males) and age and mating/calving experience (females) (Cassinello 1995). Specifically, mature females may increase their hierarchical status during mating and after parturition, whereas calf weaning may lead to a decrease of maternal rank (Cassinello 1995). Gregariousness of Barbary sheep is characterized by the segregation of sexes outside the rut. Six group types can be distinguished in the wild: solitary, nursery (females, calves, and juveniles), mixed, all male, all female, and all juveniles (Gray and Simpson 1982). Group leadership is ascribed to females when adults of both sexes are present, while the composition and size of groups may vary depending on the season (Gray and Simpson 1982; Cassinello 2000; Cassinello et al. 2004). Barbary sheep usually forms large groups (Anadón et al. 2018), nursery groups being normally the largest, and may consist of several tens of individuals (up to 80 individuals have been recorded in Sierra Espuña, Cassinello 2000). Normally, however, most groups do not exceed 10 individuals (Fig. 3).

Behavioral conflicts between mothers and their calves have been analyzed in captivity by Cassinello (1997b). Mothers that conceive in the next mating season accelerate the weaning time of their current offspring; also, high-ranked mothers wean their male calves earlier. Aggressive behaviors were recorded during the resumption of the mother's sexual activity, but not during the weaning time.

The spatial association between individuals has been analyzed in captivity, and it is mainly determined by kinship, resembling group composition in free-ranging conditions, i.e., females stay with offspring and bachelor males (Cassinello and Calabuig 2008).

Mating behavior

Barbary sheep are polygynous, seasonal breeders and nonterritorial, with a mating season occurring from September to November, although in

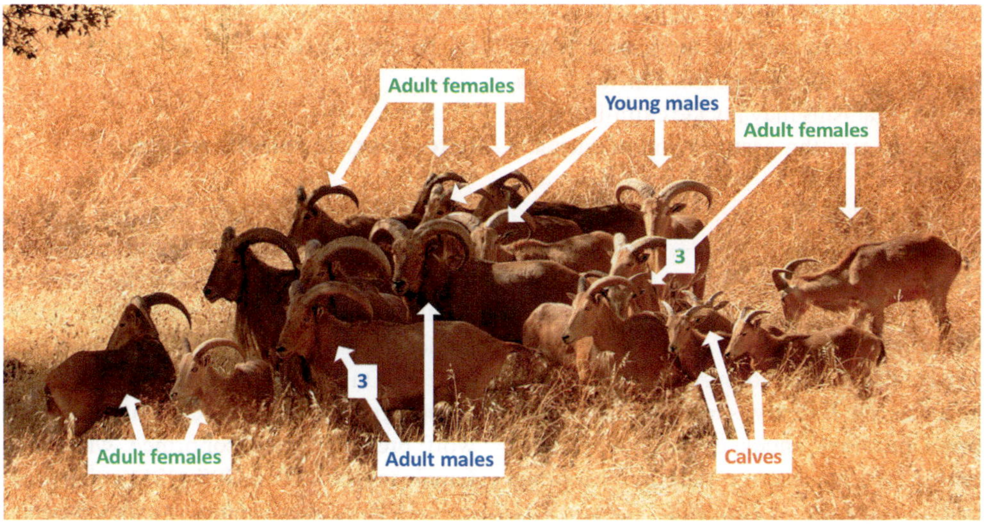

Fig. 3 Barbary sheep group; the differentiation among sex and age classes is based on horn and body size (photograph by Andrés Ríos/Ungulata research team)

captivity and under favorable conditions they can reproduce year-round (Alados and Vericad 1993; Cassinello 1997a).

Activity

Based on a 2-year camera trap survey, the daily activity pattern of Barbary sheep can be described as bimodal (morning-evening), a pattern most frequently found also in other ungulates (Gančević et al. 2016b).

Parasites and Diseases

Parasites

Barbary sheep population of Sierra Espuña suffered from a sarcoptic mange epidemic in 1991 that caused a population decline of 86% until 1995, though their recovery was equally fast (González-Candela et al. 2004). At the start of this epidemic, population density was at its highest levels since the population introduction and likely exceeded the habitat capacity, so that the animals began to share peripheral areas with small domestic ruminants infected with *Sarcoptes scabiei* (Ambiental 1993). These risk factors likely caused the rapid spread of sarcoptic mange in the Barbary sheep population. During the epidemic regression phase the infection rate was higher in males (22%) than in females (17%) and young animals (5%). Males over 5 years of age were the most affected, followed by offspring of both sexes. Few animals had superficial mange lesions (7%), while most individuals (72%) had lesions of moderate severity. The most common lesions were located on the neck, head and back. Since 1999 there have been no recorded cases of sarcoptic mange in free ranging Barbary sheep, although the disease was present in other species (González-Candela et al. 2004).

Feces analysis from 22 free-ranging Barbary sheep on Mt. Mosor in Croatia showed the presence of endoparasites (*Trichostrongylus, Strongyloides, Cooperia*) but no ectoparasites (P. Gančević, unpublished data).

Samples obtained and analyzed from a captive population in Spain (EEZA-CSIC, Almeria, Spain) showed that 87.5% of animals were parasitized with multiple infections, such as *Teladorsagia circumcincta* and *Camelostrongylus mentulatus* in the abomasum, *Trichostrongylus colubriformis* and *Trichostrongylus vitrinus* in the small intestine, and *Skrjabinema ovis* in the large intestine (Mayo et al. 2013).

Cho et al. (2006) found *Balantidium coli* within the lymphatic ducts of the gastric lymph node and abdominal submucosa. On rare occasions, these organisms may invade extra-intestinal organs, in this case the gastric lymph nodes and abomasum.

Infectious diseases

Despite being considered an invasive ungulate outside its native range (North Africa), little information exists on Barbary sheep as a pathogen reservoir. González-Candela et al. (2009) showed that neither free nor captive Barbary sheep had antibodies against *Brucella melitensis*, *Chlamydophila abortus* or BVDV-BDV. However, compared to other ungulates in Spain, Barbary sheep have a high prevalence of antibodies against *Mycobacterium bovis* (free 50%: captive 8%), very high prevalence of antibodies against *M. avium* subsp. *paratuberculosis* (free 19%; captive 56%) and intermediate prevalence of antibodies against *Salmonella* spp. (free 13%; captive 0%) or *Toxoplasma gondii* (free 2%; captive 24%).

Finally, the presence of bluetongue virus antibodies was detected in one of 22 individuals in the Guadiana Valley (Ruiz-Fons et al. 2008), and two seropositive individuals of the bluetongue virus were detected in a sample of Barbary sheep ($n = 3$) examined between 2008 and 2010 (Lorca-Oró et al. 2014).

Population Ecology

Population dynamics

Available literature lacks specific information on population dynamics. With respect to population abundance, in Southeastern Spain up to 2000 individuals are present (Cassinello et al. 2004), and around 140 individuals are present in Croatia (Lazarus et al. 2019). Population densities have been obtained from Spanish exotic populations, reaching an average of 2.6 individuals/km^2 on the island of La Palma (Canary Islands) in the late 1980s (Ornistudio 1992), and 5.5 individuals/km^2 in Sierra de Espuña (Murcia, Spain) (Fernández-Olalla et al. 2016).

Interspecific relationships

In Sierra Espuña, Spain, both the golden eagle (*Aquila chrysaetos*) and feral dogs prey on young Barbary sheep (J. García, pers. comm.), but in general in Spain there are very few natural predators, and adults are threatened only by hunters. On Mt. Mosor, Croatia, the main Barbary sheep predator is the grey wolf (*Canis lupus*); wolf predation was confirmed when a 4-year female tagged with a GPS-GSM collar was killed only 11 days after capture (predation occurred at night at 23:45) (Gančević et al. 2016a).

In Africa, the main competitors are livestock, given the small number of wild herbivores that exploit the same niche. Exceptions are Cuvier's gazelle (*Gazella cuvieri*) and Nubian ibex (*Capra nubiana*), which might be potential Barbary sheep competitors when living in sympatry and food resources are scarce. With respect to exotic Barbary sheep populations, no compelling studies have been carried out. In Spain, they may compete for resources with another exotic species, the European mouflon, when they live in sympatry (Sicilia et al. 2011), though Barbary sheep feeding habits differentiate from the native species, such as red deer and Iberian wild goat (Miranda et al. 2012). Ferretti and Mori (2020) reviewed available literature on interactions between alien and native ungulate species and found only one study in which Barbary sheep was found dominant.

Available empirical information indicates that Barbary sheep could be considered invasive on La Palma Island, Spain (Cassinello 2015, 2018), where they may negatively affect the endemic flora (Rodríguez Piñero and Rodríguez Luengo 1992).

Although Barbary sheep was considered a potential threat to the native Iberian wild goat in the Southeastern Iberian peninsula (Acevedo et al. 2007), recent empirical data on Iberian wild goat expansion throughout regions already occupied by Barbary sheep (Eguía et al. 2015; Anadón et al. 2018), and Barbary sheep trophic ecology

(Miranda et al. 2012; Velamazán et al. 2017) suggest a neutral if not positive effect on the host ecosystem, due to its grazing habits (Cassinello 2018).

Effects of climate change

Recent climate changes caused lower rainfall regimes and higher mean annual temperatures in Southeastern Spain (e.g., Puigdefábregas and Mendizábal 2004). These lead to strong desertification which may cause significant habitat changes and, in turn favor the expansion of a desert ungulate such as the Barbary sheep.

Conservation Status

Being a non-native species, most European Barbary sheep populations have no special conservation status and are hunted. Only in Czech Republic, the Barbary sheep population is protected and thus not hunted (Bartoš et al. 2010).

In its native habitat, the IUCN lists Barbary sheep as Vulnerable C1 at the species level (Cassinello et al. 2008), i.e., the population size is estimated to be less than 10,000 mature individuals with an estimated continuing decline of at least 10% within 10 years or three generations, whichever is longer (up to a maximum of 100 years in the future). The decline of Barbary sheep in its native habitat is believed to be mainly the result of poaching and habitat degradation, along with competition with domestic livestock (Alados and Shackleton 1997). Reintroduction programs are currently being in place only in Tunisia (DGF and IUCN 2017).

The current status of the six recognized subspecies is variable, with some subspecies being likely on the verge of extinction (Cassinello 2013).

Management

Native Barbary sheep populations are generally threatened due to a combination of overgrazing by domestic livestock, deforestation, habitat destruction, and poaching (see Shackleton 1997). The general conservation programs of each country where the species occurs were summarized by Shackleton (1997). It is worth mentioning the Strategic Action Plan currently being implemented in Tunisia (DGF and IUCN 2017), where five main actions have been put forward: (1) habitat conservation, (2) poaching eradication, (3) monitoring and data collection, (4) awareness and valuation, and (5) sustainable financing. It is advisable to promote similar protection and management programs in other North African countries. However, the actual status of Barbary sheep populations in its native distribution area is mostly unknown or uncertain (Cassinello et al. 2008; Cassinello 2013, 2015), requiring proper surveys to update the knowledge about actual distribution and abundance of the species, which would allow to develop and implement targeted management policies. Unfortunately, a series of logistical and political issues make it particularly difficult to obtain field data and carry out proper protection actions in native areas (see also Alados and Shackleton 1997).

Due to hunting interests, Barbary sheep was introduced with much success to the USA and Spain (Ogren 1965; Ortuño and De la Peña 1979; Christian 1980), where populations expanded very rapidly (Cassinello 1998). The introduction of a species into territories outside its natural range should be managed carefully, in order to control its expansion and to minimize its potential threat to native species. However, empirical evidence seemingly shows a neutral effect of Barbary sheep on host ecosystems, with the exception of La Palma Island (Canary Islands), where it threatens the Macaronesian flora (Cassinello 2015, 2018).

Exotic Barbary sheep populations have been, and still are, a source of conflict between different stakeholders. Hunting communities support and promote their presence, while shepherds and farmers want them eradicated. Indeed, high Barbary sheep densities in Sierra Espuña and surrounding areas have facilitated sheep presence on farmlands (Cassinello 2000), and farmers made large investments to build fences that would keep Barbary sheep off crops. Furthermore, the management of the species in Spain has changed in recent years, as legislations classified it as an invasive exotic species. However, exotic Barbary sheep populations that were established before the Spanish Law 42/2007 on the Natural Heritage and Biodiversity was approved, in December 2007, are allowed to be sustainably managed through hunting (basically, the populations present in the province of Murcia), whereas in recently colonized areas, they should be controlled and eventually eradicated (Jefatura del Estado 2018). Current scientific evidence does not demonstrate harmful effects of Barbary sheep on the Iberian environment despite its exotic origin. Currently, the populations present in the Murcia Region and other adjacent areas can be managed by hunting, although uncontrolled expansion must be avoided. Today, however, Barbary sheep is still present in the Spanish List of Invasive Exotic Species (see Cassinello 2018).

The illegal introduction of Barbary sheep to Croatia has caused much controversy between hunters and the environmental community. Hunters perceive the introduced animals as an attractive game species and a potential source of revenue. Environmentalists stress the possible negative effects of introduced exotic species on the native flora and fauna and call for the application of existing laws that prescribes complete removal of introduced animals (Official Gazette 2018).

Barbary sheep are not hunted (Bartoš et al. 2010) in Czech Republic, regardless of the fact that they are not native, since they are listed as vulnerable according to IUCN (Cassinello et al. 2008).

Future Challenges for Research and Management

It is urgent to investigate the genetic characteristics and profile of the Barbary sheep populations, which have been assigned to the six subspecies based solely on geographical distribution and morphology. This will allow to properly classify subspecies and identify the most valuable genetic units and populations, and in turn, plan the necessary conservation actions.

With respect to exotic Barbary sheep populations, these have been shown to have a great dispersal capacity and to be able to survive in virtually any areas resembling their native habitats (e.g., Cassinello 2015). However, some exotic populations have not expanded much since their release, such as in northern Italy and Croatia (Mori et al. 2017; Lazarus et al. 2019). Therefore, close monitoring of old and new exotic populations is necessary to prevent undesirable expansions or their presence in areas with protected flora.

Finally, although empirical evidence has suggested little to no negative effects of exotic Barbary sheep populations on the host environments, studies and experiments should be conducted to elucidate this aspect. New releases in non-native environments should be avoided, whereas Barbary sheep present on islands or in areas with threatened flora should be eradicated or strictly managed and controlled.

Acknowledgments The authors would like to express their gratitude to an anonymous reviewer and to Luca Corlatti for suggestions and improvement of this chapter. They also thank Andrés Ríos and the EEZA-CSIC for providing the photographs and Jean-Marc Cugnasse for providing the information about the French population.

References

Abaigar T, Domene MA, Cassinello J (2012) Characterization of the estrous cycle and reproductive traits of the aoudad (*Ammotragus lervia*) in captivity. Theriogenology 77:1759–1766

Acevedo P, Cassinello J, Hortal J, Gortazar C (2007) Invasive exotic aoudad (*Ammotragus lervia*) as a major threat to native Iberian ibex (*Capra pyrenaica*): a habitat suitability model approach. Divers Distrib 13:587–597

Alados C, Shackleton DM (1997) Regional summary. In: Shackleton DM (ed) Wild sheep and goats and their relatives: status survey and conservation action plan for Caprinae. IUCN/SSC Caprinae Specialist Group, Gland, pp 47–48

Alados CL, Vericad JR (1993) Aoudad, *Ammotragus lervia* from Western-Sahara. International Studbook. Boletín del Instituto de Estudios Almerienses, Ciencias 11/12:65–91

Ambiental (1993) Estudios básicos para el Plan de Ordenación de los Recursos Naturales de Sierra Espuña. Consejería de Medio Ambiente de la Comunidad Autónoma de Murcia, Murcia, 690pp

Anadón JD, Pérez-García JM, Pérez I, Royo J, Sánchez-Zapata JA (2018) Disentangling the effects of habitat, connectivity and interspecific competition in the range expansion of exotic and native ungulates. Landsc Ecol 33:597–608

ARMAN (1991) Informe sobre el Parque Natural de la Sierra de Espuña. Unpublished report, Agencia Regional del Medio Ambiente y la Naturaleza (ARMAN)

Bartoš L, Kotrba R, Pintíř J (2010) Ungulate and their management in the Czech Republic. In: Apollonio M, Andersen R, Putman R (eds) European ungulates and their management in the 21st century. Cambridge University Press, Cambridge, pp 243–261

Ben Mimoun J, Nouira S (2015) Food habits of the aoudad *Ammotragus lervia* in the Bou Hedma mountains, Tunisia. S Afr J Sci 111:1–5

Benirschke K (1971) Chimerism, mosaicism and hybrids. In: de Grouchy J, Ebling FJG, Henderson IW (eds) Human genetics. Excerpta Medica, Amsterdam, pp 212–231

Blyth E (1840) A summary monograph of the species of the genus *Ovis*. Proc Zool Soc London 8:12–13

Brown D (1988) Components of lifetime reproductive success. In: Clutton-Brock TH (ed) Reproductive success. The University of Chicago Press, Chicago, pp 439–453

Bunch TD, Rogers A, Foote WC (1977) G-band and transferrin analysis of aoudad-goat hybrids. J Hered 68:210–212

Cassinello J (1995) Factors modifying female social ranks in *Ammotragus*. Appl Anim Behav Sci 45:175–180

Cassinello J (1996) High ranking females bias their investment in favour of male calves in captive *Ammotragus lervia*. Behav Ecol Sociobiol 38:417–424

Cassinello J (1997a) High levels of inbreeding in captive *Ammotragus lervia* (Bovidae, Artiodactyla): effects on phenotypic variables. Can J Zool 75:1707–1713

Cassinello J (1997b) Mother-offspring conflict in the Saharan arrui. Relation to weaning and mother's sexual activity. Ethology 103:127–137

Cassinello J (1998) *Ammotragus lervia*: a review on systematics, biology, ecology and distribution. Ann Zool Fenn 35:149–162

Cassinello J (2000) *Ammotragus* free-ranging population in the south-east of Spain: a necessary first account. Biodivers Conserv 9:887–900

Cassinello J (2013) *Ammotragus lervia*. In: Kingdon JS, Hoffmann M (eds) Mammals of Africa. Vol VI. Pigs, hippopotamuses, chevrotain, giraffes, deer and bovids. Bloomsbury Publishing, London, pp 595–599

Cassinello J (2015) *Ammotragus lervia* (aoudad). In: Invasive Species Compendium. CAB International, Wallingford. www.cabi.org/isc

Cassinello J (2018) Invasive species misconception and mismanagement: the paradoxical case of an alien ungulate in Spain. Conserv Lett 11:e12440

Cassinello J, Alados CL (1996) Female reproductive success in captive *Ammotragus lervia* (Bovidae, Artiodactyla). Study of its components and effects of hierarchy and inbreeding. J Zool Lond 239:141–153

Cassinello J, Calabuig G (2008) Spatial association in a highly inbred ungulate population: evidence of fine-scale kin recognition. Ethology 114:124–132

Cassinello J, Gomendio M (1996) Adaptive variation in litter size and sex ratio at birth in a sexually dimorphic ungulate. Proc R Soc Lond B Biol Sci 263:1461–1466

Cassinello J, Serrano E, Calabuig G, Pérez JM (2004) Range expansion of an exotic ungulate (*Ammotragus lervia*) in southern Spain: ecological and conservation concerns. Biodivers Conserv 13:851–866

Cassinello J, Cuzin F, Jdeidi T, Masseti M, Nader I, de Smet K (2008) *Ammotragus lervia*. In: IUCN Red List of Threatened Species. www.iucnredlist.org

Cho HS, Shin SS, Park NY (2006) Balantidiasis in the gastric lymph nodes of Barbary sheep (*Ammotragus lervia*): an incidental finding. J Vet Sci 7:207–209

Christian TT (1980) Development and economic role of aoudad hunting in a cattle ranching operation. In: Simpson CD (ed) Symposium on ecology and management of Barbary sheep. Texas Tech University Press, Lubbock, pp 104–105

Cugnasse JM, Rigaux P (2021) *Ammotragus lervia* (Pallas, 1777) Mouflon à manchettes. In: Savouré-Soubelet A, Arthur C, Aulagnier S, Body G, Callou C, Haffner P, Marchandeau S, Moutou F & Saint-Andrieux C (eds) Atlas des Mammifères Sauvages de France volume 2 : Ongulés et Lagomorphes. Muséum national d'Histoire naturelle, Paris, pp 142–145

Curtain CC, Fudenberg HH (1973) Evolution of the immunoglobulin antigens in the Ruminantia. Biochem Genet 8:301–308

Damm GR, Franco N (2014) CIC Caprinae atlas of the world. CIC International Council for Game and Wildlife Conservation, Budakeszi, in cooperation with Rowland Ward Publications, Johannesburg

DGF and IUCN (2017) Stratégie et plan d'action pour la conservation du mouflon à manchettes (*Ammotragus lervia*) en Tunisie 2018–2027. IUCN/DGF, Malaga, 56pp

Dickinson TG, Simpson CD (1980) Home range movements, and topographic selection of Barbary sheep in the Guadalupe Mountains, New Mexico. In: Simpson CD (ed) Symposium on ecology and management of Barbary sheep. Texas Tech University Press, Lubbock, pp 78–86

Eguía S, Martínez-Noguera E, Botella F, Pascual R, Giménez-Casalduero A, Sánchez-Zapata JA (2015) Evolución del área de distribución del arruí (*Ammotragus lervia* Pallas 1777) y la cabra montés (*Capra pirenaica* Schinz, 1838) en la Región de Murcia. V Congreso de la Naturaleza de la Región de Murcia, 2015. Asociación de Naturalistas del Sureste, Murcia

Fernández-Olalla M, Martínez-Jáuregui M, Perea R, Velamazán M, San Miguel A (2016) Threat or opportunity? Browsing preferences and potential impact of *Ammotragus lervia* on woody plants of a Mediterranean protected area. J Arid Environ 129:9–15

Ferretti F, Mori E (2020) Displacement interference between wild ungulate species: does it occur? Ethol Ecol Evol 32:2–15

Gančević P, Šprem N, Jerina K (2016a) Space use and activity patterns of introduced Barbary sheep (*Ammotragus lervia*) in Southern Dinarides, Croatia. In: Hadjisterkotis E (ed) Abstract book of 6th world congress on mountain ungulates and 5th international symposium on mouflon, Nicosia, 41pp

Gančević P, Jerina K, Safner T, Šprem N (2016b) Space use and activity patterns of Barbary sheep in Mosor Mountain. In: Jelaska S (ed) Book of abstracts, of 2nd Croatian symposium on invasive species, Zagreb, 43pp

Geist V (1971) Mountain sheep. A study in behaviour and evolution. The University of Chicago Press, Chicago

González-Candela M, León-Vizcaíno L, Cubero-Pablo MJ (2004) Population effects of sarcoptic mange in Barbary sheep (*Ammotragus lervia*) from Sierra Espuña Regional Park, Spain. J Wildl Dis 40:456–465

González-Candela M, Serrano E, Martinez-Carrasco C, Martín-Atance P, Cubero MJ, Alonso F, Leon L (2009) Coinfection is an important factor in epidemiological studies: the first serosurvey of the aoudad (*Ammotragus lervia*). Eur J Clin Microbiol Infect Dis 28:481–489

Gray AP (1972) Mammalian hybrids. A check list with bibliography. Commonwealth Agricultural Bureau, Slough

Gray GG, Simpson CD (1980) *Ammotragus lervia*. Mamm Species 144:1–7

Gray GG, Simpson CD (1982) Group dynamics of free-ranging Barbary sheep in Texas. J Wildl Manag 46:1096–1101

Hassanin A, Ropiquet A, Couloux A, Cruaud C (2009) Evolution of the mitochondrial genome in mammals living at high altitude: new insights from a study of the tribe Caprini (Bovidae, Antilopinae). J Mol Evol 68:293–310

Jefatura del Estado (2018) Ley 7/2018, de 20 de julio, de modificación de la Ley 42/2007, de 13 de diciembre, del Patrimonio Natural y de la Biodiversidad. BOE 176:73196–73202

Johnston DS (1980) Habitat utilization and daily activities of Barbary sheep. In: Simpson CD (ed) Symposium on ecology and management of Barbary sheep. Texas Tech University Press, Lubbock, pp 51–58

Kavčić K, Gančević P, Šprem N (2020) Morphological analysis of the aoudad: the introduced population is well adapted to the Mediterranean habitat. J Cent Eur Agric 21:553–564

Lazarus M, Gančević P, Orct T, Barišić D, Jerina K, Šprem N (2019) Barbary sheep tissues as bioindicators of radionuclide and stabile element contamination in Croatia: exposure assessment for consumers. Environ Sci Pollut Res 26(14):14521–14533

Lorca-Oró C, López-Olvera JR, Ruiz-Fons F, Acevedo P, García-Bocanegra I, Oleaga A, Gortázar C, Pujols J (2014) Long-term dynamics of bluetongue virus in wild ruminants: relationship with outbreaks in livestock in Spain, 2006–2011. PLoS One 9:e100027

Manwell C, Baker CMA (1975) *Ammotragus lervia*: progenitor of the domesticated sheep or specialized offshoot of caprine evolution? Experientia 31:1370–1371

Mayo E, Ortiz J, Martínez-Carrasco C, Garijo MM, Espeso G, Hervías S, Ruiz de Ybáñez MR (2013) First description of gastrointestinal nematodes of Barbary sheep (*Ammotragus lervia*): the case of *Camelostrongylus mentulatus* as a paradigm of phylogenic and specific relationship between the parasite and its ancient host. Vet Res Commun 37:209–215

Mereu P, Palici di Suni M, Manca L, Masala B (2008) Complete nucleotide mtDNA sequence of Barbary sheep (*Ammotragus lervia*). DNA Seq 19:241–245

Miranda M, Sicilia M, Bartolomé J, Molina-Alcaide E, Gálvez-Bravo L, Cassinello J (2012) Contrasting feeding patterns of native red deer and two exotic ungulates in a Mediterranean ecosystem. Wildl Res 39:171–182

Moore NW, Halnan CR, McKee JJ, Watson JI (1981) Studies on hybridization between a Barbary ram (*Ammotragus lervia*) and domestic ewes (*Ovis aries*) and nanny goats (*Capra hircus*). J Reprod Fertil 61:79–82

Mori E, Mazza G, Saggiomo L, Sommese A, Esattore B (2017) Strangers coming from the Sahara: an update of the worldwide distribution, potential impacts and conservation opportunities of alien aoudad. Ann Zool Fenn 54:373–386

Nadler CF, Hoffmann RS, Woolf A (1974) G-band patterns, chromosomal homologies, and evolutionary relationships among wild sheep, goats, and aoudads (Mammalia, Artiodactyla). Experientia 30:744–746

Official Gazette (2018) Act on the prevention of the intake and spread of foreign and invasive foreign species and their management. Republic of Croatia 15/18

Ogren H (1965) Barbary sheep. New Mexico Department of Game and Fish Bulletin 13. New Mexico Department of Game and Fish, Santa Fe, 117pp

Ornistudio SL (1992) Seguimiento del Arrui en el Parque Nacional de la Caldera de Taburiente (La Palma, Islas Canarias). Unpublished report, Parque Nacional de La Caldera de Taburiente, La Palma

Ortuño F, De la Peña J (1979) Reservas y cotos nacionales de caza. 4. Región mediterránea. Colección Naturaleza Española, 5. Editorial Incafo, Madrid

Pallas PS (1777) Antilope lervia. Spicilegia Zool 12:12

Pascual-Rico R, Pérez-García JM, Sebastián-González E, Botella F, Giménez A, Eguía S, Sánchez-Zapata JA (2018) Is diversionary feeding a useful tool to avoid human-ungulate conflicts? A case study with the aoudad. Eur J Wildl Res 64:67

Puigdefábregas J, Mendizábal T (2004) Prospects for desertification impacts in Southern Europe. In: Marquina A (ed) Environmental challenges in the Mediterranean 2000–50. Kluwer Academic, Dordrecht, pp 155–172

Rodríguez Piñero JC, Rodríguez Luengo JL (1992) Autumn food-habits of the Barbary sheep (*Ammotragus lervia* Pallas 1777) on La Palma Island (Canary Islands). Mammalia 56:385–392

Ruiz-Fons F, Reyes-García AR, Alcaide V, Gortázar C (2008) Spatial and temporal evolution of bluetongue virus in wild ruminants, Spain. Emerg Infect Dis 14:951–953

Safner T, Stipoljev S, Stuhne T, Svetličić I, Galov A, Gančević P, Cassinello J, Šprem N (2018) A first insight into mitochondrial control region diversity of

Barbary sheep (*Ammotragus lervia*) introduced in Europe. In: Jelaska S (ed) Book of abstracts of the 3rd Croatian symposium on invasive species, Zagreb, 67pp
Schaffer WM, Reed CA (1972) The co-evolution of social behaviour and cranial morphology in sheep and goats (Bovidae, Caprini). Fieldiana 61:1–62
Schmitt J (1963) *Ammotragus lervia* Pallas, Mähnenschaf oder Mähnenziege? Z Säugetierk 28:7–12
Shackleton DM (ed) (1997) Wild sheep and goats and their relatives: status survey and conservation action plan for Caprinae. IUCN/SSC Caprinae Specialist Group, Gland
Sicilia M, Miranda M, Cassinello J (2011) Interspecific behaviour in temperate ungulates: an alien adult male associates with a group of non-conspecifics. Belg J Zool 41:56–58
Simpson CD, Krysl LJ, Hampy DB, Gray GG (1978) The Barbary sheep: a threat to desert bighorn survival. Desert Bighorn Counc Trans 22:26–31
Simpson CD, Krysl LJ, Dickinson TG (1980) Food habits of Barbary sheep in the Guadalupe Mountains, New Mexico. In: Simpson CD (ed) Symposium on ecology and management of Barbary sheep. Texas Tech University Press, Lubbock, pp 87–91
Solbert AGK (1980) Social organization and behaviour of aoudad (*Ammotragus lervia* Pallas) in Texas. In: Simpson CD (ed) Symposium on ecology and management of Barbary sheep. Texas Tech University Press, Lubbock, pp 66–72
Stuhne T (2018) The origin analysis of Croatian Barbary sheep (*Ammotragus lervia* Pallas, 1777) population using mitochondrial DNA sequences. Master's thesis, University of Zagreb, Faculty of Agriculture. https://urn.nsk.hr/urn:nbn:hr:204:363243. (in Croatian with English abstract)
Tolić I (2005) Barbary sheep (*Ammotragus lervia*) morphological, biological and ecological properties. Sumar List 3–4:177–181. (in Croatian with English abstract)
Vaufrey R (1955) Prehistorie de l'Afrique. I. Le Maghreb. Publications de l'Institut des Hautes Etudes de Tunis, vol 4. Masson, Paris, pp 1–435
Velamazán M, San Miguel A, Escribano R, Perea R (2017) Threatened woody flora as an ecological indicator of large herbivore introductions. Biodivers Conserv 26:917–930
Williams GC (1979) The question of adaptive sex ratio in outcrossed vertebrates. Proc R Soc Lond B Biol Sci 205:567–580

Further Reading

Dixon A, Jones D (eds) (1988) Conservation and biology of desert antelopes. Christopher Helm Ltd, Kent
Grubb P (2005) Artiodactyla. In: Wilson DE, Reeder DM (eds) Mammal species of the world. A taxonomic and geographic reference, 3rd edn. Johns Hopkins University Press, Baltimore, pp 637–722
Meester J, Setzer HW (eds) (1970) The mammals of Africa: an identification manual. Smithsonian Institution Press, Washington, DC
Shackleton DM (ed) (1997) Wild sheep and goats and their relatives: status survey and conservation action plan for Caprinae. IUCN/SSC Caprinae Specialist Group, Gland

Alpine Ibex *Capra ibex* Linnaeus, 1758 16

Alice Brambilla, Bruno Bassano, Iris Biebach, Kurt Bollmann, Lukas Keller, Carole Toïgo, and Achaz von Hardenberg

Contents

A. Brambilla (✉)
Department of Evolutionary Biology and Environmental Studies, University of Zurich, Zurich, Switzerland

Gran Paradiso National Park, Alpine Wildlife Research Center, Torino, Italy
e-mail: alicebrambilla1@gmail.com

B. Bassano
Gran Paradiso National Park, Alpine Wildlife Research Center, Torino, Italy
e-mail: bruno.bassano@pngp.it

I. Biebach · L. Keller
Department of Evolutionary Biology and Environmental Studies, University of Zurich, Zurich, Switzerland
e-mail: iris.biebach@ieu.uzh.ch; lukas.keller@ieu.uzh.ch

K. Bollmann
Biodiversity and Conservation Biology, Swiss Federal Research Institute WSL, Birmensdorf, Switzerland
e-mail: kurt.bollmann@wsl.ch

C. Toïgo
Office Français de la Biodiversité, Unité Ongulés Sauvages, Gières, France
e-mail: carole.toigo@oncfs.gouv.fr

A. von Hardenberg
Conservation Biology Research Group, Department of Biological Sciences, University of Chester, Chester, UK
e-mail: a.vonhardenberg@chester.ac.uk

© Springer Nature Switzerland AG 2022

L. Corlatti, F. E. Zachos (eds.), *Terrestrial Cetartiodactyla*, Handbook of the Mammals of Europe,
https://doi.org/10.1007/978-3-030-24475-0_32

Common Names

English	Alpine ibex
German	Alpensteinbock
French	Bouquetin des Alpes
Spanish	Ibice de los Alpes
Italian	Stambecco alpino
Russian	Альпийский козерог

Taxonomy, Systematics and Paleontology

The Alpine ibex (*Capra ibex*, Fig. 1) is a wild mountain ungulate belonging to the order Cetartiodactyla, suborder Ruminantia, family Bovidae, subfamily Caprinae, tribe Caprini, and genus *Capra*. The genus *Capra* is a young genus that appeared about 5–6 mya according to fossils and estimates from molecular data (Manceau et al. 1999; Vrba and Schaller 2000; Ropiquet and Hassanin 2006). Fossil data suggest that *Capra* first appeared in Central Asia (Pilgrim 1947) and from there species radiation rapidly occurred in the Plio-Pleistocene (Hartl et al. 1990; Manceau et al. 1999). The currently accepted classifications based on phylogenetic analyses (Sokolov 1979; Pidancier et al. 2006; Kazanskaya et al. 2007; Gebremedhin et al. 2009) identify nine wild-living species: Bezoar goat or wild goat, *Capra aegagrus*; Alpine ibex, *Capra ibex*; Siberian ibex, *Capra sibirica*; Nubian ibex, *Capra nubiana*; Walia ibex, *Capra walie*; Spanish or Iberian wild goat, *Capra pyrenaica*; Markhor, *Capra falconeri*; Kuban or Western tur, *Capra caucasica*; and Dagestan or Eastern tur, *Capra cylindricornis*. Furthermore, the genus *Capra* includes the domestic goat, *Capra hircus*. Mitochondrial and nuclear Y-chromosome DNA analyses suggest that Alpine ibex is the sister species of the geographically close Iberian wild goat (Pidancier et al. 2006). The most likely scenario for the radiation of the genus *Capra* in Europe

Fig. 1 (**a**) A fully grown male Alpine ibex (*c.* 12 years old) courting an adult female (>8 years old). (**b**) A subadult male Alpine ibex (*c.* 7 years old) with two kids of unknown sex (6 months old, in the foreground) and a very young female (*c.* 1.5 years old, in the background). Male and female Alpine ibex usually live separate during the whole year except for the rutting season in December–January (photographs by Dante Alpe)

suggests one wave of immigration from central Asia around 300 kya (Cregut-Bonnoure 1992). In that period, the ancestor of the species reached its largest geographic distribution also outside of the Alpine region, to which it was restricted at the end of the Riss glaciation as a result of forest recolonization of low altitude areas. The colonization of the Iberian Peninsula probably occurred between the Riss and Würm glaciations (about 126–115 kya) and lead to the geographic separation and speciation of *Capra ibex* and *Capra pyrenaica*.

Current Distribution

In the Middle Ages, Alpine ibex occurred across all of the European Alps but intensive hunting following the development of firearms almost brought the species to extinction (Grodinsky and Stüwe 1987). The first signs of the decline of the species were identified in the seventeenth century, and in the following centuries, it gradually disappeared from the Alpine arc except for the region around the Gran Paradiso massif along the Italian-French border. At the beginning of the nineteenth century, there were no more than a hundred individuals left in that area (Grodinsky and Stüwe 1987). Hunting was prohibited by the Piedmontese government in 1821, but the population really started to recover only after the Gran Paradiso area was declared a royal hunting reserve in 1856 by the King Vittorio Emanuele II (Passerin d'Entrèves 2000). Thanks to the active protection of the species in the reserve, the Gran Paradiso population increased to approximately 3000 individuals by the beginning of the twentieth century. A further strong numerical reduction occurred during the Second World War with no more than 600 animals left (Gran Paradiso National Park Archives Unpublished).

In the first half of the twentieth century, around 90 ibex were brought from the Gran Paradiso population into captive-breeding programs in two Swiss wildlife parks, Peter & Paul (St. Gallen) and Interlaken Harder (Stüwe and Scribner 1989). About half of these animals were illegally captured and smuggled across the Italian-Swiss border (Giacometti 2006). Initially based on this captive-breeding population, conservation programs over the past century have reintroduced the species first to Switzerland and then to all the other countries of the Alps (Tosi et al. 1986; Wiersema and Gauthier 1990; Stüwe and Nievergelt 1991; Giacometti 1991). Some restocking and translocations continue to be carried out with the goal of genetic reinforcement of reintroduced populations referred to as colonies below.

Nowadays, the species is present again across the entire Alpine arc (Fig. 2). However, the distribution is quite fragmented and not all suitable habitat is occupied. A total number of 180 Alpine

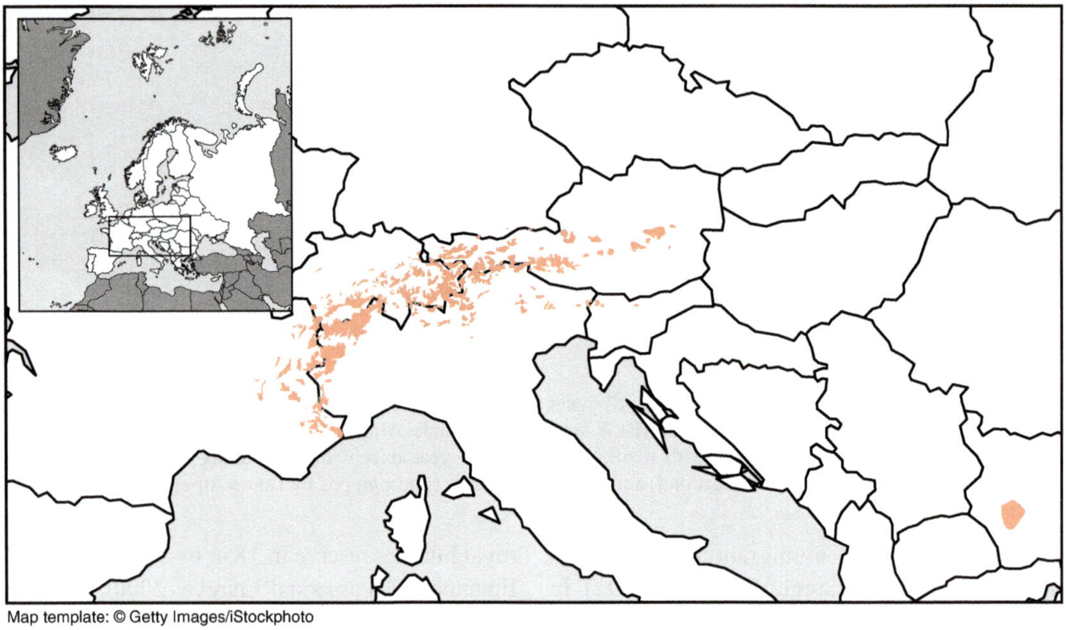

Fig. 2 Distribution map of Alpine ibex in Europe. Distribution is modified from the IUCN Red List of Threatened Species (Version 2017–2) based on data from the authors (Map template: © Copyright Getty Images/iStockphoto)

ibex colonies are distributed over the Alps and an overall estimate of about 52,000 individuals was assessed in the years 2008–2017 (Brambilla et al. 2020). Alpine ibex populations are present (from West to East with respective approximate numbers of individuals) in: France (9400); Italy (16,400); Switzerland and Liechtenstein (18,000); Germany (500); Austria (7400); and Slovenia (300). An additional ibex population can be found in Bulgaria, a region far away from the species' native range, where it has been introduced around 1980.

Description

Size and Morphology

Alpine ibex is a sexually dimorphic and polygynous species with remarkable, sabre shaped, backward-curving horns in both sexes (Fig. 1). The body is compact with a short, broad head, a strong neck, and relatively short and sturdy legs due to shorter metapodials than in other ungulate species (Couturier 1962). Males are larger and heavier than females. The average body mass of fully grown males is 75–95 kg with a maximum up to 130 kg, whereas that of females ranges on average between 35 and 45 kg, exceptionally reaching 60 kg (Couturier 1962; Ratti 1981; Loison et al. 1999; Lüps et al. 2007; Bergeron et al. 2008, 2010, see also Fig. 6b). Differences in body length and shoulder height between sexes are about 22%, and for thoracic circumference, about 30% (Giacometti et al. 1997). A male ibex grows up to 163 cm in length and to a shoulder height of 94–95 cm (Couturier 1962; Ratti 1981). Females are smaller with a body length of 127–139 cm (Couturier 1962; Ratti 1981) and a maximum height of 81–82 cm.

Alpine ibex are remarkably sure-footed and agile, which enables them to climb steep cliffs and rocky ledges. The deeply split hooves of the stocky legs with the rubber-like toe balls and sharp edges of the hooves give the animals an extraordinary climbing ability and security (Schaller 1977). Front legs have interdigital glands. The subcaudal glands impart the male with a goat-like smell (Parrini et al. 2009).

Pelage

Alpine ibex are well adapted to low temperatures and have a short, straight strong coat. The coat is quite uniform in appearance with brownish gray hair over most of the body and a pale abdomen. The species moults only once a year in spring when it replaces the winter coat (Nievergelt and Zingg 1986). In autumn, Alpine ibex complement this coat with longer top hairs and a dense, woolly undercoat. Coats vary in colour with sex and age from brownish gray to reddish chestnut brown in summer to yellowish brown-gray in winter. In particular, old males >12 years turn to a more gray and blurry appearance. Newborn kids appear brighter, pale brown. The abdomen, the rump patch, the hairy part of the tail bottom as well as the caudal part of the tarsus are pale, in particular in females and kids. The upper side of the tail is black. Adult male Alpine ibex generally have a chin beard (Couturier 1962; Meile et al. 2003; Damm and Nicolás 2014).

Head and Dentition

The frontal bone (*Os frontale*) and skull bones (*Os parietale*) are strongly convex shaped with two distinct bone cones for the horns (Fig. 3). The tear pit is missing, and the nasal bone has a contact zone with the premaxilla.

Two teeth generations: milk teeth 0.0.3.0/3.1.3.0; permanent dentition 0.0.3.3/3.1.3.3. The exchange of milk teeth starts in the second year of life and ends in the fourth year at approximately 40 months of age (Ratti and Habermehl 1977).

Horns and Age Determination

Newborn kids are hornless. At 1 month, the tips of the growing horns are palpable, and 20–25 mm long at 2 months (Ratti and Habermehl 1977). Horns of both sexes grow lifelong and develop an annual ring during each winter (Fig. 4). An individual's age can be determined from the number of rings, which are best visible on the medial side of the horns. Annual rings are easier to identify on male's horns. Horns of males, reaching up to 100 cm in length, are substantially larger and thicker than those of females which have a maximum length of 35 cm. Longer horns have exceptionally been reported for both sexes (Couturier 1962; Ratti and Habermehl 1977). The axis of male horns forms an angle of 40°–60° and the horn's cross section is rectangular while it is oval in females (Couturier 1962). Horn growth is age dependent with the longest increment (annulus) in the second year of life (Nievergelt 1966), decreasing with age afterwards (Ratti and Habermehl 1977; von Hardenberg et al. 2004; Willisch et al. 2015). Age-specific increments strongly depend on environmental conditions such as climate and forage availability in spring (Giacometti et al. 2002; von Hardenberg et al. 2004; Büntgen et al. 2014). A male's fully grown horn pair can reach a mass of 5–6 kg, while in females they weigh 0.2–0.3 kg (Couturier 1962). Males also form numerous nodes on the frontal side (up to three per year) starting at the second annulus. The nodes are only marginally developed in females. Age- and sex-specific body and horn growth tables from different populations can be found in Couturier (1962), Ratti (1981), Nievergelt and Zingg (1986), Giacometti et al. (1997), Giacometti et al. (2002), von Hardenberg et al. (2004), Bergeron et al. (2008), Toïgo et al. (1999, 2013), Büntgen et al. (2014), and Willisch et al. (2015).

Physiology

Alpine ibex can be found up to 3500 m above sea level. They evolved specific adaptations to the harsh and strongly seasonal high-altitude habitat they live in, such as a high hematocrit and hemoglobin levels. Alpine ibex hematocrit (% of packed cells volume – PCV) equals 50.45% ± 4.38 (mean ± sd, Cook et al. 1986) and is higher than that measured in other ungulates such as Iberian wild goat (44.35% ± 8.9, Pérez et al. 1999) and mouflon, *Ovis gmelini musimon* (41.3% ± 4.8, Peinado et al. 1999). Similar differences can be found for hemoglobin (g/dl mean ± sd): Alpine

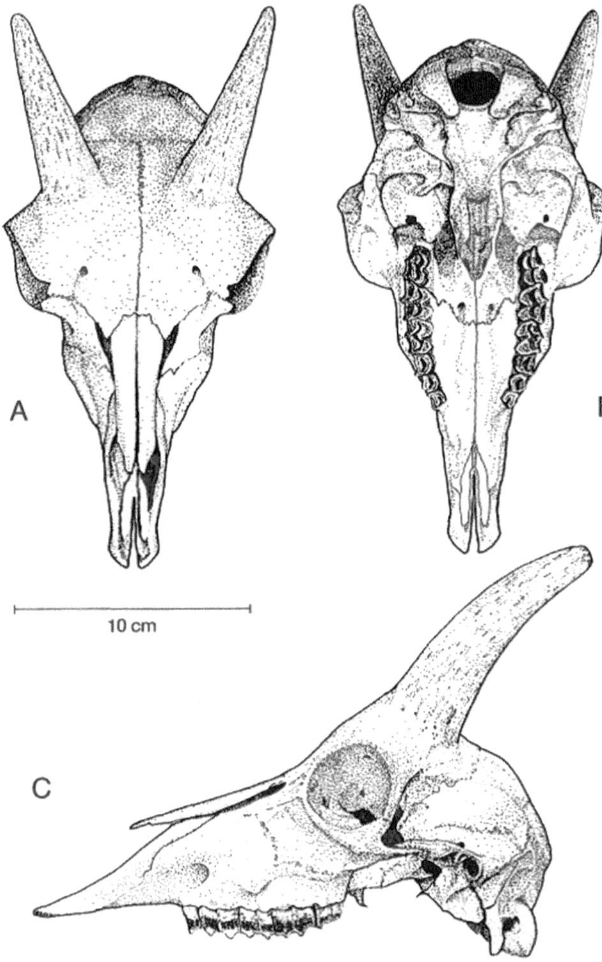

Fig. 3 Skull of an adult *Capra ibex* female. View from dorsal (**A**), ventral (**B**), and lateral (**C**). (From Nievergelt and Zingg (1986))

ibex 17.87 ± 0.93 (Cook et al. 1986); Iberian wild goat 13.59 ± 3.33 (Pérez et al. 1999); and mouflon 15.4 ± 1.7 (Peinado et al. 1999). All these values were measured in captive animals.

Blood and leukocyte values obtained by hematological analysis of wild Alpine ibex are reported in Table 1 (Gauthier et al. 1994).

Despite being less adapted to snowy environment than other mountain ungulates (e.g., Alpine chamois *Rupicapra rupicapra rupicapra*), Alpine ibex evolved several adaptations to the harsh Alpine climate such as a compact body shape, a dark and insulating fur during winter, as well as the ability to accumulate large amounts of body fat during summer and to save energy during winter. Indeed, Alpine ibex metabolism follows a seasonal cycle with many physiological and behavioral adaptations serving to reduce energy loss during winter (Signer et al. 2011). Heart rate is maximum during summer months with values around 100 bpm (beats per minute) in June and minimum during winter decreasing down to 40–50 bpm in January–February, corresponding to a 60% reduction (Signer et al. 2011). Breath and metabolic rates are expected to change accordingly. Body core temperature also varies during the year with lower values registered in winter (38–38.2 °C) and peak values in summer (39.5 °C) although the range of variation is lower (3% reduction) than that of heart rate. The

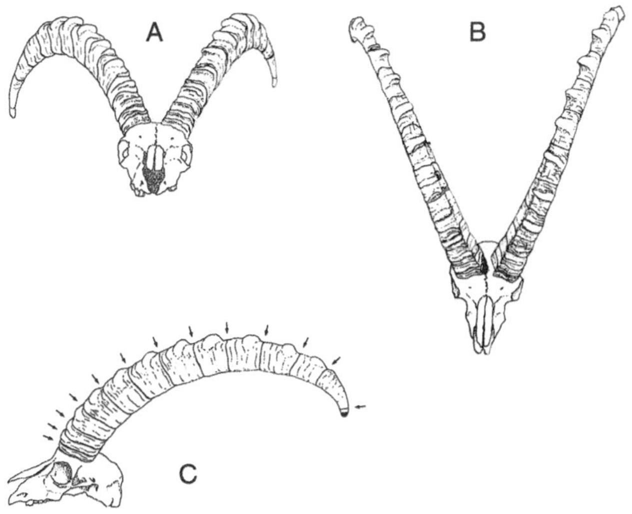

Fig. 4 Skull with horns of a *Capra ibex* male of 13.5 years. View from frontal (**A**), dorsal (**B**), and lateral (**C**). (From Nievergelt and Zingg (1986))

Table 1 Hematologic values obtained from Alpine ibex ($N = 138$) captured in Vanoise National Park (Gauthier et al. 1994)

Parameter		Mean ± sd	Min – Max
Erythrocytes	$10^6/mm^3$	13.6 ± 3.0	6.5–25.2
Leukocytes	$10^3/mm^3$	9.5 ± 5.7	3.2–29.0
Neutrophiles	%	35.9 ± 13.4	6–87
Eosinophiles	%	8.7 ± 7.7	0–40
Basophiles	%	1.5 ± 1.7	0–8
Lymphocyte	%	50.4 ± 13.9	7–82
Monocyte	%	3.5 ± 3.0	0–18

locomotory activity is also strongly reduced during winter (see "Habitat and Diet"). Finally, during winter, Alpine ibex bask in the sun in the morning and actively use exogenous heat to increase the body temperature (Signer et al. 2011).

Thanks to these adaptations, Alpine ibex spend less than half the energy in winter than in summer (Signer et al. 2011). This allows them to better survive the winter by relying almost exclusively on the fat accumulated during spring/summer months when nutritious vegetation is abundant. During winter, food intake is extremely reduced because the vegetation is dry and fibrous, low in protein, and often covered by snow. Moreover, males voluntarily reduce forage intake during the rut (December–January) because of time budget constraints generated by mating-related activities (Brivio et al. 2010). At the beginning of spring, the metabolic rate and body temperature of ibex increase in concert with the start of the vegetation period and the growth of the gut and visceral organs in preparation for processing the large amount of food available during summer (Signer et al. 2011). The high availability of resources in spring allows male ibex to invest energy in secondary sexual characters such as horns which growth pauses during winter. However, most of the studies on Alpine ibex physiology have been conducted on males, and only sparse information is available about ecophysiological adpatations of females to the environment.

Genetics

Chromosomes

Alpine ibex have $2n = 60$ chromosomes.

Reintroduction History and Its Consequence for Genetics

The near extinction of Alpine ibex in the nineteenth century created a severe bottleneck with less than 100 individuals remaining in the Gran Paradiso population (Stüwe and Nievergelt 1991). When individuals were taken from Gran Paradiso to found new populations either in the wild or in wildlife parks in Switzerland, a second bottleneck occurred. Individuals from these new founded populations in the wild or from the breeding program in Swiss wildlife parks were then reintroduced to several locations in the wild, which led to a third bottleneck. Those wild populations, in turn, provided founding animals for multiple other populations colonizing the Alps, which created a fourth bottleneck for these populations. Hence, all extant Alpine ibex populations have experienced one to four population bottlenecks (Biebach and Keller 2009).

Each bottleneck led to a loss of genetic diversity because the founders carried only a subset of the genetic diversity of the source populations. Small population sizes and isolation – the lack of genetic exchange with other populations – resulted in an additional decrease of genetic diversity for at least the first few generations after founding. The reintroduction history of Alpine ibex with the serial bottlenecks and low population sizes resulted not only in low genetic diversity but also in high inbreeding levels.

Genetic Diversity

Genetic diversity of Alpine ibex populations is very low (Biebach and Keller 2009) and comparable to other populations that experienced severe reductions in size or descend from a small number of founders, such as Scandinavian wolves (Kardos et al. 2018). Genetic diversity is also lower than that of related ibex species: while the closely related Iberian wild goat exhibits only slightly higher genetic diversity, the Siberian ibex and the bezoar goat, the ancestor of the domestic goat, are genetically multiple times more diverse than the Alpine ibex (Grossen et al. 2018 and Christine Grossen, pers. comm.).

As expected from the demographic history, genetic diversity also differs among Alpine ibex populations (Biebach and Keller 2012). The ancestral Gran Paradiso population retained the highest genetic diversity, whereas reintroduced populations have 15–25% less genetic diversity (measured as expected heterozygosity from 13,341 genetic markers (SNPs), Grossen et al. 2018). Among reintroduced populations, the amount of genetic diversity depends on the reintroduction history of each population. Genetic diversity is lower and, the more bottlenecks a population experienced, the fewer individuals and source populations were represented in the founder stock (Biebach and Keller 2012).

The genes of the major histocompatibility complex (MHC) play a central role in the immune response of vertebrates (Klein and Figueroa 1986), and many studies have shown the benefits for disease resistance of possessing numerous genetic variants at the MHC (Grossen et al. 2018; Westerdahl et al. 2011). Accordingly, MHC genes in vertebrates are among the genes with the highest diversity. Alpine ibex, however, have very low genetic diversity at this gene complex (Grossen et al. 2014). Exon 2 of the DRB gene in the MHC has only 2 variants (i.e., alleles), in the Alpine ibex, while the domestic goat and its wild ancestor, the bezoar goat, have 44 different alleles at this exon (Grossen et al. 2014). The low genetic diversity at the MHC is known to be disadvantageous for Alpine ibex. Individuals with lower heterozygosity at the MHC have a lower body mass and a higher risk of showing symptoms of keratoconjunctivitis, an ocular infection that can lead to blindness, during an outbreak of the disease (Brambilla et al. 2018).

Genetic Structure

The reintroduction history is the main determinant of the current genetic structure among Alpine ibex populations, even though more than one hundred years have passed since its beginning (Biebach and Keller 2009). Small populations are subject to strong genetic drift, random changes of frequencies of gene variants over generations, which – over time – leads to genetic differentiation among populations. Genetic structure analysis, as described in Biebach and Keller (2009), revealed that genetic drift has produced five separate genetic

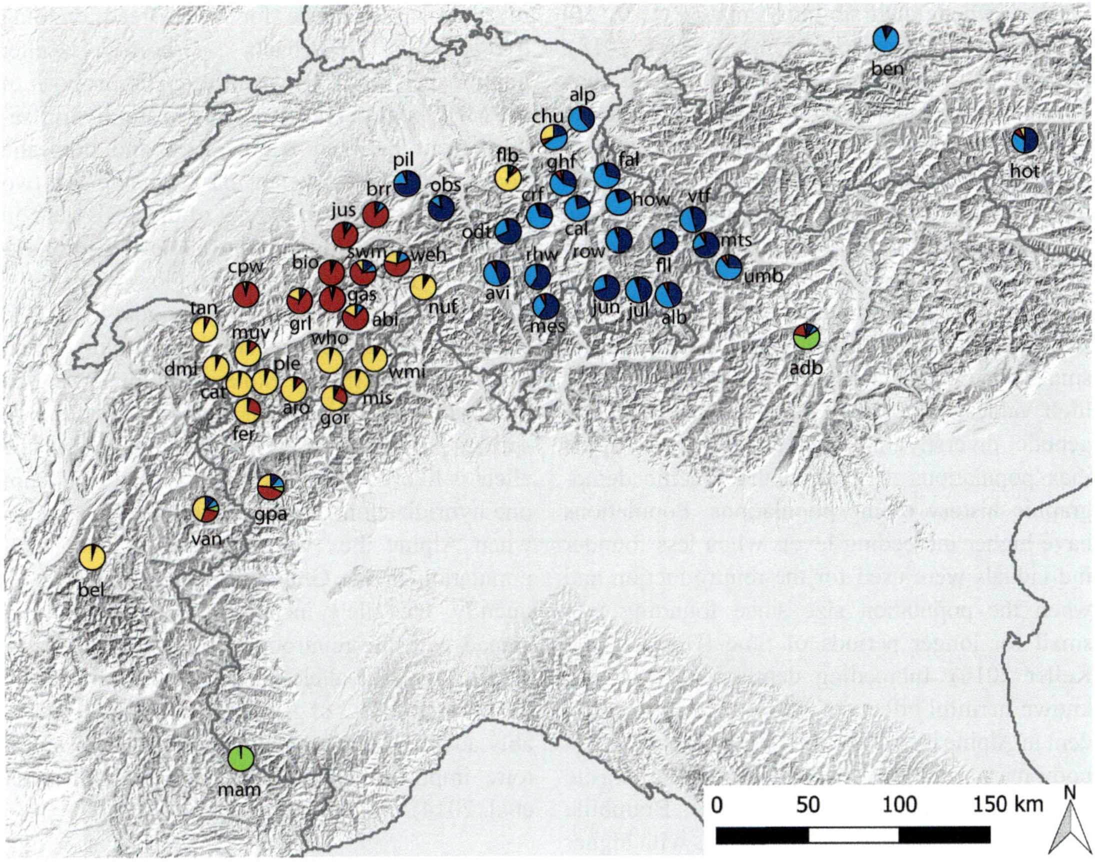

Fig. 5 Genetic structure of 50 Alpine ibex populations across the Alps. The segments show the proportion of each population belonging to one of five genetic clusters. Two genetic clusters belong to the Piz Albris population and its descendants (dark blue and light blue). The remaining three clusters belong to the following populations together with their descendants: Brienzer Rothorn (red), Mont Pleureur (yellow), and Mercantour/Alpe Marittime (green). Population abbreviations: *abi* Aletsch-Bietschhorn, *adb* Adamello-Brenta, *alb* Albris, *alp* Alpstein, *aro* Arolla, *avi* Adula-Vial, *bel* Belledonne, *ben* Benediktenwand, *bio* Bire-Öschinen, *brr* Brienzer Rothorn, *cal* Calanda, *cat* Catogne, *chu* Churfirsten, *cpw* Cape au Moine-Pierreuse-Wittenberg, *crf* Crap da Flem, *dmi* Dents du Midi, *fal* Falknis, *fer* Ferret, *flb* Fluebrig, *fll* Flüela, *gas* Gastern, *ghf* Graue Hörner-Foostock, *gor* Gornergrat, *gpa* Gran Paradiso, *grl* Gross Lohner, *hot* Hohe Tauern, *how* Hochwang, *jul* Julier Süd-Val Bever, *jun* Julier Nord, *jus* Justistal, *mam* Mercantour-Alpe Marittime, *mes* Mesocco, *mis* Mischabel, *mts* Macun-Terza-Sesvenna, *muv* Muveran, *nuf* Nufenen, *obs* Oberbauenstock, *odt* Oberalp-Düssi-Tödi, *pil* Pilatus, *ple* Pleureur, *rhw* Rheinwald, *row* Rothorn-Weissfluh, *swm* Schwarzmönch, *tan* Tanay, *umb* Umbrail, *van* Vanoise, *vtf* Vereina-Tasna-Fergen Seetal, *weh* Wetterhorn, *who* Weisshorn, *wmi* Weissmies

clusters of Alpine ibex populations, consisting of reintroduced populations founded between 1920 and 1930 and their descendant populations (Fig. 5). Geographically nearby populations tend to belong to the same genetic cluster not because of migration among populations but because founder individuals for a new population were usually taken from nearby populations. As expected from the fact that all populations ultimately derive from Gran Paradiso, the Gran Paradiso population belongs to all five genetic clusters (Fig. 5). Populations with mixed ancestry belong to genetic clusters in proportion to the origin of the founder individuals (e.g., Schwarzmönch in Fig. 5).

Inbreeding

Inbreeding levels of Alpine ibex populations are on average above the level of inbreeding that results when half-siblings mate (f>0.125) (Biebach and Keller 2010; Grossen et al. 2018). Levels of inbreeding vary among Alpine ibex populations, ranging from values that are lower than those found in offspring from first cousins (f<0.0625, e.g., Gran Paradiso f = 0.05) to values that are higher than those found in offspring from full-siblings (f>0.25, e.g., Mercantour/Alpi Marittime f=0.36). Note that this does not necessarily imply that full-sib matings are taking place. Instead, inbreeding accumulates over time in small populations and can thus result in such high values (e.g., Ewing et al. 2008). Just like genetic diversity, inbreeding levels of Alpine ibex populations depend on the specific demographic history of the populations. Populations have higher inbreeding levels when less founder individuals were used for the reintroduction and when the population size since founding was small for longer periods of time (Biebach and Keller 2010). Inbreeding depression, the well-known harmful effects of inbreeding, is also evident in Alpine ibex. Inbred individuals have lower body mass, smaller horns, and increased parasite burden (von Hardenberg et al. 2007; Brambilla et al. 2015). Additionally, populations with higher average inbreeding levels have lower population growth rates (Bozzuto et al. 2019).

Hybridization

The genus *Capra* is a relatively young genus and thus all *Capra* species are capable to hybridize and produce fertile offspring in captivity (Couturier 1962). In nature, hybridization among *Capra* species still occurs when *Capra* species inhabit the same range (for a review, see Iacolina et al. 2019). Additionally, genetic analysis suggests ancestral hybridization events among *Capra* species (Pidancier et al. 2006; Kazanskaya et al. 2007). The domestic goat is the only *Capra* species occurring in the same region as free-ranging Alpine ibex. Ongoing hybridization has been reported (Giacometti et al. 2004). Where it has been detected, hybrids were culled to prevent genetic introgression into Alpine ibex. A genetic survey of many present-day ibex populations revealed no evidence for widespread ongoing hybridization (Biebach, Grossen, Keller unpublished data). However, genetic analysis of the MHC region revealed a successful hybridization event between Alpine ibex and domestic goats in the more distant past. One of the two MHC alleles found at exon 2 of the DRB gene in Alpine ibex is identical to one allele of the domestic goat. The chromosomal region around the MHC of Alpine ibex carrying the goat-like allele shows the signature of recent introgression and no signs of the alternative explanation for sharing an allele at the MHC between species, ancient trans-species polymorphism (Grossen et al. 2014). The allele is likely introgressed into Alpine ibex from one hybridization event a few hundred years ago when Alpine ibex were reduced to one small population in the Gran Paradiso region. Subsequently, the allele increased in frequency and spread with the reintroductions across the Alps. The introgressed allele critically increases genetic diversity at MHC of Alpine ibex and is presumably adaptive as a broader MHC sequence repertoire improves the immune response (Grossen et al. 2014).

Conservation Genetics

Alpine ibex have the IUCN Red List status of least concern. However, they carry the burden from their population history of repeated bottlenecks that might pose a threat in the future. The low genetic diversity of Alpine ibex indicates low adaptive potential for forthcoming environmental changes, and the harmful effects of inbreeding will increase over time unless inbreeding is reduced.

Inbreeding can be decreased and genetic diversity enhanced by restocking populations with individuals from distantly related Alpine ibex populations with relatively high genetic diversity. Some translocations to reduce inbreeding and increase genetic diversity have already been carried out (e.g., Mercantour/Alpi Marittime, Graue Hörner, Pilatus). For those populations, the assignment to genetic clusters (Fig. 5) will likely change as the newly introduced genes spread in the populations.

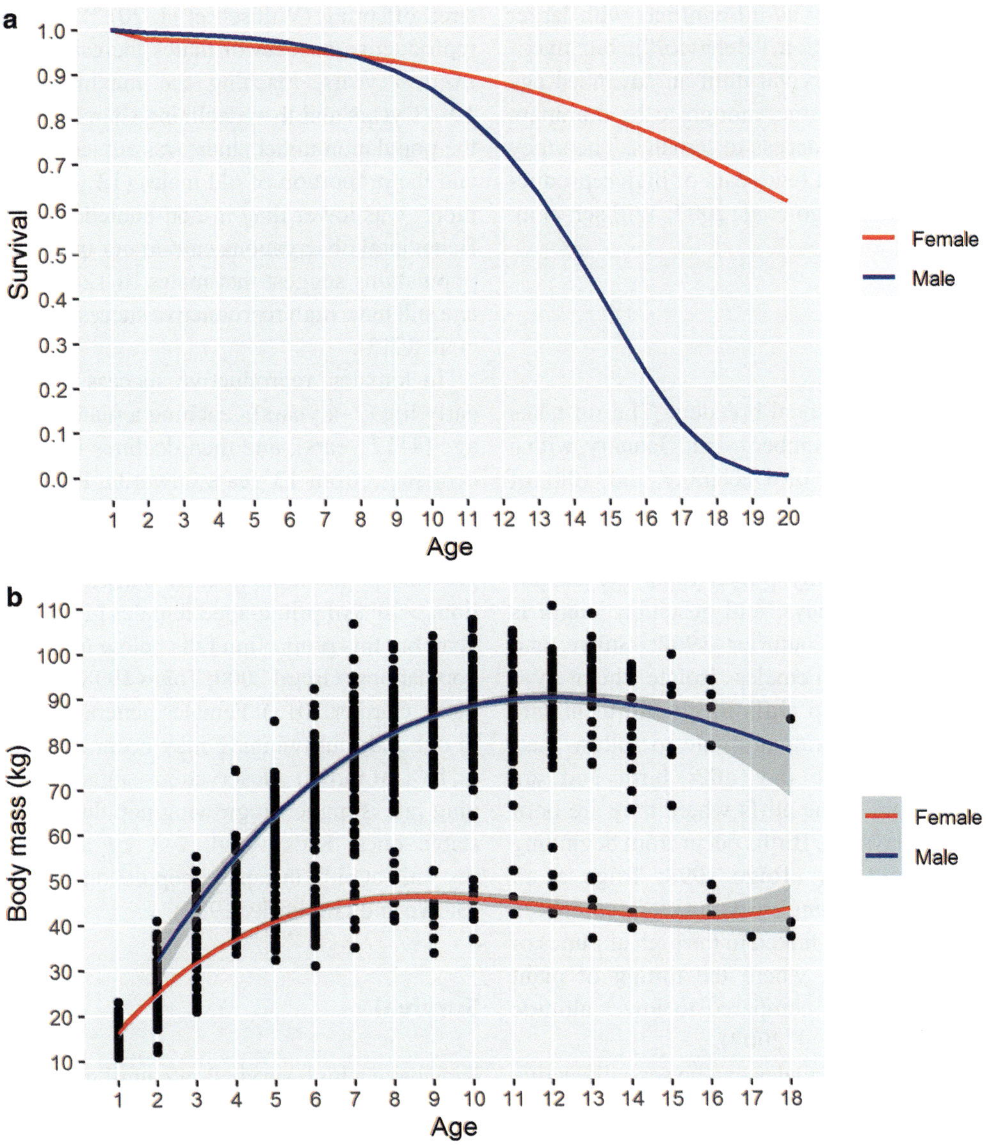

Fig. 6 (**a**) Age-related survival rates of male and female Alpine ibex (modified from Toïgo et al. 2007); (**b**) Growth curves of body mass in male and female Alpine ibex. Data for males ($N = 577$) were collected in Gran Paradiso National Park (Italy) from 2000 to 2018 via a platform scale baited with salt and standardized at the 1st of August. Data for females ($N = 310$) were collected in the Belledonne population (France) during captures that occurred in May

Life History

Growth

Growth curves of male and female ibex are peculiar of the species and mirror the sex-specific survival pattern (see section "Survival"): while maximum body size in females is reached at the age of around 5 years (Toïgo et al. 2002), male Alpine ibex keep growing until the age of 10–12 years when they reach asymptotic mass (Fig. 6b). Such unusual patterns highlight the fact that male ibex have developed a very conservative life-history strategy: young males invest little in the rut. They avoid conflicts with larger individuals (see section "Behavior"), but maximize growth and survival until an advanced age when they will be large enough to be dominant and have privileged access to females. Then they wear out following a few years of high reproductive investment (Toïgo et al. 2007; Willisch et al. 2012).

Reproduction

Alpine ibex are seasonal breeders. The rut takes place from early December to mid-January, with a peak between 15th of December and 10th of January (Couturier 1962; Willisch and Neuhaus 2009, 2010; Apollonio et al. 2013). Alpine ibex females are seasonally polyestrous, the estrous cycle averages 20 days, and gestation length is about 170 days (Couturier 1962; Stüwe and Grodinsky 1987). Females isolate themselves and select very steep cliffs to give birth in safe areas. Alpine ibex kids are able to follow their mother already one day after birth and are extremely agile in the cliffs where they are born as early as 3 days old. Births occur from beginning of June to mid-July (Toïgo 1998; Toïgo et al. 2002). This late timing of parturition, compared to other ungulates, is linked to the high altitudes of Alpine ibex ranges, where the timing of plant green-up is delayed compared to lower altitude regions (Pettorelli et al. 2007).

Both males and females can be sexually active as young as at 1.5 years of age but age at first reproduction is highly variable. In females, it strongly depends on population density: they breed for the first time at 2 years of age in growing populations, while age at primiparity is delayed to 3 or even 4 years old in stable populations (Toïgo et al. 2002). Age at first reproduction is variable also in males. Despite being fertile from the second year of life, less than 5% of males reproduce at 2 years of age because older and bigger males which are dominant prevent the younger males to have access to oestrus females (Willisch et al. 2012).

Reproductive success depends on age in both sexes. The only study reporting genetic data on sired offspring (Willisch et al. 2012) showed that reproductive success of males increases with age from 6 years, reaching the maximum around 10–11 years and then declining sharply. However, the population under study was subject to hunting and the proportion of old males (12 years old and more) was lower than in non-hunted populations. Behavioral observations carried out in non-hunted populations suggest that males of 12–13 years of age still have high reproductive success (Apollonio et al. 2013).

In females, reproductive success increases in early life (2–4 years), reaching a plateau in prime age (4–12 years), and then declines in the senescent stage from 13 years onwards. Females may produce a kid yearly until the age of 16 years, but very rarely later on (e.g., Toïgo et al. 2002). In growing populations in early phases of colonization, >90% of prime-aged females give birth each year, but this proportion falls below 60% in stable populations (Girard 2000; Toïgo 1998; Largo et al. 2008; Garnier 2013). Females generally give birth to one kid, but twinning may occur at a low rate (<10% of births). Observations suggest that twinning rate is higher in growing populations than in stable ones. Kids weigh 2–3 kg at birth, and sex-ratio at birth in captive populations is unbiased (Stüwe and Grodinsky 1987).

Survival

Kids have a high survival rate until the beginning of winter. Juvenile survival during the first winter is usually <0.50 but highly variable (from 0.20 to 0.80) according to weather conditions (Jacobson et al. 2004; Pettorelli et al. 2007). However, little detailed information on juvenile survival is available because of the difficulty of capturing and marking kids in this species.

Like reproductive performance, survival rates are also strongly affected by age in both sexes (Fig. 6a). Female survival is very high from 1 to 8 years of age (>0.95) in all populations where this parameter has been estimated, regardless of the environmental conditions, and mirrors the classical pattern reported for other ungulate females, with a slow and regular decline afterwards (Girard et al. 1999; Toïgo et al. 2007; Largo et al. 2008; Garnier 2013). Males instead show a completely atypical pattern compared to other large herbivore species. Having survived the first winter, they survive at a very high rate until 10–12 years of age, and then express a very sharp decline in survival rate (Toïgo et al. 2007). The unequal survival rates of males and females lead to a sex ratio biased towards females in the oldest age classes (i.e., >11 years). Adult survival of both sexes is buffered against variation in population density (Toïgo et al. 2007) but may be strongly affected when winter snow cover is particularly deep (Jacobson et al. 2004; Lima and Berryman 2006). Longevity is around 13–14 up to 15–17 years for males and 16–17 years for females, but it is not rare to observe males aged 17 years old and females up to 22 years old.

Habitat and Diet

Habitat Selection and Movement

With two exceptions (Jura Mountains in Switzerland and Bulgaria), Alpine ibex habitat is restricted to the biogeographic delineation of the Alps. The species mostly lives in open habitats above the upper treeline ecotone, in the alpine and subnivale zone between 1600 and 3200 m above sea level (Nievergelt and Zingg 1986). Alpine meadows and pastures characterize this craggy terrain, where south-exposed, steep and rocky slopes are relatively well represented. Alpine ibex can be found on both, siliceous and carbonaceous substrates. In total, 44 vegetation types were described across the species' distribution range (Wiersema 1989), half of them are characteristic of thermophile and xerophile conditions, which translate mainly to xeric alpine grasslands and rupicolous and scree communities. Although the current distribution of Alpine ibex is strongly influenced by its reintroduction history (Biebach and Keller 2009), the majority of the large Alpine ibex populations can be found in regions with an inner-alpine continental, dry climate (Wiersema 1989). This climate seems to positively influence the performance of Alpine ibex populations as suggested by Couturier (1962) and Nievergelt (1966) who showed that the amount of rainfall during the gestation period is inversely related to the species' reproductive success and to population growth rate (Grøtan et al. 2008; Bozzuto et al. 2019). Depending on the season and the region, Alpine ibex occasionally inhabit forests as long as they are gappy, sunny, and interspersed with rocks (Hofmann and Nievergelt 1972).

Compared to other ungulates, Alpine ibex show high site fidelity and a conservative use of space (Parrini et al. 2003). They tend to be philopatric from year to year, prefer slopes of 30–45° inclination, and use small caves and overhangs for shelter (Nievergelt 1966). Predator avoidance, thermoregulation, and access to food play a central role in habitat selection and space use. Hence, Alpine ibex show seasonal migration between summer and winter home ranges. Males cover a larger altitude range than females and can be found at lower altitudes in spring (Herfindal et al. 2019) and higher altitudes in summer (Abderhalden 2005), although this pattern largely depends on site topography. Average winter-summer dislocation distances are larger for males than for females (Abderhalden 2005), strongly varying among individuals and among populations (from tens of meters to several kilometers). Home range sizes are inversely related to snow cover in both sexes (Parrini et al. 2003; Grignolio et al. 2004) and tend to be smallest in winter, largest during summer and autumn, and intermediate in spring. Grignolio et al. (2004) showed that in Gran Paradiso, the mean overlap between female annual home ranges of consecutive years was 82% and the activity centers differed less than 200 m. In males, home range overlap ranged between 79% and 100%, and activity centers were at most 900 m apart from one year to the next (Parrini et al. 2003).

The marked age- and sex-specific social and spatial segregation of this species can be explained by the differences in body size among males and between males and females and by behavioral differences (Villaret et al. 1997; Ruckstuhl and Neuhaus 2001; Neuhaus and Ruckstuhl 2002; Brivio et al. 2010, 2014). Adult males tend to feed on relatively low inclination pasturelands where they find large amounts of biomass (Villaret et al. 1997). Due to their large size, the rumen of males is apt to intake and process larger quantity of fiber-rich biomass than that of females. Similar differences are also found among males of different age classes that differ in body size and energy metabolism. Younger males behave and use areas more similar to females than to older males (Grignolio et al. 2007a). As females are smaller and more subject to predation risk, they occupy different habitat than males for most of the year, generally staying close to rock cliffs and precipitous rocky terrain (Grignolio et al. 2019). Preference for safe and easy-to-escape areas is further increased during summer for breeding females.

Both sexes prefer south to south-west exposed habitat in winter, which is first free of snow due to solar radiation and avalanches (Nievergelt 1966), whereas summer preferences vary with sex and location (Girard 2000; Abderhalden 2005). In spring, Alpine ibex migrate to lower altitudes with snow melt and the onset of plant growth that provides access to high-quality vegetation after the rigors of the winter with restricted access to food (Wiersema 1983). Spring is a critical period for Alpine ibex to secure enough energy for growth and reproduction (Giacometti et al. 2002; Büntgen et al. 2014). As ambient daily temperatures increase in spring and summer and snow melts at higher altitudes, Alpine ibex follow the green wave upwards and move to the alpine pastures (Parrini et al. 2003; Grignolio et al. 2004). In autumn, they use the alpine pastures in relation to environmental conditions, in particular weather and the first onset of snow (Nievergelt 1966; Büntgen et al. 2017).

Feeding

Ibex are herbivores and intermediate feeders (sensu Hofmann 1989). Due to their large body mass, they have large rumens and are able to slowly process large quantities of food. The slow digestion increases the digestive efficiency, which enables Alpine ibex to live on large amounts of food even if of low quality (Illius and Gordon 1992). Due to their pronounced sexual dimorphism, males and females also differ in their digestive efficiency and thus in their dietary needs: females have a lower digestive efficiency than males and thus need forage of higher quality with a lower C/N (Carbon to Nitrogen) ratio (Geist 1974).

Females and males also show different foraging patterns during summer. Male Alpine ibex have foraging peaks in the early morning and late afternoon (Nievergelt 1966) while females, both lactating and non-lactating, can have more than six foraging peaks per day during daylight hours (Neuhaus and Ruckstuhl 2002). Some studies also report occasional nocturnal feeding in summer (Rauch 1937; Schnitter 1962) and autumn (Ten Houte de Lange 1978).

As inhabitants of open landscapes, Alpine ibex rely mainly on grass and herbs in all seasons, but woody plants and cryptogams in low quantities are also eaten (Couturier 1962). The ratio of grass to dicotyledonous herbs is 2.4, and as such considerably higher than in Alpine chamois (1.1); red deer, *Cervus elaphus* (1.2), and roe deer, *Capreolus capreolus* (0.9) (Frei 1972). In an alpine, silicaceous study area between 1900 and 3000 m above sea level, the diet of various ibex groups mainly composed of females, kids, and young males comprised 60% grass species, 38% herbs, and 2% woody plants (Ten Houte de Lange 1978). Out of 200 available plant species, 99 species were eaten. However, almost 80% of the total food intake consisted of 25 plant species only. Strongly preferred were the following species: *Anthoxanthum odoratum*, *Bupleurum stellatum*, *Carex curvula*, *Carex sempervirens*, *Festuca rubra*, *Festuca violacea*, *Poa alpina*, and *Senecio incanus*. In a nearby colony inhabiting alpine grasslands on calcerous soils, the preferred species were *Carex sempervirens*, *Festuca rubra*, *Helianthemum nummularium*, *Poa alpina*, *Sesleria caerulea*, *Valeriana montana*, and *Valeriana supina*. During winter and in areas where the extent of Alpine grasslands between the timberline and the mountain ridges is small, the ratio of woody plants in the nutrition is slightly higher (Couturier 1962; Nievergelt and Zingg 1986). However, male and female diet differs due to different energetic needs related to body size and lactation. Herfindal et al. (2019) have shown that males have a higher use of grasslands than females throughout the year.

Similar to other ruminants, Alpine ibex show a strong attraction for salt and minerals of natural or artificial origin. They have been documented standing on the sheer face of dams in several areas where they lick the stonework to obtain mineral salts (Biancardi and Minetti 2017).

Behavior

Social Behavior

Alpine ibex are gregarious and live in groups for most of the year. Typically, male and female Alpine ibex live in separate groups outside the rut (Bon et al. 2001). Females tend to have more stable social bonds than males, have smaller home ranges, and are more philopatric (Villaret and Bon 1995; Marchand et al. 2017). Female groups are typically composed of 5–10 females with or without kids (Villaret and Bon 1998). Two- to 3-year-old males move often between male and female groups. Typical group sizes of males range between 2 and 16 individuals, although it is possible to observe groups of more than 50 males. Typical group size also depends on population density and is generally larger in spring and early summer than during winter (Toïgo et al. 1995, 1996).

During spring and summer, males form highly connected and cohesive social networks with a fission fusion dynamic: group composition changes frequently but all individuals often come into contact (Marchand et al. 2017). Young males often group with older males. However, despite the high rate of fission-fusion events, subgrouping patterns are evident, and age seems to be a key factor influencing group composition with individuals of similar age more often grouping together. This may be partly related to differences in time budget and synchrony as it is the case for sexual segregation, but it is probably also due to social reasons. Young males need to develop their fighting skills by interactions with individuals of the same age class and need to avoid risky interactions with older males.

Mating Behavior

Alpine ibex have a serially polygynous mating system: males can mate with several females during a reproductive season but they only can court one female at a time. During the rutting season, males join groups of females on slopes relatively free from snow (Villaret et al. 1997) and court, follow, and monopolize access to them until they are sexually receptive. Courting of females is mainly done by large, dominant males that attempt to prevent other males from mating. Along with this "tending" tactic achieved by dominant, i.e., old males, younger subordinate males developed an alternative "coursing" tactic. Coursing young males try to achieve temporary access to a female when they run away from the tending, dominant male (Willisch and Neuhaus 2009; Apollonio et al. 2013). Although reproductive success is biased toward old dominant males that use the tending tactic, reproduction by young or very young individuals has also been confirmed with genetic analyses (Willisch et al. 2012). The coursing tactic appears to be low-cost compared to tending and can also be more suitable in snowy winters (Apollonio et al. 2013)

Dominance in males is established via agonistic interactions that occur throughout the year except in late winter immediately after the end of the rutting season (Toïgo, pers. obs.). Age and body mass are the most important factors shaping the dominance hierarchy (Bergeron et al. 2010). Most of the interactions among males occur during spring when males start to group in meadows with fresh, green vegetation. Interaction rates decrease in summer as males spend more time foraging and increase again before the rutting season. At the time of the rut, however, most of the hierarchy is already established and contests are most frequent among males of similar age-body size (Willisch and Neuhaus 2009; Bergeron et al. 2010; Willisch and Neuhaus 2010). As the rutting season occurs in winter with high snow cover and reduced forage, the early establishment of hierarchies possibly allows energy savings during the rut.

Communication

Male and female ibex vocalizations consist of a single short and sharp whistle. Generally, they are produced singularly but may also be repeated at short intervals. Whistles are used as alarm-stress calls but may also have other social functions as they are sometimes produced in absence of danger during group interactions. Bleating is used as a call between females and kids (Couturier 1962).

Activity

The high degree of dimorphism in body size between males and females and among males of different ages is also associated with physiological differences that lead to different activity budgets all year round (Tettamanti and Viblanc 2014). Adult males, young males, and females alternate feeding and resting periods at different time scale. Males feed and rest for longer than females that alternate more frequently between feeding and ruminating. Young males show an intermediate cycle between fully grown males and females (Ruckstuhl and Neuhaus 2001; Neuhaus and Ruckstuhl 2002). Such different feeding cycles lead to an asynchrony in the activity rhythms of males and females providing a mechanistic explanation for the sexual segregation that arises even when both sexes occupy the same feeding sites (Ruckstuhl and Neuhaus 2002).

Antipredatory Strategy

Flight initiation distances from approaching potential predators are shorter in Alpine ibex than in other sympatric ungulates. This is due to the species' antipredator strategy to live close to escape terrain such as cliffs that can be quickly reached in case of perceived danger (Grignolio et al. 2007a, b). Flight initiation distances are shorter in adult males than in females as the latter are more susceptible to predation risk. Currently, however, predation risk is generally low for the species due to the low densities of large carnivores in most parts of the distribution (Grignolio et al. 2019). Once on a cliff, ibex are out of reach for terrestrial predators such as wolf *Canis lupus* and females can better defend themselves and their kids from raptors such as golden eagle *Aquila chrysaetos* that occasionally depredate kids.

Parental Care

Prior to parturition, ibex females isolate themselves to give birth in safe steep and rugged terrain (Grignolio et al. 2007b). Within just a few days, the neonates are able to follow their mothers who join other females with kids, forming nurseries. Parental care is provided exclusively by females, which lactate from 3 to 5 months although kids stay with their mothers at least until the following spring.

Parasites and Diseases

Endoparasites

Common endoparasites found at subclinical levels in Alpine ibex include gastrointestinal and bronchopulmonary nematodes. The two dominant gastrointestinal nematode species are *Teladorsagia circumcincta* and *Marshallagia marshalli*, accounting together for 69% of the total parasite burden (Zaffaroni et al. 2000). Other species frequently observed are: *Nematodirus filicollis*, *Nematodirus oiratianus*, *Ostertagia lyrata*, and *Ostertagia ostertagi*. Symptoms of these infections mainly consist of reduced appetite, reduced food intake, and reduced digestive efficiency (Gunn and Irvine 2003) thus leading to physical deterioration. Bronchopulmonary nematodes are mostly of the *Protostrongylus* and *Muellerius* group and may result in parenchymal lesions and act as predisposing factors for multifactorial pneumonia (Cassini et al. 2015).

Infectious Diseases

Alpine ibex can be affected by several other infectious diseases. A list of some of the pathogens often detected by serological analysis (Tolari et al. 1987; Gennero et al. 1993; Hars and Gauthier 1994; Marreros et al. 2011) is presented in Table 2.

As often in wildlife populations, it is difficult to relate the influence of these infections to effects on population dynamics, as the pathogen-host dynamic is the result of the interaction between complex behavioral, physiological, and genetic mechanisms. Nonetheless, some diseases have been associated with occasional population crashes, such as: sarcoptic mange (caused by *Sarcoptes scabiei*), pneumonia (different pathogens: *Mycoplasma agalactiae*, *Mycoplasma* spp. *Pasteurella/Mannheimia haemolytica*, Herpes virus), infectious keratoconjunctivitis (*Mycoplasma conjunctivae*), and contagious ecthyma (*Parapoxvirus*) (De Danieli and Sarasa 2015). The diseases that affect survival or, as zoonoses, can be transmitted to humans are briefly described hereafter.

Sarcoptic Mange

The consequences of sarcoptic mange outbreaks in the Alpine ibex populations of the Italian Eastern Alpine arc were in many cases dramatic, similar to those recorded in Iberian wild goat (León-Vizcaíno et al. 1999): the mortality reached up to 80–90% and in some cases lead to local extinctions (León-Vizcaíno et al. 1999; Carmignola et al. 2006; Rossi et al. 2007). The etiological agent, *Sarcoptes scabiei*, is the same that causes mange in Alpine chamois (Lavin et al. 2000; Rossi et al. 2007) and has relevant affinity with the *Sarcoptes* species affecting domestic ruminants, domestic goat in particular. Several authors provided evidence for domestic-wild transmission (Gortázar et al. 2007). The mortality is often due to the severe self-intoxication caused by tissue destruction resulting from the proliferation of the parasite (Rossi et al. 2007).

Table 2 Serological prevalence of some typical wildlife diseases recorded in Alpine ibex. Prevalence values: N = absence 0%; L = low prevalence <5%; M = medium prevalence 5–25%; H = high prevalence >25%. Values obtained from: Gennero et al. 1993; Hars and Gauthier 1994; Tolari et al. 1987; Marreros et al. 2011; Bassano unpublished

Pathogen agent	Prevalence
Mycobacterium avium subsp. *paratuberculosis*	L
Brucella abortus/melitensis	L/H
Brucella ovis	L
Neospora caninum	N/L
Toxoplasma gondii	L
Mycoplasma agalactiae	L
Salmonella abortusovis / abortusequi	H/L
Leptospira spp.	L
Besnoitia spp.	L
Pestivirus	L
Infectious Bovine Rhinotracheitis virus	L
Bluetongue virus	N
Maedi Visna/CAEV	N

Respiratory Diseases

In some outbreaks of respiratory diseases, mortality can be high, especially in high density populations: population crashes were described both in France (Demeautis 1982; Delorme 2008) and in the Italian Alps (Dotta 2009; Bassano unpublished). The origin of the respiratory diseases in the Alpine ibex is still not clear. In other mountain ungulates, the infection is frequently caused by pathogens derived directly from domestic ruminants (e.g., in bighorn sheep, *Ovis canadensis*: *Pasteurella haemolytica*, Foreyt et al. 1994; *Mannheimia haemolytica*, Lawrence et al. 2010; *Mycoplasma ovipneumoniae*, Cassirer et al. 2018). In these cases, livestock herd management is very important for reducing the risk of contamination of wild populations (Richomme et al. 2006).

Infectious Keratoconjunctivitis

Although Alpine ibex are not very sensitive to infectious keratoconjunctivitis (caused by *Mycoplasma conjunctivae*), in some situations high mortality can result, often caused by falls linked to bilateral loss of vision, mainly during the winter season, and by starvation of blind animals (Belloy et al. 2003). Mortality can also be a consequence of brain lesions associated with ocular infection (Bassano et al. 1994). Ocular infection comes from direct contact with infected and often asymptomatic sheep *Ovis aries* or goats (Giacometti et al. 1998; Belloy et al. 2003) or by flies that act as vectors between the species (Degiorgis et al. 1999). Recently, it has been demonstrated that the pathogen may also be present in healthy Alpine ibex, which, in this case, would serve as reservoirs (Ryser-Degiorgis et al. 2008).

Brucellosis

Zoonoses represent an important management issue for wild ungulates, as they often result in drastic interventions, as in the case of the *Brucella melitensis* infection in Alpine ibex (see below). Brucellosis is a major zoonosis of worldwide public health and economic importance, and the disease is mainly caused by *Brucella abortus* and, more rarely, by *Brucella melitensis* where cattle are sympatric with infected sheep or goats (Mick et al. 2014).

The presence of *Brucella melitensis* in mountain ungulates, until a few years ago, was absolutely rare and characterized by very low prevalence, <2% (in Alpine chamois: Garin-Bastuji et al. 1990; Richomme et al. 2006; in Alpine ibex: Ferroglio et al. 1998). The recent outbreak in the Bargy massif (France), discovered in 2012, changed this perception. Brucellosis was found in the population of Alpine ibex in Bargy with a prevalence never described before (seroprevalence nearly reached 45%, Mick et al. 2014; Marchand et al. 2017). This serious zoonosis outbreak, originating from infected cattle, led to drastic measures of intervention (culling of half of the population). The unusual sensitivity of Alpine ibex in Bargy to Brucellosis is not understood yet, and several hyptoheses are being investigated. One of them may be related to the reintroduction history of this population, with few individuals reintroduced in the 1970s in an almost closed and isolated massif, leading therefore to the potential loss of genetic variation conferring resistance (Anses 2016; Quéméré et al. 2020). Although this link has not been fully demonstrated, it highlights the importance of conservation genetic studies in support of management decisions, to prevent economic interests to dominate over conservation interests.

Contagious Ecthyma

Contagious ecthyma is a widespread and common disease in many wild ungulates (Samuel et al. 1975; Nandi et al. 2011). Although this disease is localized to single parts of the body (predominantly to the lips, tongue, foot, and breast), it can affect kid survival, reducing daily milk intake, and exposing them to malnutrition and to diseases caused by ubiquitous pathogens that become fatal (especially acute pneumonia). In Rocky Mountain bighorn sheep, contagious ecthyma was described as an important cause of lamb mortality (Goldstein et al. 2005). This disease, caused by a *Parapox-Orf virus* (Scagliarini et al. 2011), was described in Alpine ibex a long time ago (Couturier 1962). Nevertheless, to date, no effect on population dynamics has been described. This disease has a chronic course, remains for a long time inside the population, and can be transmitted by domestic goats and sheep. Some progressive declines of Alpine ibex populations, which are related to the decrease of kid survival, may be due to this disease, often difficult to diagnose, although clinically evident. However, the potential effects of this and other diseases that affect kid survival are difficult to quantify because the small carcasses of kids are quickly removed by opportunistic scavengers and are thus difficult to detect.

Population Ecology

Population Dynamics

The dynamics of Alpine ibex populations differ according to the time since population establishment: recently founded populations show a typical exponential growth pattern (Loison et al. 2002), while more established populations typically show regular oscillations around the carrying capacity, mostly driven by density dependent and meteorological factors (von Hardenberg et al. 2000).

In the Gran Paradiso population, more than 80% of the variability in Alpine ibex population growth until the mid 1980s could be explained by the interaction between population density and average snow depth, a proxy of the winter climatic conditions (Jacobson et al. 2004). Snow depth, together with precipitation and temperature, have been identified as major limiting factors of yearly population growth rate also in the Swiss National Park (Sæther et al. 2002) and in other 26 reintroduced populations in Switzerland, for which changes in population sizes were shown to be influenced by climatic variables of the previous winter (33% of the populations), but also by climatic variables of the winter before the last one (23%), of the previous summer (21%) and the summer before the last one (23%) (Grøtan et al. 2008). These strong effects of climatic variability on population growth suggest that Alpine ibex might be particularly sensitive to global climate change. In particular, declining snow cover during winter and spring in the Alps (Beniston 1997), while favoring adult winter survival, has changed the timing of plant phenology causing a mismatch between the timing of ibex births and the peak in primary production of alpine pastures. Changes in the phenology of pastures have been associated with declining kid survival in Alpine ibex as well as in bighorn sheep and mountain goats, *Oreamnos americanus* (Pettorelli et al. 2007). In agreement with these results, Mignatti et al. (2012) showed that the relationship between both kid survival and weaning success and snow depth is nonlinear, with a maximum at average snow depth and lower survival and weaning success both at lower and higher snow depths. Concordant with these results, warm spring or summer temperatures have been found to negatively affect ibex survival in both sexes as well as female reproductive success in the Belledonne population (Toïgo et al. 2016). Climate change appears to have changed also the altitudinal distribution of the Alpine ibex: between 1991 and 2013, ibex hunting locations shifted in altitude by an average of 135 m in the Canton Graubünden, Switzerland (Büntgen et al. 2017), indicating a behavioral response of the species to warmer autumn conditions. Similarly, Herfindal et al. (2019) found that Alpine ibex of both sexes shift towards higher altitudes during warmer springs and summers. While it is not clear so far if these climate-induced changes in the altitudinal distribution have negative impacts on the overall population performance, Mason et al. (2017) showed that Alpine ibex shifting to higher altitudes to thermoregulate in summer have a lower quality diet due to the use of less productive pastures but do not compensate by increasing feeding time. Büntgen et al. (2014) found a contrary signal in males of eight Swiss colonies with better horn growth performance in years with a climate-induced early onset of vegetation growth between 1978 and 2011. In the Belledonne population, on the other hand, warm springs and summers have been found to have negative effects on individual performance (horn length, survival, and reproductive success, Toïgo et al. 2016, 2017). These contrasting results could be linked to the altitudinal range of the study sites. In Belledonne, ibex already occupy the whole altitudinal range of the massif and cannot move upwards with increasing temperature, as did Alpine ibex in Swiss populations.

Interspecific Interactions

Alpine ibex live in sympatry with Alpine chamois over the whole species range but evidence for direct interspecific resource competition affecting Alpine ibex population growth is lacking. This is also supported by studies from the Swiss National Park where Alpine ibex population growth rate was not influenced by chamois population size. However, dietary overlap between Alpine ibex, chamois, and red deer was high, and red deer are colonizing ranges that were previously occupied only by chamois and Alpine ibex (Anderwald et al. 2015). A recent study by Herfindal et al. (2019) suggests that high densities of red deer and chamois may affect ibex spatial behavior, pushing them upward and toward more marginal areas. Altitudinal range and habitat use of female Alpine ibex seems to be more affected than that of males by chamois abundance while competition with red deer seems to affect both sexes similarly (Herfindal et al. 2019).

Interactions of Alpine ibex with domestic sheep and goat are frequent, particularly during summer (Richomme et al. 2006). No study so far has addressed direct competition for food between Alpine ibex and livestock, but the decrease of resources available for ibex due to domestic herbivores grazing is suspected to be high. The presence of domestic ungulates in Alpine pastures has been shown to increase the risk of inter-species pathogen transmission (Ryser-Degiorgis et al. 2002; Mick et al. 2014). Moreover, the presence of domestic goats may increase the risk of hybridization with Alpine ibex (Grossen et al. 2014; Iacolina et al. 2019).

Conservation Status

The Alpine ibex is included in the Bern Convention (Convention on the Conservation of European Wildlife and Natural Habitats, Appendix III – Protected Fauna Species, 1979) and in the European Directive 43/92/CEE "Habitat," Annex V (Updated with Directive 97/62/CE, 27 Oct 1997). The IUCN classifies Alpine ibex as a species of Least Concern (LC) "in view of its wide distribution, presumed large population, and because it is not declining at nearly the rate required to qualify for listing in a threatened category." However, the IUCN status also declares that "the species needs conservation action to prevent future decline" (Toïgo et al. 2020).

Management

In most parts of the species distribution, Alpine ibex habitat has experienced significant changes in the last decades because of a widespread abandonment of agriculturally used mountain pastures. These areas have naturally reforested at low to medium altitude causing a reduction of the spring habitat of Alpine ibex. On the other hand, resources for wild ungulates may be limited in areas with high densities of livestock. Currently, there is no clear evidence of the effects of these opposing trends on the population dynamics of the species. However, changes in Alpine ibex range use are observed in areas with evident land-use alterations.

Human-induced disturbance is increasing in the Alps (e.g., recreation and outdoor sports, flights of helicopter, and hang-gliders) and may cause changes in the activity patterns of Alpine ibex (Ingold 2005; Brambilla and Brivio 2018). The economic value of this iconic ungulate species is very high for Alpine tourism and the appeal of protected areas (GPNP unpublished data). Conflicts with humans are limited and were generally related to the impact of browsing on rejuvenating forests and on mountain pastures in the past. In recent times, concerns arose about the possible spreading of zoonoses (e.g., the case of *Brucella melitensis* in the Bargy massif, Mick et al. 2014).

Although the Alpine ibex is a protected species, there is no common, coordinated management, and conservation strategy among the different countries. Here, we provide a summary of the protection and hunting laws for each country. **France:** Alpine ibex is protected, and hunting is not permitted on the whole French territory. Alpine ibex is included in the List of Endangered Species (French Game and Wildlife Office – Office National de la Chasse et de la Faune Sauvage ONCFS), and there was a National Action Plan for the reintroduction of the species: "Strategie de reintroduction des bouquetins en France 2000–2015." **Italy:** Alpine ibex is protected, and hunting is not permitted except in the autonomous province of Bolzano, where some individuals are culled for containment of sarcoptic mange. **Switzerland:** Alpine ibex is a protected species (LCP, 20 June 1986 on the Hunting and Protection of Wild Mammals and Birds) with a special decree (ORES 922.27, 30 April 1990) that regulates its management and hunting. Hunting started in 1977 with a current annual bag of about 1000 individuals. Annual bags are based on annual spring censuses, which are conducted by the cantons for each colony. **Austria:** In Austria, ibex are hunted since 1978. Hunting regulations are different for each of the five states that host ibex colonies (Styria, Carinthia, Salzburg,

Vorarlberg, and Tyrol) and for each hunting district. **Germany:** Alpine ibex occur only in few small colonies in Bavaria. Ibex are protected and not a game species with regular hunting season but some animals are culled to reduce the browsing impact in forests. **Slovenia:** In Slovenia, the Alpine ibex is a game species, hunted in the colony of Bovec and Brana. In the colony of Triglav, ibex affected by sarcoptic mange were culled for sanitary reasons.

Future Challenges for Research and Management

The biggest issues for the conservation of Alpine ibex in a period of rapidly changing environmental conditions in the Alps appear to be the isolation of the populations in combination with low recolonization rates, low genetic diversity, and high inbreeding levels. Low genetic diversity is disadvantageous as it lowers the adaptive potential of populations to a changing environment and their resilience in the face of climate change and resistance to disease outbreaks. Therefore, to ensure the long-term viability of ibex populations, genetic diversity needs to be enhanced and inbreeding reduced. However, the basis for any management plan is the availability of detailed data on the dynamics and genetics of each population. A coordinate collection of census data in a standardized format across different populations and countries is thus an essential requirement to plan balanced conservation and management actions.

Furthermore, other important points to address in future research are:

- Study the effects of climate change on population dynamics, both through its direct effects on ibex (thermoregulation, activity budget, habitat selection, reproductive success, and survival) and through its indirect effects (forage quantity and quality, parasite infections)
- Improve the knowledge about sex-specific differences in life history, behavior, and physiology
- Study competition between domestic species, chamois, red deer, and Alpine ibex and the effects on population dynamics both in terms of disease transmission as well as resource competition and disturbance
- Identify traits and genes that are involved in inbreeding depression and how they affect population growth rate
- Explore diseases, e.g., factors for susceptibility to keratoconjunctivitis, brucellosis, or sarcoptic mange
- Study the effect of large predators (e.g., wolves) on population dynamics before/after the recolonization of mountain habitats by the predator in the near future
- Investigate the effect of hunting on the demography and viability of ibex populations
- Investigate how the expansion of recreational activities (e.g., climbing, paragliding) may impact behavior and demography of ibex populations.

Acknowledgments We would like to thank Dante Alpe for kindly providing pictures of Alpine ibex and Peter Neuhaus and Luca Corlatti for their useful suggestions to improve this chapter.

References

Abderhalden W (2005) Raumnutzung und sexuelle Segregation beim Alpensteinbock *Capra ibex ibex*. Nationalpark-Forschung in der Schweiz 92. Albert-Ludwigs-Universität Freiburg im Breisgau, 184 pp

Anderwald P, Herfindal I, Haller RM et al (2015) Influence of migratory ungulate management on competitive interactions with resident species in a protected area. Ecosphere 6(11):1–18

Anses (2016) AVIS de l'Agence nationale de sécurité sanitaire de l'alimentation, de l'environnement et du travail relatif à l'évaluation approfondie et réactualisée de mesures de maîtrise du foyer de brucellose chez les bouquetins du Bargy. Saisine n° 2016-SA-0229 https://www.anses.fr/fr/system/files/SABA2016SA0229.pdf

Apollonio M, Brivio F, Rossi I et al (2013) Consequences of snowy winters on male mating strategies and reproduction in a mountain ungulate. Behav Process 98:44–50

Bassano B, Bollo E, Peracino V et al (1994) Brain lesions associated with infectious kerato-conjunctivitis in chamois and Alpine ibex. Ibex J Mount Ecol 2:17–22

Belloy L, Janovsky M, Vilei EM et al (2003) Molecular epidemiology of *Mycoplasma conjunctivae* in Caprinae: transmission across species in natural outbreaks. Appl Environ Microb 69(4):1913–1919

Beniston M (1997) Variations of snow depth and duration in the Swiss Alps over the last 50 years: links to changes in large-scale climatic forcings. In: Diaz HF,

Beniston M, Bradley RS (eds) Climatic change at high elevation sites. Springer, Dordrecht

Bergeron P, Festa-Bianchet M, von Hardenberg A et al (2008) Heterogeneity in male horn growth and longevity in a highly sexually dimorphic ungulate. Oikos 117:77–82

Bergeron P, Grignolio S, Apollonio M et al (2010) Secondary sexual characters signal fighting ability and determine social rank in Alpine ibex (*Capra ibex*). Behav Ecol Sociobiol 64(8):1299–1307

Biancardi CM, Minetti AE (2017) Gradient limits and safety factor of Alpine ibex (*Capra ibex*) locomotion. Hystrix 28(1):1–5

Biebach I, Keller LF (2009) A strong genetic footprint of the re-introduction history of Alpine ibex (*Capra ibex ibex*). Mol Ecol 18:5046–5058

Biebach I, Keller LF (2010) Inbreeding in reintroduced populations: the effects of early reintroduction history and contemporary processes. Conserv Genet 11:527–538

Biebach I, Keller LF (2012) Genetic variation depends more on admixture than number of founders in reintroduced Alpine ibex populations. Biol Conserv 147:197–203

Bon R, Rideau C, Villaret JC, Joachim J (2001) Segregation is not only a matter of sex in Alpine ibex, *Capra ibex ibex*. Anim Behav 62(3):495–504

Bozzuto C, Biebach I, Muff S et al (2019) Inbreeding reduces long-term growth of Alpine ibex populations. Nat Ecol Evol 3(9):1359–1364

Brambilla A, Brivio F (2018) Assessing the effects of helicopter disturbance in a mountain ungulate on different time scales. Mamm Biol 90:30–37

Brambilla A, Biebach I, Bassano B et al (2015) Direct and indirect causal effects of heterozygosity on fitness-related traits in Alpine ibex. P R Soc B 282(1798): 20141873

Brambilla A, Keller L, Bassano B et al (2018) Heterozygosity-fitness correlation at the major histocompatibility complex despite low variation in Alpine ibex (*Capra ibex*). Evol Appl 11(5):631–644

Brambilla A, Von Hardenberg A, Nelli L, Bassano B (2020) Distribution, status, and recent population dynamics of Alpine ibex *Capra ibex* in Europe. Mammal Rev 50(3):267–277

Brivio F, Grignolio S, Apollonio M (2010) To feed or not to feed? Testing different hypotheses on rut-induced hypophagia in a mountain ungulate. Ethology 116(5): 406–415

Brivio F, Grignolio S, Brambilla A et al (2014) Intra-sexual variability in feeding behavior of a mountain ungulate: size matters. Behav Ecol Sociobiol 68(10):1649–1660

Büntgen U, Liebhold A, Jenny H et al (2014) European springtime temperature synchronises ibex horn growth across the eastern Swiss Alps. Ecol Lett 17:303–313

Büntgen U, Greuter L, Bollmann K et al (2017) Elevational range shifts in four mountain ungulate species from the Swiss Alps. Ecosphere 8(4):e01761

Carmignola G, Stefani P, Gerstgrasser L (2006) 6° Rapporto rogna sarcoptica. Provincia autonoma di Bolzano, Ufficio caccia e pesca

Cassini R, Párraga MA, Signorini M et al (2015) Lungworms in Alpine ibex (*Capra ibex*) in the eastern Alps, Italy: an ecological approach. Vet Parasitol 214(1–2): 132–138

Cassirer EF, Manlove KR, Almberg ES et al (2018) Pneumonia in bighorn sheep: risk and resilience. J Wildl Manag 82:32–45

Cook RA, Roskop ML, Bowerman DL (1986) Hematologic and serum chemistry values for captive Alpine ibex, *Capra ibex ibex*. J Zoo Anim Med 17(2):65–68

Couturier MAJ (1962) Le Bouquetin des Alpes: *Capra aegagrus ibex ibex*. L. Arthaud, Grenoble

Cregut-Bonnoure E (1992) Dynamics of bovid migration in western Europe during the middle and late Pleistocene. Cour Forschungsinst Senck 153:177–185

Damm GR, Nicolás F (2014) The CIC Caprinae atlas of the world. CIC International Council for Game and Wildlife Conservation, Budakeszi and Rowland Ward Publications RSA, Johannesburg

De Danieli C, Sarasa M (2015) Population estimates, density dependence and the risk of disease outbreaks in the Alpine ibex *Capra ibex*. Anim Biodivers Conserv 38(1):101–119

Degiorgis MP, Obrecht E, Ryser A et al (1999) The possible role of eye-frequenting flies in the transmission of *Mycoplasma conjuntivae*. Bull Soc Entomol Suisse 72:189–194

Delorme M (2008) Recensement des populations de bouquetin des Alpes dans le Parc national de la Vanoise – Été 2008. Parc National de la Vanoise

Demeautis G (1982) Enzooties 1974–1976: pleuropneumonies des chamois et bouquetins en Vanoise. In Grande faune sauvage des montagnes d'Europe et ecosystemes d'altitude (pp. 217–245). Centre Pyrénéen de Biologie et Anthropologie des Montagnes

Dotta R (2009) Piano di Programmazione per la Gestione degli Ungulati per il quinquennio 2009–2014. Comprensorio Alpino Torino 4 Valli di Lanzo, Ceronda e Casternone

Ewing SR, Nager RG, Nicoll MAC et al (2008) Inbreeding and loss of genetic variation in a reintroduced population of Mauritius Kestrel. Conserv Biol 22:395–404

Ferroglio E, Tolari F, Bollo E et al (1998) Isolation of *Brucella melitensis* from Alpine Ibex. J Wildlife Dis 34(2):400–402

Foreyt WJ, Snipes KP, Kasten RW (1994) Fatal pneumonia following inoculation of healthy Bighorn sheep with *Pasteurella haemolytica* from healthy domestic sheep. J Wildl Dis 30(2):137–145

Frei P (1972) Ökologische Untersuchungen an Huftieren im Val Trupchun. 2. Jahresbericht Arbeitsgruppe Wildforschung 32–33

Garin-Bastuji B, Oudar J, Richard Y et al (1990) Isolation of *B. melitensis* Biovar 3 from a chamois (*Rupicapra rupicapra*) in the Southern French Alps. J Wildl Dis 26:116–118

Garnier A (2013) Conséquences des pathologies sur la dynamique des populations d'ongulés sauvages: exemple du bouquetin des Alpes dans le Parc National de la Vanoise. Ecole Pratique des Hautes Etudes

Gauthier D, Prud'homme C, Garnier F (1994) Normes sanguines chez le Bouquetin des Alpes: hématologie et diagnostic de gestation. Trav Sci Parc Natl Vanoise 18:47–52

Gebremedhin B, Ficetola GF, Naderi S et al (2009) Combining genetic and ecological data to assess the conservation status of the endangered Ethiopian walia ibex. Anim Conserv 12:89–100

Geist V (1974) Relationship of social evolution and ecology in ungulates. Am Zool 14:205–220

Gennero MS, Meneguz PG, Mandola ML et al (1993) Indagine sierologica sui ruminanti selvatici in Piemonte. Atti Soc Ital Sci Vet 1979–1980 47:979–983

Giacometti M (1991) Beitrag zur Ansiedlungsdynamik und aktuellen Verbreitung des Alpensteinbockes (*Capra i. ibex* L.) im Alpenraum. Z Jagdwiss 37(3): 157–173

Giacometti M (2006) Von Königen und Wilderern: Die Rettung und Wiederansiedlung des Alpensteinbockes. Salm Verlag Wohlen

Giacometti M, Bassano B, Peracino V et al (1997) The constitution of the alpine ibex (*Capra i. ibex* L.) in relation to sex, age, area of origin, and season in Graubünden (Switzerland) and in the Parco Nazionale Gran Paradiso (Italy). Z Jagdwiss 43:24–34

Giacometti M, Nicolet J, Frey J et al (1998) Susceptibility of Alpine ibex to conjunctivitis caused by inoculation of a sheep-strain of *Mycoplasma conjunctivae*. Vet Microbiol 61:279–288

Giacometti M, Willing R, Defila C (2002) Ambient temperature in spring affects horn growth in male Alpine ibexes. J Mammal 83:245–251

Giacometti M, Roganti R, De Tann D et al (2004) Alpine ibex *Capra ibex ibex* x domestic goat *C. aegagrus domestica* hybrids in a restricted area of southern Switzerland. Wildl Biol 10:137–143

Girard I, ToÏgo C, Gaillard JM, Gauthier D, Martinot JP (1999) Patron de Survie Chez le Bouquetin des Alpes (*Capra ibex ibex*) dans le Parc National de la Vanoise. Revue d'éologie

Girard I (2000) Dynamique des populations et expansion géographique du bouquetin des Alpes (*Capra ibex ibex*, L.) dans le Parc national de la Vanoise. Thèse de l'université de Savoie. 229 pp

Goldstein EJ, Millspaugh JJ, Washburn BE et al (2005) Relationships among fecal lungworm loads, fecal glucocorticoid metabolites, and lamb recruitment in free-ranging Rocky Moutain Bighorn sheep. J Wildl Dis 41(2):416–425

Gortázar C, Ferroglio E, Höfle U et al (2007) Diseases shared between wildlife and livestock: a European perspective. Eur J Wildlife Res 53:241

Grignolio S, Rossi I, Bassano B et al (2004) Seasonal variations of spatial behavior in female Alpine ibex (*Capra ibex ibex*) in relation to climatic conditions and age. Ethol Ecol Evol 16:255–264

Grignolio S, Rossi I, Bassano B et al (2007a) Predation risk as a factor affecting sexual segregation in Alpine ibex. J Mammal 88(6):1488–1497

Grignolio S, Rossi I, Bertolotto E et al (2007b) Influence of the kid on space use and habitat selection of female Alpine ibex. J Wildl Manage 71(3):713–719

Grignolio S, Brivio F, Sica N, Apollonio M (2019) Sexual differences in the behavioural response to a variation in predation risk. Ethology 125(9):603–612

Grodinsky C, Stüwe M (1987) With lots of help alpine ibex return to their mountains. Smithsonian 18:68–77

Grossen C, Keller L, Biebach I et al (2014) Introgression from domestic goat generated variation at the major histocompatibility complex of Alpine ibex. PLoS Genet

Grossen C, Biebach I, Angelone-Alasaad S et al (2018) Population genomics analyses of European ibex species show lower diversity and higher inbreeding in reintroduced populations. Evol Appl 11:123–139

Grøtan V, Saether BE, Filli F et al (2008) Effects of climate on population fluctuations of ibex. Glob Change Biol 14(2):218–228

Gunn A, Irvine RJ (2003) Subclinical parasitism and ruminant foraging strategies: a review. Wildl Soc Bull 31:117–126

Hars J, Gauthier D (1994) Pathologie du bouquetin des Alpes: bilan sanitaire des populations françaises. Travaux scientifiques du Parc national de la Vanoise 18:53–98

Hartl GB, Burger H, Willing R et al (1990) On the biochemical systematics of the Caprini and the Rupicaprini. Biochem Syst Ecol 18(2–3):175–182

Herfindal I, Anderwald P, Filli F et al (2019) Climate, competition and weather conditions drive vertical displacement and habitat use of an alpine ungulate in a highly topographic landscape. Landsc Ecol 34(11): 2523–2539

Hofmann RR (1989) Evolutionary steps of ecophysiological adaptation and diversification of ruminants: a comparative view of their digestive system. Oecologia 78:443–457

Hofmann A, Nievergelt B (1972) Das jahreszeitliche Verteilungsmuster und der Äsungsdruck von Alpensteinbock, Gemse, Rothirsch und Reh in einem begrenzten Gebiet im Oberengadin. Z Jagdwiss 18: 185–212

Iacolina L, Corlatti L, Buzan E et al (2019) Hybridisation in European ungulates: an overview of the current status, causes, and consequences. Mamm Rev 49(1): 45–59

Illius AW, Gordon IJ (1992) Modelling the nutritional ecology of ungulate herbivores: evolution of body size and competitive interactions. Oecologia 89(3): 428–434

Ingold P (2005) Freizeitaktivitäten im Lebensraum der Alpentiere: Konfliktbereiche zwischen Mensch und Tier mit einem Ratgeber für die Praxis. Haupt, Bern

Jacobson AR, Provenzale A, von Hardenberg A et al (2004) Climate forcing and density dependence in a mountain ungulate population. Ecology 85:1598–1610

Kardos M, Akesson M, Fountain T et al (2018) Genomic consequences of intensive inbreeding in an isolated wolf population. Nat Ecol Evol 2:124–131

Kazanskaya EY, Kuznetsova MV, Danilkin AA (2007) Phylogenetic reconstructions in the genus *Capra* (Bovidae, Artiodactyla) based on the mitochondrial DNA analysis. Russ J Genet 43(2):181–189

Klein J, Figueroa F (1986) Evolution of the major histocompatibility complex. Crit Rev Immunol 6:295–386

Largo E, Gaillard JM, Festa-Bianchet M et al (2008) Can ground counts reliably monitor ibex *Capra ibex* populations? Wildl Biol 14:489–499

Lavin S, Ruiz-Bascaran M, Marco I et al (2000) Experimental infection of chamois (*Rupicapra pyrenaica parva*) with *Sarcoptes scabiei* derived from naturally infected goats. J Vet Med 47:693–699

Lawrence PK, Shanthalingam S, Dassanayake RP et al (2010) Transmission of *Mannheimia haemolytica* from domestic sheep (*Ovis aries*) to bighorn sheep (*Ovis canadensis*): unequivocal demonstration with green fluorescent protein-tagged organism. J Wildl Dis 46(3):706–717

León-Vizcaíno L, Ruíz de Ybáñez MR, Cubero MJ et al (1999) Sarcoptic mange in Spanish ibex from Spain, 1999. J Wildl Dis 35(4):647–659

Lima M, Berryman A (2006) Predicting nonlinear and nonadditive effects of climate: the Alpine ibex revisited. Clim Res 32(2):129–135

Loison A, Gaillard JM, Pelabon C et al (1999) What factors shape sexual size dimorphism in ungulates? Evol Ecol Res 1:611–633

Loison A, Toïgo C, Appolinaire J et al (2002) Demographic processes in colonizing populations of isard (*Rupicapra pyrenaica*) and ibex (*Capra ibex*). J Zool 256(2):199–205

Lüps P, Blöchlinger B, Schmid P et al (2007) Ontogenese und Variabilität verschiedener Körpermerkmale des Steinwildes *Capra i. ibex* im Berner Oberland (Schweizer Alpen). Beiträge zur Jagd- und Wildforschung 32:495–510

Manceau V, Després L, Bouvet J et al (1999) Systematics of the genus *Capra* inferred from mitochondrial DNA sequence data. Mol Phylogenet Evol 13(3):504–510

Marchand P, Freycon P, Herbaux JP et al (2017) Socio-spatial structure explains marked variation in brucellosis seroprevalence in an Alpine ibex population. Sci Rep 7(1):15592

Marreros N, Hussy D, Albini S et al (2011) Epizootiologic investigation of selected abortive agents in free-ranging alpine ibex (*Capra ibex ibex*) in Switzerland. J Wildl Dis 47(3):530–543

Mason TH, Brivio F, Stephens PA et al (2017) The behavioral trade-off between thermoregulation and foraging in a heat-sensitive species. Behav Ecol 28:908–918

Meile P, Giacometti M, Ratti P (2003) Der Steinbock: Biologie und Jagd. Salm Verlag, Bern

Mick V, Le Carrou G, Corde Y et al (2014) *Brucella melitensis* in France: persistence in wildlife and probable spillover from Alpine ibex to domestic animals. PLoS One 9(4):e94168

Mignatti A, Casagrandi R, Provenzale A et al (2012) Sex-and age-structured models for Alpine ibex *Capra ibex ibex* population dynamics. Wildl Biol 18(3):318–332

Nandi S, De UK, Chowdhury S (2011) Current status of contagious ecthyma or orf disease in goat and sheep—a global perspective. Small Rumin Res 96(2–3):73–82

Neuhaus P, Ruckstuhl KE (2002) Foraging behaviour in Alpine ibex (*Capra ibex*): consequences of reproductive status, body size, age and sex. Ethol Ecol Evol 14(4):373–381

Nievergelt B (1966) Der Alpensteinbock (*Capra ibex* L.) in seinem Lebensraum. Ein ökologischer Vergleich. Mammalia depicta. Paul Parey Hamburg und Berlin

Nievergelt B, Zingg R (1986) *Capra ibex* Linnaeus, 1758 – Steinbock. In: Niethammer J, Krapp F (eds) Handbuch der Säugetiere Europas. Aula-Verlag, Wiesbaden

Parrini F, Grignolio S, Luccarini S et al (2003) Spatial behavior of adult male Alpine ibex *Capra ibex ibex* in the Gran Paradiso National Park, Italy. Acta Theriol 48:411–423

Parrini F, Cain JW, Krausman PR (2009) *Capra ibex* (Artiodactyla: Bovidae). Mamm Species 830:1–12

Passerin d'Entrèves P (2000) Le Chasses Royales in Valle d'Aosta (1850–1919). Umberto Allemandi, Torino

Peinado VI, Celdrán JF, Palomeque J (1999) Basic hematological values in some wild ruminants in captivity. Comp Biochem Physiol A 124(2):199–203

Pérez JM, Granados JE, González FJ et al (1999) Hematologic parameters of the Spanish ibex (*Capra pyrenaica*). J Zoo Wildl Med 30(4):550–554

Pettorelli N, Pelletier F, von Hardenberg A et al (2007) Early onset of vegetation growth vs rapid green-up: impacts on juvenile mountain ungulates. Ecology 88:381–390

Pidancier N, Jordan S, Luikart G et al (2006) Evolutionary history of the genus *Capra* (Mammalia, Artiodactyla): discordance between mitochondrial DNA and Y-chromosome phylogenies. Mol Phylogenet Evol 40(3):739–749

Pilgrim GE (1947) The evolution of the buffaloes, oxen, sheep and goats. J Linn Soc 41:272–286

Quéméré E, Rossi S, Petit E et al (2020) Genetic epidemiology of the Alpine ibex reservoir of persistent and virulent brucellosis outbreak. Scient Rep 10:1–10

Ratti P (1981) Zur Hege des Steinwildes im Kanton Graubünden. Z Jagdwiss 27:41–57

Ratti P, Habermehl KH (1977) Untersuchungen zur Altersschätzung und Altersbestimmung beim Alpensteinbock (*Capra ibex ibex*) im Kanton Graubünden. Z Jagdwiss 23:188–213

Rauch A (1937) Der Steinbock wieder in den Alpen. Orell Füssli, Zürich

Richomme C, Gauthier D, Fromont E (2006) Contact rates and exposure to inter-species disease transmission in mountain ungulates. Epidemiol Infect 134:21–30

Ropiquet A, Hassanin A (2006) Hybrid origin of the Pliocene ancestor of wild goats. Mol Phylogenet Evol 41(2):395–404

Rossi L, Fraquelli C, Vesco U et al (2007) Descriptive epidemiology of a scabies epidemic in chamois in the Dolomite Alps, Italy. Eur J Wildl Res 53:131–141

Ruckstuhl KE, Neuhaus P (2001) Behavioral synchrony in ibex groups: effects of age, sex and habitat. Behavior 138(8):1033–1046

Ruckstuhl KE, Neuhaus P (2002) Sexual segregation in ungulates: a comparative test of three hypotheses. Biol Rev 77(1):77–96

Ryser-Degiorgis MP, Ingold P, Tenhu H et al (2002) Encounters between Alpine ibex, Alpine chamois and domestic sheep in the Swiss Alps. Hystrix 13(1–2):1–11

Ryser-Degiorgis MP, Bischof DF, Marreros N et al (2008) Detection of *Mycoplasma conjunctivae* in the eyes of healthy, free-ranging Alpine ibex: possible involvement of Alpine ibex as carriers for the main causing agent of infectious keratoconjunctivitis in wild Caprinae. Vet Microbiol 134(3–4):368–374

Sæther BE, Engen S, Filli F et al (2002) Stochastic population dynamics of an introduced Swiss population of the ibex. Ecology 83(12):3457–3465

Samuel WM, Chalmers GA, Stelfox JG et al (1975) Contagious ecthyma in bighorn sheep and mountain goat in western Canada. J Wildl Dis 11(1):26–31

Scagliarini A, Vaccari F, Turrini F et al (2011) Parapoxvirus infections of red deer, Italy. Emerg Infect Dis 17(4):684–687

Schaller GB (1977) Mountain monarchs. Wild sheep and goats of the Himalaya. University of Chicago Press, Chicago

Schnitter M (1962) Beobachtungen zur Nahrungsaufnahme beim Steinwild, *Capra ibex* L. Verhandl. d. Schweiz. Naturf Ges Scuol 142:110–112

Signer C, Ruf T, Arnold W (2011) Hypometabolism and basking: the strategies of Alpine ibex to endure harsh over-wintering conditions. Funct Ecol 25(3):537–547

Sokolov VE (1979) *Sistematika mlekopitayushchikh* (Systematics of mammals). Vysshaya Shkola, Moscow

Stüwe M, Grodinsky C (1987) Reproductive biology of captive alpine ibex (*Capra i. ibex*). Zoo Biol 6:331–339

Stüwe M, Nievergelt B (1991) Recovery of alpine ibex from near extinction: the result of effective protection, captive breeding, and reintroductions. Appl Anim Behav Sci 29(1–4):379–387

Stüwe M, Scribner KT (1989) Low genetic variability in reintroduced alpine ibex (*Capra ibex ibex*) populations. J Mammal 70(2):370–373

Ten Houte de Lange SM (1978) Zur Futterwahl des Alpensteinbocks (*Capra ibex* L.). Z Jagdwiss 24:113–138

Tettamanti F, Viblanc VA (2014) Influences of mating group composition on the behavioral time-budget of male and female Alpine ibex (*Capra ibex*) during the rut. PLoS One 9(1):e86004

Toïgo C (1998) Stratégies biodémographiques et sélection sexuelle chez le Bouquetin des Alpes (*Capra ibex ibex*). Université Claude Bernard – Lyon 1, Lyon, p 180

Toïgo C, Michallet J, Blanc D (1995) Structure des groupes chez le bouquetin des Alpes (*Capra ibex ibex*) dans le massif de Belledonne-Sept-Laux (Isère). Gibier Faune Sauvage 12:133–146

Toïgo C, Gillard J-M, Michallet J (1996) La taille des groupes: un bioindicateur de l'effectif des populations de bouquetin des Alpes (*Capra ibex ibex*)? Mammmalia 60:463–472

Toïgo C, Gaillard J-M, Michallet J (1999) Cohort affects growth of males but not females in Alpine ibex (*Capra ibex ibex*). J Mammal 80:1021–1027

Toïgo C, Gaillard J-M, Gauthier D et al (2002) Female reproductive success and costs in an alpine capital breeder under contrasting environments. Ecoscience 9:427–433

Toïgo C, Gaillard JM, Festa-Bianchet M et al (2007) Sex-and age-specific survival of the highly dimorphic Alpine ibex: evidence for a conservative life-history tactic. J Anim Ecol 76(4):679–686

Toïgo C, Gaillard JM, Loison A (2013) Alpine ibex males grow large horns at no survival cost for most of their lifetime. Oecologia 173:1261–1269

Toïgo C, Blanc D, Couilloud F et al (2016) Réchauffement climatique et performance chez le bouquetin des Alpes. L'exemple de la population de Belledonne. Faune Sauvage 312:27–31

Toïgo C, Blanc D, Couilloud F et al (2017) Climate change and biodemographic performance in a mountain ungulate – warm temperatures are bad for Alpine ibex (*Capra ibex*). IUGB Congress, Monpellier

Toïgo C, Brambilla A, Grignolio S, Pedrotti L (2020) *Capra ibex*. The IUCN Red List of Threatened Species e. T42397A161916377. Downloaded on 24 February 2021

Tolari F, Meneguz PG, De Meneghi D et al (1987) Indagini siero epidemiologiche su stambecchi, camosci e ovini presenti nel Parco Naturale Argentera. In: Proceedings of the International Conference "Lo Stambecco delle Alpi: realtà attuale e prospettive", pp 83–92

Tosi G, Scherini G, Apollonio M et al (1986) Modello di valutazione ambientale per la reintroduzione dello stambecco (*Capra ibex ibex*). Supplemento alle ricerche di biologia della selvaggina 77:5–77

Villaret JC, Bon R (1995) Social and spatial segregation in Alpine ibex (*Capra ibex*) in Bargy, French Alps. Ethology 101(4):291–300

Villaret JC, Bon R (1998) Sociality and relationships in Alpine ibex (*Capra ibex*). Rev Ecol (Terre Vie) 53:153–170

Villaret JC, Bon R, Rivet A (1997) Sexual segregation of habitat by the Alpine ibex in the French Alps. J Mammal 78(4):1273–1281

von Hardenberg A, Bassano B, Peracino V et al (2000) Preliminary results on the temporal variability of the Alpine ibex population in the Gran Paradiso National Park. J Mt Ecol 5:201–210

von Hardenberg A, Bassano B, Arranz MPZ et al (2004) Horn growth but not asymmetry heralds the onset of senescence in male Alpine ibex (*Capra ibex*). J Zool 263:425–432

von Hardenberg A, Bassano B, Festa-Bianchet M et al (2007) Age-dependent genetic effects on a secondary sexual trait in male Alpine ibex, *Capra ibex*. Mol Ecol 16(9):1969–1980

Vrba ES, Schaller GB (2000) Phylogeny of Bovidae (Mammalia) based on behavior, glands and skull morphology. In: Vrba ES, Shaller GB (eds) Antelopes, Deer, and relatives: fossil record, behavioral ecology, systematics, and conservation. Yale University Press, New Haven, CT, pp 203–222

Westerdahl H, Asghar M, Hasselquist D, Bensch S (2011) Quantitative disease resistance: to better understand

parasite-mediated selection on major histocompatibility complex. P R Soc B 279(1728):577–584
Wiersema G (1983) Project ibex – ibex habitat inventory and mapping in the European Alps. Mt Res Dev 3:303–305
Wiersema G (1989) Climate and vegetation characteristics of ibex habitats in the European Alps. Mt Res Dev 9:119–128
Wiersema G, Gauthier D (1990) Status and aspects of reintroduction and management of the Ibex in the Alps. Trav Sci Parc Natl Vanoise 1(7):235–252
Willisch CS, Neuhaus P (2009) Alternative mating tactics and their impact on survival in adult male Alpine ibex (*Capra ibex ibex*). J Mammal 90(6):1421–1430
Willisch CS, Neuhaus P (2010) Social dominance and conflict reduction in rutting male Alpine ibex, *Capra ibex*. Behav Ecol 21(2):372–380
Willisch CS, Biebach I, Koller U et al (2012) Male reproductive pattern in a polygynous ungulate with a slow life-history: the role of age, social status and alternative mating tactics. Evol Ecol 26(1):187–206
Willisch CS, Biebach I, Marreros N et al (2015) Horn growth and reproduction in a long-lived male mammal: no compensation for poor early-life horn growth. Evol Biol 42:1–11
Zaffaroni E, Manfredi MT, Citterio C et al (2000) Host specificity of abomasal nematodes in free ranging alpine ruminants. Vet Parasitol 90(3):221–230

Iberian Wild Goat *Capra pyrenaica* Schinz, 1838

17

Ricardo García-González, Juan Herrero, Pelayo Acevedo, Mari Cruz Arnal, and Daniel Fernández de Luco

Contents

R. García-González
Instituto Pirenaico de Ecología (IPE-CSIC), Jaca, Spain
e-mail: ricardogarciagonzalez@gmail.com

J. Herrero (✉)
Technical School, University of Zaragoza, Huesca, Spain
e-mail: herreroj@unizar.es

P. Acevedo
Instituto de Investigación en Recursos Cinegéticos, IREC (UCLM-CSIC-JCCM), Ciudad Real, Spain
e-mail: pelayo.acevedo@uclm.es

M. C. Arnal · D. Fernández de Luco
Departamento de Patología Animal, Facultad de Veterinaria, Universidad de Zaragoza, Zaragoza, Spain
e-mail: maricruz@unizar.es; luco@unizar.es

© Springer Nature Switzerland AG 2022

L. Corlatti, F. E. Zachos (eds.), *Terrestrial Cetartiodactyla*, Handbook of the Mammals of Europe,
https://doi.org/10.1007/978-3-030-24475-0_33

Common Names

English	Iberian wild goat
German	Iberischer Steinbock
French	Bouquetin Ibérique
Spanish	Cabra montés
Italian	Stambecco iberico
Russian	Пиренейский козёл

The name of Spanish ibex to designate *Capra pyrenaica* occurs in the scientific literature, and some have defended its appropriateness (Sarasa et al. 2012); however, in our opinion, the designation is inappropriate. Etymological, historical, biological, and human societal arguments support the use of "Iberian wild goat" as common name (García-González et al. 2021). *C. pyrenaica* is not an ibex. Adult males have lyre-shaped and typically smooth horns (Fig. 1), which do not correspond to the ibex morphotype, which have scimitar-shaped and knotted horns (Pidancier et al. 2006). In addition, *C. pyrenaica* is an Iberian endemic, and not an exclusively Spanish species. The probable genetic proximity between *C. ibex* and *C. pyrenaica* (Manceau et al. 1999) does not necessarily imply a proximity of morphotypes (Pidancier et al. 2006), as is the case, for example, between *C. falconeri* and *C. aegagrus* (Zvychaynaya 2010).

Taxonomy, Systematics and Paleontology

The original description of the species by Schinz (1838) was based on a male specimen from the Zürich Museum and on skins and drawings from Carl F. Bruch. Schinz never saw a live *C. pyrenaica*, and, in his publication, he noted that they were extremely rare in the Pyrenees.

Fig. 1 Iberian wild goats: adult male, adult female, and young (photograph © Alberto Portero)

Ten years later, Schimper (1848) described a new species of *Capra* for Iberia, *C. hispanica*, relying on specimens collected in an expedition to Sierra Nevada (Southern Iberia) in 1847. Thus, until the twentieth century, taxonomically, two species of wild goat occurred in Iberia.

In an influential article, Cabrera (1911) modified substantially the taxonomy of the Iberian wild goat. Cabrera combined into a single species, *C. pyrenaica*, the two species described initially by Schinz, *C. pyrenaica*, and Schimper, *C. hispanica*, which he designated as subspecies, and he described two new subspecies: *lusitanica* and *victoriae*. Therefore, the current taxonomical classification at the intraspecific level accepted by the IUCN (Shackleton and Lovari 1997) is *Capra pyrenaica* Schinz, 1838, with four subspecies: *C. p. pyrenaica* Schinz, 1838 (extinct); *C. p. hispanica* Schimper, 1848; *C. p. lusitanica* Schlegel, 1872 (extinct), and *C. p. victoriae* Cabrera, 1911.

Terra typica of the nominotypical subspecies *C. p. pyrenaica* was found in the Pyrenees, although it was extremely rare there during the twentieth century (García-González and Herrero 1999). The last Pyrenean wild goat died in January 2000 (Fernández de Luco et al. 2000). *C. p. hispanica* occurs throughout the mountains and steep slopes in Southern and Eastern Iberia and, recently, has expanded its range rapidly (see "Current Distribution"). *C. p. lusitanica* inhabited Northern Portugal and, like the Pyrenean wild goat, was rare. Barboza du Bocage (1857) wrote the first detailed description of the Portuguese subspecies. It became extinct at the beginning of the twentieth century. *C. p. victoriae* lives in the mountains of the center and northwest of the Iberian Peninsula, and Cabrera (1911) defined it as an intermediate form between *pyrenaica* and *hispanica*.

Cabrera (1911, 1914) based the subspecies definition on a few specimens and on highly variable characters, i.e., the pattern of the black hair on the male winter dress and the shape of the horns (Schaller 1977). Couturier (1962) and Clouet (1980) questioned that classification. The subspecies definition of Cabrera was also rejected by Camerano (1917) who defended the recognition of the two original forms that had the greatest divergence of characteristics in Iberia, *C. pyrenaica* and *C. hispanica*, which he felt should be considered distinct species. Hybridization between those species led to the other geographic varieties (*victoriae*, *lusitanica*).

Although some have identified environmental differences in the ecological niches of the two living subspecies, *victoriae* and *hispanica* (Acevedo and Real 2011), recent morphological (García-González 2012) and molecular studies (Angelone-Alasaad et al. 2017; Ureña et al. 2018) have indicated that the two share a close kinship and a clear genetic distance from the extinct Pyrenean subspecies. *C. p. pyrenaica* is genetically and morphologically as far from the *victoriae-hispanica* group, as it is from the Alpine ibex *C. ibex* (Manceau et al. 1999; Ureña et al. 2018). Thus, it is logical to return to the original proposal of Schinz (1838) and Schimper (1848), supported by Camerano (1917); namely, two species in Iberia, the extinct *C. pyrenaica* and *C. hispanica*.

The oldest fossil records of *Capra* in Eurasia were found in Fonelas P-1 (Southeastern Iberia); they date to the beginning of the Pleistocene (2.0 mya) and belong to *Capra baetica* (Arribas and Garrido 2008). Other ancient fossils of *Capra* such as *C. alba* from the Early Pleistocene (1.3–1.0 mya) have been found in Venta Micena (Moyà-Solà 1987) and Quibás (Carlos Calero et al. 2006), also in Southeastern Iberia. Findings of *Capra* sp. from the Middle Pleistocene are much less abundant (García-González 2011). In Iberia and in Southern France, *Capra* records from the Upper Pleistocene are numerous, many of which assigned to *C. pyrenaica* (Sauqué et al. 2016).

The relationship between the ancient goats from the Early and Middle Pleistocene and the extant and the Upper Pleistocene *C. pyrenaica* is unknown. Possibly, those ancient Iberian goats became extinct; however, that *Capra* had been present on the Iberian Peninsula continuously throughout the Pleistocene is highly probable (Couturier 1962; Montoya et al. 2001; García-González 2011), with episodes of introgressions with the neighboring taxon, *C. ibex* from the Alps (see "Genetics").

The double-wave migratory hypothesis of Crégut-Bonnoure (1992, 2006) posits that the ancestors of *C. ibex* from the Alps, of the type *C. camburgensis*, arrived in France in a first migratory wave that came from the Near East at the end of the Middle Pleistocene. In a second migratory wave, a common ancestor of *C. pyrenaica* and the complex *caucasica/cylindricornis* arrived to the Massif Central in France during the Eemian (130–115 kya). That ancestor, *C. caucasica praepyrenaica*, evolved to become *C. pyrenaica,* which spread into Southern France and reached the Pyrenees in the Magdalenian (17–12 kya) without interbreeding with *C. ibex*. Following that hypothesis, *C. pyrenaica* colonized Iberia (Cabrera 1911) through the Pyrenees after 18 kya. Recently, the validity of that theory has been questioned. *C. pyrenaica* has been present in Iberia for more than 40,000 years (Sauqué et al. 2016); there is high variability in the differential morphological characters proposed by Crégut-Bonnoure (Magniez 2009); and kinship between *C. ibex* and *C. pyrenaica* is likely and supported by molecular (Pidancier et al. 2006; Ureña et al. 2018) and morphological studies (García-González 2012).

Alternatively, the single-wave migratory theory, mainly based on studies of molecular genetics (Manceau et al. 1999; Ureña et al. 2018), posits that *C. ibex* and *C. pyrenaica* are a monophyletic clade. They should have come from a common ancestor related to *C. camburgensis* also coming from the Middle East. Nevertheless, most studies suggest the time of divergence between *C. ibex* and *C. pyrenaica* did not coincide with the age of *C. camburgensis*. Manceau et al. (1999) estimated the divergence time of the two species to be between 2.5 and 0.5 mya, which is congruent with the estimates of Lalueza-Fox et al. (2005) and Pérez et al. (2014), who estimated the divergence time to be 0.6 and 0.72 mya, respectively. Ureña et al. (2018) analyzed fossil and recent material of *C. pyrenaica* and estimated the divergence time to be only 90–50 kya.

Overall, the actual systematic position of the Iberian goat is yet to be determined; however, it seems likely that *C. pyrenaica* was already present in the Iberian Peninsula and differentiated in the Late Pleistocene. Introgression by hybridization seems to be common in the evolution of the *Capra* genus (Ropiquet and Hassanin 2006), and *C. pyrenaica* might have a polyphyletic origin through the intervention of hybridization processes with *C. ibex* and other ancient *Capra* taxa.

Current Distribution

Cabrera (1911) first reported that the species occurred in most of the mountainous areas of Iberia until the nineteenth century. High hunting pressure and landscape changes, including those related to the increase of extensive livestock farming, caused a remarkable reduction in population size and distribution range in the nineteenth century and the first half of the twentieth century, and the species only survived in some isolated populations (e.g.,

Alados 1997). The status of the species remained the same until the 1960s when the population of Sierra de Cazorla was the only one whose distribution was clearly defined (Rodríguez de la Zubía 1969; De la Cerda and De la Peña 1971). Also in the 1960s, the establishment of Game Reserves in Spain has been fundamental to understand the population recovery of wildlife in general and game species in particular (Pita Fernández et al. 2012). From the 1960s onward, the species was translocated to several locations, mainly using individuals from Sierra de Cazorla and Tortosa and Beseit Game Reserve for the subspecies *C. p. hispanica* and from Sierra de Gredos and the Batuecas Game Reserve for the subspecies *C. p. victoriae*. This significantly expanded the species' range to the current distribution (Acevedo and Cassinello 2009a; see also "Management"). More recently, several studies have assessed the species' distribution throughout its range (e.g., Alados 1997) and in particular its expansion (González et al. 2013).

The first exhaustive revision on the species distribution and status identified more than 50 populations (Pérez et al. 2002). In the 1990s, the distribution range generally increased, but with some heterogeneity among populations (Acevedo et al. 2007a; Cano-Manuel et al. 2010). Information on the species presence in Spain was compiled at the beginning of this century (Granados et al. 2002). In this period, the species returned to Portugal after an escape from Spain (Moço et al. 2006). Acevedo and Cassinello (2009a) updated the information on its complete Iberian range. Since the 1980s, the species started the recolonization of the Pyrenees (Herrero et al. 2013b) in Spain. New reinforcements on the French Pyrenees created two new populations (Herrero et al. 2020). All these populations are expanding.

Currently, the species occupies most of the mountain ranges in Eastern Iberia and the Central Mountain Range and is particularly abundant in Sierra de Guadarrama, Sierra de Gredos, Batuecas Game Reserve, and the surrounding areas. In Northern Spain, the population of the Cantabrian Mountains is well established in Riaño Game Reserve and Ancares Game Reserve, and the species has expanded across the Northern Portuguese border (Fonseca et al. 2017b). According to recent data, the current distribution of the Iberian wild goat is shown in Fig. 2 (Acevedo and Cassinello 2009a; Fonseca et al. 2017a). In most nuclei, there has been a generalized expansion of the species' range. Currently, the species occupies 926 10 ×

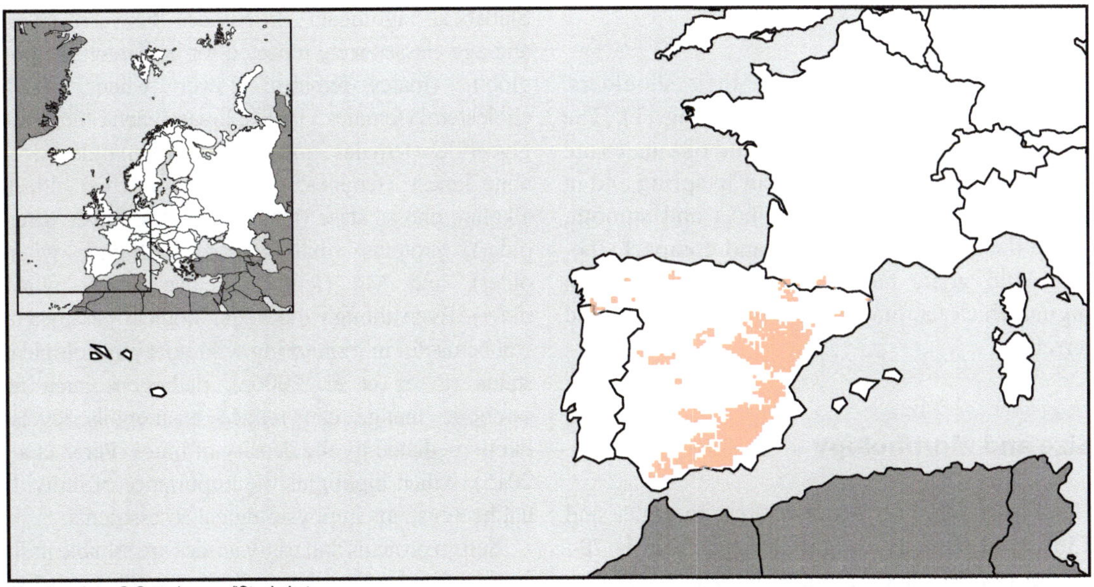

Map template: © Getty Images/iStockphoto

Fig. 2 Distribution of Iberian wild goat (Map template: © Copyright Getty Images/iStockphoto)

10 km UTM squares, of which 47.8% have been occupied since 2008. Population densities range from <1 to >30 individuals km^{-2} (Escós and Alados 1988; Escós et al. 1994; Pérez et al. 1994, 2002; Granados et al. 2001a, 2004; Torres et al. 2014; Refoyo et al. 2014; Prada et al. 2019).

The methods used to count animals and estimate population numbers include block count from trails or vantage points and distance sampling. Although a proper estimate of the total number of Iberian wild goat has not been conducted, current assessments suggest that it is well over 100,000 animals. Some examples in Spain are Sierra Nevada (15,000), Sierra de Gredos (8,000), Iberian System (more than 50,000), Serranía de Ronda and Sierra de Grazalema (4,000), Sierra de Cazorla (4,000), Sierra Tejeda y Almijara (2,500), Sierras de Antequera (2,000), and Sierra Morena (2,000); for Portugal and Spain, Peneda Geres National Park and Baixa Limia and Serra do Xurés Natural Park and surroundings (over 600 individuals); and for the Pyrenees, French Pyrenees (about 400 individuals) and Spanish Pyrenees (over 400) (Herrero et al. 2020).

Description

Pelage

Males have a black patch on their shoulders, which increases in size with age (Fig. 1). The young have brown-reddish fur and old males are brown or dark gray. Molts occur in spring and in winter. The summer coat is short and smooth, while the winter coat is long and compact. Iberian wild goats have elastic tissues between digits, which facilitate their movement in rugged terrain.

Size and Morphology

Head-body length is about 140 cm in males and 130 cm in females. Height at the withers is 70–90 cm in males and 75 cm in females (Table 1). Females grow more rapidly than males until 3–4 years of age, and males grow quickly until 5 or 6 years old and reached the maximum value at about 9 years of age. Males reach 80 kg and females 46 kg (for a review, see Fandos and Vigal 1988, and Granados et al. 2001a). Central and northern animals are larger than the southern ones (Granados et al. 1997).

Iberian wild goat exhibits strong sexual size dimorphism, with males being much larger than females (Fandos and Vigal 1993). Both sexes have horns that have a circular section and rings, with each ring representing 1 year of age. Environmental factors influence the size of the rings (Fandos 1995). Male horns reach up to 1 m in length.

Dentition

0.0.3.3 / 3.1.3.3 (Vigal and Machordom 1987).

Physiology

The physiology of Iberian wild goat is similar to that of most temperate European mountain ungulates. Pérez et al. (2003a) reported hematological and biochemical reference values for free-ranging, physically restrained, and clinically normal animals, which have provided key diagnostic information. Statistical significant differences between sexes and age classes were reported for hematocrit/hemoglobin (males>females; lower when older), cholesterol (females>males; lower when older), triglycerides (females>males; lower when older), creatine kinase (females>males; lower when older), alkaline phosphatase (males>females; lower when older), proteins (males>females; higher when older), and Mg (females>males; lower when older). Even though the hematological parameters can be useful in monitoring wild goat physiological status (Pérez et al. 2006a), pathogens such as sarcoptic mange can reduce eosinophils levels, likely mediated by the density of mites (Pérez et al. 2015), which highlights the importance of individual health status in physiological assessments.

Serum proteins and triglycerides are reliable indicators of body condition in Iberian wild goats (Serrano et al. 2008). Body condition was also studied from kidney fat stores (Serrano et al. 2011; see also

Table 1 Live body mass in kg, body size and horn size in cm from several Iberian wild goat populations. When more than two individuals were measured, the mean, sample size (in brackets) and standard deviation are given. M = males; F = females

Pyrenees	Sex	Body mass	Body length	Cross height	Horn length	Basal horn perimeter
Schinz (1838)	M		154–155	87–83	77–46	
	F		129	67		
Cabrera (1911)	M				74.6 (10) 12	25 (6) 14
Cabrera (1914)	M		148	75	86	26
	F				26.8	14
García-González et al. (unpublished)	M	(70)[a]			60.2 (18) 15.8	23.9 (15) 1.9
	F	41.5–55	138.5–133	76–83	22.3 (5) 3.7	12.6 (4) 0.9
Sierra de Gredos						
Cabrera (1911)	M				70.6 (10) 6.5	23.7 (10) 2.8
Cabrera (1914)	M		135.5	70	73.2 (3) 8.3	24.4 (3) 2.1
	F				16.5	10
Fandos and Vigal (1988)[b]	M	58.1 (23) 8.1			74.5 (4) 3.4	
	F	36.1 (17) 4.5			24.3 (12) 3.8	
García-González et al. (unpublished)	M				49.6 (24) 10	21.2 (22) 1.5
	F				19.4 (27) 5.2	10.2 (27) 0.8
Sierra de Cazorla						
Fandos and Vigal (1988)[b]	M	50.9 (45) 9			67.1 (8) 3.7	
	F	30.1 (29) 4.4			16.1 (18) 2	
Fandos (1991)	M	56	132.7	81		
	F	31	112.8	69.5		
García-González et al. (unpublished)	M				53.3 (44) 1.3	19.7 (44) 1.9
	F				15.3 (33) 2.2	8.9 (33) 0.6
Sierra Nevada						
Granados et al. (1997)	M	50.4 (123) 11.9	108.6 (100) 11.3	79.3 (100) 5.7	47.5 (137) 16.1	20.7 (138) 2.2
	F	31.3 (73) 5.2	96.9 (62) 9	69 (62) 4.3	13.9 (72) 2.7	9.7 (71) 0.83
Cabrera (1911)	M				62.9 (8) 7	21.6 (7) 1.9
Cabrera (1914)	M		121	65.5	73.3	18.9
Portugal						
França (1917)	M		142–138	73–74	36–44.5	23.5
	F		120 (3) 19.2	67 (4) 3.2	18.2 (4) 7	10.5 (3) 1.7

[a]Estimated from Fandos et al. (1989) equations
[b]Live body mass: males >5 years; females >4 years. Horn length: males >9 years; females >8 years

Santos et al. 2013). Body condition varies seasonally; fat stores are highest in the warmest months and lowest in the coldest months, although with sex- and age-specific differences (Serrano et al. 2011). Males have higher reserves than females; however, in winter, body condition decreases more in males than in females, and goatlings need one season more than young or adults to restore their reserves. In addition, population density has negative effects on fat reserves, especially in winter, possibly owing to intraspecific competition when food resources are limited (Serrano et al. 2011).

Genetics

Chromosomes

Iberian wild goat has $2n = 60$ chromosomes.

Genetic Diversity

Molecular studies in *Capra* genus have produced contradictory results that differ based on the technique used (Pidancier et al. 2006), although all agree that the kinship between Iberian wild goat and Alpine ibex is close (Manceau et al. 1999; Kazanskaya et al. 2007; Zvychaynaya 2010; Bibi et al. 2012; Hassanin et al. 2012).

Few comparative studies are available on the molecular genetics of current Iberian populations, including the Alpine ibex. An early study based on the cytochrome b gene of mitochondrial DNA (mtDNA) found that the Pyrenean wild goat differentiated clearly from the other Iberian populations, with a similar genetic distance between these (4.9%) and Alpine ibex (5.3%) (Manceau et al. 1999). In addition to the Pyrenean lineage, this study identified two evolutionarily significant units (ESU): one group that included the original populations in Northern Iberia (Sierra de Gredos, Tortosa and Beseit Game Reserve) and Muela de Cortes Game Reserve and another from the South (Sierra Nevada National Park, Sierra de Cazorla, and Ronda-Grazalema).

Ten haplotypes were identified in *Capra* populations from Southern Iberia when studying the variability of the cytochrome b gene of mtDNA, seven of which occurred in the Sierra Nevada population (Márquez et al. 2002). Manceau et al. (1999) also detected a high genetic diversity in that population, which probably did not experience severe bottlenecks. Based on Marquez et al. (2002) and unpublished data, three ESU have been identified (Pérez et al. 2002): *C. p. pyrenaica* in the Pyrenees, *C. p. victoriae* in Sierra de Gredos and surrounding areas, and *C. p. hispanica* in Southeastern Iberia which are, in principle, in agreement with the subspecies recognized by Cabrera (1914). More recent studies suggest that the molecular genetic differences between *C. p. victoriae* and *C. p. hispanica* are not consistent. Using microsatellite markers, Angelone-Alasaad et al. (2017) found that the genetic divergence between two populations (East Iberian System and Sierra Nevada) of the same subspecies, *C. p. hispanica,* was greater (31.5%) than the divergence with a population (Sierra de Gredos, 24.8%) belonging to another subspecies, *C. p. victoriae.* Using ancient mtDNA techniques and next-generation sequencing technologies, three major clades of *Capra* were identified in Western Europe (Ureña et al. 2018): *C. ibex*, *C. p. pyrenaica*, and the group comprising the subspecies *hispanica* and *victoriae.* This genetic structure indicates the distinctiveness of the Pyrenean wild goat from the other Iberian wild goats and suggests that this group is an ESU.

Hybridization

Hybridization between domestic goat *C. hircus* and *C. pyrenaica* in captivity or artificial conditions is well-known (Fernández-Arias et al. 1999; Alasaad et al. 2012). Under natural conditions, there are well-founded suspicions of its occurrence, although it is not frequent, possibly because of physiological and behavioral barriers. Placental incompatibility, immunological rejection, and differences in gestation length (162 ± 2 days in Iberian wild goat and 150 days in domestic goat) are potential limiting factors (Fernández-Arias et al. 2001).

Recently, in several isolated populations of Iberian wild goat in Southern Iberia, Angelone et al. (2018) found an MHC allele characteristic of the domestic goat, which suggests past hybridization. Moreover, Cardoso et al. (2021) have detected eight Iberian wild goats from Tortosa-Beceite population with signs of domestic goat introgression. Given numerous official and not-official translocations among different regions, introgression between subspecies of *C. pyrenaica* might have occurred (Manceau et al. 1999; Acevedo and Cassinello 2009a).

Life History

Growth

Average body mass at birth is 2.2 kg (Fandos 1989). Asymptotic body mass is reached at 3–4 years of age in females and at *c.* 9 years of age in males, leading to strong sexual size dimorphism (cf. Description). Compensatory horn growth does not occur in male Iberian wild goats (Carvalho et al. 2017). A reduction in horn length was documented, probably caused by harvest practice (Pérez et al. 2011), and the synergistic effect of harvest practice and habitat changes (Carvalho et al. 2020).

Reproduction

Typically, the number of goatlings per female is one, but twins are frequent (Rodríguez de La Zubía 1969; Fandos 1989). After parturition, adult females tend to stay alone in sheltered areas, protected from predators, mainly golden eagle *Aquila chrysaetos*. After that, mothers gather and care for their kids while sharing vigilance.

Females typically reach sexual maturity at 30 months of age, when they weigh at least 24 kg (Fandos 1989, 1991), but sexual maturity can occasionally occur earlier (Granados et al. 2001a). In captivity, they reach sexual maturity at *c.* 12–16 months (Fernández-Arias et al. 1997). Gestation lasts 23–24 weeks (Fandos 1991). Rut peaks at the beginning of December, and parturition starts in the second half of April and peaks in the second half of May in Southern Iberia (Fandos 1988b, 1989; Alados and Escós 1988, 1996).

Information on male breeding success is not available.

Survival

Sex ratio at birth can be biased toward males both in the wild (Escós 1988) and in captivity (Fernández-Arias et al. 1997). Mortality rate is high among young animals up to 3 years of age, stable until 8 years of age in males and 10–11 years of age in females, and high in both sexes thereafter (Fandos 1991). Males and females can reach 14 and 19–22 years of age, respectively (Fandos 1991).

Habitat and Diet

Habitat Use and Selection

At the species range scale, the Iberian wild goat is mostly associated with forested areas that have high levels of insolation and outcrops (Acevedo and Real 2011). Those results are consistent with studies performed with a biogeographical perspective at regional scales (Acevedo et al. 2007a, b; Acevedo and Cassinello 2009b) and also in semiarid environments (Lucas et al. 2016).

At the population scale, the Iberian wild goat uses forested landscapes on high-slope areas (Acevedo et al. 2011a). Males occur at higher elevations than females, probably because of the sexual differences in nutritional requirements (see below; Gonzalez 1982; Escós and Alados 1992b). In autumn, individuals select sunny slopes at dusk and dawn (in spring, at dusk only); in summer, however, they prefer northern slopes (Escós and Alados 1992b). In spring and summer, wild goats in the Sierra Nevada National Park occupy northeast- and northwest-facing slopes, while in autumn and winter they dwell on slopes facing south or southeast. In the Sierra de Cazorla population, due to milder climatic conditions, there are no significant differences in the seasonal variation of selected exposures (Escós and Alados 1992b). Climate severity influences seasonal patterns of habitat use within the species' distribution range; extreme weather conditions occur both in summer and in winter. In addition to topography, food quality is an important factor influencing the seasonal patterns of the species' habitat use. For instance, partial vertical migrations are frequent in summer, owing to higher nutritional quality of the vegetation at high elevation pastures (Gonzalez 1982; Escós 1988; Fandos and Martínez 1988). Furthermore, seasonal variations in dietary quality (Acevedo et al. 2011a) and body condition (Carvalho et al. 2015) suggested that primary productivity influences the species' habitat use.

Sex, habitat availability, and season influence the home range size of Iberian wild goat at the individual scale. Females have smaller home ranges than males, and ranges are larger in spring than in autumn (e.g., Viana et al. 2018). In Sierra de Cazorla, the average home range for females was 0.81 km^2 in spring and 0.25 km^2 in autumn, whereas in males the home range was 4.28 km^2 in spring and 1.05 km^2 in autumn (Escós and Alados 1992a). In addition, habitat suitability and population density negatively correlates with home range (Escós and Alados 1988; Viana et al. 2018). All animals exhibited rut area fidelity, but females only showed fidelity to the area used in spring. During the rut, males can show dispersal movements of over 7 km (Escós and Alados 1992a).

Diet

Members of the *Capra* genus classify as browsers. In *C. pyrenaica* several studies showed high browse proportions in their diet, 61% and 72% in Sierra de Cazorla (Martínez et al. 1985; Cuartas et al. 1996, respectively), 71% in Tortosa and Beseit Game Reserve (Martínez 1994), and 83% in Sierra Tejeda (Martínez 1988). Holm oak leaves *Quercus rotundifolia* and *Q. ilex* were one of the main browse species (García-González and Cuartas 1992a). Food availability, however, is one of the main factors that influence diet selection (Ellis et al. 1976), and, in open habitats or alpine pastures, herbaceous vegetation predominates in the diet of Iberian wild goats: 57.5% grasses in Sierra Nevada (Martínez 2002) and 80% grasses in Sierra de Gredos (Martínez and Martinez 1987). Therefore, the species can be more precisely classified as an intermediate feeder based on Hofmann's (1989) feeding type classification or more accurately as an intermediate feeder with preference for browsing based on Van Soest (1994).

Moço et al. (2013) described different feeding behaviors at two neighboring areas in Northern Portugal (grazer in Gerês but browser in Amarela). Wild goats from Sierra de Gredos had greater tooth wear of those in Sierra de Cazorla, possibly due to the increased consumption of grasses, richer in silica content (Fandos et al. 1993).

As an intermediate feeder, the Iberian wild goat exhibits high forage selectivity and tries to avoid consuming fibrous forage (Hofmann 1989). Foraging is opportunistic with pronounced seasonal differences in diet, reflecting changes in forage quality. For instance, in Sierra de Cazorla, the consumption of woody species is higher in September than in February and in May, as the end of summer is a dry and poor resource season in Mediterranean ecosystems. The highest proportion of forbs in the diet is in May, and consumption of grasses is similar in May and February. Number of plant species, diet diversity, and dietary quality are higher in May than in September or February (Cuartas and García-González 1992). In forest habitats, goats can cause serious damage to preferred plants if population density is high (Perea et al. 2015).

Intrinsic factors such as sex-age class can also affect diet selection. In Sierra de Cazorla, adult males ate more browse (77%) than herbs (23%), the young ate more herbs (67%) than browse (33%), and adult females ate similar proportions of both (52% vs 48%) (Alados and Escós 1987).

Dietary Overlap with Other Ungulates

In an area in Sierra de Cazorla used simultaneously by Iberian wild goat, red deer *Cervus elaphus,* and fallow deer *Dama dama*, Iberian wild goat and red deer had similar diets and a high overlap index (74%), which increased in winter when food resources were reduced (García-González and Cuartas 1992b). Fallow deer had a different diet, which included a higher proportion of grasses and rather low overlap indexes with wild goat (34%) and red deer (42%). The coexistence of the three intermediate feeders might have been based either on a divergence of habitat (red deer) or on a deviation of the trophic niche (fallow deer).

Some studies have investigated the diet of Iberian wild goat and domestic goats that shared the same habitat. In Sierra de Cazorla, domestic goats had an almost exclusively woody diet, but the Iberian wild goat exhibited a more eclectic diet, which reflected available vegetation (Cuartas and García-González 1992). In the Pre-Pyrenees, feral goats showed highly selective feeding behavior, which was considered "a feature that distinguished them from their domestic counterparts and approach them to wild goats" (Aldezabal and Garin 2000).

Behavior

Social Behavior

Males and females segregate most of the year, except during the rutting season (Granados et al. 2001a). Sexual segregation reflects differences in sexual size dimorphism and thus sex-specific ecological requirements that lead to different habitat use and selection at local scales (Alados 1985). This pattern, however, appears somewhat simplistic because mixed groups can occur throughout the year in some populations, except in August (e.g., Alados and Escós 1996). Several hypotheses have been proposed to explain the differences between populations (see Acevedo and Cassinello 2009a), but no empirical studies have been carried out so far to disentangle the processes underlying the differences in segregation patterns.

As in many other ungulates, the mother-kid pair is the basic unit of organization (Alados and Escós 1996). However, as a gregarious species, other forms of social units occur, including groups of females with kids, groups of adult males, and groups of yearlings (males or females) (Alados 1985). In the Southern Iberian populations, mixed groups (adult or young) are common. Granados et al. (2001a) found that 44.5% of the individuals observed in Andalusian populations formed mixed groups, 22.4% were groups of females with kids, and 20.2% were all-male groups, although the proportions vary seasonally. Density affects positively population proportion of mixed groups. In Sierra de Cazorla, a high-density population, the proportion of individuals in mixed groups was larger, greater than 40% throughout the year, and was highest (up to 80%) during the rut. However, in Sierra Nevada National Park, a low-density population, mixed groups were only common during the rut (Alados 1985; Alados and Escós 1996).

Granados et al. (2001a) reported that the average group size in Andalusian populations was 5.2 individuals, larger than the average sizes of all-male groups (4.9), female groups (3.2), and female with kid groups (4.2) but smaller than the average size of mixed groups (6.9). In addition, population density positively influences group size. Outside the rut, Alados and Escós (1996) reported an average group size of 3.7 individuals in Sierra Nevada National Park and 3.0 in Sierra de Cazorla, which are intermediate between the group sizes to those reported by Nievergelt (1974) for the Alpine ibex (7.6) and the walia ibex *C. walie* (1.5). Open habitats favor large groups, while closed habitats favor small groups (Alados 1985, 1986b).

Activity

The time spent in different activities (eating, moving, and resting) varies daily and seasonally. In winter, activities that involve movements occur around midday hours (Alados 1986c). In summer, goats are most active during night hours and at dawn and dusk. In winter, displacement movement occurs throughout the day; in summer, however, they move at night (Fandos 1988a). Males spend less time eating than females, and males interact more with males than with females, especially during rut. The foraging behavior of subadult males is more similar to that of females than to that of adult males (Alados 1986c).

Mating Behavior

The Iberian wild goat is a polygynous species. Males establish a hierarchy during rut, which dictates access to estrus females (Alados 1986a; Álvarez 1990). Alados (1984, 1986a) and Fandos (1991) described Iberian wild goat rutting behavior. Resources allocated to reproduction in polygenic species are biased toward females because males only invest significantly in reproduction in those years in which the social hierarchy should be established (Granados et al. 2001a).

Parasites and Diseases

Wild goat can suffer many pathologies, but sarcoptic mange appears to be the only pathology that can cause severe mortality.

Ectoparasites

Sarcoptic mange caused by *Sarcoptes scabiei* is the main epizootic disease that affects Iberian wild goat population dynamics (Acevedo and Cassinello 2009a). Several outbreaks of sarcoptic mange have occurred, with different population crashes. The first *foci* were in the late 1980s and 1990s in the Sierra de Cazorla, where a population reduction of 81% occurred (Fandos 1991; León-

Vizcaíno et al. 1999), Sierra Nevada National Park (Pérez et al. 1997), and Sierra Mágina (Palomares and Ruiz Martínez 1993). At the beginning of the twentieth century, mange affected Iberian wild goat in Muela de Cortes Game Reserve (Sánchez-Isarria et al. 2008). In 2014, an outbreak occurred in Tortosa and Beseit Game Reserve (Mentaberre et al. 2015).

Sporadically, mild infestations by *Demodex* sp. have occurred in Northeastern Iberia (Revilla 2012). Several ticks affect the species: *Dermacentor marginatus*, *D. reticulatus*, *Haemaphysalis sulcata*, *H. punctata*, *Hyalomma marginatum marginatum*, *H. lusitanicum*, *Ixodes ricinus*, *I. ventalloi*, *Rhipicephalus bursa*, and *R. sanguineus* (Hueli and Díaz 1989; Antón et al. 2002; García-Moreno et al. 2009; Varela-Castro et al. 2018). Less frequent and lesser-known ectoparasites are lice such as *Bovicola crassipes*, *B. alpine*, and *Linognathus stenopsis* (Antón et al. 2002).

Endoparasites

Parasites found in the respiratory system are *Oestrus caucasicus* larvae, known as nasal myiasis (Pérez et al. 1996; Antón et al. 2002), as well as *Cystocaulus ocreatus*, *Dictyocaulus filaria*, *Neostrongylus* sp., *Muellerius capillaris*, and *Protostrongylus* sp. that colonize the bronchi and lungs (Antón et al. 2002; Alasaad et al. 2009).

Digestive abomasum helminths are *Haemonchus contortus*, *Marshallagia marshalli*, *M. occidentalis*, *Ostertagia lyrata*, *O. ostertagi*, *Teladorsagia circumcincta*, *T. davtiani*, and *T. trifurcata*. In the small intestine, *Nematodirus davtiani*, *N. abnormalis*, *N. filicollis*, *N. oiratianus*, *N. spathiger*, *Trichostrongylus axei*, *T. capricola*, *T. colubriformis*, and *T. vitrinus* (Rossi et al. 1992; Lavín et al. 1997; Pérez et al. 2003b, 2006b; Revilla 2012) occur. In addition, in the large intestine *Chabertia ovina*, *Oesophagostomum venulosum*, *Skrjabinema* sp., and *Trichuris ovis* occur (Rossi et al. 1992; García-Moreno et al. 2009; Revilla 2012).

Described coccidious are *Eimeria arloingi*, *E. aspheronica*, *E. bakuensis*, *E. capraovina*, *E. caprina*, *E. christenseni*, *E. faurei*, *E. jolchijevi*, *E. gonzalezi*, *E. hirci*, *E. ninakohlykimovae*, *E. ovina*, and *E. parva* (Cordero del Campillo et al. 1994; Antón et al. 2002; García-Moreno et al. 2009).

Larval form of cestodosis as hydatid cyst (*Echinococcus granulosus*), *Coenurus cerebralis* (*Taenia multiceps*), and *Cysticercus tenuicollis* (*Taenia hydatigena*) has been detected in the Iberian wild goat (Antón et al. 2002; Oleaga-Pérez et al. 2003; Pérez et al. 2006b; Revilla et al. 2007). Other cestodes include *Moniezia expansa* and *M. benedeni* (Antón et al. 2002).

Fasciola hepatica and *Dicrocoelium dendriticum* exhibit low egg production in the feces of Iberian wild goat (Antón et al. 2002; Alasaad et al. 2008; Refoyo et al. 2016).

Various serological studies have shown antibodies against *Sarcocystis* sp. (Granados et al. 2001b; Antón et al. 2002; Luzón et al. 2008; Santiago-Moreno et al. 2010), *Babesia* sp. (Ferrer et al. 1998; García-Moreno et al. 2009), *Toxoplasma gondii*, and *Neospora caninum* (Gauss et al. 2006; Almería et al. 2007; García-Bocanegra et al. 2012). There was no detection of seroprevalence against *Besnoitia* sp. in Iberian wild goat (Gutiérrez-Expósito et al. 2016).

Infectious Diseases

Brucellosis has low seroprevalence: 0.5% (Muñoz et al. 2010), 0.9% (Antón et al. 2002), 3.6% (Carvajal et al. 2007), and 6% (León-Vizcaíno et al. 1994). Identification of *B. melitensis* occurred in three populations (Cubero et al. 2002; León-Vizcaíno et al. 2009; Muñoz et al. 2010).

Tuberculosis is a disease under study (Fernández de Luco and Arnal 2002; Mentaberre et al. 2010; Revilla 2012), and only one adult female with the disease has been detected (Cubero et al. 2002).

The detection of salmonellosis by *S. enterica* serovar Enteritidis occurred in one clinical case with septicemia (Navarro-González et al. 2014). Other subspecies have been isolated from asymptomatic animals: *S. enterica* serovar Enteritidis

(Navarro-González et al. 2014), *S. arizonae* (González-Candela et al. 2006), *S. typhimurium*, and *S. paratyphi* C (Cubero et al. 2002).

Infectious keratoconjunctivitis caused by *Mycoplasma conjunctivae* is less severe among free-ranging animals (Cubero et al. 2002; Arnal et al. 2009; Revilla 2012) than it is among captive animals, which can be affected by sever outbreaks (Fernández-Aguilar et al. 2017).

Serological studies have been conducted to investigate bacterial infections such as contagious agalactia (León-Vizcaíno et al. 1994; Cubero et al. 2002), Q fever, and borreliosis (Santiago-Moreno et al. 2011; Candela et al. 2017).

Sporadically, isolated bacteria have been isolated: *Arcanobacterium pyogenes*, *Corynebacterium pseudotuberculosis, E. coli, E. coli* O157: H7, *Erysipelothrix rhusiopathiae*, *Fusobacterium necrophorum*, *Klebsiella* sp., *Mannheimia haemolytica*, *Mycobacterium avium paratuberculosis*, *Pasteurella multocida*, *Pseudomonas* sp., *Staphylococcus* spp., *Streptococcus* sp., and *Streptococcus* β-hemolytic (Cubero et al. 2002; Revilla et al. 2007; Revilla 2012; Colom-Cadena et al. 2014; Navarro-González et al. 2015; Arnal et al. 2016; Pizzato et al. 2017; Varela-Castro et al. 2017).

Viral diseases are present but their impact on wild populations is unknown. Some examples are as follows: contagious ecthyma (Cubero et al. 2002; Revilla 2012; Camacho et al. 2017), bluetongue (García et al. 2009; Lorca-Oró et al. 2011; Santiago-Moreno et al. 2011; Revilla 2012), small ruminant lentivirus (Carvajal et al. 2007; García-Moreno et al. 2009; Martín-Atance and León-Vizcaíno 2009; Santiago-Moreno et al. 2011; Revilla 2012), and pestivirus (Astorga-Márquez et al. 2014, Fernández-Sirera et al. 2011).

Neoplasia

Several neoplasms have been observed in Iberian wild goat: KIT-positive gastrointestinal stromal tumors (Velarde et al. 2008), pheochromocytoma, cutaneous horn, intestinal leiomyoma, thyroid carcinoma tumor (Arnal et al. 2006), and disseminated melanoma (Arnal and Fernández de Luco 2017).

Population Ecology

Population Dynamics

Multiple factors can regulate the dynamics of Iberian wild goat populations. Populations typically show a density-dependent growth pattern, where density mainly affects reproduction and goatling survival but does not have a significant effect on adult survival and fecundity (Escós et al. 1994; Escós and Alados 1998), at least in the range of densities considered in these studies. Escós and Alados (1988) reported an adult survival rate of 0.87, a breeding success (i.e., kids-to-adult females ratio in summer) of 0.27, and an annual growth rate of 0.98 for the Iberian wild goat population in Cazorla Game Reserve (11 goats km^{-2}). Similarly, Escós et al. (1994) reported an adult survival rate of >0.92, a young survival of 0.49, a breeding success of 0.38, and an annual growth rate of 1.05 in the Sierra Nevada National Park (2 goats km^{-2}).

In general, mortality due to predation appears to be low: predation by golden eagle and red fox *Vulpes vulpes* has been reported only occasionally (Fandos 1991). Large predators such as wolf *Canis lupus* are absent in most of the species' range. Mortality caused by legal and illegal hunting is the main limiting factor, at least in populations not affected by diseases. Typically, game hunting focuses on specific age and sex classes (old males are preferred) which can alter the population structure. The selective removal of large-horned animals might contribute to a decrease in horn size in the population (Pérez et al. 2011). Hunting bags have increased significantly in the last decade, and, currently, the number of Iberian wild goats that are hunted each year is over 11,000 (Garrido et al. 2019); however, detailed information on the regulatory effects of hunting on Iberian wild goat population dynamics is missing.

Over the recent years, most populations showed an increasing trend. New populations have been established thanks to translocations (Moço et al. 2006; Refoyo et al. 2014), but most of the growth, both demographic and geographic, has occurred through natural expansions (Lucas et al. 2016). Other populations have merged, creating a very dynamic situation of subpopulations as part of metapopulations. Sarcoptic mange has a high demographic impact, with sporadic outbreaks with mortality rates around 20% and extreme outbreaks with mortality rates up to 81%, as in Cazorla at the end of the 1980s (Fandos 1991; León-Vizcaíno et al. 1999).

Intra- and Interspecific Competition

A mechanism to avoid competition for resources is asynchrony between age-class groups in the daily activity patterns of Iberian wild goat (Alados and Escós 1987). Interspecific relationships with phytophagous insects, better described as amensalism or predation rather than as pure competition, reduce insect abundance (Zamora and Gómez 1993; Gómez and González-Megías 2002). Besides, there is overlap with the (macro) ecological requirements of native, but reintroduced, red deer (Acevedo and Cassinello 2009b), exotic aoudad *Ammotragus lervia* (Acevedo et al. 2007b), and domestic goats (Acevedo et al. 2007a), showing a high potential of these species to interact with wild goat. One Pyrenean population lives in sympatry with feral goats (>900 individuals) (Herrero et al. 2013a, b), an anomalous situation that should be monitored closely and managed to prevent competition and hybridization.

Global Change

Species distribution modeling suggests that species environmental suitability would increase in the near future (Real et al. 2013). The effects of changes in land use (mainly related with renaturalization processes, i.e., from cultures to woodlands) on the species' distribution might increase the species' range and abundance in Andalusia (Acevedo et al. 2011b). At a local scale, demographic data indicated increases in the species' range and population size in recent decades, which is expected to continue in the future years (Cano-Manuel et al. 2010, González et al. 2013; Gortázar et al. 2000). Such an increase can be also attributed, at least partly, to changes in land use. In general, the species' population size and range has expanded and the process continues nowadays (Lucas et al. 2016). Currently, the densities of some populations suggest the need for population control or hunting, rather than restocking, because negative effects of population overabundance have been reported (Perea et al. 2015). In the future, however, stochastic events and prolonged droughts might have significant effects on population dynamics, mainly due to reduction in availability of food resources, emergence of new pathogens, alterations in phenology, etc. (Cano-Manuel et al. 2010).

Conservation Status

Iberian wild goat is listed in Appendix III of the Bern Convention and in Annex V of the EU Habitats and Species Directive. *C. p. victoriae* occurs in the Central Mountain Range, Cantabrian Mountains, and the Pyrenees. *C. p. hispanica* occurs in Southern and Eastern Iberia. The species occurs in a number of protected areas; however, most of the species range is outside of protected areas.

The IUCN lists Iberian wild goat as Least Concern (LC) (Herrero et al. 2020). The species is abundant, and its range and population have expanded because of numerous important socioeconomic changes such as law enforcement, rural abandonment, and translocations (reintroductions, conservation introductions, reinforcements). Game Reserves and protected areas have played a crucial role in favoring the species' recovery.

Management

Early Conservation and Outbreaks

Management has influenced the contemporary history of Iberian wild goat, which began with a conservation program in 1905 to preserve the remaining populations (Alados 1997). Hunting regulations and changes in land use have helped improving the conservation status of the species in the last decades of the twentieth century (e.g., Fandos et al. 2010; Acevedo et al. 2011b). At the end of the twentieth century, some mange outbreaks occurred (see "Population Dynamics" section) and regions implemented management programs for the species. The aim was to preserve some isolated populations, and perception of sarcoptic mange as the main threat promoted some programs aimed at improving the control of this disease at the population level. For instance, in Andalusia, a regional strategy for the conservation of the species includes a monitoring program (distribution and abundance) and a reference regional field station of Iberian wild goat. The latter consists in an enclosure for maintaining a small population, intensively monitored at the population and health levels. Habitat management, mainly consisting in the creation of pastures and in the elimination of shrubs, occurs in some of those enclosures and public Game Reserves. Animal translocations are an extensively used management practice, either to reintroduce the species in areas where it had been extirpated or, lately, to remove animals in order to control population size and minimize negative effects of overabundance.

Translocations and Escapes

Since the nineteenth century, Iberian wild goat has been translocated to formerly occupied areas as well as to new areas (e.g., Pérez et al. 2002; Acevedo and Cassinello 2009a; Prada and Herrero 2013; Cardoso et al. 2021; see also Fig. 2). It is remarkable that a limited number of populations have acted as sources for the reintroduction programs (see also "Current Distribution" section). In the twentieth century, unintentional escapes from enclosures (Moço et al. 2006; Herrero et al. 2013b) or the deliberate creation of new populations (Peral 1993; Refoyo et al. 2014) have occurred. Administrations and hunters promoted translocations mainly for hunting purposes. The genetic characteristics of most of the populations are not well-known (but see Manceau et al. 1999), and understanding the impact of these translocations is important for assessing the genetic status of the species (Cardoso et al. 2021). For instance, all current *C. p. victoriae* populations come from individuals from the Sierra de Gredos population, which experienced a strong bottleneck at the end of the twentieth century (Acevedo and Cassinello 2009a; Prada and Herrero 2013). Extreme population declines might increase the loss of genetic variability, which has already occurred in some source populations.

Trophy Hunting

The history of the conservation and management of the Iberian wild goat, an exclusively Spanish hunting trophy for more than a century, has been largely affected by trophy hunting during the twentieth century. For instance, trophy hunting stopped the efforts to recover the Pyrenean subspecies, which once occupied the entire Pyrenees (García-González and Herrero 1999). The fact of considering *C. p. pyrenaica* a subspecies rather than a species (as it was defined, originally, in the nineteenth century) also reduced the recovery efforts (Gippoliti et al. 2018). In many rural areas in Spain, Iberian wild goat is an important public and private economic resource, and Spain is the only country that allows hunting, while the species is protected in France and Portugal. In Spain, Iberian wild goat management is mainly based on trophy hunt, although culling has become more common in some populations.

Overabundance

Damage control is one of the key aspects of Iberian wild goat management, even if scientific information is not available and technical information is limited. Given the recent numerical increase of goat populations, management strategies should be adapted to a very different scenario in which the species has become overabundant, with subsequent damages to human activities (e.g., agriculture, natural environments, car accidents). Perea et al. (2015) examined woody vegetation to assess the ecological sustainability 25 years after the species reintroduction to the Sierra de Guadarrama. Almost one quarter of the woody species exhibited unsustainable levels of browsing, with a low level of natural regeneration; a 50% reduction in current population density (i.e., 47 goats km^{-2}) is needed to increase the probability of successful regeneration to nearly 60%. In fragmented habitats that have agricultural land near shelter areas (Lucas et al. 2016), Iberian wild goats tend to feed mainly on cereal crops, almond trees *Prunus dulcis*, and olive trees *Olea europaea*. Such conflicts influence hunting quotas and thus hunting pressure in those areas. Population trends in most nuclei suggest that quotas are insufficient to regulate populations, and, consequently, an increase of damages caused by wild goat is expected (e.g., Escós et al. 2008; Marco et al. 2011). However, culling is currently undertaken to decrease population densities and avoid the undesirable effects of overabundant populations (e.g., Refoyo et al. 2014).

Future Challenges for Research and Management

The Iberian wild goat has experienced an incredible expansion in range and abundance over the last 30 years, which has changed its conservation status and, accordingly, conservation and management needs. Consequently, the research, conservation, and management priorities should be as follows:

- Characterize the genetics of all populations; this will help sound decision-making aimed at increasing genetic diversity and/or at prioritizing local adaptations. Future studies should be based on large sample sizes from all populations and should be combined with complete genome sequencing: this will provide a much clearer understanding of the taxonomy and phylogeography of Iberian wild goat.
- A comprehensive revision of the species' taxonomic status that integrates genetic, paleontological, and morphological data. This information will be needed to elucidate the taxonomic status of *C. pyrenaica*.
- Monitor all populations: assessing demography, health, damages (forest, agriculture and car crashes), and biometrics. Special attention should be given to sarcoptic mange, avoiding selective culling of affected animals, which has proved to be expensive and ineffective (Meneguz et al. 1996). Coordinated large-scale monitoring should lead to data-based population management that can meet multiple objectives: increase, decrease, or stabilize populations. Adapt monitoring methods to suit specific habitats and population characteristics.
- Promote autochthonous wild mountain ungulates and prevent the expansion of introduced species with potential to interact with Iberian wild goat, such as mouflon and aoudad.
- Surveys of domestic ungulate health status is a priority because domestic goats probably are the origin of several mange and other disease outbreaks (Cassinello and Acevedo 2007; Astorga-Márquez et al. 2014).
- Eradicate feral goats to avoid issues of hybridization with Iberian wild goat (Moço et al. 2014; Herrero et al. 2013a).
- Studies of the basic biology of the species are lacking for most populations. Dispersal, home range, etc. remain poorly known.
- Evaluate the need or opportunity for further reintroductions and reinforcements in small populations.

Acknowledgments We would like to thank Luca Corlatti, who reviewed the draft of this text.

References

Acevedo P, Cassinello J (2009a) Biology, ecology and status of Iberian ibex *Capra pyrenaica*: a critical review and research prospectus. Mammal Rev 39:17–32

Acevedo P, Cassinello J (2009b) Human-induced range expansion of wild ungulates causes niche overlap between previously allopatric species: red deer and Iberian ibex in mountainous regions of southern Spain. Ann Zool Fenn 46:39–50

Acevedo P, Real R (2011) Biogeographical differences between the two *Capra pyrenaica* subspecies, *C. p. victoriae* and *C. p. hispanica*, inhabiting the Iberian Peninsula: implications for conservation. Ecol Model 222:814–823

Acevedo P, Cassinello J, Gortázar C (2007a) The Iberian ibex is under an expansion trend but displaced to suboptimal habitats by the presence of extensive goat livestock in Central Spain. Biodivers Conserv 16:3361–3376

Acevedo P, Cassinello J, Hortal J, Gortázar C (2007b) Invasive exotic aoudad (*Ammotragus lervia*) as a major threat to native Iberian ibex (*Capra pyrenaica*):

a habitat suitability model approach. Divers Distrib 13:587–597

Acevedo P, Farfán MA, Márquez AL, Delibes-Mateos M, Real R, Vargas JM (2011a) Past, present and future of wild ungulates in relation to changes in land use. Landsc Ecol 26:19–31

Acevedo P, Santos JPV, Real R, Gortázar C (2011b) Evaluación del estado de la población de cabra montés de los Montes de Toledo: relaciones con el ciervo. Pirineos 166:9–27

Alados CL (1984) Etograma de la cabra montés (*Capra pyrenaica*) y comparación con otras especies. Doñana, Acta Vertebrata 11(2): 289–309.

Alados CL (1985) Group size and composition of the Spanish ibex (*Capra pyrenaica* Schinz) in the sierras of Cazorla and Segura. In: Lovari S (ed) The biology and Management of Mountain Ungulates. Croom Helm, London/Sydney/Dover/New Hampshire, pp 134–147

Alados CL (1986a) Aggressive behaviour, sexual strategies and their relation to age in male Spanish ibex (*Capra pyrenaica*). Behav Process 12:145–158

Alados CL (1986b) Spatial structure in groups of Spanish ibex (*Capra pyrenaica*). Biol Behav 11:167–185

Alados CL (1986c) Time distribution of activities in the Spanish ibex, *Capra pyrenaica*. Biol Behav 11:70–82

Alados CL (1997) Status and distribution of Caprinae by region. Spain. In: Schackleton DM (ed) Wild sheep and goats and their relatives. IUCN, Cambridge, pp 125–130

Alados CL, Escós J (1987) Relationship between movement rate agonistic displacements and forage availability in Spanish ibexes (*Capra pyrenaica*). Biol Behav 12:245–255

Alados CL, Escós J (1988) Parturition dates and mother-kid behavior in Spanish ibex (*Capra pyrenaica*) in Spain. J Mammal 69(1):172–175

Alados CL, Escós J (1996) Ecología y Comportamiento de La Cabra Montés y Consideraciones para su Gestión. CSIC, Madrid

Alasaad S, Granados JE, Cano-Manuel FJ, Meana A, Zhu ZQ, Pérez JM (2008) Epidemiology of fasciolosis affecting Iberian ibex (*Capra pyrenaica*) in southern Spain. Parasitol Res 102:751–755

Alasaad S, Morrondo P, Dacal-Rivas V, Soriguer RC, Granados JE, Serrano E, Zhu XQ, Rossi L, Pérez JM (2009) Bronchopulmonary nematode infection of *Capra pyrenaica* in the Sierra Nevada massif, Spain. Vet Parasitol 164:340–343

Alasaad S, Biebach I, Grossen C, Soriguer RC, Pérez JM, Keller LF (2012) Microsatellite-based genotyping of MHC class II DRB1 gene in Iberian and Alpine ibex. Eur J Wildl Res 58:743–748

Aldezabal A, Garin I (2000) Browsing preference of feral goats (*Capra hircus* L.) in a Mediterranean mountain scrubland. J Arid Environ 44:133–142

Almería S, Vidal D, Ferrer D, Pabón M, Fernández-de-Mera MI, Ruiz-Fons F, Alzaga V, Marco I, Calvete C, Lavín S, Gortázar C, López-Gatius F, Dubey JP (2007) Seroprevalence of *Neospora caninum* in non-carnivorous wildlife from Spain. Vet Parasitol 143:21–28

Álvarez F (1990) Horns and fighting in male Spanish ibex, *Capra pyrenaica*. J Mammal 71(4):608–616

Angelone S, Jowers MJ, Molinar Min AR, Fandos P, Prieto P, Pasquetti M, Cano-Manuel FJ, Mentaberre G, Olvera JRL, An R-B, Espinosa J, Pérez JM, Soriguer RC, Rossi L, Granados JE (2018) Hidden MHC genetic diversity in the Iberian ibex (*Capra pyrenaica*). BMC Genet 19:28

Angelone-Alasaad S, Biebach I, Pérez JM, Soriguer RC, Granados JE (2017) Molecular analyses reveal unexpected genetic structure in Iberian ibex populations. PLoS One 12(81):e0170827

Antón JM, Boticario D, Granados JE, Habela MA, Márquez FJ, Meana A, Meneguz PG, Peña J, Pérez J, Pérez JM, Pérez MC, Rossi L, Ruiz-Martínez I, Sangenis J (2002) Estudio de las enfermedades parasitarias de la cabra montés. In: Pérez JM (ed) Distribución, genética y estatus sanitario de las poblaciones andaluzas de cabra montés. Universidad de Jaén-Consejería de Medio Ambiente de la Junta de Andalucía, Jaén, pp 117–153

Arnal MC, Fernández de Luco D (2017) Melanoma diseminado y feocromocitoma en un macho montés *Capra pyrenaica hispanica*. 35èmes Rencontres du Groupe d'Étude sur l'Eco-pathologie de la Faune Sauvage de Montagne (GEEFSM). Cofrentes, Valencia, 1–4 June 2017, Abstracts

Arnal MC, Fernández de Luco D, Ferroglio E, Formisano F, Gortázar C, Höfle U, Martín MP, Meneguz PG, Revilla M, Rossi L, Vicente J (2006) Neoplasias en rumiantes silvestres. 24émes Rencontres du Groupe d'Étude sur l'Eco-pathologie de la Faune Sauvage de Montagne (GEEFSM). Roquetes, Tarragona, 9–11 June 2006, Abstracts

Arnal MC, Revilla M, Corrales JC, Martínez D, Sánchez A, Herrero J, Contreras A, Fernández de Luco D (2009) Queratoconjuntivitis infecciosa por *M. conjunctivae* en sarrio (*Rupicapra p. pyrenaica*) y en cabra montés (*Capra pyrenaica*) en Aragón. XXI Reunión de la Sociedad Española de Anatomía Patológica Veterinaria, Lugo, 24–26 June 2009, Abstracts, p 61

Arnal MC, Martínez-Durán D, Revilla M, Pérez V, Garrido J, Fernández de Luco D (2016) Paratuberculosis y queratoconjuntivitis en un macho montés *Capra pyrenaica* en Teruel (Aragón). 34èmes Rencontres du Groupe d'Étude sur l'Eco-pathologie de la Faune Sauvage de Montagne (GEEFSM). Pont de Camps. Laruns, Valleéd'Ossau. Béarn, 29 September–2 October 2016, Abstracts, p 14

Arribas A, Garrido G (2008) Los representantes más antiguos del género *Capra* (Bovidae, Artiodactyla, Mammalia) en el registro euroasiático (Fonelas P-1, Cuenca de Guadix, España). In: Arribas A (ed) Vertebrados del Plioceno superior terminal en el suroeste de Europa: Fonelas P-1 y el Proyecto Fonelas. Instituto Geológico y Minero de España, Madrid, pp 461–473

Astorga-Márquez RJ, Carvajal A, Maldonado A, Gordon SV, Salas R, Gómez-Guillamón F, Sánchez-Baro A, López-Sebastián A, Santiago-Moreno J (2014) Influence of cohabitation between domestic goat (*Capra aegagrus hircus*) and Iberian ibex (*Capra pyrenaica* hispanica) on seroprevalence of infectious diseases. Eur J Wildl Res 60:387–390

Barboza du Bocage JV (1857) Memoria sobre uma especie nova do genero Capra L. A Cabra-Montez da Serra do Gerez, em Portugal. Memorias da Academia Real das Ciencias de Lisboa 2:1–20

Bibi F, Vrba E, Fack F (2012) A new African fossil caprin and a combined molecular and morphological bayesian phylogenetic analysis of caprini (Mammalia: Bovidae). J Evol Biol 25:1843–1854

Cabrera Á (1911) The subspecies of the Spanish ibex. Proc Zool Soc London 66:963–977

Cabrera A (1914) Fauna ibérica: Mamíferos. Museo Nacional de Ciencias Naturales, Madrid

Camacho L, Gómez-Guillamón F, Risalde MA, González D, Zorrilla I, García-Bocanegra I (2017) Primer caso de ectima contagioso en cabra montés (*Capra pyrenaica hispanica*) en libertad. 35èmes Rencontres du Groupe d'Étude sur l'Eco-pathologie de la Faune Sauvage de Montagne (GEEFSM). Cofrentes, Valencia, 1–4 June 2017, Abstracts, p 85

Camerano L (1917) Contributo allo studio degli Stambecchi Iberici. Bollettino dei Musei di Zoologia ed Anatomia comparata della R. Università di Torino 32:1–30

Candela MG, Caballol A, Atance PM (2017) Wide exposure to *Coxiella burnetii* in ruminant and feline species living in a natural environment: zoonoses in a human–livestock–wildlife interface. Epidemiol Infect 145:478–481

Cano-Manuel FJ, Granados JE, Aspizua R, Barea JM, Sánchez FJ, Henares IL, Navarro FJ (2010) Cambio global y cabra montés (*Capra pyrenaica*) en Sierra Nevada, España. Galemys 22:433–445

Cardoso TF, Luigi-Sierra MG, Castelló A, Cabrera B, Noce A, Mármol-Sánchez E, García-González R, Fernández-Arias A, Alabart JL, López-Olvera JR, Mentaberre G, Granados-Torres JE, Cardells-Peris J, Molina A, Sánchez A, Clop A, Amills M (2021) Assessing the levels of intraspecific admixture and interspecific hybridization in Iberian wild goats (*Capra pyrenaica*). Evol Appl 14:2618–2634

Carlos Calero JA, Van der Made J, Mancheño MA, Montoya P, Romero G (2006) *Capra alba* Moyà-Solà, 1987 del Pleistoceno Inferior de la Sierra de Quibas (Albanilla, Murcia, España). Estud Geol 62:571–578

Carvajal A, Santiago-Moreno J, Toledano-Díaz A, Gómez A, Pulido-Pastor A, López-Sebastián A (2007) Estudio sanitario de la cabra montés (*Capra pyrenaica*) de la Reserva Andaluza de Caza de Tejeda y Almijara. II Congreso Internacional Género *Capra* en Europa. Granada, 20–23 November 2007, Abstracts, pp 58–59

Carvalho J, Granados JE, López-Olvera JR, Cano-Manuel FJ, Pérez JM, Fandos P, Soriguer RC, Velarde R, Fonseca C, Ráez A, Espinosa J, Pettorelli N, Serrano E (2015) Sarcoptic mange breaks up bottom-up regulation of body condition in a large herbivore population. Parasit Vectors 8:1–9

Carvalho J, Eizaguirre O, Pérez JM, Mentaberre G, Lavín S, Fandos P, Olmo JR, Olivé-Boix X, Torres RT, Fonseca C, Pettorelli N, Serrano E (2017) Evidence for phenotypic plasticity but not for compensatory horn growth in male Iberian ibex. Mamm Biol 87:101–106

Carvalho J, Büntgen U, Pettorelli N, Mentaberre G, Olivé-Boix X, Eizaguirre O, Pérez JM, Fandos P, Torres RT, Lavín S, Fonseca C, Serrano E (2020) Habitat and harvesting practices influence horn growth of male ibex. J Wildl Manag 84:1–15

Cassinello J, Acevedo P (2007) Capra pyrenaica Schinz, 1838. Ficha Libro Rojo. In: Palomo LJ, Gisbert J, Blanco C (eds) Atlas y Libro Rojo de los Mamíferos Terrestres de España. Dirección General para la Biodiversidad-SECEM-SECEMU, Madrid, pp 458–459

Clouet M (1980) Note sur la systématique du bouquetin-d'Espagne. Bull Soc Hist Nat de Toulouse 115:269–277

Colom-Cadena A, Velarde R, Salinas J, Borge C, García-Bocanegra I, Serrano E, Gass D, Bach E, Casas-Díaz E, López-Olvera JR, Lavín S, León-Vizcaíno L, Mentaberre G (2014) Management of a caseous lymphadenitis outbreak in a new Iberian ibex (*Capra pyrenaica*) stock reservoir. Acta Vet Scand 56:1–11

Cordero del Campillo M, Castañón-Ordóñez L, Reguera-Feo A (1994) Índice catálogo de zooparásitos ibéricos, 2nd edn. Secretariado de Publicaciones de la Universidad de León, León

Couturier AJ (1962) Le bouquetin des Alpes (*Capra aegagrus ibex ibex* L.). Ed. by the author, Grenoble

Crégut-Bonnoure E (1992) Intérêtbiostratigraphique de la morphologiedentaire de *Capra* (Mammalia, Bovidae). Ann Zool Fenn 28:273–290

Crégut-Bonnoure E (2006) European Ovibovini, Ovini and Caprini (Caprinae, Mammalia) from the Plio-Pleistocene: new interpretations. Cour Forsch-Inst Senckenberg 256:139–158

Cuartas P, García-González R (1992) *Quercus ilex* browse utilization by wild and domestic ungulates in Cazorla Sierra. Vegetatio 99–100:317–330

Cuartas P, García-González R, Alados CL, Escós J (1996) Selección de dieta y comportamiento alimentario. In: Alados CL, Escós J (eds) Ecología y comportamiento alimentario de la cabra montés. Consideraciones para su gestión. Monografías del Museo de Ciencias Naturales CSIC, Madrid, pp 151–198

Cubero MJ, González M, León L (2002) Enfermedades infecciosas de las poblaciones de cabra montés en Andalucía. In: Pérez JM (ed) Distribución, genética y estatus sanitario de las poblaciones andaluzas de cabra montés. Universidad de Jaén and Consejería de Medio Ambiente de la Junta de Andalucía, Jaén, pp 196–254

De la Cerda JM, De la Peña J (1971) La cabra montés española. Montes 159:219–227

Ellis JE, Wiens JA, Rodell CF, Anway JC (1976) A conceptual model of diet selection as an ecosystem process. J Theor Biol 60:93–108

Escós J (1988) Estudio sobre la ecología y etología de la cabra montés (*Capra pyrenaica hispanica* Schimper 1848) de las Sierras de Cazorla y Segura y Sierra Nevada oriental. PhD Thesis, Universidad de Granada, Granada

Escós J, Alados CL (1988) Estimating mountain ungulate density in sierras de Cazorla y Segura. Mammalia 52(3):425–428

Escós J, Alados CL (1992a) The home-range of the Spanish ibex in spring and fall in two ranges. Mammalia 56(1):57–63

Escós J, Alados CL (1992b) Habitat preference of Spanish ibex and other ungulates in sierras de Cazorla y Segura (Spain). Mammalia 56(3):393–406

Escós J, Alados CL (1998) Parámetros demográficos como base de una gestión cinegética. Foresta 1:21–23

Escós J, Alados CL, Emlen JM (1994) An application of the stage-projection model with density dependent fecundity to the population dynamics of Spanish ibex. Can J Zool 72:731–737

Escós JM, Alados CL, Pulido A, Romera J, González-Sánchez N, Martínez F (2008) Estimating population trends using population viability analyses for the conservation of *Capra pyrenaica*. Acta Theriol 53:275–286

Fandos P (1988a) Différences saisonnières dans la répartition des activités quotidiennes du bouquetin, *Capra pyrenaica* de Cazorla. Mammalia 52:3–10

Fandos P (1988b) Croissance et development foetal du bouquetin de Cazorla (Espagne). Mammalia 51(4): 579–585

Fandos P (1989) Reproductive strategies in female Spanish ibex (*Capra pyrenaica*). J Zool (Lond) 218(2):339–343

Fandos P (1991) La cabra montés (*Capra pyrenaica*) en el Parque Natural de Cazorla, Segura y Las Villas. Colección Técnica. Icona, Madrid. 176 p

Fandos P (1995) Factors affecting horn growth in male Spanish ibex (*Capra pyrenaica*). Mammalia 59(2): 229–235

Fandos P, Martínez T (1988) Variaciones en la agregación y distribución de la cabra montés (*Capra pyrenaica* Schinz, 1838) detectadas con un muestreo de excrementos. Doñana Acta Vertebrata 15:133–140

Fandos P, Vigal CR (1988) Body weight and horn length in relation to age of the Spanish wild goat. Acta Theriol 33:339–344

Fandos P, Vigal CR (1993) Sexual dimorphism in size of the skull of Spanish ibex *Capra pyrenaica*. Acta Theriol 38(1):103–111

Fandos P, Vigal CR, Fernández-López JM (1989) Weight estimation of Spanish ibex, *Capra pyrenaica,* and chamois, *Rupicapra rupicapra* (Mammalia, Bovidae). Zeitschrift für Säugetierkunde 54:239–242

Fandos P, Orueta JF, Aranda Y (1993) Tooth wear and its relation to kind of food: the repercussion on age criteria in *Capra pyrenaica*. Acta Theriol 38:93–102

Fandos P, Arcenegui P, Lora MA, Buron D, Granados JE, Cadenas R (2010) Evolución demográfica de la cabra montés en Andalucía en los últimos 100 años. Galemys 22:347–358

Fernández de Luco D, Arnal MC (2002) Health status study on a population of Iberian wild goat (*Capra pyrenaica*) from Northeastern Spain. In: Third world conference on mountain ungulates, Zaragoza, 10–15 June 2002

Fernández de Luco D, Seijas JM, García-González R, Gonzalo M, Fernández M, Fernández JG, Serodio HC, Ferreira D, Arnal MC, Herrero J, Clavero JL. (2000) The last bucardo (*Capra pyrenaica pyrenaica*) in the Pyrenees died from politraumatism after being squashed under a falling fir (*Abies alba*). In: Fourth European wildlife disease association, Zaragoza, 20–23 September 2000

Fernández-Aguilar X, Cabezón O, Granados JE, Frey J, Serrano E, Velarde R, Cano-Manuel FJ, Mentaberre G, Ráez-Bravo A, Fandos P, López-Olvera JR (2017) Postepizootic persistence of asymptomatic *Mycoplasma conjunctivae* infection in Iberian ibex. Appl Environ Microbiol 83(15):1–15

Fernández-Arias A, Folch J, Alabart JL (1997) Reproductive characteristics of Spanish ibex *Capra pyrenaica hispanica* in captivity. In: Second world conference on mountain ungulates. Abstract Volume, Aosta, 5–7th May 1997. Regione Autonoma Valle de Aosta/Parco Nazionale del Gran Paradiso, pp 70–71

Fernández-Arias A, Alabart JL, Folch J, Beckers JF (1999) Interspecies pregnancy of Spanish ibex (*Capra pyrenaica*) fetus in domestic goat (*Capra hircus*) recipients induces abnormally high plasmatic levels of pregnancy-associated glycoprotein. Theriogenology 51:1419–1430

Fernández-Arias A, Roche A, Alberio R, Alabart JL, Folch J (2001) Use of hybrids as recipients in interspecies embryo transfer in the *Capra* genus. Theriogenology 55:383

Fernández-Sirera L, Cabezón O, Rossi L, Meneguz PG, Rosell R, Casas-Díaz E, Lavín S, Marco I (2011) Investigations of pestivirus infection in wild Caprinae in Europe. Vet Rec 169:15

Ferrer D, Castellà J, Gutiérrez JF, Lavín S, Marco I (1998) Seroprevalence of *Babesia ovis* in Spanish ibex (*Capra pyrenaica*) in Catalonia, Northeastern Spain. Vet Parasitol 75:93–98

Fonseca C, Carvalho J, Santos JPV, Torres RT, Bencatel J (2017a) Artiodáctilos (Cetartiodactyla). In: Bencatel J, Alvares F, Moura AE, Barbosa AM (eds) Atlas de Mamíferos de Portugal. Universidade de Évora, Évora

Fonseca C, Migueis D, Fernandes T, Carvalho H, Loureiro A, Carvalho J, Torres RT (2017b) The return of the Iberian wild goat *Capra pyrenaica* to Portugal: from reintroduction to recolonization. J Nat Conserv 38:56–61

França C (1917) Le Bouquetin du Gerez (*Capra Lusitanica*). Notes sur une espèce éteinte. Arquivo da Universidade de Lisboa 6:19–54, +10 planches

García I, Napp S, Casal J, Perea A, Allepuz A, Alba A, Carbonero A, Arenas A (2009) Bluetongue epidemiology in wild ruminants from southern Spain. Eur J Wildl Res 55:173–178

García-Bocanegra I, Cabezón O, Pabón M, Gómez-Guillamón F, Arenas A, Alcaide E, Salas-Vega R, Dubey JP, Almería S (2012) Prevalence of *Toxoplasma gondii* and *Neospora caninum* antibodies in Spanish ibex (*Capra pyrenaica hispanica*). Vet J 191:257–260

García-González R (2011) Elementos para una filogeografía de la cabra montés ibérica (*Capra pyrenaica* Schinz, 1838). Pirineos 166:87–122

García-González R (2012) New Holocene *Capra pyrenaica* (Mammalia, Artiodactyla, Bovidae) skulls from the southern Pyrénées. Comptes Rendues Palevol 11:241–249

García-González R, Cuartas P (1992a) Feeding strategies of Spanish wild goat in the Cazorla sierra (Spain). In: Spitz F, Janeau G, Gonzalez G, Aulagnier S (eds) Ongulés/Ungulates 91. SFEPM-IRGM, Paris/Toulouse, pp 167–170

García-González R, Cuartas P (1992b) Food habits of *Capra pyrenaica*, *Cervus elaphus* and *Dama dama* in the Cazorla sierra (Spain). Mammalia 56:195–202

García-González R, Herrero J (1999) El Bucardo de los Pirineos: historia de una extinción. Galemys 11:17–26

García-González R, Herrero J, Nores C (2021) The names of the southwestern European goats: is Iberian ibex the best common name for *Capra pyrenaica*? Animal Biodiversity and Conservation 44(1):1–31

García-Moreno AM, Vega FG, Moreno AM, Gragera-Slikker A, Habela MA (2009) Epidemiological surveillance of infectious contagious processes in Iberian wild goat (*Capra pyrenaica victoriae*) in Cáceres (Extremadura, Spain). In: V World conference on Mountain Ungulates, Granada, 10–14 November 2009, Abstracts, p 256

Garrido JL, Gortázar C, Ferreres J (2019) Las especies cinegéticas españolas en el siglo XXI. Edición propia, Wroclaw, 116 p

Gauss CB, Dubey JP, Vidal D, Cabezón O, Ruiz-Fons F, Vicente J, Marco I, Lavín S, Gortázar C, Almería S (2006) Prevalence of *Toxoplasma gondii* antibodies in red deer (*Cervus elaphus*) and other wild ruminants from Spain. Vet Parasitol 136:193–200

Gippoliti S, Cotterill FPD, Groves CP, Zinner D (2018) Poor taxonomy and genetic rescue are possible co-agents of silent extinction and biogeographic homogenization among ungulate mammals. Biogeographia 33:41–54

Gómez JM, González-Megías A (2002) Asymmetrical interactions between ungulates and phytophagous insects: being different matters. Ecology 83:203–211

Gonzalez G (1982) Eco-ethologie du bouquetin en Sierra de Gredos. Acta Biol Montana 1:177–215

González J, Herrero J, Prada C, Marco J (2013) Changes in wild ungulate populations in Aragon, Spain between 2001 and 2010. Galemys 25:51–59

González-Candela M, Cubero-Pablo MJ, Martín-Atance M, León-Vizcaíno L (2006) Potential pathogens carried by Spanish ibex (*Capra pyrenaica hispanica*) in southern Spain. J Wildl Dis 42:325–334

Gortázar C, Herrero J, Villafuerte R, Marco J (2000) Historical examination of the distribution of large mammals in Aragón, Northeastern Spain. Mammalia 61:411–422

Granados JE, Pérez JM, Soriguer C, Fandos P, Ruiz-Martínez I (1997) On the biometry of the Spanish ibex, *Capra pyrenaica*, from Sierra Nevada (southern Spain). Folia Zool 46(1):9–14

Granados JE, Fandos P, Márquez FJ, Soriguer RC, Chirosa M, Pérez JM (2001a) Allometric growth in the Spanish ibex, *Capra pyrenaica*. Folia Zool 50(3):234–238

Granados JE, Pérez JM, Márquez FJ, Serrano E, Soriguer RC, Fandos P (2001b) La cabra montés (*Capra pyrenaica*, Schinz 1838). Galemys 13:3–37

Granados JE, Soriguer RC, Pérez JM, Fandos P, García-Santiago JG (2002) *Capra pyrenaica* Schinz, 1848. In: Palomo LJ, Gisbert J (eds) Atlas de los mamíferos terrestres de España. Dirección General de Conservación de la Naturaleza-SECEM-SECEMU, Madrid, pp 366–368

Granados JE, Cano-Manuel J, Castillo A, Pérez JM, González FJ, Fandos P, Serrano E, Soriguer JR (2004) La gestión de la cabra montés en Andalucía. Quercus 226:16–21

Gutiérrez-Expósito D, Arnal MC, Martínez-Durán D, Regidor-Cerrillo J, Revilla M, Fernández-de-Luco D, Jiménez-Meléndez A, Calero-Bernal R, Habela MÁ, García-Bocanegra I, Arenas-Montes A, Ortega-Mora LM, Álvarez-García G (2016) The role of wild ruminants as reservoirs of *Besnoitia besnoiti* infection in cattle. Vet Parasitol 223:7–13

Hassanin A, Delsuc F, Ropiquet A, Hammer C, Jansen van Vuuren B, Matthee C, Ruiz-Garcia M, Catzeflis F, Areskoug V, Nguyen TT, Couloux A (2012) Pattern and timing of diversification of Cetartiodactyla (Mammalia, Laurasiatheria), as revealed by a comprehensive analysis of mitochondrial genomes. C R Biol 335:32–50

Herrero J, Fernández-Arberas O, Prada C, García-Serrano A (2013a) Size and density of a feral goat *Capra hircus* population in a protected area in the Pyrenees. Wildl Biol Pract 9(3):1–6

Herrero J, Fernández-Arberas O, Prada C, García-Serrano A, García-González R (2013b) An escaped herd of Iberian wild goat (*Capra pyrenaica*, Schinz 1838, Bovidae) begins the re-colonization of the Pyrenees. Mammalia 77(4):403–407

Herrero J, Acevedo P, Arnal MC, Fernández de Luco D, Fonseca C, García-González R, Pérez JM, Sourp E (2020) *Capra pyrenaica*. In: The IUCN Red List of Threatened Species e.T3798A170192604

Hofmann RR (1989) Evolutionary steps of ecophysiological adaptation and diversification of ruminants: a comparative view of their digestive system. Oecologia 78:443–457

Hueli LE, Díaz V (1989) Ixódidos (Acarina:Ixodidae) parásitos de *Capra pyrenaica hispanica* en la provincia de Granada. In: VI Congreso Nacional y I Congreso Ibérico de Parasitología, Cáceres, 25–29 Septembre 1989, Abstracts, p 223

Kazanskaya E, Kuznetsova M, Danilkin A (2007) Phylogenetic reconstructions in the genus *Capra* (Bovidae, Artiodactyla) based on the mitochondrial DNA analysis. Russ J Genet 43:181–189

Lalueza-Fox C, Castresana J, Sampietro L, Marquès-Bonet T, Alcover JA, Bertranpetit J (2005) Molecular dating of caprines using ancient DNA sequences of *Myotragus balearicus*, an extinct endemic Balearic mammal. BMC Evol Biol 5:70

Lavín S, Marco I, Rossi L, Meneguz PG, Viñas L (1997) Haemonchosis in Spanish ibex. J Wildl Dis 33:656–659

León-Vizcaíno L, De Meneghi D, Meneguz PG, Rosati S, Rossi L (1994) Investigaciones serológicas sobre enfermedades infecciosas de la cabra montés (*Capra pyrenaica*) en el Parque Natural Sierra de las Nieves, Ronda-Málaga (E): Resultados y consideraciones preliminares. Actas del I Congreso Internacional del Género *Capra* en Europa, Ronda, pp 219–222. Abstracts

León-Vizcaíno L, Ruiz de Ybáñez MR, Cubero MJ, Ortiz JM, Espinosa J, Pérez L, Simón MA, Alonso F (1999) Sarcoptic mange in Spanish ibex from Spain. J Wildl Dis 35:647–659

León-Vizcaíno L, Garrido-Abellán F, González-Candela M, Pérez-Béjar L, Cubero-Pablo MJ, Perales-Flores A, Pacheco I, Martín-Atance P (2009) Ineficacia de los rumiantes silvestres como reservorio de *Brucella* en el medio natural andaluz (Sierras Béticas). In: V World conference on Mountain Ungulates, Granada, 10–14 November 2009, Abstracts, p 303

Lorca-Oró C, Pujols J, Arenas A, Gómez-Guillamón F, Zorrilla I, Domingo M, Arenas-Montes A, Ruano MJ, García-Bocanegra I (2011) Epidemiological surveillance of bluetongue virus serotypes 1, 4 and 8 in Spanish ibex (*Capra pyrenaica hispanica*) in southern Spain. Vet Microbiol 149:230–235

Lucas P, Herrero J, Fernández-Arberas O, Prada C, García-Serrano A, Saiz H, Alados CL (2016) Modelling the habitat of a wild ungulate in a Mediterranean semi-arid environment of South-West Europe: small cliffs as key predictors for the presence of Iberian wild goat. J Arid Environ 129:56–63

Luzón M, Santiago-Moreno J, Meana A, Toledano-Díaz A, Pulido-Pastor A, Gómez-Brunet A, López-Sebastián A (2008) Parasitism and horn quality in male Spanish ibex (*Capra pyrenaica hispanica*) from Andalucía based on coprological analysis and muscle biopsy. Span J Agric Res 6:353–361

Magniez P (2009) Nouvelles données sur le genre *Capra* Linné, 1758 (Mammalia, Bovidae) du Pléistocènesupérieur de la grotte Tournal (Bize-Minervois, France): Implications biochronologiques et évolutives. Quaternaire 20:509–525

Manceau V, Crampe JP, Boursot P, Taberlet P (1999) Identification of evolutionary significant units in the Spanish wild goat, *Capra pyrenaica* (Mammalia, Artiodactyla). Anim Conserv 2:33–39

Marco J, Herrero J, Escudero MA, Fernández-Arberas O, Ferreres J, García-Serrano A, Giménez-Anaya A, Labarta JL, Monrabal L, Prada C (2011) Veinte años de seguimiento poblacional de ungulados en Aragón. Pirineos 166:135–153

Márquez FJ, Pérez JM, Granados JE, Soriguer R, Fandos P (2002) Estudios genéticos de las poblaciones de cabra montés. In: Pérez JM (coord) Distribución, genética y estatus sanitario de las poblaciones andaluzas de cabra montés. Universidad de Jaén. Junta de Andalucía, Consejería de Medio Ambiente, Sevilla, Jaén, pp 257–274

Martín-Atance P, León-Vizcaíno L (2009) Evaluación serológica de agentes contagiosos asociados al síndrome de mortalidad perinatal congénita en los rumiantes de la Serranía de Cuenca. In: V World conference on Mountain Ungulates, Granada, 10–14 November 2009, Abstracts, p 315

Martínez T (1988) Données sur l'alimentation de bouquetin d'Espagne (*Capra pyrenaica*) dans la Sierra de Tejeda (Granada). Mammalia 52:284–285

Martínez T (1994) Dieta estacional de la cabra montés (*Capra pyrenaica*) en los Puertos de Tortosa y Beceite. Ecología 8:373–380

Martínez T (2002) Summer feeding strategy of Spanish ibex *Capra pyrenaica* and domestic sheep *Ovis aries* in South-Eastern Spain. Acta Theriol 47:479–490

Martínez T, Martínez E (1987) Diet of Spanish wild goat, *Capra pyrenaica*, in spring and summer at the Sierra de Gredos, Spain. Mammalia 51:547–557

Martínez T, Martínez E, Fandos P (1985) Composition of the food of the Spanish wild goat in sierras de Cazorla and Segura, Spain. Acta Theriol 30:461–494

Meneguz PG, Rossi L, Sommavilla GM, de Martin P, Rodolfi M (1996) Sulla più temibile parassitosi della fauna alpina: la rogna sarcoptica del camoscio. Large Anim Rev 2(3):75–83

Mentaberre G, Serrano E, Velarde R, Marco I, Lavín S, Mateos A, de Juan L, Domínguez L, Olivé X, Romeva J (2010) Absence of TB in Iberian ibex (*Capra pyrenaica*) in a high-risk area. Vet Rec 166:700

Mentaberre G, Velarde R, García-Readigòs MA, Serrano E, Jovaní J, Olivé-Boix X, López-Olvera JR, Castillo R, Ruiz-Olmo J, Lavín S (2015) Nuevo brote de sarna sarcóptica en la cabra montés (*Capra pyrenaica*). In: 33èmes Rencontres du Groupe d'Étude sur l'Ecopathologie de la Faune Sauvage de Montagne (GEEFSM). 21–24 May 2015, Balme, Valli di Lanzo, Abstracts, p 15

Moço G, Guerreiro M, Ferreira AF, Rebelo A, Loureiro A, Petrucci-Fonseca F, Pérez JM (2006) The ibex *Capra pyrenaica* returns to its former Portuguese range. Oryx 40:351–354

Moço G, Serrano E, Guerreiro M, Ferreira AF, Petrucci-Fonseca F, Maia MJ, Soriguer RC, Pérez JM (2013) Seasonal dietary shifts and selection of Iberian ibex *Capra pyrenaica* Schinz, 1838 in Peneda-Gerês National Park (PGNP, Portugal). Galemys 25:13–28

Moço G, Serrano E, Guerreiro M, Ferrerira AF, Petrucci-Fonseca F, Santana D, Maia MJ, Soriguer RC, Pérez JM (2014) Does livestock influence the diet of Iberian Ibex *Capra pyrenaica* in the Peneda-Gerês National Park (Portugal). Mammalia 78:393–399

Montoya P, Alberdi MT, Barbadillo LJ, van der Made J, Morales J, Murelaga X, Penalver E, Robles F, Ruiz Bustos A, Sanchez A (2001) A very diverse lower Pleistocene faunistic assemblage from sierra de Quibas (province of Murcia, Spain). Comptes Rendus de l'Academie des Sciences – Series IIA 332:387–393

Moyà-Solà S (1987) Los bóvidos (Artiodactyla, Mammalia) del yacimiento del Pleistoceno inferior de Venta Micena (Orce, Granada, España). Paleontologia i evolució 1:181–236

Muñoz PM, Boadella M, Arnal MC, de Miguel MJ, Revilla M, Martínez D, Vicente J, Acevedo P, Oleaga A, Ruiz-Fons F, Marín CM, Prieto JM, de la Fuente J, Barral M, Barberán M, Fernández de Luco D, Blasco JM, Gortázar C (2010) Spatial distribution and risk factors of brucellosis in Iberian wild ungulates. BMC Infect Dis 10:46

Navarro-González N, Velarde R, Porrero MC, Mentaberre G, Serrano E, Mateos A, Domínguez L, Lavín S (2014) Lack of evidence of spill-over of *Salmonella enterica* between cattle and sympatric Iberian ibex (*Capra pyrenaica*) from a protected area in Catalonia, NE Spain. Transbound Emerg Dis 61:378–384

Navarro-González N, Porrero MC, Mentaberre G, Serrano E, Mateos A, Cabal A, Domínguez L, Lavín S (2015) *Escherichia coli* O157:H7 in wild boars (*Sus scrofa*) and Iberian ibex (*Capra pyrenaica*) sharing pastures with free-ranging livestock in a natural environment in Spain. Vet Q 35:102–106

Nievergelt B (1974) A comparison of rutting behaviour and grouping in the Ethiopian and Alpine ibex. In: Geist V, Walther F (eds) The behaviour of ungulates and its relation to management. IUCN Publications, Morges, pp 324–340

Oleaga-Pérez A, Pérez-Sánchez R, Ramajo-Martín V (2003) Parasitosis en las dehesas salmantinas: Reservorios silvestres. Albéitar 66:8–10

Palomares F, Ruiz Martinez I (1993) Status and conservation perspectives for the Spanish ibex population (*Capra pyrenaica* Schinz, 1838) of Sierra Mágina Natural Park. Zeitschrift für Jagdwissenschaft 39(2):87–94

Peral JC (1993) La cabra montés vuelve a Riaño. Trofeo 272:70–76

Perea R, Perea-García-Calvo R, Diaz-Ambrona CG, San Miguel A (2015) The reintroduction of a flagship ungulate *Capra pyrenaica*: assessing sustainability by surveying woody vegetation. Biol Conserv 181:9–17

Pérez JM, Granados JE, Soriguer RC (1994) Population dynamic of the Spanish ibex *Capra pyrenaica* in Sierra Nevada Natural Park (southern Spain). Acta Theriol 39(3):289–294

Pérez JM, Granados JE, Soriguer RC, Ruiz-Martínez I (1996) Prevalence and seasonality of *Oestrus caucasicus* Grunin, 1948 (Diptera: Oestridae) parasitizing the Spanish ibex, *Capra pyrenaica* (Mammalia: Artiodactyla). J Parasitol 82:233–236

Pérez JM, Ruiz-Martínez I, Granados JE, Soriguer RC, Fandos P (1997) The dynamic of sarcoptic mange in the ibex population of Sierra Nevada in Spain – influence of climatic factors. J Wildlife Res 2:86–89

Pérez JM, Granados JE, Soriguer RC, Fandos P, Marquez FJ, Crampe JP (2002) Distribution, status and conservation problems of the Spanish ibex, *Capra pyrenaica* (Mammalia: Artiodactyla). Mammal Rev 32:26–39

Pérez JM, González FJ, Granados JE, Pérez MC, Fandos P, Soriguer RC, Serrano E (2003a) Hematologic and biochemical reference intervals for Spanish ibex. J Wildl Dis 39:209–215

Pérez JM, Granados JE, Pérez MP, Márquez FJ, Ferroglio E, Rossi L (2003b) A survey of the gastrointestinal nematodes of Spanish ibex (*Capra pyrenaica*) in a high mountain habitat. J Parasitol 89:315–318

Pérez JM, González FJ, Serrano E, Granados JE, Fandos P, Carro F, Soriguer RC (2006a) Is blood collected from shot Iberian ibex (*Capra pyrenaica*) useful for monitoring their physiological status? Eur J Wildl Res 52(2):125–131

Pérez JM, Meneguz PG, Dematteis A, Rossi L, Serrano E (2006b) Parasites and conservation biology: the 'ibex-ecosystem'. Biodivers Conserv 15:2033–2047

Pérez JM, Serrano E, González-Candela M, León-Vizcaíno L, Barberá GG, de Simón MA, Fandos P, Granados JE, Soriguer RC, Festa-Bianchet M (2011) Reduced horn size in two wild trophy-hunted species of Caprinae. Wildl Biol 17:102–112

Pérez T, González I, Essler SE, Fernández M, Domínguez A (2014) The shared mitochondrial genome of *Rupicapra pyrenaica ornata* and *Rupicapra rupicapra cartusiana*: old remains of a common past. Mol Phylogenet Evol 79:375–379

Pérez JM, Serrano E, Soriguer RC, González FJ, Sarasa M, Granados JE, Cano-Manuel FJ, Cuenca R, Fandos P (2015) Distinguishing disease effects from environmental effects in a mountain ungulate: seasonal variation in body weight, hematology, and serum chemistry among Iberian ibex (*Capra pyrenaica*) affected by sarcoptic mange. J Wildl Dis 51:148–156

Pidancier N, Jordan S, Luikart G, Taberlet P (2006) Evolutionary history of the genus *Capra* (Mammalia, Artiodactyla): discordance between mitochondrial DNA and Y-chromosome phylogenies. Mol Phylogenet Evol 40:739–749

Pita Fernández M, Casas Bargueno S, Herrero J, Prada C, García Post R (2012) Game reserves in Spain: the public management of hunting. For Syst 21(3):398–404

Pizzato F, Ortega N, Serrano E, Salinas J, Castillo R, López-Olvera JR, Lavín S, Mentaberre G (2017) Estudio sobre paratuberculosis en cabra montés (*Capra pyrenaica*). In: 35èmes Rencontres du Groupe d'Étude sur l'Eco-pathologie de la Faune Sauvage de Montagne (GEEFSM), Cofrentes, Valencia (Spain), 1–4 June 2017, Abstracts, p 121

Prada C, Herrero J (2013) Viabilidad del hábitat y del reforzamiento poblacional de *Capra pyrenaica* en la Reserva Nacional de Caza de Os Ancares. Pirineos 168:59–76

Prada C, Herrero J, García-Serrano A, Fernández-Arberas O, Gómez C (2019) Estimanting Iberian wild goat abundance in a large rugged forest habitat. Pirineos 174:1–7

Real R, Romero D, Olivero J, Estrada A, Márquez AL (2013) Estimating how inflated or obscured effects of climate affect forecasted species distribution. PLOS One 8(1):e53646

Refoyo P, Olmedo C, Polo I, Fandos P, Muñoz B (2014) Demographic trends of a reintroduced Iberian ibex *Capra pyrenaica victoriae* population in Central Spain. Mammalia 79(2):139–145

Refoyo P, Olmedo C, Barba M, Muñoz B (2016) Parasite load in the Iberian ibex, *Capra pyrenaica victoriae*. Arxius de Miscel·lània Zoològica 14:108–113

Revilla M (2012) Estudio sanitario de la cabra montés *Capra pyrenaica hispanica* en Aragón. PhD Thesis, Universidad de Zaragoza, Zaragoza

Revilla M, Arnal MC, Fernández de Luco D (2007) Seguimiento sanitario de la cabra montesa *Capra pyrenaica* en Aragón. In: II Congreso Internacional Género *Capra* en Europa, Granada, 20–23 November 2007, Abstracts, pp 69–70

Rodríguez de la Zubía M (1969) La cabra montés de Sierra Nevada. Ministerio de Agricultura, Madrid
Ropiquet A, Hassanin A (2006) Hybrid origin of the Pliocene ancestor of wild goats. Mol Phylogenet Evol 41:395–404
Rossi L, Marco-Sánchez I, Meneguz PG (1992) Nematodes of the digestive tract of *Capra pyrenaica*: preliminary contribution. In: Actas del I Congreso Internacional del Género *Capra* en Europa, 20–22 October 1992, Ronda, Abstracts, pp 249–251
Sánchez-Isarria MA, Hermoso J, Theureau J, Casanova G, Burgui JM, Sanchos G, Arévalo PRS (2008) Metodología empleada en la estrategia del control de la sarna sarcóptica en la cabra montés de la Reserva Valenciana de Caza de la Muela de Cortes entre los años 2002–2007 (Valencia). In: Granados Torres JEC-MLJ, Fandos París P, Cadenas del Llano Aguilar R (eds) Tendencias actuales en el Estudio y Conservación de los Caprinos Europeos. II Congreso Internacional del género *Capra* en Europa, Granada 2007, pp 255–268
Santiago-Moreno J, Luzón M, Coloma MA, Pulido-Pastor-A, Gómez-Guillamón F, Salas-Vega R, Toledano-Díaz A, López-Sebastián A (2010) Relationship between Iberian ibex (*Capra pyrenaica*) sperm quality and level of parasitism. Eur J Wildl Res 56(4):605–611
Santiago-Moreno J, Carvajal A, Astorga RJ, Coloma MÁ, Toledano-Díaz A, Gómez-Guillamón F, Salas-Vega R, López-Sebastián A (2011) Potential impact of diseases transmissible by sperm on the establishment of Iberian ibex (*Capra pyrenaica*) genome resource banks. Eur J Wildl Res 57:211–216
Santos JPV, Fernández-de-Mera IG, Acevedo P, Boadella M, Fierro Y, Vicente J, Gortázar C (2013) Optimizing the sampling effort to evaluate body condition in ungulates: a case study on red deer. Ecol Indic 30:65–71
Sarasa M, Alasaad S, Pérez JM (2012) Common names of species, the curious case of *Capra pyrenaica* and the concomitant steps towards the 'wild-to-domestic' transformation of a flagship species and its vernacular names. Biodivers Conserv 21:1–12
Sauqué V, García-González R, Cuenca-Bescós G (2016) A late Pleistocene (MIS3) ungulate mammal assemblage (Los Rincones, Zaragoza, Spain) in the Eurosiberian–Mediterranean boundary. Hist Biol 28:358–389
Schaller GB (1977) Mountain monarchs. Wild sheep and goats of the Himalaya. University of Chicago Press, Chicago/London
Schimper WP (1848) Note sur une troisième espèce de Bouquetin en Europe (*Capra hispanica*). Comptes Rendus hebdomadaires séances de l'Académie des Sciences 26:318–320
Schinz HR (1838) Bemerkungen über die Arten der wilden Ziegen, besonders mit Beziehung auf den Sibirischen Steinbock, den Steinbock der Alpen und den Steinbock der Pyrenaeen. Neue Denkschriften der allgemeinen Schweizerischen Gesellschaft für die gesammten Naturwissenschaften 2:1–26
Serrano E, González FJ, Granados JE, Moço G, Fandos P, Soriguer RC, Pérez JM (2008) The use of total serum proteins and triglycerides for monitoring body condition in the Iberian wild goat (*Capra pyrenaica*). J Zoo Wildl Med 39:646–649
Serrano E, Granados JE, Sarasa M, González FJ, Fandos P, Soriguer RC, Pérez JM (2011) The effects of winter severity and population density on body stores in the Iberian wild goat (*Capra pyrenaica*) in a highly seasonal mountain environment. Eur J Wildl Res 57(1):45–55
Shackleton DM, Lovari S (1997) Classification adopted for the Caprinae survey. In: Shackleton DM (ed) Wild sheep and goats and their relatives. Status survey and conservation action plan for Caprinae. IUCN/SSC Caprinae Specialist Group, Gland/Cambridge, pp 9–14
Torres R, Herrero J, Prada C, García-Serrano A, Fernández-Arberas O, García-Post R (2014) Estimating the population density of Iberian wild goat *Capra pyrenaica* and mouflon *Ovis aries* in a Mediterranean forest environment. For Syst 23(1):36–43
Ureña I, Ersmark E, Samaniego JA, Galindo-Pellicena MA, Crégut-Bonnoure E, Bolívar H, Gómez-Olivencia A, Rios-Garaizar J, Garate D, Dalén L, Arsuaga JL, Valdiosera CE (2018) Unraveling the genetic history of the European wild goats. Quat Sci Rev 185:189–198
Van Soest PJ (1994) Nutritional ecology of the ruminant, 2nd edn. Cornell University Press, Ithaca/London
Varela-Castro L, Lara-Vergara J, Ortega N, Salinas J, Colom-Cadena A, Lavín S, Tizzani P, Velarde R, Serrano E, Mentaberre G (2017) Endemic caseous lymphadenitis in a wild Caprinae population. Vet Rec 180:405
Varela-Castro L, Zuddas C, Ortega N, Serrano E, Salinas J, Castellà J, Castillo-Contreras R, Carvalho J, Lavín S, Mentaberre G (2018) On the possible role of ticks in the eco-epidemiology of *Coxiella burnetii* in a Mediterranean ecosystem. Ticks Tick-Borne Dis 9:687–694
Velarde R, Mentaberre G, Sánchez J, Marco I, Lavín S (2008) KIT-positive gastrointestinal stromal tumours in two Spanish ibex (*Capra pyrenaica hispanica*). Vet J 177:445–447
Viana DS, Granados JE, Fandos P, Pérez JM, Cano-Manuel FJ, Burón D, Fandos G, Aguado MÁP, Figuerola J, Soriguer RC (2018) Linking seasonal home range size with habitat selection and movement in a mountain ungulate. Mov Ecol 6:1
Vigal CR, Machordom A (1987) Dental and skull anomalies in the Spanish wild goat, *Capra pyrenaica* Schinz, 1838. Zeitschrift für Säugetierkunde 52:38–50
Zamora R, Gómez JM (1993) Vertebrate herbivores as predators of insect herbivores: an asymmetrical interaction mediated by size differences. Oikos 66:223–228
Zvychaynaya EY (2010) Genetic differentiation of wild goats (genus *Capra*) based on the analysis of mitochondrial gene cytochrome *b* and fragment of nuclear gene DRY. Galemys 22(Special issue):255–276

Caucasian Tur *Capra caucasica* Güldenstaedt et Pallas, 1783

18

Paul Weinberg and Bejan Lortkipanidze

Contents

P. Weinberg (✉)
North Ossetian State Nature Reserve, Alagir, Russian Federation
e-mail: tu_r@rambler.ru; tur31161@gmail.com

B. Lortkipanidze
NACRES – Centre for Biodiversity Conservation & Research, Tbilisi, Georgia
e-mail: bejan.lortkipanidze@nacres.org

© Springer Nature Switzerland AG 2022

L. Corlatti, F. E. Zachos (eds.), *Terrestrial Cetartiodactyla*, Handbook of the Mammals of Europe,
https://doi.org/10.1007/978-3-030-24475-0_36

Common Names

English	Eastern tur, Western tur
German	Ostkaukasischer Steinbock, Westkaukasischer Steinbock
French	Tur occidental, Bouquetin du Caucase, Bouquetin oriental, Tur du Caucase
Spanish	Tur occidental, Tur oriental, Tur del Este
Italian	Tur occidentale, stambecco del Caucaso occidentale, Tur orientale, stambecco del Caucaso orientale
Russian	дагестанский (восточнокавказский) тур, западнокавказский (кубанский) тур, кавказский тур

Taxonomy, Systematics and Paleontology

Tur is a mountain ungulate (family Bovidae, subfamily Caprinae) endemic to the Greater Caucasus. In the nineteenth century, several tur morphotypes were described which were treated as different species. From west to east these were: *Capra dinniki* Satunin 1905, *Capra severtzovi* Menzbier 1887, *Capra caucasica* Güldenstaedt et Pallas 1783, and *Capra cylindricornis* Blyth 1841, the first three living in the West Caucasus in a comparatively narrow stripe merely some 300 km long. Already by the end of the nineteenth century, it became clear that the two first morphotypes were fully sympatric, occurring in the same groups, and there was no reproductive isolation between them. Having studied morphological peculiarities of tur, Tsalkin (1955) concluded that *C. severtzovi* (different spellings can be found: *severtzowi*, *sewertzowi*) and *C. dinniki* were not species but merely different morphs whose proportions in the population might depend upon climatic conditions (Romashin 2001). Taxonomic position of *C. caucasica*, located in Balkaria, eastward of Mt. Elbrus, remained unclear, but Tsalkin (1955) considered it to be the same Western or Kuban tur, being explicitly different from the Eastern or Daghestan tur (*C. cylindricornis*). Opinion about two separate tur species was later supported by Nasimovich (1950) and Heptner (Heptner et al. 1961) who thought that the population occurring between Baksan and Balkar Cherek Rivers, on the Northern side of the Greater Caucasus, and Inguri and Rioni Rivers, on the Southern side, was partly mixed and, possibly, partly hybrid. Sokolov (1959) suggested one tur species (*Capra caucasica*) with three subspecies: Western (*C. c. severtzovi*), Central Caucasian (*C. c. caucasica*), and Eastern (*C. c. cylindricornis*). Some zoologists (e.g., Ellerman and Morrison-Scott 1966) relied upon Lydekker (1913), who considered Western tur a subspecies of Alpine ibex *C. ibex*, but also chose the name *C. caucasica* for the Eastern tur. This name had been usually applied to the Western tur, regardless of its taxonomic status. Schaller (1977) agreed with Lydekker, but suggested the name *C. cylindricornis* for the Eastern tur. This opinion is essentially based upon external resemblance of adult males of the *dinniki* morph (Western tur) with Alpine and Asiatic ibex *Capra sibirica*: long drooping beard, scimitar-curved horns with distinct transverse ridges. However, the shape of horn cores and cranial features testify to closer affinity of all tur (Heptner et al. 1961; Veinberg 1993; Fig. 1).

Reasons exist why specialists cannot reach consensus on tur taxonomy. Firstly, the occurrence of population eastward of Mt. Elbrus, combining morphological features of Western and Eastern tur. Secondly, some features show displacement along the Greater Caucasus, e.g., weakening of spiraling twist of horn sheaths in North Ossetia, when compared with Daghestan (Fig. 2). The latter served as a starting point to suggest the existence of clineal variation in tur, caused by geographic gradient in solar radiation and climate along the Greater Caucasus (Aiunts and Kolomyts 1986).

Supporters of existence of two species declared the Central Caucasian population mixed and partly hybrid in the overlap zone of the two species (Heptner et al. 1961), while supporters of one species regarded it either a separate subspecies (Chlaidze 1975; Sokolov 1959; Tembotov 1974) or a fragment in a cline (Aiunts and Kolomyts 1986). It seems that the last two opinions did not consider the scale of the distribution range. With respect to the proposed Central Caucasian subspecies, Sokolov (and his later supporters) did not pay attention to the very small distribution of this subspecies which, according to Sokolov's own descriptions, covered merely some 80 × 40 km range between Mts. Elbrus and Shkhara (Sokolov 1959; Sokolov and Tembotov 1993). An isolated, probably relict subspecies with such a small range may exist, but hardly a non-isolated, parapatric ungulate subspecies tucked in between two other subspecies. Clineal geographic variation of several traits (Weinberg et al. 2010) also does not agree with lumping Caucasian tur into one species with three subspecies (Sokolov 1959; Tembotov 1974), the middle one of them occurring exactly in the steep and fluctuating part of the cline. Multiple and correlated clineal variations in an ungulate within a limited range (750 km long and up to 80 km wide) can hardly be explained by geographic dynamics of environmental factors. The shape of the cline is also very telling (short western and long eastern sloping parts with a steep and fluctuating middle part eastward of Mt. Elbrus), suggesting that this is, in fact, a pseudo-cline or a secondary cline created by a secondary contact and subsequent hybridization (Mayr 1968). Since there is one steep part of the cline, contact of just two primary taxa may have

Fig. 1 Fully grown Western tur male (upper left, photograph by S. Trepet); fully grown Eastern tur male (right, photograph by G. Darchiashvili); Western tur female with kid (lower left, photograph by S. Trepet)

occurred, initially separated by one geographic barrier in the Central Caucasus. The most probable barrier would be a mighty glaciation center which was pulsating during the Pleistocene in the area including Mts. Elbrus in the west and Kazbek in the east, partly surviving up to now (Gerasimov and Markov 1939; Kotlyakov and Krenke 1980; Milanovsky 1966), and situated where the steep and fluctuating part of the cline occurs. This glaciation center could have periodically separated the all-Caucasus tur population into two and created conditions for the evolution of two taxa: the Eastern and Western tur (Weinberg et al. 2010). It could also have acted not as a physical barrier, but rather created unfavorable conditions which prevented gene flow (Gavashelishvili et al. 2018).

Pulsating glaciation can also explain hybridization between the two initial taxa because the barrier was not constant and, probably, periods of isolation were insufficiently long. Populations joined periodically (as now), hybridized, and did not evolve mechanisms of effective reproductive isolation for generating distinct species. Thus, Western and Eastern tur can be regarded as semi-species at best (Weinberg et al. 2010).

To date, modern methods have failed to clarify the situation. Chromosomal numbers are the same in all *Capra* (Kuliev and Mamedov 1974; Nadler and Lay 1975). Gene sequence studies also provided controversial results showing completely different female and male lineages for the same species, leading to conclusions that almost all *Capra* species are of a hybrid origin (e.g., Kazanskaya 2007; Manceau et al. 1999; Pidancier et al. 2006). For example, it had been suggested that Western tur was a hybrid of the Eastern tur and the wild goat *Capra aegagrus* (Kazanskaya 2007). This suggestion has been later considered erroneous (Kazanskaya pers. comm.).

Fossil remnants are almost of no help to clarify tur taxonomy. Fossil tur remnants have been briefly mentioned from Late Pleistocene deposits from the Azykh Cave in Western Azerbaijan,

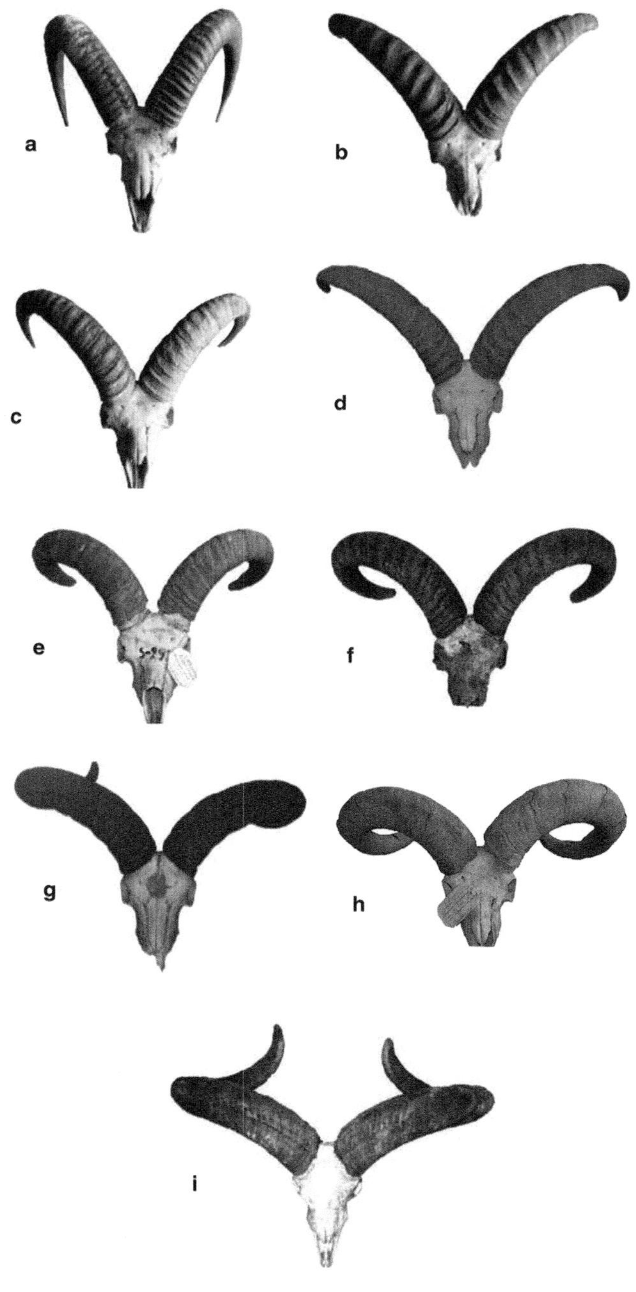

Fig. 2 Horns of adult Caucasian tur males from west to east. (**a**) Tchugush Mt. massif, Caucasus Nature Reserve; (**b**) Teberda Valley, 120 km eastward; (**c**) Malka Valley, 195 km eastward; (**d**) Baksan Valley, 200 km eastward; (**e**) Chegem Valley, 235 km eastward; (**f**) Bezengi Valley, 245 km eastward; (**g**) Alagir Valley, 315 km eastward; (**h**) Avar Koisu Valley, 500 km eastward; (**i**) Zaqatala Valley, 550 km eastward

Caucasus Minor (Aliev 1969; Gadjiyev 1977), and Tsopi Cave in Eastern Georgia (Vekua 1967). As tur are believed to have originated in the Greater Caucasus, their occurrence in the Caucasus Minor is considered unlikely (Baryshnikov 1987; Vereshchagin 1959). All Pleistocenic findings (no older records are available) from various caves of both sides of the West and Central Caucasus are defined as *C. caucasica*, except a single late Pleistocenic fossil from the Southern side of the Central Caucasus belonging to the *ibex* group (Baryshnikov 1987). Descriptions do not define fossils as belonging to Western or Eastern tur.

Summing up, in this chapter we treat *C. caucasica* and *C. cylindricornis* as a single species with the name *Capra caucasica.*

Current Distribution

Tur is distributed along the Greater Caucasus mountain chain (Fig. 3). The eastern limit is situated in the Babadagh Mountain massif (40°56′ N, 48°31′ E) in Azerbaijan (Kuliev 1981; Vereshchagin 1938). Presently, the westernmost area still harboring tur is Abago Mt. in Tchugush Mt. massif (*c.* 43°53′ N, 40°08′ E) (Kotov 1968; Nasimovich 1949; Trepet 2014). The total length of the contemporary range is about 750 km, the maximal widths occurring in the river basins of Avar Koisu and Andi Koisu Rivers in Daghestan (up to 65 km, Magomedov et al. 2001), and close to the Elbrus Mt. massif (up to 80 km, Kopaliani and Gurielidze 2009; Zalikhanov 1967; Akkiev pers. comm.). The distribution is most narrow (c. 12 km) in North Ossetia, where the five parallel ranges constituting the Northern side of the Greater Caucasus (Forest, Pasture, Rocky, Side, and Main Ranges) are compressed, and animals have been historically absent from the Main (Watershed) Range (Veinberg 2000). The distribution has changed little since the nineteenth century when it was slightly wider than today, encompassing peripheral mountain ranges (Veinberg 2000; Vereshchagin 1959). Currently, tur is almost absent from large southern spurs of the Main (Watershed) range (Southern side of the Greater Caucasus), such as Kodori and Svaneti Ranges in Georgia (Gavashelishvili 2004; Kopaliani and Gurielidze 2009). Because of steepness and narrowness of the Southern side of the Greater Caucasus, especially its eastern part, tur distribution there is naturally rather narrow. The situation is better on the Northern side, which is considerably wider and where animals usually inhabit the crystalline Side Range as well and sometimes even the limestone Rocky Range, the latter mainly in Kabardino-Balkaria and partly in North Ossetia (Akkiev 2018; Veinberg 2000; Zalikhanov 1967).

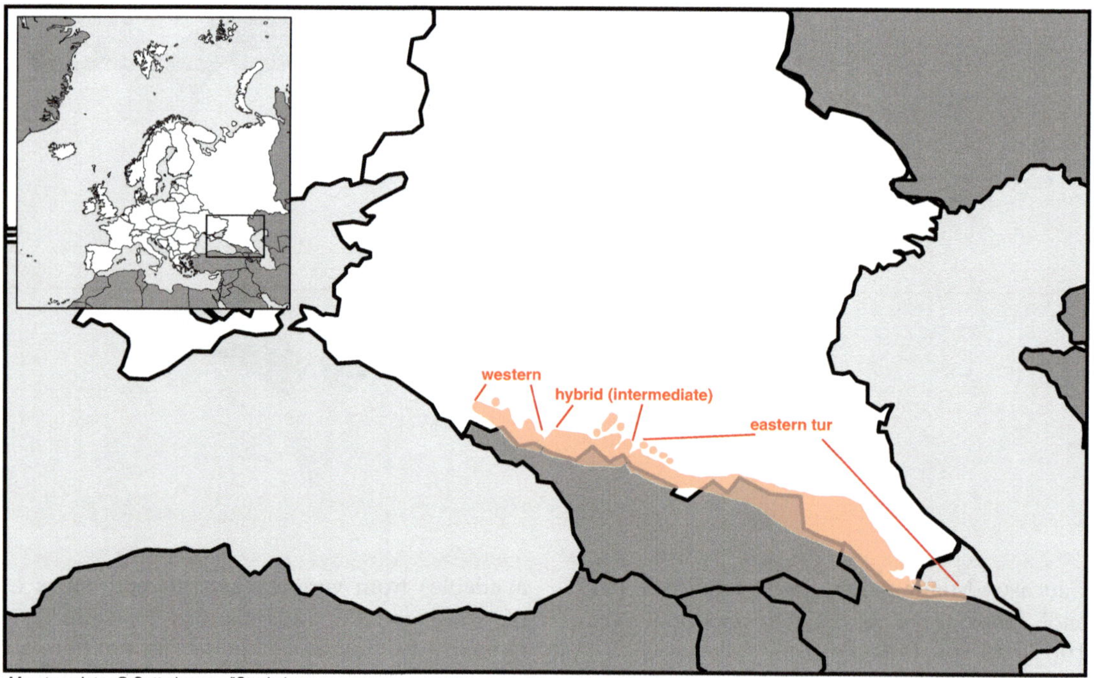

Map template: © Getty Images/iStockphoto

Fig. 3 Distribution of Western and Eastern tur. Distribution is modified from the IUCN Red List of Threatened Species (Version 2017-2) based on data from the authors and from Akkiev (2018) and Trepet (2014) (Map template: © Copyright Getty Images/iStockphoto)

Tur inhabit Georgia, Azerbaijan, and several administrative territories within the Russian Federation. From west to east: Krasnodarsky Kray, autonomous republics Adygea, Karachay-Cherkessia, Kabardino-Balkaria, North Ossetia-Alania, Ingushetia, Chechnya, and Daghestan (Weinberg et al. 1997). Since tur distribution is almost continuous, there are only few isolated populations, and population abundance estimates are available for countries, data are presented in the "Population Ecology" section (Population dynamics).

Description

Size and Morphology

Caucasian tur cannot be characterized as a uniform morphotype because of considerable geographic variability, especially in adult males. Tur, as all *Capra*, have pronounced sexual dimorphism. Females are smaller than males, their body mass being about 1/2 of adult male body mass (Table 1); females have small horns and, usually, no beard.

In general, measures and body mass of tur decline from west to east (Table 1). In Daghestan, however, animals may be considerably larger than shown in Table 1, with body length reaching 138 cm in adult females and 190 cm in 12–16-year-old males (Abdurakhmanov 1973).

In females, the mass of the axial part of skeleton is slightly lower (48.7%) than that of the appendicular part (51.3%), and bones of the hind limbs are heavier than those of the forelimbs. Peculiarities of the appendicular skeleton characterize tur as a climber, well-adapted to rugged, precipitous terrain (Kuliev 1981).

Tur females have two mammae; mammaries enlarge visibly during lactation. Tur have a subcaudal skin gland with characteristic smell (Pocock 1910; Sokolov 1959; Sokolov and Tembotov 1993).

Pelage

Summer coat of all tur is short and coarse. In the West Caucasus, summer coat in males and females is rather bright-colored, ranging from reddish grey to reddish chestnut. Underparts are whitish or yellowish gray. The tail is dark brown, as are the stripes along the frontal surface of the legs and the beard in males. In winter, regardless of age and sex, tur coloration varies from grayish brown to dirty white (Heptner et al. 1961; Kotov 1968).

The main difference in pelage between the Eastern tur and the Western tur is the very dark winter coloration of adult males in the former. Their coat is homogenously dark brown from the 3rd year of age; only 2–4-year-old males still display barely visible darker stripes on the legs (Dinnik 1910; Veinberg 1993). The grayish or dirty-yellow rump patch in all age and sex classes is very small and usually covered by the tail (Abdurakhmanov 1977; Veinberg 1984, 1993). Beard in adult Western tur is usually longer than 12 cm, and it can reach up to 18 cm. Beard of adult Eastern tur male is ≤12 cm in winter coat and, when stretched forward, never extends beyond the chin. Being short, coarse, and pointed forward, it differs from the long and drooping beards of most other *Capra*, including

Table 1 Basic measurements of adult tur (females ≥3 years, males ≥6 years, $\frac{mean}{min-max}$)

	Area					
	Caucasus NR (Kotov 1968)		N. Ossetia (Weinberg unpubl. data)		Daghestan (Magomedov et al. 2001)	
Dimensions	♀	♂	♀	♂	♀	♂
Body length (in cm)	$\frac{150}{136-176}$	$\frac{179}{159-196}$	$\frac{134}{114-140}$	$\frac{158}{146-178}$	$\frac{97}{85-119}$	$\frac{132}{98-170}$
Shoulder height (in cm)	$\frac{87}{78-92}$	$\frac{106}{101-110}$	$\frac{79}{72-84}$	$\frac{90}{85-99}$	$\frac{76}{71-80}$	$\frac{92}{86-98}$
Hindfoot length (in cm)			$\frac{31}{30-32}$	$\frac{34}{31-36}$	$\frac{31}{27-34}$	$\frac{36}{32-37}$
Full body mass (in kg) (autumn, pre-rut)	58–71	$\frac{143}{124-155}$			$\frac{55}{42-68}$	$\frac{115}{93-148}$

the Western tur. In summer coat, the beard is hardly noticeable (Veinberg 1993; Weinberg et al. 2010). Females rarely have beard but, when present, it appears wispy (Fig. 1). The beard is small in yearling males, quite conspicuous in 2-year-old males and fully developed in 4–5-year-old males in winter pelage (Fig. 1).

A single molt occurs in different age and sex classes from March to August (Veinberg 1984). In North Ossetia in warm years, females, yearlings, and juveniles stop shedding their winter coat by the beginning of May; in cold years they molt a month later, before parturition. Males 2–4 years of age shed their winter coat about 1 month later, older males do so 2 months later. Older individuals of both sex molt later than mature individuals (Veinberg 1984).

Skull and Dentition

The skull of Western tur male differs from that of the Eastern tur: firstly, the coronal suture is almost straight in the former, but projects forward at an angle in the latter (Tsalkin 1955; Heptner et al. 1961). Secondly, the highest point of the skull between the horn cores is situated closer to the back of the cores in adult Eastern tur, while in the Western tur it lies near the line connecting the front parts of the cores (Tsalkin 1955; Heptner et al. 1961; Weinberg et al. 2010). Thirdly, the parietal shows a depression in the eastern animals which is lacking in western ones (Tsalkin 1955). Western tur male horn cores lack faint spiralling twist evident in the Eastern tur, and nasals almost reach the line connecting centers of the orbits (Weinberg et al. 2010).

Dental Formula

0.0.3.3 / 3.1.3.3, total 32 (Heptner et al. 1961). Development of permanent dentition allows age determination in the first 5 years, since every summer, starting from the second one, a pair of incisors and then a pair of canines replace milk teeth. However, eruption of permanent dentition may delay in small undergrown specimens. One such yearling female also had one upper canine, normally lacking in caprines (Weinberg unpublished data).

Horns

Horns grow throughout the animal's life with cessations in the cold period of each year, approximately between October and April, and annual increments are much bigger in males than in females. Horns of juveniles are straight or slightly curved and oval in cross section. Horns of Western tur males diverge at an angle of no less than 60° (Sokolov 1959). Horn cores and sheaths of males older than 6–7 years are subtriangular in cross section and scimitar-curved, and the radius of the curve is large. Horn tips are usually pointed down-inward (Fig. 2a). The sheaths are massive and thick. In younger males, each annual segment usually has two transverse notches on the frontal surface, as in ibexes. These notches become less conspicuous with age. According to measurements taken by Tsalkin (1955), the length of 2–5th segments is on an average 36 cm, circumference at base in males older than 6 years is on average 29 cm, and the mass of the sheaths is 1–1.7 kg. Horn length of adult Eastern tur females usually reaches up to 20–22 cm, while in males it reaches 70–90 cm, with *c.* 30 cm of base circumference at 10–15 years of age (Abdurakhmanov 1973; Tsalkin 1955; Veinberg 1984). Maximal annual increment (≤16.5 cm) occurs in the second year in males, but reduces sharply after that, and is merely some 4 cm by the age of 10 years, or even less (Veinberg 1984). In yearling males, horns are triangular in cross section at base, and curve up, out, and then back. From the 3rd year, transverse wave-like notches (usually 2–3 on each annual segment) are prominent on frontal surface. The spiraling twist develops with age, and horn tips become distinctly pointed upward by 6–7 years forming about 3/4 of an open spiral curl, or even a full curl, in Daghestan and Azerbaijan (Figs. 1 and 2f–i). This shape usually does not depend upon individual rate of growth; horns of a 7-year-old male can be below the average size and significantly shorter than those of a 5-year-old one, but develop the "adult" 3/4 curl nevertheless (Veinberg 1993). Horns diverge widely, at ≥60° in males and *c.* 45° in females (Sokolov 1959). Average mass of horn sheath of a 14-year-old male is 2.45 kg (Abdurakhmanov 1973).

Anectodal evidence suggests that an area on the southern side on the border between Georgia (Lagodekhi Nature Reserve) and Azerbaijan (Zaqatala NR) harbors very large forest-dwelling males whose horns reach extraordinary length of 128 cm. A skull of a male from Lagodekhi NR, possibly killed by an avalanche, had huge horns, the longest reaching 125 cm (Weinberg unpublished data).

Physiology

The heart, lung, and liver indices (i.e., the proportion of the mass of the specific organ to the full body mass) are usually smaller in Daghestan tur males than in females, indicating that the latter may be less adapted to high altitudes (Abdurakhmanov 1973). Females indeed live at lower elevations than adult males (Veinberg 1984). In this respect, much bigger horn cores may offer still more advantage to males than to females, as they supposedly generate blood (Korzhuyev et al. 1978). Tur from Daghestan display high oxygen concentration in blood: 14.34 million erythrocytes/1 mm^3 and 12.4 g% of hemoglobin (Bulatova 1962). Oxygen concentration is slightly lower in Teberda NR, while both indices increase with age and are almost constantly higher in adult males than in females. This also seemingly indicates better adaptation to higher elevations in adult males, particularly in summer (Bobyr' and Semyonov 2008).

Body temperature is 40.0–41.6°C in juveniles and 38.5–41.0°C in adults. Pulse and breathing rates (per min.) also diminish with age (Bobyr' and Semyonov 2008).

Summer forage of females contains 1.3 times more nutrients (protein, fat, cellulose, and nitrogen-free substances) than that of males, likely because females prefer the shady slopes of northern exposures, where vegetation is richer in nutrients than on sunny slopes (Magomedov et al. 2001). In summer, relative food intake (per kg of the body mass) of young animals exceeds 1.5 times that of adults; this difference increases in winter, because juveniles do not lower intensity of their feeding, as older animals do (Magomedov et al. 2001).

Genetics

Chromosomes

Eastern tur have $2n = 60$ chromosomes (Kuliev and Mamedov 1974).

Genetic Diversity

There are no data on genetic diversity of tur, but since local tur populations usually are not isolated from each other and have not gone through recent bottlenecks, it may be assumed that diversity has not diminished in the recent history.

Hybridization

Eastern tur crossbreed with wild goat *C. aegagrus* and domestic goat *C. hircus* in captivity producing fertile offspring (Pfitzenmayer 1915; Sarkisov 1953). Information on hybridization of both species in the wild is indefinite (Pfitzenmayer 1915) and has not been confirmed by subsequent research (Heptner et al. 1961; Weinberg 1999). However, genetic studies revealed that certain Eastern tur specimens from Daghestan belong to the female lineage of local wild goat (Pidancier et al. 2006), suggesting that at least some limited hybridization might have taken place there. Other studies (Kazanskaya 2007; Kazanskaya et al. 2007) proposed that Western tur belonged to female lineage of the wild goat and to male lineage of the Eastern tur, thus being completely hybrid.

Life History

Growth

After a gestation of 160–165 days (Chlaidze 1967; Janashvili 1977; Vereshchagin 1938) generally one young is born between the end of May and August (Abdurakhmanov 1973; Chlaidze 1967; Veinberg 1984). The average full body mass of four neonates in Daghestan was 3.4 kg, and their body length was 38 cm (Magomedov et al. 2001). In the West Caucasus, neonates weigh about 4 kg (Bobyr' and Semyonov 2008; Zalikhanov 1967). Male and female juveniles in their first winter do not differ in size, and in North Ossetia their body length is 95 (87–100) cm (Weinberg 2002c).

Growth is slow and shows significant individual variation (cf. Table 1). Comparison of all data indicates that females reach full size and mass by 5–6 years of age, while in males growth slows down after 6–7 years, though they continue gaining mass until later years (Abdurakhmanov 1973; Bobyr' and Semyonov 2008; Kotov 1968; Magomedov et al. 2001; Weinberg 2002a; Zalikhanov 1967). Cranial sutures fuse by 6–7 years in males (Vereshchagin 1938). Complete dentition is achieved at 5 years. Teeth are heavily worn by the age of 12 (Vereshchagin 1938), and after this age animals usually start losing teeth (Weinberg unpublished data).

Reproduction

The rut normally lasts from mid-November to the end of December (Bobyr' and Semyonov 2008; Chlaidze 1967; Kotov 1968; Kurashvili et al. 1981; NACRES 2017; Vereshchagin 1938) but may delay in Kabardino-Balkaria and North Ossetia, lasting until beginning of January (Veinberg 1984; Zalikhanov 1967). Rut is often triggered by a cold spell (Abdurakhmanov 1977; Veinberg 1984), but later the rutting intensity can be hampered by cold weather, and the length of the mating season can be prolonged (Abdurakhmanov 1977; Enukidze 1965; Veinberg 1984).

In concordance with the times of rut, parturition occurs earlier on the warmer Southern side of the Greater Caucasus and is latest on the Central part of North Caucasus (Abdurakhmanov 1973; Kotov 1968; Kuliev 1981; Magomedov et al. 2001; Veinberg 1984; Vereshchagin 1938; Zalikhanov 1967). In Teberda Reserve, the latest births occurred at the beginning of August (Ekvtimishvili 1953b).

Twinning is exceptionally rare in the West and Central Caucasus, and in North Ossetia less than 0.1% of all females with newborns nursed two kids (Veinberg 1984), while in Daghestan 3–4% of females that gave birth had twins (Magomedov et al. 2001). Proportion of barren females varies throughout the range: from about 3–7% in the West Caucasus (Kotov 1968) and Kabardino-Balkaria (Zalikhanov 1967) to 40% in North Ossetia and Azerbaijan in some years (Veinberg 1984; Vereshchagin 1938). The proportion of barren females seemingly correlates with the snow depth of the previous winter, or even of the winter before (Weinberg 2002a). The average proportion of kids present in the population 1–2 months after parturition is just 13% in the West Caucasus and in Kabardino-Balkaria (Bobyr' and Semyonov 2008; Kotov 1968; Romashin 2001; Zalikhanov 1967), about 16% in North Ossetia (Veinberg 1984), and over 20% in Daghestan (Magomedov et al. 2001) and Azerbaijan (Kuliev 1981). This proportion correlates positively with the productivity of subalpine and alpine pastures in the West Caucasus (Romashin 2001).

Juvenile/female ratio in August–September is much more informative than juvenile proportion. In Daghestan, this index correlates inversely with population density and average group size and correlates positively with proportion of accessible summer pastures and slopes of southern exposures, which serve as main winter pastures (Magomedov et al. 2001). In North Ossetia, juvenile/female ratio correlates negatively with the snow depth of the previous winter. Juvenile survival (yearling/female to juvenile/female ratio of the previous year) also showed similar correlation, but less strong (Weinberg 2002a). Yearlings usually make only about 5–9% of the population (Bobyr' and Semyonov 2008; Kotov 1968; Romashin 2001; Veinberg 1984; Zalikhanov 1967).

Yearling tur females from Daghestan (Heptner et al. 1961) and 3-year-old males (Janashvili 1950) already mate in captivity. In the wild, females reach sexual maturity at 2 years of age in East Caucasus (Abdurakhmanov 1977; Vereshchagin 1938), though in Daghestan only a small proportion of 3-year-old females have kids (Magomedov et al. 2001). In West Caucasus, females usually give first birth at 4 years; out of 22 captured 3-year-old females only 1 was pregnant (Kotov 1968). Males participate in the rut from the age of 4–5 years in Kabardino-Balkaria (Zalikhanov 1967).

Direct information on age at primiparity and pregnancy rate of female tur is not available for the most part of the distribution. Pregnancy and lactation, however, are energetically expensive, and they might influence horn growth increments (cf. Miura et al. 1987 for Japanese serow *Capricornis crispus*). In turn, this suggests that the relative length of horn annuli might offer some insights into these life history traits in female tur (Weinberg 2002b). Age at primiparity is seemingly 4 years in North Ossetia, though some individual heterogeneity occurs, and females in good conditions might give birth already at the age of 2. The proportion of barren females appears lowest between 8 and 12 years, after which their proportion in the population grows (Weinberg 2002b). Some females, however, can still reproduce after 16 years of age (Magomedov et al. 2001, Weinberg 2002b). Annual horn segments in females older than 2 years indicated that 69% of them might have given birth, which is slightly higher than the proportion of observed females with kids in summers (56%). This difference most likely owes to death of newborns and to birth of stillborns (Weinberg 2002b).

Sex and Age Structure, Survival

Sex ratio varies throughout the distribution range and over the years. Because male and female habitat preferences are different, a robust value of sex ratio can be obtained only during the rut, and in Azerbaijan this ratio was significantly in favor of females, both within protected (0.61:1) and unprotected areas (0.44:1) (Guliev et al. 2009). Extraordinarily biased sex ratio (1:15–20) has been reported for the Khevsureti area in Georgia (Kopaliani and Gurielidze 2009), though no explanation was provided as to the reasons for such figures. In North Ossetian Nature Reserve, sex ratio was male-biased in 1979–1987 (1.02:1) and fully adult males made up *c.* 13% of the population during the rut (Weinberg 2002a). After 1987, the situation got worse: sex ratio was permanently in favor of females, dropping to some 0.2:1 in 1993/1994, and adult males made up just 4.7% of the population. Both figures, especially the very low proportion of fully mature males, indicate a high poaching level (Weinberg 2002a).

In Daghestan, mortality is very high during the 1st year of life and reaches *c.* 40% (Abdurakhmanov 1977; Magomedov et al. 2001). Annual mortality gets increasingly sex-specific with age: at 3 years it is 9.5% for both males and females, whereas by the age of 14 it is 32% and 14%, respectively (Magomedov et al. 2001). Longevity is 15–16 years (Heptner et al. 1961; Magomedov et al. 2001), but some animals can live up to 21–23 years (Vereshchagin 1938; Weinberg 2002b; Zalikhanov 1967). In North Ossetia, only females exceeded the age of 16 years (Weinberg 2002b).

Habitat and Diet

Habitat Selection and Movement

Tur distribution and seasonal movements depend largely upon vegetation and, hence, climatic conditions which differ significantly over the tur range. Annual precipitation is generally higher on the southern side of the Main Range: 644–688 mm in North Ossetia and Daghestan and 1394 mm in Azerbaijan. Precipitation occurs mostly in the second half of spring and in summer (Gasanov 1990; Malikova 1960), and it increases from east to west, reaching about 1200 mm on the Northern side and over 2000 mm on the Southern side of the West Caucasus (Akatov et al. 1990).

Tur live at 1000–4000 m of elevation, in forest, subalpine, alpine, and subnival habitats (Abdurakhmanov 1977; Ekvtimishvili 1952a; Veinberg 1984; Vereshchagin 1938). However, areas >3500 m a.s.l. are visited only in summer and rarely even then (Veinberg 1984). Animals avoid thick forests on gentle slopes but are often found in open forests growing on steep, precipitous slopes. These areas are used not only in winter, but also during snow-free periods, sometimes all year round, especially by females with kids and by young males (Veinberg 1984). Some populations live in broad-leaved forests composed of beech *Fagus orientalis*, hornbeam *Carpinus caucasica*, and oak *Quercus macranthera* on the Southern side of the East Caucasus and are separated from the subalpine zone (Janashvili 1963; Markov 1938; Vereshchagin 1938). On average, over 1/3 of tur live in the forest throughout the year in Georgia (Ekvtimishvili 1952a). High levels of human disturbance and occupation of subalpine and alpine zones by domestic livestock might have forced these animals to move into the forest (Chlaidze 1967; Markov 1938; NACRES 2017). The proportion of animals dwelling below the timberline decreases westward, because animals tend to avoid closed tall spruce *Picea abies* and fir *Abies nordmanniana* stands, widespread in the West Caucasus, but totally absent in the Central and East Caucasus where pine *Pinus kochiana* and birch *Betula litwinowii* dominate, forming more open grassy stands (Bobyr' and Semyonov 2008; Kotov 1968; Magomedov et al. 2001; Trepet 2014; Veinberg 1984). Long periods of cold and rainy weather cause subalpine and alpine-dwelling females with kids to seek shelter in forests even in summer (Veinberg 1984).

Choice of slopes with different exposures also varies over the distribution range. Exposure selection is most consistent in winter when the majority of animals (up to 90%) all over the Caucasus concentrate on sunny slopes (Aiunts and Kolomyts 1986; Bobyr' and Semyonov 2008; Kotov 1968; Magomedov et al. 2001; Veinberg 1984; Zalikhanov 1967). This can be explained by a much higher snow cover above timberline on north slopes than on south slopes (ratio 1:8) in West Caucasus (Kotov 1968). Similarly, in spring tur try to find early grass on sunny slopes. However, in summer and autumn, the choice of exposure may differ significantly, because animals seek more nutritious vegetation. In the less humid and more insulated East and Central Caucasus, vegetation dries quickly, and fresh vegetation can be more easily found on shady slopes (Aiunts and Kolomyts 1986; Magomedov et al. 2001; Veinberg 1984: Zalikhanov 1967). In the over-humid West Caucasus, and possibly on the humid Southern side of the whole Greater Caucasus, fresh vegetation mainly occurs on south slopes which are preferred by animals (Bobyr' and Semyonov 2008; Gavashelishvili et al. 2018; Kotov 1968). Approaching autumn, the last snowfields with fresh vegetation around can be found on shady slopes (Bobyr' and Semyonov 2008). However, in summer and autumn, preference of specific exposures is not as consistent as in winter.

Male and female tur select different habitats. In North Ossetia, females use forests more intensively than adult males. Altitudinal distribution of females and adult males fully overlaps only during the rut. From February to April, over 80% of females, kids, yearlings, and young males dwell in the forest, while over 50% of adult males remain in subalpine and alpine areas. In summer, above timberline, over 80% of adult males feed on gentle slopes, while about 63% of females with yearlings and kids graze on cliffs (Veinberg 1984). In Daghestan, adult males are more numerous on medium-steep slopes (26–46°), while females appear less selective to steepness of the slopes (Magomedov et al. 2001). In general, females usually live in closed habitats (more forested or more precipitous and rugged), while adult males prefer more open habitats. The habitat differences described above, combined with differences in diurnal activity, generate ecological and, often, spatial segregation between females with offspring and adult males (Veinberg 1984).

Seasonal migrations rarely exceed 5 km distances in North Ossetia and Daghestan (Abdurakhmanov 1977; Veinberg 1984). Two regular migrations from Azerbaijan to Daghestan occur across the Watershed Range: one in winter, when snow is more abundant on the southern side, and the other in summer, in order to avoid biting insects (Vereshchagin 1938). In Kabardino-Balkaria, however, adult males may cover some 20–25 km during pre-rut migrations, from the Side Range to much less snowy Rocky Range where females are present (Zalikhanov 1967). In general, animals move northward, away from the snowy Main (Watershed) Range. This regularity is common in Northern Caucasus (Aiunts and Kolomyts 1986; Veinberg 1984, Weinberg 2002a). Tur perform mainly vertical migrations in West Caucasus, with limited changes in geographical distribution over the year (Kotov 1968; Trepet 2014). Seasonal migrations of adult tur males usually exceed those of other age and sex classes (Veinberg 1984).

Diurnal movements are most evident in summer and above timberline (Abdurakhmanov 1973;

Veinberg 1984). Their direction depends upon the location of feeding areas and resting sites. Usually the latter occur at higher elevations above the subalpine meadows in North Ossetia, and animals move downslope in the evenings to move back upward in the mornings (Veinberg 1984). The scarcity of vegetation above 3,200 m reduces the amount of forage, but tur may move to higher elevations either to avoid daytime summer heat or to escape from biting insects, predation, or hunting (Gavashelishvili et al. 2018). In Kabardino-Balkaria and, especially, Daghestan, ridges are less rocky, and animals may spend the day below their feeding sites, in shady canyons, so the direction of diurnal movements may be opposite to that described above (Aiunts and Kolomyts 1986; Vereshchagin 1938). Typically, males move up to 2000 m horizontally and 1000 m vertically during the day, whereas females move *c.* 500 m horizontally and 300 m vertically (Chlaidze 1975; Magomedov et al. 2001; Veinberg 1984).

Population Density

Average population density varies from 0.15 to 17 animals/km^2 (Kotov 1968; Kuliev 1981; Magomedov et al. 2001; Veinberg 1984).

Feeding

Tur are mixed feeders. Practically all plant species occurring in the habitat are eaten, but only about 60 constitute the main diet, and some 20 of these are preferred (Magomedov et al. 2001). Grasses dominate in the autumn and winter diet throughout the range, while forbs prevail in spring and summer (Kotov 1968; Magomedov et al. 2001) and make on average no less than 30% of the rumen contents in all seasons in Daghestan. In winter, animals browse more than in summer. *Euonymus*, pine, dog rose *Rosa* sp., and willow *Salix* sp. are preferred browse all over the range (Abdurakhmanov 1973; Romashin 2001; Veinberg 1984). In snowy winters, particularly in February and March, animals may feed on pine needles (Veinberg 1984). On the southern side, trees, shrubs, and low shrubs are essential in winter (Ekvtimishvili 1953a; Kuliev 1981). Lichens and especially mushrooms are seldom eaten (Zalikhanov 1967). Tur also eat different parts of the plants in different seasons. In May and June, for example, animals feed on whole plants, while in July–August, they select buds and blossoms, i.e., the parts with highest nutritional value (Kotov 1968).

Impact on Vegetation

In Daghestan, tur use 29.0–51.3% of the phytomass in different types of plant communities above timberline (Magomedov et al. 2001). They can change the vegetation composition and lower the productivity of feeding sites. In areas where tur were the only large grazers, productivity of plant communities within enclosures protected from tur rose approximately twofold already in the second year, and new grass species emerged (Magomedov et al. 2001). In Teberda, tur may consume over 70% of phytomass in their most important wintering sites (Bobyr' and Semyonov 2008).

Mineral Feeding

Animals often visit natural salt-licks (mineral springs, clay, soil etc.). Tur use salt-licks mostly at the end of spring–beginning of summer (Bobyr' and Semyonov 2008; Kotov 1968; Romashin 2001; Zalikhanov 1967). Daghestan tur eat soil impregnated with natural minerals. Animals lick minerals deposited on the surface of cliffs and drink from mineral springs (Vereshchagin 1938). Otherwise they seldom drink (Zalikhanov 1967), apparently obtaining sufficient water from fresh grass and snow. Kids start salt-licking before the age of 1 month (Kotov 1968).

Behavior

Social Behavior

Four types of groups exist: adult male groups, sometimes including younger males; young male groups; female groups including young males; and mixed groups with at least one adult male and one adult female, apart from other animals. These latter groups are rare outside the rut. Based on the behavior during the rut and association with females outside the rut, only males older than 6–7 years can be regarded as fully adult in non-hunted populations (Veinberg 1984). In North Ossetia, in years of high density, the average size of mixed groups was about 10 animals during the rut. Adult male groups were the largest ones outside the rut (with *c.* 12 individuals on average). They split in November, just prior to rut, when adult males start roaming in search of females, and reunite in January or February. Female groups are the most stable ones, and their size fluctuates little throughout the year, averaging seven individuals. Young male groups can occur throughout the year but are rarely observed and small, probably because most young males are members of adult male or female groups. The overall mean group size increases from *c.* 6 individuals in November–January to *c.* 10 individuals in May–July (Veinberg 1984), this regularity being common in most parts of the range, because animals disperse over smaller accessible plots of pastures in snowy winters (Bobyr' and Semyonov 2008). The average group size also depends on topography, being 2.5 times larger in more rolling and gentle valleys than in more rugged areas, despite generally lower population density is found in the former (Veinberg 1984). Therefore, in less precipitous Azerbaijan, average group size was about 78 (Kuliev 1981). Group size is also negatively affected by human disturbance (Gavashelishvili et al. 2018).

Females form communities with well-defined home ranges covering 4–6 km^2. Groups are not stable within these communities; they join and split, even daily. Home ranges of adult males are considerably larger, about 25–30 km^2 due to their active roaming during the pre-rut and rut. Adult male groups are philopatric, using perennial trails, resting sites, and feeding areas (Veinberg 1984). Animals display full knowledge of topography within their home ranges. In case of danger, they do not hesitate but run across precipitous terrain, in each case using a fixed and familiar escape route. When individually known animals were met in precipitous terrain outside their home ranges, they demonstrated lack of knowledge of local topography and inability to choose instantly correct escape routes used by local animals (Weinberg unpublished data).

Mating Behavior

Courtship is generally displayed only during the rut, although young, 3–4-year-old males may attempt to court females already in spring and summer. Rutting males keep their tails up and folded over the back. Males older than 5 years do so constantly; 3-year-olds raise their tails only in absence of older males and only for short periods. Courtship generally resembles guarding and cautious approaching. Low-stretch (the neck and head are stretched and are in line with the dorsal spine) forms the basis of adult male behavior while courting females, and most other courtship patterns (kick and tongue flicking) are performed by the male while low-stretching (Bobyr' and Semyonov 2008; Kotov 1968; Veinberg 1984). This posture seemingly makes the most intimidating feature of adult males – their large horns – less conspicuous. The intensity of courtship is much higher in adults (3 displays/h) than in 3–6-year-old males (2 displays/h) (Veinberg 1984). Unlike cautious "tending" tactic of adult males, young ones usually do not guard and court females; on the contrary they try to mate with unreceptive females, chase and harass them using a "coursing" tactic. Young male may attack a female if she tries to drive him away. Females escape from *c.* 1/3 courtship attempts of adult and about 1/2 attempts of young males and respond aggressively to *c.* 9.5% attempts of adult and *c.* 15.5% of young males, clearly showing preference for adult males. Young males court yearlings and kids much more often than adult males do. They probably have more chance to breed only after the main rutting period, when adult males are exhausted but some females still remain unfertilized. However, no research with genetic analyses had been conducted on success of different mating tactics: "tending" of adult males vs "coursing" of young males. No successful matings have been recorded in North Ossetia (Veinberg 1984), and only one copulating pair had been observed in Teberda, already after sunset (Bobyr' and Semyonov 2008). Observations on tur (Veinberg 1984) and mating Asiatic (Siberian) ibex, that took place just after sunrise (Fedosenko et al. 1992), indicate that presence of other animals (besides the courting pair) irritates female and prevents copulation.

Agonistic Behavior

Adult males display agonistic behavior almost exclusively during the rut. Aggressive interactions take place mainly between animals of different age classes. A young male typically experiences about 0.8 aggressions/h from adult males, while adult male receives only 0.2 from his own class (Veinberg 1984). Adult dominant tur males never expel other males from rutting groups. Aggressive contests are usually based on indirect threats, such as broadside and demonstrating horns. The only exceptions to this rule are fights that are rare in North Ossetia (Veinberg 1984), contrary to Daghestan (Abdurakhmanov 1977), Kabardino-Balkaria (Zalikhanov 1967), and Azerbaijan (Vereshchagin 1938). Fights take place between equally sized males and are preceded by aggressive displays. They are extremely fierce and are not ritualized. Clashing from a bipedal position is followed by horn wrestling. Animals lock horns and sometimes roll down the slope together. Blows are delivered to the head and body. Males may fight in head-to-tail position and butt each other on the belly from below. The tail is kept down during a fight. The loser leaves the group, followed, but not chased, by the winner, who then rejoins the group (Veinberg 1984). One lethal fight had been observed in Teberda Nature Reserve (Bobyr' and Semyonov 2008).

During the rut, females display aggression only as a response to courtship. They avoid adult males outside the rut and are dominated even by 2–3-year-old males. This might possibly be one reason why females tend to separate from young males at the beginning of the birth period. One–two months after the parturition period, juvenile/female ratio correlates inversely with the presence of young male(s) in small (2–5 females) female groups and also with the number of females in the group (Weinberg 2004). Females are often aggressive toward other females' kids. Female aggression is primarily intrasexual (2/3 of all events) and is mainly based on direct forms, like butting. This behavior is probably related with late pregnancy, because it is usually observed in April and May (Veinberg 1984). No dominance-submission relationships occur between adult males and females (Veinberg 1984).

Activity

In summer, tur feed at night, even if undisturbed, and by the end of dawn animals start moving back to their resting sites. Adult males are practically inactive from 0800 h until 1600 h in nice weather, whereas females' activity never ceases completely (Bobyr' and Semyonov 2008; Kotov 1968; Veinberg 1984, etc.). Activity shifts to a more nocturnal cycle when some form of anthropogenic impact occurs, such as livestock pasturing or hunting (Chlaidze 1975; Kotov 1968; Zalikhanov 1967; NACRES 2017). In Azerbaijan, where shepherds seldom hunt, animals seem to be less disturbed by presence of livestock and often choose daytime resting sites close to livestock camps. During the rest of the year, animals are active in daylight hours (Veinberg 1984).

Feeding Behavior

Animals paw through snow (up to 30 cm deep), dig out underground parts of plants, but also bend branches with their forelegs; rear up on their hind legs to reach leaves, twigs, or pine needles; and even climb slanting trees, though are less skilled in this respect than wild goats (Veinberg 1984; Weinberg 1999).

Parental Care

Usually females are solitary just prior to giving birth, stay in rugged terrain and avoid other females for about a weak after parturition, and tend to dwell below timberline or just above it (Kotov 1968; Zalikhanov 1967). After parturition, females sometimes hide their neonates, leaving them alone for 2–3 days, periodically visiting to nurse (Abdurakhmanov 1977; Kotov 1968). Even after this hiding phase is over, the female with her kid may avoid other conspecifics for another 3–7 days (Zalikhanov 1967). Kids suckle in head-to-tail position. Weaning begins in September, and females not only step away but also kick and rarely even butt their kids to prevent suckling (Veinberg 1984). However, kids continue suckling till December although they begin grazing already in July. A female found dead in January had milk in her udder (Veinberg 1984). Sometimes even a yearling may suckle, supposedly when the female has lost her neonate (Zalikhanov 1967). Female tur can adopt an orphan kid and take care of it together with her own kid (Enukidze 1965).

Communication

Vocalization is poor. The alarm signal is a sharp and hissing whistle (Abdurakhmanov 1977; Veinberg 1984; Vereshchagin 1938). Females and kids bleat to each other (Dinnik 1910; Zalikhanov 1967).

Marking is rare (32 observations) and performed by all age and sex classes though mostly by adult males (2/3 of all observations) who mark only during the rut (Veinberg 1984). Males de-bark trunks or bigger branches using their horns and then rub against the bare places with the postcornual area, sniffing periodically at the mark, that may be spread over some 70 cm. In North Ossetia numerous old and new marks are characteristic of wintering areas and most often occur on pines. Marks of adult males have durable scent (Veinberg 1984) although no specific postcornual glands are known in tur (Sokolov 1973; Sokolov and Tembotov 1993). Young males mark during the rut and outside it; females do so only outside the rut. They mark twigs and do not de-bark them. On the whole, marking evidently belongs to agonistic behavior. It is not connected with territoriality, as tur are not territorial (Veinberg 1984).

Parasites and Diseases

Ectoparasites

Mange, caused by the tick *Acarus siro*, occurred in Azerbaijan (Vereshchagin 1938), Daghestan (Magomedov et al. 2001), and Kabardino-Balkaria (Zalikhanov 1967). Animals suffer from lice *Mallophaga*, ticks *Haemaphysalis sulcata*, *H. warburtoni*, *Rhipicephalus turanicus*, and larvae of gadfly *Oestrus turanicus* and *Oestrus caucasicus* (Abdurakhmanov 1973; Asadov 1959; Bobyr' and Semyonov 2008; Ekvtimishvili 1952b; Kotov 1968; Rukhlyadev 1964; Vereshchagin 1938; Zakariyev 1980).

Endoparasites

The helminth fauna of tur is comparatively poor due to the severe environment of the highlands (Asadov 1959; Zakariyev 1980). Fifteen examined animals from Azerbaijan had no worms at all (Vereshchagin 1938). Nevertheless, known tur parasites are tapeworms *Coenurus cerebralis* (*Taenia multiceps*) and *Cysticercus tenuicollis* (*Taenia hydatigena*), larvae of *Oestrus caucasicus*, flukes *Dicrocoelium lanceatum* and *Fasciola hepatica*, *Echinococcus granulosus*, 29 species of nematodes (Abdurakhmanov 1973; Akkiyev 1997; Asadov 1959; Bobyr' and Semyonov 2008; Ekvtimishvili 1952b; Rukhlyadev 1964; Vereshchagin 1938; Zakariyev 1980). Zalikhanov (1967) reports that about 65% of hunted animals in Kabardino-Balkaria had sick lungs. In Georgia, 100% of tur population had Trichostrongylidae and Protostrongylidae nematodes (Chlaidze 1967). Parasites of the following genera were described for tur in Georgia: *Cysticercus*, *Chabertia*, *Oesophagostomum*, *Marshallagia*, *Muellerius*, *Cystocaulus*, *Neostrongylus*, *Gongylonema*, *Nematodirus*, *Ostertagia*, *Protostrongylus*, *Rinadia*, *Trichocephalus*, *Haemonchus*, *Linguatula*, *Skryabinagia*, and *Skryabinema* (Chlaidze 1967; Enukidze 1965).

Tur reportedly suffered from foot-and-mouth disease in Kabardino-Balkaria and, perhaps, in Daghestan (Zalikhanov 1967). Some unknown plague reduced tur numbers in Central and Eastern Caucasus by the end of the nineteenth century, killing about 2/3 of tur population in Baksan Valley (Elbrus Mt. massif: Zalikhanov 1967).

Population Ecology

Population Dynamics

Tur numbers were low at the end of the nineteenth–beginning of the twentieth century. Conservation status remained bad until after the Second World War. A period of population growth started after the Second World War and lasted approximately till the mid-1980s (Heptner et al. 1961; Weinberg et al. 1997). In the late 1960s and early 1970s total numbers of tur were calculated to be approximately 48,000 animals (Ravkin 1975). These data appear roughly adequate, except those for Krasnodar Krai, including the Caucasus Nature Reserve (NR), where 26,000 and 16,000 animals had been calculated, respectively (Table 2), but they were most likely overestimated. These figures have been revised lately, and maximum numbers turned out to be approximately 7,000 in the Caucasus NR (Romashin 2001; Trepet 2014), so that the total number for the North Caucasus had been hardly over 35,000 in the 1960s–1970s (Table 2).

This increase was followed by a decrease that was not simultaneous throughout the range (Table 2). In Georgia, the numeric decrease began in the mid-1970s (Eriashvili 1990). In Russia, it started a decade later and continued throughout the 1990s: it was most drastic in Western Caucasus and North Ossetia and very evident in the nature reserves (Caucasian, Teberda and North Ossetian NRs) where numbers dropped two or even three times (Romashin 2001; Trepet 2014; Weinberg 2002a). Several major factors may have caused this decrease. Firstly, the extremely snowy winter of 1986–1987, which triggered catastrophic decrease of reproduction that lasted several years, and secondly, subsequent outbreak of poaching due to several ethnical conflicts in the Caucasus, disintegration of the Soviet Union, collapse of law enforcement system, and appearance of refugees (Weinberg 2002a). The situation seemed better in Daghestan, and tur numbers there had been estimated at 18,000–20,000 in the second half of the 1990s (Magomedov et al. 2001). However, tur population in Daghestan is now estimated to be about 15,000–16,000 (Babaev et al. 2017). Likewise, too optimistic estimates were presented for Azerbaijan, 13,700 tur in 1993 (Guliev 2000).

Lately, tur population has stabilized and totals some 35,000 individuals, of which Western tur makes only up to 4000 animals (Caucasus NR and maybe Svaneti area in Georgia), and the overall trend is positive (Table 2). It should be noted that the relative numbers of Eastern and, especially, Western tur depend upon taxonomic views of the surveyors and reviewers. However, Western tur indeed makes a small proportion of the totals. Many local tur populations, particularly in Azerbaijan (mostly on the Southern side) and Daghestan, Russia (Northern side of the Greater

Table 2 Dynamics of tur numbers (Western tur/Eastern tur)

Year	Country			Total
	Russia	Georgia	Azerbaijan	
1960s–1970s	16,000/25,000[a]	500/4,000[b]	7000[c]	35,000
1980s				12,000/18,000–20,000[d]
1990s		2,500/2,800[e]	13,700[f]	
2003–2013	3,000/24,000[g]	1,000/5,900[h]	6,000[i]	4,000/35,000

[a]Abdurakhmanov (1977), Ravkin (1975), Veinberg (1984), Zalikhanov (1967), Romashin (2001), Trepet 2014; [b]Chlaidze (1967); [c]Alekperov et al. (1976); [d]Weinberg et al. (1997); [e]NACRES (1996) [f]Guliev (2000); [g]Trepet (2014), Bobyr' and Semyonov (2008), Akkiyev and Pkhitikov (2007); [h]NACRES (2003), Gurielidze et al. (2013), Kopaliani and Gurielidze (2009); [i]Guliev et al. (2009)

Caucasus) are transboundary (see above), so it seems likely that double counts occur on both slopes of the Main Range.

The population growth potential of tur is low due to rare twinning and low yearling proportion (see "Life History"); thus recovery after a decrease is usually slow and may take several years even under full protection, as it happened in the Caucasus NR after disintegration of the USSR (Trepet 2014).

Inter-specific Interactions

Tur are sympatric with red deer *Cervus elaphus* in Eastern and Western Caucasus, but, since red deer usually avoid rugged areas and are much less numerous above timberline, competition is minimal (Popkova 1968; Veinberg 1984). Wild goat is restricted to Daghestan, Chechnya, and North-Eastern Georgia (Heptner et al. 1961; Weinberg 1999) and shares forest habitats with the Eastern tur, outnumbering tur in these habitats by 6–7 times, but rarely occurs above the timberline (Weinberg 1999). Tur males were observed joining wild goat herds during the rut, but wild goat females were evidently frightened and avoided them (Weinberg 1999). Chamois *Rupicapra rupicapra* are more common in the West Caucasus and on the south slope of the Watershed Range in East Caucasus and often share habitat with tur, competing for food (Heptner et al. 1961; Popkova 1968). Usually, occurrence and altitudinal distribution of chamois may depend upon the presence of tur and not vice versa (Popkova 1968; Veinberg 1984). In general, the spectrum of habitats used by chamois depends upon the occurrence of competing ungulate species, like tur, red deer, etc. (Veinberg 1983). Tur sometimes mixed with chamois groups and stayed together for a while at natural salt-licks in Svaneti (Gabliani 1930). Livestock, sheep in particular, are the main tur competitors outside strict nature reserves. Winter densities of tur population decrease from *c.* 30–35 animals/km^2 in areas without intensive shepherding to merely 12–15 animals/km^2 in areas where >80% of wintering pastures of Daghestan tur are used by sheep in summers (Magomedov et al. 2001).

Natural Mortality and Impact of Predators

Snow avalanches are the main cause of mortality in Western and Central Caucasus, where they are responsible for more than a half of adult tur mortality (Bobyr' and Semyonov 2008; Veinberg 2000) and may kill about 4% of the population (Bobyr' and Semyonov 2008; Kotov 1968; Zalikhanov 1967). In North Ossetia they killed mostly adult males who usually stayed above the timberline in winters after the end of the rut (Veinberg 1984). In the less snowy Daghestan, they account for *c.* 35% of tur annual natural mortality (Abdurakhmanov 1973).

Wolves *Canis lupus* are a greater threat in less rugged terrain. In a valley of Kabardino-Balkaria, a pack of 5 wolves killed 31 tur in one winter (Zalikhanov 1967), while in Teberda, tur remnants are found in 46% of wolf droppings throughout the year, the proportion rising to 70% in winter–spring, and annual loss was estimated at *c.* 4% of the population (Bobyr' 1997). This figure seems too high, because combined with similar amount of mortality from avalanches (see above) would make 8% of the population, equaling or even surpassing the proportion of yearlings (see "Reproduction: Life History"), and would thus totally prevent population growth. In the Caucasus Reserve, the proportion of tur remnants in wolf droppings is considerably lower (23%: Kotov 1968), probably due to abundance of red deer, the main wolf prey (Alexandrov 1968). No clear correlation was found between numbers of wolf and population density and juvenile proportion of tur in Caucasus NR (Trepet 2014). Wolves account for 20% of natural tur mortality in Daghestan (Abdurakhmanov 1973). By contrast, only two successful wolf hunts are known in the steep areas of North Ossetian Reserve, and even wolf tracks are rarely seen within tur habitat, whereas lynx *Lynx lynx* is a regular predator on tur there, killing mostly kids (about 50% of all prey) (Veinberg 2000). All over the Caucasus, lynx hunts tur throughout the year, concentrating mainly on young animals and females (Kotov 1968; Zalikhanov 1967); its impact, however, may be considered negligible because of its low density (Bobyr' and Semyonov 2008). In Georgia, both wolves and lynx predate on tur (Chlaidze 1967, 1975; Enukidze 1965), but in Lagodekhi the highest tur proportion in wolf diet was merely 2%, possibly because of abundant presence of red deer (Gurielidze 2004). Bears *Ursus arctos* predate on females with kids and sometimes try to prey on adult males (Chlaidze 1967; B. Lortkipanidze pers. obs.). In North Ossetia, tur remnants are found in bear feces but, as bears constantly inspect avalanches and dig out dead animals, it is impossible to estimate bears' direct hunting impact upon tur (Weinberg, unpublished data).

Golden eagles *Aquila chrysaetos* hunt neonatal tur (Abdurakhmanov 1977; Veinberg 1984; Zalikhanov 1967), causing 5–7% of their mortality (Abdurakhmanov 1973). All four bearded vulture *Gypaetus barbatus* nests examined in the North Ossetian Reserve contained many hooves and fragments of skulls of newborn tur kids (Veinberg et al. 1983), but whether they hunt actively or collect already dead newborns is unclear. In Russian literature, bearded vulture is traditionally regarded as a real predator on juvenile mountain ungulates (e.g., Abdurakhmanov 1973; Magomedov et al. 2001) though direct evidence is lacking, in contrast with golden eagle whose attacks on tur and wild goat kids had been witnessed (Magomedov et al. 2001; Weinberg 1999).

The rest of adult tur mortality is due to poaching (legal hunting being very limited). In Teberda, this factor rose from 1.4% in 1986–1990 to 21.6% of total mortality in 1996–2000 (Bobyr' and Semyonov 2008). This 20% additional poaching-caused mortality not only eliminated natural increase, measured by proportion of yearlings in the population (up to 9% of yearlings in the whole population, see section "Life History: Reproduction"), but could diminish population by 10% annually. All researchers agree that poaching is the main anthropogenic threat to tur all over the Caucasus.

Effects of Climate Change

Glaciation area in the Greater Caucasus shrunk by 794 km^2 or 36% between 1910 and 1976 (Panov 1993). This process continues in the twenty-first century, and it causes displacement of vertical zonation, constantly rising the timberline and the upper limit of the periglacial zone. However, to date the impact of these processes on tur is unknown.

Conservation Status

Western tur *Capra caucasica* is listed as Endangered A2ad ver 3.1 in the IUCN Red List (Weinberg 2008a), while the much more numerous Eastern tur *Capra cylindricornis* is listed as Near Threatened ver 3.1 (Weinberg 2008b). The same categories have been suggested during the latest re-assessment of the status (Weinberg 2020; Lortkipanidze and Weinberg 2020). Western and Eastern tur are also listed in the Bern Convention as European endemics, respectively, as Endangered and Vulnerable (Preliminary European Red List of Vertebrates, Vol. 2). Caucasian tur is also listed in the CITES Appendix II.

Caucasian tur, treated as a single species (*Capra caucasica*), was regarded rare in the USSR but was not included in Red Data Books of the USSR and Russian Federation. Eastern tur occurs in several nature reserves and national parks: Zaqatala, Ilisu, Shakhdagh (Azerbaijan), Lagodekhi, Tusheti, Kazbegi (Georgia), North Ossetian, and Kabardino-Balkarian (Russia: Weinberg et al. 1997). Western tur is protected in just one nature reserve: Caucasus NR (Russia). Both tur species are included in the Red List of Georgia. Eastern tur is Vulnerable and Western tur is Critically Endangered (Ordinance of the Government of Georgia on Adoption of Georgian "Red List" 2014). Hunting of both species is prohibited in Georgia.

Management

Habitat Management and Anthropogenic Threats

The conservation status of all tur populations depends crucially on the intensity of anthropogenic pressure, including poaching, competition with livestock, industrial activities (mining, road construction, and alike) and recreation activities, such as mountaineering and tourism. In the 1940s–1980s, competition with livestock and industrial land use were the greatest threats to tur, causing destruction of its habitat. By the end of the 1980s, numbers of livestock decreased drastically and industrial projects were stopped or postponed, although the Zaramag Hydropower Station in North Ossetia and the Transcaucasian Highway running through North Ossetia to Georgia both had been completed after collapse of the Soviet Union, possibly causing an impact on tur. Industrial infrastructures usually are built in the valley bottoms and may 1) fragment tur habitat and 2) facilitate human access to areas which had been hardly accessible before, especially in winter. The most controversial projects that might directly affect tur highland habitat are possibly ski resorts.

Habitat management primarily consists in control of sustainable seasonal highland pasture use by domestic sheep. In general, the number of livestock driven to the highlands for summer had decreased already after the Russian Revolution of 1917. Since the collapse of the Soviet Union, distant shepherding has shrunk considerably in certain areas, e.g., North Ossetia, but grew in others, e.g., in Daghestan and in Azerbaijan, where sheep density is extremely high on summer highland pastures. Thus, even though tur pastures are currently almost sheep-free in North Ossetia, they may be overpopulated in other areas. However, negative impact of competition with livestock upon tur populations might be exaggerated, because in Azerbaijan, where sheep summer density is very high, tur tolerate their presence and pasture in vicinity of shepherd camps. This might be explained by lower poaching pressure by shepherds due to considerably weaker tur hunting traditions in Azerbaijan.

Poaching intensified in the 1990s due to political instability and large quantities of illegal modern firearms obtained by civilians as a result of military conflicts in the Caucasus. The situation improved considerably after the year 2000. Legal hunting had always been very limited, and tur will probably remain subject to low pressure of recreational hunting.

Introductions

In 1888, 20 individuals of *C. caucasica* and *C. cylindricornis* were introduced into a large enclosure (300 ha) in Borzhomi valley in Central Georgia, Caucasus Minor (Pfitzenmayer 1915). After some time, animals had been released into the wild, and this exotic population grew to about 500 free-ranging individuals by 1917 (Heptner et al. 1961; Vereshchagin 1938) but was exterminated during the Russian Revolution and the Civil War in 1917–1921 (Heptner et al. 1961; Sokolov and Syroechkovski 1990). In 1934 and 1941, once again, several Eastern tur were bred in an enclosure in Borzhomi, but this project was stopped because of the Second World War (Sokolov and Syroechkovski 1990). No examples of successful reintroduction or introduction into the wild are known.

Abundance Estimation

Assessment of population size is based on direct counts of animals along transects or from fixed vantage points, usually during snow-free periods. Counts are either complete over the distribution area or carried out in sampling plots, and data are extrapolated over the whole distribution area (Abdurakhmanov 1973; Magomedov et al. 2001; Zalikhanov 1967).

Conflicts with Humans

There is no conflict between tur and humans because tur cannot, in any way, overcompete livestock or create negative impact on human activities. Tourism, mountaineering, or other forms of recreation may, in fact, favor tur populations, if they do not destroy habitat, as they likely diminish the impact of poaching. In general, tur easily and quickly get accustomed to presence of humans (hiking routes, camps, noise, etc.), if there is no direct threat (Veinberg 1984).

Economic Value

Tur meat is valued by hunters and hides had been used in thc past for making saddles, though this practice is not in use anymore. Wine drinking vessels are produced from male tur horns, set into silver, and can be quite expensive. However, in general, the main direct economic value of tur lies in the revenue deriving from issued hunting licenses and from fees of hunting services.

Hunting Laws and Regulations

In Azerbaijan and Russia, all morphs or subspecies of tur are harvested, despite being listed in IUCN Red List. In Russia, hunting quota should be 3–5% of the total population. Considering total numbers (27,000), about 1300 animals could be legally harvested annually. In reality, the number of issued hunting licenses is much lower. The general belief is that poaching is significantly higher than legal hunting. In Georgia, due to being Red Listed, tur hunt is prohibited.

Future Challenges for Research and Management

Clarifying the taxonomy and evolution of tur should be a priority in future research. Tur relations with sympatric mountain ungulates, such as chamois and especially wild (bezoar) goat, are important for understanding the species habitat selection, thus for planning proper conservation and management actions.

With respect to management, the main problem is lessening the impact of poaching and, accordingly, increasing legal hunting quotas, especially for local highlanders. The latter would help local communities to understand and approve measures of nature conservation in the Caucasus. Right now there is no such understanding, and the general attitude toward nature conservation is rather lightminded if not entirely hostile. Poaching is locally considered just a minor and even partly justified misbehavior, being essential to traditional lifestyle.

Standardized counting methods are necessary for proper management; although they have been operatively defined, they are not yet approved by most zoologists and game managers and are not recommended for general use.

Acknowledgments We would like to thank Luca Corlatti for careful editing.

References

Abdurakhmanov MG (1973) Ecology, protection, and use of tur in Daghestan. Candidate dissertation thesis. Izdaniye Dagestanskogo Universiteta (Daghestan University), Makhachkala. (in Russian)

Abdurakhmanov MG (1977) Caucasian tur. In: Kaletsky AA (ed) Kopytnye (Ungulates). Lesnaya Promyshlennost', Moscow, pp 186–200. (in Russian)

Aiunts KR, Kolomyts EG (1986) Morphological index of variability in Caucasian tur and its connection with highland climate. Trudy Vysokogornogo geofizicheskogo instituta (Proc Highland Geophys Inst) 64:3–21. (in Russian)

Akatov VV, Golgofskaya KY, Gorcharuk LG, Durov VV, Kipiani VV, Kudaktin AN, Nemtsev AS, Pridnya MV, Til'ba PA, Topilina VG (1990) Caucasus Nature Reserve. In: Sokolov VY, Syroyeckovsky YY (eds) Zapovedniki SSSR. Zapovedniki Kavkaza (Nature reserves of the USSR. Nature reserves of the Caucasus). Mysl', Moscow, pp 69–100. (in Russian)

Akkiev MI (2018) Evaluation of potential leopard habitats in Kabardino-Balkaria. In: Semyonov UA (ed) Vosstanovleniye leoparda na Kavkaze. Tovarishchestvo nauchnykh izdaniy KMK, Moscow, pp 165–212. (in Russian)

Akkiyev MI (1997) Fauna of helminths of Caucasian tur as a factor influencing ecological status of the species within Kabardin-Balkarian Nature Reserve. In: Trudy Regional'noi hauchno-prakticheskoi konferencii "Biosfera i chelovek" (Proceedings of Regional scientific-practical conference "Biosphere and man"). Maykop, pp 35–41. (in Russian)

Akkiyev AM, Pkhitikov AB (2007) Ungulates as hunting objects: contemporary status, problems of conservation and use (Central Caucasus, Kabardino-Balkarian Republic). In: Rozhnov VV, Tembotova FA (eds) Mlekopitayushchiye gornykh territoriy. Materialy mezhdunarodnoi konferencii (Mammals of mountain territories. Materials of International Conference). Moscow, pp 11–17. (in Russian)

Alekperov KM, Yerofeeva IK, Rakhmatulina IK (1976) Contemporary status of certain mammal species of Azerbaijan. In: Sokolov VY (ed) Redkiye mlekopitayushchiye fauny SSSR (Rare mammals of the USSR fauna). Nauka, Moscow, pp 28–34. (in Russian)

Alexandrov VN (1968) Ecology of the Caucasian Red Deer. Trudy Kavkazskogo zapovednika (Proc Cauc Nat Reserve) 4:95–200. (in Russian)

Aliev SD (1969) Fauna of the Palaeolitic Azykh encampment. Candidate dissertation thesis. Zoologicheskiy Institut, Baku. (in Russian).

Asadov SM (1959) Helminth fauna of the Daghestan tur (*Capra cylindricornis* Blyth) in Azerbaijan. Izvestiya Akademii Nauk Azerbaijanskoi SSR, seriya biologicheskikh i sel'skokhozyaistvennykh nauk (News Acad Sci Azerb SSR) 3:33–44. (in Azerbaijani, Russian summary)

Babaev EA, Yarovenko YA, Magomedov UM, Gamodova MK, Yarovenko AY (2017) Population status of ungulates of Daghestan. Trudy Dagestanskogo gosudarstvennogo pedagogicheskogo instituta (Proc Daghestan State Pedagogical Inst) 3(11):44–50. (in Russian)

Baryshnikov GF (1987) Mammals of the Caucasus of the early Palaeolitic. Trudy Zoologicheskogo Instituta Akademii Nauk SSSR (Proc Zool Inst USSR Acad Sci) 168:3–20. (in Russian, English summary)

Bobyr' GYa (1997) On studying impact of wolf upon the population of West-Caucasian tur in Teberda reserve. In: Khokhlov AN (ed) Nauchnoye naslediye N.Ya. Dinnika i ego rol' v razvitii sovremennogo yestestvoznaniya (Scientific legacy of N.Ya. Dinnik and his role in development of modern natural history). Stavropol', pp 28–37. (in Russian)

Bobyr' KG, Semyonov UA (2008) West Caucasian tur in Teberda Nature Reserve (ecology, physiology, diseases). Servisshkola, Stavropol. (in Russian)

Bulatova NJ (1962) Peculiarities of blood in highland animals. Trudy Instituta Morfologii Zhivotnykh (Proc Inst Morphol Anim) 41:11–46. (in Russian)

Chlaidze Z (1967) Materials for study of Caucasian Tur (*Capra cylindricornis* Blyth). In: Japaridze N, Kokhia S (eds) Masalebi sakartvelos paunisatvis (Materials for Georgian Fauna), vol II. Metsniereba, Tbilisi, pp 200–236. (in Georgian)

Chlaidze Z (1975) Some data on Tur ecology in Caucasus of Ratcha, Svaneti and Abkhazia. Sakartvelos SSR metsnierebata akademiis zoologiis institutis shromebi (Proc Inst Zool Acad Sci Georgian SSR) IV:481–521. (in Georgian)

Dinnik NY (1910) Mammals of the Caucasus. Zapiski Kavkazskogo Otdela Imperatorskogo Russkogo Geograficheskogo Obshchestva (Notes Caucasian Sect Russian Geogr Soc) 27:1–247. (in Russian)

Ekvtimishvili ZS (1952a) Vertical distribution of certain ungulate species on the southern slope of the Greater Caucasus. Soobshcheniya Akademii Nauk Gruzinskoi SSR (Commun Acad Sci Georgian SSR) 13(8):476–484. (in Georgian, Russian summary)

Ekvtimishvili ZS (1952b) Severtsov's tur suffering from *Coenurus cerebralis* in Georgia. Soobshcheniya Akademii Nauk Gruzinskoi SSR (Commun Acad Sci Georgian SSR) 13(9):541–544. (in Russian)

Ekvtimishvili ZS (1953a) Materials on foraging of Daghestan tur (*Capra cylindricornis* Blyth). Trudy Instituta Zoologii Akademii Nauk Gruzinskoi SSR (Proc Inst Zool Acad Sci Georgian SSR) 11:15–23. (in Georgian, Russian summary)

Ekvtimishvili ZS (1953b) Vertical distribution of tur and chamois during different seasons in Teberda NR. Trudy Instituta Zoologii Akademii Nauk Gruzinskoi SSR (Proc Acad Sci Georgian SSR) 14(3):169–186. (in Georgian, Russian summary)

Ellerman J, Morrison-Scott T (1966) Checklist of Palaearctic and Indian Mammals 1758 to 1946, 2nd edn. British Museum (Natural History), London

Enukidze G (1965) Some aspects of ecology and protection of tur and chamois in Lagodekhi State reserve. In: Gulisashvili V (ed) Sakartvelos bunebis datsva (Protection of Georgia's nature), vol I. Metsniereba, Tbilisi, pp 37–54. (in Georgian)

Eriashvili IG (1990) Long-term data on dynamics of numbers of tur and red deer in Lagodekhi reserve. In: Filonov KP (ed) Prirodnyye zapovedniki SSSR, ikh nastoyashcheye i budushcheye (Nature reserves of the USSR, their present and future). Goskomitet SSSR po okhrane prirody, Novgorod, pp 337–340. (in Russian)

Fedosenko AK, Veinberg PI, Skuratov NI (1992) Siberian ibex in the Western Sayan. In: Fedosenko AK (ed) Okhrana i izucheniye redkikh i ischezayushchikh vidov zhyvotnykh v zapovednikakh (Protection and studies of rare and endangered animals in nature reserves). Tsentral'naya Nauchno-issledovatel'skaya Laboratoriya Okhotnichyego Khozyaistva, Moscow, pp 29–44. (in Russian)

Gabliani E (1930) Tur in Svanetian Mountains. Sakhelgami, Tbilisi. (in Georgian)

Gadjiyev DV (1977) Quaternary changes in species composition of ungulates of Azerbaijan under impact of anthropogenic factors. Trudy zapovednikov Azerbaijana (Proc Nat Reserves Azerb) 4:108–128. (in Russian)

Gasanov KN (1990) Zakatala nature reserve. In: Sokolov VY, Syroyeckovsky YY (eds) Zapovedniki SSSR. Zapovedniki Kavkaza (Nature reserves of the USSR. Nature reserves of the Caucasus). Mysl', Moscow. (in Russian)

Gavashelishvili A (2004) Habitat selection by East Caucasian tur (*Capra cylindricornis*). Biol Conserv 120:391–398

Gavashelishvili A, Yarovenko YA, Babayev EA, Mikeladze G, Gurielidze Z, Dekanoidze D, Kerdikoshvili N, Ninua L, Paposhvili N (2018) Modeling the distribution and abundance of eastern tur (*Capra cylindricornis*) in the Caucasus. J Mammal 99(4):885–897

Gerasimov BP, Markov KK (1939) Glaciation period on the territory of the USSR. Trudy Instituta geografii Akademii Nauk SSSR (Proc Inst Geogr Acad Sci USSR) 33:1–464. (in Russian)

Guliev SM (2000) Daghestan tur – *Capra cylindricornis* Blyth. In: Gadjiyev DV, Pakhmatulina IK (eds) Zhivotny mir Azerbaijana, tom III. Pozvonochnye (Animal world of Azerbaijan, vol. III. Vertebrates). Elm, Baku, pp 610–613. (in Russian)

Guliev SM, Weinberg PJ, Askerov E (2009) Daghestan tur (*Capra cylindricornis* Blyth) conservation strategy in Azerbaijan. In: Zazanashvili N, Mallon D (eds) Status and protection of globally threatened species in the Caucasus. CEPF, WWF: Contour Ltd., Tbilisi, pp 53–61

Gurielidze Z (2004) Red Deer: natural mechanisms of population regulation. Proc Inst Zool, Tbilisi XXII:314–324

Gurielidze Z, Gavashelishvili A, Javakhishvili Z, Murtskhvaladze M, Ninua L, Dekanoidze D, Rajebashvili G, Kerdikoshvili N, Paposhvili N, Modebadze M, Kopaliani N (2013) Surveys in Georgia for monitoring wild animals. Ecology Institute of Ilia State University / Ministry of Environment and Natural Resources of Georgia (Project #225 06.08.2013), Tbilisi. Unpublished Report. (in Georgian)

Heptner VG, Nasimovich AA, Bannikov AG (1961) Mammals of the Soviet Union. Artiodactyls and Perissodactyls, vol 1. Vysshaya Shkola, Moscow. (in Russian)

Janashvili A (1950) Tur. Library of Tbilisi Zoo. Sakhelgami, Tbilisi. (in Georgian)

Janashvili A (1963) Animal world of Georgia, vol III. Sakartvelos SSR metsnierebata akademii, Tbilisi. (in Georgian)

Janashvili A (1977) Ungulates of Georgia. Sabchota sakartvelo, Tbilisi. (in Georgian)

Kazanskaya EY (2007) Origin of the West Caucasian tur (*Capra caucasica*, Bovidae, Artiodactyla). In: Rozhnov VV (ed) Teriofauna Rossii i sopredel'nykh territoriy (Mammal fauna of Russia and adjoining territories). Tovarishchestvo nauchnykh publikatsiy KMK, Moscow, p 179. (in Russian)

Kazanskaya EY, Kuznetsova MV, Danilkin AA (2007) Phylogenetic reconstructions in the Genus *Capra* (Bovidae, Artiodactyla) based on the mitochondrial DNA analysis. Russian J Genet 43(2):181–189

Kopaliani N, Gurielidze Z (2009) Status of Tur in Georgia and conservation action plan. In: Zazanashvili N, Mallon D (eds) Status and protection of globally threatened species in the Caucasus. CEPF, WWF: Contour Ltd., Tbilisi, pp 61–69

Korzhuyev PA, Timoshevskaya TV, Kadyrova IA, Spasskaya GK (1978) Horns as an adaptation of Daghestan tur to highland conditions. Ekologiya (Ecology) 2:100–102. (in Russian)

Kotlyakov VM, Krenke AN (1980) Contemporary glaciation and climate. In: Gerasimov IP, Bravar I, Dresh Z (eds) Al'py – Kavkaz (The Alps – The Caucasus). Nauka, Moscow, pp 157–177. (in Russian)

Kotov VA (1968) Kuban tur, its ecology and economical significance. Trudy Kavkazskogo Gosudarstvennogo Zapovednika (Proc Cauc Nat Reserve) 10:201–293. (in Russian)

Kuliev SM (1981) The wild goat and Daghestan tur of Azerbaijan. Candidate dissertation thesis. Vsesoyuznyi nauchno-issledovatel'skii institut okhrany prirody, Moscow. (in Russian)

Kuliev GK, Mamedov TO (1974) Studies of the chromosomal set of the wild goat (*Capra aegagrus* Erxleben) and Daghestan tur (*C. c. cylindricornis* Blyth). Izvestiya Akademii Nauk Azerbaijanskoi SSR, seriya biologicheskaya (News Acad Sci Azerb SSR Biol Ser) 5–6:88–93. (in Russian)

Kurashvili B, Kokhia S, Arabuli A, Rodonaia T, Claidze Z, Kapanadze A, Matsaberidze G (1981) Atlas of Georgian mammals. Institute of Zoology. Metsniereba, Tbilisi. (in Georgian and Russian)

Lortkipanidze B, Weinberg P (2020) *Capra cylindricornis*. The IUCN red list of threatened species 2020, e.T3795A91287269 [Internet; downloaded 19 Sept 2020]

Lydekker R (1913) Catalogue of the ungulate mammals in the British Museum (Natural history), vol 1. British Museum, London, pp 1–250

Magomedov M-RD, Akhmedov EG, Yarovenko YA (2001) Daghestan tur. Populational and trophic acpects of ecology. Nauka, Moscow. (in Russian)

Malikova VF (ed) (1960) Agroclimatic reference book on the North Ossetian ASSR. Gidrometeoizdat, Leningrad. (in Russian)

Manceau V, Després L, Bouvet J, Taberlet P (1999) Systematics of the genus *Capra* inferred from mitochondrial DNA sequence data. Mol Phylogenet Evol 13:504–510

Markov YL (1938) Game animals of the Lagodekhi nature reserve. Gruzinskoye otdeleniye Akademii Nauk SSSR, Tbilisi. (in Russian)

Mayr E (1968) Animal species and evolution. Mir, Moscow. (in Russian)

Milanovsky EE (1966) Main questions in history of glaciation of the central Caucasus. In: Khain VG (ed) Problemy geologii i paleogeografii antropogena (Problems of geology and paleogeography of Anthropogenic). Izdatel'stvo Moskovskogo universiteta, Moscow, pp 5–49. (in Russian)

Miura S, Kita J, Sugimura M (1987) Horn growth and reproductive history in female Japanese serow. J Mammal 68:826–836

NACRES (1996) Georgian biodiversity country study report. Unpublished report, Tbilisi

NACRES (2003) South Caucasus animal data base (Amphibia, Reptilia Aves, Mammalia). Unpublished report, Tbilisi. (in Russian)

NACRES (2017) Eastern tur (*Capra cylindricornis*) monitoring in Lagodekhi protected area. Unpublished Final Report to Caucasus Nature Fund, Tbilisi

Nadler C, Lay D (1975) Chromosomes of some Asian wild sheep (*Ovis*) and goats (*Capra*). Chromosome Inf Serv 18:28–31

Nasimovich AA (1949) Essay on ecology of the West-Caucasian tur. Trudy Kavkazskogo zapovednika (Proc Cauc Nat Reserve) 3:61–67. (in Russian)

Nasimovich AA (1950) On organization of nature reserve of highland wildlife of the Central Caucasus. Okhrana prirody (Nat Protect) 2:72–79. (in Russian)

Ordinance of the Government of Georgia on Adoption of Georgian "Red List" (Sakartvelos mtavrobis dadgenileba sakartvelos "tsiteli nuskhis" damtkicebis shesakheb) (2014) Document number 190. Sakartvelos sakanondeblo matsne Publisher, Tbilisi. Registration code 360060000.10.003.017809, [Internet; cited 2018 Aug 21] www.matsne.gov.ge/ka/document/view/2256983. Downloaded 21 Aug 2018

Panov VD (1993) Evolution of contemporary glaciation of the Caucasus. Gidrometeoizdat, St-Petersburg. (in Russian)

Pfitzenmayer EV (1915) Some interesting bovid mongrels from Transcaucasia. Izvestiya Kavkazskogo Muzeya (Proc Cauc Museum) 8:251–266. (in Russian)

Pidancier N, Jordan S, Luikart G, Taberlet P (2006) Evolutionary history of the genus *Capra* (Mammalia, Artiodactyla): discordance between mitochondrial DNA and Y-chromosome phylogenies. Mol Phylogenet Evol 40:739–749

Pocock RJ (1910) On the specialized cutaneous glands of ruminants. Proc Zool Soc Lond 1910:840–986

Popkova IF (1968) Chamois on the southern slopes of the main Caucasus range. Trudy Teberdinskogo Gosudarstvennogo Zapovednika (Proc Teberda Nat Reserve) 7:160–211. (in Russian)

Ravkin YS (1975) Resources of the wild ungulates in the North Caucasus and anthropogenic impact upon them. In: Sokolov VY (ed) Kopytnye fauny SSSR (Ungulates of the USSR fauna). Nauka, Moscow, pp 17–18. (in Russian)

Romashin AV (2001) Ecological-populational analysis of highland ungulates of the West Caucasus and their sustainable use. SPP, Sochi. (in Russian)

Rukhlyadev DP (1964) Helminth fauna of wild ungulates of the Crimea and the Caucasus in ecological and zoogeographical view. Izdatel'stvo Saratovskogo Universiteta, Saratov. (in Russian)

Sarkisov AA (1953) On bovid crossbreeds. Priroda (Nature) 42:113. (in Russian)

Schaller GB (1977) Mountain monarchs. Wild sheep and goats of the Himalaya. University of Chicago Press, Chicago

Sokolov II (1959) USSR fauna. Mammals (Artiodactyla and Perissodactyla). Trudy Zoologicheskogo Instituta Akademii Nauk SSSR (Proc Zool Inst Acad Sci USSR) 14:1–295. (in Russian)

Sokolov VE (1973) Skin cover of mammals. Nauka, Moscow. (in Russian)

Sokolov VE, Syroechkovski EE (eds) (1990) Nature reserves of the Caucasus. Mysl', Moscow. (in Russian)

Sokolov VE, Tembotov AK (1993) Vertebrates of the Caucasus. Mammals: Ungulates. Nauka, Moscow. (in Russian)

Tembotov AK (1974) On geographic variation of tur (*Capra caucasica* Güld.) in the North Caucasus. In: Tembotov AK (ed) Fauna, ekologiya i okhrana zhivotnykh Severnogo Kavkaza (Fauna, ecology and protection of animals of the North Caucasus). Izdatel'stvo Kabardino-Balkarskogo Universiteta, Nal'chik, pp 118–131. (in Russian)

Trepet SA (2014) Ungulates of the North-West Caucasus. Contemporary status and mechanisms of stability of populations. Kubanskoye knizhnoye izdatel'stvo, Krasnodar. (in Russian)

Tsalkin VI (1955) Variability and systematics of tur of the West Caucasus (Caprinae, Artiodactyla). Byulleten Moskovskogo Obshshestva Ispytatelei Prirody, Otdel biologicheskii (Bull Mosc Nat Soc Biol Sect) 60:17–33. (in Russian)

Veinberg PI (1983) Chamois habitats depending upon presence of competing species. In: Sokolov VYe (ed) Redkiye vidy mlekopitayushchikh SSSR i ikh okhrana. Materialy III Vsesoyuznoi konferencii (Rare mammal species of the USSR and their protection. Materials of the 3rd All-union conference). Moscow, pp 160–161. (in Russian)

Veinberg PI (1984) Daghestan tur. Nauka, Moscow. (in Russian)

Veinberg PI (1993) Analysis of horn shape and coat coloration in *Capra* (Artiodactyla). Byulleten Moskovskogo Obshchestva Ispytatelei Prirody, Otdel biologicheskii (Bull Mosc Nat Soc Biol Sect) 98:3–14. (in Russian, English summary)

Veinberg PI (2000) East-Caucasian tur (*Capra cylindricornis* Blyth, 1841). In: Amirkhanov AM (ed) Zhyvotnyi mir Respubliki Severnaya Osetiya–Alania (Animal world of Republic North Ossetia-Alania). Proekt Press, Vladikavkaz, pp 256–258. (in Russian)

Veinberg PI, Komarov YE, Lipkovich AD (1983) Materials on nesting biology of lammergeier in the Central Caucasus. In: Flint VE (ed) Okhrana khishchnykh ptits (Conservation of birds of prey). Nauka, Moscow, pp 101–105. (in Russian)

Vekua AK (1967) New representative of Lagomorphidae from the Palaeolitic of Georgia. Soobshcheniya Akademii Nauk Gruzinskoi SSR (Commun Acad Sci Georgian SSR) 45:139–143. (in Russian)

Vereshchagin HK (1938) Daghestan tur (*Capra cylindriricornis* Blyth) in Azerbaijan. Trudy Zoologicheskogo Instituta Akademii Nauk Azerbaijanskoi SSR (Proc Zool Inst Acad Sci Azerb SSR) 9:1–70. (in Russian)

Vereshchagin HK (1959) Mammals of the Caucasus. Izdatel'stvo Akademii Nauk SSR, Moscow. (in Russian)

Weinberg PJ (1999) The present state and biology of the wild goat (*Capra aegagrus* Erxleben) Dagestan population. Byulleten Moskovskogo Obshchestva Ispytatelei Prirody, Otdel biologicheskii (Bull Mosc Nat Soc Biol Sect) 104:12–21. (in Russian, English summary)

Weinberg PJ (2002a) Long-term dynamics of numbers, age and sex structure of the Daghestan tur (*Capra cylindricornis* Blyth, 1841) population in North Ossetian Nature Reserve. Byulleten Moskovskogo Obshchestva Ispytatelei Prirody, Otdel biologicheskii (Bull Mosc Nat Soc Biol Sect) 107(2):14–22. (in Russian, English summary)

Weinberg PJ (2002b) Annual horn segments in female Daghestan tur (*Capra cylindricornis* Blyth, 1841) as possible indicators of reproduction. Byulleten Moskovskogo Obshchestva Ispytatelei Prirody, Otdel biologicheskii (Bull Mosc Nat Soc Biol Sect) 107(6):3–8. (in Russian, English summary)

Weinberg PJ (2002c) *Capra cylindricornis*. Mammalian species no 695, pp 1–9

Weinberg PJ (2004) Dependence of juvenile index upon size and composition of female groups in Daghestan tur (*Capra cylindricornis* Blyth). Byulleten Moskovskogo Obshchestva Ispytatelei Prirody, Otdel biologicheskii (Bull Mosc Nat Soc Biol Sect) 109(5):63–65. (in Russian, English summary)

Weinberg P (2008a) *Capra caucasica*. The IUCN red list of threatened species 2008: e.T3794A10088217 [Internet; downloaded 28 Jun 2018]

Weinberg P (2008b) *Capra cylindricornis* (errata version published in 2016). The IUCN red list of threatened species 2008: e.T3795A91283066 [Internet; downloaded 28 Jun 2018]

Weinberg P (2020) *Capra caucasica*. The IUCN red list of threatened species 2020, e.T3794A22143809 [Internet; downloaded 19 Sept 2020]

Weinberg PI, Fedosenko AK, Arabuli AB, Myslenkov AI, Romashin AV, Voloshina IV, Zheleznov NK (1997) The Commonwealth of Independent States (former USSR). In: Shackleton DM (ed) Wild sheep and goats and their relatives. International Union for Conservation of Nature, Gland/Cambridge, UK, pp 173–193

Weinberg PJ, Akkiev MI, Buchukuri RG (2010) Clineal variation in Caucasian Tur and its taxonomic relevance. Galemys 22(especial):375–394

Zakariyev AY (1980) Wild ungulates of the Caucasus as bearers of parasitic worms. In: Sokolov VE (ed) Kopytnyye fauny SSSR (Ungulates of the USSR fauna). Nauka, Moscow, pp 159–160. (in Russian)

Zalikhanov MC (1967) Tur in Kabardino-Balkaria. Kabardino-Balkarskoye Knizhnoye Izdatel'stvo, Nal'chik. (in Russian)

Wild Goat *Capra aegagrus* Erxleben, 1777

19

Paul Weinberg and Sandro Lovari

Contents

P. Weinberg (✉)
North Ossetian State Nature Reserve, Alagir, Russian Federation
e-mail: tu_r@rambler.ru; tur31161@gmail.com

S. Lovari
Research Unit of Behavioural Ecology, Ethology and Wildlife Management, Department of Life Sciences, University of Siena, Siena, Italy
e-mail: sandro.lovari@gmail.com

© Springer Nature Switzerland AG 2022

L. Corlatti, F. E. Zachos (eds.), *Terrestrial Cetartiodactyla*, Handbook of the Mammals of Europe,
https://doi.org/10.1007/978-3-030-24475-0_31

Common Names

English	Wild goat, Bezoar goat
German	Bezoarziege
French	Chèvre égagre, Bouc sauvage
Spanish	Cabra bezoar
Italian	Capra selvatica
Russian	безоаровый козел

Taxonomy, Systematics and Paleontology

The wild goat (Fig. 1a, b) was described from Daghestan, North-East Caucasus, which thus harbors the typical subspecies *Capra aegagrus aegagrus* Erxleben, 1777. Some authors preferred the name *Capra hircus* for both wild goat and domestic goat, regarding the former as a wild form of the latter (e.g., Ellerman and Morrison-Scott 1966). Later, several other scientific names were introduced, some of which became subspecific (e.g., *Capra blythi* Hume, 1875). Usually, three indisputably wild subspecies are being recognized: the already mentioned *C. a. aegagrus* inhabiting the Greater and Minor Caucasus, Asia Minor, and Iran; *C. a. turcmenicus* Zalkin, 1950, in Kopet-Dagh Mountains and Large and Small Balkhan in Turkmenistan, and adjoining Turkmen-Khorasan Mountains in Iran; *C. a. blythi* Hume, 1875 in India and Pakistan (Danilkin 2005; Heptner et al. 1961; Sokolov 1959). Some zoologists do not recognize *C. a. turcmenicus*, but add *C. a. chialtanensis* Lydekker, 1913 (Schaller 1977; Shackleton and Lovari 1997). However, subspecies are often considered not valid (e.g., Schaller 1977; Valdez 2011), all being quite similar in appearance (e.g., Weinberg 2001), and differing only in adult male body size and horn length (e.g., Tsalkin 1950).

Fossil remnants of the wild goat were found in Acheul deposits (probably mid-Pleistocenic) of the Azykh Cave in Karabakh, and in Mousterian layers (end of Pleistocene) in Tsopi Cave, south-east Georgia (both Caucasus Minor) (Baryshnikov 1987). Nevertheless, Vereshchagin (1959) suggested that the wild goat might have reached the Greater Caucasus during one of more arid periods of Pleistocene, or even Pliocene. However, absence of pronounced morphological differences between the isolated North Caucasian and southern wild goat populations (see section "Description") and also the age of known fossils suggests a late appearance of the wild goat at the Greater Caucasus.

Fig. 1 (**a**) Adult wild goat male during the rut (photograph © A. Malkhasyan); (**b**) wild goat female with juveniles during the rut (photograph © A. Malkhasyan)

Current Distribution

Distribution of the wild goat is probably the largest one of all *Capra*, stretching from the south-east Pakistan to the south-west Turkey and from the Arabian Sea to the North Caucasus. Of this, the species' range in Europe makes only a small fraction. Currently, and according to earlier descriptions by Dinnik (1910), the wild goat range in the Greater Caucasus occurs only on the north slope of the mountain chain. In the beginning of the twentieth century, wild goats reportedly inhabited the eastern part of the Table (Stolovaya) Mountain massif (N 42°50′, E 44°51′) of the limestone Rocky Range quite close to Vladikavkaz, but later on this species disappeared from this area, as well as from all of the headwaters of Assa R. (Batkhiyev 1989). No evidence exists that the wild goat has ever occurred westwards of Terek R., and wild goat remnants were not found in North Ossetian shrines, famous for their wild artiodactyl skull collections dating as early as the end of the eighteenth century (Dinnik 1910; Vereshchagin 1959). Although wild goat vanished from the Rocky Range in Ingushetia, it still survives on this range in the neighboring Chechnya (Lukarevsky 2018). Currently the westernmost area inhabited by the wild goat is the headwaters of Chanty-Argun R. in Chechnya (Batkhiyev 1980, 1989) and farther more upstream – Khevsureti area of Georgia (approx. N 42°37′, E 45°07′) (Fig. 2). Most Georgian sources (Arabuli 1989; Kapanadze 1978) mention wild goat distribution only in the more easternly Tusheti area (basins of Tusheti and Pirikiti Alazani Rivers which are the headwaters of Andi Koisu R.), but a recent survey found a small wild goat population in Khevsureti (NACRES 2013). From Tusheti its distribution stretches eastwards, encompassing the Side Range and the north slope of the Main (Watershed) Range into Daghestan, where the main part of the species distribution in the Greater Caucasus occurs (Fig. 2) (Magomedov et al. 2014; Weinberg 2001). There this species inhabits the upper halves of Andi Koisu and Avar Koisu Riverbasins, and in general, distribution follows the fluvial net (Fig. 2) because the wild goat does not stick to the highlands (as tur *Capra caucasica* does), but it dwells in the lower steep forested slopes of the canyons (Weinberg 1999). There is also an isolated distribution patch, or more precisely two, or maybe even three, separate patches on the right bank of the lower flow of Andi Koisu R. near Tlokh and Karata villages (N 42°38′ E 46°22′ and N 42°36′ E 46°16′) (Magomedov et al. 2014). More easternly Kazikumukh Koisu and Karakoisu Riverbasins are almost completely forestless and therefore lack wild goat (Dinnik 1910; Magomedov et al. 2014). According to Dinnik (1910), the wild goat used to inhabit some areas within the Samur Riverbasin. Indeed, a completely isolated wild goat population has been found recently in a forested area near Khoredj Village in Chiragchai Riverbasin (approx. N 41° 43′ E 47° 50′) (Fig. 2),

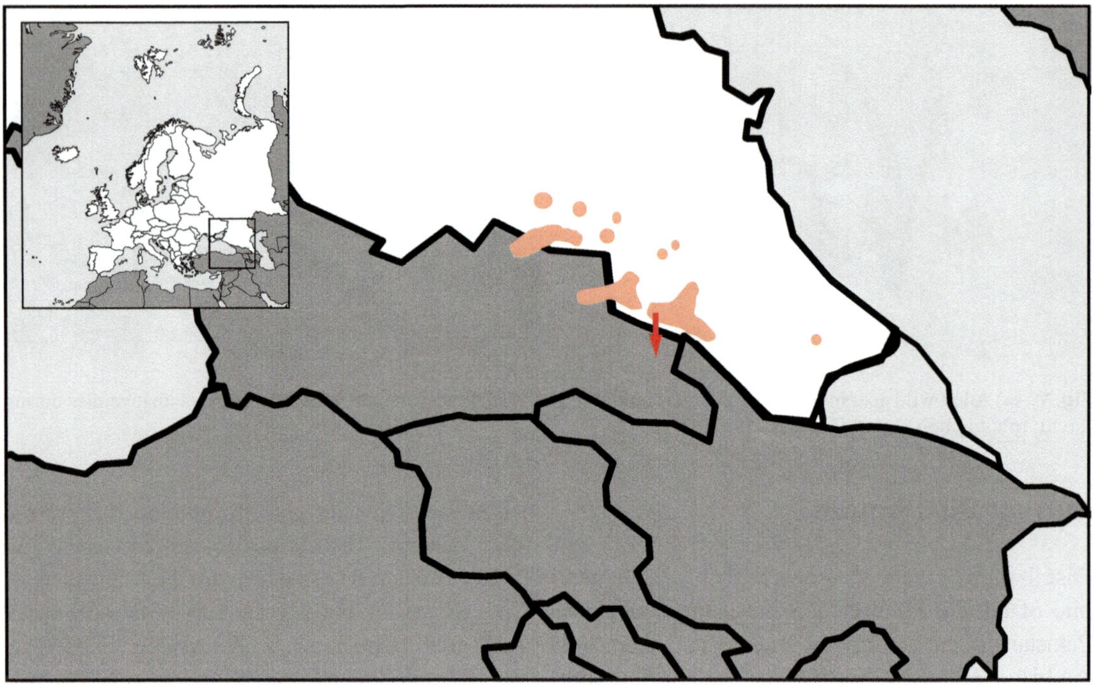

Fig. 2 Distribution of the wild goat in the Greater Caucasus. Distribution is based on the IUCN Red List of Threatened Species, version 2017-2. Arrow indicates migration of wild goat males over the Main Range into Lagodekhi NR on the south slope in Georgia (Map template: © Copyright Getty Images/iStockphoto)

more than 100 km eastwards of its nearest occurrence in the headwaters of Avar Koisu Riverbasin (Babaev et al. 2017). Such small isolated population undoubtedly is a remnant of a once much wider and more continuous range of the wild goat. Probably more thorough surveys will produce more locations in this latter area. On the other hand, the previously indicated population on the southern slopes of Mount Babadagh, Azerbaijan (Vereshchagin 1959) does not exist, and probably did not exist in the nearest past (Weinberg, unpublished data).

In 2016 and 2017, camera traps placed in Lagodekhi Nature Reserve on the south slope of the Main Range in East Georgia, just near the border with Azerbaijan and Russia (Fig. 2), photographed wild goat males in autumn (Nugzar Zazanashvili, pers. comm.).

During the last couple of centuries, at least, the wild goat range in the Greater Caucasus has been completely isolated from the nearest areas inhabited by the species in the Caucasus Minor and separated from them by almost 200 km of Kura R. and Alazani R. valleys. Nevertheless, Caucasus Minor was the place from where the wild goat migrated to the Greater Caucasus during some drier epoch, possibly along several ridges connecting both mountain systems, e.g., the Surami Range (Vereshchagin 1959) or the Trialeti Range. At the beginning of the twentieth century, a small wild goat population still survived near Borzhomi, at the south base of the Trialeti Range (Dinnik 1910; Vereshchagin 1959).

Description

Morphology

The wild goat, as all *Capra* spp., is dimorphic (Fig. 1). Males are larger and heavier than females. They have long horns up to 110, or even 130 cm (Heptner et al. 1961), while adult female horns may reach 38 cm (Sokolov 1959), both being longer than those of tur. Horns are scimitar-shaped, curved along a single plane, without evident twisting, but tips of adult male horns are often pointed inwards or outwards. Horn sheaths of males are drop-shaped in cross-section with a distinct frontal keel which is much less pronounced in females (Weinberg 1993). They can reach over 10 cm width at the base in males and hardly more than 3 cm in females. Horns diverge at an angle of 30–40° (Sokolov 1959), grow throughout life with cessations in the cold period of each year, and annual increments are much bigger in males than in females. Adult males have a long and drooping beard up to 20 cm long, while in females the beard is rare and wispy, if at all present (Heptner et al. 1961).

Skulls of younger males are distinctly convex in the center of the forehead, just below the frontal keels of the horn sheaths, and from aside this bulge looks like a base for those keels, below which, just above the nasal bones, the profile is bent-in. In adult males, this bulging is a bit less pronounced because of much wider horn cores, but very characteristic bent-in bridge of the nose differing from all other *Capra* species is still distinct (Fig. 3).

Pelage and Age Determination in the Field

Wild goats are one the most richly colored members of the genus, with a most intricate coloration pattern (Fig. 1). Kids are gray in summer, grayish-yellow in winter. They have dark stripes on the face (along the forehead and stretching from eyes to nose), and on the frontal surfaces of the legs. The underside of the body and the inner sides of the legs are white or dirty-white in all animals, regardless of sex, age, and season. Even in winter, kids' horns are hardly longer than the ears. Yearling and adult females are sandy-chestnut or fawny in summer coat and grayish-fawny in winter. In addition to head and leg stripes, they have also a narrow stripe along the spine in all seasons. Yearling females are smaller than adults, and their horns are only slightly longer than ears. Yearling males are redder or darker than females in summer, with the same pattern of stripes. They turn grayer in winter and display a vague flank stripe. Other stripes are the same as in the summer coat. The beard is inconspicuous. They are smaller than adult females. In winter, the horns of yearling males are twice as long as the ears, more sharply bent and wider at the base than in females. Young 2–3-year-old males are fawny-brown in summer; the stripes on the face are indistinct, while those on the legs are more distinct and contrasting than in the previous classes. In winter, these males are grayish-fawny, while the head gets darker. They display narrow and distinct stripes on the legs, along the spine, and broad and less definite stripes on the flanks and across the shoulders. The flank and shoulder stripes are lacking in the previous classes, as is a small but noticeable beard. Young 4–6-year-old males are grayish-fawny in summer, with stripes on the legs, flanks, and along the spine. In winter, they are sandy-gray, with black stripes. There is a black mane on the neck and the withers. The head and the

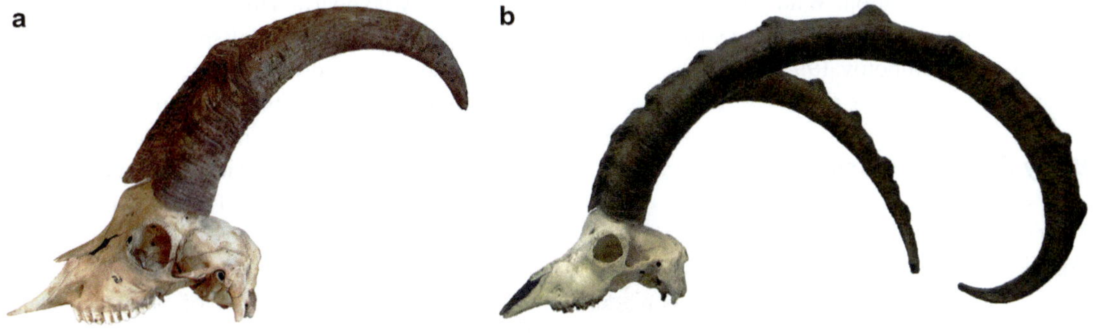

Fig. 3 Skulls of 2-year-old (left, photograph © P. Weinberg) and fully adult (right, photograph © N. Nasrulayev) wild goat males

conspicuous beard are black in all seasons. Adult males are sandy-gray in summer, with leg, flank, shoulder, and dorsal stripes. They are silvery-gray in winter, with a contrasting pattern of connected black stripes on the legs, shoulders, flanks, along the spine (mane), and throat. The chest and the head are black too (Weinberg 2001).

Thus, the coloration pattern develops with age in males, adding new stripes, and getting more contrasted. Adult wild goat males, older than 6 years, in winter coat are the most contrastingly colored and conspicuous ones of all classes (Weinberg 1993). On the whole, wild goats of the North Caucasus do not differ in coat coloration from Turkmen (collection of Zool. Institute in St. Petersburg) and even Sind (Schaller and Laurie 1974) conspecifics (Weinberg 1993, 2001).

Age of males can be easily determined by annual horn sheath segments, and often it is possible to tell the age of a male in the field to an approximation of a year thanks to knobs on the front keel of the horns, which develop from the second year on and are situated at the boundaries of the annual segments. These knobs usually become indistinguishable from 8th to 9th segments on (Fig. 3) (Weinberg 2001).

In summer, animals practically lack underwool. Winter pelage has 35–127 mm long guard hairs and much underwool, up to 70 mm long and *c.* 11–30 μm thick (Sokolov and Tembotov 1993).

Mammae

Females have two mammae; mammaries enlarge visibly during lactation. Wild goats, as all *Capra*, have a subcaudal skin gland (Sokolov 1959; Sokolov and Tembotov 1993).

Body Size and Mass

Data on body size and live mass of the wild goat, especially from the Greater Caucasus, are very scarce. The species in general is considered smaller than Caucasian tur or Alpine ibex *Capra ibex*. Dinnik (1910) states that body length of an adult male can reach 150 cm, rump height about 100 cm, while shoulder height is a few cm less. Tail is up to 20 cm long (with terminal hairs), ears 10–11 cm (Sokolov and Tembotov 1993). Body size and live mass are provided only for specimens from the Caucasus Minor and these show that males weigh just 35–38 kg, while females weigh 26–30 kg (Dahl' 1951). Heptner et al. (1961) suggested that adult males can be larger and heavier, reaching 70–80 kg. In Daghestan, an adult wild goat male was seen joining a mixed tur group during the rut and trying to court tur females (Weinberg 2001): it did not seem any smaller or less massive than tur males.

Dentition

Permanent dentition is: 0.0.3.3/3.1.3.3, total 32, as in all *Capra* spp. (Heptner et al. 1961). Age during the first 5 years can be determined by the growth of three pairs of lower incisors and one pair of canines. One pair of adult teeth grows every summer, starting from the second summer to the fifth one.

Physiology

No data available.

Genetics

The karyotype of the wild goat has $2n = 60$ (FN $= 60$) chromosomes (Kuliev and Mamedov 1974). The wild goat and the domestic goat cross in captivity producing fertile offspring (Heptner et al. 1961) and supposedly do so in the wild as well (Pfitzenmayer 1915), although there are no recent data on such events, e.g., in Daghestan (Weinberg 2001). They are close genetically (Manceau et al. 1999; Pidancier et al. 2006). The wild goat and the Eastern tur also cross in captivity producing fertile offspring (Pfitzenmayer 1915; Sarkisov 1953). Information on hybridization of both species in the wild is vague (Pfitzenmayer 1915) and has not been confirmed by subsequent research (Heptner et al. 1961; Weinberg 1999).

Life History

Growth

Newborns, captured almost immediately after birth, weighed 3.2–3.5 kg. By the age of 2 years (third summer) these animals weighed 17–20 kg on average (Magomedov et al. 2014).

Reproduction

Magomedov et al. (2014) observed yearling females with newborns, suggesting that juvenile females at the age of 7–8 months participate in the rut and can get pregnant. Yearling males already may lift their tails during the rut, indicating their sexual activity, as adult males of all *Capra* do, but in healthy populations with natural sex ratio and adult male proportion, males can effectively participate in the rut only from their 5th to 6th year of age.

Batkhiyev (1989), Dinnik (1910), and Ekvtimishvili (1954) estimate the beginning of the rut in the North Caucasus earlier than subsequent research has suggested (Weinberg 2001; Magomedov et al. 2014). According to Ekvtimishvili (1954), the times of rut and parturition could vary significantly over the years, possibly due to different weather conditions (Table 1). On 1995, in Avar Koisu riverbasin (Daghestan), males started courting females on December 17th. Magomedov et al. (2014) state that the rut peaks in the first decade of January. Pregnancy lasts about 5–5.5 months. The birth season probably lasts from mid-June to mid-July (Weinberg 2001; Magomedov et al. 2014) (Table 1). Late reproduction may be triggered by the thermophilicity of the species, since the North Caucasus is the northernmost limit of its range, and newborns might not tolerate low temperatures and occasional snowfalls in spring. On the other hand, the long vegetative period lasting from April to September in montane forests may allow late parturition, since kids have enough time for growing up and gaining body mass until winter.

Twins are common in wild goats, and triplets may also occur (Magomedov et al. 2014). Magomedov et al. (2014) reported that up to 45% of females with offspring had twins, and Ekvtimishvili (1954) stated that in some years twins were observed in most cases. Some goats manage to keep both kids until the next spring (Weinberg 2001), but on the whole, proportion of females with twins diminished by ¼ – ⅓ during the period from June to January (Magomedov et al. 2014). Juvenile/female ratio was 1.25:1 in Daghestan on June-August, while yearling/female ratio was 0.68:1 (Weinberg 2001). However, these indices may change over the years, probably due to differing weather conditions, e.g., amount of snow in the winter, because Magomedov et al. (2014) present slightly lower totals for juvenile/female ratio – 0.85:1. Generally, twinning is typical of the wild goat all over the species range (Khorozyan et al. 2009; Korshunov 1994; Kuliev 1981; Talibov et al. 2009; Schaller 1977). It is characteristic of taxa inhabiting arid and warm environments with low and unpredictable vegetation production and

Table 1 Rutting and birth seasons of the wild goat in different parts of the species range

Location	Rutting season	Birth season
Daghestan[a, b, c]	November to December Mid-December to January	May to June Mid-June to mid-July June 10 to mid-July
Chechnya[d]	Mid-December to beginning of January	
Tusheti (Georgian part of Andi Koisu riverbasin)[e]	Mid-November to mid-January (varying through the years)	Mid-April to mid-June (varying through the years)
Caucasus Minor (Armenia & Azerbaijan)[f, g]	November to December	End of April to May
Kopet Dagh (Turkmenistan)[h]	November to mid-December	End of March to early May

Sources: [a]Dinnik (1910); [b]Weinberg (2001); [c]Magomedov et al. (2014); [d]Batkhiyev (1989); [e]Ekvtimishvili (1954); [f]Dahl' (1951); [g]Kuliev (1981); [h]Korshunov (1994)

compensates them for losses in droughts (Schaller 1977). *Capra* populations of boreal mountains – including East Caucasian tur in Daghestan – give birth to singlets, as a rule (Abdurakhmanov 1973; Fandos 1989; Kotov 1968; Magomedov et al. 2001; Veinberg 1984). The generally accepted explanation is that twins are smaller at birth, grow slower, and have less chance to survive harsh winters. Therefore, the essentially thermo- and xerophilic wild goat is peculiar in having retained high twinning rate in the boreal mountains of Daghestan. High fecundity may possibly be due to living in montane forests with a long vegetative period and comparatively mild winters.

Population Structure and Survival

In natural, unharvested *Capra* populations, sex ratio is usually close to 1, although adult males make more than 10% of the encountered animals during the rut (Kotov 1968; Schaller 1977; Veinberg 1984). These indices were respectively only 0.71:1 and 4.8% in Avar Koisu R. canyon (Weinberg 1999). A similar figure of sex ratio was presented by Magomedov et al. (2014). Low number of adult males, shown by winter counts in Avar Koisu canyon, is most certainly due to poaching, as highlanders always try to shoot big males. It is worth noting that yearling males slightly outnumbered yearling females during the rut and outside it (Weinberg 2001).

After being very high in juveniles, mortality then lowers during the 2nd–4th years of life and starts rising after that and, in general, is higher in males than in females, and may differ in various areas, depending upon intensity of poaching (Magomedov et al. 2014).

Habitat and Diet

Habitat Selection and Movement

Within the wild goat range in Daghestan, valley bottoms lie at 1000–1500 m a.s.l., and the ridges rise up to 3000, at places to 4000 m a.s.l. They are often composed of shale, so the ridges are usually broad and quite rolling, whereas the valleys of the rivers and bigger brooks are narrow and canyon-like. However, the upper ends of the valleys are wider and less steep, and thus less appropriate for the wild goat. At the altitude of 1500 m a.s.l., the mean monthly temperature is about −6 °C in January and +16 °C in July, and the total precipitation about 600–750 mm. Most of the latter (up to 60%) occurs on spring and early summer (Himmelreich 1967).

Montainous forests are composed of pine *Pinus kochiana* = *P. sosnowskyi* and birch *Betula litwinovii*, with admixture of aspen *Populus tremula*, beech *Fagus orientalis*, hornbeam *Carpinus caucasica*, lime-tree *Tilia cordata* and oak *Quercus macranthera*. The forest floor is of grasses, sedges, and herbs. *Spiraea crenata* shrubbery usually covers openings on sunny slopes. Precipitous southern slopes near valley bottoms often harbor small patches of xerophitic open stands of oak, pine, and tree-like juniper *Juniperus communis* = *Juniperus polycarpus* up to 6 m high in various combinations with understorey of *Berberis vulgaris*, *Paliurus spina-christi*, *Rosa* spp., and *Spiraea*. Spiky cushions of *Tragacantha* spp. are also characteristic of such stands (Weinberg 2001).

Wild goats inhabit only montane forests (not shrubbery) from the valley bottoms to the timberline (2600–2700 m a.s.l.). The forest zone is represented by narrow stripes (up to 1.5–2 km wide) along rivers and bigger brooks, due to the steepness of the slopes. Therefore, the wild goat distribution generally follows the fluvial net. Animals prefer steep, precipitous slopes and avoid tall, dense stands on gentle slopes or tree-less areas.

Observations in different seasons indicate that animals do not perform altitudinal migrations, remaining within the same, relatively narrow, forest zone throughout the year (Weinberg 2001; Magomedov et al. 2014). Even in summer, wild goats have been only twice observed above the timberline (Weinberg 2001). However, Magomedov et al. (2014) state that males forage in subalpine zone, and that their vertical distribution is generally higher than that of females in summer. Local hunters say that occasionally males cross some high ridges in search of females during the rut. In fact, wild goat males may appear on the south slope of the Main Range which the wild goat does not inhabit permanently (see section "Current Distribution"), suggesting that populations inhabiting neighboring riverbasins communicate with each other, at least via migratory males. Although it occurs on high mountains in the Greater Caucasus, the wild goat leads a life typical of conspecific populations inhabiting low mountains or hills, such as in some areas of Turkmenistan (Heptner 1956) or Caucasus Minor (Khorozyan et al. 2009).

Selection of southern slopes is somewhat higher in the snowy season (Magomedov et al. 2014). Sexual differences in ecological distribution, so evident for the East Caucasian tur (Veinberg 1984; Magomedov et al. 2001), are weak in the wild goat due to peculiarities of its habitat, but do exist. In spring and summer, 84.1% of adult males ($N = 44$) and only 20.0% of females ($N = 90$) were sighted closer to the timberline than to the valley bottoms. Males avoid places where timberline has been lowered by man. Human settlements and agricultural land have often replaced the upper part of the forest zone, and merely a narrow stripe 300–500 m wide has remained along the bottom of the valley. Such forests are inhabited by females with young but not by adult males, who mainly visit them during the rut. Thus, a partial ecological and occasionally even spatial separation of adult males from females with offspring exists (Weinberg 2001). In summer, females dwell on the northern slopes more often than adult males, but this difference may not be significant (Magomedov et al. 2014).

Feeding

In Chechnya, the wild goat diet includes 160 plant species. Seasonal selection is rather distinct. In spring, animals choose mostly herbs and certain grasses, like *Festuca* and *Poa*. During late summer and autumn, wild goats consume various Fabaceae and other herbs, while in winter their diet gets poorer and contains lots of shrubs, tree branches, and even bark, as well as dry grasses (Tochiev and Batkhiyev 1980; Batkhiyev 1989). Magomedov et al. (2014) described diet and foraging of captured animals, which may differ from that of wild ones. They state that on natural pastures animals prefer herbs like Fabaceae, but grasses, sedges, and trees and shrubs make the largest proportion in forage, as herbs are being quickly consumed.

Due to the huge species range (see section "Current Distribution"), its habitats vary throughout the range from lowland semi-deserts to highland alpine zone, with steppe and all sorts of forest and shrubbery in between. Spatial movements and diet vary accordingly.

Behavior

Foraging Behavior and Activity

Foraging behavior of the wild goat is quite various, mainly due to their ability to climb trees. Their climbing ability outperforms that of tur (Veinberg 1984). Even massive adult males can climb oaks and junipers and feed in their canopy.

In summer, animals are mostly inactive from 9 to 10 h till 16–17, avoiding open tree- and shrubless areas. Often they cross the openings so quickly that an observer has barely time to estimate age and sex classes (Weinberg 1999). During the rut, animals are active all day long, especially males, which roam in search of females, like all *Capra* do (Weinberg 1999). This behavior is used by poachers who often wait for adult males near the paths used by them (Batkhiyev 1989).

Mating Behavior

The courtship repertoire does not differ from that of other *Capra* species (Fedosenko et al. 1992; Schaller and Laurie 1974; Veinberg 1984) and include guarding, low-stretch, tongue-flick, kick, naso-nasal and naso-genital contacts, twist, jerk, urination (or ejaculation) on muzzle (Weinberg 2001).

Social Behavior and Protection of the Newborns

Females are met on cliffs during the lambing season and after it while kids are small. Sometimes, they concentrate in such lambing and nursing sites as, for example, on a cliff wall with numerous ledges and clumps of trees. A hiding phase is characteristic of the wild goat, unlike tur. The kids (supposedly 3–5 days old) are very agile. Equally active tur kids always follow their mothers, whereas wild goat females leave their young not only when feeding, but also in case of danger. A female with a kid, found by a dog, escaped to the cliffs, reappeared without her offspring, and evidently tried to lead people away from the hidden kid. Of 13 observed females having newborns, 8 left them in some secure place (Weinberg 2001). Females and kids bleat, while the alert and danger signal is an abrupt snorting, not a hissing whistle, as in tur.

Since females are secretive around parturition, individuals of other sex/age classes seem more numerous then, yearlings in particular. Most yearlings tend to associate with females in winter and spring, whereas only some do so during the birth season. Others roam in twos and threes looking independent, without adult females in the vicinity (Weinberg 2001).

Four types of groups can be distinguished in wild goats, as well as in other *Capra* species: adult male (older than 5 years) groups that may include young males too; young male groups; female groups occasionally including young males; mixed groups with adult animals of both sex (Weinberg 2001). The first three types are met all year round, but the last one almost exclusively during the rut, due to the fact that adult males seldom associate with females outside the rutting period. Small percentages of 2–3-year-old males and 4–5-year-old ones have been observed in female groups, on spring and summer. Average group size in Daghestan (3.3) is almost similar to that in Chechnya (Batkhiyev 1989) being four to five times lower than in conspecific populations on the Caucasus Minor (Kuliev 1981) and Kopet-Dagh (Korshunov 1994), at close population densities. Single females with their offspring of the last 2 years make 38.2% of all female group sightings ($N = 136$) in Daghestan. Outside the rut, adult males tend to stay in bigger groups. Yearly dynamics of overall group size is inconspicuous. Adult males follow the common pattern of all *Capra* being gregarious outside the rut and almost solitary during it, when not associated with females (Weinberg 1999, 2001).

Parasites and Diseases

Just four helminth species have been found in the digestive tract and respiratory system of wild goats in Daghestan: *Moniezia expansa*, *Chabertia ovina*, *Nematodirus junctispicularis*, and *Trichostrongylus vitrinus*. Paucity of helminth fauna may be caused by absence of possible intermediary hosts (Zakariev 1982). However, the parasite faunas of conspecific populations in the Caucasus Minor (Grigoryan 1949) and sympatric Caucasian tur in Daghestan (▶ Chap. 18, "Caucasian Tur *Capra caucasica* Güldenstaedt et Pallas, 1783," by Weinberg and Lortkipanidze, this volume) are much richer. Further studies are required.

Population Ecology

Population Dynamics and Density

Being residents of forested precipitous slopes in the Greater Caucasus (see section "Habitat and Diet"), wild goats defy methods of direct counting commonly applied to mountain ungulates (Weinberg 2001). Methods used for forest-dwelling ungulates, like transect counts of tracks in the snowy period, are also impossible on precipitous and steep slopes of river canyons. In the Greater Caucasus, estimates were often based on information provided by state institutions of hunting control and were nothing more than rough estimates (e.g., Ravkin 1975; Tochiev 1975). Data from the Daghestan part of the species range have been more detailed, especially for 1990s–2010s, as they were based on actual counts performed by zoologists (Weinberg 2001; Magomedov et al. 2014) (Table 2). Data from Chechnya are poor,

Table 2 Dynamics of wild goat numbers in the Greater Caucasus

Year	Area (country)			Total
	Chechnya (Russia)	Daghestan (Russia)	Khevsureti and Tusheti (Georgia)	
1970s	250[a]–600[b] 350–360[c]	550[a]	200–300[d]	1000[a]
1980s		1000[e]	300[f]	1300
1990s		1500[g]		1500
1998–2000s		2500–2600[h]	100[i]	2600–2700
2010s		1200–1500[j]	180[k]–310[l]	1200–1500

Sources: [a]Ravkin (1975); [b]Tochiev (1975); [c]Batkhiyev (1980); [d]Kapanadze (1978); [e]Prilutskaya and Pishvanov (1989); [f]Arabuli (1989); [g]Weinberg (2001); [h]Magomedov et al. (2014); [i]Mallon et al. (2007); [j]Babaev et al. (2017); [k]NACRES (2010); [l]Ilia State University (2013)

and there are even suggestions that the species is close to extinction or has already vanished from there, and even from Georgia (e.g., Magomedov et al. 2014). Daghestan harbors the bulk of the wild goat population on the Greater Caucasus, but the human impact is growing there and numbers are sinking (Babaev et al. 2017). In 1990s, local villagers unanimously stated that wild goats used to be noticeably more abundant 10–15 years before. However, wild goats are really hard to see and therefore seem to be rare. In fact, they may be common within their range, and the relatively restricted size of the population could be a result of the limited range and habitat.

Population density varies from 0.7 to 0.9/km^2 in winter in Chechnya (Batkhiyev 1980) to average 3.9/km^2 (1.1–11.0/km^2) in summer in Daghestan (Magomedov et al. 2014).

Competition with other Ungulates

Wild goats are sympatric with red deer *Cervus elaphus*, East Caucasian tur, chamois *Rupicapra rupicapra*, roe deer *Capreolus capreolus*, and wild boar *Sus scrofa*. Red deer usually avoid precipitous areas and most of the animals migrate over the Watershed Range to Azerbaijan and Georgia for winter, thus its competition with wild goat is insubstantial (Magomedov et al. 2014). Chamois are more common on the north slope of the Watershed Range in Daghestan and Tusheti where they mostly dwell above timberline and rarely occur in forested precipitous habitats situated far from the ridge. They are seldom noticed in areas harboring wild goat. However, both species are sympatric on the Rocky Range in Chechnya, where chamois is much more numerous than the wild goat, but habitat separation of the two species has not been described for this area (Lukarevsky 2018). However, as on the Watershed Range, most probably the chamois occupies all of the altitudinal range while the wild goat dwells in the forest, as it happens in the Greater Caucasus. Roe deer are rare in wild goat habitats, while wild boar have a completely different diet and usually do not favor precipitous slopes (Magomedov et al. 2014).

Tur are the main, most numerous ungulates in montainous Daghestan and live mostly above the timberline, but there are places where they dwell in the forest not only in winter, but all year round. Then they are usually seen in the upper parts of the forest zone and prefer more open, precipitous sites or more humid slopes. Arid patches of dry open oak stands with admixture of juniper and understorey of *Berberis*, *Paliurus* and *Spiraea* situated on lower sections of sunny slopes near valley bottoms, were the only fragments of forest zone inhabited by the wild goat in Daghestan, which tur did not visit (Weinberg 1999). Nevertheless, both *Capra* species sometimes feed at close distances from each other on grasslands (Tochiev and Batkhiyev 1980), but especially below timberline in winter (Weinberg 2001). Wild goats outnumber tur seven times in the forest zone of Daghestan (Weinberg 2001), but the ratio is opposite above timberline (Magomedov et al. 2014).

Effects of Climate Change

The glacier area in the Greater Caucasus has shrunk by 794 km^2 or 36% in 1910–1976 (Panov 1993). This process is continuing in the twenty-first century. The timberline and the upper limit of the periglacial zone seem to be rising constantly. However, overgrowing of formerly timber-less areas may sometimes depend more on diminishing anthropogenic impact than on climatic changes. Theoretically, global warming should favor wild goat.

Conservation Status

IUCN categories were not used in the USSR, thus the wild goat was included as "low-numbered (rare)" in the first edition of Red Data Book of the USSR (Borodin et al. 1978), "diminishing" in the second edition (Borodin et al. 1984) and as "endangered" in the Red Data Book of the Russian Federation (Eliseev et al. 1983). Different assessment is caused by the fact that in the USSR the wild goat used to be quite common in the Caucasus Minor (Azerbaijan and Armenia) and in Turkmenistan, much more numerous than in the Greater Caucasus (Russian Federation). In the first Red Data Book of the new Russian Federation (after the disintegration of the Soviet Union), the wild goat was regarded as "diminishing and being on the periphery of its distribution" (Weinberg 2000). The species was previously listed as "Vulnerable" in the IUCN Red List (Weinberg et al. 2008) but has since been downgraded to "Near threatened" (Weinberg and Ambarli 2020). It is also listed in Appendix III of CITES (as *Capra hircus aegagrus*).

There are one federal and two local sanctuaries (*zakaznik*) within the wild goat range in Daghestan, i.e., Tlyarata, Kosob-Keleb, and Bezhta. Altogether they cover a territory of approx. 1500 km^2 and are aimed primarily at tur and wild goat protection. Unfortunately, they do not fulfill their task, because, unlike the nature reserves, they do not, in reality, protect the landscape. Human settlements and land-use are allowed on their territories. It would be natural to expect a further decrease of population under such circumstances. A nature reserve was being proposed, but presently it seems that a huge federal sanctuary may be organized on an area of approx. 3000 km^2, incorporating the above-mentioned local sanctuaries, to protect potential leopard habitat. Unfortunately, even a federal sanctuary does not protect the landscape, but according to the current Russian law, a nature reserve needs land without human population and that cannot be practically found within the wild goat range. Any institution of lower ranking than a strict nature reserve will be no more effective than the already existing ones, and even the nature reserve status cannot provide quick favorable results. Only the ongoing emigration of highlanders to the lowlands may offer positive changes to the wild goat status in Daghestan.

The wild goat is listed in the Red List of Georgia as Critically Endangered (Ordinance of the Government of Georgia on Adoption of Georgian "Red List" 2014). Consequently hunting wild goat is totally prohibited in that country. Wild goat is being protected in Tusheti Protected Areas, which encompasses the Tusheti Strict Nature Reserve, Tusheti National Park and Tusheti Protected Landscape.

Management

Counting Methods

Counting methods in the Greater Caucasus are based on direct counts of animals along transects or from fixed observation points, usually during the snowy period of the year. Animals are counted on some plots, and the data are extrapolated over the rest of the area (Magomedov et al. 2014; Weinberg 2001).

Anthropogenic Impact

Wild goats live side by side with man in Daghestan and are being constantly hunted, or better said poached. This proximity is not seasonal and is not restricted to professional groups (e.g., shepherds, as it happens with almost all wild *Capra* populations). On the contrary, it is permanent and with settled people, because villages and farmsteads (on average situated no more than 5–6 km away from one another, along the valleys), agricultural land, paths and roads, main resources of timber and firewood, all occur within narrow stripes of montane forests which are the wild goat habitat. Despite emigration to plains and lowlands, highland Daghestan still displays high human population density unparalleled in the rest of the North Caucasus. Existence of *Capra* species under such circumstances is difficult. It became possible, presumably, thanks to good shelter provided by montane forests and their quick natural reforestation. It is unlikely that the species could have survived in such a close neighborhood with man in the thin, open stands typical of the main, more southern and arid part of the range – Caucasus Minor, Kopet-Dagh, Iran, and Sind. The arid forest itself could hardly resist anthropogenic impact and seldom it does. Mere fragments of open stands still survive within pine-birch forests in Daghestan, but they probably covered most of the country when the species penetrated there. Later, deforestation in lower and drier parts of Daghestan probably forced the species out from there. Climate turned cooler and more humid; it changed the forests, but wild goats managed to adapt to these changes. The anthropogenic pressure can be responsible for certain peculiarities in this species biology in Daghestan, namely secretiveness, preference for closed forest, and, on the whole, avoiding open places. Unlike tur, which under human pressure usually escape to higher and less accessible areas, wild goats do not have this opportunity and remain within their habitat. Therefore the survival strategy of this species is typical of forest ungulates. The anthropogenic influence changed a good deal during the last decades. Emigration from highlands began after the Second World War and is still going on. Abandoned villages and farmlands are overgrown with forest, and wild goats live among ruins in some cases (Weinberg 2001). Timber felling reduced in the 1970s and 1980s, but increased again in the beginning of the 1990s. It is illegal, but presently is the main and sometimes the only substantial source of income for villagers. Tall stands on gentle slopes suffer most, so the wild goat habitat remains comparatively less harmed, but, if the felling will continue, the situation may get worse.

Another serious threat is exploitation of water resources in montainous Daghestan. Several hydropower stations have already been built, and presumably there are plans for constructing more.

Poaching is the main threat and has become quite intensive during the last years. The wild goat is, in fact, the most often hunted big game within its range. It has been always hunted (shot, or caught with snares and traps) all year round, despite being listed in the Red Data Books with all the accompanying prohibitions. If previously poachers used old and battered rifles from the Second World War or even the Russian Civil War (1918–1921), shotguns and small-bore rifles, now they have equally illegal automatic rifles of the newest models. Local enthusiasts of nature conservation try to influence people through Muslim priests, but without any obvious results (Weinberg 2001).

Natural Predators

There are some predators which present potential threats to wild goats: wolf *Canis lupus*, golden jackal *Canis aureus* and red fox *Vulpes vulpes*, the last two a danger only for newborns, lynx *Lynx lynx*, leopard *Panthera pardus* and brown bear *Ursus arctos*. Seemingly, the bear looks almost the least dangerous of all, but one of the authors witnessed a bear chasing and actually catching a 2-week-old wild goat kid in Armenia. Nevertheless, the brown bear and the wolf can hardly be considered the principal predators influencing the wild goat population in the Greater Caucasus, bearing in mind that wild goat is essentially a cliff dweller. The lynx, as well as the leopard, can successfully hunt wild goats, but, due to their low numbers or, in the case of leopard, sporadic occurrence within the wild goat distribution in the Greater Caucasus, their impact upon wild goat populations can hardly be important.

Golden eagles *Aquila chrysaetos* and Bearded vultures *Gypaetus barbatus* inspect rocky massifs all day long, in search of newborns. They try to hunt grown-up kids, too. In August 1990, an eagle was observed hunting a wild goat kid in Daghestan. The bird repeatedly tore it off from a cliff wall, but finally the kid managed to escape in the forest (Weinberg 2001). Snow avalanches have not been indicated as a cause of death to the wild goat due to its dwelling on the lower forested slopes in the snowy season.

Economic Value

At the moment, the economic value of the wild goat is minimal, being red-listed and thus not hunted legally. Consequently, there are no laws and regulations organizing its use in any way, and it can be regarded maybe just as a potential object of ecological tourism. If goat populations could become so numerous to stand e.g., trophy hunting, their economic value could quickly grow.

Introductions

A failed attempt to introduce wild goats to the High Tatra mountains in Slovakia (then Austrian-Hungarian Empire) before the First World War has been anecdotally reported by Turček and Hickey (1951). In 1953 there was another introduction of the wild goat to the Czech territory, in a densely populated hilly area. The goats were intended as "game" animals, and during the 1960s the herd was supplemented with new animals imported from the former USSR. Since 1994, the whole herd has been gradually captured and moved to the Vřísek game preserve, where the goats have lived ever since. Local zoologists considered those animals not pure-bred, but hybrids with domestic goat (L. Bartoš, pers. comm.) and recent photographs substantiate this view (https://liberec.rozhlas.cz/jedine-misto-v-republice-kde-zije-koza-bezoarova-je-nedaleko-ceske-lipy-6004893). The wild goat is not listed in the mammal fauna of Czech Republic.

Future Challenges for Research and Management

The main conservation issue is lessening the impact of poaching and, accordingly, increasing legal hunting quotas, especially for local highlanders. The latter would help highlanders to back nature conservation efforts in the Caucasus. Right now there is no such understanding, and the general attitude to nature conservation is rather shallow, if not entirely hostile. Poaching is considered a minor and even partly justified misbehavior, being a part of the local lifestyle.

References

Abdurakhmanov MG (1973) Ecology, protection, and use of tur in Daghestan. Candidate dissertation thesis, Izdaniye Dagestanskogo Universiteta, Makhachkala. (in Russian)

Arabuli AB (1989) Contemporary status of rare ungulate species of Georgia. In: Sablina TB (ed) Ekologiya, morfologiya, ispol'zivaniye i okhrana dikikh kopytnykh, chast' 1 (Ecology, morphology, use and protection of wild ungulates, part 1). All-Union Theriological Society Publication, Moscow, pp 28–29. (in Russian)

Babaev EA, Yarovenko YuA, Magomedov UM, Gamodova MKh, Yarovenko AYu (2017) Population status of ungulates of Daghestan. Izvestiya Dagestanskogo gosudarstvennogo pedagogicheskogo instituta. Yestesvennye i tochnye nauki (Communications of Daghestan State Pedagogical Institute. Natural and precise sciences) 3:44–50. (in Russian)

Baryshnikov GF (1987) Mammals of the Caucasus of the early Palaeolitic. Trudy Zoologicheskogo Instituta Akademii Nauk SSSR (Proceedings of Zoological Institute of the USSR Academy of Sciences) 168:3–20. (in Russian, English summary)

Batkhiyev AM (1980) Distribution and numbers of wild goat in Chechen-Ingushetia. In: Tembotov AK (ed) Fauna, ekologiya i okhrana zhivotnykh Severnogo Kavkaza (Fauna, ecology and conservation of animals of the North Caucasus). Kabardino-Balkarskiy gosuniversitet, Nal'chik, pp 46–56. (in Russian)

Batkhiyev AM (1989) Wild goat in montane Chechen-Ingushetia. In: Anisimov VI (ed) Priroda i khozyaystvo Checheno-Ingushskoi ASSR (Nature and economy of Chechen-Ingush ASSR). Checheno-Ingushskoye knizhnoye izdatel'stvo, Grozny, pp 103–109. (in Russian)

Borodin AM et al (eds) (1978) USSR Red Data Book, 1st edn. Lesnaya Promyshlennost', Moscow. (in Russian)

Borodin AM et al (eds) (1984) USSR Red Data Book, vol 1, 2nd edn. Lesnaya Promyshlennost', Moscow. (in Russian)

Dahl' SK (1951) Data on biology, distribution and numerical ratios in wild goat herds at the Urts Range. Izvestiya Akademii Nauk Armyanskoi SSR. Seriya biologicheskikh i sel'skokhozyaystvennykh nauk (Procedings of Academy of Sciences of Armenian SSR. Biological and agricultural science series) 4:33–40. (in Russian)

Danilkin AA (2005) Bovids. KMK, Moscow. (in Russian)

Dinnik NYa (1910) Mammals of the Caucasus. Zapiski Kavkazskogo Otdela Imperatorskogo Russkogo Geograficheskogo Obshchestva (Notes of Caucasian Section of the Imperial Russian Geographic Society) 27:1–247. (in Russian)

Ekvtimishvili Z (1954) Materials for study on reproduction of bezoar goat (*Capra aegagrus* Erxl.). Trudy Instituta Zoologii Akademii Nauk Gruzinskoi SSR (Proccedings of Institute of Zoology of Academy of Sciences of Georgian SSR) XIII:93–97. (in Georgian, Russian summary)

Eliseev NV et al (eds) (1983) RSFSR Red Data Book (animals). Rossel'khozizdat, Moscow. (in Russian)

Ellerman J, Morrison-Scott T (1966) Checklist of Palaearctic and Indian mammals 1758 to 1946, 2nd edn. British Museum (Natural History), London

Fandos P (1989) Reproductive strategies in female Spanish ibex (*Capra pyrenaica*). J Zool Lond 218:339–344

Fedosenko AK, Veinberg PI, Skuratov NI (1992) Siberian ibex in the Western Sayan. In: Fedosenko AK (ed) Okhrana i izucheniye redkikh i ischezayushchikh vidov zhyvotnykh v zapovednikakh (Protection and studies of rare and endangered animal species in nature reserves). Tsentral'naya Nauchno-issledovatel'skaya Laboratoriya okhotnichyego khozyaistva, Moscow, pp 29–44. (in Russian)

Grigoryan GA (1949) Parasitic worms of wild goats of Armenia. Trudy NII veterinarii Armyanskoy SSR (Proceedings of Research Institute for Veterinary of Armenian SSR) 6:151–158. (in Russian)

Heptner VG (1956) Vertebrate animal fauna of Badkhyz. Turkmenskaya Academiya Nauk, Ashkhabad. (in Russian)

Heptner VG, Nasimovich AA, Bannikov AG (1961) Mammals of the Soviet Union. Artiodactyls and Perissodactyls, vol 1. Vysshaya Shkola, Moscow. (in Russian)

Himmelreich VA (1967) Geography of the Daghestan ASSR. Daguchpedgiz, Makhachkala. (in Russian)

Ilia State University (2013) Wildlife census in the framework of national biodiversity monitoring in Georgia. Unpublished report, Tbilisi. (in Georgian)

Kapanadze AD (1978) Structure and numbers of populations of highland ungulates of the Caucasus. In: Panteleev PA (ed) II s'yezd Vsesoyuznogo teriologicheskogo obshchestva (2nd congress of All-Union Theriological Society). Nauka, Moscow, pp 137–138. (in Russian)

Khorozyan IG, Weinberg PI, Malkhasyan AG (2009) Conservation strategy for Armenian mouflon (*Ovis [orientalis] gmelini* Blyth) and bezoar goat (*Capra aegagrus* Erxleben) in Armenia. In: Zazanashvili N, Mallon D (eds) Status and protection of globally threatened species in the Caucasus. CEPF, WWF, Countour Ltd., Tbilisi, pp 46–53

Korshunov VM (1994) Ecology of the bearded goat (*Capra aegagrus* Erxleben, 1777) in Turkmenistan. Monogr Biol 72:231–246

Kotov VA (1968) Kuban tur, its ecology and economical significance. Trudy Kavkazskogo Gosudarstvennogo Zapovednika (Proceedings of Caucasus State Reserve) 10:201–293. (in Russian)

Kuliev SM (1981) The wild goat and Daghestan tur of Azerbaijan. Candidate dissertation thesis, Vsesoyuznyi nauchno-issledovatel'skii institut okhrany prirody, Moscow. (in Russian)

Kuliev GK, Mamedov TO (1974) Studies of the chromosomal set of the wild goat (*Capra aegagrus* Erxleben) and Daghestan tur (*C. c. cylindricornis* Blyth). Izvestiya Akademii Nauk Azerbaijanskoi SSR, seriya biologicheskaya (Proceedings of Academy of Sciences of Azerbaijan SSR, biological series) 5–6:88–93. (in Russian)

Lukarevsky VS (2018) Field surveys in Chechen Republic, Republic of Daghestan and Republic North Ossetia-Alania in order to evaluate habitats of leopard and work out suggestions on current species range for restoration of wild population. Unpublished report to ANO "Center for Nature of the Caucasus" (in Russian)

Magomedov M-RD, Akhmedov EG, Yarovenko YuA (2001) Daghestan tur. Populational and trophic aspects of ecology. Nauka, Moscow. (in Russian)

Magomedov M-RD, Akhmedov EG, Yarovenko YuA, Nasrulayev NI (2014) Wild goat in Daghestan: organization of the population and peculiarities of ecology. KMK, Moscow. (in Russian)

Mallon D, Weinberg P, Kopaliani N (2007) Status of the prey species of the leopard in the Caucasus. Cat News, Special Issue 2:22–27

Manceau V, Boursot P, Despres L, Taberlet P (1999) Systematics of genus *Capra* as inferred by mitochondrial DNA sequence data. Mol Phylogenet Evol 13:504–510

NACRES (2010) Report on biodiversity assessment of Tusheti protected areas complex, UNDP/GEF project "Catalyzing financial sustainability of Georgia's protected areas system" (00070382), Testing site-level revenue generation mechanisms in Tusheti PA. Unpublished report, Tbilisi

NACRES (2013) Report on assessment of Leopard and its prey species in Khevsureti. WWF Caucasus program office project. Unpublished report, Tbilisi. (in Georgian)

Ordinance of the Government of Georgia on Adoption of Georgian "Red List" (Sakartvelos mtavrobis dadgenileba sakartvelos "tsitelinuskhis" damtkicebis shesakheb) (2014) Document number 190. Registration code 360060000.10.003.017809, [Internet; cited 2018 Aug 21]. Sakartvelos sakanondeblo matsne Publisher, Tbilisi. www.matsne.gov.ge/ka/document/view/2256983. Downloaded on 21 August 2018

Panov VD (1993) Evolution of contemporary glaciation of the Caucasus. Gidrometeoizdat, St-Petersburg. (in Russian)

Pfitzenmayer EV (1915) Some interesting bovid mongrels from Transcaucasia. Izvestiya Kavkazskogo Muzeya (Proceedings of the Caucasus Muzeum) 8:251–266. (in Russian)

Pidancier N, Jordan S, Luikart G, Taberlet P (2006) Evolutionary history of the genus *Capra* (Mammalia, Artiodactyla): discordance between mitochondrial DNA and Y-chromosome phylogenies. Mol Phylogenet Evol 40:739–749

Prilutskaya LI, Pishvanov YuV (1989) Range and numbers of the wild goat in Daghestan. In: Sablina TB (ed) Ekologiya, morfologiya, ispol'zovaniye i okhrana dikikh kopytnykh, chast' 2 (Ecology, morphology, use and protection of wild ungulates, part 2). All-Union Theriological Society Publication, Moscow, pp 243–245. (in Russian)

Ravkin YeS (1975) Resources of the wild ungulates in the North Caucasus and anthropogenic impact upon them. In: Sokolov VYe (ed) Kopytnye fauny SSSR (Ungulates of the USSR fauna). Nauka, Moscow, pp 17–18. (in Russian)

Sarkisov AA (1953) On bovid crossbreeds. Priroda (Nature) 42:113. (in Russian)

Schaller GB (1977) Mountain monarchs. Wild sheep and goats of the Himalaya. University of Chicago Press, Chicago

Schaller GB, Laurie A (1974) Courtship behaviour of the wild goat. Z Säugetierkd 39:115–127

Shackleton DM, Lovari S (1997) Classification adopted for the Caprinae survey. In: Shackleton DM, The IUCN/SSC Caprinae Special Group (eds) Wild sheep and goats and their relatives, status survey and conservation action plan for Caprinae. IUCN, Gland/Cambridge, UK, pp 9–14

Sokolov II (1959) USSR fauna. Mammals (Artiodactyla and Perissodactyla). Trudy Zoologicheskogo Instituta Akademii Nauk SSSR (Proceedings of Zoological Institute of Academy of Sciences of the USSR) 14:1–295. (in Russian)

Sokolov VE, Tembotov AK (1993) Vertebrates of the Caucasus. Mammals: ungulates. Nauka, Moscow. (in Russian)

Talibov TH, Weinberg PI, Mammadov IB, Mammadov EN, Talibov ST (2009) Conservation strategy of the Asiatic mouflon (*Ovis [orientalis] gmelini* Blyth) and bezoar goat (*Capra aegagrus* Erxleben) in Azerbaijan. In: Zazanashvili N, Mallon D (eds) Status and protection of globally threatened species in the Caucasus. CEPF, WWF, Countour Ltd., Tbilisi, Georgia, pp 46–53

Tochiev TYu (1975) Rare ungulate species of Chechen-Ingushetia and their conservation. In: Sokolov VYe (ed) Kopytnye fauny SSSR (Ungulates of the USSR fauna). Nauka, Moscow, pp 284–285. (in Russian)

Tochiev TYu, Batkhiyev AM (1980) On foraging of wild goat in Chechen-Ingush ASSR. In: Tembotov AK (ed) Fauna, ekologiya i okhrana zhivotnykh Severnogo Kavkaza (Fauna, ecology and conservation of animals of the North Caucasus). Kabardino-Balkarskiy gosuniversitet, Nal'chik, pp 57–64. (in Russian)

Tsalkin VI (1950) New subspecies of the wild goat from Turkmenistan (*Capra aegagrus turcmenicus* subsp. nova). Doklady Akademii Nauk SSSR. Novaya seriya (Reports of USSR Academy of Sciences. New Series) 70:323–326. (in Russian)

Turček FJ, Hickey JJ (1951) Effect of introductions on two game populations in Czechoslovakia. J Wildl Manag 15:113–114

Valdez R (2011) Wild goat *Capra aegagrus*. In: Wilson RD, Mittermeier RA (eds) Handbook of the mammals of the world. Vol. 2 Hoofed mammals. Lynx Ediciones, Barcelona, p 917

Veinberg PI (1984) Daghestan tur. Nauka, Moscow. (in Russian)

Vereshchagin HK (1959) Mammals of the Caucasus. Izdatel'stvo Akademii Nauk SSR, Moscow. (in Russian)

Weinberg PJ (1993) Analysis of horn shape and coat coloration in *Capra* (Artiodactyla). Byulleten Moskovskogo Obshchestva Ispytatelei Prirody, Otdel biologicheskii (Bulletin of Moscow Naturalist Society, Biological Section) 98:3–14. (in Russian, English summary)

Weinberg PJ (1999) The present state and biology of the wild goat (*Capra aegagrus* Erxleben) Dagestan population. Byulleten Moskovskogo Obshchestva Ispytatelei Prirody, Otdel biologicheskii (Bulletin of Moscow Naturalist Society, Biological Section) 104:12–21. (in Russian, English summary)

Weinberg PJ (2000) Wild goat. In: Danilov-Danilyan VI (ed) Krasnaya kniga Rossii (Red Data Book of Russia). Astrel', Moscow, pp 717–718. (in Russian)

Weinberg PJ (2001) On the status and biology of the wild goat in Daghestan (Russia). J Mt Ecol 6:31–40

Weinberg P, Ambarli H (2020) *Capra aegagrus*. The IUCN Red List of Threatened Spcies: e. T3786A22145942. Downloaded on 29 July 2022

Weinberg P, Jdeidi T, Masseti M, Nader I, de Smet K, Cuzin F (2008) *Capra aegagrus*. The IUCN Red List of Threatened Species 2008: e.T3786A10076632. Downloaded on 04 September 2018

Zakariyev AYa (1982) To ecology of the wild goat in Daghestan. In: Panteleev PA (ed) Mlekopitayushchiye. III s'yezd Vsesoyuznogo teriologicheskogo obshchestva (Mammals. III congress of All-Union theriological society). Akademiya Nauk SSSR, Moscow, pp 202–203. (in Russian)

Mediterranean Feral Goat *Capra hircus* Linnaeus, 1758 — 20

Haritakis Papaioannou and Sandro Lovari

Contents

H. Papaioannou (✉)
Department of Biological Applications and Technology, School of Health Studies, University of Ioannina, Ioannina, Greece
e-mail: agriogido@hotmail.com

S. Lovari
Research Unit of Behavioural Ecology, Ethology and Wildlife Management, Department of Life Sciences, University of Siena, Siena, Italy
e-mail: sandro.lovari@gmail.com

© Springer Nature Switzerland AG 2022
L. Corlatti, F. E. Zachos (eds.), *Terrestrial Cetartiodactyla*, Handbook of the Mammals of Europe,
https://doi.org/10.1007/978-3-030-24475-0_161

Common Names

English	Feral goat
German	Hausziege (verwildert)
French	Chèvre férale
Spanish	Cabra asilvestrada
Italian	Capra rinselvatichita
Russian	Одичавшая коза

Taxonomy, Systematics and Paleontology

The feral goats of the Mediterranean islands have all been introduced by humans (Groves 1989; Vigne 1994; Manceau et al. 1999; Kahila Bar-Gal et al. 2002; Masseti 2009). In fact, fossils of the wild goat are absent from the original late Pleistocene fauna of Southern Europe (Schultze-Westrum 1963; Payne 1968; Clutton-Brock 1981; Masseti 1981; Davis 1984, 1987; Helmer 1989 in Masseti 2009). Some of these island populations may be ancient, but translocations of domestic goats have continued through time. As the "wild" phenotype tends to be genetically dominant over the domestic phenotypes, several of the older populations appear relatively homogeneous and may look as wild goats (Lorenz-Liburnau 1899; Couturier 1959; Schultze-Westrum 1963; Clutton-Brock 1981; Groves 1989) in spite of the introgression of domestic genes.

Domesticated goats descend mainly from the wild goat. The latest suggestion made by the International Commission for Zoological Nomenclature (2003: OPINION 2027; Case 3010) has been to retain the name *Capra hircus* for all domesticated goats and the name *Capra aegagrus* for the wild goat. As all the goats in the Mediterranean islands were introduced by humans, thus undergoing at least some domestication, we have used the specific name of *Capra hircus* for them.

Feral goats currently exist in six Mediterranean islands – excluding other recorded introductions implemented during the last few centuries, but presently extinct, as mentioned below. All of them show a variable number of morphological features similar to those of the Asiatic wild or Bezoar goat (*Capra aegagrus* Erxleben, 1777) (cf. Fig. 1).

Cretan goat, classified as *Capra hircus cretica* Schinz, 1838 (*Capra hircus cretensis* Lorenz, 1899) in Crete Island, Greece. Youra goat, classified as *Capra hircus jourensis* Ivrea, 1899 or *Capra hircus dorcas* Reichenow, 1888, in Youra Island (Sporades, Aegean Sea, Greece). Antinimilos goat classified as *Capra hircus pictus* Erhard, 1858 in Antimilos (Erimomilos) island (Cyclades, Aegean Sea, Greece). Majorcan Goat classified as *Capra hircus* var. *majorcan* in Majorca Island (Western Mediterranean, Spain). Montecristo goat, in the islet of Montecristo (Tuscan archipelago, Northern Tyrrhenian Sea, Italy). Eastern Atlantic Ocean goat in the island of Bugio (Desertas islands, off the southeastern coast of Madeira, Portugal) (Fig. 2).

Cretan and Aegean Feral Goats

It has been reported that the Cretan goat (*Capra hircus cretica* Schinz, 1838) has descended from a "primitive" domestic goat, which, after its introduction by humans to Crete during the Aceramic Neolithic period (7.000 BCE), became – and remain today – feral (Kahila Bar-Gal et al. 2002). In addition, the anthropochorus origin of the Cretan goat and those of other Mediterranean islands has been assumed in the past by several authors (Schultze-Westrum 1963; Papageorgiou 1974; Groves 1989; Masseti 1981; Logan et al. 1994 and others). Most likely, semi-wild goats were initially introduced to the island and released as a food source. Later, introgressions with domestic goats, especially females, have influenced its genotype (Horwitz and Bar-Gal 2006) or might be of mixed origin from two wild goat populations (from Taurus and Zagros mountains) (Geskos 2011). The Greek name for all feral goats with phenotype similarities to Bezoar, is "Agriokatisko" (Hatzisarantos & Kannelis 1955).

There have been numerous records regarding the presence of the Cretan goat (local name "Agrimi") in Crete Island from ancient times till the first half of the twentieth century (Hatzisarantos and Kannelis 1955). At the end of the nineteenth century, the Cretan goat formed free-ranging populations in all three of the highest mountain regions of Crete

Fig. 1 Adult male Cretan goat in Theodorou island (left, photograph by and courtesy of Anastasios Sakoulis); female Cretan goat at Mt. Parnitha (right, photograph by and courtesy of Konstantinos Fikas)

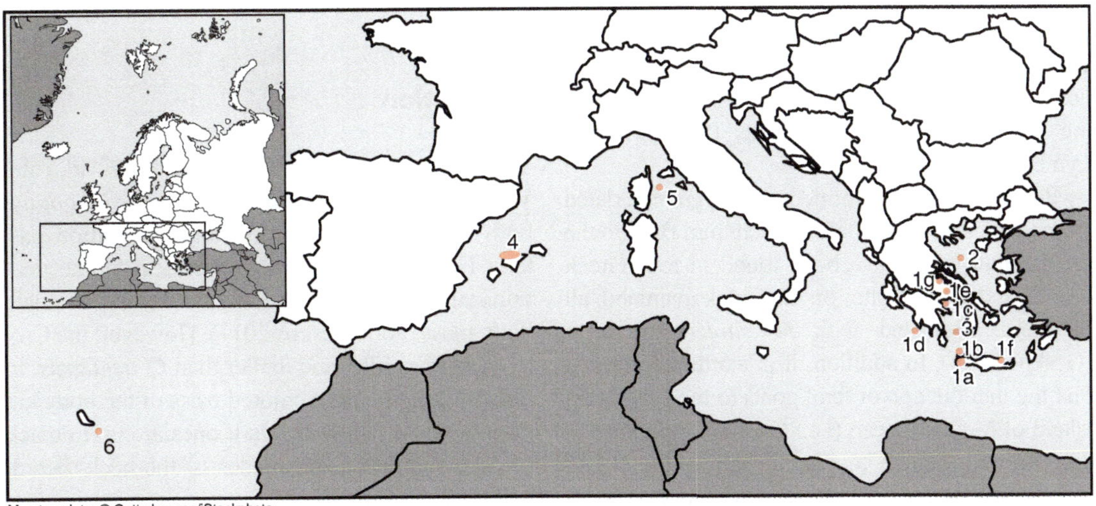

Fig. 2 Distribution map of Mediterranean feral goats: (1) Cretan goat (a) Lefka ori mountains, (b) Theodorou island, (c) Moni island, (d) Sapientza island, (e) mt. Parnitha, (f) Agii Pandes island, (g) Atalandi island, (2) Youra goat, (3) Antimilos goat, (4) Majorkan goat, (5) Montecristo goat, and (6) Eastern Atlantic ocean goat (Map template: © Copyright Getty Images/iStockphoto)

(Lefka Ori, Idi, and Dikti mountains), but it was already extinct from Idi mountain in 1913 and its last record in Dikti mountain is going back to 1904 (Sfougaris 1995; Geskos 2013).

Two additional insular feral goat populations occur in the Aegean Sea, in the islands of Youra and Antimilos (Erimomilos), where crossbreeding started between local feral goats – perhaps from ancient stock – and contemporary domestic goats. The Youra goat has been classified as *C. h. dorcas* Reichenow, 1888 (or *C. h. jourensis* Ivrea, 1899), whereas the Antimilos goat as *C. h. pictus* Erchard, 1858 (Sfougaris 1995; Adamakopoulos et al. 1997; Sfougaris and Lymberakis 2009; Papaioannou 2010).

Archeological evidence of the introduction of goats to Youra Island dates back to the Mesolithic period and it is, according precise C^{14} dates, the most ancient introduction of *C. aegagrus* by humans in the northern Aegean Sea (Masseti 2003a; Masseti and Trantalidou 2002; Trantalidou 2003) or perhaps even in the Mediterranean islands – excluding Cyprus. The extant goats are more similar to the Neolithic goats than to the goats present on the island in Upper Mesolithic times (Trantalidou 2003; Masseti 2009). According to the phenotypical patterns of the Antimilos goat, this population might have been introduced to the island in Neolithic times (Masseti 2009), but, as partly it looks like domestic goats (it has homonymous horns), it was perhaps introduced sometime between ~350 and 450 or 600 years ago from Crete for hunting purposes (Lorenz-Liburnau 1899; Schultze-Westrum 1963).

Another feral goat population existed till recently in Samothrace Island, Northern Aegean Sea, Greece, classified as *C. h. pictus* (Lorenz-Liburnau 1899; Schultze-Westrum 1963), but it went extinct on the second half of the twentieth century (Sfougaris 1994, 1995; Sfougaris and Lymberakis 2009). It had horns curved outward, which suggested crossbreeding with modern domestic goats (Sfougaris and Lymberakis 2009).

In general fossils identified as *C. aegagrus*, dated from the middle of the ninth millennium BCE to the sixth century CE, have been found in four Greek islands and in a few sites on the Greek mainland, all of them associated with *H. sapiens* presence (Geskos 2013). In addition, it is worth mentioning that the introduction of feral goats to the far eastern island of Mediterranean (i.e., Cyprus) took place at the end of the ninth to second half of the eighth millennium (Masseti 2009). However, as goat specimens from Shillourokambos excavation site (in Cyprus) were identical in bone size and shape to the continental *C. aegagrus* form – although introduced and captive bred – the goat from this site (Shillourokambos) has been termed as "pre-domestic" (Vigne and Buitenhuis 1999 in Vigne 2001).

Current Distribution

The only one long-established free-ranging population of Cretan goat is currently restricted to an area of approximately 150 km^2 in the south-west corner of Crete, in Samaria – Lefka Ori National Park (Sfougaris 1995; Adamakopoulos et al. 1997; Sfougaris and Lymberakis 2009; Geskos 2013; Papaioannou 2010; Geskos 2011) (Fig. 2). Its population size, ranging from 750 to 1.380 individuals during the period 2004–2007 (Geskos 2009, 2011), is currently totaling about 1.000–1.500 individuals (Papaioannou 2016; Samaria – Lefka ori mountains National Park Managing Authority, pers. comm.).

The Youra goat (*Capra hircus dorcas* Reichenow, 1888/*C. h. jourensis* Ivrea, 1899) is currently forming a population of 200 individuals in Youra island in Northern Sporades (Sfougaris and Lymberakis 2009), whereas the Antimilos goat (*Capra hircus pictus* Erhard, 1858) numbers currently some 300 individuals in Antimilos island of Cyclades and it has crossbred with domestic goats (Sfougaris 1994, 1995; Sfougaris and Lymberakis 2009) (Fig. 2).

Description

To some extent, the Cretan goat looks like the Wild Bezoar goat (*Capra aegagrus aegagus*) both in morphology and general coloration pattern. The only trait that may indicate a "domestication stage" is the relatively small adult male/female body mass ratio (Geskos 2011). However, the Cretan goat is smaller and darker than *C. aegagrus*. In older males, the light colored parts of the body are gray to white, while the dark ones are up to black (dorsal, shoulders, undersides of the neck, beard, upper tail, lower ribs and front of the sternum, and legs). The horns of the male are curved back to one level and strongly curved at the point. In each annual ring, they have a bulge on the anterior outer edge of the horn (Paragamian 1994). The male mass is up to 50 kg, the length of the horns up to 81 cm (the largest size recorded by Lorenz-Liburnau 1899), and the height (up to the shoulder) is 50 cm; the corresponding values for the females are 17 kg, 18 cm, and 45 cm. In the introduced population of Theodorou islet (and the populations derived from it), the animals are even smaller. The males do not exceed 40 kg (average mass in 6 years is 30.6 kg and 8 years is 35.2 kg), while the females do not exceed 25 kg (average mass at 6 years is

19.94 kg and at 8 years is 23.3 kg) (Papageorgiou 1974; Husband et al. 1986). The Antimilos goat looks very much like the Cretan goat, with outwards curving horn tips (Sfougaris 1995).

Life History

The male Cretan goat reaches reproductive maturity in 4–5 years, while females in 2–3 years. Their lifespan is 10–15 years. The mating period of the only long-established free-ranging population of Cretan goats (Lefka Ori mountains in Crete) starts at the end of October and ends at the beginning of November. Adult males are first seen associated with females at that time. Mixed adult herds form till mid-November. The mating period continues at least till December for the subadults of both sexes (Geskos 2011). In Theodorou island, females reproduce from their third year of age, whereas males mate only after they are 8 years old (Husband and Davis 1984). The mating period is in late October to early November. They give birth to 1 or 2 young in April–May. Cretan, Antimilos, and Youra goats mate earlier (Couturier 1959; Schultze-Westrum 1963; Husband and Davis 1984). In Lefka Ori mountains, almost every female gives birth each year (Stauffer 1986; Hablutzel 1990), while in Theodorou island only one-third of females gives birth (Husband and Davis 1984) due to overpopulation and poor living conditions. In the Lefka Ori mountains, in areas below the subalpine zone (Samaria gorge), the observed mean rate of twinning/year was 13.4% (Geskos 2009). Regarding the Antimilos and Youra goat populations, mating occurs on August–September and June–July respectively (Sfougaris 1995).

Habitat and Diet

The Cretan goat uses areas of steep terrain and occupies elevations between 600 and 2000 m a.s.l. It also prefers rocky areas with more than 50% vegetation cover and slopes greater than 30% (Stauffer 1986). The Cretan goat habitat in Samaria gorge (Lefka Ori mountains) consists of *Pinus brutia, Cupressus sempervirens, Acer creticum, Quercus coccifera, Quercus ilex, Pistacia lentiscus,* and *Phillyrea media* (Sfougaris 1995; Schultze-Westrum 1963).

Behavior

As to the Cretan goat on Lefka Ori mountains, the herd type recorded most often is the family group: adult (≥3 years of age) female with male and female offspring up to 2 years of age (Nicholson and Husband 1992; Geskos 2009). Adult females occur always in small herds (Geskos 2009). The largest aggregation observed has been of 34 animals of all sex and age classes, excluding adult males (Geskos 2009).

Diurnal behavior of the Cretan goat has been studied in the introduced small insular population of Theodorou Island by Nicholson and Husband (1992). Adult Cretan goats exhibit a crepuscular feeding pattern similar to that found in other ungulates. Males spend less time than females in feeding during the day. Apparently, males do not reduce feeding time during the rut, and a decline in the condition of dominant males has not been observed. Group size does not differ greatly from those reported for other species of *Capra*, but the Cretan goat exhibits a smaller average group size than any other species of *Capra*. The Cretan goat appears to associate without regard to sex. It has clear social-dominance relationships and these relationships are ordered by age and sex, with old males dominant over all others. The dominance structure remains intact during the breeding season and there is lack of aggression of males to other males during the rut. There seems to be a hierarchy within groups. The physically larger males dominate the smaller ones, while all males from 1 year and older dominate all females, regardless of age and size (Husband and Davis 1984).

Conservation Status

Goats found on Mediterranean islands are generally recognized as introduced by humans (Shackleton 1997; Wilson and Reeder 2005), and genetic and archaeozoological studies confirm it (e.g., Groves

1989; Vigne 1994; Manceau et al. 1999; Kahila Bar-Gal et al. 2002). Thus, they cannot be included in any category of IUCN (Weinberg and Ambarli 2020). However, formerly the Cretan goat was included in the VU D1+2 IUCN category (Shackleton 1997), as – at that time – it was classified as a subspecies (*Capra aegagrus cretica* Schinz, 1838) of the wild goat (*Capra aegagrus*). The Cretan goat is included, as *Capra aegagrus cretica*, in the Appendix II of the Bern convention and in the Appendix II and IV of the 92/43 EEC Directive on the conservation of natural habitats and of wild fauna and flora. All feral goats in Greek islands (Crete, Youra, and Antimilos) are classified as Endangered (E) in the Red Data Book of Threatened Animals of Greece (Legakis and Marangou 2009).

Management

An introduced population of Cretan goats exists on the small island of Theodorou (Agii Theodori) (68 ha), close to the north coast of the western part of Crete (Adamakopoulos et al. 1997; Sfougaris 1995; Sfougaris and Lymberakis 2009; Papaioannou 2010). It was established during the first half of the twentieth century, as a result of introductions with a few individuals from Lefka Ori mountains, totaling in 2006 around 100 individuals, close to the carrying capacity of the area (Husband and Davis 1984; Sfougaris 1995; Papaioannou 2010). A few additional introduced populations – all originating from Theodorou island – exist on Moni island (160 ha) near Aegina and Sapientza Island (880 ha) offshore of the south-western Peloponnese, totaling around 200 individuals in 2009 (Sfougaris and Lymberakis 2009; Papaioannou 2010).

A further population on the Agii Pandes Island (30 ha) off northern Crete and another one on Atalandi Island (180 ha) are considered as crossbreeds derived by founders from Theodorou Island and domestic goats (Sfougaris 1995; Sfougaris and Lymberakis 2009; Papaioannou 2010), whereas another crossbred population on Dia Island (Crete) was exterminated sometime between 1997 and 2004 (Papaioannou 2010).

As to the mainland, the only one free-ranging population of Cretan goats occurs on mt. Parnes (Parnitha), close to Athens. It comes from introduced founders probably from Theodorou island (Crete) some decades ago (1963?) and apparently it has not crossbred with domestic goats yet (Sfougaris 1995; Sfougaris and Lymberakis 2009; Papaioannou 2010, 2016).

The main threat affecting conservation of these Mediterranean feral goats is crossbreeding with free-ranging domestic goats, for example, on Lefka Ori mountains (Adamakopoulos et al. 1997; Sfougaris and Lymberakis 2009; Geskos 2009; Papaioannou 2010).

A further conservation problem is inbreeding which inevitably takes place in the Youra goat and Antimilos goat populations as well in all introduced populations of Cretan goats (Sfougaris and Lymperakis 2009), as most of them originated from only three pairs from Lefka Ori mountains, introduced to Theodorou islet on 1928, 1937, and 1945 (Husband and Davis 1984; Papaioannou 2010). Low genetic variation, lack of polymorphism and heterozygosity, high inbreeding coefficients, and low kid production was found in the Cretan feral goat population of Theodorou Island (Logan 1989).

Poaching for illegal meat trade (Sfougaris and Lymperakis 2009; Geskos 2009), free access to road use in Cretan goat habitats which increase poaching incidents, and food competition with livestock are some more conservation problems (Adamakopoulos et al. 1997; Geskos 2009; Papaioannou 2010). Regarding the Youra goat, Antimilos goat and some introduced Cretan goat populations, lack of water resources and food – mainly in summer – are additional problems (Sfougaris and Lymperakis 2009).

Most of the "natural" distribution range of the Cretan goat in Crete is within the borders of the Samaria (Lefka ori) National Park, an area included in the Natura 2000 network as well. All the area of the Youra goat distribution range is included in the National Sea Park of Allonisos and Northern Sporades, whereas that of the Antimilos goat is included in the Natura 2000 network (Sfougaris and Lymperakis 2009).

A number of small populations of feral goats (less than 100 individuals each) are kept in private areas (for hunting or recreation) and in most of the

State-Controlled Hunting Reserves (Papaioannou 2010). It is believed that most of these animals come from the Theodorou Cretan goat population (Husband and Davis 1984), but actually their origin is unknown (Papaioannou 2010).

According to the Greek national law, hunting of the Cretan goat in natural ecosystems has been officially forbidden since 1969, according to presidential decree 86/69 and its modifications. Very low quotas of all feral goats – including the Cretan goat – may be shot within the State-Controlled Hunting Reserves or within the state and private Game Enclosures (Papaioannou 2010). Monitoring projects are undertaken by the Management authority of the Lefka Ori – National Park in Samaria. Beside the above mentioned ones, no other specific measure has been undertaken regarding the free-ranging Cretan goat population in the Lefka Ori mountains.

Reports on feral goat relationships with predators are absent. However, the Golden eagle is permanently present in the goat distribution range in Crete and the wolf – after a long absence – has been recolonizing mt. Parnitha, near Athens (Iliopoulos 2018), where the only free-ranging Cretan goat population in mainland roams. There are no data on relationships between the Cretan goat and the wolf.

Supplementary feeding is implemented in a few small, insular populations of feral goats (Husband et al. 1986; Sfougaris 1995; Papaioannou 2010).

Future Challenges for Research and Management

Some conservation measures regarding the Cretan goat, proposed by Adamakopoulos et al. (1997), still remain on paper; these were: (1) Establish a strict control program to eliminate domestic goats from Samaria (Lefka ori) National Park to remove the threat of crossbreeding with the Cretan goat. Reducing or eliminating the herds of domestic goats from the surrounding areas will prevent their immigration into the Park; (2) Improving antipoaching measures by regularly patrolling the park during winter, and by allowing no further increases to access; (3) Impose a greater control on tourism development in the Park. These should include banning the development of major tourist facilities in the mountain regions of the Park and its surroundings, and strictly controlling visitor movements; (4) The population should be regularly counted and the area surveyed throughout the Lefka Ori mountains, paying special attention to the degree of crossbreeding and human encroachment; (5) Overall, it may be desirable to eradicate domestic × feral goat crossbreeds. Unfortunately, F1 crossbreeds are difficult to distinguish from Cretan goats in both look and behavior; (6) Determine the degree of crossbreeding of feral goats found on the various islands, to help local conservation management decisions.

Majorcan Goat

The Majorcan goat (*Capra hircus* var. *majorcan*) is basically located in the Sierra de Tramontana, a mountainous chain that runs along the Northwestern fringe of Mallorca Island, Spain, and in the mountains of Artá, in the Northeast of this island (Fig. 2). The goats occupy all habitats, from sea level to the highest peaks, including coastal cliffs, pine, and holm oak forests, up to the most inaccessible rocky areas (Adrover et al. 2017). Considering that the colonization of the island by humans took place between 2350 and 2150 BCE – as there is no evidence of previous human presence (Alcover 2010) – the arrival of the first goats is assumed to have occurred between 2300 and 2050 BCE (Seguí et al. 2005).

The Majorcan goat is slightly smaller than the domestic goat. The males and females weigh an average of 50 kg and 32 kg, respectively. The height at the withers is 70 cm in males and 56 cm in females (Seguí et al. 2005). It is well characterized morphologically and genetically (Seguí et al. 2005) and its phenotypic description for the purposes of management and regulatory use is contained in the National Decree 91/2006, regarding the regulation of caprine populations, the management of the hunting exploitation of the Majorcan goat, and the modification of the relative technical plans (Adrover et al. 2017).

The Majorcan goat is characterized genetically and phenotypically, showing both its old origin and the effects of anthropic management (Adrover et al. 2017).

In 2007, the population size of the Majorcan goat was 1.500–2.000 individuals, out of a total of 10.000 ind. including crossbreeds and feral descendants of modern domestic goats (Giannatos et al. 2007).

Several groups of Majorcan goats have been protected in fenced areas and captive bred (Giannatos et al. 2007). Since 2006, hunting of purebred individuals has been banned, whereas hunting of crossbreeds and feral domestic goats is encouraged. Trophy hunting of Majorcan goats will be allowed again after populations have reached target levels (according to the National Decree 91/2006).

Selective hunting should be used to remove the crossbreeds from the population (Palomo and Gisbert 2002). Conservation efforts should be implemented to prevent poaching, capturing alive animals for trade, and illegal hunting of crossbreeds for commercial purposes (Giannatos et al. 2007). A normative improvement taking into account the ecological and biodiversity significance of the Majorcan goat is a local priority, together with hunting directives appropriate to avoid crossbreeding, overpopulation, vegetation damage, and control of domestic goats to recover the Majorcan goat population, which shows great potential as a hunting resource (Seguí et al. 2005).

Montecristo Goat

A population of feral goats with some individuals showing a few phenotypical characters identical to those of wild goats, currently occurs on the islet of Montecristo (10.39 km^2, Tuscan archipelago, Northern Tyrrhenian Sea, Italy) (Toschi 1953, 1965 in Masseti 2009) (Fig. 2). Domestic goats must have been introduced to this tiny island not only in ancient times (Spagnesi et al. 1986), but introductions continued probably up to the first decade of the second half of the last century (Masseti 2002, 2003b in Masseti 2009) and it is confirmed by the remarkable morphological and genetic variability of these goats (Spagnesi et al. 1986; Randi et al. 1990).

Montecristo Island is a Nature Reserve. Culling for research and population control is conducted periodically, as the Montecristo goat exerts a strong pressure on the island vegetation (Randi et al. 1990; Tosi and Lovari 1997). Culling is based on phenotypic (pelage) criteria (Spagnesi et al. 1986).

During 2010–2014, this population has been drastically reduced upon the implementation of actions aimed at restoring the ecological balance in this island, mainly focused on the eradication of the local population of black rats, *Rattus rattus* (L., 1758), using poisoned pellets (Masseti 2016). The present minimum numbers of Montecristo goats are consistent with those in previous years, i.e., some 200 goats (https://www.restoconlife.eu/en/2017/03/30/2016-montecristo-goat-census).

Other historical populations of feral goats in Mediterranean islands were those living on Asinara (Italy), Tavolara (Italy), Lampedusa (Italy), and La Galite (Tunisia) which went extinct between the nineteenth and the twentieth centuries (Masseti 1997; Masseti and Zava 2002 in Masseti 2009). Furthermore, the islands of Capraia, Caprara, San Pietro, and Linosa also had their feral goat populations, now extinct (Sommier 1908; Toschi 1953; De Beaux 1955; Masseti and Zava 2002 in Masseti 2009).

Eastern Atlantic Feral Goat

In the Eastern Atlantic islands, goats of "pre-Hispanic" phenotypes still occur today on Bugio in the Desertas islands off the southeastern coast of Madeira (Masseti 2009) (Fig. 2). There are records of their existence during the previous century in Deserta Grande, Gran Canaria, Fuerteventura, in an islet off Lanzarote (Cook and Yalden 1980; Nogales et al. 1992; Muzio 1925 in Masseti 2009), and in the far past in the islands of Lanzarote, Gran Canaria, Tenerife, La Gomera, El Hierro, and especially Fuerteventura dated between 1000 and 1700 BCE (Castillo et al. 1996; Meco Cabrera et al. 1982 in Masseti 2009).

On the Eastern Atlantic islands – as in the Mediterranean ones – the natural occurrence of goats has not been proven by any fossil record and the oldest goat populations, imported onto these islands, probably are linked with the arrival of the earliest human inhabitants around 1000–2000 BCE (Rodríguez-Piñero and Rodríguez-Luengo 1993 in Masseti 2009), characterized as having a "pre-Hispanic" origin due to their phenotypic characters (Masseti 2009).

The population of the Eastern Atlantic goat of Bugio numbered no more than 60 goats early in this century (Masseti 2009). Most likely, these goats were imported to the island from the nearby island of La Palma, in historical times, possibly during the fifteenth century (Pena and Cabral 1997 in Masseti 2009).

References

Adamakopoulos T, Hablutzel C, Hatzirvassanis V (1997) Status and distribution of Caprinae in Greece. In: Shackleton DM (ed) Wild sheep and goats and their relatives, status survey and conservation action plan for Caprinae. IUCN/SSC Caprinae Specialist Group, Cambridge, pp 104–108

Adrover AB, Campaner BC, Gallego JIR (2017) The hunting of the Mallorcan wild goat in the context of cynegetic tourism. Eria 2017-2:278–282

Alcover JA (2010) Dia D, Hora H: El començament de la transformació ecològica antròpica dels sistemes naturals de les Balears. In: Mayol J, Muntaner L, Rullan O (eds) Homenatge a Bartomeu Barceló i Pons, Geògraf. Lleonard Muntaner, Palma, pp 89–108

BZN (2003) International Commission on Zoological Nomenclature, Opinion 2027 (Case 3010). Usage of 17 specific names based on wild species which are pre-dated by or contemporary with those based on domestic animals (Lepidoptera, Osteichthyes, Mammalia): conserved. Bull Zool Nomencl 60:81–84

Castillo C, López M, Martí M, Rando JC (1996) La paleontologia de vertebrados en Canarias. Rev Esp Paleont 15:237–247

Clutton-Brock J (1981) Domesticated animals from early times. Heinemann and British Museum (Natural History), London

Cook LM, Yalden DW (1980) A note on the diet of feral cats on Deserta Grande. Bocagiana 52:1–4

Couturier MJ (1959) Statut actuel des représentants du genre *Capra* dans le bassin Méditerranéen. In: Animaux et Végétaux rares de la Région Mediterranéenne. Réunion Technique. IUCN, Athens/Bruxelles, pp 12–19

Davis SJM (1984) Khirokitia and its mammal remains. A neolithic Noah's Ark. In: Le Brun A (ed) Fouilles récentes à Khirokitia (Chypre) 1977–1981. Mémoire, vol 41. Recherche sur les Civilisations, Paris, pp 147–162

Davis SJM (1987) The archaeology of animals. Routledge, Oxon

De Beaux O (1955) Posizioni sistematica degli stambecchi e capre selvatiche viventi (*Capra* Linneo 1758) e loro distribuzione geografica. Atti Accad Sci Lett 12:1–110

Geskos A (2009) Agrimi on Lefca Ori, Creta: population status, threats and conservation. Newsletter of the IUCN/SSC Caprinae Specialist Group 9:2–5

Geskos A (2011) The behavior, ecology and conservation of the agrimi (*Capra* sp.) in the Samaria National Park, Crete, Greece. PhD thesis, School of Science and the Environment, The Manchester Metropolitan University

Geskos A (2013) Past and present distribution of the genus *Capra* in Greece. Acta Theriol 58:1–11

Giannatos G, Herreto J, Lovari S (2007) *Capra hircus*. In: IUCN 2013. IUCN Red List of Threatened Species. Version 2013.1. http://www.iucnredlist.org. Downloaded on 09 December 2013

Groves CP (1989) Feral mammals of the Mediterranean islands: documents of early domestication. In: Clutton-Brock J (ed) The walking larder. Unwin Hyman, London, pp 46–58

Hablutzel C (1990) Population size, distribution and habitat selection of the agrimi (*Capra aegagrus cretica*) in the White Mountains of Crete, Greece. University of Zurich-Irchel, Switzerland

Hatzisarantos H, Kannelis A (1955) Chamois and wild goats in Greece. To Vouno Athens 187:142–164. (in Greek)

Helmer D (1989) Le dévelopement de la domestication au Proche-Orient da 9500 à 7500B.P.: les nouvelles donneés d'El Kown et de ras Shamra. Paléorient 15(1):111–121

Horwitz LK, Bar-Gal G (2006) The origin and genetic status of insular caprines in the eastern Mediterranean: a case study of free-ranging coat (*Capra aegagrus cretica*) on Crete. Hum Evol 21(2):123–138

Husband TP, Davis PB (1984) Ecology and behaviour of the Cretan agrimi. Can J Zool 63:411–420

Husband TP, Davis PB, Brown JH (1986) Population measurements of the Cretan agrimi. J Mammal 67:757–759

Iliopoulos Y (2018) Conservation status of the wolf in Greece, conflict issues and mitigation. Technical report. (in Greek). pp 79

Kahila Bar-Gal G, Smith P, Tchernov E, Greeblatt C, Ducos P, Gardeisen A, Horwits KL (2002) Genetic evidence for the origin of the agrimi goat (*Capra aegagrus cretica*). J Zool 256:369–377

Legakis A, Marangou P (2009) The Red Book of Endangered Animals of Greece. Hellenic Zoological Society, Athens

Logan GT (1989) A study of the conservation biology of the Cretan wild goat. MSc thesis, University of Rhode Island

Logan GT, Brown JH, Husband TP, Nicholson MC (1994) Conservation biology of the Cretan agrimi. Biol Gallo-hellenica 22:241–246

Lorenz-Liburnau L (1899) Die Wildziegen der griechischen Inseln und ihre Beziehungen zu anderen Ziegenformen. Wiss Mitti Bosnien Herzegovina 6:851–886

Lymberakis P, Vardakis M, Nikolakakis M, Georgiakakis P, Giannakakis A (2009) Development of a monitoring study of the population and ecological requirements of Cretan goat (*Capra aegagrus cretica*) in the Lefka ori mountains with subsequent management proposals. National Forest Management Authority of Samaria. Technical report. Hania. pp 26

Manceau V, Després L, Bouvet J, Taberlet P (1999) Systematics of the genus *Capra* inferred from mitochondrial DNA sequence data. Mol Phylogenet Evol 13:504–510

Masseti M (1981) La capra selvatica nel Mediterraneo. L'Universo LXI:177–218

Masseti M (1997) The prehistorical diffusion of the Asiatic mouflon, *Ovis gmelini* Blyth, 1841, and of the bezoar goat, *Capra aegagrus* Erxleben, 1777, in the Mediterranean region beyond their natural distributions. In: Hadjisterkotis E (ed) Proceedings of the 2nd international symposium on Mediterranean mouflon, 'The Mediterranean mouflon: management, genetics and conservation'. Game Fund, Ministry of Interior, Nicosia, pp 12–27

Masseti M (2002) Uomini e (non solo) topi. Gli animali domestici e la fauna antropocora. Firenze University Press/University of Florence, Florence, Italy

Masseti M (2003a) Holocene endemic and non endemic mammals of the Aegean islands. In: Kotjabopoulou E, Hamilakis Y, Halstead P, Gamble C, Elefanti P (eds) Zooarchaeology in Greece. Recent advances, British School at Athens studies, vol 9. British School at Athens, London, pp 47–57

Masseti M (2003b) Fauna toscana. Galliformi non migratori, Lagormorfi e Artiodattili. Agenzia Regionale per lo Sviluppo e l'Innovazione nel settore Agricolo-forestale (ARSIA)/Regione, Toscana, Firenze

Masseti M (2009) The wild goats *Capra aegagrus* Erxleben, 1777 of the Mediterranean Sea and the Eastern Atlantic Ocean islands. Mammal Rev 39(2):141–157

Masseti M (2016) The wild goat, *Capra aegagrus* Erxleben, 1777, of the Island of Montecristo (Northern Tyrrhenian Sea, Italy): does it still exist? Mammalia 80:125–141

Masseti M, Trantalidou K (2002) Boars and goats in a Mesolithic fishing community: Youra (Northern Sporades, Greece), a case study. Poster presented at 9th ICAZ meeting, Durham

Masseti M, Zava B (2002) Nineteenth century wild ungulates (Mammalia, Artiodactyla) of the island of Lampedusa (Pelagian archipelago, Italy). Biogeographia 23:199–215

Meco Cabrera J, Hernández F, Sánchez Velázquez D (1982) La cueva de Villaverde (Fuerteventura) y su mastologia (Nota preliminary). In: Libro homenaje a Jesús Arencibia. Servicio de Publicaciones de la Escuela Universitaria de Formación del Profesorado de E.G.B. de Las Palmas, Las Palmas de Gran Canaria, pp 187–194

Muzio C (1925) Isole africane. Mundus, Monografia 49a. Casa Editrice Sonzogno, Milano

Nicholson MC, Husband TP (1992) Diurnal behavior of the agrimi, *Capra aegagrus*. J Mammal 73:135–142

Nogales M, Marrero M, Hernández EC (1992) Efectos de las cabras cimarronas (*Capra hircus*) en la flora endémica de los pinares de Pajonales, Ojeda e Inagua (Gran Canaria). Bot Macarones 19(20):79–86

Palomo LJ, Gisbert J (2002) Atlas de los mamiferos terrestres de Espana. Direccion General de Conservation de la Naturaleza-SECEM-SECEMU, Madrid

Papageorgiou N (1974) Population energy relationships of the agrimi (*Capra aegagrus cretica*) on Theodorou Island, Greece. Unpublished PhD dissertation, Michigan State University, East Lansing

Papaioannou H (2010) Ungulates and their management in Greece. In: Apollonio M, Andersen R, Putman R (eds) European ungulates and their management in the 21st century. Cambridge University Press, Cambridge, pp 540–562

Papaioannou H (2016) The Balkan Chamois (*Rupicapra rupucapra balcanica* Bolkay, 1925) in Greece. PhD thesis, Department of Environmental and Natural Resources, University of Patras (in Greek)

Paragamian K (1994) The Agrimi (*Capra aegagrus cretica* Schintz, 1838). Management plan for the Samaria National Park. Forest Research Institute, Chania. (in Greek)

Payne S (1968) The origins of domestic sheep and goats: a reconsideration in the light of the fossil evidence. Proc Prehist Soc 34:368–384

Pena A, Cabral J (1997) Região autónoma da Madeira. Roteiros da Natureza. Temas & Debates, Lisbon

Randi E, Tosi G, Toso S, Lorenzini R, Fusco G (1990) Genetic variability and conservation problems in Alpine ibex, domestic and feral goat populations (genus *Capra*). Z Säugetierk 55:413–420

Rodríguez-Piñero JC, Rodríguez-Luengo JL (1993) The effect of herbivores on the endemic Canary flora. Bol Mus Mun Funchal Sup 2:265–271

Schultze-Westrum T (1963) Die Wildziegen der ägäischen Inseln. Säugetierk Mitt 11:145–182

Seguí B, Payeras L, Ramis D, Martínez A, Delgado JV, Quiroz J (2005) The Majorcan wild goat: origin, genetics, morphology, ecological notes and taxonomic implications. Boll Soc Hist Nat Balears 48:121–151

Sfougaris A (1994) Wild goats (*Capra aegagrus*) in Greece: distribution and status. Biol Gallo-hellenica 22:233–240

Sfougaris A (1995) The distribution, ecology and management of goats *Capra aegagrus* in Greece. An outline. Carpinae, 8/9:5–9

Sfougaris A, Lymperakis P (2009) *Capra aegagrus* Erxleben, 1777 (Aegagros, Agriokatsiko). In: Legakis A, Marangou P (eds) The Red Book of Endangered Animals of Greece. Hellenic Zoological Society, Athens, pp 373–375

Shackleton DM, The IUCN/SSC Caprinae Special Group (eds) (1997) Wild sheep and goats and their relatives, status survey and conservation action plan for Caprinae. IUCN, Gland/Cambridge, UK, pp 390 +vii

Sommier S (1908) Le Isole Pelagie. Lampedusa, Linosa e Lampione e la loro flora con un elenco completo delle piante di Pantelleria. Stabilimento Pella, Luigi Chiti successore, Florence

Spagnesi M, Cagnolaro L, Perco F, Scala C (1986) La Capra di Montecristo (*Capra aegagrus hircus* Linnaeus, 1758). In: Ricerche di Biologia della Selvaggina, vol 76. Istituto Nazionale per la Fauna Selvatica, Bologna, pp 1–148

Stauffer C (1986) Observations of Agrimi. In: Nievergelt B, Stocker J (eds) Field Course for Ethologists and Wildlife Biologists. Lefka Ori (White Mountains), Western Crete, 11 September- 2 October 1985). Department of Ethology and Wildlife Research, Institute of Zoology, University of Zurich-Irchel. Zurich, Switzerland, pp 43–55

Toschi A (1953) Note sui vertebrati dell'Isola di Montecristo. Ric Zool Appl Caccia 23:1–52

Toschi A (1965) Fauna d'Italia. Mammalia. Lagomorpha, Rodentia, Carnivora, Ungulata, Cetacea. Edizioni Calderini, Bologna

Tosi G, Lovari S (1997) Status and distribution of Caprinae in Italy. In: Shackleton DM, The IUCN/SSC Caprinae Special Group (eds) Wild sheep and goats and their relatives, status survey and conservation action plan for Caprinae. IUCN, Gland/Cambridge, UK, pp 111–118

Trantalidou K (2003) Faunal remains from the earliest strata of the cave of Cyclope, Youra. In: Galanidou N, Perlès C (eds) The Greek Mesolithic problems and perspectives. British School at Athens studies, vol 10. British School at Athens, Athens, pp 143–179

Vigne JD (1994) Les transferts anciens de mammiferes en Europe occidentale: histoires, mecanismes et implications dans les sciences de l'homme et les sciences de la vie. Colloq Hist Connaissanccs Zool 5:15–37

Vigne JD (2001) Large mammals of early Aceramic Neolithic Cyprus: preliminary results from Parekklisha Shillourokambos. In: London G, Swiny S (eds) The earliest prehistory of Cyprus. From colonization to exploitation. Archaeological reports, vol 5. American Schools of Oriental Research, Boston, pp 55–60

Weinberg P, Ambarli H (2020) *Capra aegagrus. The IUCN Red List of Threatened Species* 2020: e. T3786A22145942. Downloaded on 15 March 2021

Wilson DE, Reeder DM (2005) Mammal species of the world. Johns Hopkins University Press, Baltimore

Mouflon *Ovis gmelini* Blyth, 1841

21

Mathieu Garel, Pascal Marchand, Gilles Bourgoin, Julián Santiago-Moreno, Elodie Portanier, Holger Piegert, Eleftherios Hadjisterkotis, and Jean-Marc Cugnasse

Contents

M. Garel (✉)
Office Français de la Biodiversité, Direction de la Recherche et de l'Appui Scientifique, Service Anthropisation et Fonctionnement des Écosystèmes Terrestres, Gières, France
e-mail: mathieu.garel@ofb.gouv.fr

P. Marchand
Office Français de la Biodiversité, Direction de la Recherche et de l'Appui Scientifique, Service Anthropisation et Fonctionnement des Écosystèmes Terrestres, Juvignac, France
e-mail: pascal.marchand@ofb.gouv.fr

G. Bourgoin
CNRS, Laboratoire de Biométrie et Biologie Evolutive UMR 5558, Université de Lyon, Université Lyon 1, Villeurbanne, France

VetAgro Sup, Université de Lyon, Marcy l'Etoile, France
e-mail: gilles.bourgoin@vetagro-sup.fr

J. Santiago-Moreno
Department of Animal Reproduction, INIA-CSIC, Madrid, Spain
e-mail: moreno@inia.csic.es

E. Portanier
CNRS, Laboratoire de Biométrie et Biologie Evolutive UMR 5558, Université de Lyon, Université Lyon 1, Villeurbanne, France

CNRS, Station Biologique de Roscoff, UMR 7144, Sorbonne Université, Roscoff, France

H. Piegert
Thale, Germany

E. Hadjisterkotis
Ministry of Agriculture, Rural Development and Environment, Agricultural Research Institute, Nicosia, Cyprus
e-mail: ehadjisterkotis@ari.gov.cy

J.-M. Cugnasse
Vignes Basses, Brousse le Château, France

© Springer Nature Switzerland AG 2022

L. Corlatti, F. E. Zachos (eds.), *Terrestrial Cetartiodactyla*, Handbook of the Mammals of Europe,
https://doi.org/10.1007/978-3-030-24475-0_34

Common Names

English	Mouflon
German	Mufflon
French	Mouflon
Spanish	Muflón
Italian	Muflone
Russian	Муфлóн

In Cyprus, the Greek name for this species of sheep is Αγρινό.

Taxonomy and Systematics

The genus *Ovis* constitutes one of the most complex mammalian genera owing to its evolution and systematic. During the last centuries, wild sheep taxonomy was revised many times based on different criteria (morphologic, genetic, and biogeographic, e.g., Valdez 1982; Festa-Bianchet 2000; Hiendleder et al. 2002; Rezaei et al. 2010). Wild sheep taxonomy has been complicated by the fact that the different species of the genus *Ovis* can crossbreed and produce fertile offspring (Valdez et al. 1978; e.g., in the Asiatic mouflon/Urial *Ovis vignei* hybrid zone; Fig. 2), leading some authorities to consider, for instance, Asiatic mouflon and Urial as the same species (*Ovis orientalis*). A basic difference among classifications lies in the number of species recognized. No definitive consensus was achieved, but most classifications distinguish at least three morphological types (Bunch et al. 2006) and six wild species: Pachyceriforms of Siberia and North America, *Ovis canadensis* (bighorn sheep, $2n = 54$), *Ovis dalli* (Dall sheep, $2n = 54$), and *Ovis nivicola* (snow sheep, $2n = 52$); Argaliforms of Central Asia, *Ovis ammon* (Argali, $2n = 56$); Moufloniforms of Eurasia, *Ovis vignei* (Urial, $2n = 58$), and *Ovis gmelini* (Asiatic mouflon, $2n = 54$). Note that the name *Ovis orientalis* has often been used/recommended for Asiatic mouflon (Shackleton and IUCN/SSC Caprinae Specialist Group 1997), but as it seems to refer to a hybrid of Alborz red sheep (Valdez 1982), the name is unusable and may enter into homonymy (Groves and Grubb 2011; Hadjisterkotis and Lovari 2016).

The scientific denomination of mouflon present in Mediterranean islands (Cyprus, Corsica, and Sardinia), and subsequently introduced into continental Europe during the eighteenth century (Weller 2001; Fig. 1), is probably among the most controversial (e.g., Cugnasse 1994; Gentry et al. 2004; Rezaei et al. 2010; Guerrini et al. 2015). However, archaeological (Poplin 1979; Vigne 1992; Groves 1989) and genetic evidence (e.g., Chessa et al. 2009; Rezaei et al. 2010; Demirci et al. 2013; Guerrini et al. 2015; Mereu et al. 2019; Portanier et al. 2022) that has now accumulated gives strong support for ranking them as subspecies of Asiatic mouflon *O. gmelini* (see also subsection "Phylogeny and Phylogeography"). Most recent genetic advances, based on microsatellite and mitochondrial DNA, separate into two independent subspecies the Corsican/Sardinian *O. g. musimon* and the Cyprus populations *O. g. ophion* which had a separate evolutionary history (Chessa et al. 2009; Guerrini et al. 2015, 2021; Sanna et al. 2015; Satta et al. 2021; Portanier et al. 2022). It can also be mentioned that Cugnasse (1994) suggested to differentiate Corsican and Sardinian mouflon by using *O. g. musimon* var. *corsicana* and *O. g. musimon* var. *musimon,* respectively, to account for the demographic disconnection occurring since Neolithic between the two island populations.

Paleontology

The oldest fossil of mouflon *Ovis shantungensis* was found in central Asia (China, Nihowan) and would be dated >2 mya (Teilhard de Chardin and Piveteau 1930). During the Pleistocene, the genus *Ovis* has evolved from this ancestral area, through successive speciation events, in at least six wild species (see section "Taxonomy and Systematics"), while spreading along two major migration routes: the first passing through northeastern Asia and Bering Strait toward North America, the second one going toward Eurasia and western Europe. Nowadays, the genus *Ovis* is largely distributed over palearctic and nearctic regions (Valdez 1982).

Mouflon present in Europe originated from Asiatic mouflon that arrived about 8500 years before the common era (BCE hereafter), at the onset of the very first waves of human-mediated dispersal of livestock across the Mediterranean Basin (Zeder 2008; Vigne et al. 2014). Relics of these very first migrations include the historically genetically and morphologically preserved mouflon populations

Fig. 1 A male (left), a female and a lamb (right) of Mediterranean mouflon (photographs © Daniel Maillard – www.danielmaillard.com)

presently restricted to the islands of Cyprus, Corsica, and Sardinia (Chessa et al. 2009; Rezaei et al. 2010; Guerrini et al. 2015, 2021; Sanna et al. 2015; Mereu et al. 2019; Satta et al. 2021; Portanier et al. 2022). In these islands, Asiatic mouflon would have been introduced by Neolithic people, probably after a predomestication phase, between 4500 and 8500 years BCE as suggested by archaeological evidences in Neolithic sites (Poplin 1979; Groves 1989; Vigne 1992; Zeder 2008; Vigne et al. 2012, 2014). The domestication process is assumed to have remained primitive, likely limited to protection against predators with few interactions with humans and no morphological selection (Rezaei 2007; Zeder 2008). In addition, by returning to a feral state in mountainous areas of these scarcely populated islands, these primitive breeds survived, contrary to mainland populations, to the second migration of improved breeds (with productive traits such as wool, milk, and meat) occurring 5000–6000 years later from southwest Asia (Zeder 2008; Chessa et al. 2009; Barbato et al. 2017; Ciani et al. 2020).

Current Distribution

The native range of Asiatic mouflon includes 4 subspecies (Festa-Bianchet 2000) and extends from eastern Turkey, Armenia, southern Azerbaijan (including Nakhchivan Autonomous Republic), and northern Iraq to western Iran (Zagros mountains, southern Caucasus) for the Armenian mouflon *Ovis gmelini gmelini*, with isolated populations in central Iran for the Isfahan mouflon *Ovis gmelini isphahanica* and southern Iran for the Laristan mouflon *Ovis gmelini laristanica* (Fig. 2). It also includes an area where mouflon crossbreed with Urial *Ovis vignei* sp. The population of Anatolian mouflon *Ovis gmelini anatolica* in south-central Turkey (Konya-Bozdag area), previously isolated, is highly protected with most individuals being enclosed since 1989. Two populations (Karaman-Karadag and Ankara-Sariyar) have been created from reintroductions between 2004 and 2007 with individuals from the Konya-Bozdag area (60 and 120 founder individuals, respectively; Özüt 2009). The last population estimates were around 700 Anatolian mouflon in the Konya-Bozdag-enclosed population (after 200 individuals were removed for reintroductions), and 50–100 individuals outside the fences (Özüt 2009).

Populations created from feral individuals after the very first domestication attempts of Asiatic mouflon between the Neolithic and the Upper Paleolithic are restricted to the Mediterranean islands of Cyprus, Corsica, and Sardinia (see sections "Paleontology" and "Phylogeny and Phylogeography"; Fig. 2). In Corsica, there are two populations: one in the massif of Monte Cinto (Northwest, >900 individuals), and one in the massif of Bavella (Southeast, >200 individuals; Benedetti et al. 2019; Portanier et al. 2022). In Sardinia, the current distribution is limited to the eastern part of the island (Ogliastra, Gennardentu, Supramonte, and Albo; ~6000 individuals), Montiferru in the West, and a few managed/protected areas where the species has been introduced (Pabarile, Capo Figari, and Asinara; Apollonio et al. 2010; Puddu and Maiorano 2016; Satta et al. 2021). The Cyprus mouflon is present in the Paphos forest (North-West of the island; ~3000 individuals; Hadjisterkotis 2001) and recently extended its range toward Troodos mountains (Nicolaou et al. 2016).

Since the eighteenth century (see subsection "Phylogeny and Phylogeography"), mouflon have been introduced as a new game species in 24 European countries: Andorra, Austria, Belgium, Bulgaria, Croatia, the Czech Republic, Denmark, Finland, France, Germany, Greece, Hungary, Italy, Luxembourg, Netherlands, Poland, Romania, Serbia, Slovakia, Slovenia, Spain (including Canary islands, i.e., Tenerife, La Palma, and La Gomera), Sweden, Switzerland, and Ukraine (Fig. 2; Weller 2001; Linnell and Zachos 2011; Cassinello 2018). Mouflon were also introduced in Argentina, Chile, and the USA (California, Texas, and Hawaii; Weller 2001) and are also present in fenced areas or hunting estates in numerous countries. It has to be noted that the population from Kerguelen islands (French Southern and Antarctic Territories), of conservation concerns with endemic plants, has been eradicated in the 2010s (Terres Australes et Antarctiques Françaises 2013).

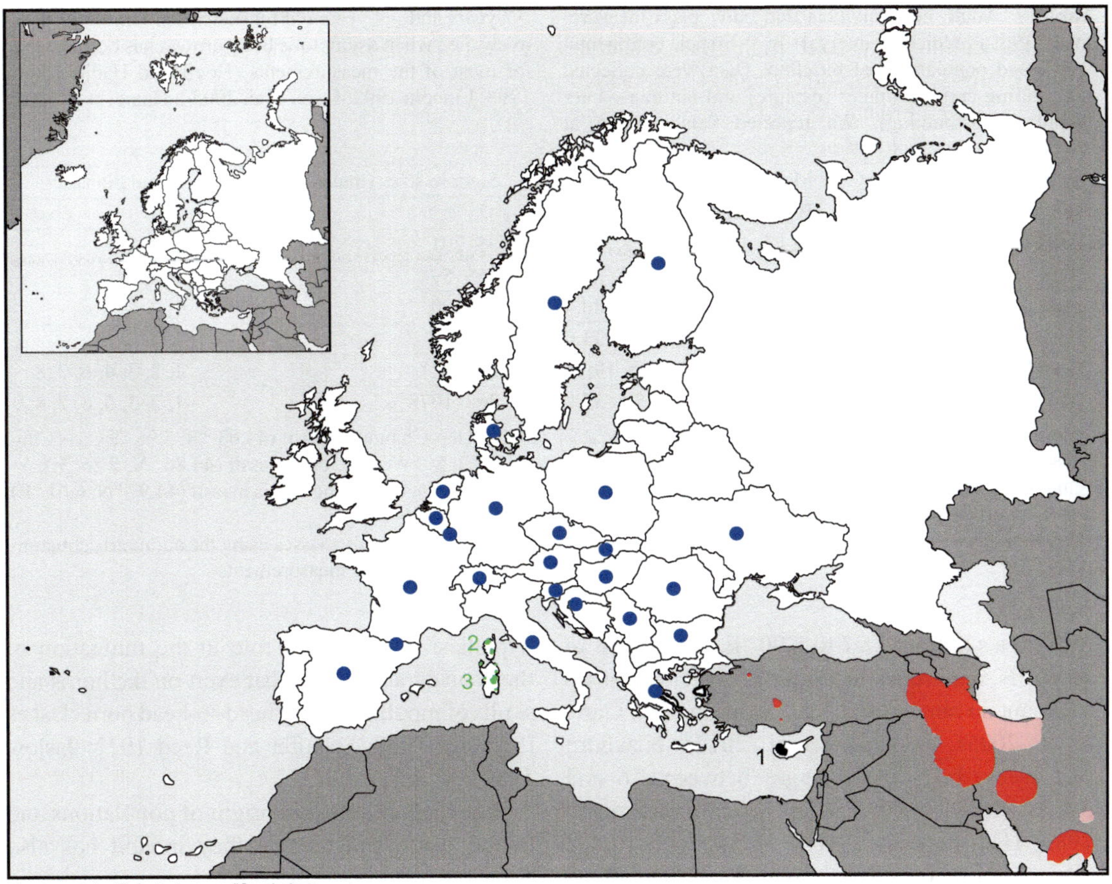

Map template: © Getty Images/iStockphoto

Fig. 2 Current distribution of mouflon in Europe. Native populations are represented in red (*Ovis gmelini gmelini*, *Ovis gmelini anatolica*, *Ovis gmelini isphahanica,* and *Ovis gmelini laristanica*; data from the International Union for Conservation of Nature, http://maps.iucnredlist.org/map.html?id=15739) or pink (area where *Ovis gmelini gmelini* and *Ovis gmelini laristanica* crossbred with Urial *Ovis vignei* sp.). Although some of these subspecies are referenced as *orientalis* on the IUCN red list, we used *gmelini* here in accordance with the section "Taxonomy and Systematics" and, e.g., Festa-Bianchet (2000). The mouflon populations on Mediterranean islands are represented in black ([1] Cyprus *Ovis gmelini ophion*) and green ([2] Corsica and [3] Sardinia *Ovis gmelini musimon*; data from Hadjisterkotis 2001; Corti et al. 2011; Puddu and Maiorano 2016). The European countries where mouflon persist after being introduced are noted with a blue point. Note that the countries where mouflon are present only in captivity are not reported on this map Map template: © Copyright Getty Images/iStockphoto

Description

Size and Morphology

The complex genetic and demographic history of island populations and of introduced populations of mouflon (see sections "Paleontology" and "Phylogeny and Phylogeography") has contributed, through genetic founding effects and/or high variability in environmental conditions encountered, to generate large inter-population variation in body and horn development. This is well illustrated by population-specific horn growth patterns (e.g., Hoefs 1982; Hoefs and Hadjisterkotis 1998; Figs. 3.5 and 7.2 in Garel et al. 2006b). To account for these inter-population differences, we summarized in Table 1 >3500 adult morphological and body mass measurements collected from known-age animals in 9 French populations spanning a large range of environments (ranging from 180 to

Table 1 Adult morphological and body mass measurements (95% prediction interval) in 9 French continental and island populations of mouflon. Data were collected both during spring-summer (capture) and autumn-winter (capture and hunting). We reported values taken at ≥7 years and ≥4 years old for males and females, respectively, i.e., when asymptotic body growth has been reached for most of the measurements (Hoefs and Hadjisterkotis 1998; Lincoln 1998; Garel et al. 2005d; Hamel et al. 2016)

Metrics	Males	Females	Sample sizes (males–females)	Populations[a]
Tail length	Bony, very short (<10 cm)		–	–
Body length (cm)	124.5–145.6	102.7–133.8	95–201	1, 2
Hind foot length (cm)	30.7–35.3	28.8–33.5	155–598	1, 3, 6
Shoulder height (cm)	70.6–86.8	61.4–80.4	97–200	1, 2
Neck circumference (cm)	30.1–48.8	20.8–33.6	99–294	1, 3
Horn length (cm)	48.6–90.3	0.5–19.1	295–121	1, 2, 3, 4, 6, 7, 8
Body mass[b] (kg)	26.1–61.8	18.8–37.7	268–1078	1, 2, 3, 5, 6, 7, 8, 9

[a]Populations: 1 – Caroux-Espinouse massif (43.63 °N, 2.97 °E); 2 – Gap-Chaudun massif (44.63 °N, 5.98 °E); 3 – Cinto massif (42.38 °N, 8.90 °E); 4 – Chartreuse massif (45.33 °N, 5.79 °E); 5 – West Vercors massif (44.86 °N, 5.26 °E); 6 – Bauges massif (45.69 °N, 6.24 °E); 7 – Queyras massif (44.76 °N, 6.88 °E); 8 – Brianconnais massif (44.97 °N, 6.61 °E); and 9 – North-East Vercors massif (45.11 °N, 5.56 °E)

[b]Combining live body masses and eviscerated body masses converted into live body masses using the allometric equation: live = $e^{0.4131} \times \text{eviscerated}^{0.9687}$ calibrated on 88 animals for which we had both measurements

2966 m a.s.l., from 2.97 to 8.90 °E, from 42.38 to 45.69 °N, and from Mediterranean/island areas up to mountain ranges) and genetic origin (e.g., Garel et al. 2005a; Portanier et al. 2017). Newborn (<1 week) body mass ranges between 1.6 and 3.4 kg (Mottl 1960; Pfeffer 1967; Briedermann 1989; Hadjisterkotis 1996b). It has to be noted that island populations might have undergone island dwarfism during their evolutionary history (e.g., shoulder height in Cyprus: males = 68.0 cm, females = 61.4 cm, n = 13-13 versus continental populations in Table 1; Pfeffer 1967; Hadjisterkotis 1993, 1996b).

Mouflon are among the smallest wild *Ovis* in the world (Valdez 1982) with a marked sexual dimorphism both in terms of size and body mass (up to ~64% during autumn), horn and coat (Table 1; see subsections "Pelage" and "Growth"). While males have large horns, females may or may not have short horns (Hoefs and Hadjisterkotis 1998; Garel et al. 2007). When present, the horns of females are thin, often asymmetric with a circular shape. In males, horns are thick, with a triangular basis and a circumference that decreases from the base to the tip (Fig. 1). They are also symmetric and show three sides with a marked fronto-nuchal edge and the fronto-orbital edge almost completely rounded off. Differences in skull design between males and females have been revealed and hypothesized to play a role in the mitigation of the considerable forces that exert on the horns and skulls of mouflon during head-to-head horn clashes (Schaffer 1968; Schaffer and Reed 1972; Jaslow 1989).

The high diversity in origin of populations and in the environments that they inhabit has also contributed to large inter-population variation in phenotypic characteristics of this species. For instance, marked differences in the proportion of horned females among populations have been documented: from 43% in the Corsican population of Bavella (only 13% in the other Corsican population of Cinto; Sanchis 2018) to values close or equal to 0%, both in island populations (Sardinia, Pfeffer 1967; S. Ciuti, pers. comm.; Cyprus, Maisels 1988; Hadjisterkotis 1993) and in most of the introduced populations (Pfeffer and Genest 1969; McClelland 1991). The fact that hornless females is the rule in most introduced populations would originate from a much larger contribution of animals coming from Sardinia than Corsica during the early phases of introduction of mouflon in continental Europe (Pfeffer and Genest 1969; Apollonio et al. 2010). Similarly, large variation in the form of horns occurs in males that can partly result, in some populations, from artificial selections against desirable or undesirable trophy characteristics (Garel et al. 2007; Schröder et al.

2016). Horns in males can be heteronymous (left horn first twists clockwise and then anticlockwise) or perverted (growth in the same plane with sickle-shaped horns), curving back behind the head toward the anterior-posterior axis of the body, with tips directed above the neck (supracervical) or at the sides of the neck (cervical; Fig. 1; Pfeffer 1967; Groves and Grubb 2011). In both cases, the basic growth pattern is converging with a relatively narrow tip-to-tip spread when compared to greatest spread as animals aged. With the notable exceptions of Cyprus and Armenian mouflon, the supracervical horn growth is rare in wild sheep (Hoefs and Hadjisterkotis 1998; Groves and Grubb 2011; Hadjisterkotis et al. 2016). At the opposite, distinct homonymous horn curl configuration also exists (left horn twists clockwise), where the tip-to-tip spread equals the widest spread with horn tips pointing outward the anterior-posterior axis of the body as animals get older. Intermediate types exist, with, for instance, horns angling more vertically, with a more parallel horn growth pattern with the tips turning inward toward the neck/face in a tight circle (Pfeffer 1967; Groves and Grubb 2011).

Pelage

The typical pelage coloration exhibited by females varies from light tan to dark brown (Valdez 1982). The dorsal patch is absent or indistinct. Lambs are similar in appearance to ewes. In males, short, coarse, and thick hair forms the outer coat (Frisina and Frisina 2000). The general color of the coat on mature rams is rufous brown or foxy red shading into chocolate brown on the head and face. A black coat is found on sides of the neck, throat, and chest with a band on the flanks and down the withers, on the outer front surfaces of the forelegs above the knees, and on the front and outer sides of the hind limbs above the hocks (Valdez 1982; Frisina and Frisina 2000). Ears are grayish with the margins and part of their interior that are white. White/grayish coat is also found as a broad band on the rear border of the saddle patch and on muzzle and chin shading into grayish rufous in the middle of the black area on the throat. Similarly, the buttocks and all under parts are white, except for a narrow black streak down the forelegs. Limbs also exhibit a streak of white on their back surface above the knees and hocks. In winter coat, the general color of the upper parts deepens and becomes more of a chestnut brown. The saddle patch on each side of body (not systematically present in some populations) lightens until in many older rams it becomes nearly white (Frisina and Frisina 2000). Both males and females display a white facial mask which shows a great inter-individual variability in size (Garel et al. 2005d, 2006a; see also subsection "Age Determination").

Age Determination

Excepted in lambs (when less than 6 months old) for which error in age determination is very unlikely (Garel et al. 2006a), horn annulus technique in males (Geist 1966; Hemming 1969) and tooth eruption and replacement of the lower incisors in both sexes (Rieck 1975; Piegert and Uloth 2005) are the only reliable criteria used to estimate age when handling animals. Tooth eruption pattern allows a reliable estimation of age until 3.5 years in autumn-winter (the hunting season for ungulates in most countries; e.g., Milner et al. 2006) during which five stages of tooth eruption can be described: no permanent incisor for lambs, two permanent incisors for yearlings, four permanent incisors for 2.5 years old, six permanent incisors for 2.5–3.5 years old, and eight permanent incisors for ≥3.5 years old. In males, the horn annulus technique (Geist 1966; Hemming 1969), which is based on counts of horn growth annuli, provides an index of age to the nearest year. Horn growth annuli correspond to drastic reduction of horn growth which mainly results from hormone-induced factors related to the reproductive cycle and from photoperiod (see subsection "Horn Growth and Reproductive Activity"). This index becomes less accurate as the animal aged and the first horn annulus is worn away.

From observations in the field, different morphological criteria have been used to estimate the age categories of mouflon (Türcke and Schmincke 1965; Pfeffer 1967; Tomiczek 1989; Boussès and Réale 1994). In males, observers have principally relied on the position of the horn tip relative to the base of the neck and eyes (the so-called circular arc theory; Piegert and Uloth 2005). In females, the size of the white facial mask, which corresponds to a whitening of the hair of the face, was considered to progress from the nostril toward the face as animal aged. From these criteria, field studies (e.g., Le Pendu et al. 1995; Ciucci et al. 1998; Cransac et al. 1998; Réale et al. 1999) generally distinguished between lambs, yearlings (1 year), and 2–3 wider age categories as animals get older and inter-individual variation in age criteria increases (e.g., 2–3, 4–6, and >6 years old; Bourgoin et al. 2018). However, except in lambs, errors in age determination are common (>20% in adult males and >50% in adult females; Boussès and Réale 1994; Garel et al. 2005d, 2006a).

Physiology

Thermoregulation

Mouflon, as all other homeothermic animals, regulate their own body temperature through internal metabolic processes. Heat stress affects significantly the heat balance, and the main thermoregulatory mechanisms are the reduction in heat production, the increase in heat loss (Johnson et al. 2003), and the molting of the pelage (Lincoln 1990). Seasonal changes in annual and daily prolactin secretion and their relationships with melatonin secretion, metabolism, and thermoregulation have been reported in mouflon. Seasonal changes have been shown to be associated with changes in day length, with highest and lowest concentrations of prolactin occurring during summer (long days) and winter (short days), respectively (Lincoln 1990; Santiago-Moreno et al. 2004). Spring molt is timely associated with rising plasma prolactin concentrations, and hair growth, that takes place between the summer and winter solstices, coincides with low-plasma concentrations of prolactin (Santiago-Moreno et al. 2004). In addition, an endogenous circannual rhythm of molting has been demonstrated in this species (Allain et al. 1994). Lastly, it has also been suggested that horns core might contribute to the heat exchanges in mouflon and play as such a thermoregulatory role in this species (Hoefs 2000).

Horn Growth and Reproductive Activity

The body mass is the prevailing factor determining horn growth throughout the first year of age (Santiago-Moreno et al. 2006). Thus, young mouflon rams, aged up to 18 months, exhibit a regular and nonseasonal development of horns through the year under captive regimen (Santiago-Moreno et al. 2000b). Afterward, in subadult (2 years old), adult (≥3 years old), and old ages (>10 years old), the rate of horn growth shows a seasonal pattern modulated by day length (Lincoln 1998) with largest monthly growth occurring in spring and summer (Santiago-Moreno et al. 2005a). The greatest growth in length occurs within the first 2–3 years of life and subsequently decreases year by year (e.g., Garel et al. 2005d).

The annual cycles of reproductive activity and horn growth follow opposite trends (Lincoln 1990), although both are mainly regulated by circulating testosterone levels and photoperiod (Toledano-Diaz et al. 2007). The onset of horn growth arrest in the mouflon coincides with maximum testosterone secretion in the premating season. The horn growth is at a minimum during the period of optimum sperm production and maximum testis and accessory sex gland activity (autumn). The rate of horn growth before the mating season may be related to springtime testosterone levels (Toledano-Diaz et al. 2007; Santiago-Moreno et al. 2012). The greater horn growth in spring involves an appreciable use of energy resources. The decrease of photoperiod in autumn appears to act as a signal to stop horn growth, allowing energy resources to be diverted toward combat and the production of better quality semen. Interestingly, the improvement of sperm quality in mouflon occurs between September and December, coinciding with the premating and mating seasons in this species. Normal sperm cell morphology is closely linked to the correct progress of spermatogenesis and is essential for optimum sperm function and fertilization capacity (Martínez-Fresneda et al. 2019). The resurgence of horn growth in spring is positively correlated with the percentage of spermatozoa with morphological abnormalities (Santiago-Moreno et al. 2003).

Reproductive Physiology

As reported in many species of large ungulates (Gaillard et al. 2000), body mass, more than age, appears to be the main factor controlling sexual maturity and reproductive performance in mouflon ewes and rams (Lincoln 1989; Santiago-Moreno et al. 2001b; Garel et al. 2005a). In females, onset of puberty is thus found to depend on a body mass threshold corresponding to ~80% of the adult body mass (Santiago-Moreno et al. 2000a, 2001b; similar threshold has also been found in other ungulates: Hewison 1996; Sand 1996) that can be reached in ewes during the first or second year of life (Mottl 1960; Lincoln 1998; Santiago-Moreno et al. 2001b). Similarly, males may initiate full spermatogenesis (male puberty) during their first year of life (Lincoln 1998). However, in males, physical maturity is also necessary for successful matings, and only rams 3 years old or more have been observed in rutting activities (Bon et al. 1992, 1995; Lincoln 1998).

During the reproductive period, ovulatory activity, assessed by measuring the plasma progesterone concentration in blood samples, reveals a mean duration of the oestrus cycle of 17 days. Ovulations are expected to be highly synchronized (Santiago-Moreno et al. 2001b) leading to highly synchronized births (80% of births may have occurred in 3–6 weeks; Bon et al. 1993a; Hadjisterkotis and Bider 1993; Garel et al. 2005a). Photoperiod is generally accepted as the primary environmental cue influencing seasonal breeding pattern, mediated via the pineal gland and melatonin secretion (Santiago-Moreno et al. 2000c). Other environmental variables, such as climate-related seasonal variation in resource availability, are also expected to shape the timing and synchrony of reproduction in large herbivores (Bunnell 1982; Rutberg 1987). Geographical variation in these factors probably contributes to explain the inter-population differences in the timing and synchrony of breeding activities reported in the literature (Pfeffer 1967; Lincoln 1989; Bon et al. 1993a; Santiago-Moreno et al. 2001b; Garel et al. 2005a). However, as a genetic basis for photoresponsiveness has also been found in both sheep (*Ovis aries*) and wild ruminants (Santiago-Moreno et al. 2001b), the genetic history of populations (see sections "Paleontology" and "Phylogeny and Phylogeography") could also contribute to the regional differences reported. In females, the existence of an endogenous rhythm of reproduction is synchronized through the circadian rhythm of melatonin secretion (Gómez-Brunet et al. 2008). Uterine regression and resumption of ovulatory activity after parturition is about 25 days, although ovulatory activity in the early postpartum period is not always associated with successful conception and pregnancy (Santiago-Moreno et al. 2001a).

Genetics

Chromosomes

$2n = 54$ (Nadler et al. 1973)

Phylogeny and Phylogeography

The Armenian mouflon *Ovis gmelini gmelini*, from western Iran and easternmost Turkey, and the Anatolian mouflon *O. g. anatolica,* endemic to central Anatolia (Fig. 2), are considered to be the most likely ancestors of all breed of domestic sheep (DNA analyses: Hiendleder et al. 2002; Bruford and Townsend 2006; Demirci et al. 2013; Sanna et al. 2015; chromosome number: Nadler et al. 1973; Valdez et al. 1978; type of hemoglobin A: Bunch et al. 1978). It is also widely acknowledged that mouflon present in Europe are the feral descendants of the very first domesticated Asiatic mouflon introduced by Neolithic people during the first waves of sheep domestication across the Mediterranean Basin (see section "Paleontology").

Some authors have initially considered mouflon present in Mediterranean islands and subsequently introduced into continental Europe (see section "Current Distribution") as belonging to a full species *O. musimon* (Nadler et al. 1973). The most recent classifications considered them either as subspecies of Asiatic mouflon *O. gmelini musimon/ophion* (e.g., Cugnasse 1994; Shackleton and IUCN/SSC Caprinae Specialist Group 1997; Festa-Bianchet 2000; Hadjisterkotis and Lovari 2016; see section "Taxonomy and Systematics") or of domestic sheep *O. aries musimon/ ophion* due to the predomestication phase that they have undergone (e.g., Wilson and Reeder 2005; International Commission on Zoological Nomenclature; Bern Convention, Table 2). However, while sheep domestication occurred ~9000 years BCE (Zeder 2008), divergence times up to 110–171 kya between sheep and mouflon haplotypes have, for instance, been reported in mitochondrial phylogeny of Sardinian and Cyprus populations (Sanna et al. 2015; Mereu et al. 2019). These populations thus did not originate from domestic sheep but have evolved separately. The same is expected for the Corsican mouflon which share a large part of their history with those from Sardinia (Satta et al. 2021; Portanier et al. 2022). This gives support for considering mouflon on Mediterranean islands as wild taxon rather than as domestic subspecies (Guerrini et al. 2015; Portanier et al. 2022). They should thus be unvaryingly referred to as *Ovis gmelini musimon/ophion*, i.e. as subspecies of the Asiatic mouflon. This taxonomic designation should help protecting the rare gene pools of these populations (Chessa et al. 2009; Guerrini et al. 2015; Hadjisterkotis and Lovari 2016; Mereu et al. 2019; Portanier et al. 2022) which have been preserved in their historical ranges from genetic introgression from domestic breeds (see subsection "Hybridization").

Cyprus mouflon occupies a place apart in the history of sheep domestication being at the start of the migration routes toward western regions which were reached by human expansion a few hundreds of years later after the first archeological evidences of domestication in the fertile crescent (Zeder 2008; Mereu et al. 2016). Mouflon were introduced 2500–4000 years later in Corsica and Sardinia (Poplin 1979; Vigne 1992; Portanier et al. 2022). This history has contributed to the genetic isolation among islands that persists since Neolithic times and the likely existence of many historical events to the present-day genetic differentiation among Cyprus and western populations (Chessa et al. 2009; Guerrini et al. 2015, 2021). In addition, it cannot be completely discarded that wild mouflon would have arrived by their own on Cyprus during the last ice age when the Mediterranean sea level was 125 m below the current sea level (Hadjisterkotis 2012). Cyprus is geographically close to the native populations (Fig. 2), and Cyprus mouflon was found to be phenotypically and genetically closely related to Anatolian individuals carrying the rare haplotype X at the D-loop region (Demirci et al. 2013; Sanna et al. 2015; Guerrini et al. 2021).

Mouflon from Corsica and Sardinia were later introduced into continental Europe (Uloth 1972; see also section "Current Distribution"). The first recorded introduction would date of 1729–1731 (Uloth 1972; Tomiczek 1989; Weller 2001) by Prince Eugen of Savoy in the game pasture of Belvedere, near Vienna, Austria. The origin of continental mouflon populations is often uncertain, and some evidences of crossbred with other domestic sheep, feral, or wild mouflon have been reported, raising caution on the genetic integrity of these populations (but see subsection "Hybridization"). Related to this uncertainty in the level of hybridization within introduced populations, Cugnasse (1994) suggested referring to them by adding the suffix "× *Ovis* sp." to the Latin name. This author also suggested referring to Mediterranean mouflon rather than European mouflon to keep track of the geographical origin of the founding populations.

Genetic Diversity

Ancestral populations seem to exhibit depleted levels of allelic richness and/or heterozygosity (e.g., Anatolian, Corsican, Sardinian and particularly Cyprus populations, Özüt 2001; Guerrini et al. 2015; Barbato et al. 2017; Satta et al. 2021; Portanier et al. 2022). In some of these populations, strong demographic bottlenecks have been reported (e.g., Pfeffer 1967; Özüt 2001; Kaya et al. 2004; Satta et al. 2021), possibly explaining this limited genetic diversity, although genetic bottlenecks have not been detected in Cyprus and Anatolian populations (Özüt 2001; Guerrini et al. 2015). On the other hand, introduced populations have been evidenced to have a higher than expected heterozygosity (Kaeuffer et al. 2007; Portanier et al. 2017) that may have contributed to their persistence despite a very low number of founder individuals (e.g., one male and one female in Kerguelen Archipelago).

Genetic diversity is an important element in conservation purposes since it can be linked to individual phenotypes and fitness (e.g., twinning rate, Kaeuffer et al. 2008; parasite resistance, Portanier et al. 2019) and thus to the adaptive potential of populations (Hedrick 2011). Several populations benefit from conservation measures such as a reintroduction programs (Corsica, Sardinia, Cyprus, and Turkey; Hadjisterkotis and Bider 1993; Özüt 2001; Berlinguer et al. 2005; Rieu 2007) to increase population sizes, favor population expansion, and enrich the genetic diversity (see section "Management"). Cloning and assisted reproductive technologies have even been considered as conservation measures (Loi et al. 2001; Hosseini et al. 2009; Hajian et al. 2011).

Hybridization

Introgression of domestic or non-native genetic material, which may occur due to hybridization between the different *Ovis* species, might be considered as a threat since it can cause phenotypic modifications (e.g., be larger, Hess et al. 2006), with possible negative effects on fitness (e.g., loss of local adaptations, Burke and Arnold 2001) and, when really high, raise questions about the definition and the delimitation of a species. Several Iranian populations have been shown to hybridize with Urial (Valdez et al. 1978; Fig. 2). Across Europe, hybridization occurs with domestic sheep, either because of contact zones between domestic and wild sheep (e.g., in Armenian and Sardinian mouflon, Bleyhl et al. 2018; Satta et al. 2021), or due to deliberate crossbreeding with primitive domestic breeds to improve fecundity or body/trophy growth (Türcke and Schmincke 1965; Uloth 1972; Lorenzini et al. 2011; Ciani et al. 2014; Schröder et al. 2016; Barbato et al. 2017). However, recent investigations in several continental and island populations show that introgression of domestic sheep into mouflon is not systematic and most often negligible, especially in historical mouflon range of island populations where populations appear overall pure (Schröder et al. 2016; Barbato et al. 2017; Iacolina et al. 2019; Mereu et al. 2019; Ciani et al. 2020). In harvested continental populations, the removal of animals with undesirable characteristics might have contributed to explain the low level of introgression detected (Schröder et al. 2016; Iacolina et al. 2019). In island populations, by occupying remote/mountainous areas, resident populations were less prone to genetic introgression from domestic sheep, and such introgressions from not locally adapted breeds were most likely counterselected by natural and sexual selection.

Population Genetics and Structuring

Genetic differentiation between mouflon populations has been shown to be relatively high (e.g., Corsica versus mainland France, Portanier 2018). Within populations, a marked sex-specific genetic structure can also be expected (e.g., in Sardinia, Satta et al. 2016, 2021, in Corsica, Portanier et al. 2022 but see Barbanera et al. 2012 in Cyprus) due to the sex-specific space-use (Marchand et al. 2015a, b, 2017) and the strong spatial segregation between sexes (e.g., Bourgoin et al. 2018). In a French-introduced population, this socio-spatial behavior has been found to reduce gene flow among females socio-spatial units (Petit et al. 1997; Portanier et al. 2017, 2018) and to contribute to the persistence of a historical genetic footprint more than 14 generations after mouflon introduction (Portanier et al. 2017). The socio-spatial behavior of ewes relies on a strong philopatry (Dubois et al. 1992, 1994; Martins et al. 2002) and a preponderant role of habitat characteristics on female movements and habitat selection (Marchand et al. 2015a, 2017; see section "Habitat and Diet"). Among habitat characteristics, both natural and anthropogenic linear landscape features, and, to a lower extent, habitats avoided during the rutting period (i.e., habitat perceived as unsafe or offering low quantity and quality resources), have been found to be resistant to gene flow (Portanier et al. 2018). Conversely, habitats selected during the rutting period (i.e., perceived as safe or offering resources) would be much more permeable. This is the case for steep slopes (used as refuge against predators) that appear to be an important determinant of landscape genetic resistance in mouflon (Portanier et al. 2018) as reported for several other wild sheep species (see Epps et al. 2007; Roffler et al. 2016).

Rams are expected to insure most of the gene flow in mouflon (Portanier et al. 2017) due to their relaxed spatial behavior (Dubois et al. 1993, 1996) and reproductive dispersal, i.e., excursions outside their home ranges during the rutting period (Martins et al. 2002; Malagnino et al. 2021), as well as the limited impacts of linear landscape features on their movements (Marchand et al. 2017). This contributes to a much lower genetic structuring in males than in females (Portanier et al. 2017).

Life History

Growth

Few studies have reported information on the prenatal growth pattern of mouflon. Santiago-Moreno et al. (2005b) found 12 body size metrics studied by transrectal ultrasonographic measurements on captive ewes to be closely related to gestational age by simple linear or allometric models. The postnatal growth pattern of mouflon has been better investigated (e.g., Hoefs 1982; Garel et al. 2007), but more in males than in females for which no reliable age criteria exist above 4 years-old (see subsection "Age Determination"). In males, growth patterns were often based upon hunting records, i.e., transversal data that may bias age-related patterns (Cam et al. 2002), or from the longitudinal monitoring of a restricted number of known-age animals experiencing captive conditions with a supplementary feeding (e.g., Lincoln 1998; Santiago-Moreno et al. 2005a). To overcome these limits, we analyzed data from free-ranging mouflon of known-age trapped during spring (before summer drought) in the Caroux-Espinouse population (for more details on the population and study site, see Garel et al. 2005a, 2007; Marchand et al. 2014a, 2015a). Mouflon show typical monomolecular patterns of growth (Gaillard et al. 1997) with males being larger and experiencing active growth in body mass ($n = 662$) and hind foot length ($n = 392$) for a longer period than females ($n = 521$ and 360, respectively). Sexual dimorphism was marked from 6 months of age onward with adult males that weighed in average 44.2% more than adult females and that had skeletal size in average 7.1% longer (see also Table 1 for data collected all over the year). Males reached their asymptotic body size and mass at 3 and 7 years-old, i.e., approximately 1 and 3 years later than females, respectively. As already mentioned for phenotypic characteristics (see subsection "Size and Morphology"), postnatal growth shows large inter-population variation in this species (e.g., Figs. 3.5 and 7.2 in Garel et al. 2006b).

Reproduction

An early onset of reproduction (~6–10 months-old) has been reported in some populations (Land 1978; Briedermann 1992; Garel et al. 2005a; Sanchis 2018) with up to 50% of female lambs pregnant (see also subsection "Reproductive Physiology"). However, in most mouflon populations, the first reproductive attempt of females is considered to arise at the second breeding season when females have benefited of two seasons of growth (age at first reproduction: 1.5 years-old; Türcke and

Schmincke 1965; Valdez 1976; Hadjisterkotis and Bider 1993; Boussès and Réale 1998; Garel et al. 2005a). High-pregnancy rates can be reached and have been found both in wild and captive populations (from six populations: median [range]: 90.6% [83.3–99%], n [range] = 106 [27–396]; Briedermann 1992; Garel et al. 2005a; Nahlik 2001) and both in primiparous yearling females (93%; Cugnasse et al. 1985) and adult multiparous ones (Garel et al. 2005a).

From studies conducted at latitudes of 34–56 °N, rutting season extends from mid-October to December in most populations (e.g., Pfeffer 1967; Bon et al. 1993a; Hadjisterkotis and Bider 1993; Garel et al. 2005a). Females give birth to lamb (s) after *c.* 148–159 days of gestation (Mottl 1960; Türcke and Schmincke 1965; Briedermann 1992), i.e., from late March to late May. Although females are commonly considered as monotocous (Türcke and Schmincke 1965; Boussès and Réale 1998; Garel et al. 2005a), twins have been observed (mostly from postmortem examination) in enclosed populations (Briedermann 1992), in free-living populations given supplemental feed (Nahlik 2001; Sanchis 2018), and in wild populations (Briedermann 1992; Hadjisterkotis and Bider 1993; Boussès and Réale 1998; Garel et al. 2005a). High twinning rates have even been reported in the wild population of the Kerguelen Archipelago (in adult females: 36.4%; Boussès and Réale 1998). In this population, the probability of twinning increased with decreasing level of heterozygosity in ewes, but this correlation could result from local genetic effects rather than general inbreeding level (Kaeuffer et al. 2008). Some authors have also posited that in such populations, females would have partly inherited their reproductive ability (e.g., of twinning) from wild ancestors crossbred with domestic sheep (Pfeffer 1967; Boussès and Réale 1998; Garel et al. 2005a) in which domestication processes had meat production and greater prolificity as an initial aim (Chessa et al. 2009). None of the primiparous females autopsied in the previous studies had twins, and the sex ratio of fetuses was statistically balanced between males and females (Garel et al. 2005a).

To our knowledge, no data have been published on the age-related reproductive success in males which require access to hardly reachable data such as the genetically reconstructed pedigree (see Coltman et al. 2002 in bighorn sheep). However, reproductive effort of many different species of ungulates is consistently higher in prime-aged than in younger males and very old age classes (Mysterud et al. 2004). We may thus expect the same marked age-structured pattern of reproductive success in mouflon rams. In females, for which the presence/absence of a lamb provides more tractable data on their breeding success, observations from ewes marked in the Caroux-Espinouse population would suggest an increase in breeding probability in very old age classes (Richard 2016), rather than the commonly expected lower and more variable reproductive performance as animals get older (Martín and Festa-Bianchet 2011). This finding would provide some support to the terminal investment hypothesis in female mouflon, i.e., when survival prospects are low (see Fig. 3), and residual reproductive value approaches zero, individuals should invest all available resources in their terminal breeding attempt (Williams 1966; Pianka and Parker 1975; Clutton-Brock 1984).

Ewes may thus express a high reproductive potential: with a breeding attempt the first year of life in some populations, high reproductive rates over their lifetime and a litter size that may sometimes exceed 1. This reproductive output can be modulated by environmental conditions encountered as suggested by the linear decrease in female productivity from continental rich environments (lamb-ewes ratios >0.8) to Mediterranean islands where resource shortage is common (lamb-ewes ratios <0.5; Ciuti et al. 2009). For instance, the wild population of Corsican mouflon in Bavella experiences among the lowest lamb-ewes ratio ever reported (0.22; Sanchis 2018); while in captivity with a supplementary feeding, >90% of adult females reproduce (G. Comiti, pers. comm.). However, whether these low lamb-ewes ratios are only a matter of low fecundity in females and/or of low survival in lambs has still been little investigated.

Surprisingly, high-reproductive investment has also been reported in some populations facing limiting conditions. In the population on Kerguelen archipelago and in Soay sheep (*Ovis aries*) on

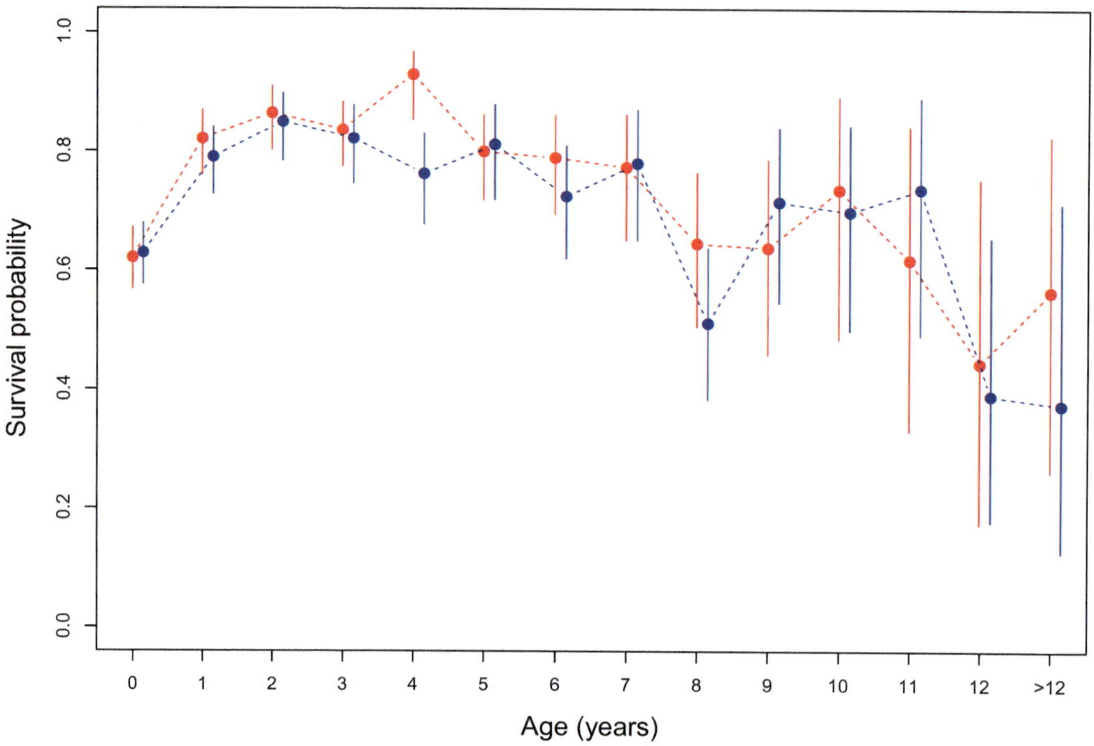

Fig. 3 Sex and age-specific survival probabilities (vertical lines indicate 95% confidence interval) of known-age mouflon (red: $n = 730$ females; blue: $n = 816$ males; 1986–2018 period) monitored by Capture-Mark-Resighting in the Caroux-Espinouse massif, France (see Garel et al. 2004). These results are an update of previous published survival analyses performed in this population during a restricted period (1986–1995) of nonlimiting conditions (Cransac et al. 1997; Dupuis et al. 2002; King and Brooks 2003). This update allows to account for density-dependent conditions encountered by this population after 1995 (Garel et al. 2007) and provides a comprehensive analysis along a gradient of density-dependent resource limitation

St Kilda archipelago (a closely related species, Chessa et al. 2009, with the same generation time, Hamel et al. 2016), females may give birth as 1 year and/or may produce twins, even under severe environmental conditions, and even at a cost of survival (Clutton-Brock et al. 1991, 1996; Boussès and Réale 1998; Tavecchia et al. 2005). Similarly, most of females in the Caroux-Espinouse population did not postpone their first breeding attempt in 40 years of monitoring, despite marked environmental changes in resources availability (Cugnasse et al. 1985; Garel et al. 2005b; Richard 2016), whereas age at primiparity is usually the first vital rate to be impacted by density-dependent food limitation in ungulates species (Bonenfant et al. 2009). Female mouflon could thus adopt in some populations a less conservative strategy than usually reported in ungulate females (Festa-Bianchet and Jorgenson 1998; Gaillard and Yoccoz 2003; Toïgo et al. 2007; Loison et al. 1999), by investing more energy in producing young and less in surviving long (Richard 2016; see subsection “Survival”).

Survival

Studies of marked individuals throughout their lifespan are also the only way to get reliable information on demographic parameters (survival and breeding probabilities) in free-ranging populations (Festa-Bianchet et al. 2017). In mouflon, very few long-term (>10 years) monitorings by capture-mark-recapture/resighting have been performed. We thus relied again on the analysis of data available from the French population of mouflon inhabiting the Caroux-Espinouse massif to illustrate age and sex-specific patterns in demographic parameters. In this population, mouflon survival of both sexes fits to the marked age-structured patterns classically reported in ungulate species with a survival improving in early life, reaching a plateau in prime age (2–7 years old), and then declining from 8 years onward (Loison et al. 1999; Gaillard et al. 2000; Fig. 3). Survival also becomes more variable as animals get older. This increase of mortality rate with age (actuarial senescence) most likely results from the decline in physiological functions with aging and is expected in any age-structured population (Hamilton 1966; Gaillard et al. 2000; Gaillard and Yoccoz 2003).

Another general rule is that the male survival should be lower and more variable than female survival (Toïgo and Gaillard 2003) in large iteroparous and highly dimorphic ungulates (sexual size dimorphism SSD $\geq$20%; Ruckstuhl and Neuhaus 2002). This is due to contrasted life-history tactics between sexes (e.g., Clutton-Brock et al. 1982; Jorgenson et al. 1997) which lead males to be more sensitive to food limitation than females (Toïgo and Gaillard 2003). Mouflon with their marked SSD (e.g., Table 1), and the limiting resources conditions experienced by the Caroux-Espinouse population (Garel et al. 2007), confirm this pattern (sex-specific survivals over all age classes are 0.751 $[0.733; 0.769]_{95\%}$ in males and 0.785 $[0.768; 0.801]_{95\%}$ in females) and illustrates possible costs caused by sexual selection in males. Although this gender difference is expected to increase with age, there is no such evidence here (Fig. 3). In accordance with this sex-specific survival pattern, maximal longevity reported for this population is lower in males (14 years) than in females (17 years). More generally, for both sexes, very few (<1%) known-age animals have been resighted or recaptured beyond 13 years-old.

Habitat and Diet

In its Caucasian original range, mouflon occupy plains, hills, lowland steppes, plateau slopes, mountain grasslands, and shrublands on dry and open slopes, with a preference for slightly rugged terrain where antipredator strategy for survival mostly relies on a high visibility (Baskin and Danell 2003; Bleyhl et al. 2018). However, as a result of its long history of introduction (see section "Current Distribution") and of an important behavioral plasticity, mouflon have been able to persist and can now be found in highly variable habitats, i.e., from coastal areas (e.g., Hawaii's Big Island) to high mountains (e.g., Alps), from flat terrains to rugged habitats, and from open areas to dense forests (e.g., pampa in Argentina versus continental forests from Eastern Europe).

Spatial Movements and Home Range

Depending on gender (males generally use larger areas as a result of reproductive dispersal during the rutting period; Dubois et al. 1993, 1996; Malagnino et al. 2021) and on habitats, the annual home range of mouflon ranges from a few hundred to a few thousands of hectares (e.g., 200–350 ha in females-males from the Caroux-Espinouse population, 950–1700 ha in females-males from the Cinto population [Corsica, France]; Marchand 2013). This space use pattern and its variation between populations are generally related to the existence of seasonal or altitudinal migration imposed by variation in resources availability and by the presence of snow cover in winter (due to the absence of interdigital membrane between digits, deep snow strongly impedes movements in mouflon). Both anthropogenic (roads, tracks, and trails) and natural (ridges, talwegs, and forest edges) linear features that mouflon can easily cross can constitute behavioral barriers for movements, delineating individual home ranges and influencing intrapopulation gene flow (Marchand et al. 2017; Portanier et al. 2018). As a result of females' philopatry and poor dispersal abilities in mouflon, a strong socio-spatial segregation is often reported within populations (Martins et al. 2002; Bourgoin et al. 2018), with the existence of several population units that use specific ranges and have little contacts and exchanges of individuals with each other (Darmon et al. 2007; Portanier et al. 2017).

Habitat Selection and Activity Pattern

Habitat selection and activity pattern of mouflon mostly depend on the trade-off between foraging resources, cover from risks imposed by potential predators (including humans, either in hunted populations or as a result of the "ghosts of predator past"; Byers 1997), and cover from adverse climatic conditions (food/cover trade-off; Lima and Dill 1990). Generally crepuscular and bimodal, the activity of mouflon can be largely modulated depending on seasonal and individual constraints (Bourgoin et al. 2008; Pipia et al. 2008). Likewise, habitat selection is characterized at the daily scale by an alternation between foraging areas used during active periods and refuge areas used during ruminating/resting periods. Once again, this pattern can be modulated depending on seasonal and/or individual constraints (Marchand et al. 2015a).

During the first months of lamb's life in spring, reproductive females generally use habitats offering both security and abundant resources, maximizing their own survival and survival of their lamb(s) while providing enough food resources to cover the increased energetic needs imposed by lactation (Hadjisterkotis and Bider 1993; Bourgoin et al. 2008; Ciuti et al. 2008, 2009; Pipia et al. 2008). During this period of abundant resources, males and nonreproductive females are highly active and focus on favorable foraging areas (Bourgoin et al. 2008; Pipia et al. 2008; Marchand et al. 2015a).

During summer, in areas where temperatures are high, mouflon are generally less active during daytime, when they select thermal refuges, and more active at nighttime when they focus on foraging areas (Bourgoin et al. 2011; Marchand et al. 2015b). Both males and females movements are thus preferentially directed toward habitats offering thermal cover when diurnal temperature overruns a given threshold (~15–17 °C in Marchand et al. 2015b; see also Sanchis 2018). In males, this selection persists during nighttime. In females, a concomitant increase in nocturnal activity rate has been reported to compensate for the drastic reduction in activity level during the hottest period of the day. This activity shift may contribute in limiting thermoregulatory costs (Bourgoin et al. 2011). Thermal cover also influences habitat preferences of males at a broader spatial scale (home range) exemplifying the importance of this factor for male fitness (Rettie and Messier 2000). Conversely, females often trade-off food and thermal cover with refuges and better conditions for lamb survival, especially at the home range scale (Marchand et al. 2015a), and their behavioral responses to hot summer conditions were thus restricted to the hottest period of the day only and by temporally shifting their feeding activity to the night. Similar results were reported in Corsica, where a site occupancy approach was used over 30,000 ha to investigate habitat selection by mouflon during summer (Sanchis 2018). Detection probability also strongly decreased with increasing temperatures and could be divided by up to 2 from 10 °C to 30 °C (Garel et al. 2005c; Cazau et al. 2011; Sanchis 2018).

During autumn and winter, foraging resources become scarce and mouflon spend most of their time foraging. In some areas, food supply is provided and seems important for successful overwintering (Heroldova et al. 2007). In areas where snow cover persists and where winters are harsh, mouflon often move to lower altitudes and use distinct areas than those used in summer (Baskin and Danell 2003; Darmon et al. 2014). Altitudinal migration, or even long-distance migration, have been consequently reported in several mountainous populations (Pfeffer and Settimo 1973; Rigaud 1985; Dubois et al. 1993; Talibov et al. 2009), but little is known on the determinants, on the proportion of individuals concerned, and on the ecological and biological consequences for mouflon populations. Collective migratory movements have been observed in Corsica (P. Benedetti, pers. comm.).

Whatever the season, these behavioral patterns can also be modulated depending on the level of human activities in natural areas (lethal, i.e., hunting, or nonlethal, such as recreational activities). For instance, human disturbance generally results in a temporal shift of activities toward nighttime and an exacerbated selection of refuge areas during daytime (Benoist et al. 2013; Marchand et al. 2014b; Sanchis 2018). However, the long-term consequences of this spatio-temporal shift related to human activities, e.g., on energetic balance, reproduction, and/or survival, remain unknown.

Feeding

Among large herbivores, mouflon have been classified as a “grazer” species based on the relative consumption of grasses and browse and on the morphology and physiology of their digestive system (Hofmann 1989). Indeed, its diet is mostly composed with grasses and forbs when these plants are available in sufficient quantity (Marchand et al. 2013). Besides, its rumen morphophysiology is particularly suitable for grass digestion and contains specific bacteria able to provide energy from cellulosis contained in grasses (Clauss et al. 2009, 2010). However, when grasses are temporally/seasonally and/or locally scarce, mouflon can feed on dicots, i.e., forbs, shrubs, fruit, and even trees (Marchand et al. 2013). As a result, >100 plant species have been reported in mouflon diet. Likewise, mouflon have been able to persist and spread in areas where grasses are uncommon (e.g., Kerguelen archipelago), calling into question the classification of mouflon as a grazer (see Clauss et al. 2010; Marchand et al. 2013). However, a strong decrease in body mass has been observed in a population facing an important loss of open (i.e., grass-rich) areas and between individuals with decreasing levels of open areas in their home ranges (Garel et al. 2007; Marchand et al. 2014b). This suggested that this mismatch between the digestives features of mouflon and their actual diets in most areas where it persists has to be interpreted owing to the introduction history of the species.

Behavior

Social Behavior

Mouflon is a gregarious species in which group size and composition are unstable (fission-fusion dynamics; Bon et al. 1990, 1993b). As a result of a strong sexual size dimorphism (Table 1) and sex-specific needs, adult males and females live in segregated groups out of the rutting season and can use different habitats depending on sex-specific seasonal needs and constraints (sexual segregation with both social and spatial components; Bourgoin et al. 2018). Sexual segregation between young males and females increases with the age of males, resulting in young males often found either in groups of females or in groups of adult males. The social component appears to be important not only for sexual segregation between the sexes but also between age groups among males (Cransac et al. 1998; Bourgoin et al. 2018).

Reproductive females generally isolate from other individuals during a few hours-days for lambing (Hadjisterkotis 1993; Marchand et al. 2021). As soon as their lamb is able to follow their mother (follower type *sensu* Lent 1974), they form groups of females and lambs that use specific habitats offering both security and abundant resources until weaning; which progressively occurs around 3 months after births.

Mating Behavior

During the rutting season, two reproductive tactics have been reported in mouflon rams, depending mostly on their age (Bon et al. 1992). Dominant old males generally tend receptive females by following them and by impeding attempts to mate of other males (tending tactic). By contrast, young subordinate males rather course females, attempting to breach the defense of dominant males (coursing tactic). Whatever their age, reproductive dispersal is often observed in males during the rutting season, i.e., important movements of males out of the home range they use the rest of the year (Malagnino et al. 2021), with consequences on gene flow (Portanier et al. 2018).

Inter- or intrasexual interactions consist in repeated behavioral postures (McClelland 1991). Interactions between males involve low-intensity agonistic displays such as foreleg kicks and twists (an animal closely approaches another one and rotates its head approximately 90° with the horn directed away from the recipient’s body). Head butting (in both sexes) and clashes (in males) predominate in intrasexual agonistic interactions, and the former behavior is regularly used to initiate the interaction in both sexes. Male-female interactions show more variation in the behaviors which initiate the encounter, but usually begin by males performing twist, flehmen (male tightly retracting its upper lip until it curls completely away from its toothless upper jaw), or low stretch (male stretches its head and neck as forward as possible). Much more behavior components are exhibited during male-male interactions as compared to female-female encounters which are often very brief.

Parasites and Diseases

Several studies on diseases of mouflon were performed not only on island and introduced populations across Europe, but also in Chile and in the sub-Antarctic Kerguelen archipelago, stressing the high diversity of both internal and external parasites mouflon may host (see Hille 2003 for a review).

Ectoparasites

Among the external parasites, some infestations by acarian and insects are sporadically reported in mouflon, including the sarcoptic mange caused by *Sarcoptes scabiei* (Bornstein et al. 2001; Poglayen et al. 2018), hypodermosis caused by larvae of the warble fly *Hypoderma diana* (Colwell 2001), and oestrosis caused by larvae of the sheep nasal bot fly *Oestrus ovis* (Moreno et al. 1999, 2015; Poglayen et al. 2018).

The most prevalent ectoparasites are the ticks belonging to the *Ixodida* order. The tick *Ixodes ricinus* was observed in most of the study areas within Europe and was often the only species observed in northern and high-elevation areas (e.g., Germany, Netherland, Poland, Czechoslovakia, Austria, and French Alps). In southern parts of Europe (Cyprus: Ioannou et al. 2011; south of France and Corsica: Grech-Angelini et al. 2016; Mercier 2016; Italia: Poglayen et al. 2018) and Ukraine (see Hille 2003 for a review), a higher diversity of tick species was observed with five to eight species per country, belonging to the genera *Haemaphysalis, Hyalomma, Rhipicephalus,* and *Dermacentor.* In the Caroux-Espinouse population, the infestation of mouflon by ticks in spring was positively correlated with the body mass of mouflon and with the level of rainfall, but negatively correlated with the mean ambient temperature (Bourgoin et al. 2014).

Ticks are of main concern as they can transmit several pathogens of veterinary and zoonotic importance. Mouflon have been shown to have direct or indirect signs (high antibody titers) of contamination with pathogens such as *Anaplasma phagocytophilum* (anaplasmosis; Stefanidesova et al. 2008; Silaghi et al. 2011; Kauffmann et al. 2017; Hornok et al. 2018), *A. ovis* (Ioannou et al. 2011), *Borrelia burgdorferi* s.l. (Lyme borreliosis; Trávnicek et al. 1999; Zeman and Januška 1999; Juřicová et al. 2000; Trávnicek et al. 2003; Štefančíková et al. 2008), *Babesia* spp. (babesiosis; Ferrer et al. 1998; Kauffmann et al. 2017), *Coxiella burnetii* (Q fever; e.g., Martinov et al. 1989; Hubalek et al. 1993; López-Olvera et al. 2009; Candela et al. 2017), and *Rickettsia* sp. (Ioannou et al. 2011; Hornok et al. 2018). Tick-borne encephalitis virus (Zeman and Januška 1999; Bagó et al. 2002) and tick paralysis (Konjevic et al. 2007) were also reported in mouflon.

Endoparasites

Necropsies of mouflon from France and Germany revealed the presence of 42 and 34 different species/morphotypes of helminths in their digestive and respiratory tracts, respectively (Gauthier et al. 2001; Hille 2003). The prevalence of lungworm infections by Protostrongylidae, belonging mostly to the *Muellerius, Cystocaulus, Neostrongylus,* and *Protostrongylus* genera, is generally high (Gauthier et al. 2001; Hille 2003; Panayotova-Pencheva 2006; Poglayen et al. 2018). The life cycle of these parasites requires a gastropod as an intermediate host, while the less prevalent *Dictyocaulus* sp., located in the trachea and the large bronchi at the adult stage, has a direct life cycle.

Regarding the parasitism of the digestive tract, the prevalence and diversity of helminth species are significant, especially for nematodes inhabiting the abomasum and the small intestine of mouflon (Gauthier et al. 2001; Hille 2003; Maerten 2014; Poglayen et al. 2018). Most of these nematodes belong to the order Strongylida, and some species have a well-recognized pathogenicity and are shared with domestic small ruminants, such as *Haemonchus contortus* (Cerutti et al. 2010). Isolation and counts of adult strongyles from mouflon hunted during autumn-winter showed an increase of the infestation with age and a negative relationship with body condition (Hille 2003; Maerten 2014). In spring, the number of eggs shed in feces is higher in young individuals, reproductive females, and individuals in poor body condition (Portanier et al. 2019; Bourgoin et al. 2021). Both candidate gene (MHC DRB1 exon 2) and neutral genetic diversity (16 microsatellites) have been found to be associated with resistance to gastrointestinal nematodes (Portanier et al. 2019).

Other helminths observed in the digestive tract of mouflon include trematodes and cestodes. The small liver fluke *Dicrocoelium dendriticum* is the most common fluke and was reported in all the European countries where parasitological studies were conducted, while the pathogenic common liver fluke *Fasciola hepatica* and the rumen fluke *Paramphistomum*/*Calicophoron* spp. are more rarely detected in mouflon (Gauthier et al. 2001; Hille 2003; Poglayen et al. 2018; Bourgoin et al. 2021). *Moniezia expansa* and *M. benedeni* are the main adult cestodes developing in the small intestine of mouflon, especially in lambs and young individuals (Hille 2003).

Intestinal infestations by the protozoan *Eimeria* sp. are highly prevalent, and often multi-infections with different species of *Eimeria* occur. The intensity of fecal oocyst excretion is generally the highest in mouflon in poor body condition and in young mouflon (Hille 2003; Bourgoin et al. 2021). More rarely described in free-ranging mouflon, infestations by *Giardia duodenalis* can be observed, especially in lambs and young individuals (Cockenpot 2013).

Mouflon can play the role of the intermediate host for different parasites. It includes *Sarcocystis* sp., with cysts located in muscles (Goldova et al. 2008; Coelho et al. 2014; Poglayen et al. 2018) and the zoonotic protozoan *Toxoplasma gondii,* with potential consequences on the reproductive success of ewes, and also risks for people consuming game meat (Gauss et al. 2006; Aubert et al. 2010; Bartova and Sedlak 2012; Gotteland et al. 2014; Verma et al. 2015; Heddergott et al. 2018; Almería et al. 2018). Larvae of cestodes can be detected during a carcass examination of mouflon such as the metacestode larval stage of *Taenia hydatigena* (*Cysticercus tenuicollis*) attached to the peritoneum in the abdominal cavity, or the hydatid cysts of *Echinococcus granulosus* in the liver and/or lungs (Hadjisterkotis 1997). The metacestode larval stage of *Taenia multiceps* (*Coenurus cerebralis*) is located in the brain or the spinal cord of the host and may cause nervous or behavioral disorders (Hille 2003; Poglayen et al. 2018).

Infectious Diseases

The infectious keratoconjunctivitis, caused by *Mycoplasma conjunctivae*, has been reported in mouflon in France (Cransac et al. 1997; Cugnasse 1997) and in the Spanish Pyrenees (Marco et al. 2009). Following epizootics, marked decreases in abundance (Garel et al. 2005c) and survival probabilities (Cransac et al. 1997) have been detected, with a constant decline for males in all age class (~−20%) and age-specific variation for females (with survival probabilities declining by up to −30–60% among lamb and young ewes).

Antibodies against other pathogens of veterinary and zoonotic importance have also been detected in mouflon, such as *Salmonella abortusovis* (abortive salmonellosis; Martín-Atance et al. 2012), *Chlamydophila abortus* (chlamydiosis; López-Olvera et al. 2009), *Mycobacterium avium* ssp. *paratuberculosis* (paratuberculosis; López-Olvera et al. 2009), caprine arthritis-encephalitis virus (Guiguen et al. 2000), bluetongue virus (Rossi et al. 2014), and Schmallenberg virus (Rossi et al. 2017).

Other diseases have been searched for in a limited number of individuals (*Leptospira* sp.,

Mycoplasma agalactiae, and *Neospora caninum*): None of the individuals tested were infected; however, this does not preclude the possibility of low prevalence of these diseases. Mouflon are also expected to be susceptible to diseases affecting ruminants in general, such as brucellosis and tuberculosis. However, the exact epidemiological role of mouflon populations in the transmission of all these infections remains to be determined.

Additional sanitary issues have been reported in some mouflon populations. For instance, the Cyprus population suffers from a large range of bone problems similar to those of domestic animals (Hadjisterkotis 1996c). The most important defects are degenerative changes and spondylosis (a.k.a. spondylosis deformans or ankylosing spondylosis). At the articular processes and facets, degenerative arthropathy is present and the reaction of the osteophytes produces ankylosis of the articulations between vertebral bodies (intervertebral parts) and between the costal facets and the head of the ribs (costovertebral parts). Disk damage with osteophytosis and sometimes degenerative changes of the vertebral body and a greater spinal curvature were also observed mainly in males throughout the thoracic vertebra, which could result from the greater strain their spinal cords endure, as in fights during the rutting season. Affections of the hoofs have also been reported in mouflon, among which the most common is the contagious foot rot disease, caused by *Dichelobacter nodosus* in association with *Fusobacterium necrophorum* (Belloy et al. 2007; Bennett et al. 2009). It is usually passed on from domestic sheep to mouflon on shared pastures. An inflammation and necrosis of the interdigital tissue among feet is observed. The animals produce a hyperplastic foot horn, show lameness, and often feed in a recumbent position (Volmer et al. 2008).

Population Ecology

Population Dynamics

In line with its fast living strategies (at least in females; see subsection "Reproduction"), mouflon have a short generation time (4.21 years; Hamel et al. 2016), may reach in some populations a very high population growth rate ($\lambda = 1.34$ in Kaeuffer et al. 2009), and show a temporal variability in prime age survival similar to that of breeding probability (Richard 2016). This suggests a lower canalization of adult survival than usually reported in other ungulates species (Gaillard and Yoccoz 2003). Variability in environmental conditions encountered during early life probably contributes to this temporal variability in demographic rates (Lindström 1999) and has been found to generate sex-specific cohort effects on phenotypic traits such as body mass (Garel et al. 2007; Hamel et al. 2016). Typically, female body mass shows less phenotypic variability among cohorts and stronger compensatory/catch-up growth than males, whereas males show more progressive changes throughout life. This supports that stronger selective pressures for rapid growth in this dimorphic species make males more vulnerable to poor environmental conditions early in life and less able to recover rapidly after a poor start.

As in any other ungulate species, population dynamics of mouflon results from complex interactions between per capita food resources (Kaeuffer et al. 2009; in Soay sheep Catchpole et al. 2000; Coulson et al. 2001), climatic conditions (Hadjisterkotis 2002; Garel et al. 2004), disease (Cransac and Hewison 1997), and, when present, predation (either natural or from hunting; Espuno 2004; Garel et al. 2005c). Age structure is also expected to shape the response of demographic rates to density and environmental stochasticity (Sæther 1997; Gaillard et al. 2000; Coulson et al. 2001). In absence of predation/hunting, mouflon populations could grow close to their maximum growth rate over a wide range of population densities (Bonenfant et al. 2009; Kaeuffer et al. 2009) and, when resources are not limiting (e.g. in populations released to a new environment), should display a typical "irruptive dynamics" (see Kaeuffer et al. 2009 in the mouflon population of the Kerguelen archipelago). In such a situation, the mouflon population quickly reaches, and may exceed, the carrying capacity of its environment, before collapsing and entering in a second phase of successive irruptions. This pattern has also been documented in Soay sheep (Clutton-Brock et al. 1991). In both species, it has been related to the high-reproductive potential of ewes (see subsection "Reproduction") in association with delayed density-dependent processes on demographic rates during the irruptive phases.

Phenotypic traits, such as body mass and trophy size, have also been found to be density-dependent and related to intraspecific competition for resources (Garel et al. 2007; in Soay sheep see Forchhammer et al. 2001). Yearly monitoring of phenotypic measurements, such as lamb body mass, could thus be used as an early warning sign of food limitation in this species, providing a quantitative basis when setting hunting quotas (Morellet et al. 2007).

Competition with Other Ungulates

Numerous populations of mouflon also share habitats with other wild or domestic ungulate species (e.g., red deer *Cervus elaphus*, roe deer *Capreolus capreolus* and axis deer *Axis axis*, Pyrenean *Rupicapra pyrenaica pyrenaica* and Alpine chamois *Rupicapra rupicapra rupicapra*, Iberian wild goat *Capra pyrenaica* and Alpine ibex *Capra ibex*, aoudad *Ammotragus lervia*, wild boar *Sus scrofa*, sheep, goat, and cow), resulting in a high potential for competition, particularly between similar-sized species (Pfeffer and Settimo 1973; Gonzales 1986; Hadjisterkotis 1993; Heroldova 1996; Bertolino et al. 2009; Darmon et al. 2012; Miranda et al. 2012; Chirichella et al. 2013; Redjadj et al. 2014; Centore et al. 2018). Although important spatial and/or diet overlaps have been reported with other species in some areas where mouflon have been introduced, contrasted results have been observed in terms of competition for native species. Species-specific adaptations to local environmental conditions (e.g., differences in activity patterns or habitat selection; Darmon et al. 2012, 2014) may contribute to coexistence without negative impacts from mouflon in most cases. However, spatial adjustments, diet overlap, and activity shifts in native species related to the presence of mouflon were also reported and interpreted as evidence for interspecific competition (Bertolino et al. 2009; Chirichella et al. 2013; Centore et al. 2018). However, to our knowledge, no study has evidenced so far a negative impact of mouflon on body mass, growth, reproduction, or survival of other ungulate species. By contrast, competition with livestock, and particularly with domestic sheep, is a key concern for mouflon (see section "Management").

Effects of Climate Change

Thermal conditions during summer have been identified as a major driver of spatial and temporal behaviors in mouflon populations inhabiting Mediterranean areas (Pipia et al. 2008; Bourgoin et al. 2011; Marchand et al. 2015b; see subsection "Habitat Selection and Activity Pattern"). However, behavioral adjustments identified did not prevent females from losing up to 15% of activity time over a 24-h period during hot summer days (Bourgoin et al. 2011). In addition, as daily temperature increases, ambient temperatures experienced by mouflon never stopped increasing despite shift in habitat selection (Marchand et al. 2015b). This suggested that this shift might be insufficient to maintain ambient temperatures around the threshold controlling thermoregulation (~15–17 °C). In line with these findings, fitness-related traits, such as body mass, and fitness components, such as lamb survival and female productivity, have been found to be impacted by summer heatwaves and droughts (Garel et al. 2004) or to be correlated with temperature-related parameters (e.g., latitude; Ciuti et al. 2009). As an example, a high lamb mortality was recorded during the 4-month period around the severe 2003 drought in the Caroux-Espinouse population (Garel et al. 2004). Adults harvested during the hunting season following this severe drought were on average 3.4 ± 0.8 kg lighter than those harvested in the three previous hunting seasons, representing a >11% and >17% mass loss in adult males and females, respectively. In Cyprus, the higher mortality in fall/early winter has also been partly associated with the low resources quality and quantity experienced by animals some months earlier during the long hot and dry Cypriot summers (Hadjisterkotis 2002). These first insights raise concerns on the long-term consequences of global warming in Mediterranean areas where mouflon populations will face in a near future more intense and more frequent heatwaves (Paeth and Hense 2005).

Conservation Status

The conservation status of mouflon is summarized in Table 2.

Management

This species remains a paradox for conservation and management (Garzón-Machado et al. 2012). On the one hand, several native populations (Özüt 2009; Hajian et al. 2011) and mouflon on Mediterranean islands are of conservation concern (Cassola 1985; Shackleton and IUCN/SSC Caprinae Specialist Group 1997; Hadjisterkotis and Lovari 2016; Portanier et al. 2022) and often benefit locally of a protection status (e.g., for Cyprus: "The Protection and Development of Game and Wild Birds Law of 1974 (39/1974), article 9," for Corsica: "Ministerial order of 1st March 2019, NOR: TREL1824291A"; see also Apollonio et al. 2010). Though not always assessed, several threats have been identified in these areas: poaching, hunting (nontarget species) and associated disturbance, habitat loss (e.g., forest encroachment), hybridization with sheep and more generally loss of genetic diversity related to population isolation, competition and sharing of pathogens with domestic herds, and development of recreational activities in natural areas (Hadjisterkotis 2001; Rieu 2007; Khorozyan et al. 2009; Talibov et al. 2009; Bleyhl et al. 2018; Satta et al. 2021; Brivio et al. 2022). In its Caucasian original range, competition with herds of several hundreds of domestic sheep, usually accompanied by several shepherds and 1–4 dogs/100 sheep, is a key issue since it might have pushed mouflon into marginal habitats (Khorozyan et al. 2009; Talibov et al. 2009; Bleyhl et al. 2018; Brivio et al. 2022).

These numerous threats justified important efforts to improve the conservation status in its

Table 2 Summary of mouflon status in international agreements for conservation of animal and plant species

List	Taxon	Status	Date	Notes
IUCN red list	*O. gmelini*	Near Threatened A2cde ver 3.1 (global)	2020-03-18	Current population trend unknown. Four subspecies included: Anatolian and Armenian mouflon lumped into *O. g. gmelini*, Isfahan mouflon *O. g. isphahanica*, Laristan mouflon *O. g. laristanica* and Cyprus mouflon *O. g. ophion*
Bern Convention	*O. aries* (*musimon*, *ammon*)	Appendix 3[a]	1979	
CITES	*O. gmelini*	Appendix 1[b]	2022-06-22	Only the population of Cyprus. No other population is included in the Appendices
Habitat directive	*O. gmelini musimon* (or *O. ammon musimon*) *O. orientalis ophion* (*or O. gmelini ophion*)	Annex 2[c] Annex 4[d]	1992	Natural populations in Corsica and Sardinia

[a]Protected Fauna species
[b]Species threatened with extinction. Trade in specimens of these species is permitted only in exceptional circumstances (e.g., scientific research)
[c]Animal and plant species of community interest whose conservation requires the designation of special areas of conservation
[d]Animal and plant species of community interest in need of strict protection

native range and in Mediterranean islands during the two last decades. As examples, between 2004 and 2007, two reintroductions of Anatolian mouflon occurred in central Turkey (Karaman-Karadag and Ankara-Sariyar) using founder individuals from the Konya-Bozdag area (Özüt 2009). In Cyprus and Corsica, some individuals have been enclosed and reproduce in captivity for reintroduction (Hadjisterkotis and Bider 1993; Rieu 2007). However, the breeding program for the Cyprus mouflon was considered a failure due to the introduction of inbred animals with reduced viability (Hadjisterkotis and Lambrou 2001). Mouflon are also one of the first wild endangered species for which cloning and assisted reproductive technologies have been considered as conservation measures (Loi et al. 2001; Hosseini et al. 2009; Hajian et al. 2011). Managing habitats is another simpler lever for mouflon conservation (Khorozyan et al. 2009; Talibov et al. 2009; Bleyhl et al. 2018). As an example, increasing grass availability, e.g., by using prescribed burning or cutting when woody plants encroach (Cazau et al. 2011), or by limiting competition with domestic herds (Bleyhl et al. 2018; Brivio et al. 2022), may help increasing forage abundance for mouflon facing habitat loss. Likewise, food supply is provided and seems important for successful overwintering in some populations where snow cover persists during winter (Nahlik 2001; Heroldova et al. 2007).

On the other hand, mouflon has been introduced with success as a game species in numerous countries (see section "Current Distribution"), allowing the development of thriving businesses based on trophy hunting (Shackleton and IUCN/SSC Caprinae Specialist Group 1997; Hofer 2002). In some populations, this selective hunting may have had detrimental effects on individual phenotypic characteristics (Garel et al. 2007). Introduced mouflon have also sometimes raised management concerns when negative impacts on ecosystems are reported (Heroldova and Homolka 2001; Michaelidou and Decker 2002). Severe impacts on commercial forests or endemic plants have been recorded when mouflon have been introduced in forested areas and sites of high economical or environmental values (e.g., Hawaii, Kerguelen, and Canary islands; Rodríguez et al. 1988; Rodríguez Luengo and Piñero 1991; Homolka and Heroldova 1992; Chapuis et al. 2001; Hess and Jacobi 2011). More generally, when grasses become scarce, mouflon can locally/seasonally cause damages on trees, artificial pastures, vineyards, orchards, and/or crop fields (e.g., in Cyprus see Hadjisterkotis and Vakanas 1997). As a result, mouflon have been classified as invasive in Hawaii and Canary islands and have been eradicated from Kerguelen archipelago during the 2010s (Giffin 1979; Garzón-Machado et al. 2012; Terres Australes et Antarctiques Françaises 2013).

Both native and introduced populations of mouflon may also play an important role in food webs and in the maintenance and/or recovery of carnivores and raptors (Poulle et al. 1997; Cugnasse and Golliot 2000; Andersen et al. 2006; Herzog 2018). Mouflon newborns and juveniles are indeed preys for the golden eagle *Aquila chrysaetos* and for the red fox *Vulpes vulpes,* but the impact of these predators on populations may be limited and may not concern adults (Hadjisterkotis 1996a; Cugnasse and Golliot 2000). More direct killings were reported by stray dogs than foxes, although foxes are also one of the major scavengers of dead mouflon (Constantinou and Hadjisterkotis 2016; Nasiadka et al. 2021). By contrast, predators such as lynx *Lynx lxnx* and wolf *Canis lupus* can play a significant role on population size and persistence. In several areas where one of these predators recovered during the last decades, marked drops in mouflon populations were recorded (Poulle et al. 1997; Andersen et al. 2006; Möckel 2017; Herzog 2018). Mouflon seem particularly sensitive to predation by large carnivores during winter as movements are impeded by snow and as group size is at its highest in restricted winter ranges.

Future Challenges for Research and Management

Further research is needed to inform managers on the extent to which populations of mouflon can adapt to the fast ongoing global changes, in particular in its native range (where knowledge is critically lacking) and more generally where populations are of conservation concerns.

Mouflon seem particularly sensitive to heat stress and droughts, with both behavioral and demographic side effects reported (Garel et al. 2004; Bourgoin et al. 2009; Marchand et al. 2015b) and a strong dependence on water availability during the hottest period in the most arid/ Mediterranean areas (Hadjisterkotis 2001). With rising temperatures and more frequent/intense droughts predicted (Paeth and Hense 2005), and numerous expected consequences on mouflon habitats, better understanding how the behavioral ecology, dynamics, and distribution of mouflon populations can be affected by global warming is an important research avenue.

Likewise, habitat loss and interactions with domestic livestock (competition, sharing of pathogens) have already been identified as major threats for numerous populations of mouflon, in particular in its native range (Hadjisterkotis 2001; Khorozyan et al. 2009; Talibov et al. 2009; Bleyhl et al. 2018; Brivio et al. 2022). Habitat fragmentation and loss of connectivity imposed by the development of human activities and associated infrastructures, the cessation of ancestral migration routes, and disturbance imposed by the development of recreational activities in natural areas may constitute other challenges for most populations of mouflon that are still to be evaluated (Marchand et al. 2014a, 2017; Bleyhl et al. 2018). As the size of domestic herds generally increases in mountainous areas (García-Martínez et al. 2009), so do the populations of other wild large herbivores that coexist with mouflon (in Europe; Apollonio et al. 2010); it is likely that questions of competition between wild and domestic species, of the impact of introduced mouflon on ecosystems, and of the sanitary issues resulting from increasing interspecific contacts will arise even more in many countries (Khorozyan et al. 2009; Talibov et al. 2009; Bleyhl et al. 2018).

Acknowledgments The authors thank Giuseppe Puddu, the Observatoire de la Grande Faune et de ses Habitats (OGFH) and Simone Ciuti for providing some of data presented in this chapter. They also warmly thank all the professionals from the Office National de la Chasse et de la Faune Sauvage (now Office Français de la Biodiversité; Thibaut Amblard, Pierre Benedetti, Thierry Chevrier, François Couilloud, Dominique Dubray, Jeanne Duhayer, Philippe Gibert, Christian Itty, Jean-Michel Jullien, Daniel Maillard, Eric Marty, Frédéric Sanchis and SD 34) and the numerous trainees and partners (GIEC du Caroux-Espinouse, GIC des Bauges, ONF, PNR des Bauges et de Corse, OEC, DREAL Corse, LDAV73, LDA2B) involved in the long-term studies performed on the Caroux-Espinouse, Bauges, and Corsican study sites. Thanks to Sebastian Piegert for translating from German to English and Luca Corlatti, Simone Ciuti, and Nicolas Morellet for the useful comments that they provided. We would like also to thank the Minister of Agriculture, Rural Development and Environment of the Republic of Cyprus (Dr. Kostas Kadis), the Senior Officer of the Ministry of Agriculture Rural Development and Environment (Dr. Marina Michaelidou Kadi), and the Director of the Agricultural Research Institute (Dr. Dora Chimonidou) for their support. JSM is supported by grant AGL2017-85753-R (AEI/FEDER, UE).

References

Allain D, Thebault R, Rougeot J, Martinet L (1994) Biology of fibre growth in mammals producing fine fibre and fur in relation to control by day length: relationship with other seasonal functions. Eur Fine Fibre Netw 2: 23–39

Almería S, Cabezón O, Paniagua J, Cano-Terriza D, Jiménez-Ruiz S, Arenas-Montes A, Dubey J, García-Bocanegra I (2018) *Toxoplasma gondii* in sympatric domestic and wild ungulates in the Mediterranean ecosystem. Parasitol Res 117:665–671

Andersen R, Linnell JDC, Solberg EJ (2006) The future role of large carnivores in terrestrial trophic interactions: the northern temperate view. In: Danell K, Bergstrom R, Duncan P, Pastor J (eds) Large herbivore ecology, ecosystem dynamics and conservation, Conservation biology. Cambridge University Press, Cambridge, UK, pp 413–448

Apollonio M, Andersen R, Putman R (2010) European ungulates and their management in the 21st century. Cambridge University Press, Cambridge, UK

Aubert D, Ajzenberg D, Richomme C, Gilot-Fromont E, Terrier ME, de Gevigney C, Game Y, Maillard D, Gibert P, Dardé ML, Villena I (2010) Molecular and biological characteristics of *Toxoplasma gondii* isolates from wildlife in France. Vet Parasitol 171:346–349

Bagó Z, Bauder B, Kolodziejek J, Nowotny N, Weissenbock H (2002) Tickborne encephalitis in a mouflon (*Ovis ammon musimon*). Vet Rec 150:218–220

Barbanera F, Guerrini M, Beccani C, Forcina G, Anayiotos P, Panayides P (2012) Conservation of endemic and threatened wildlife: molecular forensic DNA against poaching of the Cypriot mouflon (*Ovis *orientalis ophion,* Bovidae). Forensic Sci Int Genet 6: 671–675

Barbato M, Hailer F, Orozco-terWengel P, Kijas J, Mereu P, Cabras P, Mazza R, Pirastru M, Bruford MW (2017) Genomic signatures of adaptive introgression from European mouflon into domestic sheep. Sci Rep 7:7623

Bartova E, Sedlak K (2012) Toxoplasmosis in animals in the Czech Republic – the last 10 years. In: Toxoplasmosis – recent advances. IntechOpen, London

Baskin L, Danell K (2003) Mouflon – *Ovis aries*. Springer, Berlin/Heidelberg, pp 277–283

Belloy L, Giacometti M, Boujon P, Waldvogel A (2007) Detection of *Dichelobacter nodosus* in wild ungulates (*Capra ibex ibex* and *Ovis aries musimon*) and domestic sheep suffering from foot rot using a two-step polymerase chain reaction. J Wildl Dis 43:82–88

Benedetti P, Sanchis F, Garel M, Chalbos M (2019) Programme de recherche et de developpement du mouflon de Corse-synthese et résultats – 2011/2015. ONCFS, Paris

Bennett G, Hickford J, Sedcole R, Zhou H (2009) *Dichelobacter nodosus*, *Fusobacterium necrophorum* and the epidemiology of footrot. Anaerobe 15:173–176

Benoist S, Garel M, Cugnasse JM, Blanchard P (2013) Human disturbances, habitat characteristics and social environment generate sex-specific responses in vigilance of Mediterranean mouflon. PLoS One 8: e82960

Berlinguer F, Leoni GG, Bogliolo L, Bebbere D, Succu S, Rosati I, Ledda S, Naitana S (2005) In vivo and in vitro fertilizing capacity of cryopreserved European mouflon [*Ovis gmelini musimon*] spermatozoa used to restore genetically rare and isolated populations. Theriogenology 63:902–911

Bertolino S, Di Montezemolo N, Bassano B (2009) Food-niche relationships within a guild of alpine ungulates including an introduced species. J Zool 277:63–69. ISSN 1469-7998

Bleyhl B, Arakelyan M, Askerov E, Bluhm H, Gavashelishvili A, Ghasabian M, Ghoddousi A, Heidelberg A, Khorozyan I, Malkhasyan A et al (2018) Assessing niche overlap between domestic and threatened wild sheep to identify conservation priority areas. Divers Distrib 25:129–141

Bon R, Gonzales G, Im S, Badia J (1990) Seasonal grouping in female mouflons in relation to food availability. Ethology 86:224–236

Bon R, Gonzalez G, Bosch M, Cugnasse JM (1992) Ram rut-involvement in a hunted population of mouflons. Acta Theriol 37:63–71

Bon R, Dardaillon M, Estevez I (1993a) Mating and lambing periods as related to age of female mouflon. J Mammal 74:752–757

Bon R, Dubois M, Maublanc ML (1993b) Does age influence between-rams companionship in mouflon (*Ovis gmelini*). Rev Ecol (Terre Vie) 48:57–64

Bon R, Recarte JM, Gonzales G, Cugnasse JM (1995) Courtship and behavioral maturation of male mouflons. Acta Theriol 40:283–294

Bonenfant C, Gaillard JM, Coulson TH, Festa-Bianchet M, Loison A, Garel M, Loe LE, Blanchard P, Pettorelli N, Owen-Smith N, Du Toit J, Duncan P (2009) Empirical evidences of density-dependence in populations of large herbivores. Adv Ecol Res 41:313–357

Bornstein S, Morner T, Samuel WM (2001) *Sarcoptes scabiei* and sarcoptic mange. In: Parasitic diseases of wild mammals, vol 2. Iowa State University Press, Ames, pp 107–119

Bourgoin G, Garel M, Van Moorter B, Dubray D, Maillard D, Marty E, Gaillard JM (2008) Determinants of seasonal variation in activity patterns of mouflon. Can J Zool 86:1410–1418

Bourgoin G, Garel M, Dubray D, Maillard D, Gaillard JM (2009) What determines global positioning system fix success when monitoring free-ranging mouflon? Eur J Wildl Res 55:603–613

Bourgoin G, Garel M, Blanchard P, Dubray D, Maillard D, Gaillard JM (2011) Daily responses of mouflon (*Ovis gmelini musimon* × *Ovis* sp.) activity to summer climatic conditions. Can J Zool 89:765–773

Bourgoin G, Garel M, Mercier A, Massot R, Poirel MT, Itty C, Duhayer J (2014) Increased infestation of Mediterranean mouflon (*Ovis gmelini musimon* × *Ovis* sp.) by ticks in France. In: International congress on parasites of wildlife, Kruger National Park, 14–18 Sept

Bourgoin G, Portanier E, Poirel MT, Itty C, Duhayer J, Benabed S, Cockenpot A, Callait-Cardinal MP, Garel M (2021) Reproductive females and young mouflon (*Ovis gmelini musimon* × *Ovis* sp.) in poor body condition are the main spreaders of gastrointestinal parasites. Parasitology 148:809–818

Bourgoin G, Marchand P, Hewison AJM, Ruckstuhl KE, Garel M (2018) Social behaviour as a predominant driver of sexual, age-dependent and reproductive segregation in Mediterranean mouflon. Anim Behav 136:87–100

Boussès P, Réale D (1994) Valeur du masque facial comme critère d'âge des femelles de mouflons de Corse (*Ovis musimon*) dans la population de Kerguelen. Mammalia 58:301–304

Boussès P, Réale D (1998) Biology of twinning and origin of an unusually high twinning rate in an insular mouflon population. Z Saugetierkd 63:147–153

Briedermann L (1989) Beobachtungen zum Geburtsverhalten und zur Mutterfamilienbildung beim Mufflon (*Ovis ammon musimon* Schreber, 1782). Beitr Jagd- Wildforsch 16:117–126

Briedermann L (1992) Ergebnisse von Untersuchungen zur Reproduktion des Mufflons (*Ovis ammon musimon*). Z Jagdwiss 38:16–25

Brivio F, Ciuti S, Pipia A, Grignolio S, Apollonio M (2022) Livestock displace European mouflon from optimal foraging sites. Eur J Wildl Res 68:30

Bruford MW, Townsend SJ (2006) Mitochondrial DNA diversity in modern sheep. In: Documenting domestication: new genetic and archaeological paradigms. University of California Press, Berkeley, pp 306–316

Bunch T, N'guyen T, Lauvergne J (1978) Hemoglobins of the Corsico-Sardinian mouflon (*Ovis musimon*) and their implications for the origin of HbA in domestic sheep (*Ovis aries*). Ann Genet Sel Anim 10:503

Bunch TD, Wu C, Zhang YP, Wang S (2006) Phylogenetic analysis of snow sheep (*Ovis nivicola*) and closely related taxa. J Hered 97:21–30

Bunnell FL (1982) The lambing period of mountain sheep: synthesis, hypotheses, and tests. Can J Zool 60:1–14

Burke JM, Arnold ML (2001) Genetics and the fitness of hybrids. Annu Rev Genet 35:31–52

Byers JA (1997) American pronghorn: social adaptations and the ghosts of predators past. The University of Chicago Press, Chicago

Cam E, Link WA, Cooch EG, Monnat JY, Danchin E (2002) Individuals covariation in life-history traits: seeing the trees despite the forest. Am Nat 159:96–105

Candela M, Caballol A, Atance P (2017) Wide exposure to *Coxiella burnetii* in ruminant and feline species living in a natural environment: zoonoses in a human–livestock–wildlife interface. Epidemiol Infect 145:478–481

Cassinello J (2018) Detailed coverage of invasive species threatening livelihoods and the environment worldwide

Cassola F (1985) Management and conservation of the Sardinian moufflon (*Ovis musimon* Schreber): an outline. In: Lovari S (ed) The biology and management of mountain ungulates. Croom Helm, London, pp 197–203

Catchpole EA, Morgan BJT, Coulson TN, Freeman SN, Albon SD (2000) Factors influencing Soay sheep survival. Appl Stat 49:453–472

Cazau M, Garel M, Maillard D (2011) Responses of heather moorland and Mediterranean mouflon foraging to prescribed-burning and cutting. J Wildl Manag 75: 967–972

Centore L, Ugarković D, Scaravelli D, Safner T, Pandurić K, Sprem N (2018) Locomotor activity pattern of two recently introduced non-native ungulate species in a Mediterranean habitat. Folia Zool 67:6–13

Cerutti M, Citterio C, Bazzocchi C, Epis S, D'amelio S, Ferrari N, Lanfranchi P (2010) Genetic variability of *Haemonchus contortus* (Nematoda: Trichostrongyloidea) in alpine ruminant host species. J Helminthol 84:276–283

Chapuis JL, Boussès P, Pisanu B, Réale D (2001) Comparative rumen and fecal diet microhistological determinations of European mouflon. J Range Manag 54: 239–242

Chessa B, Pereira F, Arnaud F, Amorim A, Goyache F, Mainland I, Kao RR, Pemberton JM, Beraldi D, Stear MJ, Alberti A, Pittau M, Iannuzzi L, Banabazi MH, Kazwala RR, Zhang YP, Arranz JJ, Ali BA, Wang Z, Uzun M, Dione MM, Olsaker I, Holm LE, Saarma U, Ahmad S, Marzanov N, Eythors-dottir E, Holland MJ, Ajmone-Marsan P, Bruford MW, Kantanen J, Spencer TE, Palmarini M (2009) Revealing the history of sheep domestication using retrovirus integrations. Science 324:532–536

Chirichella R, Ciuti S, Apollonio M (2013) Effects of livestock and non-native mouflon on use of high-elevation pastures by Alpine chamois. Mamm Biol 78:344–350

Ciani E, Crepaldi P, Nicoloso L, Lasagna E, Sarti FM, Moioli B, Napolitano F, Carta A, Usai G, D'Andrea M et al (2014) Genome-wide analysis of Italian sheep diversity reveals a strong geographic pattern and cryptic relationships between breeds. Anim Genet 45: 256–266

Ciani E, Mastrangelo S, Da Silva A, Marroni F, Ferenčaković M, Ajmone-Marsan P, Baird H, Barbato M, Colli L, Delvento C, Dovenski T, Gorjanc G, Hall S, Hoda A, Li MH, Marković B, McEwan J, Moradi MH, Ruiz-Larrañaga O, Ruži-ć-Muslić R, Šalamon D, Simćić M, Stepanek O, Consortium E, Consortium S, Curik I, Cubric-Curik V, Lenstra JA (2020) On the origin of European sheep as revealed by the diversity of the Balkan breeds and by optimizing population-genetic analysis tools. Genet Sel Evol 52:1–14

Ciucci P, Boitani L, Ricci S (1998) Social patterns of mouflon (*Ovis gmelini*) in the northern Apennines (Italy). Mammalia 62:442–446

Ciuti S, Pipia A, Ghiandai F, Grignolio S, Apollonio M (2008) The key role of lamb presence in affecting flight response in Sardinian mouflon (*Ovis orientalis musimon*). Behav Process 77:408–412

Ciuti S, Pipia A, Grignolio S, Ghiandai F, Apollonio M (2009) Space use, habitat selection and activity patterns of female Sardinian mouflon (*Ovis orientalis musimon*) during the lambing season. Eur J Wildl Res 55:589–595

Clauss M, Fritz J, Bayer D, Hummel J, Streich WJ, Südekum KH, Hatt JM (2009) Physical characteristics of rumen contents in two small ruminants of different feeding type, the mouflon (*Ovis ammon musimon*) and the roe deer (*Capreolus capreolus*). Zoology 112: 195–205

Clauss M, Hume I, Hummel J (2010) Evolutionary adaptations of ruminants and their potential relevance for modern production systems. Animal 4:979–992

Clutton-Brock TH (1984) Reproductive effort and terminal investment in iteroparous animals. Am Nat 123: 212–229

Clutton-Brock TH, Guinness FE, Albon SD (1982) Red deer. Behaviour and ecology of two sexes. Edinburgh University Press, Edinburgh

Clutton-Brock T, Price O, Albon S, Jewell P (1991) Persistent instability and population regulation in Soay sheep. J Anim Ecol 60:593–608

Clutton-Brock TH, Stevenson IR, Marrow P, MacColl AD, Houston AI, McNamara JM (1996) Population fluctuations, reproductive costs and life-history tactics in female Soay sheep. J Anim Ecol 65:675–689

Cockenpot A (2013) Etude des facteurs de variation de l'excrétion parasitaire mesurée par analyse coproscopique chez le mouflon Méditerranéen (*Ovis gmelini musimon* × *Ovis* sp.) dans le massif du Caroux-Espinouse. Veterinary thesis, VetAgro Sup, Université de Lyon, Lyon, France, 152p

Coelho C, Vinhas B, Pires I, Gomes J, Vieira-Pinto M (2014) Detection of *Sarcocystis* spp. in large game from Portugal by histological examination. In: Trends in game meat hygiene: from forest to fork. Wageningen Academic Publishers, Wageningen, pp 15–18

Coltman DW, Festa-Bianchet M, Jorgenson JT, Strobeck C (2002) Age-dependent sexual selection in bighorn rams. Proc R Soc Lond B Biol Sci 269:165–172

Colwell DD (2001) Bot flies and warble flies (order Diptera: family Oestridae). In: Parasitic diseases of wild mammals, vol 2. Iowa State University Press, Ames, pp 46–71

Constantinou G, Hadjisterkotis E (2016) Fox predation on Cyprian mouflon and marine turtles on Cyprus. In: Hadjisterkotis E (ed) Book of abstracts, 3rd edition. 6th World conference on mountain ungulates and 5th international symposium on mouflon. Ministry of the Interior, Nicosia, pp 80–81

Corti R, Réseau Ongulés Sauvages – ONCFS, Fédération Nationale des Chasseurs (2011) Inventaire des populations françaises d'ongulés de montagne – Mise à jour 2011. ONCFS, Paris

Coulson TN, Catchpole EA, Albon SD, Morgan BJT, Pemberton JM, Clutton-Brock TH, Crawley MJ, Grenfell BT (2001) Age, sex, density, winter weather, and population crashes in Soay sheep. Science 292: 1528–1531

Cransac N, Hewison AJM (1997) Seasonal use and selection of habitat by mouflon (*Ovis gmelini*): comparison of the sexes. Behav Process 41:57–67

Cransac N, Hewison AJM, Gaillard JM, Cugnasse JM, Maublanc ML (1997) Patterns of mouflon (*Ovis gmelini*) survival under moderate environmental conditions: effects of sex, age, and epizootics. Can J Zool 75:1867–1875

Cransac N, Gérard JF, Maublanc ML, Pépin D (1998) An example of segregation between age and sex classes only weakly related to habitat use in moufon sheep (*Ovis gmelini*). J Zool 244:371–378

Cugnasse JM (1994) Révision taxinomique des mouflons des îles méditerranéennes. Mammalia 58:507–512

Cugnasse JM (1997) L'enzootie de kérato-conjonctivite chez le mouflon méditerranéen (*Ovis gmelini musimon* × *Ovis* sp.) dans le massif du Caroux-Espinouse (Hérault) à l'automne 1993. Gibier Faune Sauvage 14:569–584

Cugnasse JM, Golliot E (2000) L'aigle royal *Aquila chrysaëtos* attaque des mouflons dans l'Espinouse (Hérault). Ornithos 7:188–190

Cugnasse JM, Garcia M, Veyrac T (1985) Contribution à l'étude de la biologie de la reproduction du mouflon (*Ovis ammon musimon*), par examen post-mortem, dans le massif du Caroux-Espinouse. Bull Off Nat Chasse 89:33–35

Darmon G, Calenge C, Loison A, Maillard D, Jullien JM (2007) Social and spatial patterns determine the population structure and colonization processes in mouflon. Can J Zool 85:634–643

Darmon G, Calenge C, Loison A, Jullien JM, Maillard D, López JF (2012) Spatial distribution and habitat selection in coexisting species of mountain ungulates. Ecography 35:44–53. ISSN 1600-0587

Darmon G, Bourgoin G, Marchand P, Garel M, Dubray D, Jullien JM, Loison A (2014) Do ecologically close species shift their daily activities when in sympatry? A test on chamois in the presence of mouflon. Biol J Linn Soc 111:621–626

Demirci S, Baştanlar EK, Dağtaş ND, Pişkin E, Engin A, Özer F, Yüncü E, Doğan ŞA, Togan İ (2013) Mitochondrial DNA diversity of modern, ancient and wild sheep (*Ovis gmelinii anatolica*) from Turkey: new insights on the evolutionary history of sheep. PLoS One 8:e81952

Dubois M, Gérard JF, Maublanc ML (1992) Seasonal movements of female Corsican mouflon (*Ovis ammon*) in a Mediterranean mountain range, southern France. Behav Process 26:155–166

Dubois M, Quenette P, Bideau E, Magnac M (1993) Seasonal range use by European mouflon rams in medium altitude mountains. Acta Theriol 38:185–198

Dubois M, Bon R, Cransac N, Maublanc ML (1994) Dispersal patterns of Corsican mouflon ewes: importance of age and proximate influences. Appl Anim Behav Sci 42:29–40

Dubois M, Khazraie K, Guilhem C, Maublanc ML, Le Pendu Y (1996) Philopatry in mouflon rams during the rutting season: psycho-ethological determinism and functional consequences. Behav Process 35: 93–100

Dupuis J, Badia J, Maublanc ML, Bon R (2002) Survival and spatial fidelity of mouflon (*Ovis gmelini*): a Bayesian analysis of an age-dependent capture–recapture model. J Agric Biol Environ Stat 7:277–298

Epps CW, Wehausen JD, Bleich VC, Torres SG, Brashares JS (2007) Optimizing dispersal and corridor models using landscape genetics. J Appl Ecol 44:714–724

Espuno N (2004) Impact du Loup (*Canis lupus*) sur les ongulés sauvages et domestiques dans le massif du Mercantour. Ph.D. thesis, University Montpellier II, Sciences et Techniques du Languedoc, Montpellier

Ferrer D, Castellà J, Gutiérrez JF, Lavín S, Marco I (1998) Seroprevalence of *Babesia ovis* in mouflon sheep in Spain. J Wildl Dis 34:637–639

Festa-Bianchet M (2000) A summary of discussion on the taxonomy of mountain ungulates and its conservation implications. Workshop on Caprinae taxonomy, Ankara

Festa-Bianchet M, Jorgenson JT (1998) Selfish mothers: reproductive expenditure and resource availability in bighorn ewes. Behav Ecol 9:144–150

Festa-Bianchet M, Douhard M, Gaillard JM, Pelletier F (2017) Successes and challenges of long-term field studies of marked ungulates. J Mammal 98:612–620

Forchhammer MC, Clutton-Brock TH, Lindström J, Albon SD (2001) Climate and population density induce long-term cohort variation in a northern ungulate. J Anim Ecol 70:721–729

Frisina MR, Frisina RM (2000) Phenotype evaluation of free-ranging European mouflon (*Ovis orientalis musimon*) on Kahuku ranch, south point of Hawaii. In: Proceedings of the eleventh biennal symposium, Northern Wild Sheep and Goat Council, pp 12–17

Gaillard JM, Yoccoz N (2003) Temporal variation in survival of mammals: a case of environmental canalization? Ecology 84:3294–3306
Gaillard JM, Pontier D, Allainé D, Loison A, Hervé JC, Heizmann A (1997) Variation in growth form and precocity at birth in eutherian mammals. Proc R Soc Lond B Biol Sci 264:859–868
Gaillard JM, Festa-Bianchet M, Yoccoz NG, Loison A, Toïgo C (2000) Temporal variation in fitness components and population dynamics of large herbivores. Annu Rev Ecol Syst 31:367–393
García-Martínez A, Olaizola A, Bernués A (2009) Trajectories of evolution and drivers of change in European mountain cattle farming systems. Animal 3:152–165
Garel M, Loison A, Gaillard JM, Cugnasse JM, Maillard D (2004) The effects of a severe drought on mouflon lamb survival. Proc R Soc Lond B Biol Sci 271(Suppl): S471–S473. IF: 3.653
Garel M, Cugnasse JM, Gaillard JM, Loison A, Gibert P, Douvre P, Dubray D (2005a) Reproductive output of female mouflon (*Ovis gmelini musimon* × *Ovis* sp.): a comparative analysis. J Zool 266:65–71
Garel M, Cugnasse JM, Gaillard JM, Loison A, Santosa Y, Maublanc ML (2005b) Effect of observer experience on the monitoring of a mouflon population. Acta Theriol 50:109–114. IF: 0.459
Garel M, Cugnasse JM, Loison A, Gaillard JM, Vuiton C, Maillard D (2005c) Monitoring the abundance of mouflon in South France. Eur J Wildl Res 51:69–76
Garel M, Cugnasse JM, Maillard D (2005d) Reliability of mouflon aging using morphological characteristics from observations in the field. Wildl Biol 11:229235. IF: 0.535
Garel M, Cugnasse JM, Hewison AJM, Maillard D (2006a) Errors in age determination of mouflon in the field. Wildl Soc Bull 34:300–306. IF: 0.714
Garel M, Toïgo C, J.-M. Cugnasse, Maillard D, Gaillard JM, Hewison A, Dubray D (2006b) Long-term life history changes in a mouflon population: the role of selective harvesting. IVth world congress on mountain ungulates, Munnar
Garel M, Cugnasse JM, Maillard D, Gaillard JM, Hewison AJM, Dubray D (2007) Selective harvesting and habitat loss produce long-term life history changes in a mouflon population. Ecol Appl 17:1607–1618
Garzón-Machado V, del Arco-Aguilar MJ, Pérez-de Paz P (2012) Threat or threatened species? A paradox in conservation biology. J Nat Conserv 20:228–230
Gauss C, Dubey J, Vidal D, Cabezón O, Ruiz-Fons F, Vicente J, Marco I, Lavin S, Gortazar C, Almería S (2006) Prevalence of *Toxoplasma gondii* antibodies in red deer (*Cervus elaphus*) and other wild ruminants from Spain. Vet Parasitol 136:193–200
Gauthier D, Gibert P, Cugnasse JM (2001) Parasitic helminthofauna carried by mouflon in France: its signification for host-parasite co-evolution. In: Nahlik A, Uloth W (eds) Proceedings of the third international symposium on mouflon, Sopron
Geist V (1966) Validity of horn segment counts in aging Bighorn sheep. J Wildl Manag 30:634–635
Gentry A, Clutton-Brock J, Groves CP (2004) The naming of wild animal species and their domestic derivatives. J Archeol Sci 31:645–651
Giffin JG (1979) Ecology of the mouflon sheep on Mauna Kea. Pittman-Robertson project no. W-17-R, study no. R-III 1975–1979, State of Hawaii – Department of Land and Natural Resources – Division of Forestry and Wildlife
Goldova M, Toth S, Mojzisova J, Ciberej J, Konjevic D, Kocisova A, Slavica A et al (2008) Sarcocystosis in cloven-hoofed game in Slovak Republic. Nat Croat 17:303
Gómez-Brunet A, Santiago-Moreno J, Campo Ad, Malpaux B, Chemineau P, Tortonese DJ, Gonzalez-Bulnes A, López-Sebastián A (2008) Endogenous circannual cycles of ovarian activity and changes in prolactin and melatonin secretion in wild and domestic female sheep maintained under a long-day photoperiod. Biol Reprod 78:552–562
Gonzales G (1986) Compétition interspécifique chez les ongulés sauvages: l'isard et le mouflon dans le massif du Carlit (Pyrénées-Orientales). Bull Off Nat Chasse 100:35–39
Gotteland C, Aubert D, Gibert P, Moinet M, Klein F, Game Y, Villena I, Gilot-Fromont E (2014) Toxoplasmosis in natural populations of ungulates in France: prevalence and spatiotemporal variations. Vector Borne Zoonotic Dis 14:403–413
Grech-Angelini S, Stachurski F, Lancelot R, Boissier J, Allienne JF, Marco S, Maestrini O, Uilenberg G (2016) Ticks (Acari: Ixodidae) infesting cattle and some other domestic and wild hosts on the French Mediterranean island of Corsica. Parasit Vectors 9:582
Groves CP (1989) Feral mammals of the Mediterranean islands: documents of early domestication. In: The walking larder: patterns of domestication, pastoralism, and predation. Routledge, London, pp 46–58
Groves C, Grubb P (2011) Ungulate taxonomy. The Johns Hopkins University Press, Baltimore
Guerrini M, Forcina G, Panayides P, Lorenzini R, Garel M, Anayiotos P, Kassinis N, Barbanera F (2015) Molecular DNA identity of the mouflon of Cyprus (*Ovis orientalis ophion,* Bovidae): near Eastern origin and divergence from Western Mediterranean conspecific populations. Syst Biodivers 13:472–483
Guerrini M, Panayides P, Mofrad NN, Kassinis N, Ioannou I, Forcina G, Hadjigerou P, Barbanera F (2021) Spatial genetic structure and *Ovis* haplogroup as a tool for an adaptive conservation management of the endangered Cyprus mouflon. Zoology 148:125959
Guiguen F, Mselli-Lakhal L, Durand J, Du J, Favier C, Fornazero C, Grezel D, Balleydier S, Hausmann E, Chebloune Y (2000) Experimental infection of mouflon-domestic sheep hybrids with caprine arthritis-encephalitis virus. Am J Vet Res 61:456–461
Hadjisterkotis E (1993) The Cyprus mouflon (*Ovis gmelini ophion*). Management, conservation and evolution. Ph. D. thesis, McGill University
Hadjisterkotis E (1996a) Fox and avian predation on Cyprus mouflon *Ovis gmelini ophion*. In:

Hadjisterkotis E (ed) Abstracts of the 2nd international symposium on Mediterranean mouflon, Government Printing Office, Nicosia, p 33
Hadjisterkotis E (1996b) Herkunft, Taxonomie und neuere Entwicklung des Zyprischen Mufflons (*Ovis gmelini ophion*). Z Jagdwiss 42:104–110
Hadjisterkotis E (1996c) Vertebral pathology of the wild sheep of Cyprus. In: Botev NE (ed) Proceedings of the International Union of Game Biologists XXII, Pensoft Publishers, Sofia/St. Petersburg, 4–8 Sept 1995, pp 424–430
Hadjisterkotis E (1997) The first record of *Echinococcus granulosus* in Cyprus mouflon. In: 2nd world conference on mountain ungulates, Centro Congressi Saint-Vincent, Valle d'Aosta
Hadjisterkotis E (2001) The Cyprus mouflon, a threatened species in a biodiversity "hotspot" area. In: Nahlik A, Uloth W (eds) Proceedings of the third international symposium on mouflon, Sopron
Hadjisterkotis E (2002) Seasonal and monthly distribution of deaths of Cyprus mouflon *Ovis gmelini* ophion. In: García-Gonzales R, Herrero J, Fernandez de Luco D (eds) Third world conference mountain ungulates, Pirineos, Zaragoza, vol 157, pp 81–88
Hadjisterkotis E (2012) The arrival of elephants on the island of Cyprus and their subsequent accumulation in fossil sites. In: Elephants: ecology, behavior and conservation. Nova Science, Hauppauge, pp 49–75
Hadjisterkotis E, Bider JR (1993) Reproduction of Cyprus mouflon *Ovis gmelini ophion* in captivity and in the wild. Int Zoo Yearb 32:125–131
Hadjisterkotis E, Lambrou L (2001) The role of the Zoological Garden of Limassol in wildlife conservation. In: Abstract of the XXVth international congress of the International Union of Game Biologists I.U.G.B. and IXth international symposium Perdix, Lemessos, 3–7 Sept
Hadjisterkotis E, Lovari S (2016) Results and resolutions of the 6th world congress on mountain ungulates and the 5th international symposium on mouflon. In: Hadjisterkotis E (ed) Book of abstracts, 3rd edn. 6th World congress on mountain ungulates and 5th international symposium on mouflon, Nicosia, 28 Aug–1 Sept, pp 20–23
Hadjisterkotis E, Vakanas C (1997) Agricultural damages caused by mouflon *Ovis gmelini* ophion in Cyprus. In: Proceedings of the second international symposium on Mediterranean mouflon, Game Fund, Nicosia, pp 111–117
Hadjisterkotis E, Mereu P, Masala B (2016) A review of the nomenclatural spelling variation of the Armenian mouflon (*Ovis gmelini gmelini*) and the Cyprian mouflon (*O. g. ophion*). In: Hadjisterkotis E (ed) Book of abstracts, 3rd edn. 6th World congress on mountain ungulates and 5th international symposium on mouflon, 28 Aug–1 Sept, Nicosia, pp 48–50
Hajian M, Hosseini S, Forouzanfar M, Abedi P, Ostadhosseini S, Hosseini L, Moulavi F, Gourabi H, Shahverdi A, Dizaj AVT et al (2011) "Conservation cloning" of vulnerable Esfahan mouflon (*Ovis orientalis isphahanica*): in vitro and in vivo studies. Eur J Wildl Res 57:959–969
Hamel S, Gaillard JM, Yoccoz N, Albon S, Côté S, Craine J, Festa-Bianchet M, Garel M, Lee P, Moss C, Nussey DH, Pelletier D, Stien A, Tveraa T (2016) Cohort variation in individual body mass dissipates with age in large herbivores. Ecol Monogr 86:517–543
Hamilton WD (1966) The moulding of senescence by natural selection. J Theor Biol 12:12–45
Heddergott M, Osten-Sacken N, Steinbach P, Frantz AC (2018) Seroprevalence of *Toxoplasma gondii* in free-living European mouflon (*Ovis orientalis musimon*) hunted in central Germany. Parasite 25:21
Hedrick P (2011) Genetics of populations. Jones & Bartlett Learning, Sudbury
Hemming JE (1969) Cemental deposition, tooth succession, and horn development as criteria of age in Dall sheep. J Wildl Manag 33:552–558
Heroldova M (1996) Dietary overlap of three ungulate species in the Palava Biosphere Reserve. For Ecol Manag 88:139–142
Heroldova M, Homolka M (2001) The introduction of mouflon into forest habitats: a desirable increasing of biodiversity? In: Nahlik A, Uloth W (eds) Proceedings of the third international symposium on mouflon, Sopron, pp 37–43
Heroldova M, Homolka M, Kamler J, Koubek P, Forejtek P (2007) Foraging strategy of mouflon during the hunting season as related to food supply. Acta Vet Brno 76: 195–202
Herzog S (2018) Return of grey wolf (*Canis lupus*) to Central Europe: challenges and recommendations for future management in cultural landscapes. Ann For Res 61:203–209
Hess S, Jacobi J (2011) The history of mammal eradications in Hawaii and the United States associated islands of the Central Pacific. In: Island invasives: eradication and management. IUCN, Gland, pp 67–73
Hess S, Kawakami B Jr, Okita D, Medeiros K (2006) A preliminary assessment of mouflon abundance at the Kahuku Unit of Hawaii Volcanoes National Park. Geological Survey, Reston
Hewison AJM (1996) Variation in the fecundity of roe deer in Britain: effects of age and body weight. Acta Theriol 41:187–198
Hiendleder S, Kaupe B, Wassmuth R, Janke A (2002) Molecular analysis of wild and domestic sheep questions current nomenclature and provides evidence for domestication from two different subspecies. Proc R Soc Lond B Biol Sci 269:893904
Hille G (2003) Untersuchungen zum Endoparasitenbefall des Muffelwildes in Thüringen. Veterinary thesis, Universität Leipzig, Leipzig, Germany, 196p
Hoefs M (1982) Beitrag zur Morphometrie und Wachstumsdynamik der Schnecke des Muffelwidders (*Ovis ammon musimon,* Schreber 1782). Z Jagdwiss 28:145–162
Hoefs M (2000) The thermoregulatory potential of *Ovis* horn cores. Can J Zool 78:1419–1426
Hoefs, M. & Hadjisterkotis, E. (1998) Horn characteristics of the Cyprus mouflon. In: Peracino V, Lovari S, Bassano B (eds) Proceedings of the 2nd world

conference mountain ungulates, Parco Natl. Gran Paradiso, Saint Vincent, pp 59–67

Hofer D (2002) The Lion's share of the hunt: trophy hunting and conservation: a review of the legal Eurasian tourist hunting market and trophy trade under CITES. TRAFFIC Europe, Brussels

Hofmann RR (1989) Evolutionary steps of ecophysiological adaptation and diversification of ruminants: a comparative view of their digestive system. Oecologia 78: 443–457

Homolka M, Heroldova M (1992) Similarity of the results of stomach and faecal contents analyses in studies of the ungulate diet. Folia Zool 41:193–208

Hornok S, Sugár L, de Mera IGF, de la Fuente J, Horváth G, Kovács T, Micsutka A, Gönczi E, Flaisz B, Takács N et al (2018) Tick-and fly-borne bacteria in ungulates: the prevalence of *Anaplasma phagocytophilum*, haemoplasmas and rickettsiae in water buffalo and deer species in Central Europe, Hungary. BMC Vet Res 14:98

Hosseini S, Fazilati M, Moulavi F, Foruzanfar M, Hajian M, Abedi P, Nasiri N, Kaveh A, Shahverdi A, Hemami M et al (2009) Reproductive potential of domestic *Ovis aries* for preservation of threatened *Ovis orientalis isphahanica*: in vitro and in vivo studies. Eur J Wildl Res 55:239–246

Hubalek Z, Juticová Z, Svobodová Š, Halouzka J (1993) A serologic survey for some bacterial and viral zoonoses in game animals in the Czech Republic. J Wildl Dis 29: 604–607

Iacolina L, Corlatti L, Buzan E, Safner T, Šprem N (2019) Hybridisation in European ungulates: an overview of the current status, causes, and consequences. Mammal Rev 49:45–59

Ioannou I, Sandalakis V, Kassinis N, Chochlakis D, Papadopoulos B, Loukaides E, Tselentis Y, Psaroulaki A (2011) Tick-borne bacteria in mouflons and their ectoparasites in Cyprus. J Wildl Dis 47:300–306

Jaslow CR (1989) Sexual dimorphism of cranial suture complexity in wild sheep (*Ovis orientalis*). Zool J Linnean Soc 95:273–284

Johnson D, Ferrell C, Jenkins T (2003) The history of energetic efficiency research: where have we been and where are we going? J Anim Sci 81:E27–E38

Jorgenson JT, Festa-Bianchet M, Gaillard JM, Wishart WD (1997) Effects of age, sex, disease, and density on survival of bighorn sheep. Ecology 78:1019–1032

Juřicová Z, Halouzka J, Hubalek Z et al (2000) Prevalence of antibodies to *Borrelia burgdorferi* in game animals in South Moravia, Czech Republic. Vet Med 45:145–148

Kaeuffer R, Coltman DW, Chapuis JL, Pontier D, Réale D (2007) Unexpected heterozygosity in an island mouflon population founded by a single pair of individuals. Proc R Soc Lond B Biol Sci 274:527–533

Kaeuffer R, Réale D, Pontier D, Chapuis JL, Coltman DW (2008) Local effects of inbreeding on embryo number and consequences for genetic diversity in Kerguelen mouflon. Biol Lett 4:504–507

Kaeuffer R, Bonenfant C, Chapuis JL, Devillard S (2009) Dynamics of an introduced population of mouflon *Ovis aries* on the sub-Antarctic archipelago of Kerguelen. Ecography 33:435–442

Kauffmann M, Rehbein S, Hamel D, Lutz W, Heddergott M, Pfister K, Silaghi C (2017) *Anaplasma phagocytophilum* and *Babesia* spp. in roe deer (*Capreolus capreolus*), fallow deer (*Dama dama*) and mouflon (*Ovis musimon*) in Germany. Mol Cell Probes 31:46–54

Kaya MA, Bunch TD, Konuk M (2004) On Konya wild sheep, *Ovis orientalis anatolica,* in the Bozdag protected area. Mammalia 68:229–232

Khorozyan I, Weinberg P, Malkhasyan A (2009) Conservation strategy of Armenian mouflon (*Ovis [orientalis] gmelini* Blyth) and bezoar goat (*Capra aegagrus* Erxleben) in Armenia. In: Zazanashvili N, Mallon D (eds) Status and protection of globally threatened species in the Caucasus. CEPF Biodiversity Investments in the Caucasus Hotspot 2004-2009, Tbilisi, pp 37–45

King R, Brooks SP (2003) Survival and spatial fidelity of mouflons: the effect of location, age, and sex. J Agric Biol Environ Stat 8:486–513

Konjevic D, Janicki Z, Severin K, Stanko M, Zivicnjak T, Slavica A, Staresina V (2007) An outbreak of tick paralysis in free-ranging mouflon (*Ovis ammon musimon*). J Zoo Wildl Med 38:585–587

Land R (1978) Reproduction in young sheep: some genetic and environmental sources of variation. J Reprod Fertil 52:427–436

Le Pendu Y, Briedermann L, Gérard JF, Maublanc ML (1995) Inter-individual associations and social structure of a mouflon population (*Ovis orientalis musimon*). Behav Process 34:67–80

Lent PC (1974) Mother–infant relations in ungulates. In: Walther F (ed) The behaviour of ungulates and its relation to management. IUCN, Morges, pp 14–55

Lima SL, Dill LM (1990) Behavioral decisions made under the risk of predation: a review and prospectus. Can J Zool 68:619–640

Lincoln G (1989) Seasonal cycles in testicular activity in mouflon, Soay sheep and domesticated breeds of sheep: breeding seasons modified by domestication. Zool J Linnean Soc 95:137–147

Lincoln G (1990) Correlation with changes in horns and pelage, but not reproduction, of seasonal cycles in the secretion of prolactin in rams of wild, feral and domesticated breeds of sheep. Reproduction 90:285–296

Lincoln GA (1998) Reproductive seasonality and maturation throughout the complete life-cycle in the mouflon ram (*Ovis musimon*). Anim Reprod Sci 53:87–105

Lindström J (1999) Early development and fitness in birds and mammals. Trends Ecol Evol 14:343–348

Linnell JD, Zachos FE (2011) Status and distribution patterns of European ungulates: genetics, population history and conservation. In: Ungulate management in Europe – problems and practices. Cambridge University Press, Cambridge, UK, pp 12–53

Loi P, Ptak G, Barboni B, Fulka J Jr, Cappai P, Clinton M (2001) Genetic rescue of an endangered mammal by cross-species nuclear transfer using post-mortem somatic cells. Nat Biotechnol 19:962

Loison A, Festa-Bianchet M, Gaillard JM, Jorgenson JT, Jullien JM (1999) Age-specific survival in five populations of ungulates: evidence of senescence. Ecology 80:2539–2554

López-Olvera JR, Vidal D, Vicente J, Pérez M, Luján L, Gortázar C (2009) Serological survey of selected infectious diseases in mouflon (*Ovis aries musimon*) from south-central Spain. Eur J Wildl Res 55:75–79

Lorenzini R, Cabras P, Fanelli R, Carboni GL (2011) Wildlife molecular forensics: identification of the Sardinian mouflon using STR profiling and the Bayesian assignment test. Forensic Sci Int Genet 5:345–349

Maerten J (2014) Détermination du parasitisme de l'abomasum chez le mouflon méditerranéen (*Ovis gmelini musimon* × *Ovis* sp.) dans le massif du Caroux-Espinouse. Veterinary thesis, VetAgro Sup, Université de Lyon, Lyon, France, 128p

Maisels FG (1988) The feeding ecology of the Cyprus mouflon *Ovis orientalis gmelini* 1774, in the Paphos Forest, Cyprus. Ph.D. thesis, University of Edinburgh, Edinburgh

Malagnino A, Marchand P, Garel M, Cargnelutti B, Itty C, Chaval Y, Hewison A, Loison A, Morellet N (2021) Do reproductive constraints or experience drive age-dependent space use in two large herbivores? Anim Behav 172:121–133

Marchand P (2013) Déterminants spatio-temporels de la sélection de l'habitat chez le mouflon méditerranéen *Ovis gmelini musimon* × *Ovis* sp. Ph.D. thesis, University of Grenoble

Marchand P, Redjadj C, Garel M, Cugnasse JM, Maillard D, Loison A (2013) Are mouflon *Ovis gmelini musimon* really grazers? A review of variation in diet composition. Mammal Rev 43:275–291

Marchand P, Garel M, Bourgoin G, Dubray D, Maillard D, Loison A (2014a) Impacts of tourism and hunting on a large herbivore's spatio-temporal behavior in and around a French protected area. Biol Conserv 177:1–11

Marchand P, Garel M, Michel P, Maillard D, Loison A (2014b) Habitat-related variation in carcass mass of a large herbivore revealed by combining hunting and gps data. J Wildl Manag 78:657–670

Marchand P, Garel M, Bourgoin G, Dubray D, Maillard D, Loison A (2015a) Coupling scale-specific habitat selection and activity reveals sex-specific food/cover trade-offs in a large herbivore. Anim Behav 102:169–187

Marchand P, Garel M, Bourgoin G, Dubray D, Maillard D, Loison A (2015b) Sex-specific adjustments in habitat selection contribute to buffer mouflon against summer conditions. Behav Ecol 26:472–482

Marchand P, Garel M, Bourgoin G, Duparc A, Dubray D, Maillard D, Loison A (2017) Combining familiarity and landscape features helps break down the barriers between movements and home ranges in a non-territorial large herbivore. J Anim Ecol 86: 371–383

Marchand P, Garel M, Morellet N, Benoit L, Chaval Y, Itty C, Petit E, Cargnelutti B, Hewison AJM, Loison A (2021) A standardised biologging approach to infer parturition: an application in large herbivores across the hider-follower continuum. Methods Ecol Evol 12:1017–1030

Marco I, Mentaberre G, Ballesteros C, Bischof DF, Lavín S, Vilei EM (2009) First report of *Mycoplasma conjunctivae* from wild Caprinae with infectious keratoconjunctivitis in the Pyrenees (NE Spain). J Wildl Dis 45:238–241

Martín JG, Festa-Bianchet M (2011) Age-independent and age-dependent decreases in reproduction of females. Ecol Lett 14:576–581

Martín-Atance P, León L, Candela MG (2012) Serology as an epidemiological tool for *Salmonella abortusovis* surveillance in the wild-domestic ruminant interface. In: *Salmonella* – a diversified superbug. IntechOpen, London

Martínez-Fresneda L, O'Brien E, Velazquez R, Toledano-Díaz A, Martínez-Cáceres C, Tesfaye D, Schellander K, García-Vázquez FA, Santiago-Moreno J (2019) Seasonal variations of sperm freezability associated with testicular germinal epithelium changes in domestic (*Ovis aries*) and wild (*Ovis musimon*) rams. Reprod Fertil Dev 31(10):1545–1557

Martinov S, Pandarov S, Popov G (1989) Seroepizootology of Q fever in Bulgaria during the last five years. Eur J Epidemiol 5:425–427

Martins AG, Netto NT, Aulagnier S, Borges A, Dubois M, Vicente L, Gerard JF, Maublanc ML (2002) Population subdivision among mouflon sheep (*Ovis gmelini*) ewes and ranging behaviour of rams during the rut. J Zool 258:27–37

McClelland BE (1991) Courtship and agonistic behavior in mouflon sheep. Appl Anim Behav Sci 29:67–85

Mercier A (2016) Les maladies à tiques à l'interface faune sauvage-ruminants domestiques-homme: détermination de la prévalence des hémopathogènes vectorisés par les tiques parasitant le mouflon (*Ovis gmelini musimon* × *Ovis* sp) au sein du massif du Caroux-Espinouse. Veterinary thesis, VetAgro Sup, Université de Lyon, Lyon, France, 126p

Mereu P, Pirastru M, Barbato M, Hadjisterkotis E, Leoni G, Naitana S, Masala B, Manca L (2016) The entire mtDNA sequence of the Cyprian mouflon (*Ovis gmelini ophion*): a new method for the study of mouflon and domestic sheep evolution. In: Hadjisterkotis E (ed) Book of abstracts, 3rd edn. 6th World congress on mountain ungulates and 5th international symposium on mouflon, Nicosia, 28 Aug–1 Sept, pp 59–60

Mereu P, Pirastru M, Barbato M, Satta V, Hadjisterkotis E, Manca L, Naitana S, Leoni GG (2019) Identification of an ancestral haplotype in the mitochondrial phylogeny of the ovine haplogroup B. PeerJ 7:e7895

Michaelidou M, Decker D (2002) Challenges and opportunities facing wildlife conservation and cultural sus-

tainability in the Paphos Forest, Cyprus: historical overview and contemporary perspective. Z Jagdwiss 48:291–300
Milner JM, Bonenfant C, Mysterud A, Gaillard JM, Csanyi S, Stenseth NC (2006) Temporal and spatial development of red deer harvesting in Europe: biological and cultural factors. J Appl Ecol 43:721–734
Miranda M, Sicilia M, Bartolomé J, Molina-Alcaide E, Gálvez-Bravo L, Cassinello J (2012) Contrasting feeding patterns of native red deer and two exotic ungulates in a Mediterranean ecosystem. Wildl Res 39:171–182
Möckel R (2017) Wolf und Mufflon – ein Fallbeispiel aus dem Forst Hohenbucko (Brandenburg). Säugetierkd Inf 10(53):355–366
Morellet N, Gaillard JM, Hewison AJM, Ballon P, Boscardin Y, Duncan P, Klein F, Maillard D (2007) Indicators of ecological change: new tools for managing populations of large herbivores. J Appl Ecol 44: 634–643
Moreno V, Pérez JM, Moreno PA, Granados JE, Ruiz-Martínez I, Soriguer RC, de Simon MA (1999) Oestrid myiasis in European mouflon from Spain. J Wildl Dis 35:78–81
Moreno V, Romero-Fernández I, Marchal JA, Beltrán M, Granados JE, Habela MA, Tamadon A, Rakhshandehroo E, Sarasa M, Pérez JM et al (2015) Molecular characterization of bot flies, *Oestrus* spp. (Diptera, Oestridae), from domestic and wild Bovidae hosts. Vet Parasitol 212:473–477
Mottl S (1960) Mufloni Zver, Biologi a chov. S.Z.N, Prag
Mysterud A, Langvatn R, Stenseth NC (2004) Patterns of reproductive effort in male ungulates. J Zool 264: 209–215
Nadler CF, Korobitsina KV, Hoffmann RS, Vorontsov NN (1973) Cytogenic differentiation, geographic distribution, and domestication in Palearctic sheep (*Ovis*). Z Säugetierkd 38:109–125
Nahlik A (2001) Fecundity and survival of mouflon and factors affecting them. In: Nahlik A, Uloth W (eds) Proceedings of the third international symposium on mouflon, Sopron, pp 22–30
Nasiadka P, Wajdzik M, Skubis J (2021) A comprehensive over 100 years history of mouflon (*Ovis musimon*) in Poland: from the promising beginning in 1902 to questionable future in 2014-a case study of wildlife management history. App Ecol Environ Res 19:993–1017
Nicolaou H, Hadjisterkotis E, Papasavvas K (2016) Past and present distribution and abundance of the Cyprian mouflon. In: Hadjisterkotis E (ed) Book of abstracts, 3rd edn. 6th World congress on mountain ungulates and 5th international symposium on mouflon, Nicosia, 28 Aug–1 Sept
Özüt D (2001) Conservation genetics of Anatolian mouflon (*Ovis gmelinii anatolica*). Master's thesis, Middle East Technical University, Ankara
Özüt D (2009) Évaluation du processus d'adaptation d'une population ré-introduite de mouflon (*Ovis gmelinii anatolica*), par l'étude de sa démographie et de son écologie spatiale. Ph.D. thesis, Université Toulouse 3, Paul Sabatier
Paeth H, Hense A (2005) Mean versus extreme climate in the Mediterranean region and its sensitivity to future global warming conditions. Meteorol Z 14:329–347
Panayotova-Pencheva M (2006) New records of protostrongylid lungworms from wild ruminants in Bulgaria. Vet Med 51:477
Petit E, Aulagnier S, Bon R, Dubois M, Crouau-Roy B (1997) Genetic structure of populations of the Mediterranean mouflon (*Ovis gmelini*). J Mammal 78:459–467
Pfeffer P (1967) Le mouflon de Corse (*Ovis ammon musimon* Schreber, 1782); position systématique, écologie et éthologie comparées. Mammalia 31 (Suppl):1–262
Pfeffer P, Genest H (1969) Biologie comparée d'une population de mouflons de Corse (*Ovis ammon musimon*) du parc naturel du Caroux. Mammalia 33:165–192
Pfeffer P, Settimo R (1973) Déplacements saisonniers et compétition vitale entre mouflons, chamois et bouquetins dans la réserve du Mercantour (Alpes-Maritimes). Mammalia 37:203–219
Pianka ER, Parker WS (1975) Age-specific reproductive tactics. Am Nat 109:453–464
Piegert H, Uloth W (2005) Der Europäische Mufflon. DSV-Verlag GmbH, Edition Natur Life, Hamburg
Pipia A, Ciuti S, Grignolio S, Luchetti S, Madau R, Apollonio M (2008) Influence of sex, season, temperature and reproductive status on daily activity patterns in Sardinian mouflon (*Ovis orientalis musimon*). Behaviour 145:1723–1745
Poglayen G, Urbani L, Modugno F, Scala A, Giannetto S (2018) The Italian mouflon (*Ovis musimon*): a brief history of its parasites in the last 45 years. Res J Zool 1:2
Poplin F (1979) Origine du mouflon de Corse dans un nouvelle perspective paléontologique, par marronnage. Ann Genet Sel Anim 11:133–143
Portanier E (2018) Parasitisme et structuration génétique et spatiale: exemple chez le mouflon méditerranéen, *Ovis gmelini musimon* × *Ovis* sp. Ph.D. thesis, Université de Montpellier, Montpellier
Portanier E, Garel M, Devillard S, Marchand P, Andru J, Maillard D, Bourgoin G (2017) Introduction history overrides social factors in explaining genetic structure of females in Mediterranean mouflon. Ecol Evol 7: 9580–9591
Portanier E, Larroque J, Garel M, Marchand P, Maillard D, Bourgoin G, Devillard S (2018) Landscape genetics matches with behavioral ecology and brings new insight on the functional connectivity in Mediterranean mouflon. Landsc Ecol 33:1069–1085
Portanier E, Garel M, Devillard S, Maillard D, Poissant J, Galan M, Benabed S, Poirel MT, Duhayer J, Itty C, Bourgoin G (2019) Both candidate gene and neutral genetic diversity correlate with parasite resistance in female Mediterranean mouflon. BMC Ecol 19:1–14
Portanier E, Chevret P, Gélin P, Benedetti P, Sanchis F, Barbanera F, Kaerle C, Queney G, Bourgoin G, Devillard S, Garel M (2022) New insights into the past and recent evolutionary history of the Corsican

mouflon (*Ovis gmelini musimon*) to inform its conservation. Conserv Genet 23:91–107

Poulle ML, Carles L, Lequette B (1997) Significance of ungulates in the diet of recently settled wolves in the Mercantour mountains (southeastern France). Rev Ecol (Terre Vie) 52:357–368

Puddu G, Maiorano L (2016) Combining multiple tools to provide realistic potential distributions for the mouflon in Sardinia: species distribution models, spatial pattern analysis and circuit theory. Hystrix 27:11695

Réale D, Boussès P, Chapuis JL (1999) Nursing behaviour and mother–lamb relationships in mouflon under fluctuating population densities. Behav Process 47:81–94

Redjadj C, Darmon G, Maillard D, Chevrier T, Bastianelli D, Verheyden H, Loison A, Saïd S (2014) Intra- and interspecific differences in diet quality and composition in a large herbivore community. PLoS One 9:e84756

Rettie WJ, Messier F (2000) Hierarchical habitat selection by woodland caribou: its relationship to limiting factors. Ecography 23:466–478

Rezaei H (2007) Phylogénie moléculaire du Genre *Ovis* (Mouton et Mouflons). Implications pour la Conservation du Genre et pour l'Origine de l'Espèce Domestique. Ph.D. thesis, Université de Grenoble

Rezaei HR, Naderi S, Chintauan-Marquier IC, Taberlet P, Virk AT, Naghash HR, Rioux D, Kaboli M, Pompanon F (2010) Evolution and taxonomy of the wild species of the genus *Ovis* (Mammalia, Artiodactyla, Bovidae). Mol Phylogenet Evol 54:315–326

Richard Q (2016) Hétérogénéité individuelle, variabilité temporelle et structure spatiale comme sources de variation démographique chez les grands herbivores de montagne. Ph.D. thesis, Université Savoie Mont-Blanc, Bourget du Lac

Rieck W (1975) Muffelwildalter. Brochure de l'association des chasseurs d'Allemagne Fédérale, pp 1–10

Rieu L (2007) LIFE NATURE Conservation et extension des populations de mouflons corses en Corse – Projet LIFE03NAT/F/000099 (2003–2007). Parc Naturel Regional de Corse, Corte

Rigaud P (1985) Le mouflon (*Ovis ammon musimon*, Shreber 1782) dans le massif du Sancy (Puy de Dôme). Ecologie. Structure des populations. Régime alimentaire. Parasitisme. Veterinary Thesis, Université Claude Bernard Lyon 1, Lyon 1, 247p

Rodríguez Luengo JL, Piñero JC (1991) Autumn diet of the Corsica mouflon *Ovis ammon musimon* Shreber 1782 on Tenerife, Canary Islands. In: Bobek B, Perzanowski K, R. W.L. (eds) Global trends in wildlife management: transactions of the 18th International Union of Game Biologists Congress, Jagiellonian University, Kraków, Poland, August 1987. Świat Press, Krakow/Warszawa, pp 137–140

Rodríguez JL, Rodríguez JC, Ramos MT (1988) Autumn diet selectivity of the Corsica mouflon (*Ovis ammon musimon* Schreber, 1782) on Tenerife (Canary Islands). Mammalia 4:476–481

Roffler GH, Schwartz MK, Pilgrim KL, Talbot SL, Sage GK, Adams LG, Luikart G (2016) Identification of landscape features influencing gene flow: how useful are habitat selection models? Evol Appl 9:805–817

Rossi S, Pioz M, Beard E, Durand B, Gibert P, Gauthier D, Klein F, Maillard D, Saint-Andrieux C, Saubusse T, Hars J (2014) Bluetongue dynamics in French wildlife: exploring the driving forces. Transbound Emerg Dis 61:e12–e24

Rossi S, Viarouge C, Faure E, Gilot-Fromont E, Gache K, Gibert P, Verheyden H, Hars J, Klein F, Maillard D, Gauthier D, Game Y, Pozet F, Sailleau C, Garnier A, Zientara S, Bréard E (2017) Exposure of wildlife to the Schmallenberg virus in France (2011–2014): higher, faster, stronger (than bluetongue)! Transbound Emerg Dis 64:354–363

Ruckstuhl KE, Neuhaus P (2000) Sexual segregation in ungulates: a new approach. Behaviour 137:361–377

Ruckstuhl K, Neuhaus P (2002) Sexual segregation in ungulates: a comparative test of three hypotheses. Biol Rev 77:77–96

Rutberg AT (1987) Adaptive hypotheses of birth synchrony in ruminants: an interspecific test. Am Nat 130:692–710

Sæther BE (1997) Environmental stochasticity and population dynamics of large herbivores: a search for mechanisms. Trends Ecol Evol 12:143–149

Sanchis F (2018) Etude de l'utilisation et de la selection de l'habitat chez le mouflon de corse à différentes échelles spatiales et temporelles. Ecole Pratique des Hautes Etudes, Montpellier

Sand H (1996) Life history patterns in female moose (*Alces alces*): the relationship between age, body size, fecundity and environmental conditions. Oecologia 106: 212–220

Sanna D, Barbato M, Hadjisterkotis E, Cossu P, Decandia L, Trova S, Pirastru M, Leoni GG, Naitana S, Francalacci P et al (2015) The first mitogenome of the Cyprus mouflon (*Ovis gmelini ophion*): new insights into the phylogeny of the genus *Ovis*. PLoS One 10:e0144257

Santiago-Moreno J, Gómez-Brunet A, Gonzalez de Bulnes A, Villar D, López Sebastian A (2000a) Attainment of puberty in the European mouflon (*Ovis gmelini musimon*) and the domestic Manchega ewe (*Ovis aries*). Reprod Domest Anim 35:49–52

Santiago-Moreno J, Gonzales-Bulnes A, Gómez-Brunet A, López-Sebastian A (2000b) Effect of birth date on body weight, scrotal circumference and horn dimension growth in captive juvenile mouflon (*Ovis gmelini musimon*) rams. Game Wildl Sci 17:179–187

Santiago-Moreno J, López-Sebastian A, Gonzalez-Bulnes-A, Gómez-Brunet A, Chemineau P (2000c) Seasonal changes in ovulatory activity, plasma prolactin, and melatonin concentrations, in mouflon (*Ovis gmelini musimon*) and manchega (*Ovis aries*) ewes. Reprod Nutr Dev 40:421–430

Santiago-Moreno J, González-Bulnes A, Gómez-Brunet A, Cocero M, Del Campo A, García-García R, López-Sebastián A (2001a) Procedure for successful interspecific embryo transfer from mouflon (*Ovis gmelini musimon*) to Spanish Merino sheep (*Ovis aries*). J Zoo Wildl Med 32:336–341

Santiago-Moreno J, López-Sebastian A, Gonzalez-Bulnes-A, Gómez-Brunet A, Tortonese D (2001b) The timing of the onset of puberty, extension of the breeding season, and length of postpartum anestrus in the female mouflon (*Ovis gmelini musimon*). J Zoo Wildl Med 32: 230–235. TY – JOUR notes: 1 figure

Santiago-Moreno J, Gonzales-Bulnes A, Gómez-Brunet A, López-Sebastian A (2003) The mouflon (*Ovis gmelini musimon*): functional characterization and hunting resource [in Spanish]. Monografías INIA, Madrid, Ganadera 2, p 217. ISSN: 1577-3167

Santiago-Moreno J, López-Sebastián A, Del Campo A, González-Bulnes A, Picazo R, Gómez-Brunet A (2004) Effect of constant-release melatonin implants and prolonged exposure to a long day photoperiod on prolactin secretion and hair growth in mouflon (*Ovis gmelini musimon*). Domest Anim Endocrinol 26: 303–314

Santiago-Moreno J, Gómez-Brunet A, Toledano-Diaz A, Gonzalez-Bulnes A, Picazo RA, López-Sebastian A (2005a) Influence of age on the relationship between annual changes in horn growth rate and prolactin secretion in the European mouflon (*Ovis gmelini musimon*). Anim Reprod Sci 85:251–261

Santiago-Moreno J, Gonzales-Bulnes A, Gómez-Brunet A, Toledano-Diaz A, López-Sebastian A (2005b) Prediction of gestational age by transrectal ultrasonographic measurements in the mouflon (*Ovis gmelini musimon*). J Zoo Wildl Med 36:457–462

Santiago-Moreno J, Toledano-Díaz A, Gómez-Brunet A, López-Sebastián A (2006) Effect of constant-release implants of melatonin on horn growth in mouflon ram lambs, *Ovis gmelini musimon*. Folia Zool 55:15

Santiago-Moreno J, Gmez-Brunet A, Toledano-Daz A, Salas-Vega R, Gmez-Guillamn F, Lpez-Sebastin A (2012) Role of testosterone and photoperiod on seasonal changes in horn growth and sperm variables in the Iberian ibex: a model for polygynous wild bovids. J Endocrinol 214:155

Satta V, Lereu P, Doro M, Casula S, Casu G, Bassu G, Frongia G, Berlinguer F, Masala B, Manca L, Naitana S, Leoni GG (2016) Habitat fragmentation causes genetic differentiation of Sardinian mouflon. In: Hadjisterkotis E (ed) Book of abstracts, 2nd edn. Book of abstracts, 3rd edn. 6th World congress on mountain ungulates and 5th international symposium on mouflon, Nicosia, 28 Aug–1 Sept, pp 72–73

Satta V, Mereu P, Barbato M, Pirastru M, Bassu G, Manca L, Naitana S, Leoni GG (2021) Genetic characterization and implications for conservation of the last autochthonous Mouflon population in Europe. Sci Rep 11:14729

Schaffer WM (1968) Intraspecific combat and the evolution of the Caprini. Evolution 22:817–825

Schaffer WM, Reed CA (1972) The co-evolution of social behavior and cranial morphology in sheep and goats (Bovidae, Caprini), vol 61 (1971). Field Museum of Natural History, Chicago. https://www.biodiversitylibrary.org/item/20785

Schröder O, Lieckfeldt D, Lutz W, Rudloff C, Frölich K, Ludwig A (2016) Limited hybridization between domestic sheep and the European mouflon in Western Germany. Eur J Wildl Res 62:307–314

Shackleton DM, IUCN/SSC Caprinae Specialist Group (1997) Wild sheep and goats and their relatives: status survey and conservation action plan for Caprinae. IUCN, Gland/Cambridge, UK

Silaghi C, Hamel D, Thiel C, Pfister K, Passos LMF, Rehbein S (2011) Genetic variants of *Anaplasma phagocytophilum* in wild caprine and cervid ungulates from the Alps in Tyrol, Austria. Vector Borne Zoonotic Dis 11:355–362

Štefančíková A, Derdáková M, Škardová I, Szestáková E, Čisláková L, Kováčová D, Stanko M, Pet'ko B (2008) Some epidemiological and epizootiological aspects of Lyme borreliosis in Slovakia with the emphasis on the problems of serological diagnostics. Biologia 63: 1135–1142

Stefanidesova K, Kocianova E, Boldis V, Kostanova Z, Kanka P, Nemethova D, Spitalska E (2008) Evidence of *Anaplasma phagocytophilum* and *Rickettsia helvetica* infection in free-ranging ungulates in central Slovakia. Eur J Wildl Res 54:519–524

Talibov T, Weinberg P, Mammadov I, Mammadov E, Talibov ST (2009) Conservation strategy of Asiatic mouflon (*Ovis [orientalis] gmelini* Blyth) and bezoar goat (*Capra aegagrus* Erxleben) in Azerbaidjan. In: Zazanashvili N, Mallon D (eds) Status and protection of globally threatened species in the Caucasus. CEPF Biodiversity Investments in the Caucasus Hotspot 2004-2009, Tbilisi, pp 46–52

Tavecchia G, Coulson T, Morgan BJ, Pemberton JM, Pilkington J, Gulland F, Clutton-Brock TH (2005) Predictors of reproductive cost in female Soay sheep. J Anim Ecol 74:201–213

Teilhard de Chardin P, Piveteau J (1930) Les mammifères fossiles de Nihowan (Chine). Masson et Cie, Paris

Terres Australes et Antarctiques Françaises (2013) Bilan d'activités 2012. Terres Australes et Antarctiques Françaises, Saint Pierre

Toïgo C, Gaillard JM (2003) Causes of sex-biased adult survival in ungulates: sexual size dimorphism, mating tactic or environment harshness? Oikos 101:376–384

Toïgo C, Gaillard JM, Festa-Bianchet M, Largo E, Michallet J, Maillard D (2007) Sex- and age-specific survival of the highly dimorphic Alpine ibex: evidence for a conservative life-history tactic. J Anim Ecol 76: 679–686

Toledano-Diaz A, Santiago-Moreno J, Gómez-Brunet A, Pulido-Pastor A, Lopez-Sebastian A (2007) Horn growth related to testosterone secretion in two wild Mediterranean ruminant species: the Spanish ibex (*Capra pyrenaica hispanica*) and European mouflon (*Ovis orientalis musimon*). Anim Reprod Sci 102:300–307

Tomiczek H (1989) Altersansprache des Muffelwildes. Der Anblick 6:254–257

Trávnicek M, Halanova M, Misko J, Gurycova D, Kozuch O, Nadzamova D (1999) Presence of antibodies to some zoonoses in mouflons (*Ovis musimon*) and fallow deer (*Dama dama*) in Eastern Slovakia. Vet Med 44:215–219

Trávnicek M, Štefančíková A, Nadzamova D, Stanko M, Cisláková L, Pet'ko B, Mardzinová S, Bhide M (2003) Immunoglobulin G antibodies to *Borrelia burgdorferi* in game animals and small mammals in Eastern Slovakia. Rev Sci Tech 22:1035–1042

Türcke F, Schmincke S (1965) Das Muffelwild. Naturgeschichte. Paul Parey, Hamburg

Uloth W (1972) To the history of the distribution, introduction and cross-breeding of the Tyrrhenis mouflon in Europe and oversea. Acta Theriol 17:412–413

Valdez R (1976) Fecundity of wild sheep (*Ovis orientalis*) in Iran. J Mammal 57:762–763

Valdez R (1982) The wild sheep of the world. Wild Sheep & Goat International, Mesilla

Valdez R, Nadler C, Bunch T (1978) Evolution of wild sheep in Iran. Evolution 32:56–72

Verma SK, Su C, Dubey JP (2015) *Toxoplasma gondii* isolates from mouflon sheep (*Ovis ammon*) from Hawaii, USA. J Eukaryot Microbiol 62:141–143

Vigne JD (1992) Zooarchaeology and the biogeographical history of the mammals of Corsica and Sardinia since the last ice age. Mammal Rev 22:87–96

Vigne JD, Briois F, Zazzo A, Willcox G, Cucchi T, Thiébault S, Carrère I, Franel Y, Touquet R, Martín C et al (2012) First wave of cultivators spread to Cyprus at least 10,600 y ago. Proc Natl Acad Sci USA 109: 8445–8449

Vigne JD, Zazzo A, Cucchi T, Carrère I, Briois F, Guilaine J (2014) The transportation of mammals to Cyprus sheds light on early voyaging and boats in the Mediterranean Sea. Eurasian Prehist 10:157–176

Volmer K, Hecht W, Weiß R, Grauheding D (2008) Treatment of foot rot in free-ranging mouflon (*Ovis gmelini musimon*) populations – does it make sense? Eur J Wildl Res 54:657–665

Weller KE (2001) The status of mouflon (*Ovis musimon*) in Europe. In: Nahlik A, Uloth W (eds) Proceedings of the third international symposium on mouflon, Sopron, pp 114–140

Williams GC (1966) Natural selection, the costs of reproduction, and a refinement of Lack's principle. Am Nat 100:687–690

Wilson DE, Reeder DM (2005) Mammal species of the world: a taxonomic and geographic reference, 3rd edn. Johns Hopkins University Press, Baltimore

Zeder MA (2008) Domestication and early agriculture in the Mediterranean Basin: origins, diffusion, and impact. Proc Natl Acad Sci U S A 105:11597–11604

Zeman P, Januška J (1999) Epizootiologic background of dissimilar distribution of human cases of Lyme borreliosis and tick-borne encephalitis in a joint endemic area. Comp Immunol Microbiol Infect Dis 22:247–260

Index

© Springer Nature Switzerland AG 2022
L. Corlatti, F. E. Zachos (eds.), *Terrestrial Cetartiodactyla*, Handbook of the Mammals of Europe,
https://doi.org/10.1007/978-3-030-24475-0

GPSR Compliance

The European Union's (EU) General Product Safety Regulation (GPSR) is a set of rules that requires consumer products to be safe and our obligations to ensure this.

If you have any concerns about our products, you can contact us on ProductSafety@springernature.com

In case Publisher is established outside the EU, the EU authorized representative is:

Springer Nature Customer Service Center GmbH
Europaplatz 3
69115 Heidelberg, Germany

Batch number: 10755741

Printed by Printforce, the Netherlands